Multiphase Polymers

Multiphase Polymers

Stuart L. Cooper, EDITOR
University of Wisconsin—Madison

Gerald M. Estes, EDITOR
E. I. du Pont de Nemours & Co.

Based on a symposium sponsored by the Division of Polymer Chemistry at the 175th Meeting of the American Chemical Society, Anaheim, California, March 13–15, 1978.

ADVANCES IN CHEMISTRY SERIES **176**

AMERICAN CHEMICAL SOCIETY
WASHINGTON, D. C. 1979

Library of Congress CIP Data

Multiphase polymers.
(Advances in chemistry series; 176 ISSN 0065-2393)

Includes bibliographies and index.

1. Polymers and polymerization—Congresses.
I. Cooper, Stuart L., 1941- . II. Estes, Gerald M.
III. American Chemical Society. Division of Polymer Chemistry. IV. Series.

QD1.A355 no. 176 [QD380] 540'.8s [547'.84]
79-10972
ISBN 0-8412-0457-8 ADCSAJ 176 1-643 1979

PRINTED IN THE UNITED STATES OF AMERICA

Advances in Chemistry Series

Robert F. Gould, *Editor*

FOREWORD

ADVANCES IN CHEMISTRY SERIES was founded in 1949 by the American Chemical Society as an outlet for symposia and collections of data in special areas of topical interest that could not be accommodated in the Society's journals. It provides a medium for symposia that would otherwise be fragmented, their papers, distributed among several journals or not published at all. Papers are reviewed critically according to ACS editorial standards and receive the careful attention and processing characteristic of ACS publications. Volumes in the ADVANCES IN CHEMISTRY SERIES maintain the integrity of the symposia on which they are based; however, verbatim reproductions of previously published papers are not accepted. Papers may include reports of research as well as reviews since symposia may embrace both types of presentation.

CONTENTS

BLENDS

PREFACE

Multiphase polymers may be divided broadly into two classifications. The first group are the semicrystalline homopolymers, while a second group would include a diverse collection of block polymers, blends, and segmented elastomers. The latter systems were emphasized at a symposium on "Multicomponent Polymer Systems" organized as part of the 175th National Meeting of the American Chemical Society in Anaheim, March 13–17, 1978.

This volume contains 29 contributions presented at the three-day Anaheim symposium. They are organized into sections where one may find state-of-the-art papers on segmented copolymers, block copolymers, and blend systems. Three additional review articles were solicited and added to the meeting papers to provide nonspecialists with background material and an introduction to each section.

Segmented polyurethanes and polyesters are microphase-separated polymers that are playing an increasingly important role in thermoplastic elastomer, fiber, and adhesive technology. Although the eight papers on segmented polymers included in this volume do not describe applications, they do survey important characterization methods and structure–property relationships. Application of the techniques of electron microscopy, x-ray diffraction, thermal analysis, infrared spectroscopy, and dynamic mechanical testing may be found in the papers of this section. The contributions include experimental details and methodology that should be of value to those readers applying segmented polymer technology.

The block polymer section is headed by an excellent review paper by Mitchel Shen. Covering anionically polymerized styrene–diene block polymers primarily, the eight papers of this section explore relaxation behavior and morphology. Block polymer properties such as transition behavior, deformation characteristics, and blend effects are shown to be related both to polymer chemical structure and to microphase morphology.

Blend systems are introduced by a timely review by Donald Paul and Joel Barlow focused on blend technology. This review includes a discussion of the increasingly important topic of polymer miscibility in engineering thermoplastics. The 15 papers that follow present a variety of approaches to polymer-blend morphology study, fracture and deformation characteristics, crystallization, scattering behavior, and blending thermodynamics. Although this largest section of the volume may be less cohesive, it is a fine collection of both the polymer physics and engineering of this important class of multiphase polymer systems.

The editors wish to thank the individual contributors for their fine cooperation, which has made possible the timely publication of our symposium volume. We are also appreciative of the fine editorial assistance provided by the American Chemical Society, especially Joan Comstock and Saundra Goss.

University of Wisconsin
Madison, WI

STUART L. COOPER

E. I. duPont de Nemours & Co.

GERALD M. ESTES

February 6, 1979.

SEGMENTED COPOLYMERS

1

Morphology and Properties of Segmented Copolymers

J. W. C. VAN BOGART, A. LILAONITKUL, and S. L. COOPER

Department of Chemical Engineering, University of Wisconsin, Madison, WI 53706

The structure–property relationships of $(AB)_n$ type segmented copolymers and methods of characterization are reviewed. The two-phase microstructure exhibited by segmented elastomers, resulting from thermodynamic incompatibility of the unlike blocks, is seen to depend on segment type, segment length, segment compatibility, chemical composition, method of fabrication, and the ability of the segments to crystallize. Morphological characterization is accomplished using the techniques of small-angle x-ray and light scattering, electron microscopy, and dielectric spectroscopy. The microphase morphology is directly responsible for the unique elastomeric properties exhibited by these systems as determined by thermal analysis, dynamic mechanical testing, and stress–strain behavior.

The unique and novel properties of block copolymers have recently generated considerable interest in their synthesis, properties, and solid state morphology (*1*–*8*). Generally, these materials are synthesized by chemically combining blocks of two dissimilar homopolymers along the chain backbone. If A and B represent two homopolymers, then the possible molecular architecture includes A–B diblock structures, A–B–A triblock polymers, and $(AB)_n$ multiblock systems. The nature of the blocks and their sequential arrangement play an important role in determining block copolymer properties.

Segmented copolymers are the $(AB)_n$ type alternating block copolymers in which the blocks are relatively short and numerous. Depending on the nature of the blocks and the average segment length, the

0-8412-0457-8/79/33-176-003$07.00/0

properties of segmented copolymers may vary from those of random copolymers to thermoplastic elastomers. The former generally has been observed in systems which have either short segment lengths or similar inter- and intrasegmental secondary binding forces or both. The solid-state structure of these compatible segmented polymers is relatively homogeneous, with the copolymers displaying properties approximated by a weighted average of the homopolymer segments. However, most segmented copolymers exhibit a two-phase structure and are known as thermoplastic elastomers. At service temperatures, one of the components is viscous or rubbery (soft segment) while the other is of a glassy or semicrystalline nature (hard segment). It is now widely accepted that the unusual properties of these copolymers are directly related to their two-phase microstructure, with the hard domains acting as a reinforcing filler and multifunctional crosslink. Block copolymers of this type behave as chemically crosslinked elastomers, yet they can be processed by rapid thermoplastic-forming techniques at elevated temperature. Because of the relatively short segment length and its molecular weight distribution, microphase separation may be incomplete, suggesting impure domains and interfacial regions comprised of a mixed phase in which there is a gradient of composition. The extent of interphase mixing and how it is affected by sample fabrication methods also can control many of the important properties of segmented copolymers.

In this chapter, synthesis of segmented copolymers and the thermodynamics of phase separation will be discussed briefly. The main focus, however, summarizes recent research activities in the study of structure–property relationships of these segmented copolymers.

Synthesis of Segmented Copolymers

Segmented copolymers are usually synthesized by condensation polymerization reactions (5, 7, 9). The reaction components consist of a difunctional soft segment, the basic hard segment component, and a chain extender for the hard segment.

Soft blocks are composed of linear, dihydroxy polyethers or polyesters with molecular weights between 600 and 3000. In a typical polymerization of a thermoplastic polyurethane elastomer, the macroglycol is end capped with the full amount of aromatic diisocyanate required in the final composition. Subsequently, the end-capped prepolymer and excess diisocyanate mixture reacts further with the required stoichiometric amount of monomeric diol to complete the reaction. The diol links the prepolymer segments together while excess diol and diisocyanate form short hard-block sements, leading to the $(AB)_n$ structure illustrated in Figure 1. Block lengths in $(AB)_n$ polymers are frequently much shorter than those in anionically synthesized ABA block copolymers.

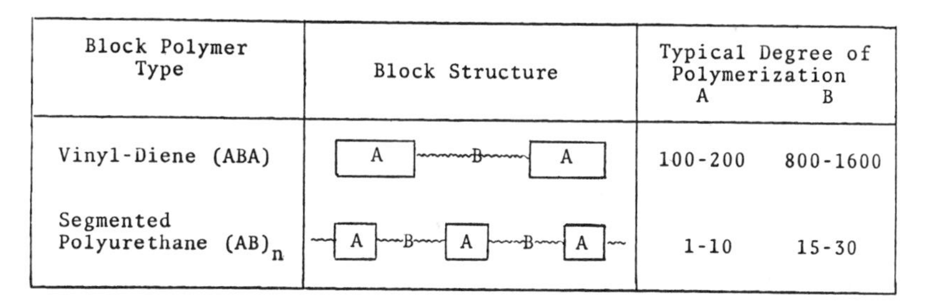

Block Polymer Type	Block Structure	Typical Degree of Polymerization A	B
Vinyl-Diene (ABA)	[A]~~~B~~~[A]	100-200	800-1600
Segmented Polyurethane $(AB)_n$	~[A]~B~[A]~B~[A]~	1-10	15-30

Figure 1. Thermoplastic elastomer molecular structures. (A) Denotes hard segment; (B) denotes soft segment.

Molecular structure can be varied by changing the chemical composition of the three reactants (macroglycol, diisocyanate, and monomeric diol) or by changing the method of polymerization. All three reactants can be simultaneously polymerized in a one-step (*10*) reaction or they can be added sequentially (*11*) after forming an isocyanate-capped prepolymer. From a theoretical standpoint, Peebles (*12, 13*) has shown that a two-step method of block polymer synthesis should lead to a narrower distribution of hard segment lengths as compared with that for a one-step synthesis, providing that reaction of the first isocyanate moiety occurs at a faster rate than that of the remaining moiety.

Chemical composition determines many molecular properties such as polarity, hydrogen bonding capability, and crystallizability of the blocks. If monomeric diols are replaced by diamines, highly polar urea linkages are formed in the hard blocks. Polyurethanes also have been synthesized with piperazine replacing the diisocyanate (*14*), thereby eliminating all possibility of hydrogen bonding. The synthesis reactions for other segmented copolymers, such as the segmented copolyesters (*15, 16, 17, 18, 19*) are analogous to those for the urethanes. In the case for the copolyesters, however, the reaction is a melt trans-esterification which produces methanol as a by-product and requires low-pressure evaporation for removal.

Thermodynamics of Phase Separation

The physics controlling the morphology of the block copolymer is simple to describe. In a material composed of units of type A and B, which have a positive heat of mixing, there is a tendency toward phase separation. The topology of the block copolymer molecules imposes restrictions on this segregation and leads to microdomain formation. From a thermodynamic point of view, there is a positive surface free energy associated with the interface between A and B domains. This serves as a driving force toward growth of the domains. As a result of the tendency of the block joints to stay in the interfacial regions, there is a loss of entropy in two ways. One is attributable to the confinement of the joints to the interface. The other has its origin in maintaining the virtually constant overall polymer density by the suppression of a vast number of polymer conformations. The equilibrium domain size and shape are a result of the balance of these three free-energy terms. Prevalent thermodynamic theories for phase separation in block copolymers are given by Krause (*20, 21, 22, 23, 24*), Helfand (*25, 26, 27, 28, 29*), Meier (*30–35*), and Le Grand (*36*).

Krause analyzed microphase separation from a strictly thermodynamic approach based on macroscopic variables. Complete phase separation with sharp boundaries between the phases was assumed. Despite the limitations of the treatment (the theory allows neither a prediction of morphology nor a calculation of microphase dimensions), the approach is very useful in demonstrating the influence of the number of blocks on phase separation. The model predicts that phase separation becomes more difficult as the number of blocks increases in a copolymer molecule of given length. It predicts easier phase separation when the molecular weight of the block copolymer increases at fixed copolymer composition and number of blocks per molecule. Furthermore, phase separation occurs more readily for a system having a 1:1 ratio of components (by volume) when the molecular chain length and number of blocks are kept constant. Phase separation also is improved by an increasing Flory–Huggins interaction parameter, χ_1, for the copolymer. Finally, all else remaining the same, a higher degree of phase separation is predicted for a copolymer system where one of the components is crystallizable. Krause's equation for the entropy change upon microphase separation, ΔS, where the "A" units are crystalline is shown below (*23*).

$$\frac{\Delta S}{k} = N_c \ln (v_A)^{v_A} (v_B)^{v_B} - 2N_c(m-1)\left[\frac{\Delta S_{dis}}{R}\right] + N_c \ln (m-1) + N_c n_A \left[\frac{\Delta S_{cryst}}{R}\right] \quad (1)$$

Here: N_c, total number of copolymer molecules in system; v_i, volume fraction of component *i; m*, number of blocks per copolymer molecule; n_A, number of "A" units per copolymer molecule; ΔS_{dis}, disorientation entropy change gain on fusion per segment; and ΔS_{cryst}, molar entropy of crystallization per "A" unit.

Meier (*30–35*) has developed criteria for the formation of domains and their size in terms of molecular and thermodynamic variables. He treated the constraints that A and B blocks must be placed in separate domains as boundary values in a diffusion problem. This model predicts the size of an assumed spherical domain in terms of the average end-to-end distance of the random flight chain for the constituent blocks (*30*). Extending this theory to consider cylindrical and lamellar domains, Meier (*31, 32*) predicts the trend of domain stability with molecular weight of one constituent block to proceed from spherical to cylindrical to lamellar.

Helfand (*25, 26, 27, 28, 29*) has formulated a statistical thermodynamic model of the microphases similar to that of Meier. This treatment, however, requires no adjustable parameters. Using the so-called mean-field-theory approach, the necessary statistics of the molecules are embodied in the solutions of modified diffusion equations. The constraint at the boundary was achieved by a narrow interface approximation which is accomplished mathematically by applying reflection boundary conditions.

Le Grand (*36*) has developed a model to account for domain formation and stability based on the change in free energy which occurs between a random mixture of block copolymer molecules and a micellar domain structure. The model also considers contributions to the free energy of the domain morphology resulting from the interfacial boundary between phases and elastic deformation of the domains.

The preceding discussion has neglected the effects of temperature on phase separation. The free energy of phase mixing, ΔG_{mix}, for a copolymer system is given in terms of the enthalpy, ΔH_{mix}, and entropy, ΔS_{mix}, of phase mixing as follows:

$$\Delta G_{mix} = \Delta H_{mix} - T\Delta S_{mix} \quad (2)$$

Generally, ΔH_{mix} and ΔS_{mix} are both positive. The sign of ΔG_{mix} will depend on the temperature, T. As a negative value for ΔG_{mix} favors phase mixing, one can see that increasing T will lower ΔG_{mix}, thus promoting a more compatible (phase mixed) system. This is a rather simple argument; however, it does demonstrate the importance of temperature as another parameter in determining the morphology of a segmented copolymer system.

Characterization of Sample Morphology

Segmented thermoplastic elastomers exhibit structural heterogeneity on the molecular, the domain, and in some cases on a larger scale involving periodic or spherulitic texture. Each level of structural organization is studied by specific methods. Molecular sequence distributions can be studied by chemical methods, such as NMR or IR spectroscopy.

Domain structures are studied directly by electron microscopy or more quantitatively by small-angle x-ray scattering (SAXS) methods which are particularly applicable because of the size range of typical domains. From the SAXS intensity curve, $I(s)$, three important parameters are obtainable which can be used to characterize the morphology of a two-phase system. These are the invariant, the interfacial thickness, and the sample's inhomogeneity length. The weighted integral of the scattering intensity,

$$\int_0^\infty s^2 I(s)\ ds, \tag{3}$$

known as the invariant, is related to the mean square of the electron density fluctuations, $\overline{\Delta\rho^2}$, for the two-phase system by the following expression (37):

$$2\pi^2 I_e \overline{\Delta\rho^2} V = \int_0^\infty s^2 I(s)\ ds \tag{4}$$

where $I_e = 7.90 \times 10^{-26}$, the Thomson-scattering constant for a free electron, V is the scattering volume, and s is equal to $4\pi \sin \theta / x$ with 2θ being the scattering angle and x is the wavelength of radiation. If a linear density gradient exists between phases,

$$\overline{\Delta\rho^2} = (\rho_2 - \rho_1)^2[\phi_1\phi_2 - \phi_3/6] \tag{5}$$

where ϕ_1 and ϕ_2 represent the respective volume fractions for the phases prior to mixing and ϕ_3 represents the fractional volume of the interfacial region (38). In this way, one can determine the interfacial volume fraction or the electron density difference between the two phases $(\rho_1 - \rho_2)$. This information is important in determining the degree of phase separation within elastomers. This analysis is confounded, however, if phases 1 and 2, rather than being pure hard- and soft-segment domains, are mixed phases with unknown volume fractions and electron densities. In this case, one can only determine a range of possible values of ρ and ϕ consistent with the experimental observations (49).

The thickness of the interfacial region, t, separating domains in a two-phase structure, assuming a linear density gradient across the interface, can be obtained from the shape of the high-angle tail of a SAXS intensity curve. Hashimoto, et al. (39) have shown that in the high-angle limit, the following relation is applicable:

$$\underset{s \to \infty}{I(s)} = \frac{K}{s^4}[1 - (t^2/12)s^2] \tag{6}$$

Here K represents the Porod constant.

If a two-phase system is randomly intersected with an infinite set of lines, the line segments lying within the phase A will have an average length, l_A. Likewise, the average length of segments lying in phase B could be defined as l_B. The reciprocal of the sum of the reciprocals of these lengths is known as the inhomogeneity length, l_p:

$$\frac{1}{l_p} = \frac{1}{l_A} + \frac{1}{l_B} \tag{7}$$

This parameter can be related to the intensity function, $I(s)$, as follows (*37*):

$$l_i = \frac{1}{2\pi V \Delta\rho^2} \int_0^\infty \frac{sI(s)}{I_e(s)} = \frac{1}{2\pi V \overline{\Delta\rho^2}} \int_0^\infty \frac{s\,I(s)}{I_e\,(s)}\,ds \tag{8}$$

The average length of all segments contained in a given phase i, l_i, can also be related to the inhomogeneity length through the following relationship:

$$l_i = \frac{\phi_i l_p}{\phi_A\,\phi_B} \tag{9}$$

Electron microscopy (*5, 40, 52, 63, 64*), on the other hand, can provide direct information on the domain structure under favorable conditions, such as when the domains are crystalline. When the samples exhibit a semicrystalline superstructure, small-angle light scattering and polarized microscopy have been used in addition to electron microscopy to study the spherulitic structure. These methods are complemented by differential scanning calorimetry, and various techniques for studying dynamic mechanical behavior which can be interpreted to give additional, if somewhat less direct, information on domain structure.

Much attention has been directed toward an elucidation of domain morphology in polyurethane-segmented copolymers. Direct evidence of a domain structure in polyurethanes was first provided by Koutsky et al. (*40*), using transmission electron microscopy. The domain structure is not as clearly observable as in most styrene–diene–styrene block copolymers because of difficulties of phase staining and the smaller domain sizes involved. A subsequent citation by Allport (*5*) also showed a similar morphology. It cannot be assumed on the basis of the evidence, however, that complete phase separation occurs. In fact, there is evidence to suggest that appreciable hydrogen bonding occurs between hard and soft blocks (*41, 42, 43*), which implies incomplete phase separation.

An illustration of the domain structure in an unstrained, segmented polyurethane proposed by Estes (*41*) is shown in Figure 2 in which the shaded areas are the hard domains. Both phases are represented as being continuous and interpenetrating. The model also presumes that phase separation is not completed, and some urethane blocks are also dispersed in the rubbery matrix. The domain size in the direction of the chain axis is given as approximately 50 Å, which agrees well with the calculated 55 Å contour length for an average hard block (*42*).

SAXS has been used to study the domain structure in segmented polyurethanes (*44–53*). Figure 3 shows SAXS intensity curves for three samples having a hard segment based on 4,4′-diphenylmethane diisocyanate (MDI) chain extended with butanediol (BD) and a polycaprolactone (PCL) soft segment (either 830 or 2000 molecular weight). Samples were compression molded, quenched, then stabilized at 125°C for 0.5 hr, followed by slow cooling. Hard-segment weight fractions for PCL 2000-178, PCL 830-123, and PCL 830-134 are 57, 53, and 61%,

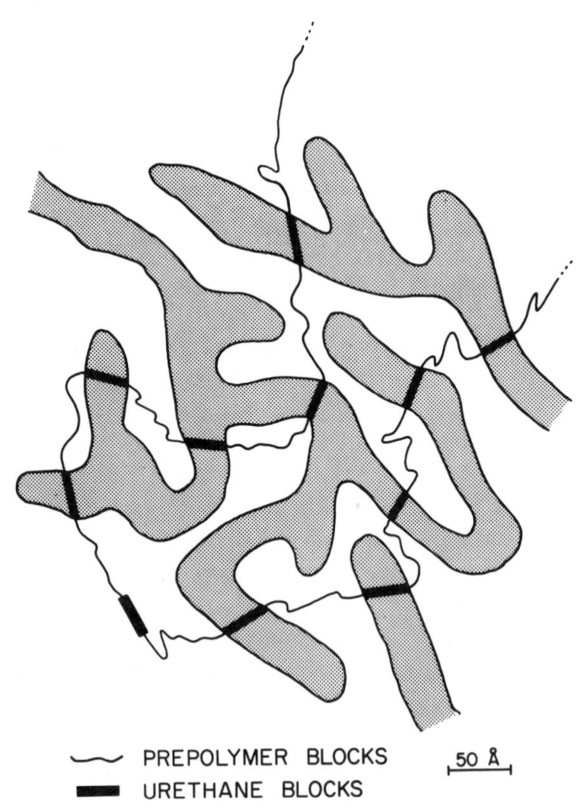

Figure 2. Representation of domain structures in a segmented copolymer

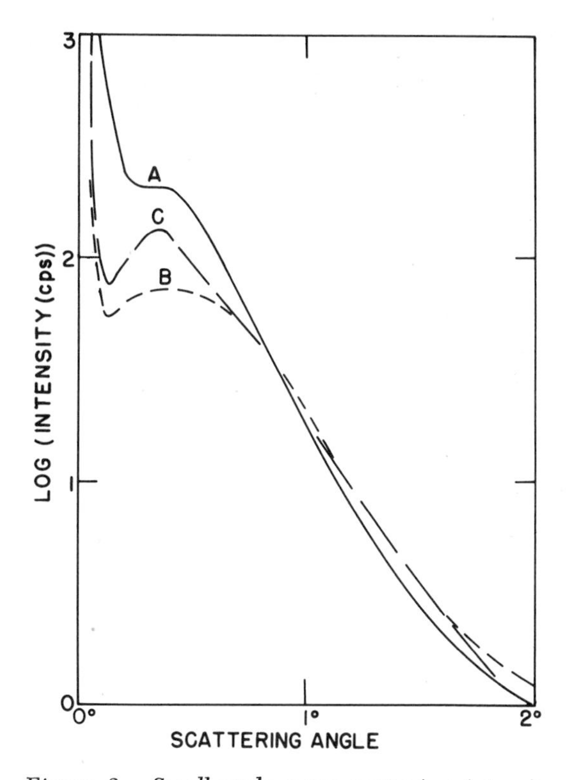

Figure 3. Small-angle x-ray scattering intensity curves demonstrating the effect of soft-segment molecular weight for three polycaprolactone polyurethanes (MDI/BD) of approximately equal hard-segment content. (A) 57% by wt, PCL 2000–178 (S); (B) 53% by wt, PCL 830–123 (S); and (C) 61% by wt, PCL 830–134 (S).

respectively. The greater scattering intensity for PCL 2000-178 is indicative of a much larger degree of microphase separation. The longer hard segments in sample PCL 830-134 also permit better phase separation relative to PCL 830-123. When the samples were annealed for 4 hr at 150°C and then slowly cooled, phase separation was improved as shown by the larger intensities of Figure 4 (*S*—stabilized; *A*—annealed). The change observed was much larger for the shorter segment length sample (830 mol wt).

Preliminary studies of the "interphase" between respective domains by Van Bogart et al. (*53*) indicate the thickness of this region, assuming a linear density gradient, is on the order of 10–20 Å for polyester and polyether urethanes (MDI-BD based). Theoretically, the interfacial thickness is inversely related to the square root of the hard segment–soft segment interaction parameter (*54*).

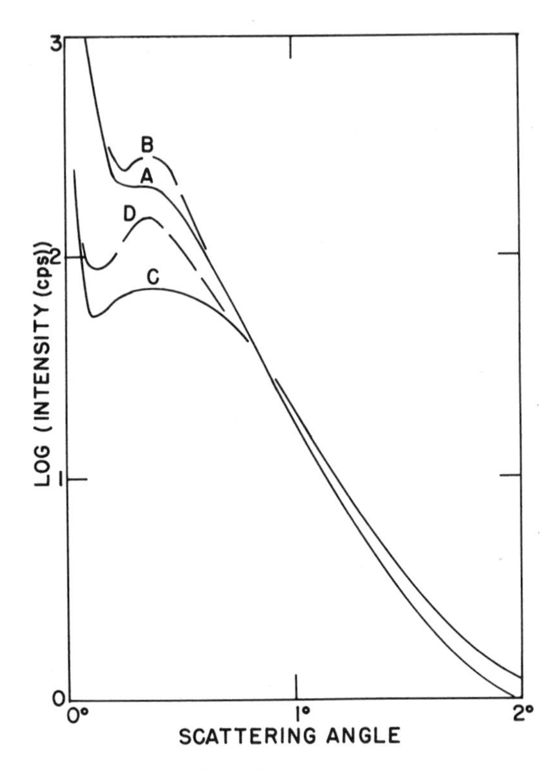

Figure 4. Small-angle x-ray scattering intensity curves showing the effects of annealing for two polycaprolactone polyurethanes (MDI/BD). (S) Stabilized at 125°C for 0.5 hr; (A) Annealed at 150°C for 4 hr. (A) PCL 2000–178 (S); (B) PCL 2000–178 (A); (C) PCL 830–134 (S); and (D) PCL 830–134 (A).

The presence of hard and soft domains in segmented polyurethanes also has been confirmed by experimental results using pulsed NMR and low-frequency dielectric measurements. Assink (*55*) recently has shown that the nuclear-magnetic, free-induction decay of these thermoplastic elastomers consists of a fast Gaussian component attributable to the glassy hard domains and a slow exponential component associated with the rubbery domains. Furthermore, the NMR technique also can be used to determine the relative amounts of material in each domain.

Dielectric relaxation study of two-phase microstructures in segmented copolymers was first attempted by North and his co-workers (*56, 57, 58, 59*). Dielectric measurements down to 10^{-5} Hz were made on MDI-based segmented polyether- and polyester-urethanes using a dc transient technique. These materials displayed large, low-frequency dielectric absorptions which were temperature dependent and characteristic of Maxwell–Wagner interfacial polarization. Both the amplitude and the relaxation frequency increased with increasing temperature. Interestingly, the dielectric loss properties were markedly reduced upon reaching the glass-transition temperature of the continuous rubbery phase. The activation energy for the low-frequency relaxation was comparable with the activation energy for bulk conductivity. The study led to the conclusion that the occluded hard domains were nonspherical with diffuse phase boundaries.

Crystallization of either segment of thermoplastic elastomers provides another mechanism of domain reinforcement. Under a suitable sample preparation method, a macroscale superstructure is usually observable. The morphology depends on several factors, such as the nature and concentration of the crystallizable component, solvent and thermal history, etc. Wilkes and Samuels (*60*) reported spherulitic morphology of a segmented piperazine polyurethane cast from chloroform whereas a similar series investigated by Cooper, et al. (*61*) showed no superstructure when using methylene chloride as solvent. Spherulites also have been observed in the segmented polyurethane–urea (*62*), segmented polycaprolactone–urethanes (*63*), and segmented polyether–ester thermoplastic elastomers (*64, 65, 66, 67, 68*). The effect of different spherulitic structures on the mechanical properties of segmented polyether–esters has been recently reported (*68*).

Thermal Analysis

The study of transition behavior by various thermoanalytical techniques (differential thermal analysis (DTA), differential scanning calorimetry (DSC), thermal expansion measurement, thermomechanical analysis) has been important to the understanding of morphology and intermolecular bonding in segmented copolymers. In some samples, as many as five transitions can be observed in a DSC thermogram, depending on the nature of the solid state structure of the sample. These include the glass-transition temperatures of each phase which appear as base line shifts, a short-range order endotherm of the hard phase attributable to storage or annealing effects, and endotherms associated with the long-range order of crystalline portions of either segment.

Typical DSC thermograms of segmented polyurethanes are shown in Figure 5 (*43*). The samples are based on MDI-BD hard segments combined with either polytetramethylene oxide (PTMO) or polytetramethylene adipate (PTMA) as soft segment. The nomenclature has been established whereby the two figures following the ET (PTMO/BD/MDI copolymer) or ES (PTMA/BD/MDI copolymer) indicate the

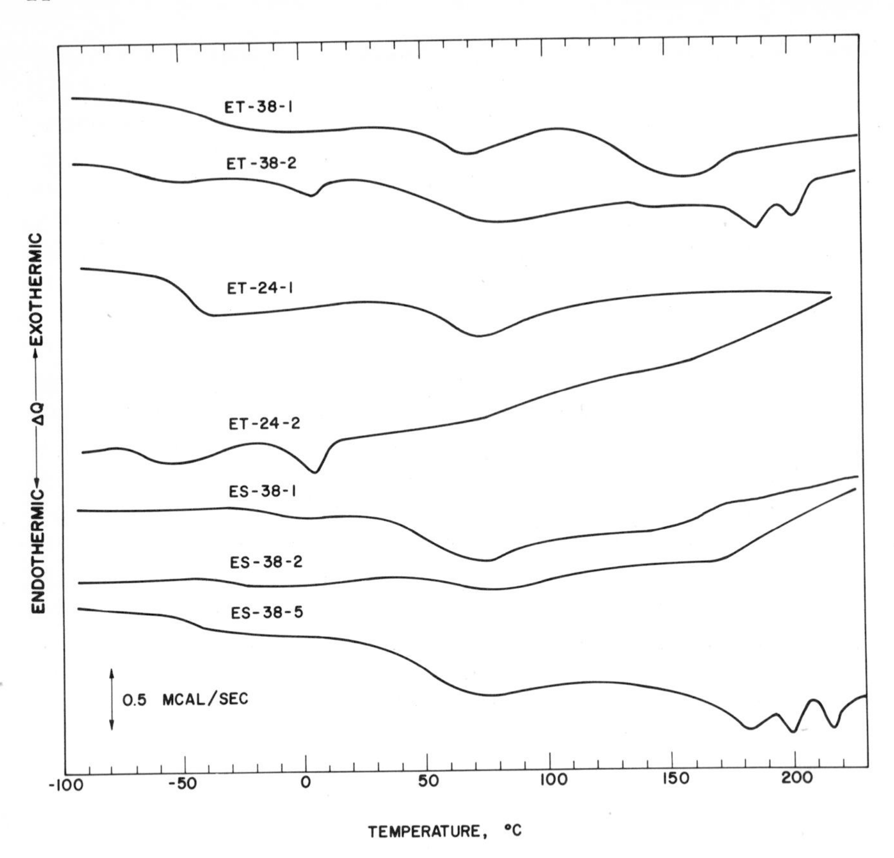

Figure 5. Differential scanning calorimetry curves for the ET-38, ET-24, and ES-38 series of polyurethanes

weight-percent diisocyanate and the soft-segment molecular weight in thousands, respectively. The shift in base line in the region of −60° to −10°C corresponds to the T_g of the soft segment. Some T_g values of interest are collected in Table I. Except for ET-38-1, ET polyurethanes having a soft-segment molecular weight of 1000 show a constant soft-segment T_g at about −44°C. In ES polyurethanes, the soft-segment T_g generally shows an increase with increasing MDI content. Both ES and ET polymers show a decrease in soft-segment T_g with increasing soft-segment molecular weight.

The variation of T_g of the soft matrix in segmented polyurethanes as a function of composition or segmental chemical structure has been monitored and used as an indicator of the degree of microphase separation. Factors influencing the phase-separation process in these MDI-based polyurethanes have been summarized by Aitken and Jeffs (69) as follows: (a) crystallization of either component, (b) the steric hindrance of the hard-segment unit in a hydrogen-bonding process, and (c) the inherent solubility between the hard and soft components. While (a) is the major factor accounting for the low T_g of the 2000-mol-wt soft-segment samples, factors (b) and (c) determine the extent of phase separation in the 1000-mol-wt soft-segment polyurethanes. However, the relatively constant T_g values observed as a function of composition exhibited by the ET elastomers of 1000-mol-wt soft segment indicates that the penetration of isolated hard segments into the soft phase is limited. The ES samples, on the other hand, have a greater tendency for the hard-segment units to be trapped in the soft matrix. This is probably because of the greater polarity of the polyester segment (*41, 43*) rather than the formation of hard segment–soft segment hydrogen bonding which also may take place in polyether–urethanes (*70*). Because of phase mixing in polyester–urethanes at 1000-mol-wt soft segment, the T_g of the elastomeric phase increases as the hard-segment content is raised.

Analogous studies on two series of segmented poly(tetramethylene oxide) (mol wt 1000) polyurethanes, having either a symmetric 2,6-toluene diisocyanate (TDI) or an asymmetric 2,4-TDI-based hard segment (butanediol chain extended), show similar results (*71*). The 2,6-TDI specimens, having crystalline hard domains which restrict phase separation, exhibit a soft segment T_g which is relatively independent of hard-segment content. The 2,4-TDI systems, on the other hand, give soft-segment T_g values which increase with increasing hard-segment content indicative of considerable phase mixing allowed by the amorphous 2,4-TDI-based hard domains.

Table I. Soft-Segment, Glass-Transition-Temperature Dependence on Hard-Segment Content for PTMO/MDI/BD(ET)- and PTMA/MDI/BD(ES)-Segmented Copolymers

Sample	*Hard Segment Content (Wt %)*	*Soft Segment T_g (°C)*
ET-38-2	50	−60
ET-38-1	46	−39
ET-35-1	42	−43
ET-31-1	37	−44
ET-28-1	31	−44
ET-24-2	30	−57
ET-24-1	25	−43
ES-38-5	54	−47
ES-38-2	54	−26
ES-38-1	48	−10
ES-35-1	41	−19
ES-31-1	37	−25
ES-28-1	31	−32
ES-24-1	26	−30

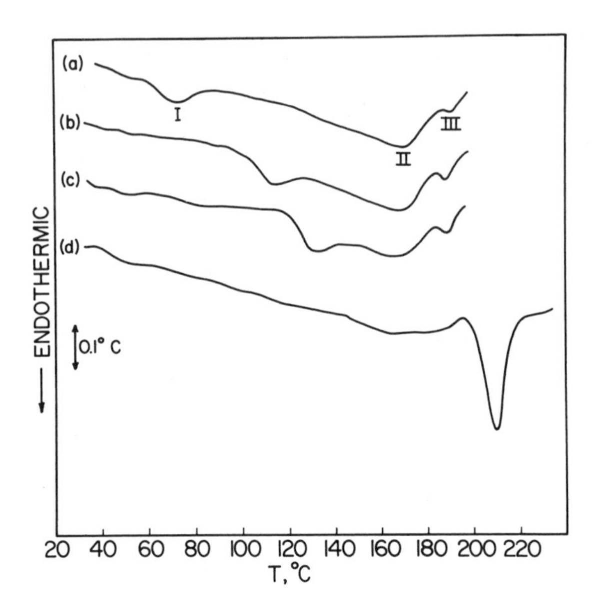

Figure 6. Effect of annealing on the differential scanning calorimetry curves for a poly(tetramethylene oxide) polyurethane (MDI/BD) (ET-38-1) containing 38% by wt MDI. Thermal treatment: (a) control; (b) 80°C, 4 hr; (c) 110°C, 4 hr; and (d) 150°C, 4.5 hr.

The DSC spectra shown in Figure 5 exhibit several endotherms associated with disordering processes which occur in the urethane domains (*43*). Early studies (*72, 73, 74*) assumed that these endotherms were attributable to hydrogen-bond disruption, e.g. an endotherm about 80°C for dissociation of hard–soft-segment hydrogen bonds and an endotherm around 150°–170°C for inter-urethane hydrogen-bond dissociation. More recent studies have shown that the intermediate DSC transitions are not attributable to hydrogen-bond dissociation (*75*). Typical DSC thermal spectra for a segmented polyether urethane are shown in Figure 6 (*76*). Transition behavior strongly depends upon thermal history. Three characteristic endothermic transitions have been observed: (I) an endotherm centered at approximately 70°C which is attributed to the disruption of domains with limited short-range order; (II) a transition at 120°–190°C which represents the dissociation of domains containing long-range order; and (III) a transition above 200°C which is attributable to the melting of microcrystallites of the hard segments. Seymour (*76*) has shown that the transition (I) can be shifted upward to merge with the higher transition (II) by annealing, the final state of order depending on the annealing history and sample composition.

Apparently, hydrogen bonding plays only a secondary role in determining the transition behavior and properties. Additional information from IR studies (*41, 43, 76, 77, 78*) indicates that it is the chain mobility, or T_g of the hard blocks, which controls hydrogen-bond dissociation rather than the opposite case. The presence of hydrogen bonds thus serves primarily to increase the overall cohesion of the hard domains, and the unusually good mechanical properties of segmented polyurethanes are instead ascribed to the microphase-separated morphology. It is of interest to mention, however, that IR studies of hydrogen bonding provide a more quantitative indication of the degree of phase mixing, through comparison of the extent of hard-segment carbonyl and NH group bonding, than afforded by DSC (*43, 47*). Furthermore, hydrogen bonding can be important in determining the arrangement of hard segments within their domains (*45, 46, 47*).

Dynamic Mechanical Property Measurements

Dynamic mechanical properties (*79*) provide information about first- and second-order transitions (T_m and T_g, respectively), phase separation, and mechanical behavior of polymers. Typical storage modulus data for several representative polymer systems (*80*) are shown in Figure 7. Below T_g the glassy state prevails with modulus values on the order of 10^{10} dyn/cm^2 for all materials. A rapid decrease in modulus is seen as the temperature is increased through the glass-transition region (above $-50°C$ for these polymers). A linear, amorphous polymer which has not been crosslinked (curve A) shows a rubbery plateau region followed by a continued rapid drop in modulus. Crosslinking (curve B) causes the modulus to stabilize with increasing temperature at about three decades below that of the glassy state. In block copolymers (curves D and E), an enhanced rubbery plateau region appears where modulus changes little with increasing temperature. Another rapid drop in modulus occurs when temperature is increased to the hard-segment transition point. In contrast, a semicrystalline polymer (curve C) maintains high modulus through the glass-transition region and up to the crystalline melting point where the structural identity of the crystallites is destroyed.

In Figure 7, curves D and E represent thermomechanical spectra for segmented, polyurethane $(AB)_n$ block copolymers. Two distinct transitions are indicated by the precipitous drops in storage moduli and the corresponding presence of two loss peaks. Ideally, for block copolymers these transitions are located at the T_m or T_g of the corresponding component homopolymers. However, sample composition, segmental length, inherent intersegment solubility, and sample preparation method have been found to influence the degree of phase separation and thereby the

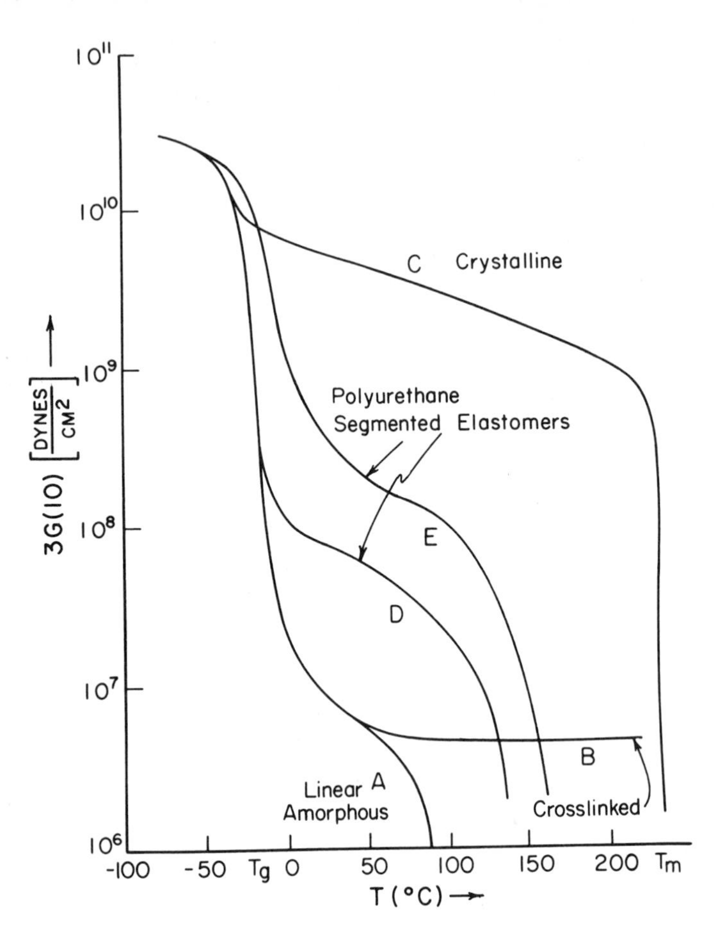

Figure 7. Storage modulus vs. temperature curves for: (A) linear amorphous polymer; (B) crosslinked polymer; (C) semi-crystalline polymer; (D) PTMA/MDI/BD-segmented copolymer (32% MDI by wt); (E) PTMA/MDI/BD-segmented copolymer (38% MDI by wt)

shape and temperature location of the dynamic-mechanical transition points. Phase mixing between domains has been indicated by a decreased slope in storage-modulus transitions and by broadened loss peaks. Some of these features are illustrated in Figure 8 for a dimethyl siloxane-polycarbonate (PDMS/BPAC) segmented copolymer system. In general, the system shows a higher degree of phase separation for samples having higher hard-segment concentration at constant PDMS block length. Also shown is that an increased hard-segment content results in an enhanced rubbery plateau. At a fixed composition ratio of PDMS to BPAC, the copolymers exhibit better phase separation and mechanical properties as the average block length increases. This latter observation is true for all segmented copolymers.

Figure 9 shows the dynamic mechanical spectra of a series of poly(tetramethylene oxide)/poly(tetramethylene terephthalate) (PTMO/PTMT) segmented copolymers (*67*). These materials reveal only one T_g and one T_m analogous to semicrystalline thermoplastics. The magnitude of both transition temperatures shifts progressively higher with increasing

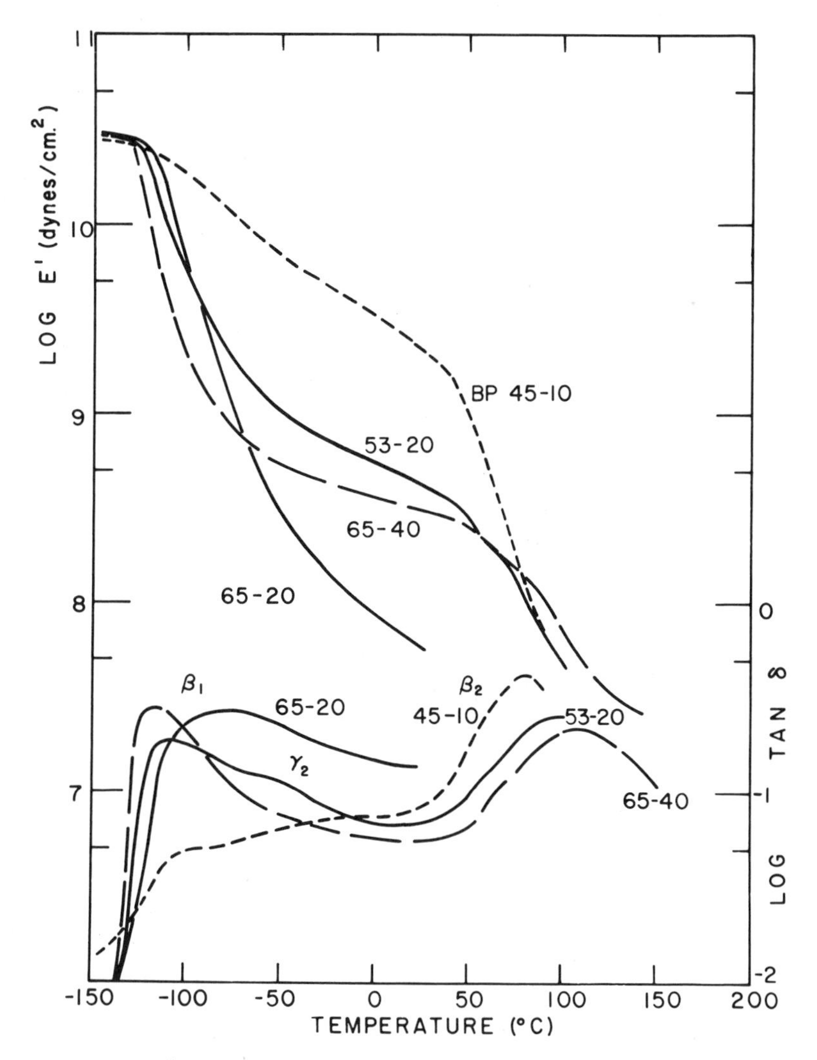

Figure 8. Effect of segment length on the dynamic mechanical properties of PDMS/BPAC segmented copolymers. (BP 45-10 contains 45% by wt PDMS with a degree of polymerization of 10 for the PDMS segment.)

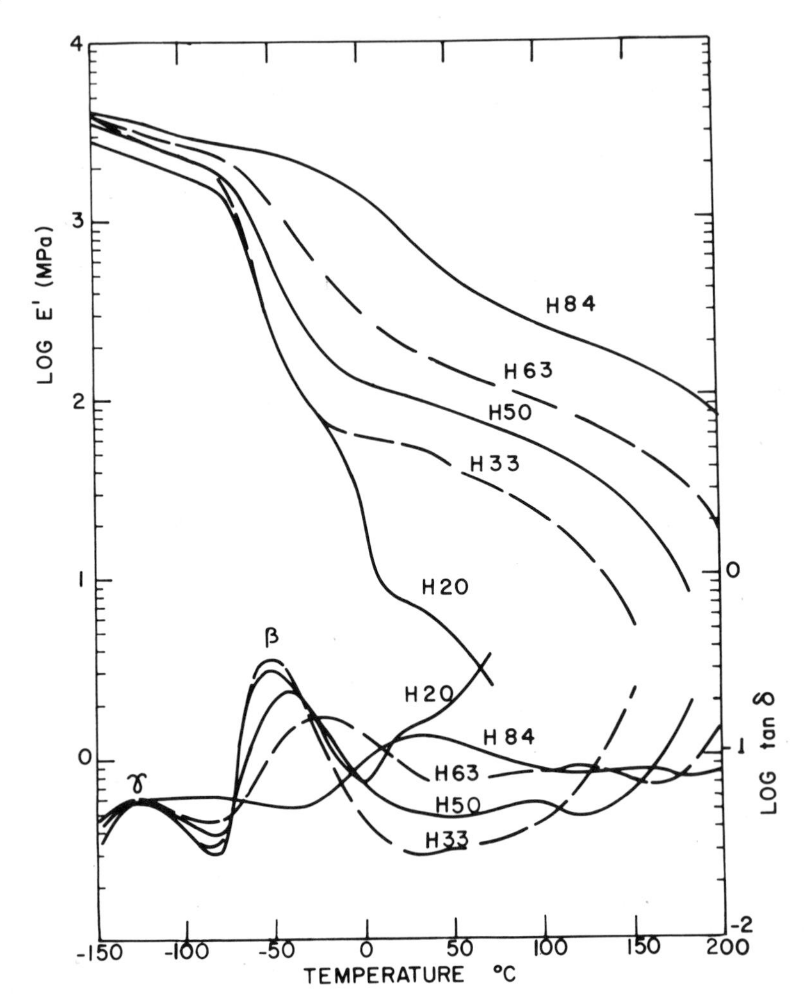

Figure 9. Effect of hard-segment content on storage modulus and loss tangent of compression-molded PTMO/PTMT-segmented copolymers

hard-segment content. Interestingly, the Gordon–Taylor equation was found to accurately model T_g behavior of these samples providing that the crystalline polyester component was not included in the definition of the hard segment used in the calculation. This suggests that uncrystallized hard segments form a relatively compatible interlamellar amorphous phase with the polyether component.

Extensive dynamic mechanical property studies have been carried out on hydrogen-bonded (*81*) and nonhydrogen-bonded (*60, 82*) polyurethanes. Several secondary relaxations were found in addition to the major hard- and soft-segment transitions. Molecular mechanisms could be assigned to each of these. A low temperature γ transition ($\sim -125°C$) was attributed to localized motion in polyether sequences. Similar γ transitions have been found in other block copolymers (*64, 65, 66, 67*). In polyurethanes with long soft segments (mol wt = 2000–5000), a soft-segment melting transition was detected. The T_g loss peak occurred at lower temperatures when soft-segment length was increased. Since the longer segments are expected to produce better-ordered and larger domains, some soft segments can exist in regions well removed from the domain interface and hard-domain interactions, so that their motion can be relatively unrestricted by the hard domains. There is also better microphase separation in these systems and therefore less hard-segment material dissolved in the soft-segment phase. Soft-segment T_g values were lower in nonhydrogen-bonded materials than in hydrogen-bonded samples with equivalent hard-segment content. Presumably this was primarily attributable to the influence of hydrogen bonding interactions with the soft segments. Since this bonding persists even above the hard-segment T_g (*41*), it is evident that the lower temperature, soft-segment T_g is not accompanied by a marked disruption of these bonds. In addition, crystallization of the hard segment in the nonhydrogen-bonded samples also serves as an additional driving force in the microphase separation process.

Stress–Strain and Ultimate Properties

T. L. Smith (*83, 84, 85, 86, 87*) has studied the relationship between segmented, copolymer ultimate properties and morphology. In general, the behavior of a strained system depends on the size and concentration of hard-segment domains (*88, 89, 90*), the strength of hard-segment aggregation, the ability of the segments to orient in the direction of stretch, and the ability of the soft segment to crystallize under strain.

Studies of two poly(tetramethylene oxide) polyurethanes (MDI-BD based) of approximately 50% by weight hard-segment content each demonstrate the effect of differences in domain size. At low strains, IR dichroism results (*91, 92*) show that hard segments in ET-38-2 (2000 mol wt soft segment) orient transverse to the direction of stress. Parallel orientation is achieved at higher stress levels. Furthermore, at a fixed strain, reorientation of the segments is time dependent. For ET-38-1 having block lengths which are 50% that of ET-38-2, however, orientation of the hard segments is always parallel with the stretch direction and much less relaxation is observed. Because of the shorter lengths, hard-segment domains for ET-38-1 are noncrystalline whereas those for ET-38-2 are crystalline, which results in the differences observed. ET-38-1 domains can be deformed readily and disrupted quickly, allowing rapid relaxation and parallel orientation unlike the crystalline ET-38-2 which tends to oppose domain disruption and reorientation.

Secant moduli (*87*) for ET-38-2 are higher than those for ET-38-1 by virtue of a greater degree of phase separation in the former. Furthermore, moduli at a fixed strain for these samples decrease with increasing temperature, a characteristic attributed to their thermoplastic nature. Samples with a urea–urethane hard segment (based on 4,4′-methylene-bis-(2-chloro-aniline)) capable of forming rigid, highly crystalline hard segments exhibit secant moduli which are basically temperature independent up to 150°C.

Stress hysteresis is prominent in segmented copolymers because of the disruption of hard segments with strain (*93, 94*). Unlike Young's modulus which depends on the rigidity and morphology of hard-segment domains, stress softening is a function of domain restructuring and ductility and the nature of the mixed hard- and soft-segment interfacial regions (*45, 87*). In segmented polyurethanes, hard-segment crystallization has been found to increase stress hysteresis, permanent set, and tensile strength. Heat build-up in polyurethanes attributable to their

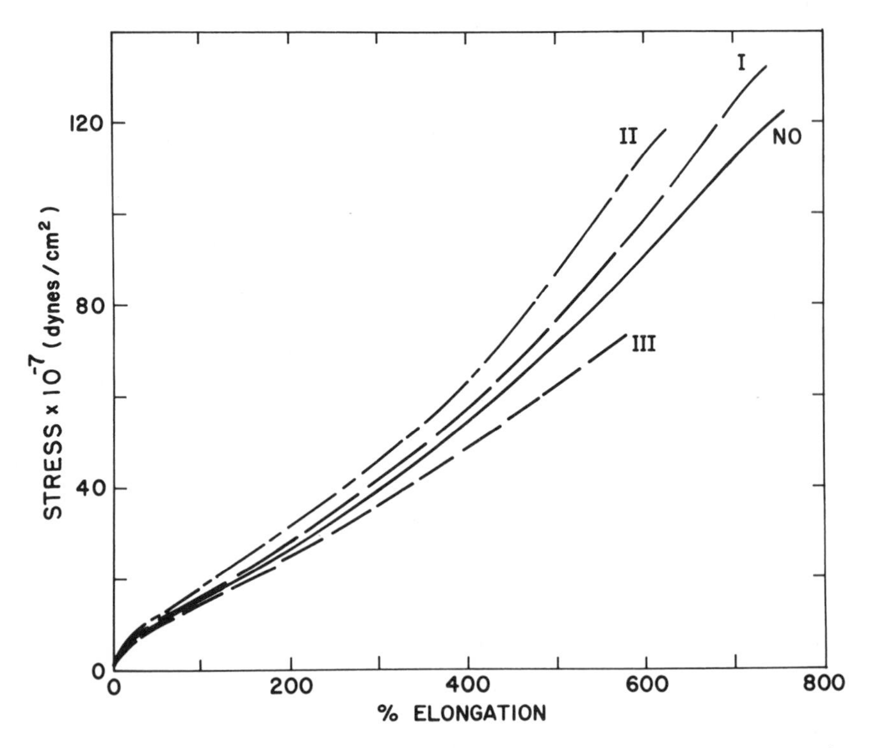

Figure 10. Effect of sample morphology on the stress–strain properties of a PTMO/PTMT elastomeric system containing 44% by wt hard segment

high hysteresis losses has limited their suitability in applications such as high-speed tires.

Studies have been conducted on poly(tetramethylene oxide)–poly-(tetramethylene terephthalate)-segmented copolymers that are identical in all respects except for their crystalline superstructure (*66, 67, 68*). Four types of structures—type I, II, and III spherulites (with their major optical axis at an angle of 45°, 90°, and 0° to the radial direction, respectively), and no spherulitic structure—were produced in one segmented polymer by varying the sample-preparation method. Figures 10 and 11 show the stress–strain and IR dichroism results for these samples, respec-

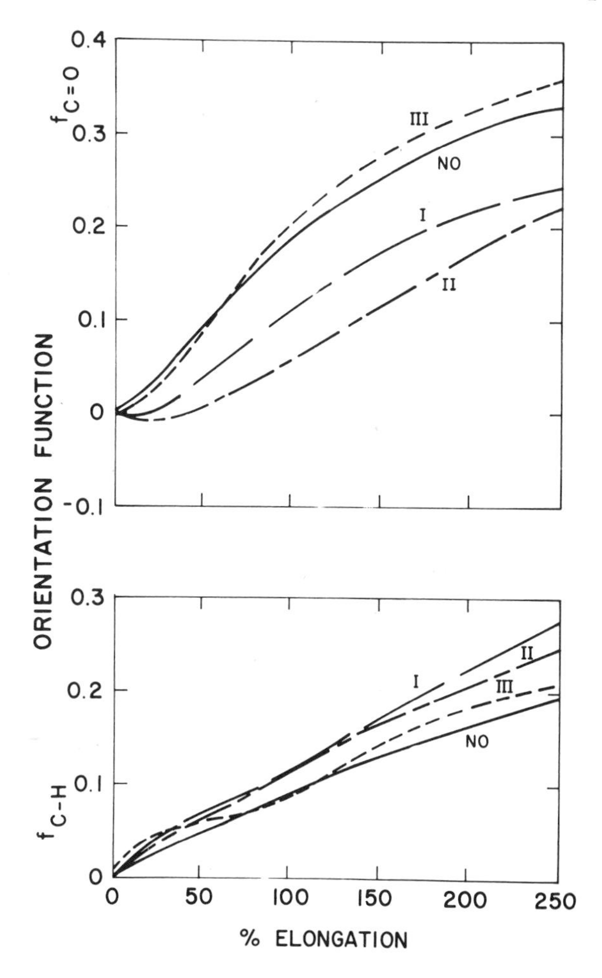

Figure 11. Hard- and soft-segment orientation function vs. elongation curves for the various morphologies exhibited by a PTMO/PTMT segmented copolymer system (44% by wt hard segment)

tively. Stresses past the yield point increased in the order of texures III, "NO," I, and II. The extent of hard-segment orientability increased in the opposite order—II, I, "NO," III. The authors thus suggest that the greater ability of hard segments to resist orientation in the stretch direction results in higher stress levels (*68*).

With regard to ultimate properties, the fracture process can be represented by three steps: initiation of microcracks or cavitation, slow crack propagation, and catastrophic failure (*83, 84, 85, 86, 87*). Dispersed phases tend to interfere with the crack propagation step, redistributing energy that would otherwise cause the cracks to reach catastrophic size. Thus, a two-phase morphology is essential to the achievement of high strength in elastomers. The presence of hard-segment domains increases energy dissipation by hysteresis and other viscoelastic mechanisms. Growing cracks can be deflected and bifurcated at phase boundaries. Upon deformation, triaxial stress fields are formed about hard-phase particles, tending to inhibit the growth of cavities. Cavities which do form can be limited to small sizes, stabilized by surface energy effects. The high modulus hard phase can also relieve stress concentrations by undergoing deformation or internal structural reorganization. At lower temperatures, strength can be raised because of the greater domain yield stresses, increasing matrix viscosity, or strain-induced crystallization effects. The relative importance of these and other reinforcement mechanisms in two-phase polymer systems depends on the type, size, and concentration of the domains or phases.

In segmented polyurethanes, strength is enhanced by long, rigid hard segments with high cohesive energy. Although hydrogen bonding can contribute to domain cohesiveness, hydrogen bonding itself is not directly responsible for high strength.

Figure 12 from Smith (*87*) shows reduced values for the strain, $\lambda_b = 1$, and true stress, $\lambda_b\sigma_b$, at break as a function of temperature for four segmented elastomers (all with poly(tetramethylene oxide) soft segments). Comparing ET-38-1 and ET-38-2, the former is superior in strength and elongation. The smaller, more numerous domains in ET-38-1 are apparently more efficient at stopping catastrophic crack growth than those for ET-38-2 (both have approximately the same hard-segment content), resulting in a larger tensile strength. Furthermore, the more readily deformable, disruptable domains in ET-38-1 permit higher elongation which increases with increasing temperature before fracture. ET-24-2 exhibits a drastic change in properties near 40°C. Below 40°C, the 2000-mol-wt soft segment partially crystallizes and strengthens the specimen (*84*) even though the lower domain concentration as compared with ET-38-1 allows a greater degree of viscous flow. (This effect is reduced at lower temperatures.) Above 40°C, however, the soft segments are no longer crystalline, and the hard-segment domains cannot retard crack

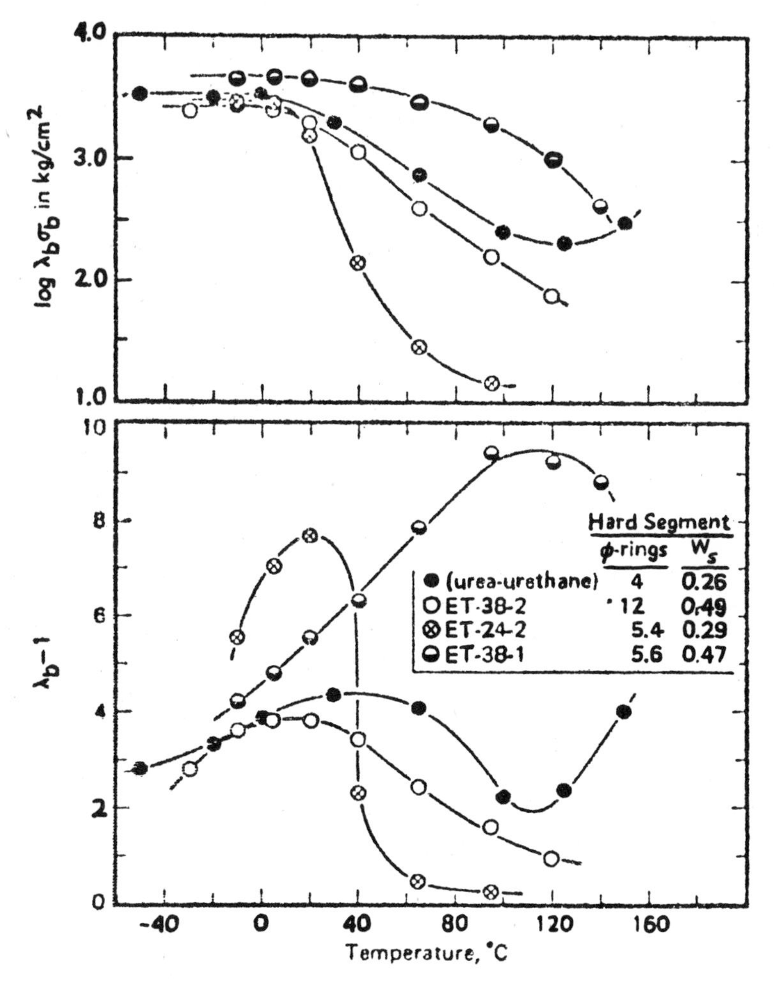

IBM Journal of Research and Development

Figure 12. Plots of log $\lambda_b\sigma_b$ (upper panel) and (λ_b − 1) (lower panel) against temperature. Data were evaluated at an extension rate of 1 min^{-1} and are for poly(urea-urethane) and polyurethane elastomers (87).

growth because of insufficient cohesive strength at these temperatures. If another hard segment characterized by high cohesive energy is used, however, as in the case of the urea–urethane specimen (similar to ET-24-2 except for the nature of the hard segment), higher tensile strengths are observed. The urea–urethane sample is unique in that elongation at break and tensile strength begin to improve above 120°C. Smith (87) offers the explanation that the hard-segment domains become more deformable above this temperature, permitting reorganization into a more fibrous structure than is possible near 100°C.

Kinetics of Microphase Separation

Recent work (*95–100*) has revealed that the morphology of segmented copolymers following thermal treatment is time dependent. As mentioned in the section on thermodynamics, raising the temperature of a polymer system induces phase mixing, as shown in Figure 13, taken from Wilkes and Emerson (*97*). Subsequent cooling causes phase separation giving the original morphology. However, because of kinetic and viscous effects, a finite amount of time is required to produce a given change in morphology.

Wilkes and Emerson (*97*) studied the time-dependent behavior of a polyester polyurethane (MDI-BD based; 40% hard segment) which was heated to 160°C for 5 min, then rapidly quenched to room temperature. To monitor changes in phase separation, SAXS intensity values (at a fixed angle) were recorded as a function of time. Furthermore, the elastic modulus and soft-segment T_g were followed with time. The results, shown in Figure 14, reveal an approximately exponential decay toward equilibrium with a good correlation between properties (T_g and modulus) and structure (inferred by SAXS intensities).

Hesketh, et al. (*99*) have performed similar time-dependent T_g studies on a number of segmented copolymers, only samples were annealed for 4 hr at various temperatures (120°, 150°, 170°, or 190°C). All samples had similar hard-segment contents ($\sim$ 50% by weight) and a poly(tetramethylene oxide) (mol wt 1000) soft segment, and differed only in hard-segment composition. Exponential decay to the steady-state

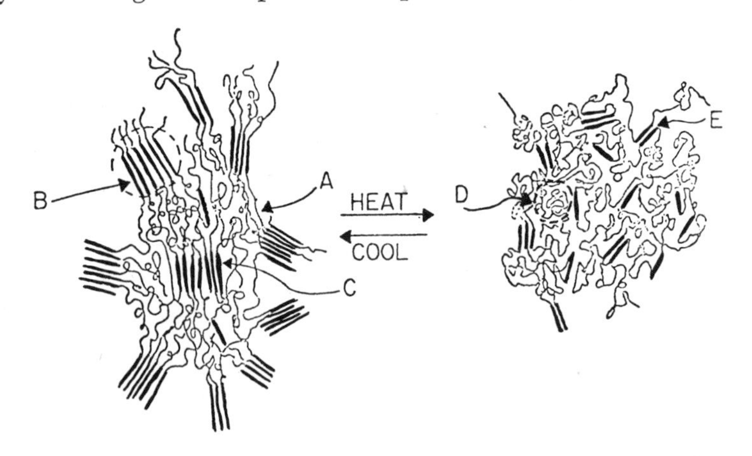

Journal of Applied Physics

Figure 13. Model depicting the morphology at both (a) a long time and (b) following heat treatment. (A) partially extended soft segment; (B) hard segment domain; (C) hard segment; (D) coiled or "relaxed" soft segment; (E) lower-order hard-segment domain (97).

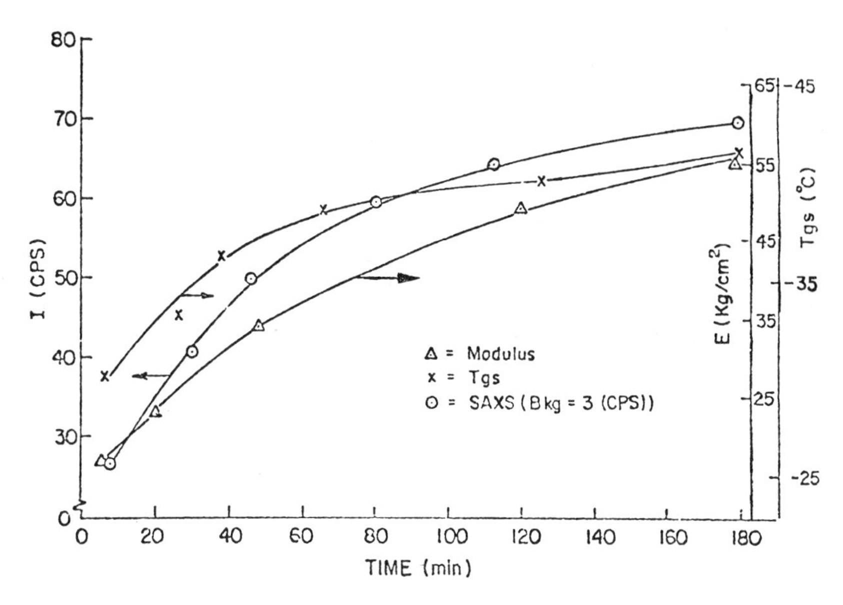

Journal of Applied Physics

Figure 14. Plots of the glass transition temperature of the soft segment phase, T_{gs}, Young's modulus E, and SAXS intensity, I(cps), vs. post-annealing time for a commercial polyester polyurethane (MDI/BD based), R53 (Hooker Chemical Company) (fixed s = 0.042 Å^{-1} for SAXS data) (97).

T_g was observed. The displacement from equilibrium at a given time, however, was greater for higher annealing temperatures. Furthermore, the more compatible the sample was at room temperature, the greater the displacement of T_g from the steady-state value. For example, the least displacement from and quickest return to equilibrium was displayed by a sample with a highly crystalline piperazine–butanediol hard segment.

Ophir, et al. (*100*) also have studied a series of polyester urethanes (poly(tetramethylene adipate) soft segment of 1000 mol wt; MDI-BD-based hard segment) with various degrees of crosslinking using a peroxide (performed at 210°C—in the phase mixed state). For well-aged samples at room temperature, the samples with lower degrees of crosslinking showed better phase separation as revealed by higher intensity SAXS curves. Upon thermal treatment and quench to room temperature, the greater displacements from steady state (as monitored by SAXS intensity at a fixed angle and storage modulus) were exhibited by the more lightly crosslinked specimens. These samples also came to steady state in a much shorter time than the more heavily crosslinked systems. Apparently the crosslinking is a domain-disruptive process (*94, 101*) which lowers rubbery modulus values and inhibits the transient response.

Acknowledgment

The authors wish to acknowledge the support of their research on segmented copolymers by the National Science Foundation Division of Materials Research Grant DMR 76-20085, the U.S. Army Research Office Materials and Metallurgy Division Grant DAA G29-76-G0041, and the donors of the Petroleum Research Fund administered by the American Chemical Society.

Literature Cited

1. Moacanin, J., Holden, G., Tschoegl, N. W., Eds., "Block Copolymers," Interscience, New York, 1969.
2. Aggarwal, S. L., Ed., "Block Polymers," Plenum, New York, 1970.
3. Molau, G. E., Ed., "Colloidal and Morphological Behavior of Block and Graft Copolymers," Plenum, New York, 1971.
4. Burke, J. J., Weiss, V., Eds., "Block and Graft Copolymers," Syracuse University, Syracuse, 1973.
5. Allport, D. C., Janes, W. H., Eds., "Block Copolymers," Wiley, New York, 1973.
6. Estes, G. M., Cooper, S. L., Tobolsky, A. V., *J. Macromol. Sci., Rev. Macromol. Chem.* (1970) **4**, 313.
7. Noshay, A., McGrath, J. E., Eds., "Block Copolymers, Overview and Critical Survey," Academic, New York, 1977.
8. Aggarwal, S. L., *Polymer* (1976) **17**(11), 938.
9. Saunders, J. H., Frisch, K. C., "Polyurethanes, Chemistry and Technology, Part I, Chemistry," Interscience, New York, 1962.
10. Schollenberger, C. S., U.S. Patent **2,871,218,** 1955.
11. Carvey, R. M., Witenhafer, D. E., Br. Patent **1,087,743,** 1965.
12. Peebles, L. H., Jr., *Macromolecules* (1974) **7,** 872.
13. Ibid. (1976) **9,** 58.
14. Harrell, L. L., Jr., *Macromolecules* (1969) **2,** 607.
15. Witsieppe, W. K., ADV. CHEM. SER. (1973) **129,** 39.
16. Hoeschele, G. K., Witsieppe, W. K., *Agnew. Makromol. Chem.* (1973) **29/30,** 267.
17. Hoeschele, G. K., *Polym. Eng. Sci.* (1974) **14,** 848.
18. Wolfe, J. R., Jr., *Rubber Chem. Technol.* (1977) **50,** 688.
19. Wolfe, J. R., Jr., *Polym. Prepr., Am. Chem. Soc., Div. Polym. Chem.* (1978) **19**(1), 5.
20. Krause, S., *J. Polym. Sci., Part A-2* (1969) **7,** 249.
21. Krause, S., *Macromolecules* (1970) **3,** 84.
22. Krause, S., "Colloidal and Morphological Behavior of Block and Graft Copolymers," G. E. Molau, Ed., p. 223, Plenum, New York, 1971.
23. Krause, S., "Block and Graft Copolymers," J. J. Burke, V. Weiss, Eds., p. 143, Syracuse University, Syracuse, 1973.
24. Krause, S., Reismiller, P. A., *J. Polym. Sci., Part A-2* (1975) **13,** 1975.
25. Helfand, E., *Polym. Prepr., Am. Chem. Soc., Div. Polym. Chem.* (1974) **15**(2), 246.
26. Helfand, E., *Polym. Sci. Technol.* (1974) **4,** 141.
27. Helfand, E., *Macromolecules* (1975) **8,** 552.
28. Helfand, E., *Acc. Chem. Res.* (1975) **8,** 295.
29. Helfand, E., Wasserman, Z., *Polym. Eng. Sci.* (1977) **17,** 582.
30. Meier, D. J., "Block Copolymers," J. Moacanin, G. Holden, N. W. Tschoegl, p. 81, Interscience, New York, 1969.

31. Meier, D. J., *Polym. Prepr., Am. Chem. Soc., Div. Polym. Chem.* (1970) **11**, 400.
32. Meier, D. J., "Block and Graft Copolymers," J. J. Burke, V. Weiss, Eds., p. 105, Syracuse University, Syracuse, 1973.
33. Meier, D. J., *Polym. Prepr., Am. Chem. Soc., Div. Polym. Chem.* (1973) **14**, 280.
34. Ibid. (1974) **15**(1), 171.
35. Meier, D. J., *J. Appl. Polym. Symp.* (1974) **24**, 67.
36. LeGrand, D. G., *Polym. Prepr., Am. Chem. Soc., Div. Polym. Chem.* (1970) **11**, 434.
37. Guinier, A., Fournet, G., "Small Angle Scattering of X-rays," Wiley, New York, 1955.
38. Vonk, C. G., *J. Appl. Crystallogr.* (1973) **6**, 81.
39. Hashimoto, T., Todo, A., Itoi, H., Kawai, H., *Macromolecules* (1977) **10**, 377.
40. Koutsky, J. A., Hien, N. V., Cooper, S. L., *J. Polym. Sci., Part B* (1970) **8**, 353.
41. Seymour, R. W., Estes, G. M., Cooper, S. L., *Macromolecules* (1970) **3**, 579.
42. Tanaka, T., Yokoyama, T., Yamaguchi, Y., *J. Polym. Sci., Part A-1* (1968) **6**, 2137.
43. Srichatrapimuk, V., M.S. Thesis, Department of Chemical Engineering, University of Wisconsin, 1976.
44. Clough, S. B., Schneider, N. S., King, A. O., *J. Macromol. Sci., Phys.* (1968) **B2**(4), 641.
45. Bonart, R., *J. Macromol. Sci., Phys.* (1968) **B2**(1), 115.
46. Bonart, R., Morbitzer, L., Hentze, G., *J. Macromol. Sci., Phys.* (1969) **B3**(2), 337.
47. Bonart, R., Morbitzer, L., Müller, E. H., *J. Macromol. Sci., Phys.* (1974) **B9**(3), 447.
48. Bonart, R., Müller, E. H., *J. Macromol. Sci., Phys.* (1974) **B10**(1), 177.
49. Ibid., **B10**(2), 345.
50. Wilkes, C. E., Yusek, C. S., *J. Macromol. Sci., Phys.* (1973) **B7**(1), 157.
51. Chang, Y. J. P., Wilkes, Garth, *J. Polym. Sci., Polym. Phys. Ed.* (1975) **13**, 455.
52. Schneider, N. S., Desper, C. R., Illinger, J. L., King, A. O., *J. Macromol. Sci., Phys.* (1975) **B11**(4), 527.
53. Van Bogart, J. W. C., West, J. C., Cooper, S. L., *Am. Chem. Soc., Div. Org. Coat. Plast. Chem., Pap.* (1977) **37**(2), 503.
54. Helfand, E., Sapse, A. M., *J. Chem. Phys.* (1975) **62**(4), 1327.
55. Assink, R. A., *J. Polym. Sci.* (1977) **15**, 59.
56. North, A. M., *J. Polym. Sci., Part C* (1975) **50**, 345.
57. North, A. M., Reid, J. C., Shortall, J. B., *Eur. Polym. J.* (1969) **5**, 565.
58. North, A. M., Reid, J. C., *Eur. Polym. J.* (1972) **8**, 1129.
59. Dev, S. B., North, A. M., Reid, J. C., in "Dielectric Properties of Polymers," F. E. Karaz, Ed., Plenum, New York, 1972.
60. Samuels, S. L., Wilkes, G. L., *J. Polym. Sci., Part C* (1973) **43**, 149.
61. Allegrezza, A. E., Jr., Seymour, R. W., Ng, H. N., Cooper, S. L., *Polymer* (1974) **15**, 433.
62. Kimura, I., Ighihara, H., Ono, H., Yoshihara, N., Nomura, S., Kawai, H., *Macromolecules* (1974) **7**, 355.
63. Chang, A. L., Thomas, E. L., *Polym. Prepr., Am. Chem. Soc., Div. Polym. Chem.* (1978) **19**(1), 32.
64. Shen, M., Mehra, U., Nieinomi, M., Koberstein, J. K., Cooper, S. L., *J. Appl. Phys.* (1974) **45**(10), 4182.
65. Seymour, R. W., Overton, J. R., Corley, L. S., *Macromolecules* (1975) **8**, 331.

66. Lilaonitkul, A., West, J. C., Cooper, S. L., *J. Macromol. Sci., Phys.* (1976) **B12**(4), 563.
67. Lilaonitkul, A., Cooper, S. L., *Rubber Chem. Technol.* (1977) **50,** 1.
68. Lilaonitkul, A., Estes, G. M., Cooper, S. L., *Polym. Prepr., Am. Chem. Soc., Div. Polym. Chem.* (1977) **18**(2), 500.
69. Aitken, R. R., Jeffs, G. M. F., *Polymer* (1977) **18,** 197.
70. Schneider, N. S., Paik Sung, C. S., *Polym. Eng. Sci.* (1977) **17**(2), 73.
71. Schneider, N. S., Paik Sung, C. S., Matton, R. W., Illinger, J. L., *Macromolecules* (1975) **8,** 62.
72. Clough, S. B., Schneider, N. S., *J. Macromol. Sci., Phys.* (1968) **B2,** 553.
73. Miller, G. W., Saunders, J. H., *J. Polym. Sci.* (1970) **A1**(8), 1923.
74. Vrouenraets, C. M. F., *Polym. Prepr., Am. Chem. Soc., Div. Polym. Chem.* (1972) **13**(1), 529.
75. Schollenberger, C. S., Hewitt, L. E., *Polym. Prepr., Am. Chem. Soc., Div. Polym. Chem.* (1978) **19**(1), 17.
76. Seymour, R. W., Cooper, S. L., *Macromolecules* (1973) **6,** 48.
77. Paik Sung, C. S., Schneider, N. S., *Macromolecules* (1975) **8**(1), 68.
78. Paik Sung, C. S., Schneider, N. S., *Macromolecules* (1977) **10**(2), 452.
79. Nielsen, L., "Mechanical Properties of Polymers," Reinhold, New York, 1961.
80. Cooper, S. L., Tobolsky, A. V., *J. Macromol. Sci.* (1970) **B4**(4), 877.
81. Huh, D. S., Cooper, S. L., *Polym. Eng. Sci.* (1971) **11**(5), 369.
82. Ng, H. N., Allegrezza, A. E., Seymour, R. W., Cooper, S. L., *Polymer* (1973) **14,** 255.
83. Smith, T. L., *Polym. Prepr., Am. Chem. Soc., Div. Polym. Chem.* (1974) **15,** 58.
84. Smith, T. L., *J. Polym. Sci., Phys.* (1974) **12,** 1825.
85. Smith, T. L., *Polym. Prepr., Am. Chem. Soc., Div. Polym. Chem.* (1976) **17,** 118.
86. Smith, T. L., *Polym. Eng. Sci.* (1977) **17**(3), 129.
87. Smith, T. L., *IBM J. Res. Dev.* (1977) **21**(2), 154.
88. Guth, E., *J. Appl. Phys.* (1945) **16,** 20.
89. Aggarwal, S. L., Livigni, R. A., Marker, L. F., Dudek, T. J., "Block and Graft Copolymers," J. J. Burke, V. Weiss, Eds., p. 157, Syracuse University, Syracuse, 1973.
90. Nielsen, L. E., *Rheol. Acta* (1974) **13,** 86.
91. Seymour, R. W., Allegrezza, A. E., Cooper, S. L., *Macromolecules* (1973) **6,** 896.
92. Seymour, R. W., Cooper, S. L., *Rubber Chem. Technol.* (1974) **47,** 19.
93. Trick, G. S., *J. Appl. Polym. Sci.* (1960) **3,** 252.
94. Cooper, S. L., Huh, D. S., Morris, W. J., *Ind. Eng. Chem., Prod. Res. Dev.* (1968) **7,** 248.
95. Wilkes, G. L., Bagrodia, S., Humphries, W., Wildnauer, R., *J. Polym. Sci., Polym. Lett. Ed.* (1975) **13,** 321.
96. Wilkes, G. L., Wildnauer, R., *J. Appl. Phys.* (1975) **46,** 4148.
97. Wilkes, G. L., Emerson, J. A., *J. Appl. Phys.* (1976) **47,** 4261.
98. Assink, R. A., Wilkes, G. L., *Polym. Eng. Sci.* (1977) **17,** 603.
99. Hesketh, T. R., Cooper, S. L., *Am. Chem. Soc., Div. Org. Coat. Plast. Chem., Pap.* (1977) **37**(2), 509.
100. Ophir, Z. H., Wilkes, G. L., *Polym. Prepr., Am. Chem. Soc., Div. Polym. Chem.* (1978) **19**(1), 26.
101. Cooper, S. L., Tobolsky, A. V., *J. Appl. Polym. Sci.* (1967) **11,** 1361.

RECEIVED June 5, 1978.

2

Morphological Studies of PCP/MDI/BDO-Based Segmented Polyurethanes

A. L. CHANG[1] and E. L. THOMAS[1]

Department of Chemical Engineering, University of Minnesota, Minneapolis, MN 55455

Polyurethane samples based on polycaprolactonediol/4,4′-diphenylmethane diisocyanate/1,4-butanediol from 1/2/1 to 1/6/5 (PCP/MDI/BDO) mole ratio were phase separated by DSC, WAXS, electron diffraction, and TEM. Soft [(PCP/MDI)$_i$] and hard [(MDI/BDO)$_j$] segment sequences can crystallize. As-reacted samples contained a more ordered, hard-segment phase while solution-cast samples contain a more ordered, soft-segment phase. The observed increase in melting point and decrease in line-width of the hard segment, WAXS reflections suggest an increase of domain size with increased hard-segment content. The fractional degree of crystallinity of the hard-segment phase is approximately constant. All solvent (DMF) cast samples except sample 1/2/1 contained a spherulitic superstructure. Bright field, defocus micrographs of solution-cast films of samples 1/5/4 and 1/6/5 exhibit a 400 A scale granularity suggestive of hard-segment-rich domains.

The interesting elastomeric properties of polyurethanes are currently attributed to the formation of microdomains (*1, 2, 3, 4*). Thermoplastic polyurethane elastomers are multiblock copolymers consisting of short, immobile, polyurethane sequences ("hard" segments) connected via long and flexible chains ("soft" segments). A variety of aliphatic and aromatic diisocyanates with diol or diamine chain extenders have been used for the hard segment with typically 1,000–3,000 molecular weight polyether or polyester polyols for the soft segment. During polymerization

[1] Current address: Department of Polymer Science and Engineering, University of Massachusetts, Amherst, MA 01002.

0-8412-0457-8/79/33-176-031$05.50/0

and solidification, the hard and soft portions of the multiblock polyurethane chain undergo microphase separation into hard and soft domains. The strong polar bonding of the hard segments causes the hard domains to act as pseudocrosslinks for the flexible soft-segment phase. Mechanical properties can then be conveniently tailored by varying the ratio of hard-to-soft phase.

The substantial work on polystyrene/polybutadiene and polystyrene/polyisoprene blends and diblock and triblock copolymer systems has lead to a general understanding of the nature of phase separation in regular block copolymer systems (*5, 6*). The additional complexities of multiblocks with variable block length as well as possible hard- and/or soft-phase crystallinity makes the morphological characterization of polyurethane systems a challenge.

In this chapter we investigate the morphology of a series of polyurethanes based on polycaprolactone polyol (PCP), diphenylmethane diisocyanate (MDI), and butanediol (BDO). Samples of as-batch-reacted and solution-cast polymers were examined by optical microscopy, transmission electron microscopy, electron and x-ray diffraction, and differential scanning calorimetry. Our interest is to provide a mapping of the size and shape of the domains (and any superstructure such as spherulites) and the degree of order as a function of the fraction of each phase present.

Chain regularity and block length as well as thermal history during and after polymerization all play important roles in determining the degree of phase separation as well as the degree of order of the soft- and hard-segment domains. Better phase separation is favored for nonpolar soft-segment systems and with longer sequence lengths of the respective hard/soft segments.

Both soft and hard domains in polyurethanes can be amorphous or partially crystalline, depending on the particular system. For partially crystalline systems, the ordered hard-segment phase is thought to be composed of fringed lamellae domains of thickness (dimension parallel to chain axis) equal to the hard-segment length and lateral width (dimensions normal to chain axis) of less than a few hundred Angstroms (*3*). Occasionally there is a spherulitic superstructure (*3, 6, 7*). Although the block length and volume fraction of each type of segment will influence the overall domain morphology, both bicontinuous (*8*) and discrete-isolated (*3, 4, 5*) domain morphologies have been proposed.

Providing the fact that a block is regular, block length is important in determining crystallinity. For polyester-based polyurethanes, Seefried et al. (*9*) have shown that a polycaprolactonediol with $\overline{M}_n \geq 3000$ was necessary for soft-segment crystallinity. Hard-segment crystallization can occur for much shorter block lengths. Harrell (*10*) prepared a systematic series of polymers with monodisperse hard-segment sizes and showed that crystallization occurred even for chains containing only a single, hard-segment repeat unit (it should be noted that Harrell's polymer has no hydrogen bonding). Interestingly, he reported that increasing the hard-segment block length did not appreciably affect the degree of order within the hard domains, though the melting point of the hard-segment phase was raised. Wilkes (*7, 11, 12*) and co-workers have characterized the morphology of these specially prepared polymers and have found both spherulitic superstructure and domain structures. Domain size apparently increased with increase in hard-segment content as evidenced by sharpening of the wide-angle x-ray (WAXS) reflections (*12*).

There are relatively few direct transmission, electron microscopy (TEM) studies of domain structures in polyurethanes (*3–7, 13, 14, 15*). No distinct micelle-lamellae platelets have been observed in the urethane systems thus studied. Instead, the domain structures which have been observed generally appear as isolated equiaxed grains 30–500 Å in diameter (*4, 13, 14, 15*). Randomly oriented fibrils with lateral dimensions of 300–600 Å have been observed in a polyether/MDI/BDO system (*3, 6*). It remains to be seen what relation the equiaxed grains (domains) and the fibrils observed by transmission electron microscopy have with the micelle-lamellae structures inferred from WAXS and SAXS (*12, 16*).

Bonart has proposed two-dimensional and three-dimensional models of MDI/BDO hard-segment (para) crystals (*17, 18, 19*). Arrangements were constructed that provide optimum hydrogen bonding. The 7.9 Å paracrystalline, hard-segment reflection is thought to arise from the hydrogen bond containing planes inclined 30° to the chain axis. Quite high temperatures and long annealing times (190°C, 12 hr) are reportedly required for significant hard-segment (MDI/BDO) crystallinity (*2*).

Several SAXS studies on MDI/BDO polyurethanes have shown a discrete small-angle maximum in the 200 Å range (*3, 12, 16, 20*). This maximum has been attributed to the average center-to-center spacing of the hard-segment domains. One would expect a systematic variation in the position of this maximum with composition, but this has not always been observed (*3, 16*). The influence of soft-segment hydrogen bond ability and hard-segment block size on the phase separation have been clearly shown in SAXS studies. For the same molecular weight polyol ($\overline{M}_n = 1000$), a 1/2/1 polyester/MDI/BDO system was single phased (i.e., compatible) whereas the 1/2/1 polyether/MDI/BDO system was phase separated (*20*).

In addition to microphase structures, MDI/BDO-based polyurethane systems have exhibited spherulitic superstructure. Characterization of the birefringence of the spherulites was used to determine the orientation of the hard-segment domains (*7*). However, because of the sensitivity of the optical anisotropy to the exact conformation of the aromatic rings as well as to the chain orientation, birefringence alone cannot determine hard-segment orientation within the spherulite (3).

The dynamic mechanical properties (torsion pendulum) of the present PCP/MDI/BDO polyurethane system have been studied both as a function of polyol molecular weight and as a function of hard-segment concentration. A series of polyurethanes with PCP/MDI/BDO mole ratio of 1/2/1 with variable $\overline{M}_n$ of the soft-segment polyol ($\overline{M}_n$ = 340, 530, 830, 1250, 2100, and 3130) exhibited a single compositional dependent T_g consistent with a compatible (noncrystalline) system for polyols of $\overline{M}_n$ of 340–2100 (e.g., 37–13 wt % hard segment) (9). For the 3130-$\overline{M}_n$ polyol, soft-segment crystallization occurred indicating a two-phase morphology. The dynamic mechanical properties for two different molecular weight polyols (830 and 2100 $\overline{M}_n$) were also studied as a function of hard-segment concentration (*21*). The lower T_g of the 830-$\overline{M}_n$ polyol system was very sensitive to the hard-segment content whereas the lower T_g of the 2100 $\overline{M}_n$-polyol system did not significantly

Table I.

Mol Ratio PCP/MDI/BDO	*Wt % MDI*	*Wt Fraction Hard Segment (%)*	T_g[a]	*52°C DMF Spherulite Size (μm)*	*As-Reacted WAXS*[b]	*52°C DMF Electron Diffraction*[b]
1/0/0	0	0	—64°C	50	XL–S	XL–S
1/1/0	11	0	—	30	XL–S	XL–S
1/2/1	19	13	−40	None	A—4.4	XL–S*
1/3/2	26	23	−40	< .5	4.6, A—4.4	XL–S*, A—4.4 (weak)
1/4/3	31	31	−32	5	4.6, 4.1, 3.75, A—4.4	XL–S*, A—4.4
1/5/4	35	38	−30	10–30	4.6, 4.1, 3.75, A—4.4	XL–S*, A—4.4
1/6/5	38	43	−30	10–30	4.6, 4.1, 3.75, A—4.4	XL–S*, A—4.4
0/1/1	74	100	+125°C	50	A—8.4, 4.9, 4.6, 4.1, 3.75	4.6, 3.9

[a] From Ref. *21* for a PCP diol of $\overline{M}_n = 2100$.

[b] The random copolymer samples are based on a PCP diol of $\overline{M}_n = 2{,}000$; XL–S: crystalline PCL reflections; XL–S*: crystalline soft segment in limited regions of sample; A–X.X: diffuse halo centered at X.X Å.

increase over the same range of hard-segment concentration (*see* Table I). This would indicate that the relative degree of phase separation was much greater for the higher molecular weight polyol system.

Experimental

The polyurethane samples were kindly supplied by F. E. Critchfield of Union Carbide Corporation. The samples were made by a one-step batch process with curing at 145°C for 16 hr. Details of the polyurethane polymerization are described elsewhere (9). The hard segment consists of 4,4′-diphenylmethane diisocyanate (MDI) and 1,4-butanediol (BDO); $(MDI/BDO)_i$, and the soft segment consists of MDI and polycaprolactone diol (PCP) with $M_n = 2000$; $(PCP/MDI)_j$.

Transmission electron micrographs and diffraction patterns were obtained using a JEOL 100CX electron microscope operated at 100 KeV. Wide-angle x-ray patterns were taken with a flat film camera using nickel-filtered copper radiation with a Phillips–Norelco generator operated at 30 KV and 20 mA. The DSC experiments were carried out using a Perkin–Elmer DSC II at a heating rate of 20°C/min. Sample size was approximately 10 mg. Calibration was done with an In standard .

Eight different polymers were studied: polycaprolactone homopolymer (PCL); the two regular pure segment copolymers, PCP/MDI and MDI/BDO, and five random copolymers with PCP/MDI/BDO mole ratios varying systematically from 1/2/1 to 1/6/5 (*see* Table I). Sample opacity increases strongly with increased hard-segment content for the five random copolymer samples. Both regular pure segment copolymers are opaque.

Two types of samples were prepared for detailed morphological examination. Sections of the as-polymerized material were examined directly by WAXS and DSC. Samples also were prepared by casting films from a solution of the polymer in dimethyl formamide (DMF) at 52°C. Solvent was allowed to slowly evaporate, and the films were dried by annealing for 670 hr at 52°C, followed by slow cooling to room temperature. Thin films suitable for electron microscopy were cast from 0.5 wt % polymer in DMF onto clean glass slides, and after 20 hr of annealing at 52°C, floated off on distilled water and mounted on 300-mesh copper grids.

Experimental Results

Differential Scanning Calorimetry. Figures 1 and 2 show 320–520 K DSC scans of the as-reacted polymer and the DMF solution cast samples at heating rate of 20°C/min. Two broad endothermic transition regions are observed—a very broad and weak transition near 350°C and a narrower, stronger transition below 500°C.

For the as-reacted polymer, the high temperature peak (circa 490 K) shifts upwards by about 10°C as the hard-segment concentration increases from 13–43%. Scans through the PCL melting range indicate

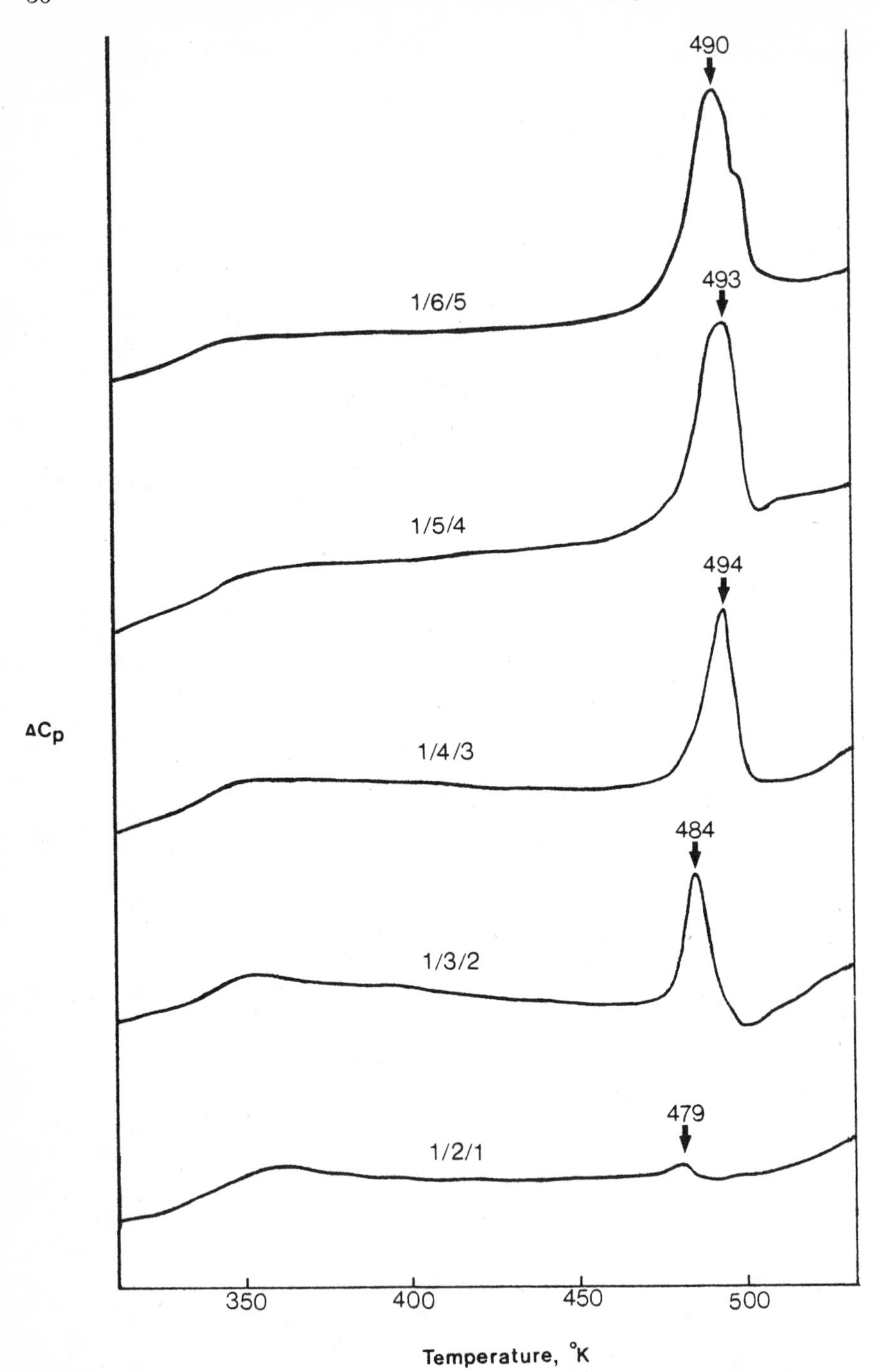

Figure 1. DSC scans at heating rate of 20°C/min of the as-reacted PCP/MDI/BDO samples over the 320–520 K range

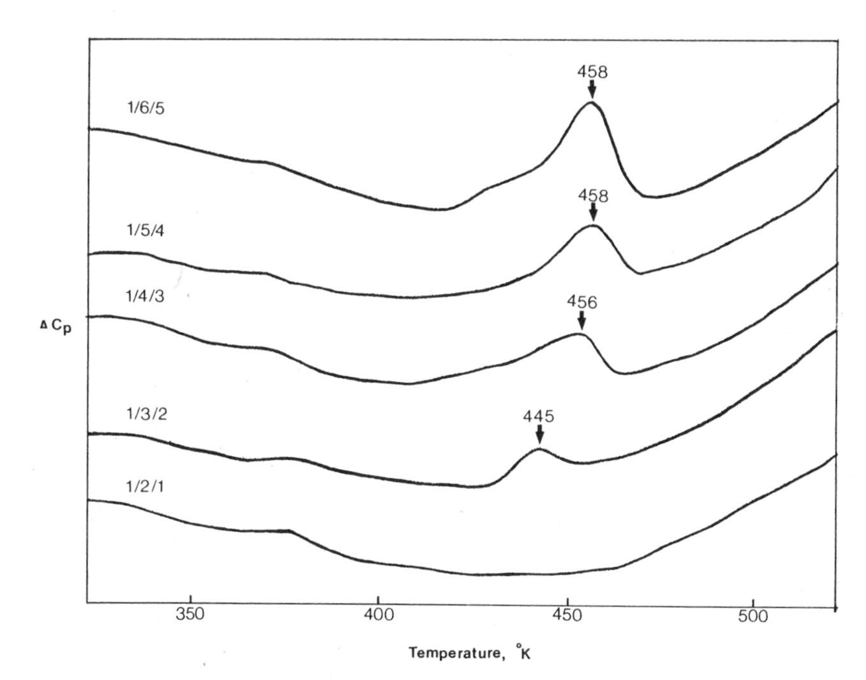

Figure 2. DSC scans at heating rate of 20°C/min of DMF solution cast PCP/MDI/BDO samples over the 320–520 K range

no significant soft-segment crystallinity is present. For the solvent-cast samples, the high temperature peak (circa 455 K) also shifts upwards with increasing hard-segment content by about 10°C. The solvent cast 1/2/1 sample has no high temperature peak but does show a small soft-segment melting endotherm at about 40°C for a scanning rate of 5°C/min. The heat of fusion of the high temperature transition increases with hard-segment content for both types of samples, with the respective DMF cast samples having much lower values than the as-reacted samples (*see* Table II).

Table II.

Mol Ratio PCP/MDI/ BDO	*Wt Fraction Hard Segment*	*As-Reacted Polymer*		*Solution Cast Polymer*	
		Δh_f *cal/g*	T_m *(K)*	Δh_f *cal/g*	T_m *(K)*
1/2/1	13	0.2	479	0	—
1/3/2	23	1.7	484	0.4	445
1/4/3	31	2.8	494	1.1	456
1/5/4	38	4.2	493	1.2	458
1/6/5	43	5.2	490	2.6	458

Diffraction. Wide-angle x-ray scattering patterns of the as-reacted polymer indicate microphase separation is occurring by the appearance of crystalline hard-segment reflections (*see* Table I and Figure 3) (*2, 3*). These reflections become sharper and stronger with increasing hard-segment content. Sample 1/2/1 with only 13% hard segment shows only one diffuse halo centered at 4.4 Å. There is no x-ray evidence of soft-segment crystallinity in any of the as-reacted random copolymer samples. Sample 1/1/0 (pure soft segment) exhibits sharp crystalline reflections which index well with the known crystal structure of PCL (*22*).

Electron diffraction of solution cast films annealed at 52°C (which is just below the melting point of PCL homopolymer) shows soft-segment

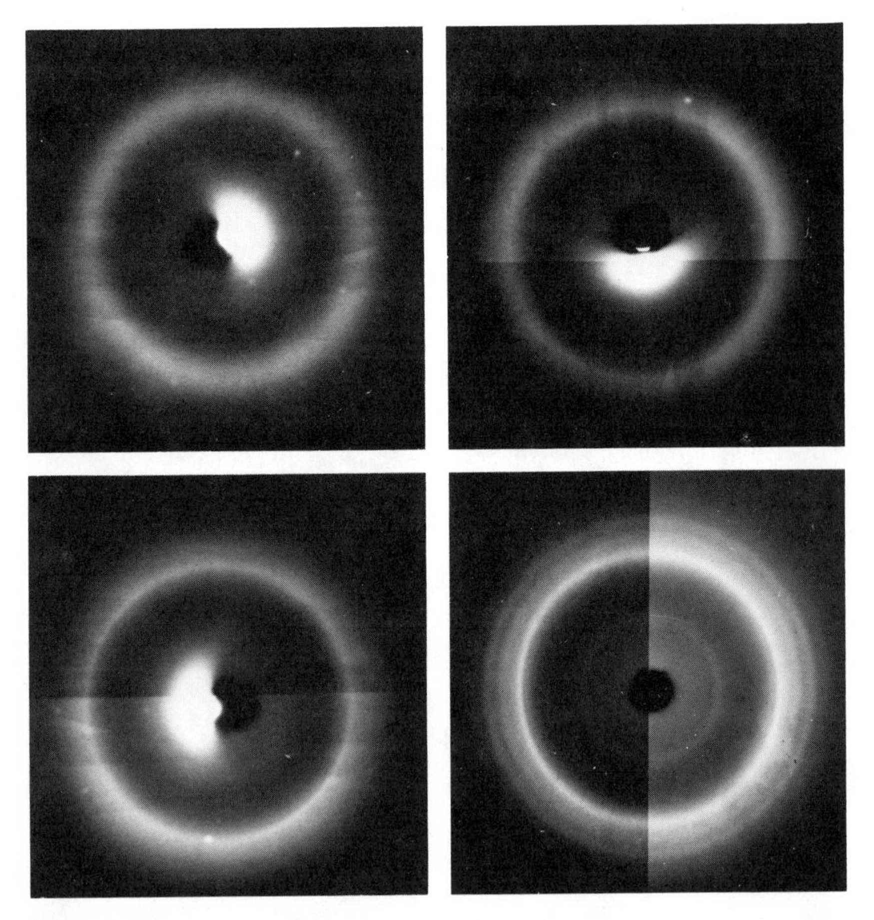

Figure 3. WAXS patterns of the as-reacted PCP/MDI/BDO polymers: (upper left) *1/2/1,* (upper right) *1/3/2,* (lower left) *1/4/3, and* (lower right) *1/6/5. All samples except upper left exhibit hard-segment crystalline reflections.*

crystallinity (*see* Figure 4). The amount of soft-segment crystallinity increases with the weight fraction of soft segment, but these crystalline soft-segment regions appear discretely (and randomly) throughout the film. It has not been possible to pin down their location within the spherulites.

No distinct hard-segment reflections are visible in solution-cast samples of the 1/2/1–1/6/5 random copolymers, instead a strong halo centered at 4.4 Å is observed inside the 4.1- and 3.7-Å PCL reflections. The pure, hard-segment sample has two broad rings centered at 4.6 and 3.9 Å.

Microscopy. The morphologies of the pure, soft-segment copolymer (PCP/MDI) and pure, hard-segment copolymer (MDI/BDO) are

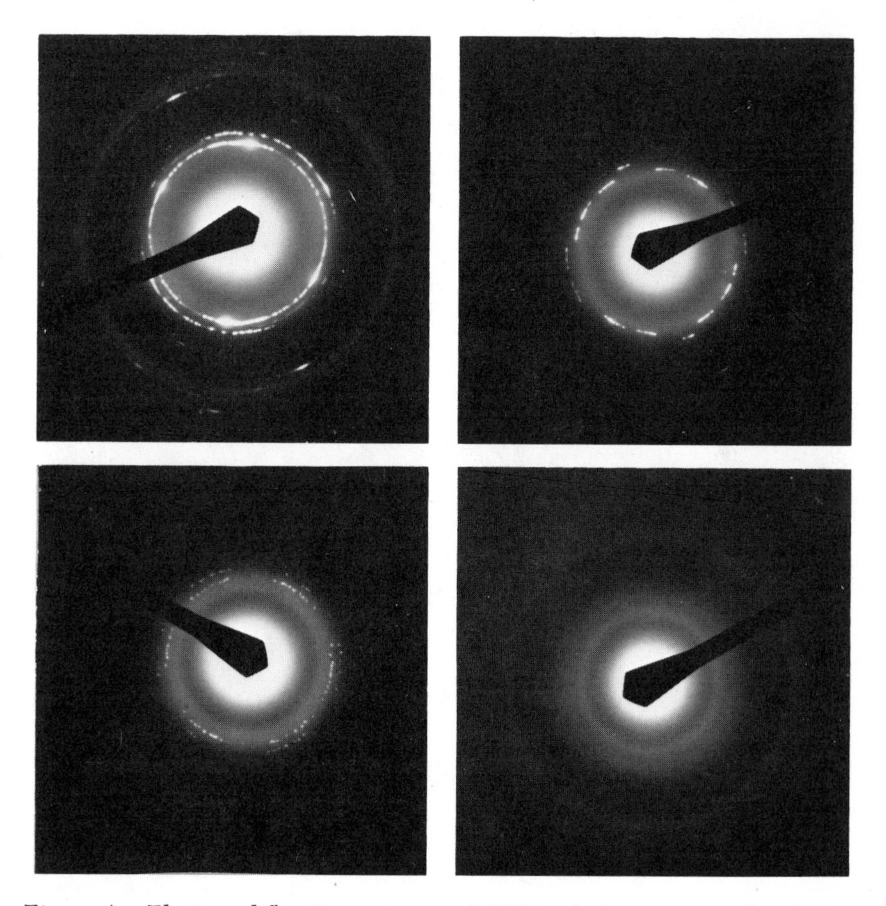

Figure 4. Electron diffraction patterns of DMF solution cast samples: (upper left) *1/3/2,* (upper right) *1/5/4,* (lower left) *1/6/5, and* (lower right) *0/1/1 (pure hard segment)*

shown in Figure 5. The observed morphology depends on the sample preparation technique, melt cast films of (PCP/MDI) and (MDI/BDO) yielding very large (50-μm diameter) nonbanded, positively birefringent (0°–90°) spherulites. Solution-cast films sometimes exhibit

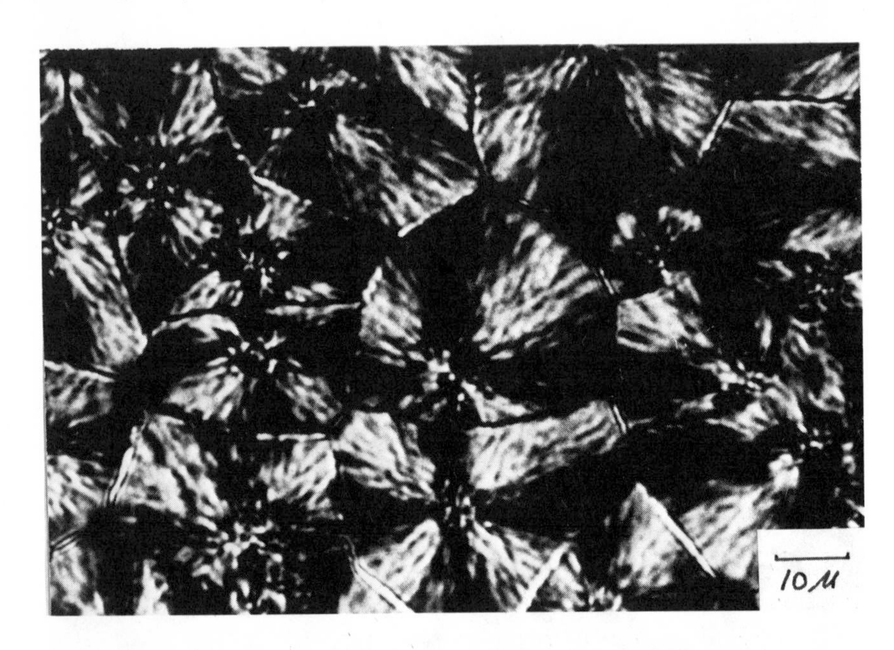

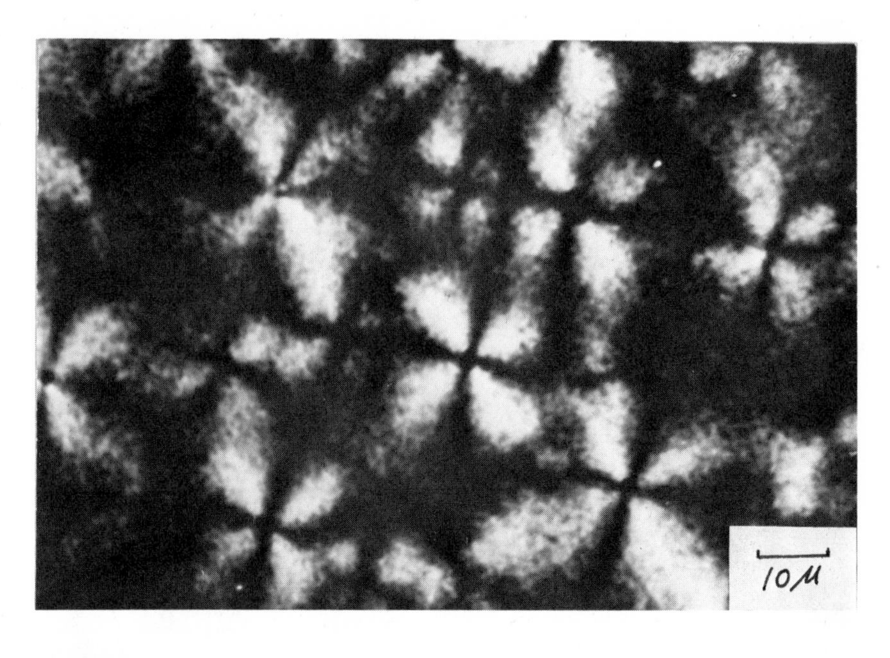

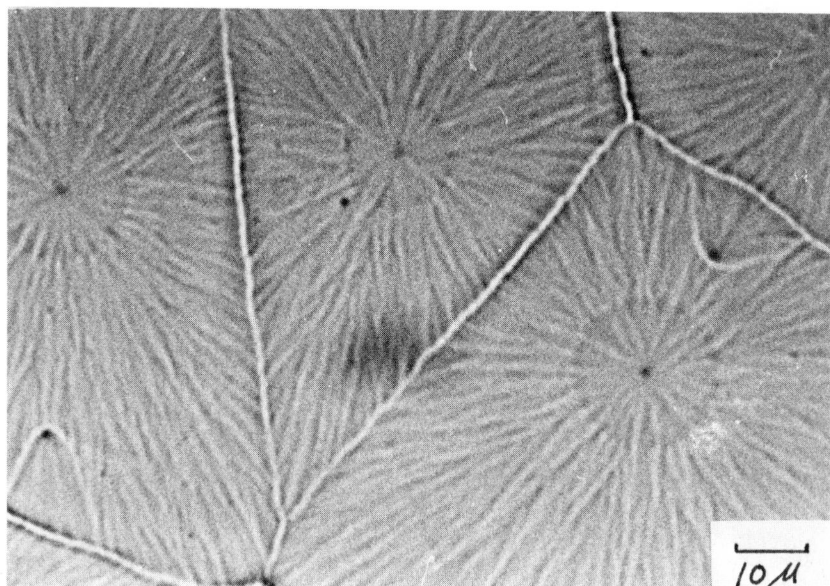

Figure 5. Optical micrographs of pure soft- and pure hard-segment copolymers: (upper left) *compression-molded soft segment,* (lower left) *DMF solution-cast soft segment,* (upper right) *compression-molded hard segment,* (lower right) *DMF solution-cast hard segment. All micrographs are for crossed polarizers except lower right.*

5μ

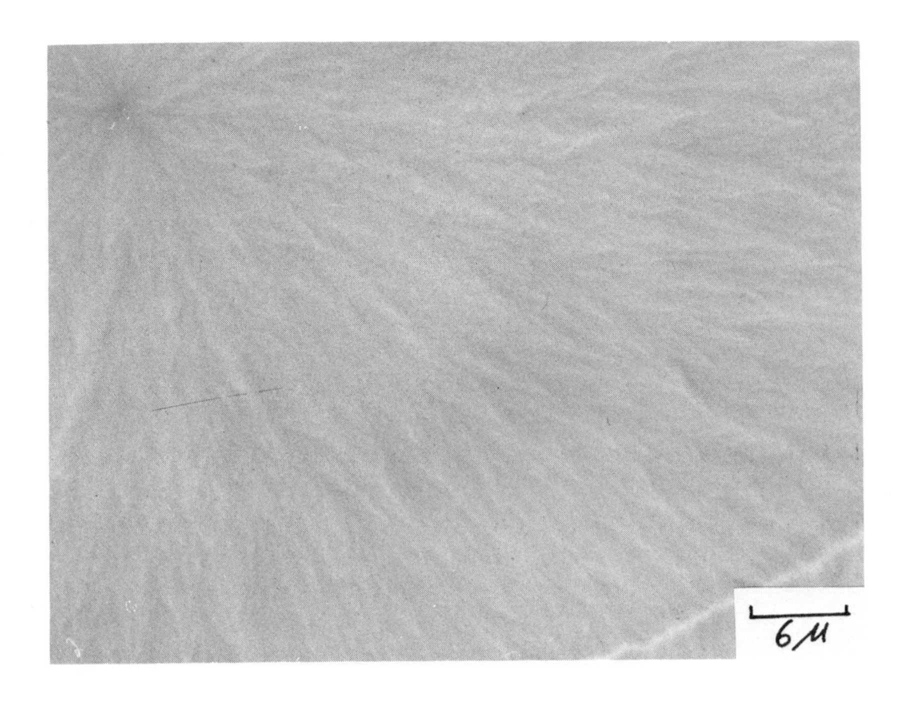
6μ

Figure 6. Transmission electron micrographs of solution-cast films of: (upper left) *pure soft-segment copolymer and* (lower left, above) *pure hard-segment copolymer*

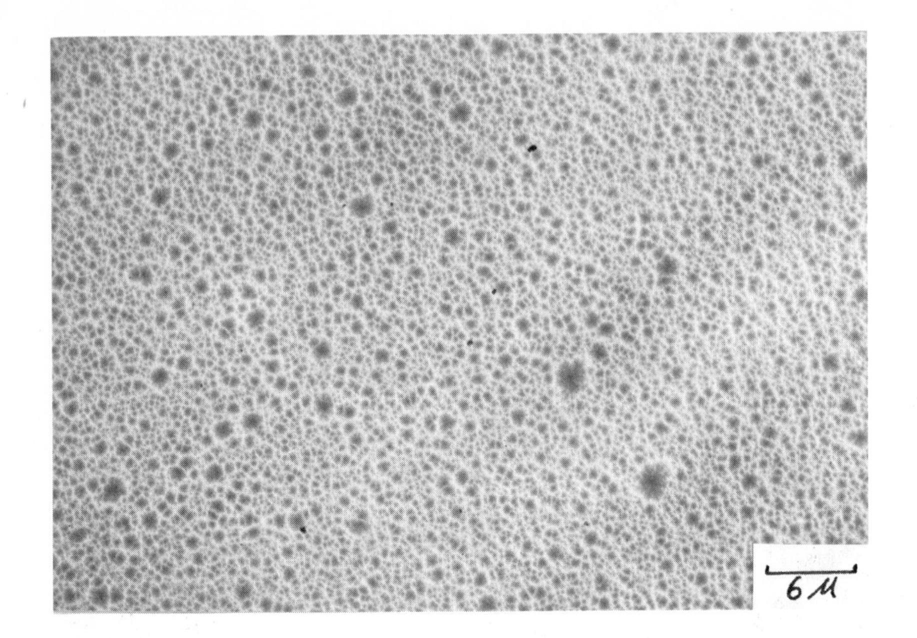
6 μ

5 μ

Figure 7. Transmission electron micrographs of solution-cast films of: (upper left) *1/3/2,* (lower left) *1/4/3, and* (above) *1/6/5. Spherulite size decreases with decreasing hard-segment content.*

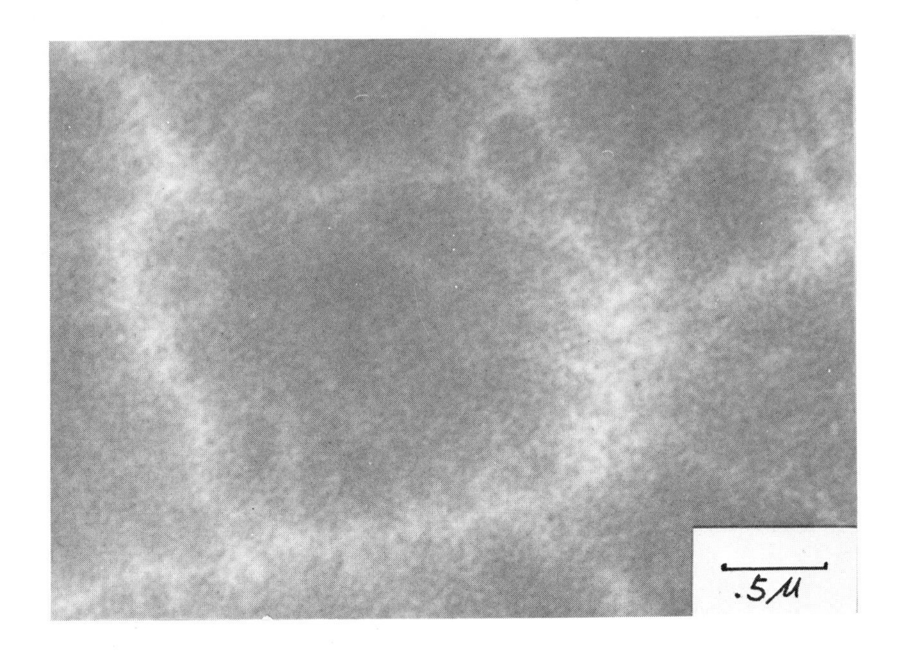

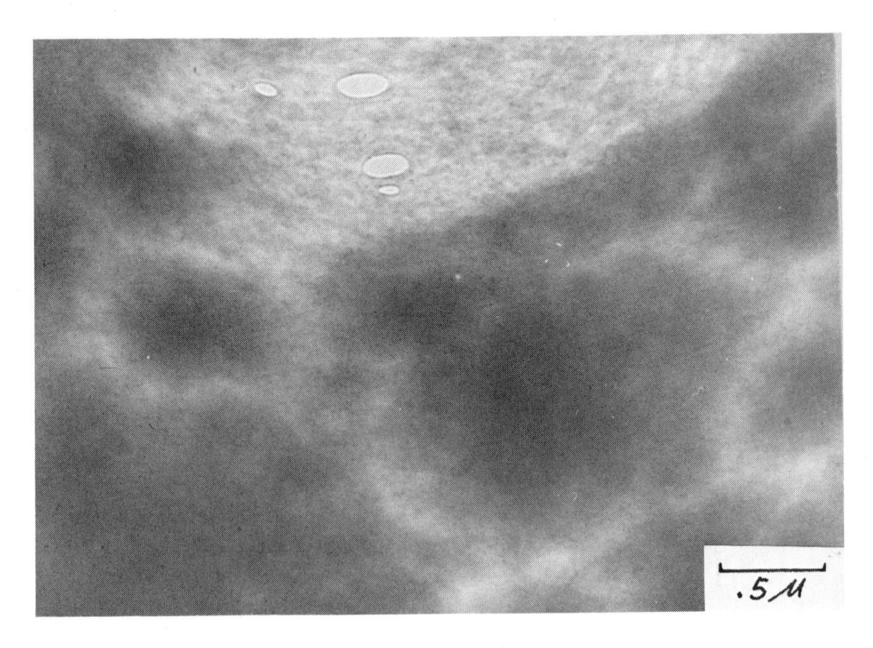

Figure 8. Bright field, defocus micrographs of solution cast films of: (a) sample 1/5/4 and (b) sample 1/6/5 (underfocus for both micrographs = 2 μm)

banded spherulites, but the optical anisotropy of the random copolymer and pure hard-segment spherulites is greatly diminished. Transmission-electron micrographs show that the detailed morphology of the pure soft and pure hard spherulites are quite different (*see* Figure 6). The pure soft-segment spherulites consist of the familiar branching, radiating lamellae, whereas the pure hard-segment spherulites exhibit a radiating, rough fibrous texture.

Figure 7 shows that spherulite size decreases markedly with decreasing hard-segment content. Interestingly, the spherulitic structure of the intermediate composition random copolymers (23%, 31%, 38%, and 43%) still resembles that of the pure hard segment even though the hard segment is the minor component. Samples 0/1/1, 1/6/5, 1/5/4, 1/4/3, and 1/3/2 all contain coarse, fibrous spherulites with rough boundaries. Sample 1/2/1, which contains the smallest amount of hard segment, exhibits no detectable superstructure.

Visualization of the microphase domain morphology proved difficult. Figure 8 shows a dark granularity on the 400-Å scale in the defocused bright-field image of samples 1/6/5 and 1/5/4. The size and amount (as well as the visibility) of this structure decreases with decreasing hard-segment content. Films of less than 38% hard segment appear rather uniform in contrast at high resolution.

Discussion

As-Reacted Samples. The thermal transition behavior of the random copolymer samples is similar to that of other previously studied MDI/BDO-based segmented polyurethanes (*23, 24*). Depending on the thermal history of the sample, up to three endothermic transitions, all believed to be associated with disordering of (para) crystalline hard-segment regions, have been observed. Sample annealing has been shown to cause the two lower endotherm peaks to shift to higher temperatures and become merged. Further annealing then causes an additional upwards shift of this combined lower peak until only a single high-temperature peak remains. The annealing behavior is interpreted in terms of the rearrangement of small, disordered hard-segment regions into more crystalline domains (*24*).

The observed weak, single low-temperature transition with a single, strong high-temperature transition suggests that ordered hard-segment domains can be directly formed in the as-reacted (and cured) polymers. The observed increase in the melting point of the random copolymers with increase in the average hard-segment length suggests an increase of domain size with increased hard-segment content. The intensity and line width of the hard-segment WAXS reflections also indicate that crystal size is increasing with increased hard-segment content.

Figure 9 shows that the heat of fusion increases approximately linearly with hard-segment content for samples containing greater than 23 wt % hard segment. The incremental heat of fusion when related to the incremental weight fraction of hard segment for random copolymers of greater than 23 wt % hard segment shows that the fractional degree of crystallinity of the hard segment phase is approximately constant (at about 47% based on an estimate of 35.5 cal/g for the pure, hard-segment heat of fusion (*29*)) (*see* Table II). The deviation from linearity of the heat of fusion data at low weight fraction hard segment may be attributable to a minimum critical hard segment length necessary for crystallization. The weight fraction of hard segment of block length equal to or greater than k consecutive hard-segment units may be estimated using Peebles (*30*) theoretical data (which assumes complete stoichiometric reaction with equal diisocyanate reactivity). From Figure 9 we see for a theoretical hard-segment distribution a block length of $k \geq 6$ is the minimum effective block length for crystalline hard segment, which is a much higher value than for Harrell's (*10*) polymer ($k \geq 1$).

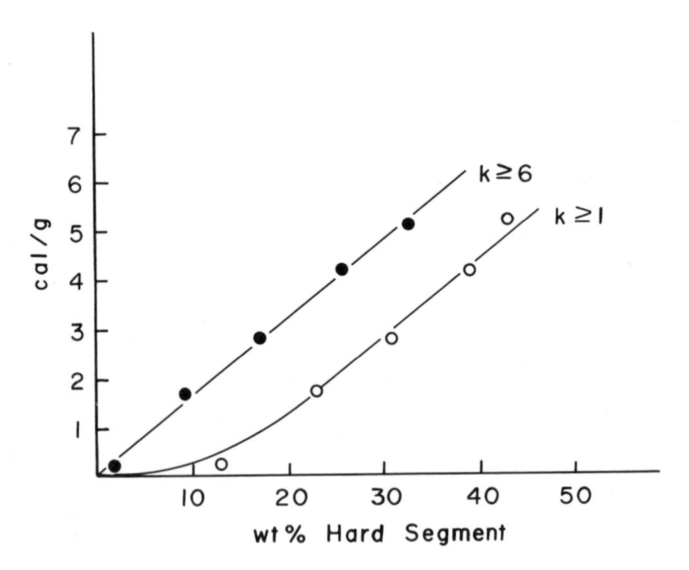

Figure 9. Plot of heat of fusion vs. weight fraction of hard segment for each random copolymer, assuming k $\geq$ *1 or* k $\geq$ *6, where* k *is the number of diisocyanate (hard-segment) units betwen two consecutive macrodiol (soft-segment) units. (Peebles* (30) *calculation of hard, block-length distribution in segmented polyurethane block copolymer is applied.)*

The present results are, however, in general agreement with those of Harrell (*10*), that is, longer hard-segment block length increases the size of hard-segment domains (hence higher T_m) but does not affect the degree of order within the hard-segment domains (hence constant, hard-segment degree of crystallinity). The absence of any soft-segment endotherm and crystalline WAXS reflections implies a noncrystalline soft-segment phase in the as-reacted samples.

Solvent-Cast Samples. The DSC scans of the DMF-solvent cast samples at heating rate of 20°C/min also show two endothermic transitions, but both peaks are shifted down in temperature in comparison with the as-reacted samples. As well, the heat of fusion of the high temperature peak is less than in the respective as-reacted sample (*see* Table II). This would indicate a lower ordering of the hard-segment phase and perhaps a poorer degree of phase separation in solution-cast films. Interestingly, sample 1/2/1 shows a small melting endotherm at approximately 40°C at heating rate of 5°C/min, indicating the presence of crystalline soft-segment regions.

The electron diffraction results support the occurrence of soft-segment crystallinity and a more disordered hard-segment phase in the solvent-cast samples. Moreover, electron diffraction indicates isolated crystalline soft-segment regions persist even in samples of up to 43% hard segment. Electron diffraction is very sensitive to small local fluctuations in the overall structure because the diffraction patterns can be obtained from regions less than 1 μm in diameter and less than .1-μm thick, whereas DSC and WAXS, of course, measure bulk polymer which yields averages over the whole sample.

The spherulite morphology of the solvent-cast samples depends strongly on the hard-segment content. The rough, fibrous-texture spherulites of the pure hard-segment copolymer are retained in the 1/6/5–1/3/2 random copolymer samples, but the average spherulite size decreases markedly with decreasing hard-segment content. The spherulite morphology is therefore controlled by the first solidifying component (T_m and T_g for pure hard segment are 232° and 125°C (*25*) and for pure soft segment 60° and −64°C (*26*)) even at quite low proportion of that component (e.g., as low as 23 wt % hard segment). Films of sample 1/2/1 are not spherulitic and appear quite uniform at high resolution even though regions of crystalline soft segment are present.

The low degree of hard-segment crystallinity in the solvent-cast samples with the added problem of loss of crystallinity from electron-beam damage prevented visualization of either the hard- or soft-segment domains by dark-field microscopy. Therefore bright-field defocus electron microscopy was used to enhance contrast between the microphases (*27*, *28*). Only for the 1/5/4 and 1/6/5 samples was a distinct domain morphology evident. Regions which appear dark in the underfocused micrograph are regions causing a greater phase change of the transmitted electrons and hence are hard-segment-rich regions. Isolated domains of 200–500 Å diameter as well as occasional short, meandering elongated domains 1000–4000 Å long and 300 Å wide appear rather uniformly distributed within the spherulites and interspherulitic regions. The rough radiating fibrous texture and the banding of the pure hard-segment spherulites are attributed to mass thickness contrast caused by variations in film thickness during growth of the solvent-cast spherulitic films. Solvent casting tends to decrease the crystallinity of the hard-segment phase as reflected in the observed lower hard-segment melting points and heats of fusion and the decreased birefringence of the spherulites. This lower degree of order in the hard-segment phase is attributed to the rapid upwards passage of the hard-segment T_m and T_g through the solvent casting temperature during the last stages of solvent evaporation, restricting crystallization of the hard-segment sequences.

Overview

DSC, WAXS, electron diffraction, and electron microscopy support a phase-separated morphology for all five of the random copolymer PCP/MDI/BDO-segmented polyurethanes studied. The slight increase observed by Seefried et al. (9) in the lower temperature T_g with hard-segment concentration is therefore likely attributable to increased reinforcement of the soft-segment matrix by an increased number of dispersed hard-segment domains.

Establishing the occurrence of phase separation by examining the crystallinity of either the hard- or soft-segment phase can be misleading since a small volume fraction of a partially crystalline phase may not be easily detected (e.g., sample 1/2/1 appears amorphous by WAXS), and/or, of course, phase separation may occur without either phase being ordered. Also the detectability of domain structures by TEM is sensitive to the domain size and the electron density difference between the phases. Even if the electron density of the phases remains constant for the various compositions, the projected electron density difference for randomly oriented domains will tend to average out unless the domain size is on the order of the sample thickness. Thus, bright field electron micrographs of (typically) 500-Å thick films enhances the visibility of larger domains while obscuring smaller domains.

Although short-segment sequences are expected to be the most compatible, the 1/2/1 random copolymer with an average hard-block-sequence length of only two units does exhibit a phase-separated morphology—as reflected for the as-reacted sample in hard-segment crystallinity and in soft-segment crystallinity in the solution-cast sample. In general, the as-reacted samples contain a more ordered, hard-segment phase than solution-cast samples, while the solution-cast samples contain a more ordered, soft-segment phase than the as-reacted samples.

This strong dependence of the order of both the soft- and hard-segment phases on the sample preparation technique may account for some of the previous disagreement on the extent of phase separation in a given polyurethane.

Acknowledgment

The authors are pleased to acknowledge suport of a grant-in-aid from the Union Carbide Corporation.

Literature Cited

1. Cooper, S. L., Tobolsky, A. V., *J. Appl. Phys.* (1966) **10**, 1837.
2. Huh, D. S., Cooper, S. L., *Polym. Eng. Sci.* (1971) **11**, 369.
3. Schneider, N. S., Desper, C. R., Illinger, J. L., King, A. O., Barr, D., *J. Macromol. Sci., Phys.* (1975) **B11**, 527.
4. Koutsky, J. A., Hein, N. V., Cooper, S. L., *J. Polym. Sci., Polym. Lett. Ed.* (1970) **8**, 353.
5. Beecher, J. F., Marker, L., Bradford, R. D., Aggarwal, S. L., *J. Polym. Sci., Part C* (1969) **26**, 117.
6. Aggarwal, S. L., *Polymer* (1976) **17**, 938.
7. Wilkes, G. L., Samuels, S. L., Crystal, R., *J. Macromol. Sci., Phys.* (1974) **B10**, 203.
8. Cooper, S. L., Seymour, R. W., West, J. C., "Encyclopedia of Polymer Science & Technology, Plastics, Resins, Rubbers, Fibers: Supplement Vol. 1—Acrylonitrile Polymers, Degradation to Vinyl Chloride," Herman F. Mark, Norbert M. Bikales, Eds., p. 521, Wiley–Interscience, New York, 1976.
9. Seefried, C. G., Jr., Koleske, J. V., Critchfield, F. E., *J. Appl. Polym. Sci.* (1975) **19**, 2493.
10. Harrell, L. L., Jr., *Macromolecules* (1969) **2**, 607.
11. Samuels, S. L., Wilkes, G. L., *J. Polym. Sci., Polym. Lett. Ed.* (1971) **9**, 761.
12. Samuels, S. L., Wilkes, G. L., *J. Polym. Sci., Polym. Symp.* (1973) **43**, 149.
13. Lagasse, R. R., *J. Appl. Polym. Sci.* (1977) **21**, 2489.
14. Kimura, I., Ishihara, H., Ono, H., Yoshihara, N., Kawai, H., IUPAC —III, *Macromol. Prepr.*, Vol. 1, 525, Boston, 1971.
15. Lagasse, R. R., Wischmann, K. B., *Am. Chem. Soc., Div. Org. Coat. Plast. Chem., Prepr.* (1977) **37**, 501.
16. Wilkes, C. E., Yusek, C. S., *J. Macro. Sci. Phys.* (1973) **B7**, 157.
17. Bonart, R., Morbitzer, L., Henze, G., *J. Macromol. Sci. Phys.* (1969) **B3**, 337.
18. Bonart, R., Morbitzer, L., Muller, E. H., *J. Macromol. Sci., Phys.* (1969) **B3**, 337.
19. Bonart, R., *Angew Makromol. Chemie* (1977) **58/59**, 259.
20. Clough, S. B., Schneider, N. S., King, A. O., *J. Macromol. Sci., Phys.* (1968) **B2**, 641.
21. Seefried, Jr., C. G., Koleske, J. V., Critchfield, F. E., *J. Appl. Polym. Sci.* (1975) **19**, 2503.

22. Chatani, Y., Okita, Y., Tadokoro, H., Yamashita, Y., *Polym. J.* (1970) **1**, 555.
23. Miller, G. W., Saunders, J. H., *J. Appl. Polym. Sci.* (1969) **13**, 1277.
24. Seymour, R. W., Cooper, S. L., *Macromolecules* (1973) **6**, 48.
25. MacKnight, W. J., Yang, M., Kajiyama, T., *Polym. Prepr., Am. Chem. Soc., Div. Polym. Chem.* (1968) **9**, 860.
26. Heijboer, J., *J. Polym. Sci., Polym. Symp.* (1968) **16C**, 3755.
27. Petermann, J., Gleiter, H., *Philos. Mag.* (1975) **31**, 929.
28. Christner, G. L., Thomas, E. L., *J. Appl. Phys.* (1977) **48**, 4063.
29. Kajiyama, T., MacKnight, W. J., *Polym. J.* (1970) **1**, 548.
30. Peebles, L. H., Jr., *Macromolecules* (1976) **9**, 58.

RECEIVED June 5, 1978.

3

Time Dependence of Mechanical Properties and Domain Formation of Linear and Crosslinked Segmented Polyurethanes

ZOHAR H. OPHIR

Polymer Materials Program, Department of Chemical Engineering, Princeton University, Princeton, NJ 08540

GARTH L. WILKES

Department of Chemical Engineering, Virginia Polytechnic Institute and State University, Blacksburg, VA 24061

A systematic series of segmented polyester-MDI polyurethanes were investigated by SAXS, DSC and stress strain measurements. The linear polymer as well as three peroxide cured samples of different crosslinking levels were utilized. Correlations were made between time-dependent changes in structure and mechanical properties following specific thermal treatments. Higher crosslinking leads to lower final domain formation and to lower modulus. Also, the crosslinked samples had their domain texture more easily disrupted at lower temperature than did the linear system. Finally, the rate of phase separation upon cooling was increased as the degree of crosslinking was decreased.

Segmented polyurethanes are thermoplastic materials that generally display elastomeric behavior—the degree of which depends on the relative amounts of "soft" and "hard" segments. The properties also depend on the extent of microphase separation and the morphological characteristics of the phases. Because of the basic thermodynamic incompatibility of these two segment types, localized microphase formation occurs leading to the well-recognized "domain" morphology. It is widely accepted that the mechanical properties of the final bulk material are

0-8412-0457-8/79/33-176-053$05.00/0

dictated by the nature and extent of the domain formation (*1–7*). In light of this, one must not always picture a domain "morphology" as shown in Figure 1a where the hard segments form a distinctly dispersed phase in the matrix comprised of the soft elastomeric segments. While this schematic may well be quite realistic, if the hard-segment composition is higher, or through different processing procedures, both the hard and soft segments could form continuous phases (Figure 1b) or finally the hard segment may become the matrix for dispersed soft segments (Figure 1c). The latter two systems are not likely to display conventional elastomeric behavior because of the interconnectivity of the hard-segment phase which, by itself, displays stiff, high modulus behavior. In view of the variation in hard-segment content within the broad class of segmented polyurethane materials (linear and crosslinked), one must not therefore expect identical structure as to specific time-dependent behavior as addressed within this chapter. Furthermore, other factors as variation in chain chemistry (of either soft or hard segment) must be recognized because of its affect on thermodynamic incompatibility as well as molecular symmetry which may lead to crystallization of one or both segment types (*2, 3, 8*). Other basic aspects which also must not be ignored when comparing segmented systems is the mode of polymerization, i.e., one- or two-step polymerization and molecular weights (and distribution) of the segments (*1, 3, 8–11*).

In recalling Figure 1, where two-phase behavior exists, the related question is "does the domain texture become instantaneously induced during processing whether it be through thermal or solvent means?" A second related question is "once established, how thermally stable (in a structural sense) is this texture?" From an applications point of view both questions have very obvious ramifications. Recent work (*7, 8, 11, 12,*

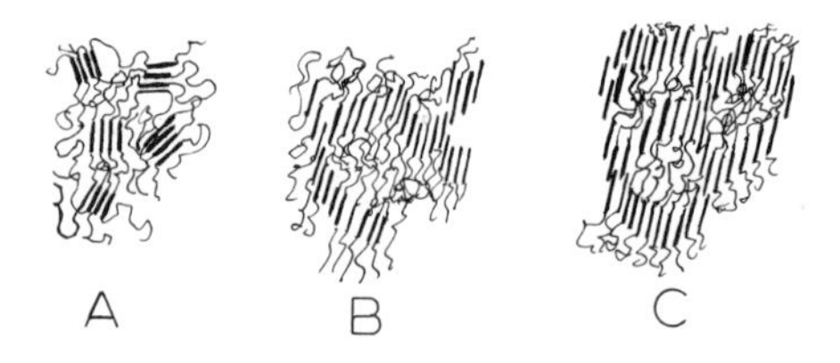

Figure 1. Molecular schematics of some possible segmented polyurethane morphologies. For convenience, good phase separation is assumed. (a) Hard-segment domains dispersed within a soft-segment matrix; (b) interconnected hard-segment phase giving rise to continuous phases of both hard and soft segments; (c) soft-segment phase is dispersed within a continuous matrix of hard segments.

13) has shown that upon heating many different types of linear segmented polyurethanes, the domains become unstable and mixing of the soft and hard segments takes place; the temperature of disruption depends on the type of polymer. Upon cooling, the hard and soft segments start to phase separate and the domains are again formed. Several methods (stress–strain, SAXS, DSC, NMR) have been applied to show that the time-dependent changes in Young's modulus, soft-segment glass transition, and the degree of phase separation are very similar in all of these experiments. These results showed that the domain formation is a thermally reversible process and that its kinetics are directly correlated with the changes in the mechanical properties with time.

The current work deals with similar time-dependent phenomena of segmented polyurethanes but focuses on a specific segmented urethane where different levels of chemical crosslinking have been induced.

Experimental

Materials. The materials used in this study were based on a single linear polyester urethane that had been crosslinked by different amounts of peroxide. The hard segments contained *p,p'*-diphenylmethyl diisocyanate (MDI) and 1,4-butanediol. The soft segment was poly(tetramethylene adipate) glycol (mol wt ~ 1100). The materials were prepared as follows: the MDI, diol, and glycol were mixed and "melt reacted" randomly to yield a polyurethane containing 30 weight percent of hard segments. Next, commercial organic peroxide was mixed into the polymer, and the films were then compression molded at 210°C for 10 minutes while curing took place. Four samples were prepared this way: ESX 2.0 (contains 2.0% peroxide by weight), ESX 1.0, ESX 0.5, and ESX 0.0 (contains no peroxide). Crosslinking was demonstrated by the lower degree of swelling in DMF as the peroxide level increased (the exact site of chemical crosslinking is not known). The well-characterized materials were kindly prepared by C. S. Schollenberger and K. Dinsbergs of the B. F. Goodrich Center, Brecksville, Ohio.

Equipment and Methods. The domain morphology and its changes with time were detected by a Kratky small angle x-ray camera, operated by a PDP 8/a computer. The x-ray source was a Siemens AG Cu40/2 tube operated at 40 kV and 25 mA. A CuKα monochromatic radiation (1.54 Å) was obtained by nickel foil filtering and the use of a pulse-height analyzer. Two types of SAXS experiments were used. The first one was a regular intensity scan with scattering angle measured at room temperature on well-aged samples. The second test consisted of measuring the scattered intensity at a fixed angle of 9.35 mrad (Bragg's Law spacing of 165 Å). The scattering functions show a maximum near this angle, and therefore changes in the intensity with time represent changes in the degree of phase separation, as discussed earlier (*12*). In the scanning experiments three sets of slits were used in each scan to cover a wide range of scattering angles, starting with a 60-μ entrance slit at 1 mrad and ending with a 150μ slit at 140 mrad. In the fixed-angle experiments a

150-μ entrance slit and 375-μ counter slit were used. In both cases the parasitic scattering was measured with the sample mounted in the non-scattering position (before the collimation system).

The intensity at the fixed angle was measured after the samples were given a thermal treatment as follows: the sample was annealed 10 minutes at a given temperature and then quenched in liquid nitrogen. (To minimize thermal degradation, only a five-minute annealing period was used at 170°C). As the sample's thickness was about 0.06 cm, it was assumed that the temperature change in the sample was rapid. Once the temperature was below the soft-segment glass transition, the morphology of the sample was fixed and did not change with time. Next the sample was heated to room temperature (24°C) within a few seconds in a water bath. At this point the annealing time was taken as zero for the time-dependent measurements. The sample was then placed in the scattering position of the camera, and the scattered intensity was measured by a detector at the fixed angle and recorded by a chart recorder as well as by a counter as a function of time.

For a two-phase system having two components with volume fractions ϕ_1 and ϕ_2 and with corresponding constant electron densities ρ_1 and ρ_2, the mean square of fluctuation in the electron density is given by:

$$\langle\rho^2\rangle = \phi_1\phi_2(\rho_1 - \rho_2)^2 \tag{1}$$

If the boundaries of the phases are not sharp, there is a negative correction term in Equation 1 (*11*). If the electron density changes linearly from one phase to the other and takes place over a transition layer of thickness E and specific surface S/V, the equation will take the form (*14*):

$$\langle\rho^2\rangle = \left(\phi_1\phi_2 - \frac{ES}{6V}\right)(\rho_1 - \rho_2)^2 \tag{2}$$

The mean square of the electron density distribution, i.e., the left-hand term above, is proportional to the invariant

$$\tilde{Q} = \int_0^\infty \tilde{I}(\theta) \cdot \theta \cdot d\theta \tag{3}$$

where θ is the scattering angle and $\tilde{I}(\theta)$ is the measured scattered intensity (smeared).

The scattering curves were analyzed by the computer program of Vonk (*14*). The major steps in this analysis are (1) parasitic scattering subtraction; (2) matching the different sections of the scattering curve in the overlapping regions and combining them into a single curve; (3) correction of wide-angle background; (4) tail fitting; (5) invariant calculation assuming an infinite slit—in this way artificial effects of desmearing are avoided; and (6) desmearing (*14*), either by the use of the measured slit-length weighing function or by assuming an infinite slit. The validity of the infinite slit assumption can be checked by comparing the two desmeared curves.

In the case of segmented urethanes it might be expected that there will always be some degree of mixing between the phases and that the boundaries between the phases are not sharp. Therefore ρ_1 and ρ_2 as well as E and S/V are not known, and the measurement of the invariant $\tilde{Q}$ cannot give us numerical values for ϕ_1 and ϕ_2. However, it is still a useful parameter for comparing materials, particularly a given material undergoing phase separation. The same argument also holds for the measurement of $\tilde{I}$ (time) at a fixed θ. This intensity cannot be used to calculate the invariant as a function of time, but rather it is a measure of the relative rate of change of the domain morphology as indicated in a previous paper (*12*) and therefore can serve as a kinetic parameter.

In this study, the mechanical property that was measured as a function of time following the thermal treatment was Young's modulus. It was measured with an Instron at room temperature using a crosshead speed of 10 cm/min. Precut dog-bone-shaped samples were thermally treated as described for the SAXS experiments. The dog-bone samples were 0.269-cm wide and 0.06-cm thick and had an effective gage length of 1 cm. Each sample was used only once to make sure that no changes in morphology were induced by the stretching.

Differential scanning calorimetry (DSC) was carried out with a DuPont Model 990 differential scanning calorimeter using sample weights of 4–6 mg with a heating rate of 20°C/min and full range sensitivity of 1.0 mcal/sec. As defined earlier by Wilkes and Wildnauer (*11*), the glass-transition temperature of the soft segment, T_{gs}, was taken as the temperature that corresponds to the initial change in slope in the heat capacity C_p upon heating from the glassy amorphous state. Values of T_{gs} determined by this method were reproducible to ± 1°C in duplicate runs. The thermal treatment was similar to those of the SAXS experiments except that the cooling was done by placing the sealed sample pan on a cold metal plate to enhance the heat transfer.

Results and Discussion

The SAXS scans of the well-aged ESX samples indicate clearly that samples with a higher crosslinking density show less phase separation (Figure 2). This data is consistent with the time-dependent results of the current work. These latter experiments show that at above 140°C, for example, the domains disrupt considerably, segment mixing takes place, and thus the SAXS intensity decreases. Since the temperature of curing was 210°C, it is clear that the crosslinking took place in the mixed or partially mixed state. Upon cooling, there is a driving force for phase separation. The mobility, however, of the crosslinked segments is restricted, and in cases where hard and soft segments are chemically bonded, they cannot separate. As a result, increasing the crosslinking density decreases the electron density difference $(\rho_1 - \rho_2)$ and also likely increases the transition-zone thickness E. This in turn causes a reduction in the invariant $<\rho^2>$ (Equation 1) and hence a reduction in the scattered intensity as measured. The above statement is directly sup-

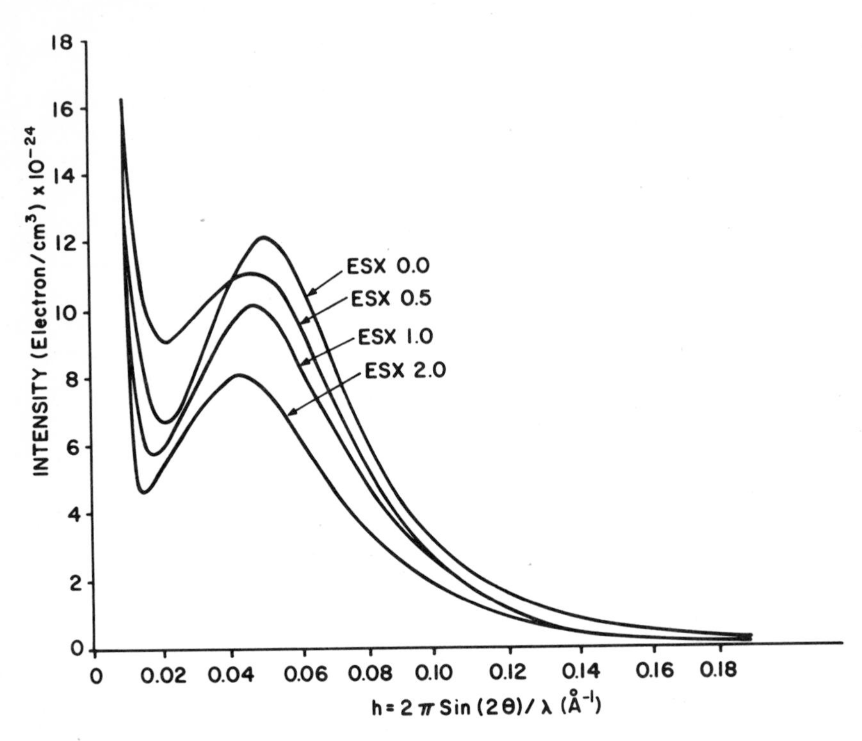

Figure 2. Desmeared SAXS of well-aged ESX samples at room temperature

Table I. Important Characterization Parameters of the ESX Series of Segmented Urethanes

Peroxide (wt %)	0.0 (ESX 0.0)	0.5 (ESX 0.5)	1.0 (ESX 1.0)	2.0 (ESX 2.0)
Swelling in DMF (%)	soluble	275	205	165
Soft-segment glass-transition temperature (°C)	−40.0	−38.5	−37.0	−35.0
Young's Modulus (MPa)	12.8	11.5	10.3	7.2
Mean square of the electron density fluctuations (mol elect/cm³)² × 10³	0.605	0.495	0.429	0.385

ported by the values of the invariant calculated from the smeared SAXS data (*see* Table I).

Values of Young's modulus as found for the well-aged ESX samples also are given in Table I. Typically for elastomeric polymers, Young's modulus increases as the degree of crosslinking increases. In the case studied here the results show an opposite trend. Similar results were reported by Kimball and Fielding-Russell (*15*) as well as by Schollenberger et al. (*1*). The explanation is as follows: the main contributor to Young's modulus in these two-phase systems is the hard domains that serve both as a filler and as physical crosslinks. In the presence of these glassy domains, the relative contribution of the chemical crosslinking (at least up to the level that was induced) is small. Therefore, increasing the crosslinking density reduces the extent of final phase separation and thus decreases Young's modulus (at room temperature). This explanation is clearly confirmed by the SAXS results.

Time-Dependence Experiments. In the fixed-angle SAXS experiments the time-dependent scattered intensity $\overline{I(t)}$ was normalized by:

$$\overline{I}(t) = \frac{I(t) - I_b}{I_o - I_b} \tag{4}$$

I_b is the background scattering which is about 10% of the total intensity and is almost constant. I_o is the initial scattered intensity of the sample before the thermal treatment. The normalization helps correct for changes in x-ray tube intensity and for differences in the sample thickness. Fixed-angle experiments were made on samples which had been annealed at the temperatures of 170°, 140°, 120°, 100°, and 80°C as described earlier. Figures 3a, 3b, and 3c show the results at the two extreme temperatures and at 120°C. From these data one observes the following. (a) Upon heating above 140°C, all of the samples apparently lose most of their phase separation. The small scattered intensity that remained can be related to local fluctuations in the electron density. A full scan measured at a temperature below the soft-segment glass-transition temperature and following thermal treatment supports this assumption (*12*). (b) Below 120°C, the higher the degree of crosslinking, the easier the domains are partially disrupted. (c) In all cases the recovery rate decreases as the crosslinking density increases.

The following explanation of this behavior is now suggested. As the SAXS data show, as the crosslinking density goes up there is less phase separation. The soft segments that remain mixed in the hard domains apparently act as plasticizers and weaken the domain structure. Crosslinking within the glassy phase itself provides only a minor contribution to its strength and at the same time prevents optimal packing and thereby

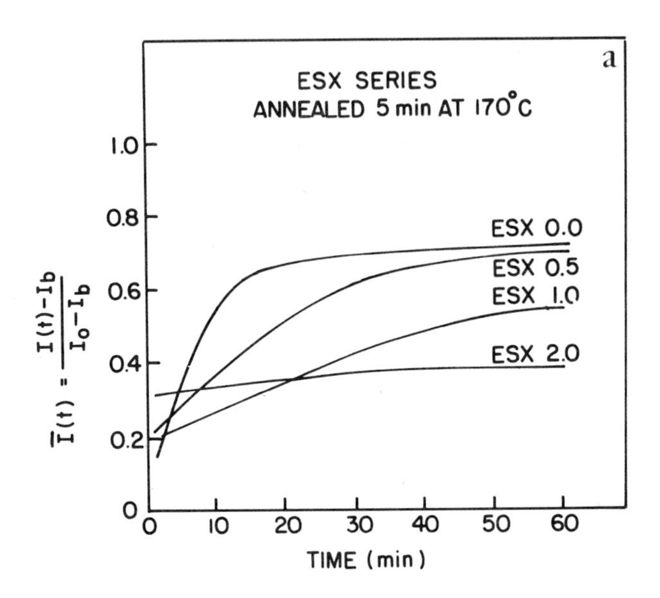
a
ESX SERIES
ANNEALED 5 min AT 170°C
$\bar{I}(t) = \frac{I(t)-I_b}{I_o-I_b}$
1.0
0.8
0.6
0.4
0.2
0
ESX 0.0
ESX 0.5
ESX 1.0
ESX 2.0
0 10 20 30 40 50 60
TIME (min)

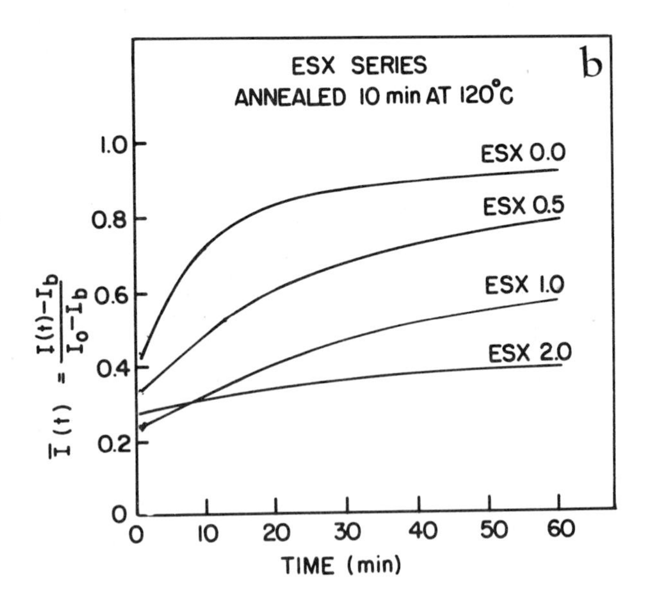
b
ESX SERIES
ANNEALED 10 min AT 120°C
$\bar{I}(t) = \frac{I(t)-I_b}{I_o-I_b}$
1.0
0.8
0.6
0.4
0.2
0
ESX 0.0
ESX 0.5
ESX 1.0
ESX 2.0
0 10 20 30 40 50 60
TIME (min)

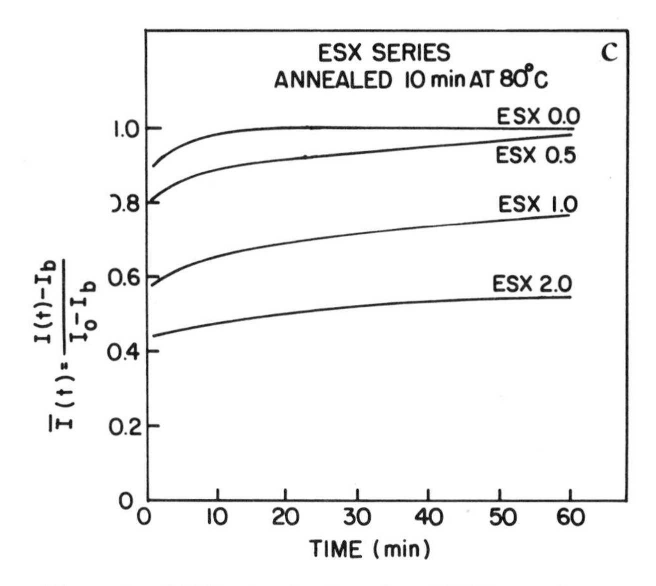

Figure 3. SAXS at a fixed angle of ESX samples at room temperature as a function of time following a thermal treatment of (a) 170°C, (b) 120°C, and (c) 80°C

possibly decreases the degree of hydrogen bonding. When the sample is heated, the effect of crosslinks within the soft phase will tend to increase the retractive elastic forces that help promote the disruption of the domains. According to the kinetic theory of rubber elasticity

$$\sigma_0 = N_v KT \cdot f(\epsilon) \tag{5}$$

where σ is the stress, N_v is the number of network segments in a unit volume (which is proportional to the crosslinking density), T is the temperature and $f(\epsilon)$ represents a strain function that will be assumed to be independent of temperature. It is noted that for a finite strain, i.e., $f(\epsilon)$ is nonzero, stress will increase in direct proportion to temperature and to degree of crosslinking. Accepting that the soft segments are indeed partially strained as a result of domain formation (*7, 11*), then one may well expect that the "internal" stress arising from strained soft segments at high temperatures could be greater for the chemically crosslinked samples. While this statement is difficult to confirm experimentally, it is likely as is the fact that the domain cohesiveness is expected to be less in the crosslinked samples because of the limitations imposed on molecular packing by the crosslink junctions. Thus, as the temperature goes up, the domains of the crosslinked sample will likely start to disrupt at lower temperatures than those in the uncrosslinked sample. Upon returning to

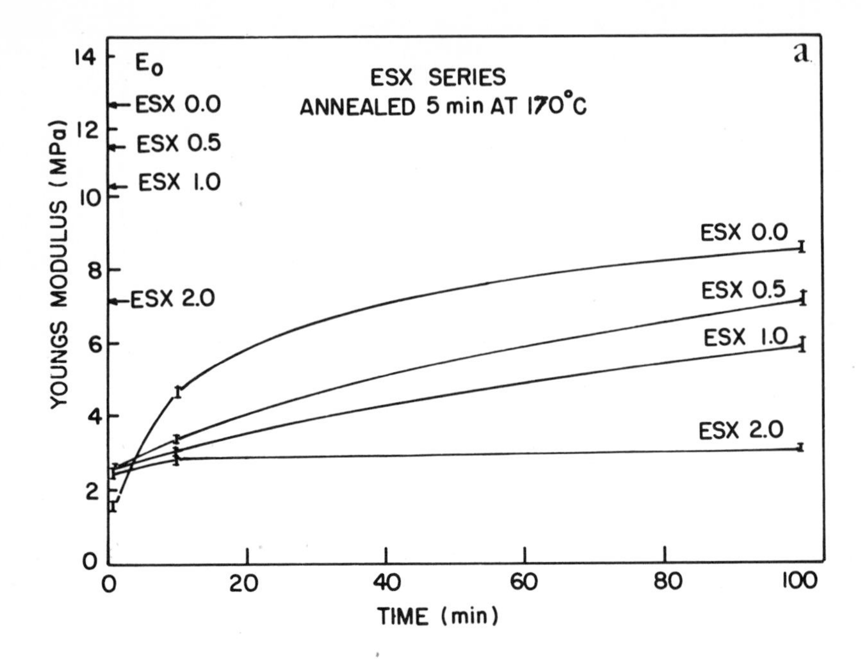
a
ESX SERIES
ANNEALED 5 min AT 170°C
E_0
ESX 0.0
ESX 0.5
ESX 1.0
ESX 2.0
ESX 0.0
ESX 0.5
ESX 1.0
ESX 2.0
YOUNGS MODULUS (MPa)
TIME (min)
0
20
40
60
80
100
2
4
6
8
10
12
14

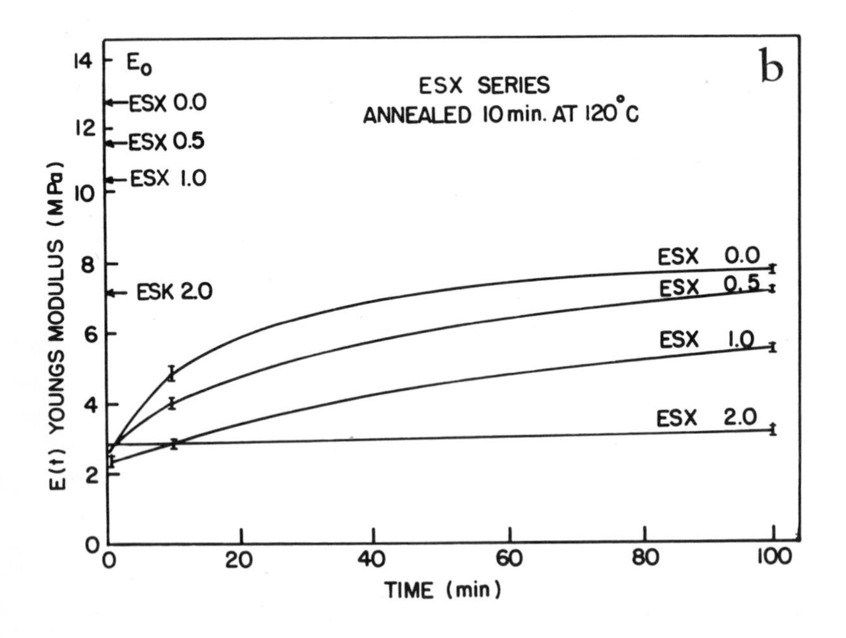
b
ESX SERIES
ANNEALED 10 min. AT 120°C
E_0
ESX 0.0
ESX 0.5
ESX 1.0
ESK 2.0
ESX 0.0
ESX 0.5
ESX 1.0
ESX 2.0
E(t) YOUNGS MODULUS (MPa)
TIME (min)
0
20
40
60
80
100
2
4
6
8
10
12
14

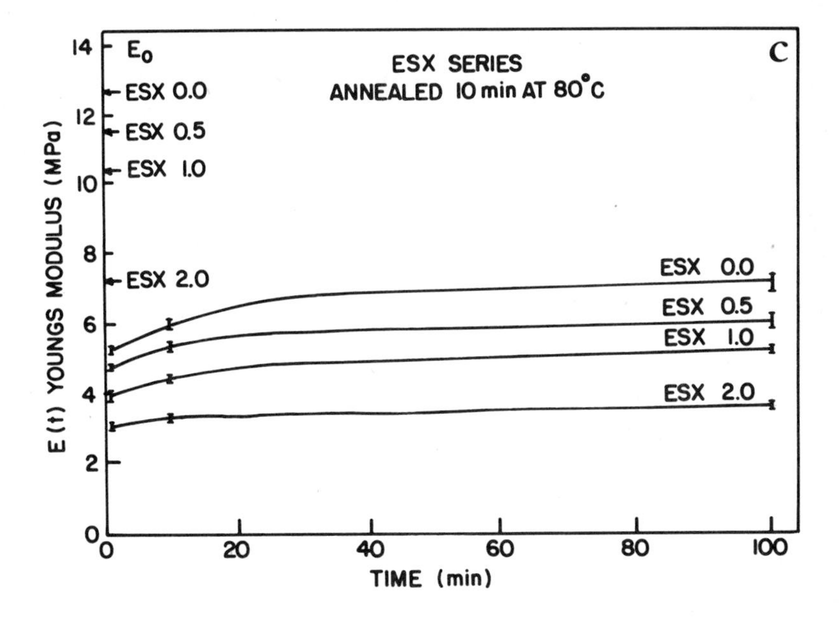

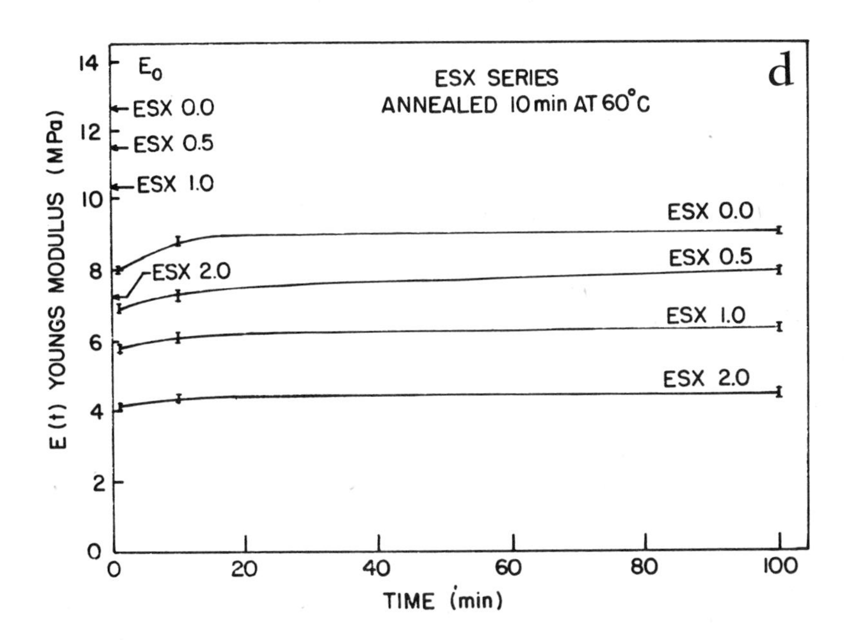

Figure 4. Young's modulus of ESX samples at room temperature as a function of time following a thermal treatment of (a) 170°C, (b) 120°C, (c) 80°C, and (d) 60°C

room temperature, subsequent to the thermal treatment, the segments start to unmix and form domains. As the crosslinking restricts the mobility of the segments, higher crosslinking density causes a lower rate of recovery.

In the stress–strain experiments, Young's modulus was measured as a function of time, following thermal treatments of 170°, 120°, 80°, and 60°C (Figures 4a, 4b, 4c, and 4d). We note that above 120°C there is strong similarity between $\bar{I}(t)$ and $E(t)$ plots (compare the normalized Young's modulus E/E_o in Figure 5 with $\bar{I}(t)$ in Figure 3a). At lower temperatures Young's modulus drops subsequent to the thermal pulse to relatively lower values than the corresponding scattered intensity values. However, the same trends with respect to temperature, crosslinking density, and recovery rates are observed. These results support the conclusions of the SAXS experiments. The difference in the behavior at low temperature is tentatively explained as follows. When the thermal treatment is above 120°C, the internal stress and softening of the hard domains is large enough to disrupt the domains and hence to reduce both the scattered intensity and Young's modulus in similar fashions. At lower temperatures, the thermal pulse can disrupt the domains only partially (depending on the crosslinking density), and therefore the changes in the scattered intensity are relatively small (especially in the uncrosslinked sample). However, this thermal pulse apparently does affect the dense

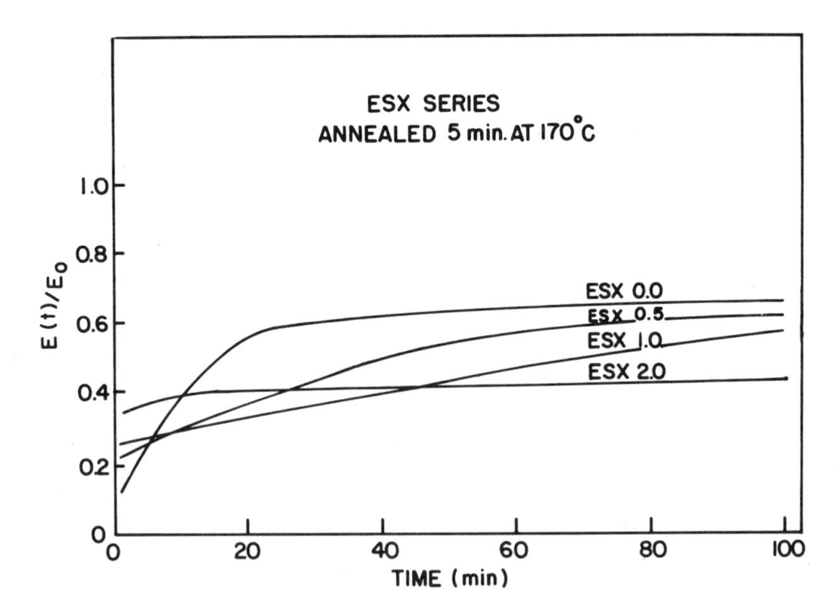

Figure 5. Young's modulus of ESX samples normalized by its initial value (E_o) measured at 24°C as a function of time following a thermal treatment of 170°C

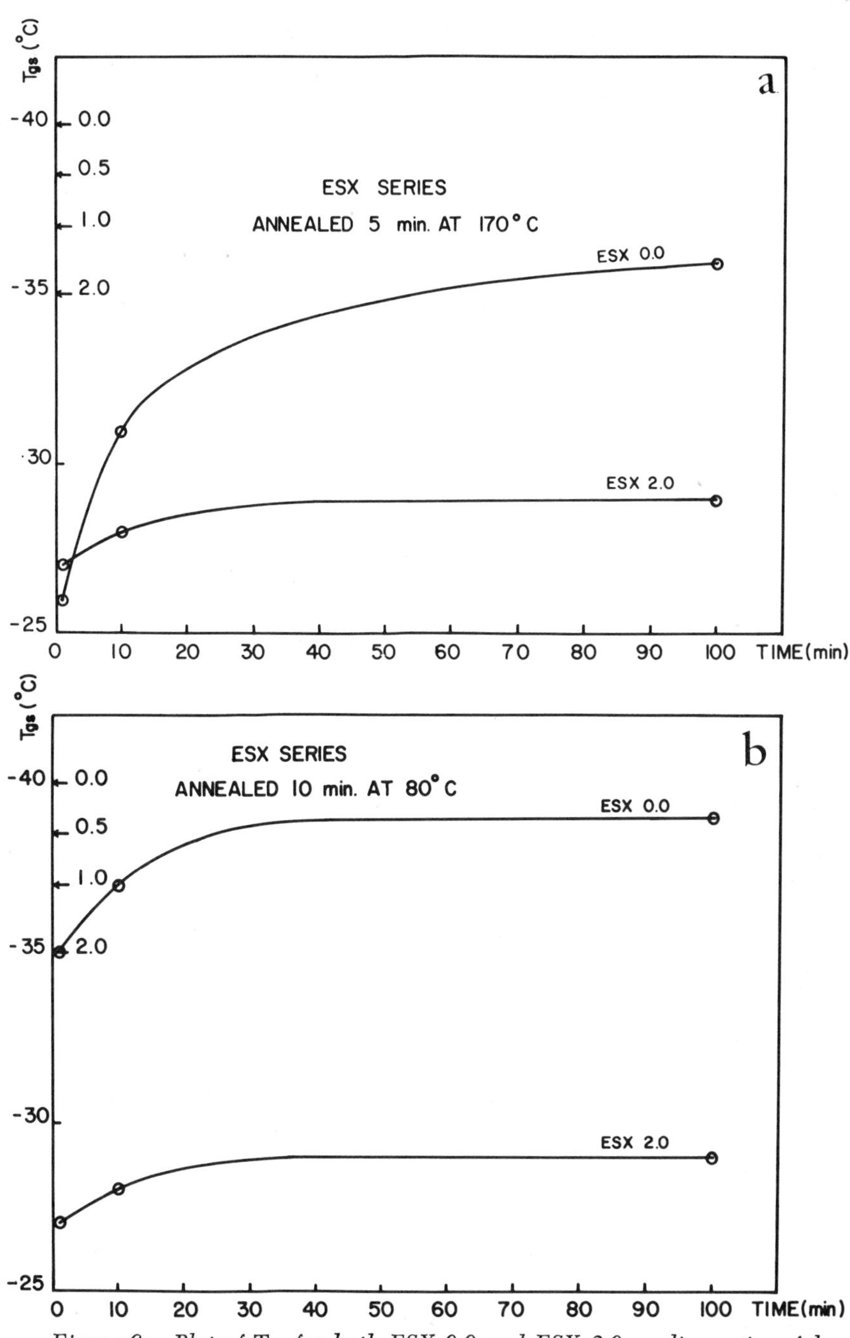

Figure 6. Plot of T_{gs} *for both ESX 0.0 and ESX 2.0 vs. linear time following thermal treatment at (a) 170°C and (b) 80°C*

packing in the glassy domains and causes them to lose their stiffness and hence to have lower modulus. The recovery of the modulus in this case may have some similarity to the physical aging phenomena observed in glassy polymers (*16, 17*). Measurement of the glass-transition temperature of the soft segment, T_{gs}, give further support to the mixing–demixing model. As shown by Wilkes et al. (*7, 11*), upon mixing with the hard segments, T_{gs} should go up because the motion of the soft segments becomes more restricted by the presence of the hard segments. In Figures 6a and 6b, T_{gs} of ESX 0.0 and ESX 2.0 are given as a function of time following a thermal treatment of 170° or 80°C. One can see the direct similarity between these figures and the corresponding SAXS and Young's modulus plots, as expected.

Conclusions and Comments

This work extends the understanding of the structure–property relationships of segmented urethane systems. An important finding is that high-temperature crosslinking of these two-phase materials can reduce their Young's modulus, thermal stability, and recovery rate. These data may well have relevance to the time-dependent behavior of network urethanes prepared by reactive molding where crosslinking is induced at high temperatures. Presently we are continuing these studies and also are noting the effects of crosslinking of the time-dependent properties when the crosslinks are induced at lower temperatures where the domain structure already has been established. This is accomplished by exposing the samples to a low-dosage gamma radiation (cobalt 60 source) at ambient temperatures. Initial results show a distinct difference in the temperature–time dependence between radiation and chemically treated samples with an equivalent average crosslinking degree (as measured by degree of swelling in DMF). These results are expected and will be published later.

Acknowledgment

The authors would like to acknowledge the partial financial support of this work obtained from both the National Science Foundation (Polymer Program), Grant No. DMR-77-05648, and the Goodyear Tire and Rubber Company.

Literature Cited

1. Schollenberger, C. S., Dinbergs, K., *J. Elastomers Plast.* (1975) **7**, 65.
2. Samuels, S. L., Wilkes, G. L., *J. Polym. Sci.* (1973) **C43**, 149.
3. Chang, Y., Wilkes, G. L., *J. Polym. Sci.* (1975) **13**, 455.

4. Schneider, N. S., Paik Sung, C. S., Matton, R. W., Illinger, J., *Macromolecules* (1975) **8**, 62.
5. Bonart, R., Morbitzer, L., Rinke, H., *Kolloid Z. Z. Polym.* (1970) **290**, 807.
6. Cooper, S. L., Tobolsky, A., *J. Appl. Polym. Sci.* (1966) **10**, 1837.
7. Wilkes, G. L., Bagrodia, S., Humphries, W., Wildnauer, R., *J. Polym. Lett.* (1975) **13**, 321.
8. Wilkes, G. L., Morra, B., Emerson, J. A., Wildnauer, R., *Proc. Int. Congr. Rheol., 7th, 1977*, p. 224.
9. Seymour, R. W., Cooper, S. L., *Rubber Chem. Technol.* (1974) **47**, 19.
10. Peebles, L. H., *Macromolecules* (1974) **7**, 872.
11. Wilkes, G. L., Wildnauer, R., *J. Appl. Phys.* (1975) **46**, 4148.
12. Wilkes, G. L., Emerson, J. A., *J. Appl. Phys.* (1976) **47**, 4261.
13. Assink, R. A., Wilkes, G. L., *Polym. Eng. Sci.* (1977) **17** ,603.
14. Vonk, C. G., *J. Appl. Crystallogr.* (1973) **6**, 81.
15. Kimball, M. E., Fielding-Russell, G. S. (1977) **18**, 777.
16. Petrie, S. E. B., *Polym. Mater., Prepr. Semin. Am. Soc. Mater.* (1973) **55.**
17. Lagasse, R., *J. Appl. Polym. Sci.,* in press.

RECEIVED June 5, 1978.

4

Surface Chemical Analysis of Segmented Polyurethanes. Fourier Transform IR Internal Reflection Studies

C. S. PAIK SUNG and C. B. HU

Department of Materials Science and Engineering, Massachusetts Institute of Technology, Cambridge, MA 02139

The chemical composition on the surfaces of two segmented polyurethanes was investigated by using Fourier transform IR internal reflection spectroscopy. The variables investigated were the difference in exposure, i.e., to air vs. to the substrate during casting procedure, and the depth from the surface. In a polyether urethane which also contains 10% by weight of polydimethyl siloxane, the air side contains a greater amount of the polyether soft segment and the silicone polymer than the substrate side, and this trend is more pronounced in a layer close to the surface. In another polyether polyurethane, only a modest increase in the relative content of the soft segment/hard segment is observed in the air side when the depth composition profile was obtained. No lateral inhomogeneity was observed in either polymer.

During the last two decades, significant progress has been made in elucidating the bulk structure of polymers, which leads to a better understanding of the structure–property relationships in polymers. By contrast, the surface structure and chemical composition of polymers have not been investigated in detail, mainly because of the lack of experimental techniques sensitive enough to provide accurate information. Many properties such as adhesion (*1*), surface crazing and crack initiation (*2*), friction (*3*), solid-state chemical reactivity (*4*), and blood compatibility (*5, 6*) of a polymer are believed to be strongly influenced by the detailed structure and the chemical composition of the surface.

0-8412-0457-8/79/33-176-069$05.00/0

Since the surface may be considerably different from the bulk, as is the case in metals and semiconductors (*7*), detailed information of the surface of polymers becomes necessary to understand and to improve surface-related properties. With the availability of new techniques such as Fourier transform IR internal reflection spectroscopy (FTIR–IRS) (*8*), electron spectroscopy for chemical application (ESCA) (*9*), ion scattering spectroscopy and secondary ion mass spectrometry (ISS/SIMS) (*10*), and Auger electron spectroscopy (AES) (*11*), now it is possible to study the surface structure and chemical composition of polymer surfaces with the accuracy required for reasonable interpretation. Among these techniques, the FTIR–IRS method provides information on a surface layer between a few tenths of a micron and a few microns while in other methods, the depth of the layer which is investigated is in the order of 10 to 100 Å without inert-gas-ion sputtering. However, by choosing a suitable sputtering condition, the depth of the layer can be increased to a few hundred angstroms or more (*12, 13*).

In this study, we have attempted to obtain a detailed, quantitative estimate of the surface chemical composition of two commercially available polyurethanes, i.e., Biomer and Avcothane, which have demonstrated a reasonable degree of blood compatibility. For example, Avcothane has been used as an intraaortic balloon pump for post-operative patients (*5*). Biomer also has been successfully used for artificial heart components in calves (*14*).

These segmented polyurethanes consist of alternating hard and soft segments. The soft segment is commonly polyether whereas the hard segment consists of an aromatic diisocyanate chain extended with a low-molecular-weight diol or diamine. Avcothane is polyether polyurethane with 10% by weight of polydimethylsiloxane (*5*). The properties of segmented polyurethanes are primarily attributable to the phase segregation of soft and hard segments, leading to the formation of hard-segment domains which are dispersed in the rubbery polyether matrix. A variety of compositional variables, such as the type of the diisocyanate and chain extender, the molecular weight of the soft segment, and the length of the hard segment, will affect the degree of phase segregation and the volume fractions of hard-segment domains in bulk (*15*). Furthermore, at or near the surface, the thermodynamic and kinetic factors can drastically alter the ratio of volume fractions of hard- to soft-segment phases, even in a given bulk composition. Since these polyurethanes are cast from solutions on mold surfaces, factors such as the thermodynamic properties of the casting solvent, the polarity of the casting mold substrate, and the rate of solvent evaporation will influence the course of phase segregation and thus the morphology of the soft-segment phase and domain phase in bulk and at the interface.

In this paper, we summarize the results obtained by using Fourier transform IR internal reflection spectroscopy. In IR internal reflection spectroscopy, the IR beam penetrates the surface of polymers between a few tenths of a micron and a few microns, depending on variables such as the reflection plate, angle of incidence, and the wavelength of the IR beam. The depth of beam penetration has been reduced by placing a thin barrier film between the trapezoidal reflection plate and the polymer under study. This barrier film was chosen because it does not absorb IR in most of the regions where polyurethanes show strong absorption peaks. By varying the thickness of the barrier film we obtained the depth profile of the concentration of IR-sensitive bands in polymer surfaces. With this method, the smallest thickness that can be studied is about 1000 Å. By using a Fourier transform IR spectrometer rather than a conventional dispersive IR instrument, we obtained higher sensitivity and accuracy because of the capability of a computer to carry out various spectral manipulations. This added advantage makes it possible to remove unwanted absorption of barrier film and to obtain difference spectra directly.

Experimental

Materials. An intraaortic pump made of Avcothane was purchased from Avco Medical Products. Avcothane is known to be a hybrid of a poly(ether urethane) (90%) and poly(dimethylsiloxane) (10%) (*16, 17, 18*). Poly(ether urethane) portion is known to be composed of MDI, butanediol, and polypropylene glycol while the poly(dimethylsiloxane) used was the diacetoxy-terminated polymer with molecular weight of approximately 54,000. Nyilas and Ward reported that Avcothane pump was made using prepolymer solutions in a 2:1 mixture of absolute THF and dioxane and by casting against stainless steel surface. They also reported that monitored by GPC, low-molecular-weight substances which are potentially leachable have been removed. Therefore, it is reasonable to assume that the surface which was analyzed does not contain oligomers reactive or otherwise. For details of the procedure, Ref. *16* can be referred to.

Solutions of Biomer were obtained from Ethicon Inc. Biomer is poly(ether polyurethane) which contains urea linkage in the hard segment according to our IR analysis (*19*). Films of Biomer were cast on clean glass plates by diluting the polymer solution in dimethyl acetylamide. The films were dried in a vacuum oven at 50°C for 24 hr. The final film thickness was around 125μ. GPC analysis showed that the content of oligomers in Biomer was also negligible. Kel-F82, which is a copolymer composed of chlorotrifluoroethylene (97%) and vinylidene fluoride (3%), was obtained from the 3M Company.

Analytical Methods. A piece of Avcothane pump or Biomer film was pressed against a KRS-5 plate which was the internal reflection plate by applying maximum pressure to insure close contact by using a sample holder provided by the Wilkes model #9 attachment. As illustrated in Chart I, the incident angle was 60° and the polymer was pressed against only one side of the KRS-5 since the IR-absorption intensity was too

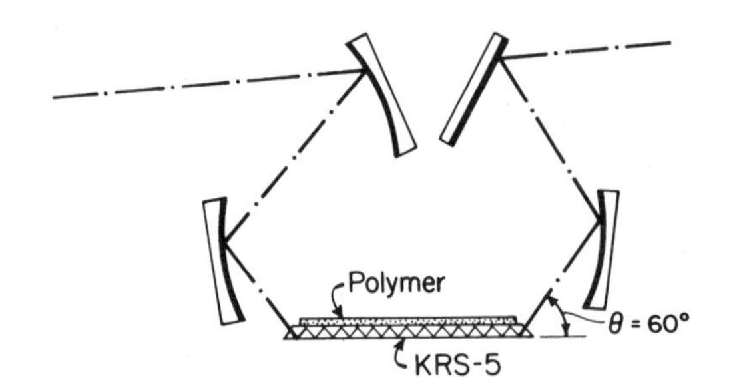

strong. The number of internally reflected beams was about 17. For Avcothane, sample covered the entire area where the IR beam penetrates. For Biomer, because of very high absorption in the ether region (~ 1110 cm^{-1}), a smaller piece of the sample was used. For studies involving barrier films to reduce the IR-beam penetration, the barrier film was cast directly from 0.1% solution in 2,5-dichlorobenzotrifluoride onto other KRS-5 plates. One of them had a dry thickness of 0.65μ and the other of 1.0 μ. The thickness was determined by monitoring the IR peak at 967 cm^{-1} (*20*). The internal reflection unit was placed in the sample compartment of the Fourier transform IR spectrometer (Digilab FTS-14). Scanning was averaged for at least 100 times to remove the noise peaks before the spectra were recorded. In the case where the barrier films of Kel-F82 was used, the spectra of Kel-F82 was subtracted from the internal reflection spectra by digital subtraction. Sometimes, we had noted an incomplete subtraction but the extent of the incompleteness was always below 10%. IR analyses were carried out during the same day with each sample in successive runs with and without barrier film. Therefore, the same area was examined at different depths.

Results

Avcothane Intraaortic Baloon Pump. The IR-reflection spectrum of the blood contacting surface (air facing side) of Avcothane with KRS-5 as the reflection plate is illustrated in Figure 1a while Figure 2a illustrates the spectrum of the substrate facing side. It is noted that the polyether polyurethane portion of Avcothane would contain diol as a chain extender rather than diamine since the characteristic urea carbonyl peaks at 1640 and 1680 cm^{-1} were absent (*16*). Inspection of Figures 1a and 2a shows that the absorbance of 3–3.5 is recorded for the most intense band at 1110 cm^{-1}. Since the IR beam is internally reflected about 17 times within the KRS-5 plate and the polymer is placed only in one side of the plate, the absorbance for each reflection can be calculated by dividing the total absorbance by 17. That amounts to an absorbance of 0.2 per each reflection for the most intense peak at 1110 cm^{-1}. Under this condition, it is reasonable to assume that Beer's law is applicable. Also, similar values of absorbance obtained for two different surfaces indicate that the degree of IR-beam penetration is similar, probably as

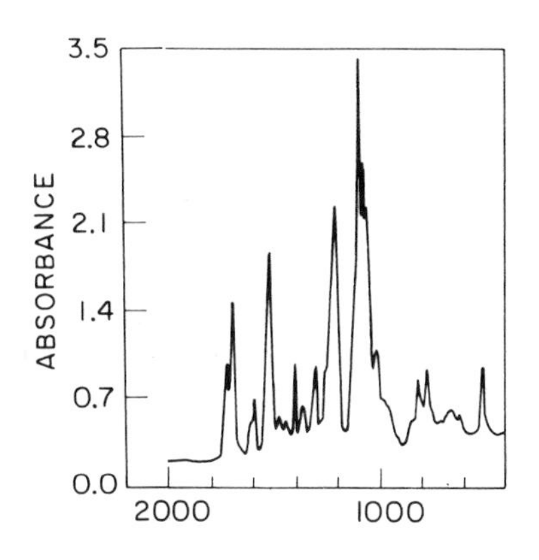

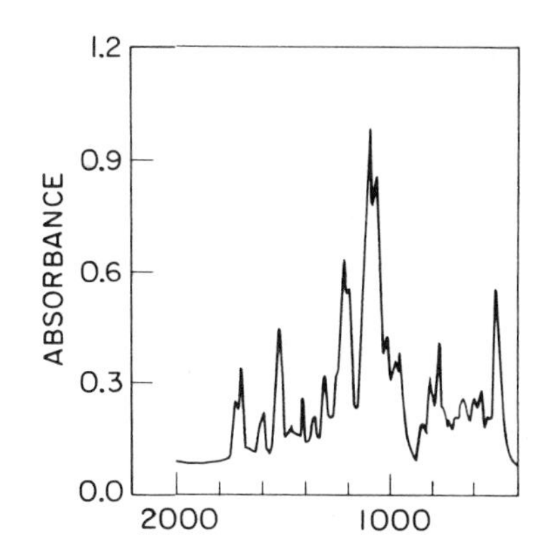

WAVENUMBER (CM^{-1})

Journal of Biomedical Materials Research

Figure 1. Difference reflection IR spectra of air surface of Avcothane intra-aortic baloon pump with KRS-5 reflection plate; (a) without a barrier film, (b) with a barrier film of thickness 0.65 μ (19)

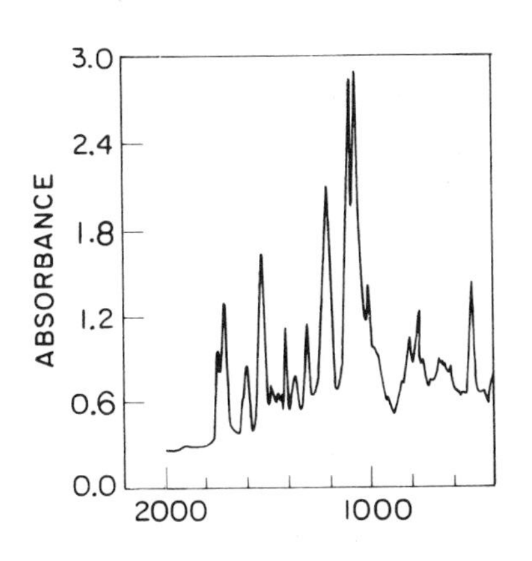

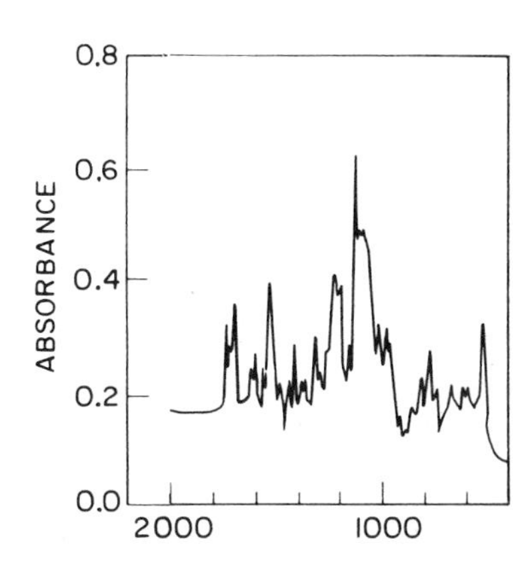

WAVENUMBER (CM^{-1})

Journal of Biomedical Materials Research

Figure 2. Difference reflection IR spectra of substrate side of Avcothane intra-aortic balloon pump with KRS-5 reflection plate; (a) without a barrier film, (b) with a barrier film of thickness 0.65 μ (19)

a result of similar contact efficiency between the KRS-5 plate and the polymer. This similarity in IR-beam penetration is a necessary condition when a quantitative comparison of the chemical composition between two different surfaces is to be made. Even though the spectra of both sides of the surfaces are basically similar, the relative concentration of hard segment, soft segment, or silicone which was calculated by absorption ratios of the characteristic peaks were found to be different. For this analysis, a peak at 1110 cm^{-1} corresponding to the C–O–C stretching (*19*) is used as an index for soft-segment concentration. Three peaks (*19*), i.e., a urethane carbonyl peak at 1710 cm^{-1}, aromatic ring at 1600 cm^{1-}, and a peak at 770 cm^{-1}, are used as representing the hard-segment concentration to verify the results. For silicone, a peak at 1020 cm^{-1} corresponding to the Si–O–Si stretching motion (*19*) and a peak at 800 cm^{-1} representing bending of Si–CH_3 (*19*) is used. The results are summarized in Table I which shows that there is a slight excess of the soft-segment concentration in the air-facing surface while silicone content is greater in the substrate surface than in the air surface. The fact that the silicone concentration was lower in the air surface which exhibited a better blood compatibility was reported earlier by Nyilas and Ward (*17*) by comparing absorption ratios near 800 cm^{-1} for silicone and ratios near 770 cm^{-1} for polyurethane with a KRS-5 plate at 45° incident angle. The depth of the IR-beam penetration in their case would be deeper by 22% than our case because of the smaller incident angle used for their studies if we assume the contact efficiency was analogous.

The depth of IR-beam penetration (*dp*) can be calculated according to the following equation proposed by Harrick (*21*):

$$dp = \frac{\lambda}{2\pi\eta_1(\sin^2\theta - \eta_{21}{}^2)^{\frac{1}{2}}}$$

where η_1 is the refractive index of the reflection plate (2.37 for KRS-5), $\eta_{21} = \eta_2/\eta_1 = 0.63$ by assuming η_2 of 1.50 for Avcothane, and θ is the effective incident angle which is 48.7. Figure 3 illustrates the changes in *dp* as a function of wave number. The depth for hard segment at 1600 and 1710 cm^{-1} is 0.9–1.0 μ, for soft segment (1110 cm^{-1}) it is 1.5 μ, and for silicone (1020 cm^{-1}) it is 1.6 μ.

Whether the depth of the IR-beam penetration predicted by this equation is really approached in an actual internal reflection experiment or not is an important factor to consider in quantitative studies. It would depend on the surface characteristics (e.g., roughness) and deformability of the polymer in addition to the surface characteristics of the reflection plate.

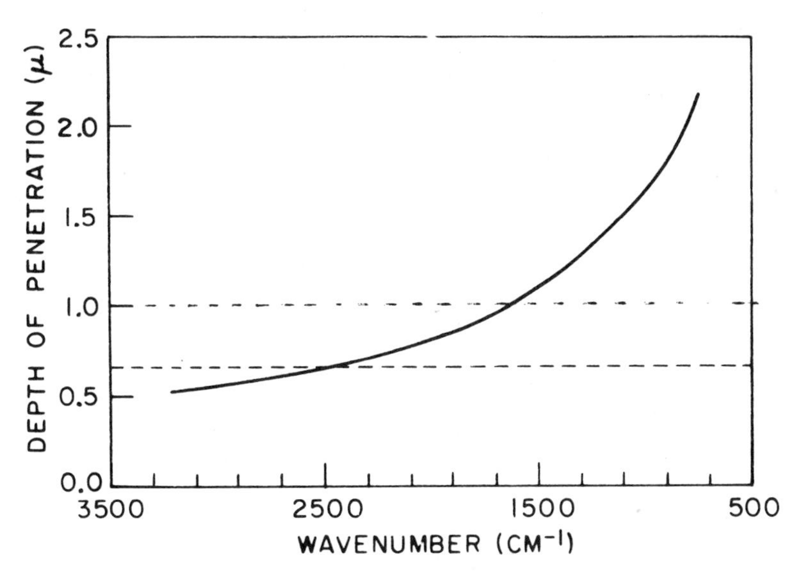

Figure 3. Depth of IR beam penetration as a function of wave number with KRS-5 plate at the incident angle of 60° (dotted lines indicate the thickness of two barrier films, respectively)

However, there is a way to assess this problem. For example, the extinction coefficient of the CH_2 group at 2960 cm^{-1} is known to be 75 L/mol cm. If one knows the concentration of CH_2 groups in the polymer, one can calculate the actual averaged depth of penetration by using the following relationship:

$$dp \text{ (experimental)} = \frac{\text{absorbance per each reflection}}{75 \times \text{(concentration of } CH_2)}$$

In Avcothane and Biomer, the exact concentration of CH_2 is not known. However, in a similar polyurethane made of 2:1:1 molar ratio of 2,4-toluene diisocyanate/ethylenediamine/polyether (mol wt 1000), our calculation shows that the actual depth can approach the predicted value if care is taken to insure close contact. Therefore, we would assume, for the sake of argument, that in Avcothane and Biomer our actual values approach the predicted values.

A question was raised as to whether there was a gradient in the concentration of the soft segment and of the silicone as one approaches the surface. To answer this question, a thin film of copolymer (Kel-F82) was cast directly on the KRS-5 plate and used as a barrier to reduce the depth of IR penetration. Kel-F82 is a copolymer consisting of chlorotrifluorethylene (97%) and vinylidene fluoride (3%) and transmits most

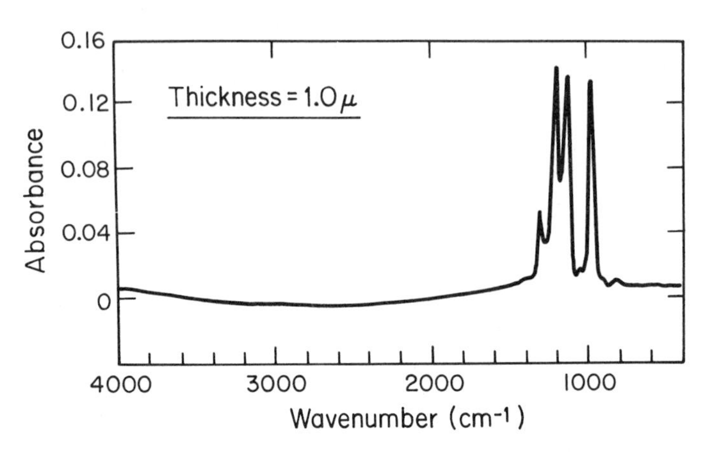

Figure 4. Transmission IR spectrum of a barrier film (Kel-F82)

of the IR beam in the range 4000–1200 cm^{-1}, as shown in Figure 4; thereby it can serve as an effective barrier film. In the range below 1200 cm^{-1}, this copolymer exhibits several absorption peaks. However, it has very little absorption at 1110 cm^{-1}, which is used for polyether, and at 1020 cm^{-1}, which is used for the silicone group. From the absorbance of the peak at 967 cm^{-1} of the transmission IR spectrum, the thickness can be calculated accurately. This technique was first developed by Sibilia who studied the surface chemical composition of bicomponent fibers (*20*).

The thickness of Kel-F82 film was found to be 0.65 μ in our case. The reduction in the depth as a consequence is illustrated in Figure 3. The reflection spectra of Avcothane when the depth of IR penetration was reduced by 0.65 μ were then obtained as the difference spectra by subtracting the contributions of the Kel-F82 and KRS-5 plate. Figure 1b represents the air-facing surface while the substrate-facing side is shown in Figure 2b. The absorption peak of NH groups at 3330 cm^{-1} was absent since the Kel-F82 film is thicker than the depth of penetration in this region as indicated in Figure 3. The rest of the absorption bands are reduced in intensity, reflecting the reduced depth of the IR-beam penetration. Sometimes the digital subtraction was incomplete. However, the extent of incompleteness was less than 10% and this error is only present in IR range below 1300 cm^{-1} where Kel-F82 absorbs substantially.

Again the relative concentration of soft segment/hard segment and silicone were calculated from Figures 1b and 2b, and the results are summarized in Table I. Now the concentration of the soft segment is increased approximately 50% in the air-facing side when compared with the case where the depth of the penetration was deeper. More significant was the fact that now the relative concentration of silicone in the air-facing surface is much greater (50%) than in the substrate side. In fact, the ratio of silicone/hard segment is almost doubled in the air surface as listed in Table I. Also a considerable increase in the ratio of $A_{800\ cm^{-1}}/A_{770\ cm^{-1}}$ is observed when the barrier film is used to reduce the depth of IR-beam penetration. These results clearly demonstrated the existence of a gradient in the chemical composition as a function of the surface depth. We also have examined several pieces of the Avcothane samples taken from various areas in the same pump to see if the lateral heterogeneity exist and found that the lateral chemical composition is homogeneous.

Biomer Films. The IR-reflection spectrum of the air-facing side of Biomer film with KRS-5 as the reflection plate is shown in Figure 5a while Figure 6a illustrates the spectrum of the glass-facing side. IR spectrum of Biomer film indicate that the Biomer is likely to contain diamine as the chain extender in the hard-segment portion since the hydrogen-bonded urea carbonyl peak at 1640 cm^{-1} is strong (*16*). Inspection of Figures 5a and 6a shows that the absorbance for the most intense band at 1110 cm^{-1} is about 1.9 for both surfaces. This indicates that the depth of IR-beam penetrations was similar for both surfaces. This value

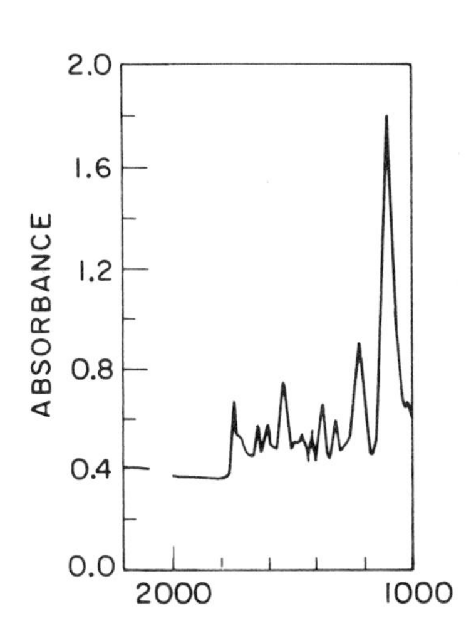

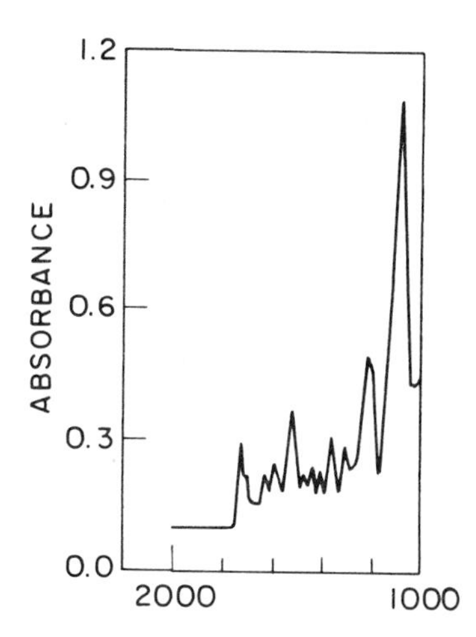

WAVENUMBER (CM^{-1})

Journal of Biomedical Materials Research

Figure 5. Difference reflection IR spectra of air surface of Biomer film with KRS-5 reflection plate; (a) without a barrier film, (b) with a barrier film of thickness 0.65 μ (19)

Table I. Absorption Ratios of Avcothane Surfaces (19)

Absorption Ratio	*Barrier Film Thickness (μ)*	*Blood Contact (Air Side) / Substrate Side*
Hard segment / Soft segment		
A 1110 cm^{-1} / A 1710 cm^{-1}	0	1.01
	0.65	1.55
A 1110 cm^{-1} / A 1600 cm^{-1}	0	1.29
	0.65	1.70
Silicone / Hard segment		
A 1020 cm^{-1} / A 1710 cm^{-1}	0	0.63
	0.65	1.43
A 1020 cm^{-1} / A 1600 cm^{-1}	0	0.81
	0.65	1.56
A 800 cm^{-1} / A 770 cm^{-1}	0	0.97
	0.65	1.58

Journal of Biomedical Materials Research

of absorbance corresponds to six reflections instead of 17 since we used a smaller size of Biomer film because of the high intensity of the polyether peak at 1110 cm^{-1}. The relative concentration of the soft and hard segments have been calculated and summarized in Table II. The result shows that the air-facing side contains approximately the same amount of

Table II. Absorption Ratios of Biomer Film Surfaces (19)

Absorption Ratio	*Barrier Film Thickness (μ)*	*Air Side/ Substrate Side*
Soft segment / Hard segment		
A 1110 cm^{-1} / A 1640 cm^{-1}	0	0.96
	0.65	1.03
	1.0	1.05
A 1110 cm^{-1} / A 1600 cm^{-1}	0	0.96
	0.65	1.01
	0.0	1.14

Journal of Biomedical Materials Research

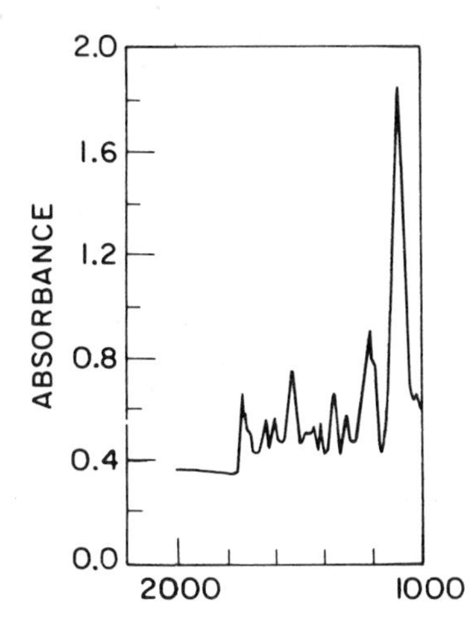

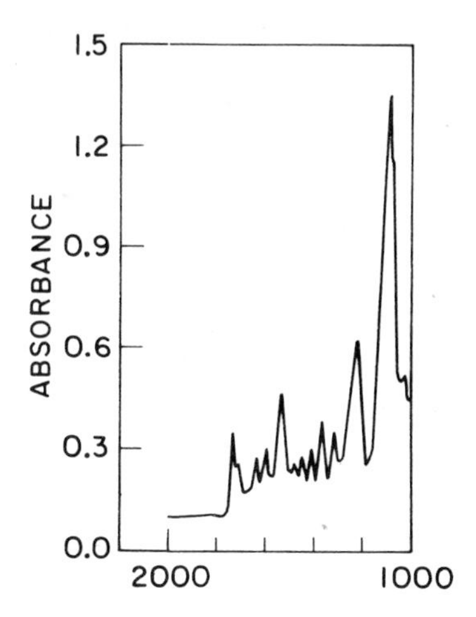

WAVENUMBER (CM^{-1})

Journal of Biomedical Materials Research

Figure 6. Difference reflection IR spectra of substrate side of Biomer film with KRS-5 reflection plate; (a) without a barrier film, (b) with a barrier film thickness 0.65 μ (19)

the soft segment as does the glass-facing side. As illustrated in Figure 8, the depth of the penetration was 1.5 μ for the soft-segment phase and about 1.0 μ for the hard-segment phase. Again the depth of IR-beam penetration was reduced by placing the Kel-F82 film between the KRS-5 plate and the Biomer film. Figures 5b and 6b show the difference spectra after the depth is reduced, reflecting the changes in chemical composition. For studies involving Kel-F82 barrier film, we used a bigger Biomer film which provided 17 reflections. The relative concentration of soft seg-

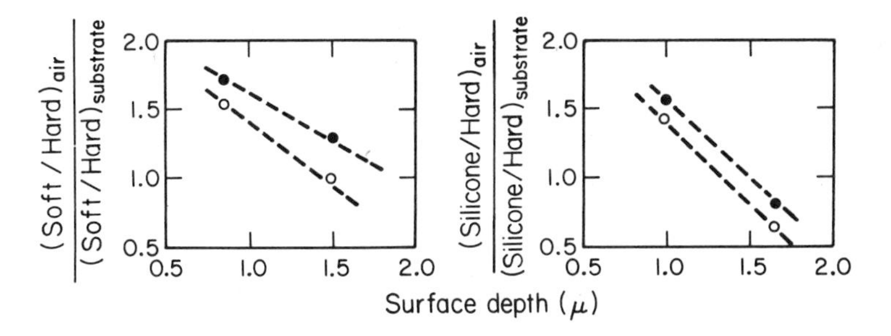

Figure 7. Surface chemical composition—depth profile of Avothane. (○) Based on 1710 cm^{-1}; (●) based on 1600 cm^{-1}. (a) The calibrated soft-segment content in air surface/substrate surface, (b) the calibrated silicone polymer content in air surface/substrate surface.

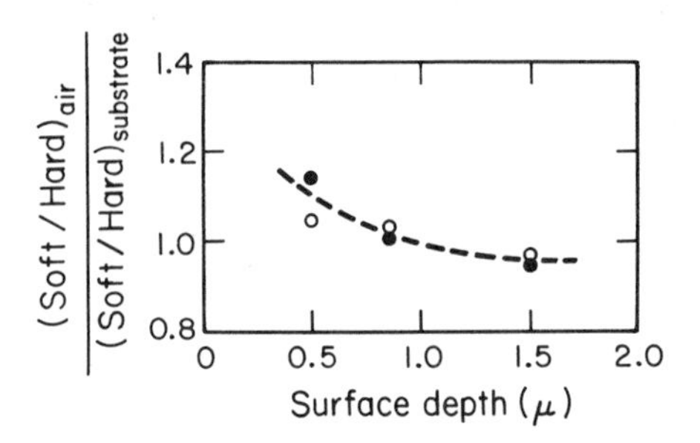

Figure 8. Surface chemical composition–depth profile of Biomer film. The calibrated soft-segment content in air surface/substrate surface. (○) Based on 1640 cm⁻¹; (●) based on 1600 cm⁻¹.

ment/hard segment, as summarized in Table II, has increased only slightly on the air side when a barrier film (0.65 μ in thickness) was used. Therefore, a thicker film (1.0 μ) was used to see if the concentration of soft segment and hard segment shows a more pronounced change on a layer which is very close to the surface. In this case, the reduction in the depth of the IR-beam penetration is illustrated in Figure 3. Specifically, now the depth of penetration for the hard-segment phase is in the order of a few hundred angstroms, and for the soft segment phase the depth is reduced to 0.65 μ. The results as listed in Table II show that even with a thicker Kel-F82 film, the relative concentration of soft segment/hard segment on the air-facing surface had increased only modestly. This finding seems to suggest that in Biomer films the concentration difference of soft- and hard-segment phases as a function of depth from the surface is less pronounced.

Discussion and Conclusion

In this study we have attempted to investigate the effective chemical composition at the surfaces of two commercially available polyurethanes, Biomer and Avcothane, with the objective of identifying chemical species which would interact with plasma proteins.

For Avcothane surfaces, the results indicate the difference in the chemical composition depending on the nature of the surface, i.e., air facing vs. substrate facing during the casting process. The air-facing side contains a greater concentration of the soft-segment phase. When the chemical composition was studied as a function of the depth for the surface, more interesting results were obtained as illustrated in Figure 7a. In the air side of Avcothane, the soft-segment concentration increased about 50% in the first 0.8-μ thick layer when compared with the 1.5-μ thick layer. A more pronounced effect was observed in regard to the silicone content as illustrated in Figure 7b. When averaging for 1.5 μ or deeper, we observed a greater amount of silicone in the substrate side than in the air side. However, when the averaging was done for a layer approximately 0.8 μ in depth, the trend was reversed, i.e., in this case the silicone content was much greater (50%) in air surface than in the substrate surface. This result indicates almost a twofold increase in the relative silicone concentration when the depth was reduced. This might suggest that the aggregation of the silicone blocks close to the surface is more pronounced than that for the hard segment. As a consequence, the hard segment in the air facing side (blood-contact side) should be less abundant immediately at the surface.

In Biomer films, only a modest increase in the relative soft-segment concentration in the air side in comparison with the substrate side was observed when one studies the depth–concentration profile. In other polyurethanes which are similar to Biomer, Lyman and co-workers recently observed that the amount of polyether in the air-facing surface is approximately the same as in the glass-facing side, even though higher concentrations of polyether were found in both surfaces than in bulk (*6*).

Acknowledgment

The authors gratefully acknowledge the financial support of the National Heart and Lung Institute under Grant #HL20079-01. Partial support to one of the authors (C.S.P.S.) from the Health Sciences Fund is also acknowledged as well as the support of the National Science Foundation for the Fourier transform IR spectrometer.

Literature Cited

1. Schonhorn, H., *Encycl. Polym. Sci. Technol.* (1970) **13**, 533.
2. Gesner, B. D., "Polymer Stabilization," W. L. Hawkins, Ed., p. 369, Wiley-Interscience, New York, 1972.
3. Lee, L. H., "Advances in Polymer Friction and Wear," Plenum, New York, 1974.
4. Wunderlich, B., "Reactions on Polymers," J. A. Moore, Ed., p. 295, Reidel, Boston, 1973.
5. Nyilas, E., Leinbach, R. C., Caulfield, J. B., Buckley, N. H., Austen, W. G., *J. Biomed. Mater. Res. Symp.* (1972) **3**, 129.
6. Lyman, D. J., Alb, D., Jr., Jackson, R., Knutson, K., *Trans. Am. Soc. Artif. Intern. Organs.* (1977) **23**, 253.
7. Blakely, J. M., "Introduction to the Properties of Crystal Surfaces," Pergamon, New York, 1973.
8. Jakobsen, R. J., "The Use of Infrared Fourier Transform Spectroscopy and ATR Techniques for the Study of Thin Surface Films," presented at the Conference on Analytical Chemistry and Applied Spectroscopy, Pittsburgh, 1975.

9. Clark, D. T., "Characterization of Metal and Polymer Surfaces," Vol. 2, p. 5, Academic, New York, 1977.
10. Sparrow, G. R., Mishmash, H. E., "Surface Analysis of Polymers and Glass by Combined ISS/SIMS," presented at the Conference on Analytical Chemistry and Applied Fields, Pittsburgh, 1977.
11. Klassen, H. E., Beaman, D. R., "Proceedings of the Eighth International Conference on X-Ray Optics and Microanalysis," held in Boston (1977) **74A.**
12. Williams, D. E., Davis, L. E., "Characterization of Metal and Polymer Surfaces," L. H. Lee, Ed., Vol. 2, p. 53, Academic, New York, 1977.
13. Sung, C. S. P., Hu, C. B., Merrill, E. W., "Surface Chemical Analysis of Segmented Polyurethane by Auger Electron Spectroscopy," *Proceedings of the Annual Technical Conference in Plastics Analysis Section of Society of Plastics Engineers* (1978) **414.**
14. Boretos, J. W., Pierce, W. S., Baier, R. E., Leroy, A. F., Donachy, H. J., *J. Biomed. Mater. Res.* (1975) **9,** 327.
15. Sung, C. S. P., Schneider, N. S., "Polymer Alloys," D. Klempner, K. C. Frisch, Eds., p. 261, Plenum, New York, 1977.
16. Ward, R. W., Jr., Nyilas, E., "Organometallic Polymers," p. 219, Academic, New York, 1978.
17. Nyilas, E., Ward, R. S., Jr., *J. Biomed. Mater. Res. Symp.* (1977) **8,** 69.
18. Nyilas, E., Ward, R. S., Jr., *Proceedings of International Symposium on Polymer Processing, August, 1978, M.I.T.*, in press.
19. Paik Sung, C. S., Hu, C. B., Merrill, E. W., Salzman, E. W., *J. Biomed. Mater. Res.*, in press.
20. Sibilia, J. P., *J. Appl. Polym. Sci.* (1973) **17,** 2911.
21. Harrick, N. J., "Internal Reflection Spectroscopy," Wiley, New York, 1967.

RECEIVED April 14, 1978.

5

Thermoplastic Polyurethane Elastomer Structure—Thermal Response Relations

C. S. SCHOLLENBERGER

B. F. Goodrich Co., Research and Development Center, 9921 Brecksville Rd., Brecksville, OH 44141

Thermoplastic polyurethane elastomers are high performance materials that have found a variety of uses including solution applications as coatings and adhesives, extrusion applications as tubing and jacketing, film applications, and molding applications, particularly flexible injection-molded automotive parts such as front ends, sight shields, fascia, etc. These polymers have a segmented structure which gives rise to heterophase morphology. This morphology is responsible for the fascinating and useful "virtually cross-linked" state of the polymers, and determines other important polymer properties. This chapter describes how thermal analysis, namely differential scanning calorimetry (DSC), can be applied to detect and investigate the contribution of the separate phases in such polymers.

The widespread use of plastic and rubber parts within vehicle passenger and engine compartments has been practiced for some time now by the automotive industry and continues to grow. However, legislation requiring specific front- and rear-end vehicle impact resistance has extended the use of both flexible plastics and rubbers to body-exterior structural members. Flexible plastic and rubber front ends, sight shields, and fascia have appeared on production model automobiles for approximately 10 years and are now well established. An important requirement of the materials used for these applications is flexibility, which enables impacted parts to distort without breaking then to completely recover their original shape undamaged.

0-8412-0457-8/79/33-176-083$05.00/0

Now another factor has appeared on the scene. It is the world-wide energy shortage which recommends the use of construction materials less dense than metal wherever practical to permit the substantial reduction of overall vehicle weight in the interest of improved fuel economy. This development combines with impact requirements to strengthen the case for much broader use of synthetic polymers in automobiles. So, the time would seem to be ripe to capitalize more fully on the use of light-weight, flexible polymeric materials in making exterior automobile-body structural members. Indeed, we already hear of the "friendly fender," the nondenting door panel, etc.

It would seem that such broadened use of flexible plastics and rubbers in exterior-body structural members will compel the conclusion that high performance materials are needed. They must not only be relatively light, flexible, serviceable at temperature extremes, and easily fabricated and finished, but they must be as strong and as tough as possible. The public will demand this. One type of flexible synthetic polymer which meets these requirements very well indeed is the thermoplastic urethane elastomers which have already proved themselves in flexible front ends, sight shields, and fascia in production model automobiles.

Thermoplastic polyurethane elastomers were invented and described almost 20 years ago (*1, 2*). They were first commercialized by the B. F. Goodrich Chemical Co. as "Estane" in 1959. In this chapter I will discuss the useful but rather complex TPU elastomers in terms of their chemical composition, chemical structure, and molecular weight, and how these factors affect their internal physical structure (i.e., morphology), nature, and properties. Specifically, I will show how differential scanning calorimetry (DSC), a method of thermal analysis, can be used to learn much about and to predict processing and performance characteristics of TPU elastomers, such as are being used currently in automotive injection-molding applications. The effects of changes in TPU chemical composition and molecular weight on thermal transitions also will be discussed.

TPU Elastomer Composition

TPU elastomers are usually made from three chemicals: (1) a diisocyanate, (2) a high-molecular-weight macroglycol, and (3) a low-molecular-weight chain extender glycol. Examples of these three compounds and their structures are shown in Figure 1.

Now, it happens that the relative amounts of the above reactant types used to make the TPU elastomer and the order in which they chemically react with each other is very important, for this determines the structure of TPU polymer chains, which must be "segmented."

Component	Typical Example	Structure
DIISOCYANATE	MDI	$OCN-C_6H_4-CH_2-C_6H_4-NCO$
MACROGLYCOL (HIGH MW, ~1000–2000)	PTAd	$HO(CH_2)_4\left[OCO(CH_2)_4CO\,O(CH_2)_4\right]_nOH$
CHAIN EXTENDER GLYCOL (LOW MW)	1,4 BDO	$HO(CH_2)_4OH$

Figure 1. Chemical components of TPU elastomer

TPU Elastomer Structure

What I mean by segmented structure in TPU elastomer chains is apparent in the schematic of Figure 2. The structure in Figure 2 shows the TPU elastomer to consist of linear polymer primary chains. These primary chains are segmented in composition. Specifically they are made up of alternating hard and soft segments which are joined end to end through strong, covalent chemical linkages.

Soft Segments. The soft segments are the linear reaction products of the diisocyanate component, e.g., MDI (diphenylmethane-*p,p'*-diisocyanate), and the macroglycol component, e.g., PTAd [poly(tetramethylene adipate) glycol], as is seen in Figure 3.

Because of the deliberate low melting or liquid nature of the macroglycol component, which has a molecular weight of about 1000–2000 and so comprises approximately 75–95% of the soft segments, the soft segments in the derived TPU chains tend to be low melting and largely amorphous. At any appreciable molecular weight, the soft segment by itself tends to be like a gum. With this explanation we may now equate the exemplary soft-segment chemical structure of Figure 3 with its schematic representation (∿∿) in Figure 2.

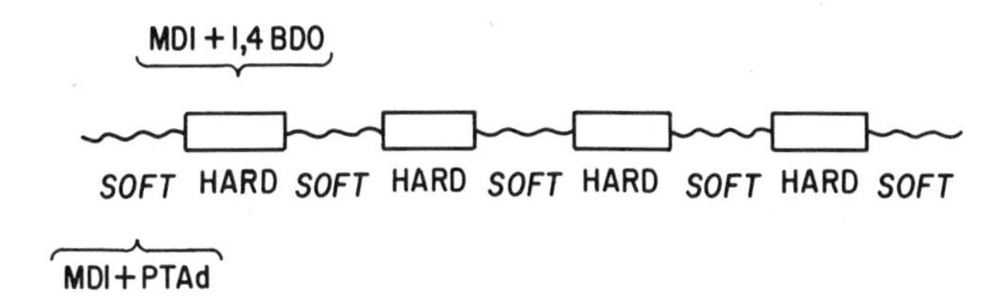

Figure 2. Schematic of TPU-elastomer polymer primary chains

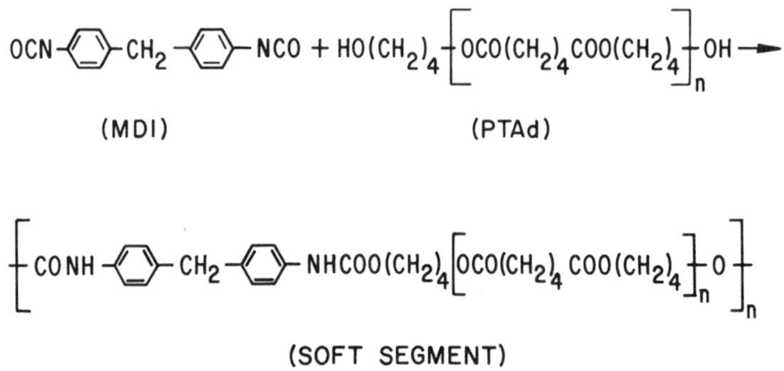

$$OCN-C_6H_4-CH_2-C_6H_4-NCO + HO(CH_2)_4\left[OCO(CH_2)_4COO(CH_2)_4\right]_n OH \rightarrow$$

(MDI) (PTAd)

$$\left[CONH-C_6H_4-CH_2-C_6H_4-NHCOO(CH_2)_4\left[OCO(CH_2)_4COO(CH_2)_4\right]_n O\right]_n$$

(SOFT SEGMENT)

Figure 3. Typical TPU-elastomer soft-segment structure

Several types of macroglycols for use in forming polyurethane soft segments are possible and practical. Examples of some of these are listed in Figure 4. The polymer chemist makes his macroglycol selection based on such considerations as the desired mechanical properties, low-temperature flexibility, environmental resistance, and economics of the derived TPU. Even by his choice of macroglycol molecular weight, the polymer chemist can further regulate such TPU elastomer properties as low-temperature flexibility.

Hard Segments. The hard segments are the linear reaction products of the diisocyanate component and the third monomer type in the TPU elastomer recipe, the small glycol-chain-extender component. In Figure 5 we see a typical TPU hard-segment structure which is formed from MDI and 1,4-BDO (1,4-butanediol).

The polymer chemist selects his diisocyanate and chain-extender-glycol components to produce high-melting hard segments in the derived TPU chains. Hard segments display some but not all of the classical characteristics of the crystalline state and for this reason have been called "paracrystalline." At any appreciable molecular weight the hard segment by itself would tend to be hard and nylonlike in character. With this explanation we may now equate the exemplary hard-segment structure of Figure 5 with its schematic representation (▭) in Figure 2.

Several types of diisocyanates (aromatic, aliphatic, cyclo aliphatic) and many different glycol-chain extenders (open-chain aliphatic, cyclo aliphatic, aromatic aliphatic) can be used to produce TPU-elastomer hard segments. In the more conventional and practical formulations only a single diisocyanate component is used to make a TPU, so the diisocyanate is common to both the hard and soft segments. The polymer chemist makes his diisocyanate and glycol-chain-extender component selections based on such considerations as desired TPU mechanical properties, upper service temperature, environmental resistance, solubility characteristics, and economics.

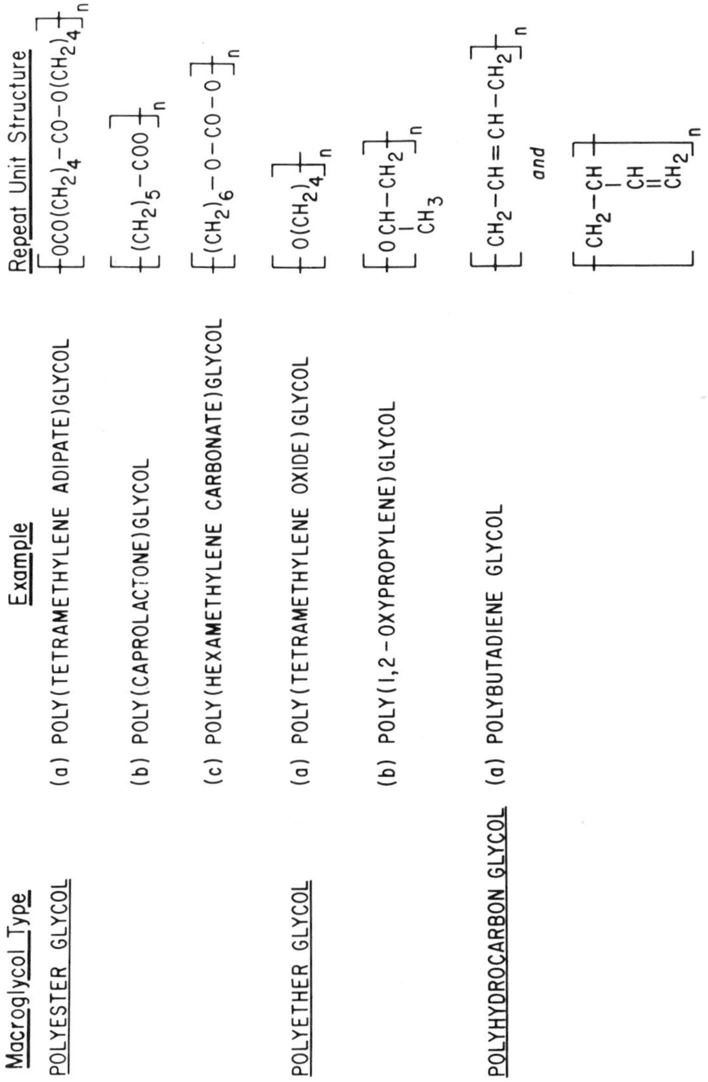

Macroglycol Type	Example	Repeat Unit Structure
POLYESTER GLYCOL	(a) POLY(TETRAMETHYLENE ADIPATE)GLYCOL	$\left[OCO(CH_2)_4-CO-O(CH_2)_4 \right]_n$
	(b) POLY(CAPROLACTONE)GLYCOL	$\left[(CH_2)_5-COO \right]_n$
	(c) POLY(HEXAMETHYLENE CARBONATE)GLYCOL	$\left[(CH_2)_6-O-CO-O \right]_n$
POLYETHER GLYCOL	(a) POLY(TETRAMETHYLENE OXIDE)GLYCOL	$\left[O(CH_2)_4 \right]_n$
	(b) POLY(1,2-OXYPROPYLENE)GLYCOL	$\left[O\underset{CH_3}{\underset{\mid}{C}}H-CH_2 \right]_n$
POLYHYDROCARBON GLYCOL	(a) POLYBUTADIENE GLYCOL	$\left[CH_2-CH=CH-CH_2 \right]_n$ *and* $\left[CH_2-\underset{\underset{CH_2}{\parallel}}{\underset{CH}{\mid}}{CH} \right]_n$

Figure 4. Some macroglycol types

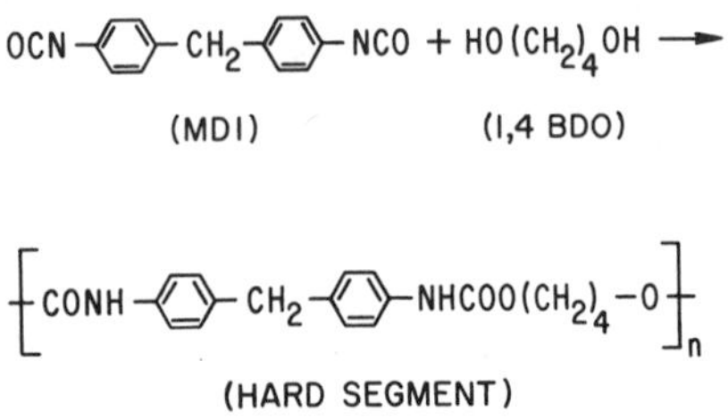

$$OCN-C_6H_4-CH_2-C_6H_4-NCO + HO(CH_2)_4OH \longrightarrow$$

(MDI) (1,4 BDO)

$$\left[CONH-C_6H_4-CH_2-C_6H_4-NHCOO(CH_2)_4-O \right]_n$$

(HARD SEGMENT)

Figure 5. Typical TPU-elastomer hard-segment structure

Just as the polymer chemist can significantly regulate TPU elastomer, low-temperature flexibility by his choice of macroglycol molecular weight and thus soft-segment length (molecular weight), he also can regulate other important TPU-elastomer properties by regulating hard-segment length (molecular weight). Such regulation is easily accomplished by varying the amounts of the diisocyanate and glycol-chain-extender components (relative to the macroglycol amount) charged in the TPU-elastomer formulation. Increasing matched amounts of the diisocyanate and glycol-chain-extender components, relative to the macroglycol amount, increase hard-segment chain length (molecular weight) and TPU-urethane concentration. This in turn increases TPU modulus (hardness, stiffness, load-bearing capacity), processing temperature, upper service temperature, etc., while reducing TPU solubility.

Polymer Chain Organization

Figure 2 depicted the TPU-elastomer primary chain schematically and, I believe, accurately. But in the solid polymer, the polymer primary chains do not really exist separately as shown in Figure 2. Rather, all evidence points to the fact that the hard segments of the TPU chains strongly attract and tend to associate with each other through urethane–urethane hydrogen bonding and aromatic π-electron attractions. As a result, the hard segments in the polymer primary chains form aggregates (domains) in the mobile soft-segment matrix and a two-phase polymer system results. This new organization of the polymer primary chains is depicted schematically in Figure 6. The separate hard-segment aggregates (domains) are seen to tie the linear polymer primary chains together in lateral fashion, in effect crosslinking them. It is also apparent that the same phenomenon extends the chains in linear fashion. The combined lateral and linear effects produce a giant crosslinked network which accounts for the elastic character of TPU elastomers.

However, these crosslinks which hold the TPU network together can be overcome by heat or by solvation by processes which more or less regenerate the polymer primary chains. This is a readily reversible process, as Figure 6 shows, and on cooling or on drying free of solvent, the

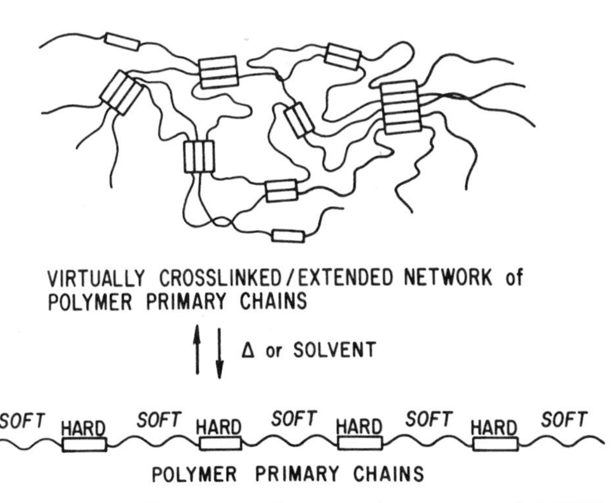

Figure 6. Schematic of chain organization in solid TPU elastomer

polymer primary chains reform the TPU-network structure in their newly fabricated form. The foregoing lability of TPU-elastomer networks leads us to refer to them as being "virtually crosslinked" (2). That is, they are crosslinked in effect but not in fact, and the hard-segment domains are the virtual crosslinks.

Hard-Segment Domain Morphology

As can be seen, hard-segment domains are a necessary structural feature of TPU elastomers since they tie the polymer primary chains together. Without these domains TPU would lack elastic character and would be gumlike in nature.

Much study of the nature of TPU hard-segment domains has been made (*3–11*), particularly since they might be expected to be crystalline but do not appear to be by conventional crystallinity tests such as wide angle x-ray diffraction (WAXD). It appears that the strong mutual attraction of the hard segments (e.g., urethane–urethane hydrogen bonding) restricts their mobility and thus their ability to organize themselves well into a crystalline lattice. Perhaps the situation is similar to that encountered in concentrated solutions of the sugars which crystallize reluctantly, again likely because of the restrictions of hydrogen bonding. The unoriented polyamides would seem to be a similar case.

The possibility also exists that the hard-segment domains are really crystalline, but the crystals are so small that they are not detected by WAXD. In this regard, small angle x-ray diffraction (SAXD) does detect the domains, so we know that they exist, that they have some order, and what their size and separation in the soft-segment matrix are. The subcrystalline state of polyurethane elastomer hard-segment domains has been referred to as the "paracrystalline state" (*4*).

Phase (Hard Segment, Soft Segment) Segregation

A final point on TPU structure. Studies have shown (*11*) that melting a TPU elastomer, as during processing, results in the remixing of its hard and soft segments. This is followed by their tendency to de-mix, or segregate, in the cooled solid polymer. It can be appreciated from Figure 6 that a TPU-elastomer sample whose hard segments have not aggregated fairly well has not developed its ultimate property potential. Often the segregation of hard and soft segments increases with time. In some TPU compositions the intermolecular forces favoring the mixed or the de-mixed states seem to largely balance each other, and de-mixing is appreciably retarded. In any case, phase segregation in TPU elastomers is a time-dependent phenomenon whose consequences cannot be overlooked in the study and use of these polymers. Thermal analysis proves to be an excellent tool for investigating the property-determining morphological state and tendencies of the TPU elastomers. In the balance of this chapter, I will demonstrate the thermal responses commonly seen in TPU by DSC, what they mean, and how they can aid in understanding and improving polymer performance.

Thermal Analysis of TPU

Differential Scanning Calorimetry (DSC). The DSC instrument heats or cools a small (10–15 mg) sample under very carefully controlled and reproducible conditions. While this is going on a very sensitive thermocouple continuously monitors changes in the heat capacity of the sample, which are recorded as a thermogram. Sample heat capacity changes whenever the temperature reaches a point where it causes a change in the organization of the molecules in the sample. Such detectable changes are called thermal responses or thermal transitions. DSC curves are plotted as in Figure 7.

In Figure 7, the verticle axis is the differential heat scale which is proportional to the heat flow (mcal/sec/in.) into or out of the TPU sample in a small aluminum pan (sample system) with respect to an identical but empty pan (reference system), as the total system (reference system, sample system) temperature is raised or lowered uniformly. The horizontal axis is the temperature scale (°C) for the actual temperature at the sample.

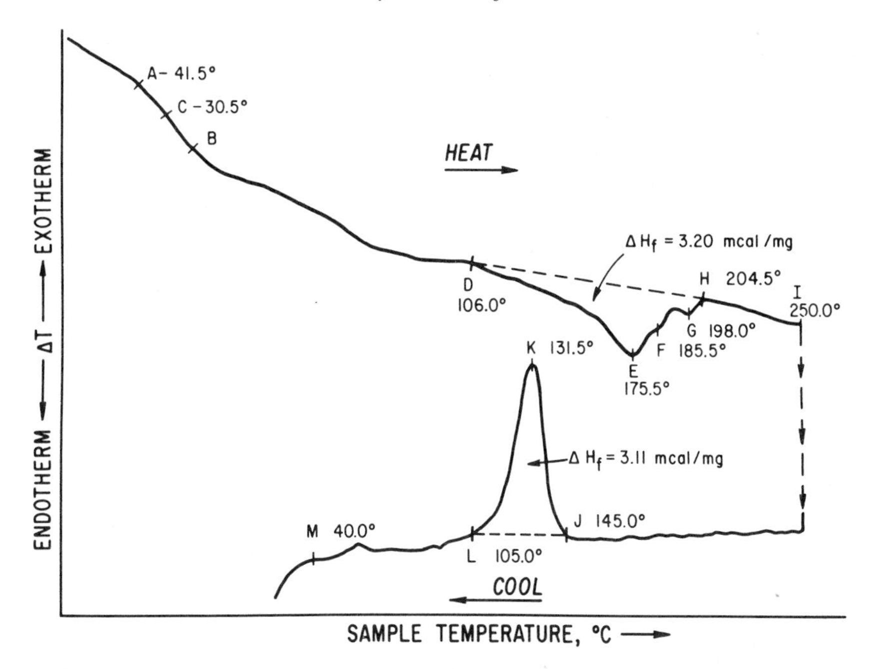

Figure 7. DSC thermogram of an injection-molding thermoplastic polyurethane elastomer

The TPU sample tested in Figure 7 was injection molded from a tough, strong, high-urethane-content injection-molding polymer which contains poly(tetramethylene adipate)–MDI soft segments (Figure 3) and 1,4-butanediol–MDI hard segments (Figure 5). Our test procedure was to: (1) rapidly cool the sample to − 120°C in our DSC instrument (duPont, Model 990); (2) heat it from − 120°C to + 250°C at 10°C/min under nitrogen; (3) immediately cool it to 25°C at the same rate.

GLASS-TRANSITION TEMPERATURE (T_g). As we slowly heat the TPU sample (top curve), an increase in its rate of heat absorption is noted at A where the slope of the upper DSC curve increases. This new rate persists until B, where it reverts to its original rate. The A–B region is the temperature range in which the TPU soft segments go from a rigid "glassy" condition to a flexible "rubbery" condition. This is called the glass-transition region of the TPU soft segments, and the change occurs when the applied thermal energy is adequate to overcome a set of relatively weak, interchain attractive forces which immobilize the TPU chain segments, allowing them to move. By convention, the midpoint C (− 30.5°C) of A–B is called the glass-transition temperature, T_g.

T_g has practical significance in TPU elastomers since it indicates the temperature at which the polymer will lose appreciable flexibility as it cools. In TPU elastomer, soft segments that produce low T_g values are desirable, for they indicate good low-temperature flexibility.

Melting Temperature (T_m) and Heat of Fusion (ΔH_f). As uniform heating of the TPU samples continues beyond B, no further interpretable thermal events are noted until D (106.0°C), where an increased rate of heat absorption again commences. This dip (endotherm) is quite pronounced and contains a series of minima at E (175.5°C), F (185.5°C), and G (198.0°C), ending at H (204.5°C). The D–H endotherm represents the melting/disruption of the total 1,4-butanediol–MDI hard-segment domains in the TPU elastomer. The multiple minimum values, E, F, and G, are called melting temperatures (T_m) and are believed to reflect the sequential melting/disruption of hard-segment domains having different degrees of organization which were developed during prior TPU thermal and processing history.

By measuring the area of the entire melting endotherm, which is bounded by D, E, and H, one can calculate the heat of fusion (ΔH_f) of the sample and express it as millicalories per milligram (mcal/mg) of sample. ΔH_f then is a measure of the "strength" of the polymer virtual crosslinks, or in other words, of the degree of molecular organization in the hard-segment domains. In Figure 7, ΔH_f is 3.20 mcal/mg.

One likes to see high T_m values for TPU hard segments, for this indicates strong virtual crosslinks which will allow the molded TPU parts to pass through paint-drying ovens and to stand in the hot sun without sag or distortion. Of course if T_m is too high, molding problems could be encountered.

Further heating of the sample to I (250°C) produces no further interpretable thermal response. And since another test, thermogravimetric analysis (TGA), which I will not discuss in this chapter, shows that the polymer sample loses a little weight (suggestive of decomposition) at approximately 250°C, we heat the sample no further but now immediately begin to cool it at 10°C/min.

Crystallization Temperature (T_c) and Heat of Crystallization (ΔH_c). Attention is now directed to the lower (cooling) curve in the Figure 7 thermogram. No thermal response is noted in the cooling sample until J (145.0°C), which is reached after about 10 minutes of cooling. Here heat suddenly evolves from the sample in a sharp exotherm, peaking at K (T_c, 131.5°C) and ending at L (105.5°C). This response represents crystallization in the sample and is ascribed to the reformation of the hard-segment domains. However, the accepted methods of detecting crystallinity do not show crystallinity in such TPU, as mentioned earlier.

By measuring the area of the entire "crystallization" exotherm which is bounded by J, K, and L, one can calculate the heat of crystallization (ΔH_c, mcal/mg) of the sample, as was done to get ΔH_f. In Figure 7,

ΔH_c is 3.11 mcal/mg. Ideally, the heat absorbed to melt the sample should equal the heat evolved when the melted sample crystallizes, so that $\Delta H_f = \Delta H_c$. But as Figure 7 shows, ΔH_c is not quite as large as ΔH_f, showing that the degree of crystallinity/order present in the sample before melting was not reestablished completely during its four-minute-crystallization exotherm. However, after standing awhile, hard-segment–soft-segment phase segregation tends to increase (*11*) and often $\Delta H_c = \Delta H_f$.

Notice also that the cooled sample started to crystallize at J (145.0°C), which is underneath the melting endotherm but beyond its center and well below H (204.5°C), the point at which melting was complete on heatup. This crystallization lag is recognized as supercooling. Too much of it means long molding cycles, distorted parts, etc. For injection molding TPU, one likes to have the crystallization exotherm (J-K-L) sharp and of about equal area and centered fairly well under a large, reasonably high-temperature melting exotherm (D-E-H). The polymer of Figure 7 is seen to approximate these requirements fairly well.

Effects of Composition and Molecular Weight

The polymer of Figure 7 is, as noted earlier, a high-urethane-content polymer whose properties, including thermal responses, make it well suited for injection-molding applications. However, Figure 7 is not representative of the thermograms encountered for all TPU, whose thermal differences are attributable primarily to polymer composition differences. This even includes polymers based on the same diisocyanate, macroglycol, and chain-extender components but containing different hard-segment concentrations. This is illustrated in Figure 8.

Figure 8 is the thermogram of a general-purpose TPU more suitable for extrusion, calendering, and solution applications than for injection molding. Like the polymer of Figure 7, it was made by melt polymerizing MDI, PTAd, and 1,4-BDO. But its hard-segment content is only

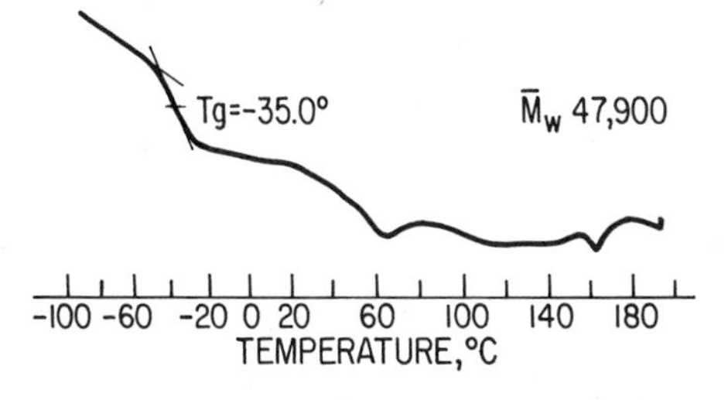

Figure 8. General-purpose thermoplastic, polyurethane elastomer thermogram

about one-half, and its macroglycol molecular weight only about one-third those of the Figure 7 polymer values, making it a less-segmented polymer. The Figure 8 thermogram was run in the same way as that of Figure 7, but the cooldown is not shown since it was essentially featureless.

Inspection of Figure 8 shows: T_g to be $-35°C$ (lower than in Figure 7); a distinct melting endotherm at approximately 67°C (absent in Figure 7); and a broad, shallow endotherm at 80°–180°C (at 106°–205°C, and much more pronounced in Figure 7) which contains a small, sharp endotherm at 165°C (at 178°, 186°, and 198°C in Figure 7).

The lower T_g value of the Figure 8 vs. the Figure 7 polymer probably reflects: a lower polymer molecular weight ($\overline{M}_w$ 48,000) and attendant free-volume effect; its lower hard-segment concentration (reduced hard-segment–soft-segment phase mixing); and its unusually long (approximately 500 days) equilibration period during 25°C shelf storage (increased hard-segment–soft-segment phase segregation).

The 67°C endotherm in Figure 8 is suspected of representing the melting of soft segment (macroglycol lengthened to high molecular weight via diisocyanate coupling) at a temperature appreciably above that of the relatively low-molecular-weight parent macroglycol itself ($\sim$ 46°C).

The broad, shallow, 80°–180°C endotherm in Figure 8 and the small sharp endotherm within it at 165°C are believed to correspond to the 106°–205°C D–E–H endotherm of Figure 7 and also to represent the thermal disruption of the (less extensive) hard segments of the Figure 8 polymer.

In addition to polymer composition, polymer molecular weight also can exert a significant effect on the thermal response of polymers including those having identical chemical makeup and composition. This can be seen in Figures 9A and 9B. The 10 polymers in these figures were all of the same composition, materials, and preparation method as the Figure 8 polymer, which reappears at the head of Figure 9A. But the molecular-weight levels of this series were varied in the range, $\overline{M}_w$ 48,000–367,000. The thermograms were all measured as in the case of the Figure 7 polymer.

Comparison of Figure 9A and 9B thermograms shows several interesting changes with increasing $\overline{M}_w$ in this polymer series. It can be seen that T_g values increase with $\overline{M}_w$ from $-35°C$ at $\overline{M}_w$ of 48,000 to $-27°C$ at $\overline{M}_w$ of 183,000, where they hold through the final member of the series, $\overline{M}_w$ of 367,000. We interpret this pattern to reflect the influence of polymer free-volume (chain end) reduction on polymer-chain segment mobility with increasing polymer $\overline{M}_w$. It appears that polymer free-volume decreases with increasing polymer molecular weight up to $\overline{M}_w$ of approxi-

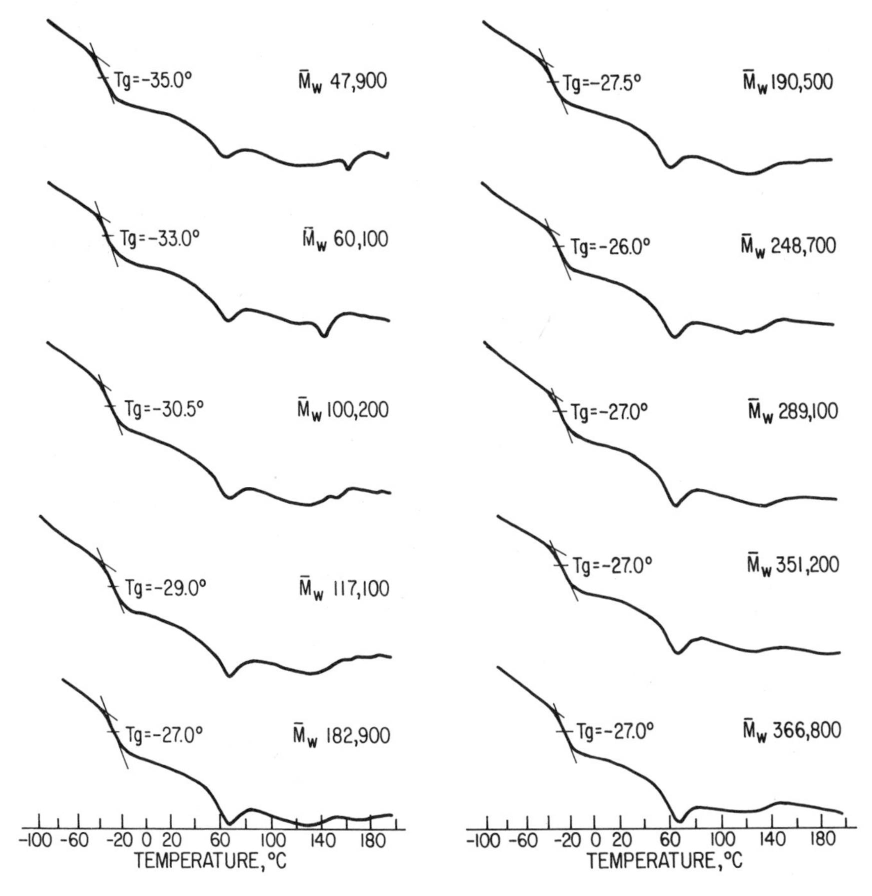

Figure 9. (a, left; *b,* right*) Effects of* $\overline{M}_w$ *on general purpose TPU elastomer thermograms*

mately 160,000, thereafter ceasing to change significantly or to further influence polymer T_g with additional $\overline{M}_w$ increase.

Notice that the 67°C endotherm shows no $\overline{M}_w$ dependency, possibly because the smallest TPU of the series ($\overline{M}_w$ of 48,000) already consists of soft segments (diisocyanate-coupled polyester chains) long enough to display the maximum temperature for this transition in this system.

The broad, shallow, 80°–180°C endotherm of the 48,000-$\overline{M}_w$ polymer (Figures 8, 9A) narrows, drawing toward the lower temperature, and also thins as polymer molecular weight increases, while the distinct 165°C endotherm at $\overline{M}_w$ of 48,000 gradually shrinks, essentially disappearing at $\overline{M}_w$ of 117,000. Hard-segment domain development would appear to be progressively impeded in the TPU as $\overline{M}_w$ increases. Again, we suggest that this is attributable to a decrease of polymer free volume and thus chain-segment mobility. Consequently, less pronounced evidence of hard-segment thermal response is apparent in the thermograms of the higher $\overline{M}_w$ polymers of this TPU series.

Other information can be obtained from thermal studies of TPU elastomers which is also helpful in understanding and improving these high performance, easily processed polymers. But it was the intention to limit this chapter to the basic DSC thermal responses which forecast strengths and weaknesses in TPU processing and performance characteristics and to indicate the parts of the TPU structure that are responsible for these thermal responses. Hopefully, this has been accomplished.

Literature Cited

1. Schollenberger, C. S., assignor to the B. F. Goodrich Company, U.S. Patent **2,871,218**, "Simulated Vulcanizates of Polyurethane Elastomers," (Appl. Dec. 1, 1955, Issued Jan. 27, 1959).
2. Schollenberger, C. S., Scott, H., Moore, G. R., "Polyurethane VC, a Virtually Crosslinked Elastomer," *Rubber World* (1958) **137**(4), 549.
3. Cooper, S. L., Tobolsky, A. V., *J. Appl. Polym. Sci.* (1966) **10**, 1837.
4. Bonart, R., *J. Macromol. Sci., Phys.* (1968) **B2**(1), 115.
5. Clough, S. B., Schneider, N. S., *J .Macromol. Sci., Phys.* (1968) **B2**(4), 553.
6. Miller, G. W., Saunders, J. H., *J. Appl. Polym. Sci.* (1969) **13**, 1277.
7. Harrell, L. L., Jr., *Macromolecules* (1969) **2**(6), 607.
8. Rausch, K. W., Farrissey, W. J., Jr., *J. Elastoplastics* (1970) **2**, 114.
9. Morbitzer, L., Hespe, H., *J. Appl. Polym. Sci.* (1972) **16**, 2697.
10. Wilkes, C. E., Yusek, C. S., *J. Macromol. Sci., Phys.* (1973) **B7**(1), 157.
11. Wilkes, G. L., Bagrodia, S., Humphries, W., Wildnauer, R., *J. Polym. Sci., Polym. Lett. Ed.* (1975) **13**, 321.

RECEIVED April 14, 1978.

6

A Dynamic Mechanical Study of Phase Segregation in Toluene Diisocyanate, Block Polyurethanes

G. A. SENICH and W. J. MAC KNIGHT

Materials Research Laboratory, Polymer Science and Engineering Department, University of Massachusetts, Amherst, MA 01003

The dynamic mechanical response of several toluene diisocyanate, polyurethane elastomers has been studied. A symmetrical hard-segment isomer and increasingly long hard and soft blocks promote phase segregation in these materials. The glass-transition temperature of the soft-segment phase is strongly influenced by the purity of the soft-segment phase and the overall degree of crystallinity. The high temperature δ relaxation is directly related to the average hard-segment length, while the plateau modulus between these relaxation regions strongly depends upon hard-segment content. Comparison of block polyurethane relaxations with corresponding loss processes of the hard- and soft-segment polymers indicates that the short block length and significant degree of intersegmental mixing contribute to poor phase organization in these block polyurethanes.

Linear polyurethane block copolymers have been objects of considerable research interest over recent years. These materials are composed of three parts, a diisocyanate, a low molecular weight chain extender (usually a diol but sometimes a diamine), and a low molecular weight rubbery polymer, frequently a polyether or polyester. The soft segment consists of the rubbery polymer while the diisocyanate and the chain extender form the hard segment. These materials are classified as thermo-

0-8412-0457-8/79/33-176-097$07.75/0

plastic elastomers since their resiliency at room temperature is characteristic of elastomers while their linear nature allows them to be processed with the ease of thermoplastics at higher temperatures.

Dynamic mechanical property studies have been successful in elucidating the molecular mechanisms underlying the mechanical behavior of these materials. Kajiyama and MacKnight (*1, 2*) proposed mechanisms for low temperature and glass-transition-region relaxations of three model hard-segment polyurethanes extended with diols containing from 2 to 10 methylene units. They observed relaxations attributable to motion of side chains, short-length main chain structures, and long main chain segments as well as crystal–crystal transitions. Dynamic mechanical investigations by Huh and Cooper (*3*) demonstrated the existence of a two-phase structure in 4,4′-diphenylmethane diisocyanate, MDI, polyurethanes, and examined the effects of composition, soft-segment molecular weight, and thermal history upon the relaxation mechanisms of this class of copolymers. Relaxations, in their notation, corresponding to a soft-segment T_g, α_a, soft-segment crystalline melting, α_c, hard-segment T_g, δ, and hard-segment melting, δ', were observed in the polyurethanes studied. They concluded that the α_a and δ relaxations were influenced by the degree of crystallinity and the nature of the domain structure. The magnitudes of the α_c, δ, and δ' relaxations were found to be related to the size of the domain structures present. Similar studies were carried out on polyurethanes synthesized with structures that eliminated hydrogen-bonding effects and which allowed blocks of controlled, molecular weight distribution to be incorporated (*4, 5*). The major conclusions from this work are that a narrow, hard-segment molecular weight distribution in the phase-segregated systems increases the hard-segment glass transition, α_h, and the modulus between the hard- and soft-segment T_g while having little effect on the soft-segment glass transition, α_s. An examination of low temperature, dynamic mechanical properties of MDI polyurethanes with various soft segments was reported by Illinger and co-workers (*6*). They found that as urethane concentration increased, α_s remained constant but crystallization of the soft segment was inhibited. Ferguson and co-workers (*7*) have examined MDI–polyether block polyurethane chains extended with 1,3-diaminopropane as a function of hard-segment content.

Systematic studies on MDI polyurethanes with polycaprolactone soft segments over a wide range of soft-segment molecular weight and hard-segment concentration have been carried out by Seefried, Koleske, and Critchfield (*8, 9*). The glass transition was observed to shift to lower temperatures as the soft-segment chain length increased, and soft-segment crystallization was found to occur for the highest molecular weight polyester sample. The T_g of lower molecular weight, soft-segment samples increased as the hard-segment content increased while the T_g of polyurethanes containing higher molecular weight soft segments was unchanged until extreme amounts of hard segments were introduced. The authors attributed the latter behavior to a high degree of phase segregation between blocks. This group also has compared MDI and toluene diisocyanate, TDI, hard segments over a more limited compositional range and studied the effects of different chain extenders on TDI polyurethane properties (*10, 11*). All TDI-containing polyurethanes studied by these workers displayed little or no phase segregation.

Schneider, Paik Sung, and co-workers have completed an extensive study of TDI polyurethanes (*12, 13, 14, 15*). This work differs from previous studies by the separate characterization of block polyurethanes prepared from the two isomers of TDI. These materials have been studied by thermomechanical analysis, TMA, DSC, x-ray scattering, and IR analysis in order to elucidate the structure of this particular class of polyurethanes. Their findings can be briefly summarized in terms of the degree of phase segregation exhibited. The glass-transition temperature of 2,4-TDI polyurethanes with polyether soft segments of about 1000 $\overline{M}_n$ showed a strong dependence on composition. Extensive hard- and soft-segment mixing was postulated in these materials. Similar polyurethanes based on the 2,6-TDI structural isomer displayed a highly ordered domain structure, indicated by a concentration invariant T_g and a strong high temperature transition attributed to melting of the easily crystallizable hard segment. An increase in soft-segment molecular weight from 1000 to 2000 induced phase segregation in the 2,4-TDI series and improved phase segregation in the 2,6-TDI series to the point that the soft segment exhibited some crystallinity. The soft-segment glass transition temperature, T_{gs}, was found to be a sensitive measure of the degree of phase segregation in these materials.

The systematic dynamic mechanical study reported here was undertaken to gain additional insight into the molecular relaxation processes which occur in TDI polyurethanes as well as to test the generality of findings developed from investigations of other chemical classes of segmented polyurethanes.

Experimental

Samples. Linear polurethane block copolymers with hard segments composed of 2,4- or 2,6-TDI, and 1,4-butanediol, BD, and either 1040 or 2060 molecular weight poly(tetramethylene oxide), PTMO, soft segments were supplied through the courtesy of N. S. Schneider of the Army Materials and Mechanics Research Center, Watertown, MA. These polymers were prepared by a two-step process, first an endcapping of the PTMO with a 5% molar excess of TDI, followed by chain extension

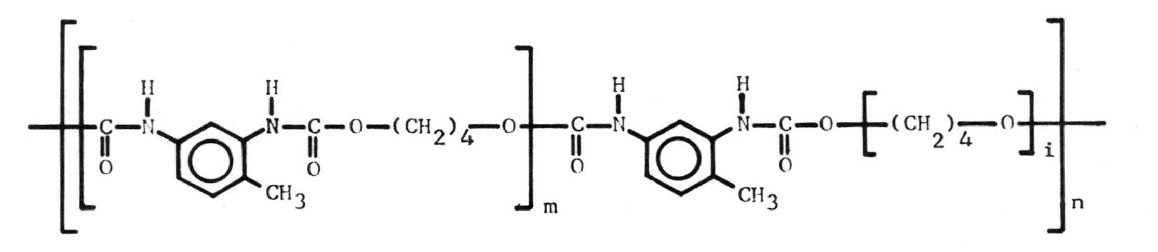

Figure 1. Structure of the block polyurethane elastomers

with BD and additional TDI. All were found to be soluble in dimethylformamide. Hard- and soft-segment model polymers also were supplied. These materials were made in a one-step condensation reaction of a 2.5% molar excess of TDI with either BD or PTMO of 1090 molecular weight. Figure 1 shows a representation of the polymers studied. The hard segment is the material contained within the brackets with subscript m. The approximate value of the index i corresponds to 14, 15, and 28 for 1040, 1090, and 2060 molecular weight PTMO, respectively. For the pure, hard-segment material the index i is, of course, zero, while for the soft segment polymer m is likewise zero. Overall polymer M_n was estimated to be on the order of 13,000–25,000. Hard- and soft-segment polymers are designated first by the TDI isomeric structure followed by either BD for the hard segment or PTMO for the soft-segment polymers. For example, 2,4-TDI/PTMO refers to the 1090 molecular weight PTMO soft segment, which contains 14 wt% 2,4-TDI. The nomenclature used throughout the discussion for the block polyurethanes first refers to the TDI isomeric structure followed by the molecular weight of PTMO in thousands and finally the hard-segment content in weight percent if a specific polymer is designated. For example, 2,6-T-2P-60 indicates a block polyurethane with 60 wt% 2,6-TDI/BD hard segments and a 2060 molecular weight PTMO soft segment.

Characterization. Film samples about 0.25-mm thick were prepared by compression molding the polyurethanes under dry nitrogen at 180°C and 3 MPa pressure. After molding for 20 min, the samples were allowed to cool slowly in the press under dry nitrogen flow. Figure 2 shows the thermal history of the 180°C-molded films. No discoloration was observed when the samples were removed from the mold. The two soft-segment polyurethanes were compression molded at 120°C. The films were allowed to stand at least one week at room temperature in a dessicator before being evaluated with the Rheovibron DDV-IIB Dynamic Visco-

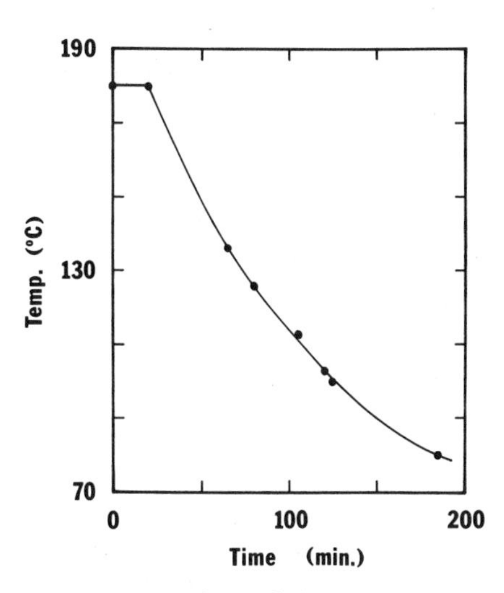

Figure 2. Thermal history of compression-molded polyurethane films

elastomer (Toyo Baldwin Co.). Samples were tested sequentially at 3.5, 11, 35, and 110 Hz while heating from − 135° to 195°C at a nominal heating rate of 1.5°C/min under a dry nitrogen atmosphere. Several samples were re-examined by the same procedure about 60 days after the first test. Apparent activation energies were determined for major relaxations from the Arrhenius dependence of frequency on inverse temperature. Least-squares lines were fitted to plots of log (frequency) vs. T^{-1} and the activation energies to the 95% confidence level determined from the slopes. Thermal studies on the compression-molded films were carried out on a Perkin–Elmer DSC-2 differential scanning calorimeter at a heating rate of 20°C/min and with a sensitivity of 5 mcal/sec.

Results

Hard- and Soft-Segment Polymers. The hard-segment polymer composed of 2,4-TDI/BD exhibited a single major α relaxation associated with the glass transition. The relaxation occurred at 108°C (11 Hz), as determined from the position of the loss modulus maximum. The storage modulus decreased from about 1.5 GPa to 50 MPa during the relaxation, and the loss of mechanical integrity and onset of flow was indicated by a rapid and unabated rise in tan(δ) as the temperature rose above T_g. The apparent activation energy, E_a, determined from an Arrhenius dependence of frequency, ω, on temperature as shown in Equation 1, was

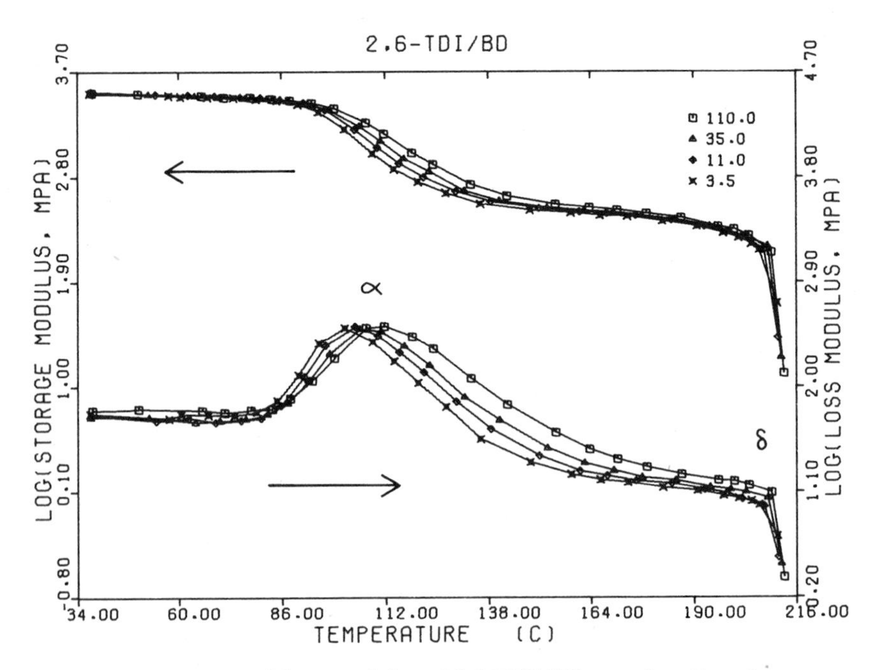

Figure 3a. Storage and loss modulus of 2,6-TDI/BD as a function of temperature at 110, 35, 11, and 3.5 Hz

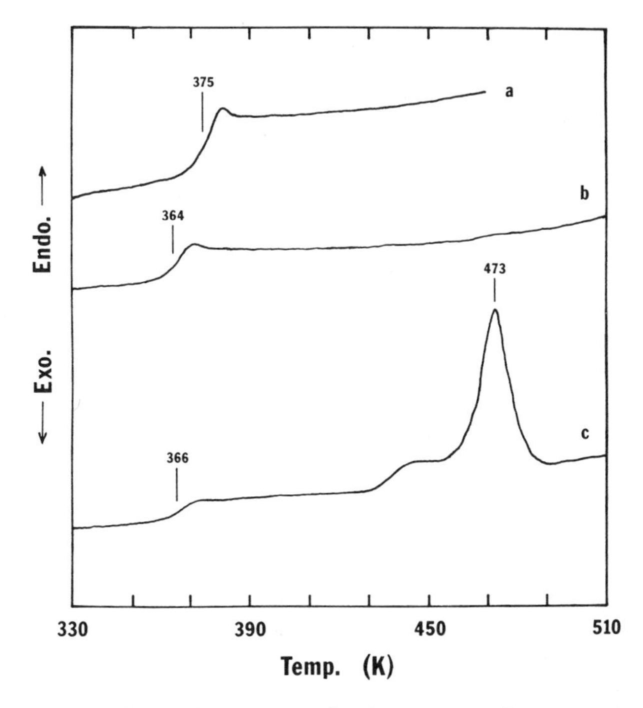

Figure 3b. DSC scans of hard-segment polymers. (a) 2,4-TDI/BD; (b) quenched 2,6-TDI/BD; (c) 2,6-TDI/BD-annealed for 1 hr at 423 K.

$$\ln(\omega) = \frac{-E_a}{RT} + C \qquad (1)$$

600 kJ/mol. The activation energy for the α relaxation agrees within experimental error with the value of 480 kJ/mol, determined by Kajiyama (*16*). No crystallinity is present in this hard-segment material, as shown by the DSC scan in Figure 3b which has no evidence of a melting endotherm.

The dynamic mechanical behavior of 2,6-TDI/BD was found to be similar to that of the 2,4-TDI/BD hard segment at low temperatures. Both showed a secondary γ relaxation at about $-115°$ to $-135°$C, with an activation energy of about 30 kJ/mol. These processes correspond to the γ_1 and/or γ_2 relaxations, involving motions of methylene sequences which were observed for model hard segments by Kajiyama and Mac-Knight (*2*) and will not be subsequently considered. The α or glass-transition relaxation for 2,6-TDI/BD occurred at 105°C (11 Hz), with an activation energy of 510 kJ/mol. Above the α-loss peak, the storage modulus remained at a level of about 300 MPa, as can be seen in Figure 3a. This phenomenon can be attributed to the presence of crystallinity in the 2,6-TDI/BD hard segment. The magnitude of the plateau in modulus was found to highly depend upon the thermal history of the sample. The results in Figure 3a represent the steady-state behavior of the fully annealed polymer. The width of the α-loss peak is about twice that of the 2,4-TDI/BD α-relaxation maximum. Broadening in the α-loss maximum by the introduction of crystallinity has been similarly observed for PET and PTFE (*17*). Crystalline structure in the 2,6-TDI/BD hard-segment polymer remained until the onset of the δ relaxation in dynamic mechanical experiments, about 212°C. The melting point, as determined from the maximum of the melting endotherm in Figure 3b, was found to be 200°C. The positions of the α relaxations for 2,4- and 2,6-TDI/BD are in good agreement with T_g values found from DSC studies, 102° and 93°C, respectively.

The model soft segment composed of 2,4-TDI/PTMO exhibited a single major α relaxation at − 55°C (11 Hz), accompanied by a drop in the level of the storage modulus of two and one half orders of magnitude. An activation energy of 200 kJ/mol was determined for this relaxation. The 2,6-TDI/PTMO soft segment displayed quite different behavior from the 2,4-TDI/PTMO polymer, as can be seen from Figure 4a. An α relaxation can be observed to occur at − 59°C (11 Hz), with an apparent activation energy of 260 kJ/mol. A subsequent α_c-loss process with a corresponding increase in the storage modulus by a factor of three can be seen to occur at about − 30°C. This second relaxation can be attributed to crystallization of the soft segment above T_g, followed by annealing and finally complete melting at 9° to 13°C. Crystallization in the material during cooling to − 135°C was inhibited by the rapid quenching process used for cooling the sample chamber. The DSC results for the 2,6-TDI/PTMO soft segment in Figure 4b confirm the interpretation of the dynamic, mechanical relaxation processes. A glass transition is indicated at − 63°C, followed by a broad crystallization exotherm centered at − 30°C and finally a melting endotherm at 10°C. The behavior of the 2,6-TDI/PTMO soft segment, containing no chain extender, is similar to that noted by Illinger (*6*) for an MDI/BD/PTMO polyurethane containing 24-wt% hard segments and 2000 $\overline{M}_n$ PTMO. The glass transition occurred at − 100°C (110 Hz), followed by crystallization, completed at − 45°C and finally melting at 30°C. The dynamic mechanical test results for the hard and soft segments are summarized in Table I.

2,4-TDI Polyurethanes. Two 2,4-T-1P samples with different hard-segment concentrations were studied and found to display a broad α-relaxation maximum, also characterized by a decline in storage modulus of about two and one half orders of magnitude, as shown in Figure 5, a plot of storage and loss modulus vs. temperature at 11 Hz. The position

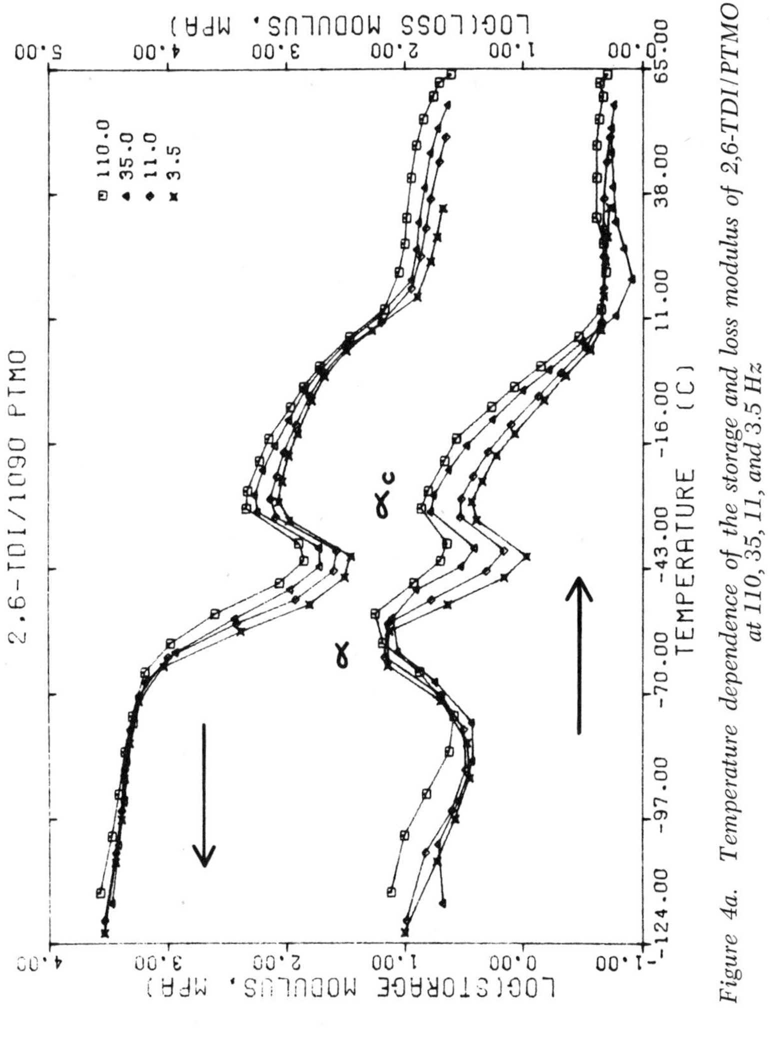

Figure 4a. Temperature dependence of the storage and loss modulus of 2,6-TDI/PTMO at 110, 35, 11, and 3.5 Hz

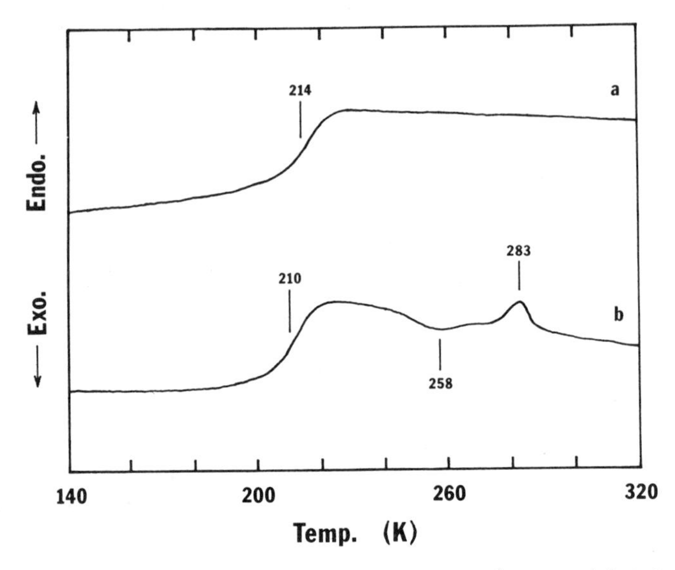

Figure 4b. DSC scans of soft-segment polymers. (a) 2,4-TDI/PTMO, (b) 2,6-TDI/PTMO.

of the α relaxation varied with hard-segment content, ranging from 22°C for 2,4-T-1P-56 to 38°C for 2,4-T-1P-60, both at 11 Hz. The activation energies found for each sample are comparable, as can be seen from Table II, a summary of 2,4-TDI polyurethane data, on the order of 290 kJ/mol. The results from dynamic mechanical studies agree with the glass-transition temperatures found by Schneider and co-workers (*12*) from a TMA study of the same polyurethanes, as can be seen from Figure 6, a comparison of both sets of data.

The dynamic mechanical response of three 2,4-T-2P samples at 11 Hz is shown in Figure 7 for three hard-segment concentrations. A low temperature relaxation maximum, α_s, in the region of − 68° to − 54°C,

Table I. Relaxations of Hard- and Soft-Segment Polymers

Sample	*Relaxation*	*Temperature*[a] *(°C)*	*Activation Energy (kJ/mol)*
2,4-TDI/BD	α	108	600 ± 140
2,6-TDI/BD	α	105	510 ± 70
2,6-TDI/BD	δ	212	—
2,4-TDI/PTMO	α	−53	200 ± 40
2,6-TDI/PTMO	α	−59	260 ± 50
2,6-TDI/PTMO	α_c	−30	—

[a] α and α_c determined from position of E''_{max}, δ from $\log(E') = 1.25$ MPa, all at 11 Hz.

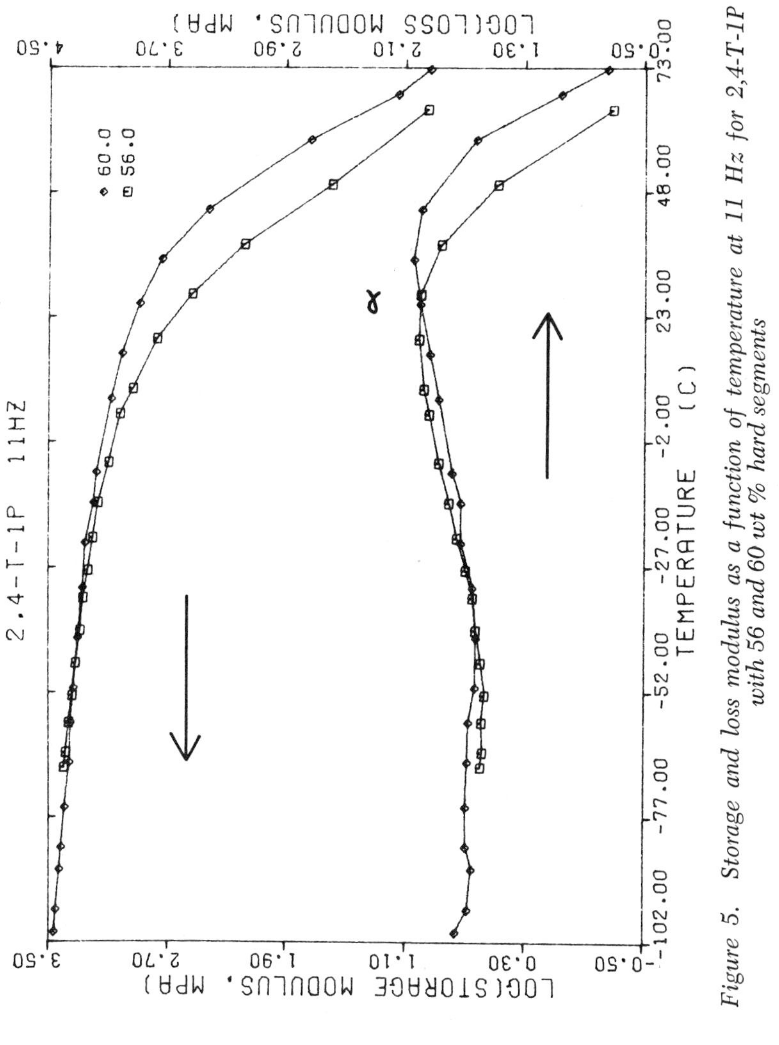

Figure 5. Storage and loss modulus as a function of temperature at 11 Hz for 2,4-T-1P with 56 and 60 wt % hard segments

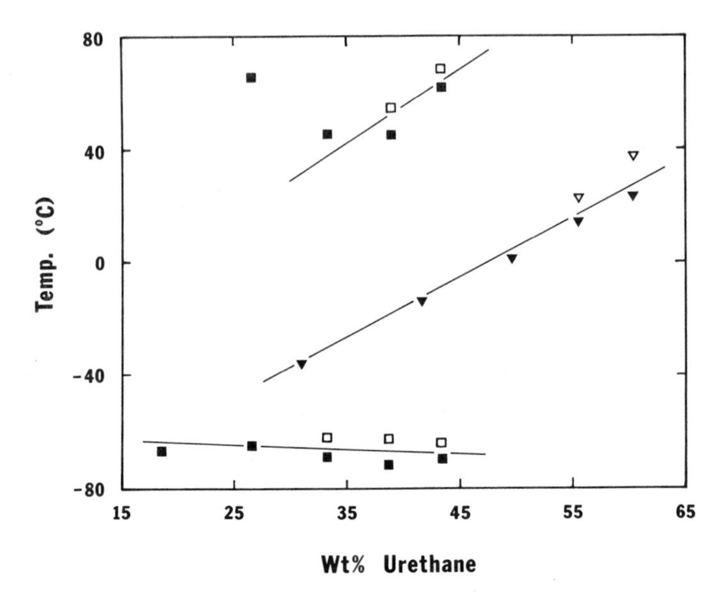

Figure 6. Comparison of 11-Hz dynamic mechanical (open points) and DSC or TMA transition temperatures (filled points, Ref. 14) of 2,4-T-1P (triangles) and 2,4-T-2P (squares) polyurethanes as a function of hard segment wt %

with an activation energy of 120 kJ/mol, is apparent for each sample. The position of the α_s-relaxation maximum, given in Table II, can be seen to decrease slightly with increasing concentration of hard segments. Again, the dynamic mechanical data compares favorably with results from DSC studies on these materials by Schneider and Paik Sung (*14*), as can be seen in Figure 6. The same small decrease in transition temperature with increasing hard-block content is evident from both determinations. The storage modulus above the α_s relaxation can be seen to highly depend upon the hard-segment content. At 0°C for

Table II. Relaxations of 2,4-TDI Block Polyurethanes

Sample	*Approximate Molar Ratio (TDI/BD/PTMO)*	*Relaxation*	*Temperature[a] (°C)*	*Activation Energy (kJ/mol)*
2,4-T-1P-56	5/4/1	α	22	320 ± 50
2,4-T-1P-60	6/5/1	α	38	280 ± 20
2,4-T-2P-33	4/3/1	α_s	−62	120 ± 20
2,4-T-2P-39	5/4/1	α_s	−63	120 ± 20
2,4-T-2P-39	5/4/1	α_h	55	260 ± 30
2,4-T-2P-43	6/5/1	α_s	−64	120 ± 20
2,4-T-2P-43	6/5/1	α_h	68	270 ± 40

[a] Determined from position of E''_{max} at 11 Hz.

example, the storage modulus of 2,4-T-2P-43 is twice that of 2,4-T-2P-39, which in turn has a modulus five times that of 2,4-T-2P-33.

On further examination of Figure 7, an α_h relaxation can be noted that becomes increasingly prominent as hard-segment content increases from 33 to 43 wt%. The α_h-loss process for 2,4-T-2P-43 is comparable in breadth to the α_s relaxation and has an apparent activation energy of about 270 kJ/mol. The position of α_h rose about 13°C as the hard-segment content increased from 39 to 43 wt%.

Several 2,4-T-2P samples were reevaluated and gave results identical within experimental error to the first determination, indicating that thermal history has a negligible effect on these polyurethanes.

2,6-TDI Polyurethanes. Three 2,6-T-1P samples of varying hard-segment content were studied and exhibited the dynamic mechanical response displayed in Figure 8. An α_s relaxation can be noted at − 64°C (11 Hz) for 2,6-T-2P-31, which progressively broadened as the quantity of hard segments incorporated into the polyurethane increased. The activation energy of the α_s process was found to decrease as hard-segment content increased, as can be noted from Table III. The level of the storage modulus above α_s was found to highly depend upon the composition of the polyurethane, increasing from 2,6-T-1P-31 to 2,6-T-1P-42 by about 250% at 0°C and by a like amount from 2,6-T-1P-42 to the next member of the series. A final point to note is the high temperature behavior of these polyurethanes. The storage modulus shows a region of rapid decline not matched by the loss modulus. This leads to a rapid rise in the loss tangent vs. temperature curves, not shown, corresponding to the onset of a softening or flow δ-relaxation process. The temperature at which the storage modulus reached a value of 1.25 MPa was taken as an arbitrary indication of the loss of mechanical integrity and melting of the hard segments of these polyurethanes. The temperatures of both α_s- and δ-relaxation processes are plotted against hard-segment content and compared with the data of Schneider and co-workers (*12*) in Figure 9. The α_s relaxation occurs at increasingly higher temperatures as hard-segment content increases while the T_{gs} determined from DSC measurements shows a slight decrease. Both sets of data show the same trend for the higher temperature, hard-segment melting transition.

The dynamic mechanical properties of four 2,6-T-2P samples containing from 19 to 43 wt% hard segments are summarized in Figure 10. A low temperature α_s relaxation is apparent at about − 70°C for all compositions examined. The transition temperatures of these loss maxima and the associated activation energies are given in Table III. A second process, the α_c relaxation, can be noted as a shoulder on the high temperature side of the α_s-loss maximum. The conclusion of this relaxation is marked by a change in slope of the loss modulus vs. temperature plots

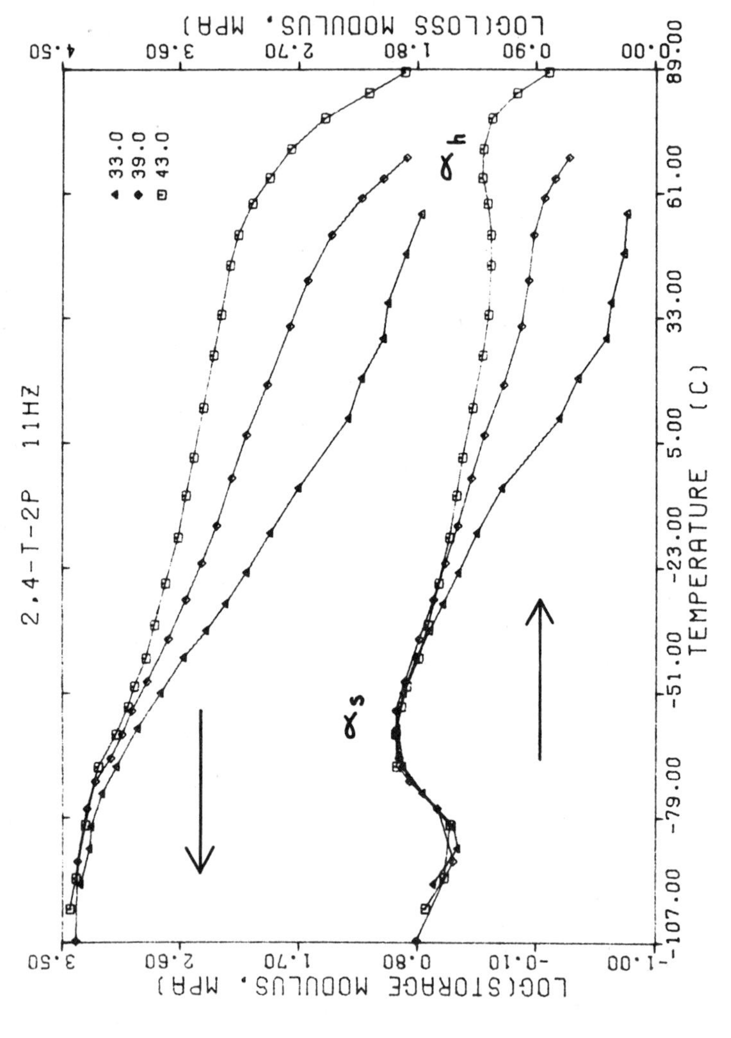

Figure 7. Temperature dependence of the storage and loss modulus of 2,4-T-2P at 11 Hz for 33, 39, and 43 wt % hard-segment concentration

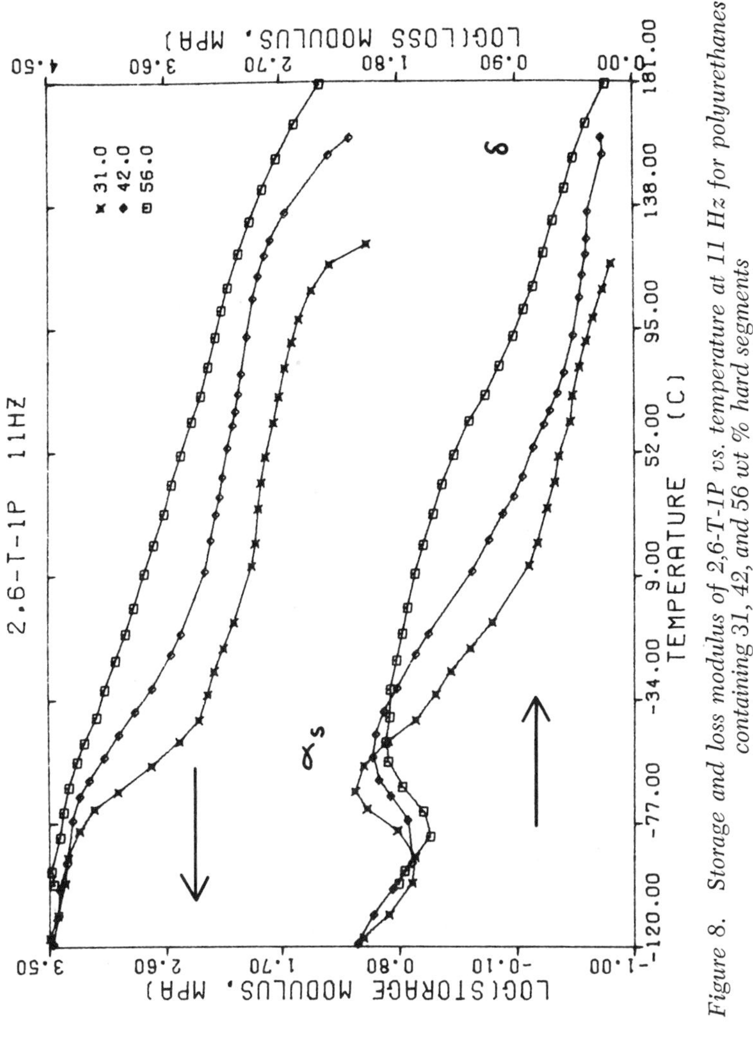

Figure 8. Storage and loss modulus of 2,6-T-1P vs. temperature at 11 Hz for polyurethanes containing 31, 42, and 56 wt % hard segments

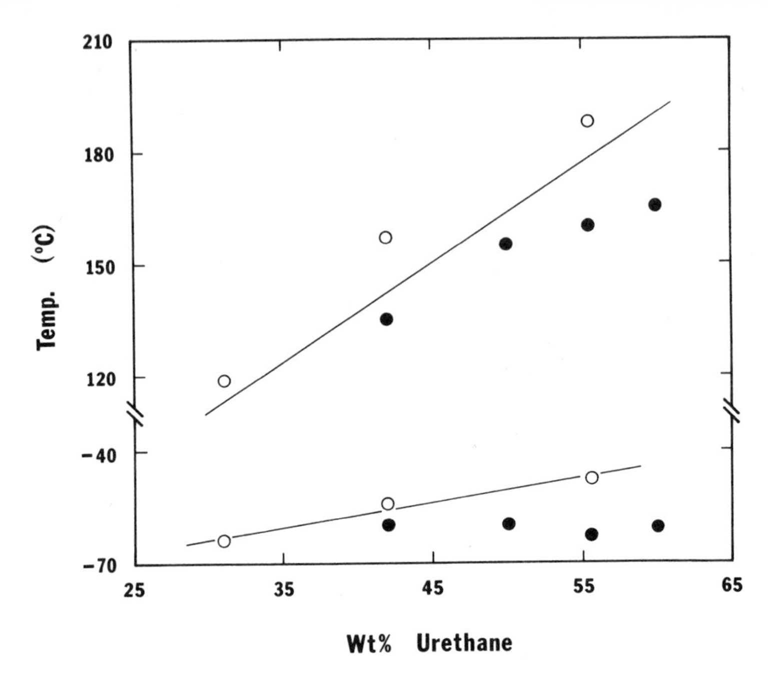

Figure 9. Comparison of 11-Hz dynamic mechanical (open points) and DSC or TMA transition temperatures (filled points, Ref. 13) *of 2,6-T-1P polyurethanes as a function of hard-segment content*

at about 2° to 7°C. This relaxation can be more clearly observed in a plot of storage and loss modulus vs. temperature at 11 Hz for two polyurethanes, 2,6-T-1P-31 and 2,6-T-2P-33, shown in Figure 11. These polymers have about the same hard-segment content and differ only in the soft-segment molecular weight. The α_c-relaxation peak can be observed in the 2,6-T-2P polyurethane as an elevation in the loss modulus over the region from about − 40° to 7°C. The storage modulus of the sample with a higher molecular-weight soft segment can be seen to be about double that of 2,6-T-1P over the same temperature range. The increase in storage modulus above α_s can be attributed to reinforcement of the soft-segment phase by crystallization above its T_g. This enhancement of strength continues up to a temperature of about 7°C, the region where soft-segment melting was found to occur in the 2,6-TDI/PTMO soft-segment polymer. The α_c relaxation is also apparent in loss tangent vs. temperature curves, not shown. The intensity of the α_s relaxation is approximately equal to that of the α_c relaxation from the tan(δ) data, and the two processes exhibit considerable overlap.

Returning to the 2,6-T-2P in Figure 10, the storage modulus above α_c and the position of the δ relaxation, the temperature at which the storage modulus reached 1.25 MPa, can be seen to depend upon hard-

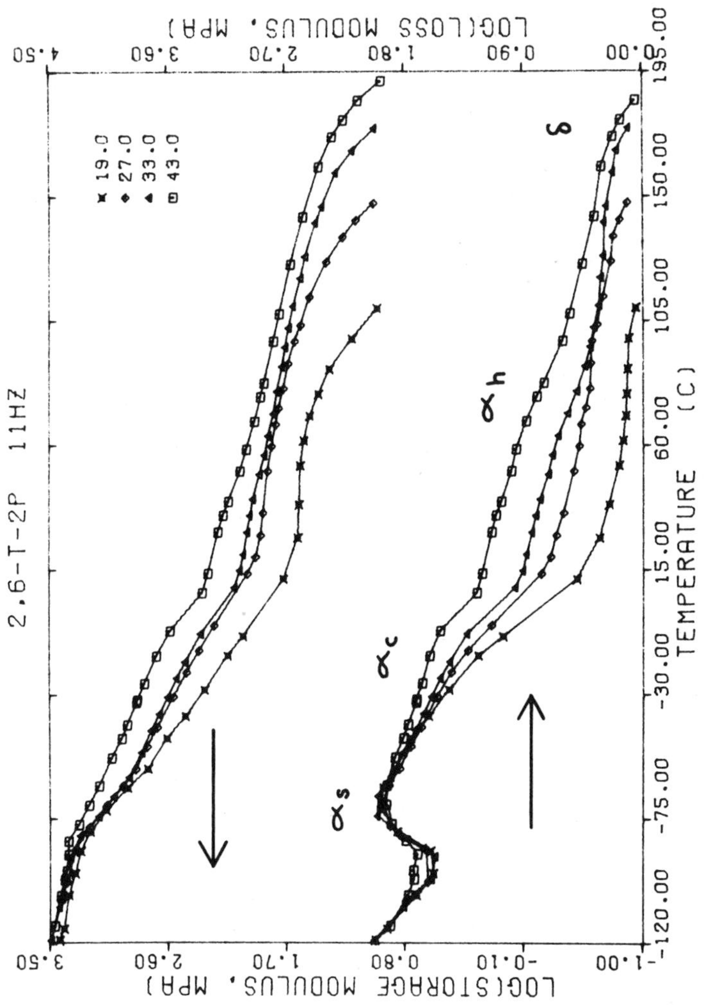

Figure 10. Temperature dependence of the storage and loss modulus of 2,6-T-2P at 11 Hz for polyurethanes containing 19, 27, 33, and 43 wt % hard blocks

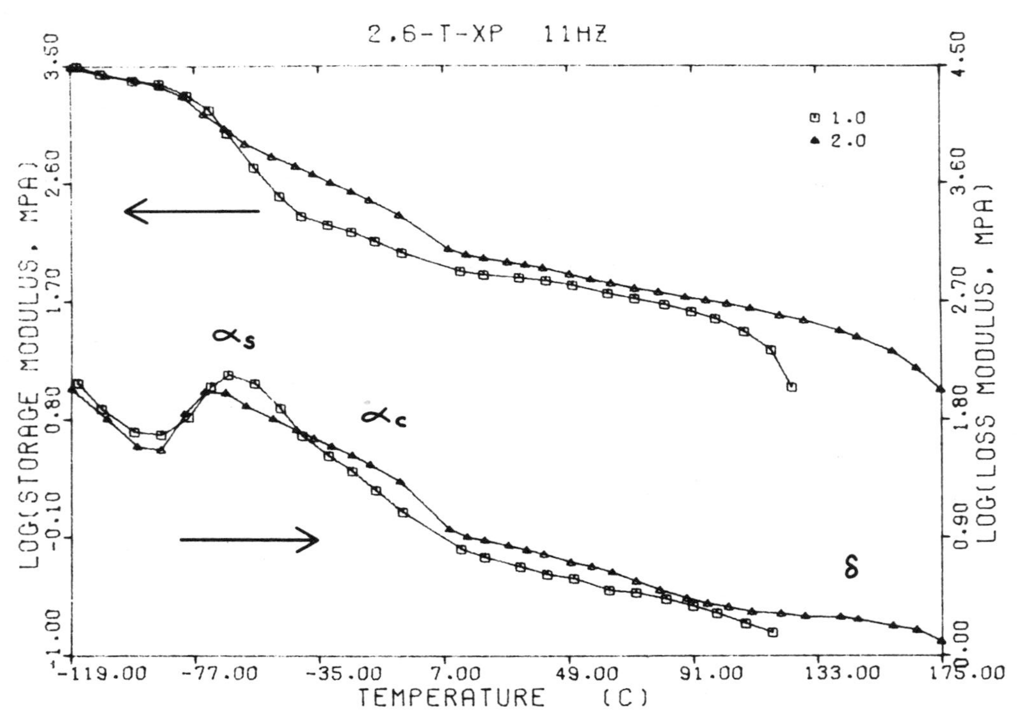

Figure 11. Comparison of the storage and loss modulus of 2,6-T-1P-31 (squares) and 2,6-T-2P-33 (triangles) as a function of temperature

Table III. Relaxations of 2,6-TDI Block Polyurethanes

Sample	*Approximate Molar Ratio (TDI/BD/PTMO)*	*Relaxation*	*Temperature*[a] *(°C)*	*Activation Energy (kJ/mol)*
2,6-T-1P-31	2/1/1	α_s	−64	170 ± 20
2,6-T-1P-31	2/1/1	δ	120	—
2,6-T-1P-42	3/2/1	α_s	−54	120 ± 40
2,6-T-1P-42	3/2/1	δ	157	—
2,6-T-1P-56	5/4/1	α_s	−48	70 ± 20
2,6-T-1P-56	5/4/1	δ	183	—
2,6-T-2P-19	2/1/1	α_s	−69	140 ± 20
2,6-T-2P-19	2/1/1	δ	97	—
2,6-T-2P-27	3/2/1	α_s	−71	180 ± 20
2,6-T-2P-27	3/2/1	α_c	−31	—
2,6-T-2P-27	3/2/1	δ	137	—
2,6-T-2P-33	4/3/1	α_s	−71	180 ± 30
2,6-T-2P-33	4/3/1	α_c	−19	—
2,6-T-2P-33	4/3/1	δ	160	—
2,6-T-2P-39	5/4/1	α_s	−72	160 ± 20
2,6-T-2P-39	5/4/1	α_c	−9	—
2,6-T-2P-39	5/4/1	δ	154	—
2,6-T-2P-43	6/5/1	α_s	−66	70 ± 20
2,6-T-2P-43	6/5/1	α_c	−8	—
2,6-T-2P-43	6/5/1	α_h	60	270 ± 90
2,6-T-2P-43	6/5/1	δ	177	—

[a] α_s determined from E''_{max}, α_c and α_h from $\tan(\delta)_{max}$, δ from $\log(E') = 1.25$ MPa, all at 11 Hz.

segment concentration in a manner similar to that observed for the 2,6-T-1P polyurethanes. The α_s- and δ-relaxation temperatures at 11 Hz are plotted as a function of hard-segment content in Figure 12 and compared with TMA and DSC results of Schneider and Paik Sung (*14*). There is good agreement between the α_s-loss process and the T_{gs} transition, and both the δ relaxation and the hard-segment melting transition show a similar trend, with the exception of 2,6-T-2P-39.

Thermal history was found to have an effect on 2,6-T-2P dynamic mechanical properties. For the two samples retested after the first experiment to 190°C, the α_s relaxation was shifted to a higher temperature after thermal treatment while the storage modulus increased by about 50% over the temperature interval from − 80° to 130°C. The δ relaxation was not affected by heating to 190°C. Table IV summarizes the dynamic mechanical data for both of the retested 2,6-T-2P polyurethanes. The apparent activation energy of the α_s relaxation increased significantly after the first determination, as can be seen from Figure 13, a plot of activation energy vs. hard-segment content.

Table IV. Relaxations of Thermally Treated 2,6-TDI Polyurethanes

Sample	*Relaxation*	*Temperature*[a] *(°C)*	*Activation Energy (kJ/mol)*
2,6-T-2P-39	α_s	−64	310 ± 40
2,6-T-2P-39	α_c	−3	—
2,6-T-2P-39	δ	154	—
2,6-T-2P-43	α_s	−64	200 ± 30
2,6-T-2P-43	α_c	−2	—
2,6-T-2P-43	α_h	71	310 ± 50
2,6-T-2P-43	δ	174	—

[a] α_s determined from E''_{max}, α_c and α_h from $\tan(\delta)_{max}$, δ from $\log(E') = 1.25$ MPa, all at 11 Hz.

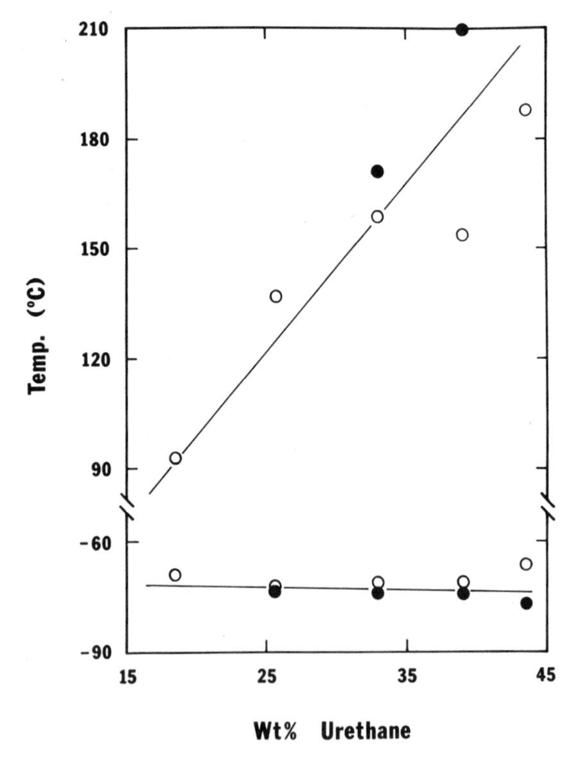

Figure 12. Comparison of 11-Hz dynamic mechanical (open points) and DSC or TMA transition temperatures (filled points, Ref. 14) of 2,6-T-2P polyurethanes as a function of hard-segment concentration

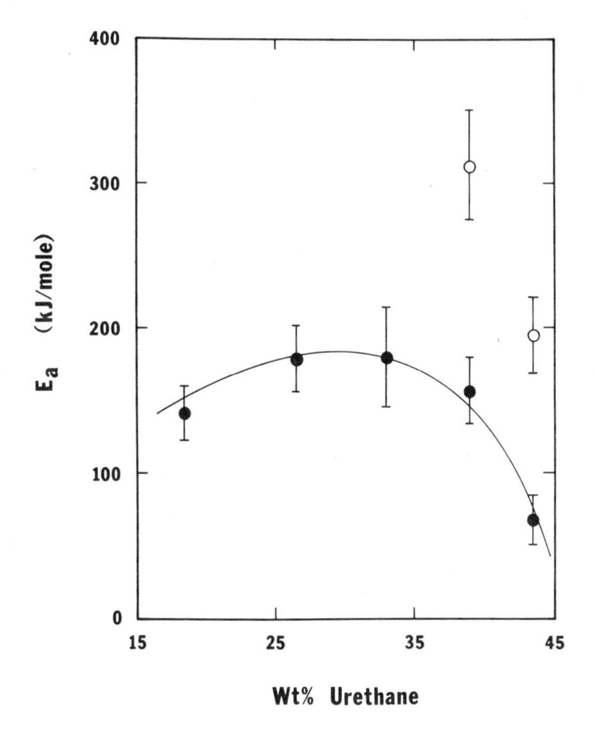

Figure 13. Apparent activation energy of the α_s relaxation in 2,6-T-2P polyurethanes as a function of hard-segment content from the first (filled points) and second determination (open points)

An additional relaxation can be detected in plots of the loss tangent as a function of temperature for 2,6-T-2P-43, not shown. The α_h relaxation occurred at 60°C (11 Hz) in the first examination of this sample. A weak α_h relaxation can be detected in the loss modulus vs. temperature plots in Figure 13, but the process is more readily apparent in the tan(δ) data. Upon retesting of this sample, the α_h dispersion was shifted to 71°C. The activation energy for the α_h relaxation did not change significantly with thermal history and was extremely weak or undetectable in 2,6-T-2P samples which contained less than 43 wt% hard segments.

Discussion

Hard- and Soft-Segment Polymers. The structure of the diisocyanate component has a significant influence on dynamic mechanical properties of both hard- and soft-segment polymers. Both 2,4- and 2,6-TDI/BD α relaxations occur at about the same temperature and have a comparable activation energy. However, the symmetrical structure of

2,6-TDI/BD facilitates crystallization of the hard-segment polymer, which has the effect of broadening the α relaxation when compared with that of the totally amorphous 2,4-TDI/BD (*see* Figure 3b). The semicrystalline nature of the 2,6-TDI hard segment provides modulus reinforcement above the α relaxation to about 212°C. This behavior can be compared with that of a MDI/BD hard segment studied by Illinger et al. (*6*), which displayed an α relaxation at 110°C (110 Hz) and a second relaxation, attributed to melting, at about 170°C (110 Hz). The symmetrical TDI hard segment maintained a higher modulus above the α relaxation to a higher temperature than did the MDI/BD hard segment studied by these workers.

The soft-segment polymers studied consisted of 1090 $\overline{M}_n$ PTMO chains, corresponding to about 75 carbon and oxygen atoms each, in an alternating arrangement with diisocyanate molecules. This material was thought to be representative of the soft segment, as the single urethane units do not function as hard segments in block polyurethanes (*18*). The glass transition for both isomers was found to be in the region − 50° to − 60°C, above the literature value for high molecular-weight PTMO, − 84°C (*19*). The elevation of the soft-segment T_g over that of the pure PTMO homopolymer can be attributed to a combination of the copolymer effect and effective crosslinking of the PTMO ether oxygens by hydrogen bonding with the diisocyanate amine group, an association which has been found to be significant for block copolyurethanes (*13, 14, 15*).

Above the α-relaxation process, the 2,4-TDI/PTMO polymer displayed a short rubbery plateau at a storage modulus of about 5 MPa while 2,6-TDI/PTMO was capable of crystallization, as evidenced by the α_c-loss process. This difference in dynamic mechanical properties demonstrates the effect of a symmetric diisocyanate structure upon soft-segment properties. As previously discussed, single urethane links can sometimes be incorporated into the soft-segment phase. The introduction of only one of these diisocyanate molecules between two long PTMO chains inhibits crystallization if the diisocyanate is asymmetric. In the case of a symmetric diisocyanate, soft-segment crystallization above T_g can readily occur. The crystals formed were found to melt about 30°C below the reported melting point for PTMO homopolymer, 37°–43°C (*19*), possibly because of disruption of the crystal structure by the bulky diisocyanate units.

The α Relaxation of 2,4-T-1P. Only the 2,4-T-1P block polyurethanes exhibited a single major α relaxation which had a strongly composition-dependent position. Similar behavior was observed by Seefried and co-workers (*10*) for polyurethanes containing a TDI/BD hard segment and a 2100-$\overline{M}_n$ polycaprolactone hard segment. The extreme dependence of the α process on hard-segment content suggests that the amount of hard urethane segments mixed with the soft segment is extensive. The location of the α relaxation cannot be predicted from the hard- and soft-segment-polymer relaxation temperatures and polyurethane composition by the use of the copolymer equation. Schneider and Paik Sung (*14*) have shown that the difference between the observed T_g and that predicted from the copolymer equation can be accounted for by considering hydrogen bonds between hard and soft segments as crosslinks which elevate the glass transition of the soft-segment phase. Hydrogen-bond crosslinking should affect the soft segment, main chain microBrownian motion through a reduction in free volume. A corresponding increase in the temperature and activation energy of the α-relaxation process is predicted with increasing crosslink density. Such effects can be observed in the dielectric relaxation data of Scott and co-workers (*20*) on a series of increasingly crosslinked natural rubber samples. An increase in the activation energy for the α process in 2,4-T-1P by about 50% over that for 2,4-TDI/PTMO can be noted. These results are consistent with the crosslinking assumption of Schneider and Paik Sung.

Estimates of the quantity of hard segments incorporated in the soft segment phase do not approach the total hard-segment content of the 2,4-T-1P polyurethanes. These materials were found to give rise to a small angle, x-ray-scattering intensity maximum which became more pronounced as the hard-segment content increased (*15*). The x-ray results indicate that these materials display some phase segregation, but no hard-segment relaxation could be detected by dynamic mechanical testing as carried out in this study because of the low modulus of these materials above the α relaxation. Additional experiments to characterize the dynamic mechanical properties of 2,4-T-1P samples supported by an elastic spring above the α-relaxation process are currently underway.

Soft-Segment Relaxations. The remaining polyurethanes studied exhibited a low temperature, α_s-relaxation much less sensitive to sample composition than the 2,4-T-1P polyurethanes. This relaxation process can be attributed to the glass transition which occurs in soft-segment domains of the phase-separated structure. The α_s relaxation of 2,4-T-2P occurred about $-63°C$ and was not a strong function of hard-segment content over the concentration range examined. An increase in soft-segment block length caused a dramatic improvement in phase separation of 2,4-TDI polyurethanes.

The increase in the α_s-relaxation temperature for 2,6-T-1P samples with increasing hard-segment content is caused by extreme broadening of the relaxation maximum attributable to increasing hard block crystallinity. These effects are similar to changes in dynamic mechanical and dielectric glass-transition relaxation processes in nylon, PET, and other semicrystalline polymers as the amount of crystallinity increases (*21*). Huh and Cooper (*3*), Seefried and co-workers (*9, 10*), and Ferguson (*7*) have observed similar effects in dynamic mechanical studies of MDI polyurethanes. The activation energy of the α_s process in 2,6-T-1P was found to decrease significantly as the hard-segment content increased, again reflecting the influence of the increasingly crystalline hard segment on soft-segment motions. Similar results have been noted for the decrease in activation energy of the PET α relaxation with increasing degree of crystallinity (*21, 22*).

The 2,6-T-2P polyurethanes exhibited the lowest temperature α_s relaxation of any of the block polyurethanes studied. A slight increase in activation energy and decrease in the temperature of the α_s relaxation occurred when the hard-segment content increased from 19 to 27 wt%. This behavior can be attributed to the increasing purity of the soft-segment phase. As the hard-segment content increases the hard-block-sequence length also increases, and the longer hard segments display a greater tendency to form a more highly organized separate phase in preference to mixing with the soft-segment material. When a well-developed, hard-segment phase has formed in 2,6-T-2P-43, the crystalline regions exert a similar broadening effect on the α_s relaxation to that noted for highly crystalline 2,6-T-1P polymers. The activation energy was found to decrease significantly and the α_s-relaxation temperature increased for the 2,6-T-2P-43 sample.

Upon repetition of the dynamic mechanical tests, the activation energy for the α_s relaxation of the two highest hard-segment content 2,6-T-2P polyurethanes showed an increase of about 250% over the first determination, as can be seen from Figure 13. The increase in activation energy and corresponding narrowing of the α_s relaxation can be understood in terms of a change in composition of the soft-segment phase on heating. Longer length hard segments are transferred from the soft-segment to the hard-segment phase during the thermal treatment. This leads to a narrower distribution of relaxation times and a corresponding narrowing of the α_s-loss process when the same samples are reevaluated. Changes in the α_s-relaxation temperature from run to run may not be significant in light of the change in breadth of the loss process.

Only the 2,6-T-2P polyurethanes exhibited a relaxation associated with crystallization and later melting of the soft segments. The position of the crystallization maximum was estimated from the location of $\tan(\delta)_{max}$ and was somewhat difficult to characterize accurately because of extreme overlapping of the α_s and α_c relaxations, as evident in Figure 10. Melting occurred over the region 2°–7°C in all samples. Phase segregation in 2,6-T-2P samples was well developed to the point that the small amount of hard-segment mixing in the soft segment resulted in freedom from restrictions of hard-segment hydrogen bonding and permitted the soft segment to crystallize above its glass transition.

Hard-Segment Relaxations. For the 2,6-TDI polyurethanes, the δ-relaxation temperature increased with increasing hard-segment content as did the α_h-relaxation temperature of 2,4-T-2P. This effect can more accurately be ascribed to an increase in hard-block length as hard-segment content increased. The longer hard-segment lengths can form a more ordered structure which melts or dissociates at higher temperatures. Peebles (*23*) has shown that the hard-block length is actually greater than that which would be predicted on a molar basis for a two-stage polyurethane reaction similar to the procedure used to synthesize the materials characterized in this study. Using the number average hard-segment length at 99% conversion and a μ value of three from Peebles' results, the equation given by Flory (*24*) can be used to obtain the melting temperature for an infinite chain (Equation 2). A plot of

$$\frac{1}{T_m} = \frac{2R}{n\Delta H_n} + \frac{1}{T_m{}^{\infty}} \tag{2}$$

the inverse of the δ relaxation temperature against the inverse of the hard-segment length, *n*, is shown in Figure 14. The δ-relaxation temperatures for both 2,6-TDI polyurethanes were found to fit a straight line with identical values of $T_m{}^{\infty}$, within experimental error, of 222°C for 2,6-T-2P and 227°C for 2,6-T-1P. The hard-segment δ relaxation depends upon chain length and is depressed over the value for an infinite chain only by the presence of end groups, in this case the soft segments which cannot enter the hard-segment crystals.

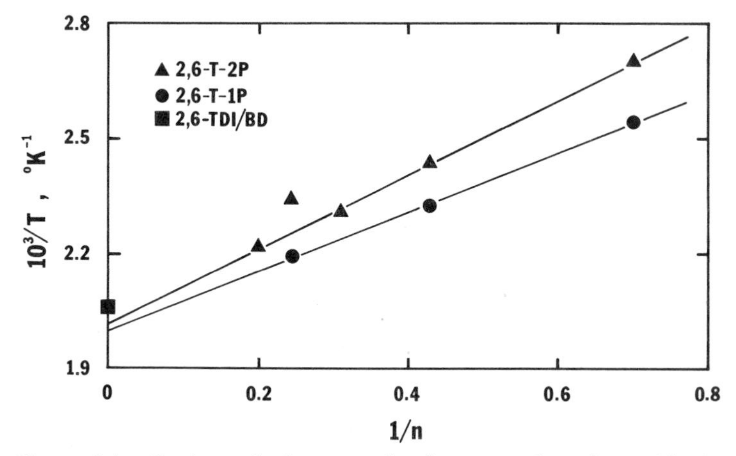

*Figure 14. Reciprocal of average hard segment length vs. 1/*T *for the 2,6-TDI-polyurethane δ relaxation*

From the slope of the lines, the heat of fusion per repeat unit, ΔH_n, was calculated as 64 J/g for 2,6-T-2P and 82 J/g for 2,6-T-1P. The hard segment in 2,6-T-1P is more highly organized as the heat of fusion per segment, the unit contained within the brackets with subscript *m* in Figure 1, is about 25% higher than the value for 2,6-T-2P. This could occur if a greater amount of soft segment is mixed in the hard-segment phase of 2,6-T-2P. Some caution should be exercised in the interpretation of the heat of fusion results, however, as the equation given by Flory applies to the melting temperature while the δ-relaxation temperature, which has no thermodynamic basis, was used in this study.

Harrell (*18*) has noted that the melting point increased with hard-block length while the degree of crystallinity did not change in a DSC study of nonhydrogen-bonding polyurethane elastomers. The degree of organization in a hard-segment polymer was found to be superior to that of a hard segment in a block polyurethane. A value of ΔH_n of about 40 J/g was determined for these materials, considerably less than for the 2,6-TDI hard segments attributable to the absence of hydrogen-bonding interactions in the piperazine-containing samples.

The δ relaxation, as defined in this study, did not exceed 185°C while the 2,6-TDI/BD hard segment had a melting temperature of about 212°C. The crystalline hard-segment domains in 2,6-TDI polyurethanes can be seen to be more poorly organized than in the hard-segment polymer because of their shorter lengths and the possibility of inclusion of soft segments into the hard-segment-domain regions. Similar results that show an increasing hard-segment melting relaxation have been reported by Ferguson and co-workers (*7*) for MDI polyurethanes of increasing hard-segment concentration.

The increase in the α_h-relaxation temperature with increasing hard-segment content for 2,4-T-2P parallels the results of Kraus and Rollman (*25*), who studied S-B-S and S-I-S block copolymers of varying block molecular weight. They found that as block length increased, the hard-segment relaxation shifted to higher temperatures while the soft-block relaxation shifted to lower temperatures. They also demonstrated that the hard-segment relaxation can be lowered by mixing of soft segment into the hard-segment phase at a given hard-block length. By using the theory of Meier (*26*), they concluded that samples which contained no compositionally pure phases could still give rise to typical block copolymer, dynamic mechanical relaxation behavior. The existence of compositionally pure phases in 2,4-T-2P polyurethanes is doubtful. Schneider and Paik Sung (*14*) have estimated that about 11–19 wt% hard segment is mixed into the soft phase in 2,4-T-2P polyurethanes. The hard-segment α_h-relaxation occurs at a much lower temperature in 2,4-T-2P polyurethanes than in the pure hard segment, indicating that a compositionally pure hard-segment phase may not be present. The existence of specific interactions of the hard and soft segment by hydrogen bonding further enhances the presumption that 2,4-T-2P polyurethanes contain no pure hard or soft phase but consist primarily of interlayer material.

The α_h relaxation in 2,6-TDI block polyurethanes could only be clearly observed for 2,6-T-2P-43. This sample contained sufficiently well-developed hard-segment domains that both an amorphous α_h and a crystalline melting δ relaxation were detectable. During the second examination of this sample, the α_h relaxation increased by about 10°C but did not change in intensity, an effect which can be attributed to the presence of a slightly greater degree of crystallinity in the hard segment after thermal treatment.

Degree of Phase Segregation. The relative degree of phase segregation of the various block polyurethanes studied can be estimated from their dynamic mechanical response. The 2,4-T-1P samples displayed the least amount of phase segregation by virtue of the highly composition-dependent α relaxation. The 2,6-T-2P polyurethanes possessed the soft-segment phase of greatest purity as evidenced by the temperature of the α_s relaxation, closest to that of the PTMO homopolymer, as well as the ability of the soft segment to crystallize above its T_g. The soft-segment domains in 2,4-T-2P and 2,6-T-1P polyurethanes are intermediate in purity, with the 2,6-T-1P samples containing a slightly greater amount of hard segment mixed into the soft segment than 2,4-T-2P, which inhibited soft-segment crystallization. The hard-segment domains in 2,4-T-2P are clearly the least organized. A large hard-segment concentration was found to be necessary in order for a well defined, α_h-relaxation process to occur. The 2,6-TDI polyurethanes contained highly ordered hard segments because of the ability of the hard blocks to crystallize. The 2,6-T-1P polyurethanes appear to be the more highly ordered of the two, as these materials exhibit a higher temperature δ relaxation for a given hard-segment length and a higher heat of fusion per segment. The degree of phase segregation in the polyurethanes studied is not nearly as well developed as in some nonpolar block copolymers of higher block molecular weight. The polyurethanes display broad relaxation processes characteristic of poorly defined phases. No pure component phases are thought to exist in these materials. From their dynamic mechanical response, the degree of mixing in each phase can be seen to depend on soft-block molecular weight and hard-segment structure.

Effects of Phase Segregation on Modulus. The magnitude of the modulus between α_s and α_h for 2,4-T-2P or δ for 2,6-TDI block polyurethanes was found to highly depend upon hard-segment content. The presence of a plateau region in modulus–temperature plots has been attributed to phase segregation in block copolymers (*27*). The hard

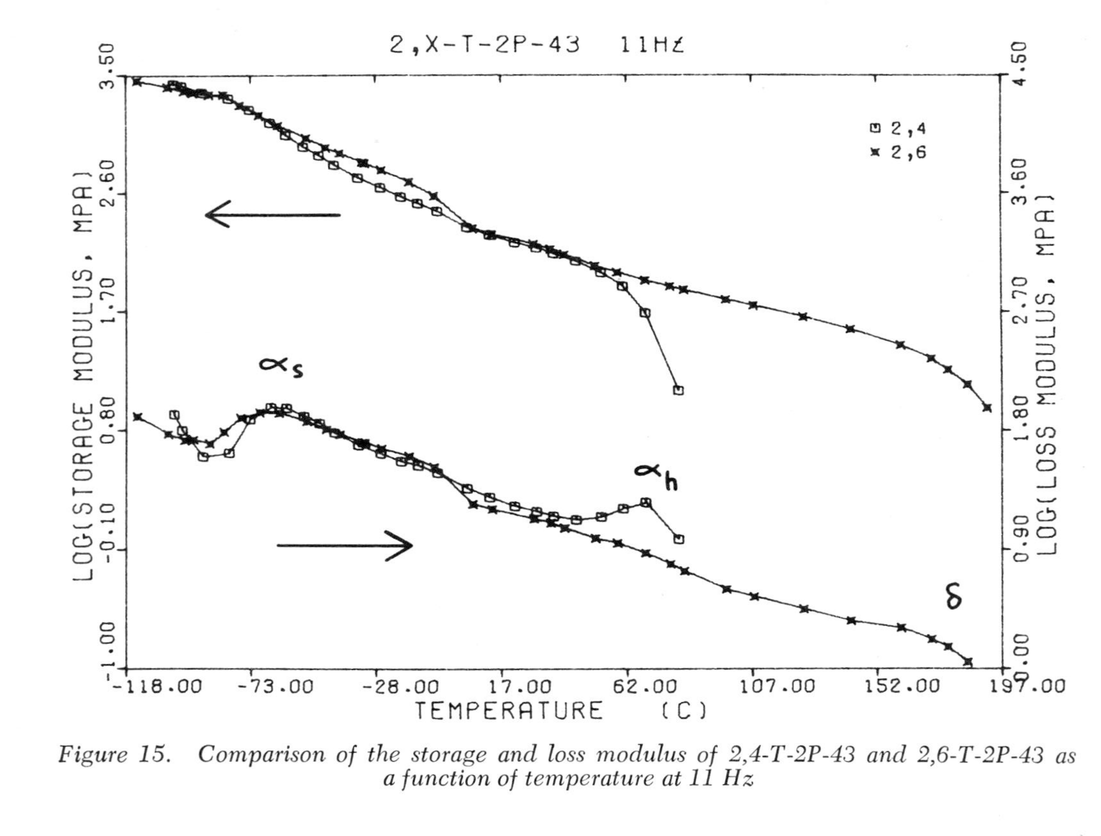

Figure 15. Comparison of the storage and loss modulus of 2,4-T-2P-43 and 2,6-T-2P-43 as a function of temperature at 11 Hz

phase can act to reinforce the plateau modulus in a manner similar to that of fillers in crosslinked elastomers. The hard phase also is thought to act as a crosslinking or anchoring region for the soft segments, a mechanism which gives rise to a plateau in modulus above the soft segment T_g. An increase in hard-segment content can induce a higher value of the plateau modulus through either of these mechanisms.

The nature of the hard-segment domains, either crystalline or amorphous, does not appear to be important to plateau modulus enhancement in TDI block polyurethanes, provided that the hard-block length is sufficient. This is illustrated in Figure 15, a modulus vs. temperature comparison of 2,4- and 2,6-T-2P-43. Both polyurethanes exhibit identical storage moduli in the region between the 2,6-T-2P-43 α_c and 2,4-T-2P-43 α_h relaxations. The difference in modulus between α_s and α_c can be accounted for by crystallinity of the soft segment in the 2,6-TDI sample. A comparison of these two polyurethanes containing 39 wt% hard segments indicates that the plateau modulus is much lower for 2,4-T-2P-39. A well developed, α_h-relaxation maximum was not present in this sample. A hard segment capable of crystallization can increase the temperature range of plateau modulus enhancement or be a more efficient method of increasing the plateau modulus, but an equal effect can be provided by a well developed, glassy urethane domain structure.

The plateau modulus in 2,6-TDI polymers was found to depend on hard-segment content rather than on hard-block length. Comparisons of the modulus levels between two 2,6-TDI samples with the same hard-segment content, such as in Figure 11, show that the plateau modulus of the 2,6-T-2P polyurethanes is only about 10% higher at temperatures above the soft-segment melting point. In contrast, for two block polyurethanes with the same hard-block length, such as 2,6-T-1P-42 and 2,6-T-2P-27, the plateau modulus of the 2,6-T-1P sample is about 250% greater.

Conclusions

The structure of the diisocyanate moiety has a significant effect on hard- and soft-segment dynamic mechanical response, which also is reflected in the properties of the block polyurethanes studied. An asymmetric 2,4-TDI hard segment displays only an amorphous α relaxation while a symmetric 2,6-TDI hard segment has a broadened glass-transition relaxation and is capable of crystallization, retaining a high modulus to the melting temperature, about 212°C. Similarly for soft-segment polymers, the relaxation associated with the glass transition occurs at low temperatures for both 2,4- and 2,6-TDI/PTMO, but the latter crystallizes above T_g and later melts at about 10°C.

The dynamic mechanical response of TDI block polyurethanes is sensitive to the degree of phase segregation in these materials. Polyurethanes composed of 2,4-T-1P contain a highly nonhomogeneous soft-segment phase, as evidenced by the compositionally dependent, α-relaxation temperature. A doubling in the soft-segment molecular weight causes phase segregation to occur in 2,4-T-2P, as a relaxation process associated with each phase is evident. The 2,6-TDI block polyurethanes studied are also phase segregated, displaying an α_s relaxtion corresponding to the soft-segment T_g and a δ relaxation, associated with hard-block melting. The α_s-relaxation temperature is affected by soft-segment-phase purity and the degree of crystallinity. The δ-relaxation temperature increases with increasing hard-block length. The level of the plateau modulus between the low- and high-temperature relaxations in phase-segregated polyurethanes strongly depends upon hard segment content.

The 2,6-T-2P samples display an additional α_c relaxation associated with crystallization of the soft segment above its glass transition, suggesting that the soft-segment phase is relatively free of intermixed hard blocks. A hard-segment glass transition was observed for the polyurethane sample in this series with the highest hard-segment content, indicating that hard-block organization also is well developed in these materials. The 2,6-T-1P polyurethanes however, contain fewer soft segments mixed in the hard-block domains than the 2,6-T-2P samples.

Comparison of the block copolyurethane relaxations with corresponding loss processes of the hard- and soft-segment polymers indicates that the organization of phases in block polyurethanes is inferior because of the short block length and significant amount of intersegmental mixing in the domains. Samples with crystallizable hard segments and the higher block length, amorphous samples are composed of domains with a sufficiently small degree of mixing that typical block copolymer, dynamic mechanical response is exhibited.

Acknowledgment

We are grateful to the Army Materials and Mechanics Research Center, Watertown, Massachusetts, for support of this research under contract No. DAAG 46-77C-0031. We are also grateful to N. Schneider for helpful discussions.

Literature Cited

1. Kajiyama, T., MacKnight, W. J., *Trans. Soc. Rheol.* (1969) **13**, 527.
2. Kajiyama, T., MacKnight, W. J., *Macromolecules* (1969) **2**, 254.
3. Huh, D. S., Cooper, S. L., *Polym. Eng. Sci.* (1971) **11**, 369.

4. Ng, H. N., Allegrezza, A. E., Seymour, R. W., Cooper, S. L., *Polymer* (1973) **14**, 255.
5. Samuels, S. L., Wilkes, G. L., *J. Polym. Sci., Part C* (1973) **43**, 149.
6. Illinger, J. L., Schneider, N. S., Karasz, F. E., *Polym. Eng. Sci.* (1972) **12**, 25.
7. Ferguson, J., Hourston, D. J., Meredith, R., Patsavoudis, D., *Eur. Polym. J.* (1972) **8**, 369.
8. Seefried, C. G., Koleske, J. V., Critchfield, F. E., *J. Appl. Polym. Sci.* (1975) **19**, 2493.
9. Seefried, C. G., Koleske, J. V., Critchfield, F. E., *J. Appl. Polym. Sci.* (1975) **19**, 2503.
10. Seefried, C. G., Koleske, J. V., Critchfield, F. E., *J. Appl. Polym. Sci.* (1975) **19**, 3185.
11. Seefried, C. G., Koleske, J. V., Critchfield, F. E., Dodd, J. L., *Polym. Eng. Sci.* (1975) **15**, 646.
12. Schneider, N. S., Paik Sung, C. S., Matton, R. W., Illinger, J. L., *Macromolecules* (1975) **8**, 62.
13. Paik Sung, C. S., Schneider, N. S., *Macromolecules* (1975) **8**, 68.
14. Schneider, N. S., Paik Sung, C. S., *Polym. Eng. Sci.* (1977) **17**, 73.
15. Paik Sung, C. S., Schneider, N. S., *Macromolecules* (1977) **10**, 452.
16. Kajiyama, T., Ph.D. Dissertation, University of Massachusetts (1969).
17. Ward, I. M., "Mechanical Properties of Solid Polymers," pp. 177, Wiley-Interscience, London, 1971.
18. Harrell, L. L., *Macromolecules* (1969) **2**, 607.
19. "Polymer Handbook," J. Brandrup, E. H. Emmergut, Eds., pp. 3–43, Wiley, New York, 1975.
20. Scott, A. H., McPherson, A. T., Curtis, H. L., *U.S. National Bureau of Standards, Journal of Research* (1933) **11**, 173.
21. McCrum, N. G., Read, B. E., Williams, G., "Anelastic and Dielectric Effects in Polymeric Solids," pp. 511, Wiley, London, 1967.
22. Saito, S., *Kolloid Z.* (1963) **189**, 116.
23. Peebles, L. H., *Macromolecules* (1974) **7**, 872.
24. Flory, P. J., "Principles of Polymer Chemistry," pp. 570, Cornell University, Ithaca, NY, 1953.
25. Kraus, G., Rollmann, K. W., *J. Polym. Sci., Part A-2* (1976) **14**, 1133.
26. Meier, D. J., *Polym. Prepr., Am. Chem. Soc., Div. Polym. Chem.* (1974) **15**(1), 171.
27. Estes, G. M., Cooper, S. L., Tobolsky, A. V., *J. Macromol. Sci., Rev. Macromol. Chem.* (1970) **C4**, 313.

RECEIVED June 5, 1978.

7

Elastomeric Polyether-Ester Block Copolymers

Properties as a Function of the Structure and Concentration of the Ester Group

JAMES R. WOLFE, JR.

Elastomer Chemicals Department, Experimental Station,
E. I. du Pont de Nemours and Co., Wilmington, DE 19898

Physical properties are related to ester-segment structure and concentration in thermoplastic polyether-ester elastomers prepared by melt transesterification of poly(tetramethylene ether) glycol with various diols and aromatic diesters. Diols used were 1,4-benzenedimethanol, 1,4-cyclohexanedimethanol, and the linear, aliphatic α,ω*-diols from ethylene glycol to 1,10-decane-diol. Esters used were terephthalate, isophthalate, 4,4′-biphenyldicarboxylate, 2,6-naphthalenedicarboxylate, and* m*-terphenyl-4,4″-dicarboxylate. Ester-segment structure was found to affect many copolymer properties including ease of synthesis, molecular weight obtained, crystallization rate, elastic recovery, and tensile and tear strengths.*

Interest in polyether–ester block copolymers that are both thermoplastic and elastomeric continues at a sustained pace (*1–9*). Most of the recent communications have dealt with the tetramethylene terephthalate/poly(tetramethylene ether) terephthalate copolymers which are continuing to find increased use in commercial applications requiring thermoplastic elastomers with superior properties.

Part I of this series explored the structure–property relationships of tetramethylene terephthalate/polyether terephthalate copolymers as a function of variations in the chemical structure, molecular weight, and concentration of the polyether units (*10*). Of the polyether monomers tested, poly(tetramethylene ether) glycol of molecular weight approximately 1000 was found to provide copolymers having the best overall combination of physical properties and ease of synthesis.

0-8412-0457-8/79/33-176-129$05.75/0

The work reported here is concerned with the syntheses and properties of polyether–ester block copolymers containing poly(tetramethylene ether) units of molecular weight of approximately 1000 as the amorphous polyether blocks and a variety of esters as the crystallizable hard segments. The purpose of this study is to correlate changes in synthesis and properties of these thermoplastic and elastomeric copolymers with changes in the concentration and nature of the ester segments, particularly the types of diol and diacid.

Experimental

Nomenclature. Poly(tetramethylene ether) glycol having a number-average molecular weight of approximately 1000 is coded as PTME glycol. The straight-chain, α,ω-diols are coded as 2G, 3G, . . . , 10G where the numerals represent the number of methylene groups between terminal hydroxyls. Thus 2G is ethylene glycol; 5G is 1,5-pentanediol; and 10G is 1,10-decanediol. 1,4-Cyclohexanedimethanol is coded as CD. Terephthalate is coded as T. CDT represents 1,4-cyclohexanedimethylene terephthalate. 4GT represents tetramethylene terephthalate.

Copolymer compositions are expressed as weight percentages of the ester units with the remainder being polyether–ester units. For instance, 40% tetramethylene terephthalate/PTME terephthalate copolymer represents a block copolymer containing 40 wt % of tetramethylene terephthalate units and by difference 60 wt % of poly(tetramethylene ether) terephthalate units.

Materials. Except where otherwise noted the materials were of commercial quality and were used as received. The PTME glycol used was obtained from E. I. du Pont de Nemours and Co. Its number-average molecular weight ranged from 975 to 1020, based on hydroxyl-number determinations (ASTM D2849-69) assuming two hydroxyls per molecule. Dimethyl terephthalate, 1,4-butanediol, and tetrabutyl titanate ("Tyzor" TBT organic titanate) were obtained from E. I. du Pont de Nemours and Co. Practical-grade 1,4-cyclohexanedimethanol and reagent-grade *trans*-1,4-cyclohexanedimethanol were obtained from Eastman Kodak Co. Dimethyl isophthalate was practical-grade material from Eastman Kodak Co. Dimethyl *m*-terphenyl-4,4″-dicarboxylate, prepared by W. H. Watson using a reported procedure (*11*), was obtained from the retained-chemical storage at the DuPont Experimental Station. Its structure was verified by elemental analysis, molecular weight determination, and melting point.

Polymer Preparation. The polyether–ester copolymers were prepared by titanate–ester-catalyzed, melt transesterification of a mixture of PTME glycol, the dimethyl ester of an aromatic diacid, and a diol present in 50–100% molar excess above the stoichiometric amount required (Figure 1). The reactions were carried out in the presence of no more than 1 wt %, based on final polymer, of an aromatic-amine or hindered-phenol antioxidant. The general procedures and equipment required have been reported in detail (*12*). The polymerization procedure consisted of adding catalyst solution to a nitrogen-blanketed mixture of the

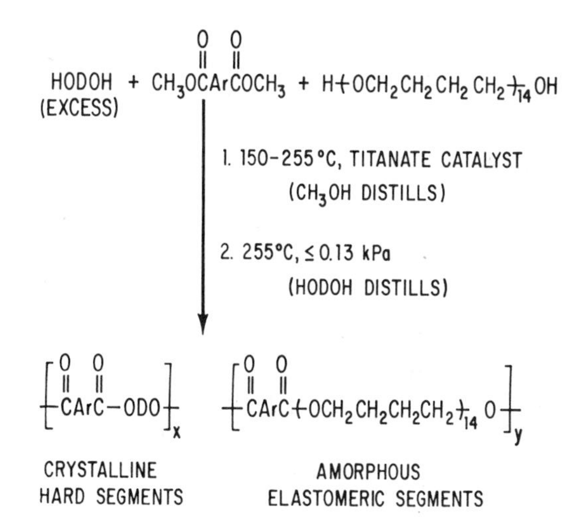

Figure 1. Synthesis and structure of polyether–ester block copolymers (D = hydrocarbon portion of diol; Ar = aromatic portion of the ester; x,y = the number of repeat units in the respective ester and polyether–ester blocks)

monomers plus antioxidant which had been heated sufficiently hot to be molten (150°–200°C), removing methanol by-product by atmospheric distillation as the stirred reaction mass was gradually heated to 255°C in 1 hr, and finally reducing the pressure below 0.13 kPa (1 Torr) and removing excess diol by distillation from the viscous reaction mass in about 30–120 min. A high surface-to-volume ratio for the viscous copolymerizing mass and efficient stirring were important to facilitate removal of excess diol in order to obtain copolymer of high molecular weight. The copolymerization was halted and the copolymer recovered when the viscosity of the molten mass ceased to increase as indicated by the stirrer torque.

Polymer compositions are based on the ratios of starting monomers on the assumption of no loss of monomers during copolymerization other than the deliberate removal of the excess diol needed to drive the copolymerization to completion. The polymerization procedures are believed to result in block copolymers composed of a random distribution of components (*12, 13*). The block structures of these copolymers are attributed to the use of PTME glycol as one of the monomers. X-ray diffraction studies of tetramethylene terephthalate/PTME terephthalate copolymers have verified the presence of crystalline tetramethylene terephthalate segments and the apparent lack of crystallinity of the PTME terephthalate segments (*14*).

Measurement Methods. Copolymer inherent viscosities (η_{inh}) are reported in L/g. They were determined at 30°C at a concentration of 1 g/L in *m*-cresol or of 5 g/L in a 60/40 (wt %) mixture of phenol and 1,1,2-trichloroethane (TCE). The latter were corrected to their expected values in *m*-cresol by the empirically determined relationship:

$$\eta_{inh}(m\text{-cresol}) = 1.28\eta_{inh}(\text{phenol/TCE}) - 0.027 \quad (1)$$

An inherent viscosity of 0.15 L/g for the tetramethylene terephthalate/PTME terephthalate copolymers corresponds to a number average molecular weight of approximately 25,000 (*12*). Melting points were determined by the use of differential scanning calorimeters (Du Pont Models 900 and 990) at heating rates of 10° or 11°C/min. The melting points reported represent the peaks of the endothermic melting-range curves.

Copolymer test specimens were prepared by pressing slabs for 0.5–3 min at temperatures approximately 20°C above the copolymer melting point, followed by cooling under pressure to room temperature in 5–10 min. Test samples were conditioned at 24°C and 50% relative humidity for at least 48 hr before testing. Stress-strain and tear-strength measurements were made on 0.6 to 0.8-mm-thick specimens. Thicker specimens yield lower values for tensile strength and elongation at break. The test methods used were:

stress at 100 and 300% at 8.5 mm/sec	ASTM D412, die C
tensile strength at 8.5 mm/sec	ASTM D412, die C
elongation at break at 8.5 mm/sec	ASTM D412, die C
permanent set at break	ASTM D412, die C
Clash–Berg torsional modulus	ASTM D1043
tear strength at 21 mm/sec	ASTM D1938
Shore hardness	ASTM D2240
compression set after 22 hr at 70°C	ASTM D395, method B.

Specimens used in the measurement of compression set and Shore hardness were prepared by stacking to a height of about 12.7 mm, discs 19 mm in diameter died out from slabs approximately 1.7–1.9 mm in thickness. Specimens used in the tear-strength test were 37 × 75 mm rectangles slit lengthwise to their center. Clash–Berg torsional-modulus results are reported as $T_{10,000}$ values, the temperature at which the apparent modulus of rigidity equals 10,000 psi (69 MPa). Permanent set at break was measured 5 min after break rather than 10 min as specified in ASTM D412.

Effects of Hard-Segment Concentration on Copolymer Properties

The concentration of crystalline ester segments in tetramethylene terephthalate/PTME terephthalate copolymers has a major effect on physical properties. Copolymers were prepared and tested which covered a range of tetramethylene terephthalate (4GT) concentrations from a low of 30% to a high of 82% (Table I).

Stress at 100% elongation increases with increasing concentration of 4GT segments. At 57.4% 4GT and above the polymers exhibit yield points which occur below 100% elongation. The occurrence of yield points in these copolymers has been attributed to disruption and orientation of the crystalline matrix of 4GT segments (*14*).

Table I. Tetramethylene Terephthalate/PTME Terephthalate Copolymers—Properties as a Function of Tetramethylene Terephthalate Content (36)

Tetramethylene Terephthalate (wt %)	*30*	*40*	*50*	*57.4*	*82*
Copolymer properties					
Inherent viscosity (L/g)	0.20	0.18	0.18	0.17	0.18
Yield strength (MPa)	—	—	—	13.1	28.4
Stress at 100% (MPa)	5.9	8.3	11.7	13.4	24.4
Tensile strength (MPa)	24.6	31.0	48.4	46.7	54.5
Elongation (%)	900	830	755	660	535
Permanent set (%)	250	335	370	365	350
Tear strength (kN/m)	13	26	48	63	108
Shore D hardness	34	44	48	54	70
Compression set (%)	60	53	52	52	53
Clash–Berg $T_{10,000}$ (°C)	−65	−52	−33	−7	> 25
Melting point (°C)	152	172	189	197	212

Polymer Preprints, ACS

Tear strength, hardness, and melting point increase with increasing content of 4GT in the copolymers. Compression set is highest at 30% 4GT content and virtually identical above 40% 4GT. Compression set resistance can be improved by annealing the copolymers at elevated temperature (*12*). None of the copolymers in this chapter were annealed prior to testing.

Resistance to low temperature stiffening as measured by the Clash–Berg test becomes increasingly poor as the concentration of 4GT is increased. The glass-transition temperature (T_g) of the amorphous phase of these crystalline-amorphous copolymers has previously been shown to increase with increasing 4GT content (*12, 14*). The increases in T_g with increased 4GT content were attributed to higher concentrations of uncrystallized 4GT units in the amorphous phase (*4, 5, 14, 15*), the increased number of crystalline tie points (*14*), and the greater reinforcement of the amorphous phase by the on-average-longer 4GT segments (*15*). Lilaonitkul, West, and Cooper have used differential scanning calorimetry measurements to relate the percentage of the polyether–ester copolymer which is crystalline to the total 4GT content of tetramethylene terephthalate/PTME terephthalate copolymers (*4, 5*). Their results indicate that for copolymers which have been compression molded and have not been annealed, approximately 35–54% of the total 4GT segments are crystallized. The percentage of 4GT segments which were found to be crystalline increased with increasing 4GT content up to about 76% 4GT in the copolymer.

Elongation at break decreases as the 4GT concentration in the copolymers increases. Permanent set at break appears to show no correlation with 4GT concentration. A considerably different picture emerges if one calculates permanent set at break as a percentage of elongation at break and plots the result vs. the concentration of 4GT in the copolymers (Figure 2). As the concentration of 4GT increases, the copolymers show less and less elastic recovery under the conditions of the set measurements. The greater the 4GT content, the less elastomeric are the block copolymers and the more closely they resemble the ester homopolymer, poly(tetramethylene terephthalate), which is a tough, rapidly crystallizing plastic.

Alkylene Terephthalate/PTME Terephthalate Copolymers

The structure of the diol in alkylene terephthalate/PTME terephthalate copolymers has an important effect on the properties of these block copolymers, as evident from the results shown in Tables II, III, and IV. The 50% tetramethylene terephthalate/PTME terephthalate copolymer prepared from 1,4-butanediol (4G) which was previously noted in Table I serves as our reference copolymer for purposes of discussing the effects of changing the structure of the crystallizable ester segments. The outstanding properties of the 4G-based copolymer are ease of synthesis, a rapid rate of crystallization from the melt, a high melting point, and excellent tensile and tear strengths.

Ethylene Terephthalate/PTME Terephtalate Copolymer. The ethylene glycol- or 2G-based copolymer (Table II) closely resembles the 4G-based copolymer in having a high melting point, even higher than the 4G copolymer, and excellent tensile and tear strengths. The 2G-based copolymer suffers from having a rather slow rate of crystallization (*8, 12*). Poly(ethylene terephthalate) homopolymer suffers from similar

Table II. 50% Alkylene Terephthalate/PTME Terephthalate

Diol	*2G*	*3G*
Copolymer properties		
Inherent viscosity (L/g)	0.13	0.17
Yield strength (MPa)	—	—
Stress at 100% (MPa)	11.4	11.7
Tensile strength (MPa)	45.5	22.8
Elongation (%)	675	660
Permanent set (%)	275	195
Tear strength (kN/m)	42	15
Shore D hardness	46	48
Compression set (%)	52	48
Clash–Berg $T_{10,000}$ (°C)	−38	−36
Melting point (°C)	224	198

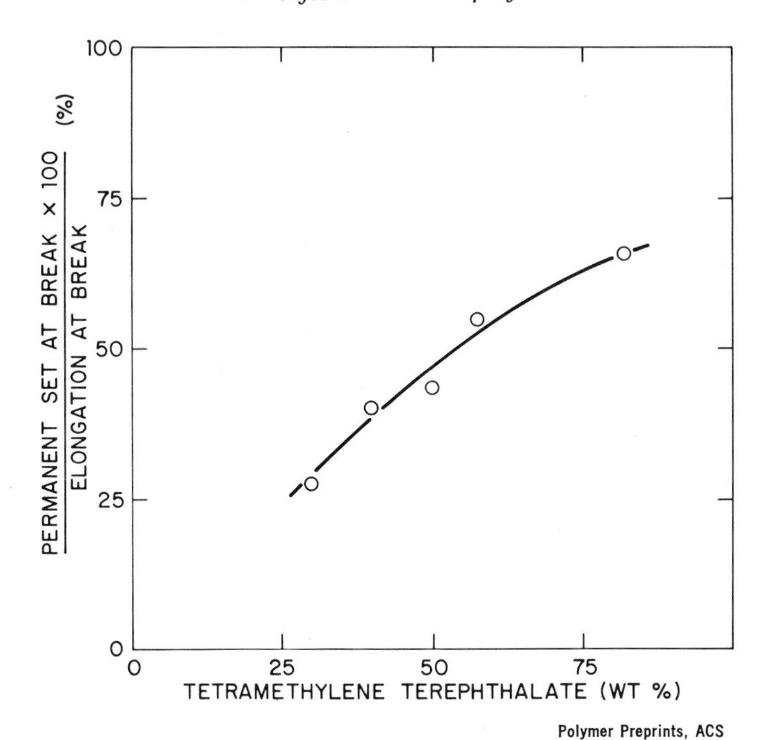

Figure 2. Permanent set at break as a percentage of elongation at break vs. the wt % of tetramethylene terephthalate segments in tetramethylene terephthalate/PTME terephthalate copolymers (36)

Copolymers—Properties as a Function of Diol Structure (36)

4G	*5G*	*6G*	*10G*
0.18	0.16	0.15	0.13
—	4.8	—	—
11.7	4.5	5.2	6.1
48.4	15.4	13.5	15.5
755	880	750	640
370	210	370	370
48	25	18	7.8
48	32	33	35
52	90	86	81
−33	−50	−53	−51
189	106	122	106

Table III. 1,4-Benzenedimethylene Terephthalate/PTME Terephthalate Copolymers—Properties as a Function of 1,4-Benzenedimethylene Terephthalate Content (36)

1,4-Benzenedimethylene Terephthalate (wt %)	*30*	*40*	*50*
Copolymer properties			
Inherent viscosity (L/g)	0.13	0.11	0.10
Stress at 100% (MPa)	5.9	9.0	11.7
Tensile strength (MPa)	19.2	21.7	21.3
Elongation (%)	628	500	380
Permanent set (%)	161	160	143
Tear strength (kN/m)	7.7	10.5	14.4
Shore D hardness	32	40	48
Compression set (%)	55	54	58
Clash–Berg $T_{10,000}$ (°C)	−61	−58	−42
Melting point (°C)	199	—	227

difficulties with crystallization rate (*16*). A slow rate of crystallization severely limits the use of injection-molding techniques to fabricate items from these polymers.

Synthesis of the 2G-based copolymer can be somewhat more difficult than synthesis of the analogous 4G-based copolymer. If the ethylene glycol and dimethyl terephthalate monomers are prereacted to form bis(2-hydroxyethyl) terephthalate, and this product is then copolymerized with poly(tetramethylene ether) glycol to form the block copolymer using tetrabutyl titanate as the transesterification catalyst, the reaction proceeds readily and copolymer of high inherent viscosity is easily obtained. If the ethylene glycol monomer is not prereacted and tetrabutyl titanate is again used as the transesterification catalyst, the copolymerization proceeds more slowly and a block copolymer of lower inherent viscosity is usually obtained.

Table IV. 1,4-Cyclohexanedimethylene Terephthalate/PTME of 1,4-Cyclohexanedimethylene

1,4-Cyclohexanedimethylene Terephthalate (wt %)
Copolymer properties
Inherent viscosity (L/g)
Stress at 100% (MPa)
Tensile strength (MPa)
Elongation (%)
Permanent set (%)
Tear strength (kN/m)
Shore D hardness
Compression set (%)
Clash–Berg $T_{10,000}$ (°C)

Trimethylene Terephthalate/PTME Terephthalate Copolymer. The properties of the 1,3-propanediol- or 3G-based copolymer are rather surprising. The melting point and the stress at 100% elongation of this copolymer are very similar to those of the 2G- and 4G-based copolymers, yet the 3G-based copolymer has much lower tensile strength at break and very low tear strength. The cause of these wide differences is unknown. Possibly related are studies carried out on oriented fibers which have shown that the crystalline regions of poly(trimethylene terephthalate) homopolymer respond quite differently to strain-induced deformation than do the crystalline regions of poly(ethylene terephthalate) or poly(tetramethylene terephthalate) homopolymers (*17*).

5G-, 6G-, and 10G-Based Copolymers. The alkylene terephthalate/PTME terephthalate co-polymers based on 1,5-pentanediol (5G), 1,6-hexanediol (6G), and 1,10-decanediol (10G) resemble each other more closely than they do those prepared from the shorter diols. The copolymers prepared from these longer diols all show, relative to those copolymers prepared from shorter diols, low melting point, low hardness, high compression set, low stress at 100% elongation, and low tensile strength. The tear strengths of the 5G-, 6G-, and 10G-based copolymers are also low relative to the 2G- and 4G-based copolymers. The 5G-, 6G-, and 10G-based copolymers excel in their resistance to stiffening at low temperatures as measured by the Clash–Berg test. It should be noted that the copolymers prepared from the longer diols contain a lower mole fraction of hard segments than do the copolymers prepared from the shorter diols.

A plot of the observed melting points of the 50% alkylene terephthalate/PTME terephthalate copolymers vs. the range of melting points of the poly(alkylene terephthalate) homopolymers as taken from the literature (*16, 18–25*) results in the linear relationship shown in Figure 3.

Terephthalate Copolymers—Properties as a Function Terephthalate Structure and Concentration

Practical CD				trans-*CD*	
20	*30*	*40*	*50*	*20*	*30*
0.12	0.12	0.11		0.13	insol
2.6	6.2	—		4.1	
5.9	11.0	8.1		6.8	
625	470	40	brittle	280	brittle
118	119	5		47	
3.2	6.0	10.2		4.6	
15	34	42		24	
75	62	53		60	
−55	−60	−53		−58	

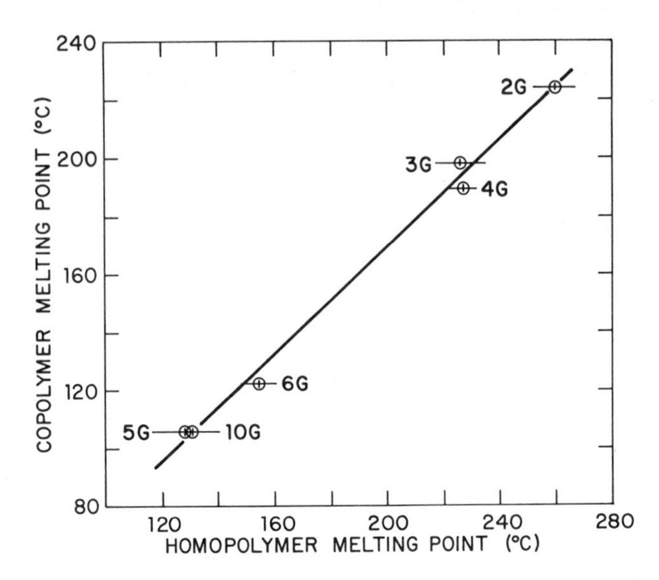

Figure 3. The melting points of 50% alkylene terephthalate/PTME terephthalate copolymers as a function of the melting points of the corresponding poly(alkylene terephthalate) homopolymers

A least-squares fit of the data using the center of the range of melting points reported for the homopolymers provided the relationship:

$$\text{mp (copolymer)} = 0.94 \text{ mp (homopolymer)} - 17 \quad (2)$$

1,4-Benzenedimethylene Terephthalate/PTME Terephthalate Copolymers. The 50% 1,4-benzenedimethylene terephthalate/PTME terephthalate co-polymer of Table III has a high stress at 100% elongation comparable to the 2G- and 4G-based copolymers of Table II but is low in inherent viscosity, tensile strength, elongation at break, and tear strength. The poor failure properties of the 1,4-benzenedimethanol-based copolymer can be attributable in part to the low molecular weight of the co-polymer as indicated by its low inherent viscosity. The tensile strengths at break of similar 4G-based copolymers have been shown to decrease as their inherent viscosities are reduced (*12*).

1,4-Cyclohexanedimethylene Terephthalate/PTME Terephthalate Copolymers. Polyether-ester copolymers based on 1,4-cyclohexanedimethanol have been prepared using both *trans*-1,4-cyclohexanedimethanol (*t*-CD) and a practical grade of 1,4-cyclohexanedimethanol (p-CD) which is reported to contain about a 68/32 mixture of *trans*- to *cis*-1,4-cyclohexanedimethanol (*26*). Gas-phase chromatography of one of our samples of p-CD indicated a 72/28 ratio of trans-to-cis isomer. By use of practical grade, 1,4-cyclohexanedimethanol, copolymers could be prepared containing 20 and 30% of 1,4-cyclohexanedimethylene terephthalate (p-CD) units whose stress at 100% elongation, tensile strength, tear strength, and Shore D hardness increase with increasing CDT content (Table IV). At 40% p-CDT content, tensile strength and elongation at break of the copolymer drop off from the values obtained with the 30% p-CDT copolymer because of phase separation in the melt during the synthesis of the 40% p-CDT copolymer. Phase separation during the synthesis of the 50% p-CDT copolymer was sufficiently severe that the resulting copolymer was a brittle solid rather than an elastomer. When *trans*-1,4-cyclohexanedimethanol is used as the diol monomer, severe phase separation during copolymerization occurs at even lower levels of CDT than when using practical-grade 1,4-cyclohexanedimethanol. The 30% *t*-CDT copolymer resembles the 50% p-CDT copolymer in that it is not elastomeric but brittle (Table IV).

Phase separation during melt copolymerization was reported in the first paper of this series, particularly with tetramethylene terephthalate/poly(propylene oxide) terephthalate copolymers prepared from the higher molecular weight polyether glycols (*10*). Phase separation occurs during the preparation of polyether–ester copolymers when the copolymerizable segments are mutually insoluble in each other. The result is a reaction mixture containing two phases, as often evidenced by the melt having an opaque rather than the usual transparent or translucent appearance. As noted in the previous paper (*10*) and as shown in Table IV, severe phase separation results in copolymers which exhibit low elongation at break and very often low tensile and tear strengths. The conclusion that the very low elongation at break of the 40% p-CDT copolymer (Table IV) is caused by phase separation during copolymerization is supported by the observation that the polymerizing melt of the 40% p-CDT copolymer was opaque while that of the 30% p-CDT copolymer was transparent.

There are several techniques available to help overcome the problem of phase separation during melt copolymerization. One technique is to add a high-boiling solvent to the polymerization melt to assist in solubilizing the copolymerizing segments. A second technique is to introduce other monomers into the copolymerization such as a second diol, a second ester (*27*), or both. The additional monomers serve to lower the concentration and melting point of any hard segment which might tend to separate out. The use of practical-grade 1,4-cyclohexanedimethanol which contains both the cis and trans isomers rather than using pure *trans*-1,4-cyclohexanedimethanol is a method of adding a second diol while simultaneously lowering the concentration of the first diol. The effectiveness of this method is evident from the results shown in Table IV. A 30%

1,4-cyclohexanedimethylene terephthalate/PTME terephthalate copolymer could be prepared using practical-grade 1,4-cyclohexanedimethanol whereas the analogous copolymer based on *trans*-1,4-cyclohexanedimethanol underwent severe phase separation in the melt during synthesis.

Alkylene Ester/PTME Ester Copolymers Other Than Terephthalate Ester

Up to this point the discussion has been concerned with alkylene terephthalate/PTME terephthalate copolymers in which the concentration of alkylene terephthalate and the chemical structure of the alkylene groups have been varied. The next section of this report is concerned with polyether–ester copolymers in which aromatic esters other than terephthalate are used in combination with PTME glycol and various diols. The objective is the same, to correlate changes in copolymer structure with changes in copolymerization results and copolymer properties. Once again the 50% tetramethylene terephthalate/PTME terephthalate copolymer (Tables I and II) with its excellent properties and relative ease of synthesis will be used as the point of reference to which the other polymers will be compared.

Alkylene Isophthalate/PTME Isophthalate Copolymers. Polyether-ester copolymers having the compositions 50% alkylene isophthalate/PTME isophthalate were prepared using as diols ethylene glycol (2G), 1,4-butanediol (4G), practical-grade 1,4-cyclohexanedimethanol (p-CD), and *trans*-1,4-cyclohexanedimethanol (*t*-CD) as shown in Table V. The 2G-based copolymer, which was prepared using bis(2-hydroxyethyl) isophthalate, had no discernible melting point as measured by differential scanning calorimetry. It exhibited low hardness, low stress at 100% elongation, and very low tensile strength. Apparently the ethylene isoph-

Table V. 50% Alkylene Isophthalate/PTME Isophthalate

Diol	*2G*
Copolymer properties	
Inherent viscosity (L/g)	0.16
Stress at 100% (MPa)	0.7
Tensile strength (MPa)	< 0.7
Elongation (%)	> 1000
Permanent set (%)	> 470
Tear strength (kN/m)	—
Shore D hardness	7
Compression set (%)	—
Clash–Berg $T_{10,000}$ (°C)	−37
Melting point (°C)	none

thalate segments did not crystallize, even though the homopolymer poly(ethylene isophthalate) is reported to have a melting point in the range 102°–240°C (*20, 23, 28*). Annealing this polyether–ester copolymer above the glass-transition temperature of poly(ethylene isophthalate) might possibly induce some crystallization of ester segments although this was not attempted.

The 50% tetramethylene isophthalate/PTME isophthalate copolymer listed under 4G in Table V exhibits outstanding tensile strength and excellent tear strength even though this copolymer is considerably lower in melting point, Shore hardness, and stress at 100% elongation than is the analogous 50% tetramethylene terephthalate/PTME terephthalate copolymer of Tables I and II. In contrast to the terephthalate-based copolymer which crystallizes and hardens in seconds, the 50% tetramethylene isophthalate/PTME isophthalate copolymer crystallizes very slowly over a period of hours. The isophthalate-based copolymer exhibits a double endotherm by differential scanning calorimetry with peaks at 85° and 112°C. The 85°C peak is slightly the larger of the two. Poly(tetramethylene isophthalate) homopolymer has reported melting points of 88°–152°C (*20, 28*).

Fifty percent 1,4-cyclohexanedimethylene isophthalate/PTME isophthalate copolymers prepared from practical-grade 1,4-cyclohexanedimethanol and from *trans*-1,4-cyclohexanedimethanol also harden slowly over a period of hours. No attempts were made to speed the rate of crystallization of these copolymers by addition of nucleating agents although it has been demonstrated that the rate of crystallization of ethylene terephthalate/PTME terephthalate copolymers, which also crystallize slowly, can be increased by nucleation.

Unlike their terephthalate analogs of Table IV, the 1,4-cyclohexanedimethylene isophthalate/PTME isophthalate copolymers show no evi-

Copolymers—Properties as a Function of Diol Structure (36)

4G	*Practical CD*	trans-*CD*
0.16	0.14	0.11
7.2	5.7	6.9
58.6	31.7	33.8
720	690	760
126	70	179
54	78	61
39	38	46
91	89	87
−31	−38	−28
85,112	147	184

Polymer Preprints, ACS

dence of phase separation during copolymerization. The difference in behavior during synthesis may be caused by the lower melting points of the isophthalate copolymers. Poly(*trans*-1,4-cyclohexanedimethylene isophthalate) homopolymer is reported to have a melting range of 190°–197°C while poly(*trans*-1,4-cyclohexanedimethylene terephthalate) has a melting range of 312°–318°C (*29*). This suggests that if the 1,4-cyclohexanedimethylene terephthalate/PTME terephthalate copolymerizations could be run at higher temperatures, phase separation in the melt might be less of a problem. Unfortunately higher polymerization temperatures would increase the rate of polymer degradation which is known to occur at a significant rate even at the melt temperatures used in the preparation of the copolymers described in this report (*13, 30*). (Two of the three curves in Figure 7 of Ref. *13* are mislabeled. The "Polytetramethylenoxid" and "Polypropylenoxid" labels should be reversed (*31*).)

Alkylene 4,4′-Biphenyldicarboxylate/PTME 4,4′-Biphenyldicarboxylate Copolymers. Tetramethylene 4,4′-biphenyldicarboxylate/PTME 4,4′-biphenyldicarboxylate copolymers containing 20 and 30% tetramethylene 4,4′-biphenyldicarboxylate were prepared without incident (Table VI). Attempts to prepare similar copolymers containing 40 and 50% tetramethylene 4,4′-biphenyldicarboxylate led to problems with phase separation in the melt during the copolymerizations.

Fifty percent 4,4′-biphenyldicarboxylate/PTME 4,4′-biphenyldicarboxylate copolymers were prepared using 1,3-propanediol (3G), 1,5-pentanediol (5G), and 1,6-hexanediol (6G) (Table VII). All were prepared without incident. The 3G-based copolymer has poor tear strength. The 5G-based copolymer has good tear strength but low tensile strength at break. The 6G-based copolymer has the best tensile strength of the three copolymers but has low tear strength.

Table VI. Tetramethylene 4,4′-Biphenyldicarboxylate/PTME 4,4′-Biphenyldicarboxylate Copolymers—Properties as a Function of Tetramethylene 4,4′-Biphenyldicarboxylate Concentration

Tetramethylene 4,4′-biphenyldicarboxylate (wt %)	*20*	*30*	*40*
Copolymer properties			
Inherent viscosity (L/g)	0.20	0.21	insol
Stress at 100% (MPa)	2.8	5.5	8.0
Tensile strength (MPa)	9.7	10.3	8.3
Elongation (%)	1050	600	200
Permanent set (%)	283	175	69
Tear strength (kN/m)	5	20	10
Shore D hardness	22	33	32
Compression set (%)	65	49	67
Clash–Berg $T_{10,000}$ (°C)	−59	−58	—

Table VII. 50% Alkylene 4,4′-Biphenyldicarboxylate/PTME 4,4′-Biphenyldicarboxylate Copolymers—Properties as a Function of Diol Structure

Diol	*3G*	*5G*	*6G*
Copolymer properties			
Inherent viscosity (L/g)	0.12	0.12	0.13
Stress at 100% (MPa)	9.8	5.5	7.8
Tensile strength (MPa)	11.0	7.6	15.2
Elongation (%)	560	490	510
Permanent set (%)	190	191	329
Tear strength (kN/m)	11	40	16
Shore D hardness	50	35	40
Compression set (%)	61	82	67
Clash–Berg $T_{10,000}$ (°C)	−40	−40	−35

Alkylene 2,6-Naphthalenedicarboxylate/PTME 2,6-Naphthalenedicarboxylate Copolymers. Fifty percent alkylene 2,6-naphthalenedicarboxylate/PTME 2,6-naphthalenedicarboxylate copolymers were prepared using each of the straight-chain, hydroxy-terminated diols from ethylene glycol (2G) to 1,10-decanediol (10G) (Table VIII). In contrast to many of the 50% alkylene terephthalate/PTME terephthalate copolymers of Table II, all of the 2,6-naphthalenedicarboxylate-based copolymers tested exhibit excellent tensile strength and tear strength regardless of the diol used or the melting point of the copolymer. As a consequence of their excellent properties, the 2,6-naphthalenedicarboxylate copolymers have been the subject of several patents (*32, 33, 34*).

Permanent set values are relatively high for the 2,6-naphthalenedicarboxylate copolymers prepared from diols containing an even number of carbon atoms but quite low for the corresponding copolymers prepared from odd-membered diols, particularly the 3G, 5G, and 7G diols. Strain-induced irreversible disruption and orientation of crystalline ester segments has been advanced as the cause of the high permanent sets of tetramethylene terephthalate/PTME terephthalate copolymers (*14*). On this basis the crystalline ester segments of the 2,6-naphthalenedicarboxylate copolymers prepared from odd-membered diols appear to be considerably more resistant to strain-induced irreversible deformation than do the crystalline segments of the corresponding copolymers prepared from even-membered diols. This in turn suggests the presence of a persistent pattern of morphological differences between the crystalline ester segments derived from the even-membered diols and those derived from the odd-membered diols.

A plot of the measured melting points of the 50% alkylene 2,6-naphthalenedicarboxylate/PTME 2,6-naphthalenedicarboxylate copolymers of Table VIII vs. the reported melting points of the corresponding

Table VIII. 50% Alkylene 2,6-Naphthalenedicarboxylate/PTME Function of Diol

Diol	*2G*	*3G*	*4G*
Copolymer properties			
Inherent viscosity (L/g)	0.11	0.17	0.15
Yield strength (MPa)	—	—	—
Stress at 100% (MPa)	6.2	7.3	10.0
Tensile strength (MPa)	35.9	54.5	51.0
Elongation (%)	460	525	660
Permanent set (%)	102	16	280
Tear strength (kN/m)	151	79	117
Shore D hardness	53	47	49
Compression set (%)	67	88	49
Clash–Berg $T_{10,000}$ (°C)	1	−15	−14
Melting point (°C)	232	168	202

poly(alkylene-2,6-naphthalenedicarboxylate) homopolymers (22) results in the linear relationship shown in Figure 4. A least squares fit of the data provided the relationship:

$$\text{mp (copolymer)} = 0.87 \text{ mp (homopolymer)} - 4.3 \quad (3)$$

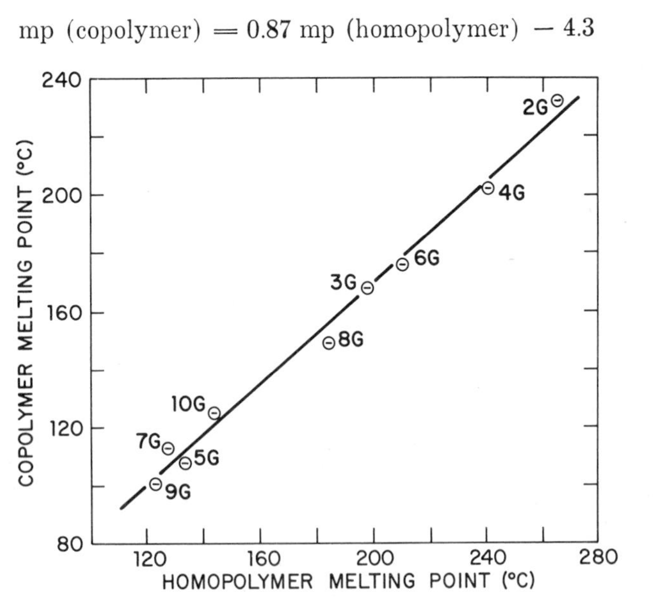

Figure 4. The melting points of 50% alkylene 2,6-naphthalenedicarboxylate/PTME 2,6-naphthalenedicarboxylate copolymers as a function of the melting points of the corresponding poly(alkylene 2,6-naphthalenedicarboxylate) homopolymers

2,6-Naphthalenedicarboxylate Copolymers—Properties as a Structure (36)

5G	6G	7G	8G	9G	10G
0.18	0.20	0.20	0.20	0.15	0.17
7.4	—	7.6	8.1	—	6.6
6.8	8.8	6.8	7.9	3.7	6.4
52.1	51.0	51.7	44.1	38.6	36.9
590	570	540	620	625	620
65	253	60	288	147	318
123	123	103	82	69	93
45	47	49	48	37	41
94	67	93	73	99	85
−5	−24	−8	−29	−34	−27
108	176	113	149	101	125

Polymer Preprints, ACS

The melting points of the 2,6-naphthalenedicarboxylate homopolymers and their corresponding copolymers containing PTME units show a distinct odd–even alternation. Polymers based on odd-membered diols are consistently somewhat lower melting than their even-membered neighbors. This type of behavior has been noted for other series of polyesters and has been attributed to "a consistent pattern of configurational and chain-packing differences between odd and even members in homologous series" (22). Thus both the melting points and the permanent sets of the alkylene 2,6-naphthalenedicarboxylate/PTME 2,6-naphthalenedicarboxylate copolymers show patterns of alteration which suggest that consistent differences in morphology exist between the crystalline ester segments derived from the even-membered diols and those derived from the odd-membered diols.

A plot of compression-set results vs. copolymer melting points for the 50% alkylene terephthalate/PTME terephthalate copolymers of Table II and the 50% alkylene 2,6-naphthalenedicarboxylate/PTME 2,6-naphthalenedicarboxylate copolymers of Table VIII indicates that compression set generally increases as copolymer melting point decreases (Figure 5). This is not surprising as the lower the melting point of the copolymer, the closer the melting point is to the 70°C temperature at which the compression set test is run and thus the more susceptible are the crystalline hard segments to distortion. This is particularly true of the copolymers melting close to 100°C. The reported melting points represent only the peaks of the endothermic melting curves as measured by differential scanning calorimetry. The melting curves often cover a range of temperatures as wide as 20°C on each side of the melting-curve peaks. Thus the lower end of the melting ranges of these low-melting

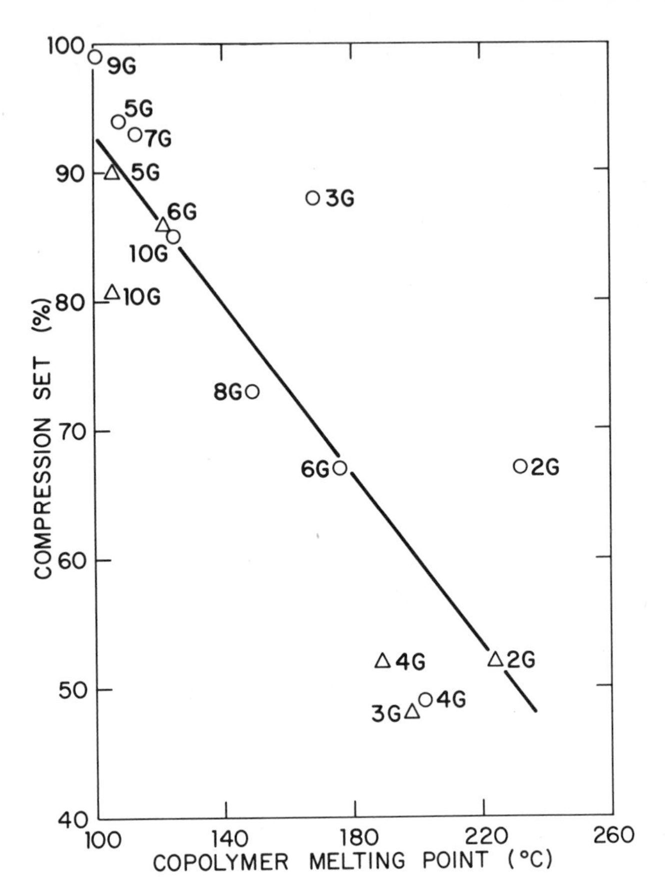

Figure 5. Copolymer compression set as a function of copolymer melting point for 50% alkylene ester/PTME ester copolymers (Δ = terephthalate; O = 2,6-naphthalene-dicarboxylate)

copolymers may lie very close to the 70°C temperature of the compression set test.

Alkylene *m*-Terphenyl-4,4″-dicarboxylate/PTME *m*-Terphenyl-4,4″-dicarboxylate Copolymers. Polyether–ester copolymers with the composition 50% alkylene *m*-terphenyl-4,4″-dicarboxylate/PTME *m*-terphenyl-4,4″-dicarboxylate were prepared using as diols 1,3-propanediol (3G) and 1,4-butanediol (4G). Both copolymers exhibit excellent tensile and tear strength as shown in Table IX. They both have very poor resistance to compression set.

After compression molding, both copolymers were initially transparent. Gradually over a period of many days the 4G-based copolymer turned opaque as if it were crystallizing (35). The 3G-based copolymer

Table IX. 50% Alkylene *m*-Terphenyl-4,4″-dicarboxylate/PTME *m*-Terphenyl-4,4″-dicarboxylate Copolymers[a]—Properties as a Function of Diol Structure (36)

Diol	*3G*	*4G*
Copolymer properties		
Inherent viscosity (L/g)	0.15	0.17
Stress at 100% (MPa)	5.7	2.8
Tensile strength (MPa)	56.2	60.3
Elongation (%)	420	390
Permanent set (%)	15	3
Tear strength (kN/m)	98	47
Shore A hardness	94	76
Shore D hardness	43	31
Compression set (%)	100+	100+
Clash–Berg $T_{10,000}$ (°C)	8	2

[a] *m*-Terphenyl-4,4″-dicarboxylate = —O—C(=O)—C₆H₄—C₆H₄—C₆H₄—C(=O)—O—.

Polymer Preprints, ACS

remained transparent over the same time span; however, that portion of the 3G-based copolymer which had been subjected to the 70°C test for compression set also turned opaque as if it had crystallized. In their transparent form neither copolymer exhibited a melting point as measured by differential scanning calorimetry.

The 50% tetramethylene *m*-terphenyl-4,4″-dicarboxylate/PTME *m*-terphenyl-4,4″-dicarboxylate copolymer was remolded. The remolded copolymer was tested three days after remolding while it was still transparent and 28 days after remolding, by which time it had turned opaque. The results are shown in Table X. The sample which was 28 days old had turned from transparent to opaque, had developed endothermic peaks at 63° and 158°C in the differential scanning calorimetry

Table X. 50% Tetramethylene *m*-Terphenyl-4,4″-dicarboxylate/PTME *m*-Terphenyl-4,4″-dicarboxylate Copolymer—Effect of Time on Physical Properties of Remolded Copolymer

Days after Remolding	*3*	*28*
Copolymer properties		
Stress at 100% (MPa)	2.2	7.9
Stress at 300% (MPa)	17.4	30.3
Tensile strength (MPa)	51.9	49.3
Elongation (%)	383	375
Permanent set (%)	3	3
Shore A hardness	74	89
Appearance	transparent	opaque

trace (Figure 6), and had shown a marked increase in Shore A hardness and in stress at 100 and 300% elongation. Apparently the 4G-based copolymer slowly crystallized over a period of time. The 158°C endothermic peak is presumably caused by crystalline tetramethylene *m*-terphenyl-4,4″-dicarboxylate blocks. Poly(tetramethylene *m*-terphenyl-4,4″-dicarboxylate) homopolymer is reported to melt at 208°C (*11*).

If the 4G-based copolymer is crystallizing slowly with time, what is holding together the newly remolded copolymer to give it such excellent tensile strength and tear strength? Possibly it is a glassy phase composed of tetramethylene *m*-terphenyl-4,4″-dicarboxylate segments. Poly(tetramethylene *m*-terphenyl-4,4″-dicarboxylate) is reported to have a glass-transition temperature of 110°C (*11*).

Other explanations which have been suggested to account for the high strength of the newly remolded copolymer are either strain-induced crystallization of the tetramethylene *m*-terphenyl-4,4″-dicarboxylate hard segments during the tensile strength and tear strength tests or the presence of microcrystalline hard segments of a size sufficiently small for the polymer to retain its transparent appearance. Strain-induced crystallization of the hard segments is considered an unlikely explanation in view of the low permanent set of the newly remolded copolymer. After removal of the strain, crystalline hard segments induced by strain would be expected to remain crystalline at least in part, since the crystalline form of the hard segment is the stable form at room temperature as indicated by the spontaneous crystallization of the 4G-based copolymer.

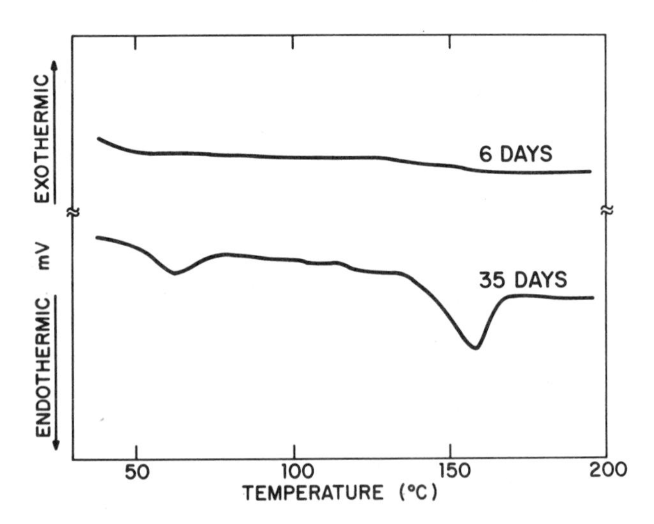

Figure 6. Differential scanning calorimetry traces of remolded 50% tetramethylene m*-terphenyl-4,4″-dicarboxylate/PTME* m*-terphenyl-4,4″-dicarboxylate copolymers taken 6 and 35 days after remolding*

Retention of strain-induced crystals would be expected to result in permanent set values for the 4G-based copolymer considerably higher than the 3% observed. The other alternative explanation, the presence of microcrystalline hard segments, appears unlikely as there is no melting endotherm in the differential scanning calorimetry trace of the newly remolded copolymer. The presence of a glassy hard-segment phase composed of alkylene *m*-terphenyl-4,4″-dicarboxylate segments seems the most likely explanation at this time for the high tensile and tear strengths of the newly remolded copolymer.

Summary and Conclusions

Elastomeric polyether–ester block copolymers were prepared by melt transesterification of poly(tetramethylene ether) glycol of molecular weight approximately 1000 with a variety of diols and esters. The ease of synthesis and the properties of these thermoplastic copolymers have been related to the chemical structure and concentration of the ester hard segments.

Among the terephthalate-based polyether–ester copolymers, those prepared using 1,4-butanediol as the diol monomer exhibit the best overall physical properties. The use of ethylene glycol as the diol monomer retards the rate of polymer formation and results in copolymers which crystallize slowly. Other aliphatic α,ω-diols yield terephthalate-based polyether–ester copolymers which are low in tensile strength and tear strength relative to the 1,4-butanediol-based copolymer. Terephthalate-based copolymers prepared with 1,4-benzenedimethanol as the diol monomer are relatively low in inherent viscosity, tensile strength, and tear strength.

When 1,4-cyclohexanedimethanol is used as the diol monomer, phase separation occurs in the polymerizing melts of terephthalate-based polyether-ester copolymers but not those of the analogous isophthalate-based copolymers. Phase separation in the melt during copolymerization, if sufficiently severe, drastically impairs the properties of the product.

All of the isophthalate-based polyether–ester copolymers which were prepared using various diols are slow to crystallize. Those copolymers that eventually do crystallize exhibit excellent tensile strength and tear strength. All of the 2,6-naphthalenedicarboxylate copolymers prepared using linear α,ω-diols exhibit excellent tensile strength and tear strength. Phase separation in the melt is encountered with 4,4′-biphenyldicarboxylate-based copolymer at the 50-wt % ester level when 1,4-butanediol is used as the diol monomer but not when 1,3-propanediol, 1,5-pentanediol, and 1,6-hexanediol are used.

Polyether–ester copolymers based on *m*-terphenyl-4,4″-dicarboxylate exhibit excellent tensile strength and tear strength despite an apparent absence of crystallinity of the ester groups. Over a period of weeks the 1,4-butanediol-based copolymer gradually crystallizes, accompanied by an increase in the modulus and hardness of the copolymer.

It is evident from this study that combining any high-melting ester component with the best available polyether segment is not sufficient to insure that the resulting block copolymer will have the properties desired of a high-strength, thermoplastic elastomer. The structure of the ester segment has an important effect on many properties including ease of synthesis, molecular weight obtained, rate of crystallization or hardening, tensile strength, tear strength, and elastic recovery.

Literature Cited

1. Noshay, A., McGrath, J. E., "Block Copolymers," pp. 313–328, 432–435, Academic, New York, 1977.
2. Kalfoglou, N. K., *J. Appl. Polym. Sci.* (1977) **21,** 543.
3. Hoeschele, G. K., *Angew. Makromol. Chem.* (1977) **58/59** (Nr 841), 299.
4. Lilaonitkul, A., Cooper, S. L., *Rubber Chem. Technol.* (1977) **50,** 1.
5. Lilaonitkul, A., West, J. C., Cooper, S. L., *J. Macromol. Sci., Phys.* (1976) **B12**(4), 563.
6. Nishi, T., Kwei, T. K., *J. Appl. Polym. Sci.* (1976) **20,** 1331.
7. Steggerda, B. F., Rijnders, R. F. R. T., *SGF Publ. (Environ. Quest., TPE, Test.)* (1976) **48,** C5.
8. Ijzermans, A. B., Pluijm, F. J., Huntjins, F. J., Repin, J. F., *Br. Polym. J.* (1975) **7,** 211.
9. West, J. C., Lilaonitkul, A., Cooper, S. L., Mehra, U., Shen, M., *Polymer Prepr.* (1974) **15**(2), 191.
10. Wolfe, J. R., Jr., *Rubber Chem. Technol.* (1977) **50,** 688.
11. Watson, W. H., U.S. Patent **3,350,354** (1967).
12. Witsiepe, W. K., Adv. Chem. Ser. (1973) **129,** 39.
13. Hoeschele, G. K., *Chimia* (1974) **28,** 544.
14. Cella, R. J., *J. Polym. Sci., Polym. Symp.* (1973) **42,** 727.
15. Seymour, R. W., Overton, J. R., Corley, L. S., *Macromolecules* (1975) **8,** 331.
16. Duling, I. N., Stearns, R. S., Scott, M. A., Scott, K. A., U.S. Patent **3,546,320** (1970).
17. Jakeways, R., Ward, I. M., Wilding, M. A., Hall, I. H., Desborough, I. J., Pass, M. G., *J. Polym. Sci., Polym. Phys. Ed.* (1975) **13,** 799.
18. Edgar, O. B., Ellery, E., *J. Chem. Soc.* (1952) 2633.
19. Dole, M., Wunderlich, B., *Makromol. Chem.* (1959) **34,** 29.
20. Korshak, V. V., Vinogradova, S. V., Belyakov, V. M., *Izv. Akad. Nauk. SSSR, Otd. Khim. Nauk.* (1957) 730.
21. Edgar, O. B., Hill, R., *J. Polym. Sci.* (1952) **8,** 1.
22. Goodman, I., "Polyesters," *in* "Encyclopedia of Polymer Science and Technology," N. M. Bikales, Ed., Vol. 11, p. 62, Wiley, New York, 1969.
23. Hill, R., Walker, E. E., *J. Polym. Sci.* (1948) **3,** 609.
24. Whinfield, J. R., *Nature* (1946) **158,** 930.
25. Flory, P. J., Bedon, H. D., Keefer, E. H., *J. Polym. Sci.* (1958) **28,** 151.
26. "1,4-Cyclohexanedimethanol," Eastman Industrial Chemicals Publication No. **N-199B,** Eastman Kodak Co., January 1976.
27. Bell, A., Kibler, C. J., Smith, J. G., U.S. Patent **3,261,812** (1966).
28. Conix, A., Van Kerpel, R., *J. Polym. Sci.* (1959) **40,** 521.
29. Kibler, C. J., Bell, A., Smith, J. G., *J. Polym. Sci., Part A* (1964) **2,** 2115.

30. Hoeschele, G. K., Witsiepe, W. K., *Angew. Makromol. Chem.* (1973) **29/30**, 267.
31. Hoeschele, G. K., private communication.
32. Wolfe, J. R., Jr., U.S. Patent **3,775,374** (1973).
33. Wolfe, J. R., Jr., U.S. Patent **3,775,375** (1973).
34. Oka, I., Tanigawa, Y., Kawase, S., Shima, T., *Jpn. Kokai* **73 96,693** (1973); *Chem. Abstr.* (1974) **80**, 134543q.
35. Buck, W. H., Cella, R. J., Gladding, E. K., Wolfe, J. R., Jr., *J. Polym. Sci., Polym. Symp.* (1974) **48**, 47.
36. Wolfe, James R., Jr., *Polym. Prepr., Am. Chem. Soc., Div. Polym. Chem.* (1978) **19**(1), 5.

RECEIVED April 14, 1978. Contribution No. 425.

8

Phase Separation and Mechanical Properties of an Amorphous Poly(ether-b-ester)

M. P. C. WATTS[1] and E. F. T. WHITE

Department of Polymer and Fibre Science, University of Manchester, Institute of Science and Technology, Manchester, England

The dynamic mechanical properties and phase separation of a series of amorphous condensation multiblock polymers, poly(oxypropylene-b-oxypropylene oxyterephthaloyl), was studied. Samples prepared from poly(oxypropylene) with a molecular weight of 3000 showed clear evidence of block phase separation from small angle x-ray scattering and two transition regions in torsional braid analysis. As the molecular weight of the poly(oxypropylene) was reduced, the stages between clear phase separation and formation of a homogeneous mixture were observed. The size and shape of the mechanical damping peaks could be controlled by the block size, the weight fractions of the two components, and the chemical composition of the components.

Polymers are often used in sound absorption and vibration damping applications (*1*), for example, in reducing hull noise in ships (*2*) and in sound-proof helmets (*3*). Their efficiency in these applications is related to the tan δ of the polymer in the range of frequencies and temperatures found in the particular application. A major limitation of conventional amorphous homopolymers is that the region of high tan δ (> 0.8) extends over only 20°–30°C, at 1 Hz, in the glass–rubber transition region (*4*). There have been many attempts to broaden this damping peak by the addition of plasticizers and fillers, but with only limited success (*5, 6*). However, the width of the peak can be readily controlled in polymer blends (*7, 8, 9*). Recently, interpenetrating networks (IPN) have been prepared that show exceptionally broad and high tan δ maxima,

[1] Current address: Materials Research Laboratory, Polymer Science and Engineering, University of Massachusetts, Amherst, MA 01003.

0-8412-0457-8/79/33-176-153$06.50/0

and they have properties in damping applications that are significantly better than other commercially available damping materials (*5*). It is thought that the broad maxima in tan δ observed in IPNs arises from a partial phase separation of the network chains such that regions of continuously varying composition are found rather than the regions of pure homopolymer normally observed in polymer blends. Each volume element with a given composition contributes to the overall properties, leading to a broad tan δ peak.

The tendency of an amorphous multicomponent polymer system to phase separate is driven by the large negative enthalpy of the phase-separated state relative to a homogeneous mixture (*11*). In terms of the Scott–Hildebrand solution theory (*10*), this negative enthalpy is related to the difference in solubility parameter of the components and their molecular weight. The tendency to phase separate is opposed by the loss in entropy caused by the reduction in chain configurations on phase separation. In principle, polymer blocks, grafts, blends, and IPNs could all show partial phase separation with a suitable choice of polymers and chain topology.

The remarkable properties of amorphous block polymers, particularly poly(styrene-b-diene), have been studied extensively in recent years. Many studies (*10, 35, 36, 37, 38*) suggest that the degree of phase separation depends on the length of, the number of, and the chemical composition of the blocks.

The aim of the present study was to prepare multiblock polymers having a suitable combination of these three variables so that the change between a polymer showing phase separation to one behaving as a homogeneous mixture could be observed in a thermoplastic. In this intermediate region, properties similar to those of IPNs might be expected. The properties of these materials also would be of interest in comparison with the semicrystalline multiblock polymers such as the segmented polyurethanes and poly(ether-b-ester)s that are available commercially.

Approach

A series of amorphous condensation multiblock polymers were prepared comprising poly(oxypropylene oxyterephthaloyl) (3IGT), a brittle glassy solid, as the hard block and poly(oxypropylene) (PPO) or poly(di(oxyethylene)oxyadipoyl) (PDEGA) as the soft block.

Polymer Preparation

The polymerization method used was a polycondensation of terephthaloyl chloride, propane 1,2-diol, and the appropriate soft block (Figure 1) in a 20% solution in dry 1,2-dichloroethane at 85°C for 48 hours.

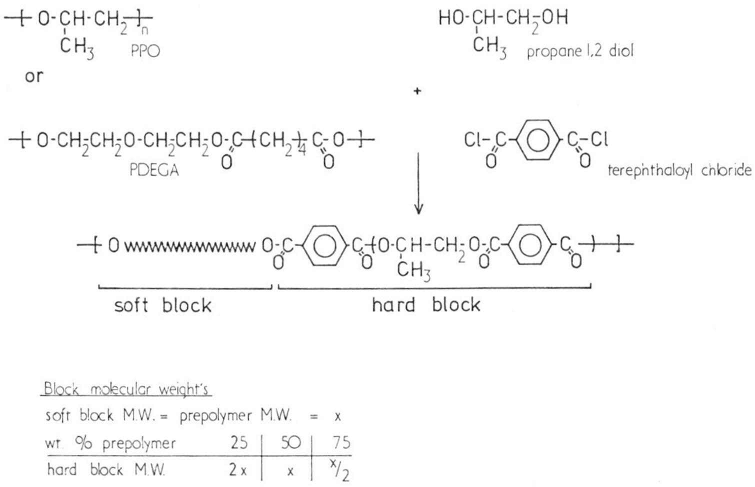

$-[O-CH(CH_3)-CH_2]_n-$ PPO

or

$-[O-CH_2CH_2O-CH_2CH_2O-C(=O)-(CH_2)_4-C(=O)-O]-$ PDEGA

$HO-CH(CH_3)-CH_2-OH$ propane 1,2 diol

+

$Cl-C(=O)-C_6H_4-C(=O)-Cl$ terephthaloyl chloride

↓

$-[O$ wwwwwwwwww $O-C(=O)-C_6H_4-C(=O)-[O-CH(CH_3)-CH_2-O-C(=O)-C_6H_4-C(=O)]-]-$

soft block | hard block

Block molecular weight's

soft block M.W. = prepolymer M.W. = x

wt % prepolymer	25	50	75
hard block M.W.	2x	x	x/2

Figure 1. Preparation of the multiblock polymers

The molar ratio of soft block to propane 1,2-diol and terephthaloyl chloride was A:B:(A + B) × 1.01. Four moles of pyridine per mole of acid chloride was added to react with the HCL produced in the polymerization. Pyridine hydrochloride was removed at the end of the reaction by washing with dilute HCl.

Multiblock polymers were prepared from PPO prepolymers with molecular weights 400, 1000, 2000, 3000, and weight fractions of PPO 25, 50, and 75%. PPO "homopolymers" were prepared by reacting PPO prepolymer with terephthaloyl chloride alone. A sample also was prepared containing 50% PDEGA of mol wt 2550.

The polymers were characterized by viscometry and GPC analysis. The viscosities were measured as 0.5 g dl^{-1} solutions in 1,1,2,2-tetrachloroethane at 25°C. Gel permeation chromatograms were measured in dimethylacetamide solution at 80°C. A typical chromatogram of 3IGT homopolymer is given in Figure 2. There is a small secondary peak at high elution volume. This secondary peak appeared in all of the polymers prepared from terephthaloyl chloride, and the elution volume of the maximum was constant, independent of the presence of PPO. This secondary peak is thought to be attributable to residual terephthalic acid and cyclic dimer and trimer that are formed in the early stages of polycondensation reactions (*15*). The simple construction shown in Figure 2 was used to remove the low-molecular-weight peak, and the chromatograms were converted to poly(styrene) molecular weight using a computer program supplied by the Department of Chemistry, University of Manchester. All results are given in Table I. The overall molecular weight of these polymers is estimated as 12,000–20,000 from viscosity

Table I. Data

Polymer	Soft Block (mol wt)	Soft Block (approx wt frac)	Wf_{PPO}	Wf_{TR}	$\bar{s}$
PPO/3IGT	400	100	0.720	0.280	
		75	0.526	0.420	2.03
		50	0.421	0.500	1.53
		25	0.215	0.637	1.17
	1000	100	0.869	0.131	
		75	0.678	0.266	1.65
		50	0.469	0.415	1.22
		25	0.226	0.584	1.08
	2000	100	0.930	0.070	
		75	0.724	0.216	1.35
		50	0.548	0.341	1.15
		25	0.314	0.511	1.06
	3000	100	0.951	0.049	
		75	0.680	0.242	1.18
		50	0.486	0.380	1.08
		25	0.255	0.544	1.03
PDEGA/3IGT	2550	100	0.944	0.059	
		50	0.55	0.329	1.11
3IGT/C	—	—	0	0.718	

[a] T_g of 3IGT homopolymer with the same molecular weight as the hard block.
[b] Single-point viscosities 0.5 g · dl^{-1} in tetrachlorethane at 25°C.

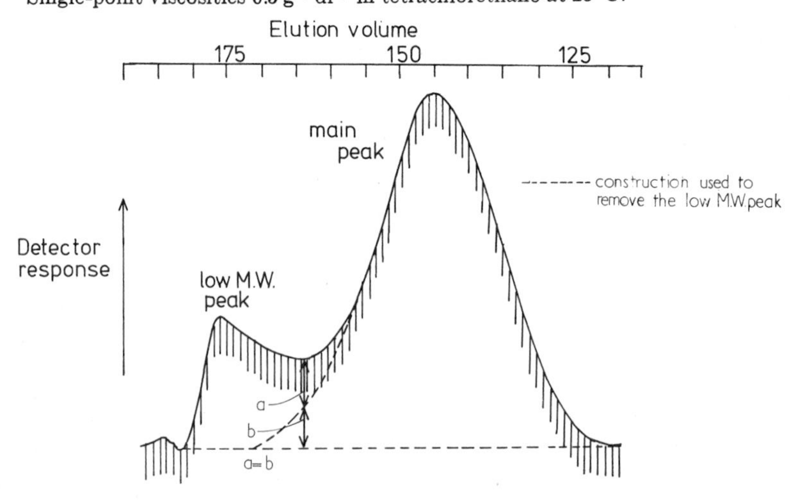

Figure 2. GPC chromatogram of 3IGT prepared using terephthaloyl chloride (3IGT/C)

for the Polymers

Hard Block			Tetrachloroethane	GPC in DMA[c]		T_g/K
$\overline{g}$	$\overline{Hn}$	$T_g/°C$[a]	0.5/dl g^{-1}[b]	($\overline{M}_n \times 10^{-3}$	$\overline{M}_w/\overline{M}_n$)	(tan δ max)
			0.401	56	2.03	251
2.04	536	8	0.615	71	2.26	271
2.96	728	13	0.548	59	2.18	283
7.39	1640	12	0.375	29	1.97	315
			0.550			229
2.59	643	9	0.842	92	1.49	248
5.54	1260	32	0.556	63	2.61	267
16.39	3607	71	0.455	54	2.28	308
			0.423	34	2.03	221
3.98	926	22	0.502	49	2.46	235
8.38	1820	46	0.490	39	2.22	260
21.84	4553	78	0.413			308
			0.425	30 (14.30)[d]	1.83 (2.99)[d]	219
7.13	1566	41	0.398	33	2.06	234
15.77	3368	68	0.494	52	2.29	—
43.02	8852	90	0.391	64	2.12	317
			0.415	31	1.74	241
10.25	2199		0.425	37	2.16	261
			0.319	51		363 (DSC)

[c] GPC in dimethyl acetamide solution at 80°C, poly(styrene) calibration.
[d] GPC in dimethyl acetamide solution at 80°C, poly(oxyethylene) calibration.

and GPC data (*16, 17*). The ratio of $\overline{M_w}/\overline{M_n}$ in terms of poly(styrene) molecular weight was 1.7/2.6; a similar range has been reported for poly(oxytetramethylene oxyterephthaloyl-co-oxysebacoyl) random copolymers (*22*).

Naming System

The multiblock polymers are distinguished by soft-block type, the molecular weight of the soft block, and the appropriate weight fraction of the soft block. Therefore, PPO 3000/50 refers to a PPO/3IGT block polymer, prepared from a PPO prepolymer with molecular weight of 3000 and containing approximately 50% by weight of PPO. PPO 3000/100 represents a PPO "homopolymer" prepared by linking a PPO prepolymer mol wt 3000 with terephthaloyl chloride. 3IGT/C refers to 3IGT prepared from terephthaloyl chloride and 3IGT/M to melt-polymerized 3IGT using dimethyl terephthalate.

Block Molecular Weights

The sequence distribution of monomer residues in copolycondensation reactions of this type has been studied in detail by Peebles (*18, 39*). If all monomers are assumed to have equal reactivity and if the reaction has gone to completion, the number-average sequence length of 3IG residues($\bar{g}$) is given by:

$$\bar{g} = \frac{[3IG]_o + [PPO]_o}{[PPO]_o} \qquad [\quad]_o = \text{initial concentration}$$

Similarly, for PPO residues:

$$\bar{s} = \frac{[3IG]_o + [PPO]_o}{[3IG]_o}$$

The sequence lengths follow a "most probable" distribution; values of $\bar{s}$ and $\bar{g}$ for each polymer are given in Table I. Because of the high molecular weight of PPO, $\bar{s}$ is generally less than two, so the molecular weight of the soft block will be taken as the molecular weight of the PPO prepolymer. The molecular weight of the hard block ($\overline{Hn}$) is given by:

$$\overline{Hn} = 206\,\bar{g} + 148$$

206 = mol wt of repeat unit

148 = mol wt of oxyterephthaloyl residue

To a first approximation, a multiblock polymer prepared from PPO with molecular weight x containing 50% PPO by weight will have a hard-block molecular weight of x; one containing 25% will have a hard-block molecular weight of $2x$; and one containing 75% PPO a hard-block of $x/2$ (Figure 1).

Melt Polymerization of 3IGT

Low-molecular-weight 3IGT homopolymers (1000–8000) were prepared by melt polycondensation (*19*) using dimethyl terephthalate, and the T_gs were measured using DSC. Molecular weights up to 4000 were measured by end-group analysis, higher molecular weights were estimated by extrapolation of the T_g/mol wt curves (*16, 17*). An attempt was made to prepare PPO 3000/50 by the same method, but the PPO formed a separate layer in the melt before high-molecular-weight polymer was formed and the attempt was abandoned.

Mechanical Testing

The room-temperature properties of the polymers varied from viscoelastic liquids for 100% PPO through soft (50% PPO) and tough (25% PPO) solids to a brittle glass (3IGT). A torsional braid pendulum was used as the dynamic mechanical-test method, because only with a supported sample could of the polymers be tested on one apparatus, and only small amounts of the polymers were available. The torsional pendulum was based on Gillham's design (*20*), with the inertia disc motion being detected by a Hall-effect device. The glass substrate consisted of two glass rovings twisted around each other by "doubling" to give a uniform bundle of 1680 filaments that acted as support for the polymer, with the ends of the glass trapped in crimp tags. The polymer was cast onto the braid as a 30% solution in a suitable solvent, and the solvent removed slowly at atmospheric pressure and under vacuum to leave a composite consisting of a polymer matrix and the glass support. The volume fraction of polymer in the braid (ϕ_p) was calculated from the weight per unit length of the braid, knowing the amount of glass present and assuming a polymer density of 1.0.

The braids were tested from $-100°C$ upwards at a heating rate of $1°C\ min^{-1}$ and a frequency of ≈ 1 Hz, using an inertial weight of $4.55 \times 10^{-5}\ kg\ m^2$. The damped sine waves were processed by hand to give log decrement and modulus. The radius of the sample was calculated from the weight per unit length, as above, with the additional assumption that the sample was a right circular cylinder and that there were no voids in the sample.

The Torsional Braid as a Composite Material

In any discussion of the results from torsional braid analysis it is important to establish what contribution the glass will make to the properties of the composite. The principal effects of the glass are illustrated in Figure 3. The modulus and log-decrement curves for a sample of poly(styrene) and a braid of poly(styrene) were cast from a toluene solution and tested using two different inertia weights. The sample of pure poly(styrene) was tested on a conventional inverted torsional pendulum (*21*). The modulus and damping curves are in fairly close agreement at low temperatures and in the first part of the transition region but beyond the maximum in log decrement, there are major differences between the curves. Most important is in the observation that at high temperatures the modulus of the composite tends to a constant limiting value. Above the glass-transition temperature, the modulus of the poly(styrene) matrix becomes very small, so this limiting

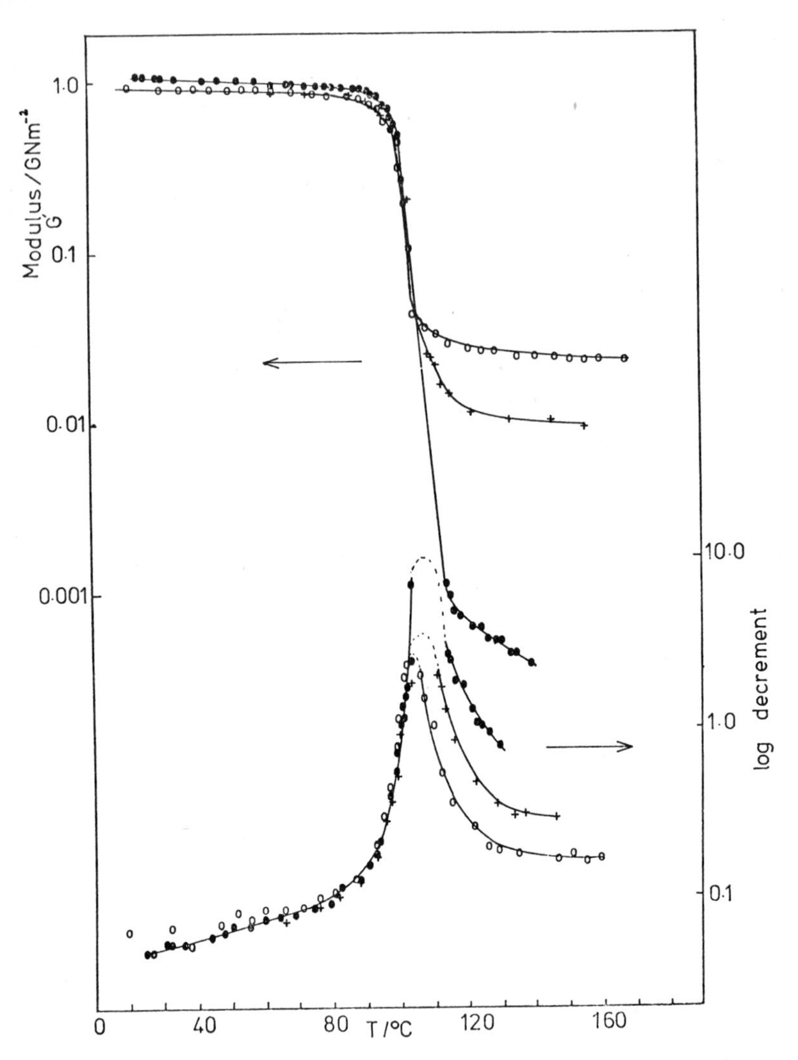

Figure 3. A comparison between the mechanical properties of poly(styrene) and a poly(styrene) braid. (●) Compression-molded sample of PS. PS braid cast from toluene, $\phi p = 0.78$. (○) I1—4.55×10^{-5} kg m^{-2}. (+) I2—8.8×10^{-6} kg m^{-2}.

modulus must correspond to the torsional rigidity of the glass trapped between the two crimp tags. The damping of the braid also tends to a low constant value at high temperatures for the same reason. The effect of the different inertia weights (I1, I2) is probably a result of the change in tension in the sample, altering the configuration of the glass. As a result of the change in the damping curves at high temperatures, the T_g as defined by the maximum in log decrement has been shifted down in temperature by about 5°C. The important conclusion of this discussion is that below the maximum in log decrement the results from the braid are comparable with a sample of pure polymer, but above the maximum the properties of the glass substrate dominate the composite braid.

The value of the modulus of the poly(styrene) braid at low temperature is smaller than the sample of pure poly(styrene) because the cross-section of the braid is rather irregular, and the approximation of an irregular cross-section by a circular one will inevitably lead to an underestimate of modulus (*21*). For this reason, the absolute values of the modulus of the multiblock polymers in Figures 4–10 are unreliable.

It should be noted that a $T_{L,L}$ would not be expected in this poly(styrene) braid sample because of its high molecular weight (approximately 100,000) (*46*). Also, this analysis of the composite structure of the braid supports Nielsen's (*47*) suggestion that $T_{L,L}$ arises solely from the residual elasticity of the glass. The torsional pendulum, braid, and $T_{L,L}$ will be discussed in more detail elsewhere (*16, 17*).

Properties of the Multiblock Polymers

Figure 4 shows the properties of PPO 3000/50 cast from methyl ethyl ketone (MEK) and toluene. The sample cast from MEK shows two equal-sized peaks in log decrement but in the sample cast from toluene, the high-temperature peak is much larger than the low-temperature peak. These variations are typical of a two-phase composite in which the morphology is controlled by the casting solvent (*21*). All of the other samples were cast from solutions in MEK. The results for all of the other samples are summarized in Figures 5, 6, 7, 8, and 9. The series of polymers prepared from PPO mol wt 400 all showed a sharp transition region similar to poly(styrene). A sample of melt polymerized 3IGT of mol wt 2750 also is shown. As the block molecular weight increases, the transition region becomes progressively broader until in PPO 3000/50 a double peak is formed. The breadth of the transition region in the samples prepared from high-molecular-weight PPO may be taken as evidence of some form of phase separation.

Small Angle X-Ray Results

Small angle x-ray pictures of melt-pressed films of PPO 3000/50 and 3000/25 cast from MEK showed a well-defined diffraction ring giving a measure of the size of the inhomogeneities in the material. A MEK-cast film of PPO 3000/50 gave identical results. A sample of PPO 2000/25 gave a diffuse scattering after three times the exposure of the PPO 3000/50, indicating the presence of irregular inhomogeneities with less difference in electron density between the scattering regions (Figure 10).

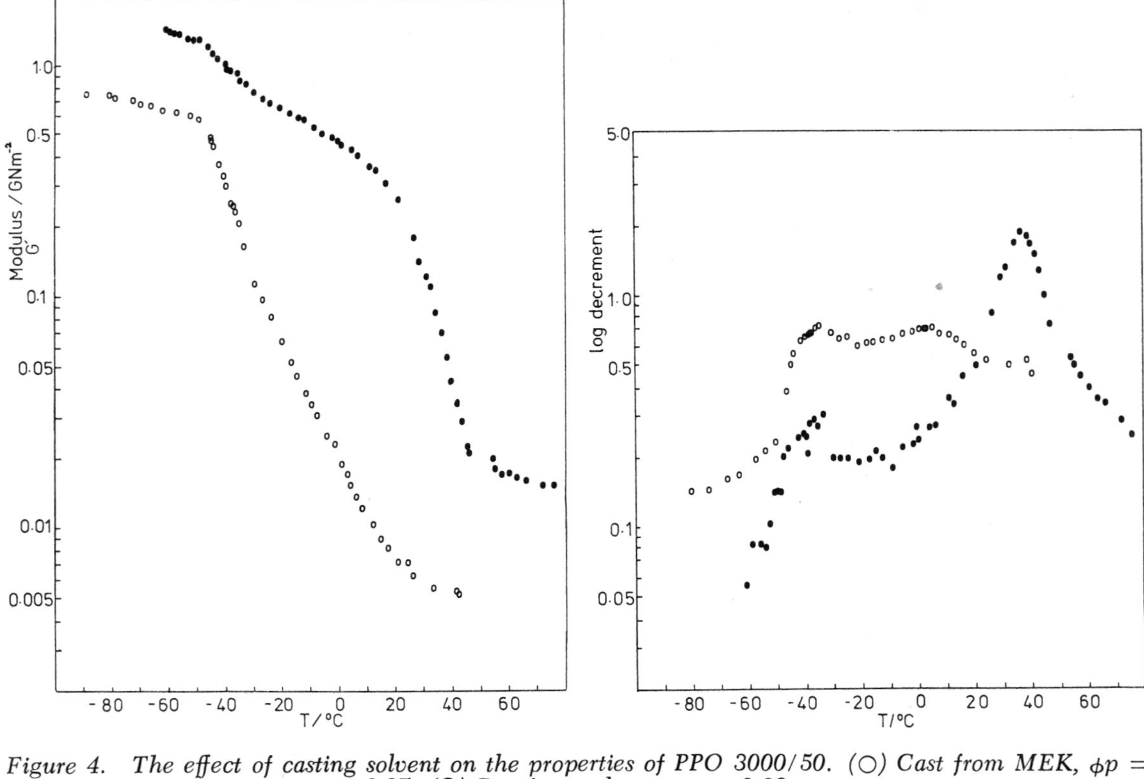

Figure 4. The effect of casting solvent on the properties of PPO 3000/50. (○) *Cast from MEK, $\phi p = 0.87$.* (●) *Cast from toluene, $\phi p = 0.66$.*

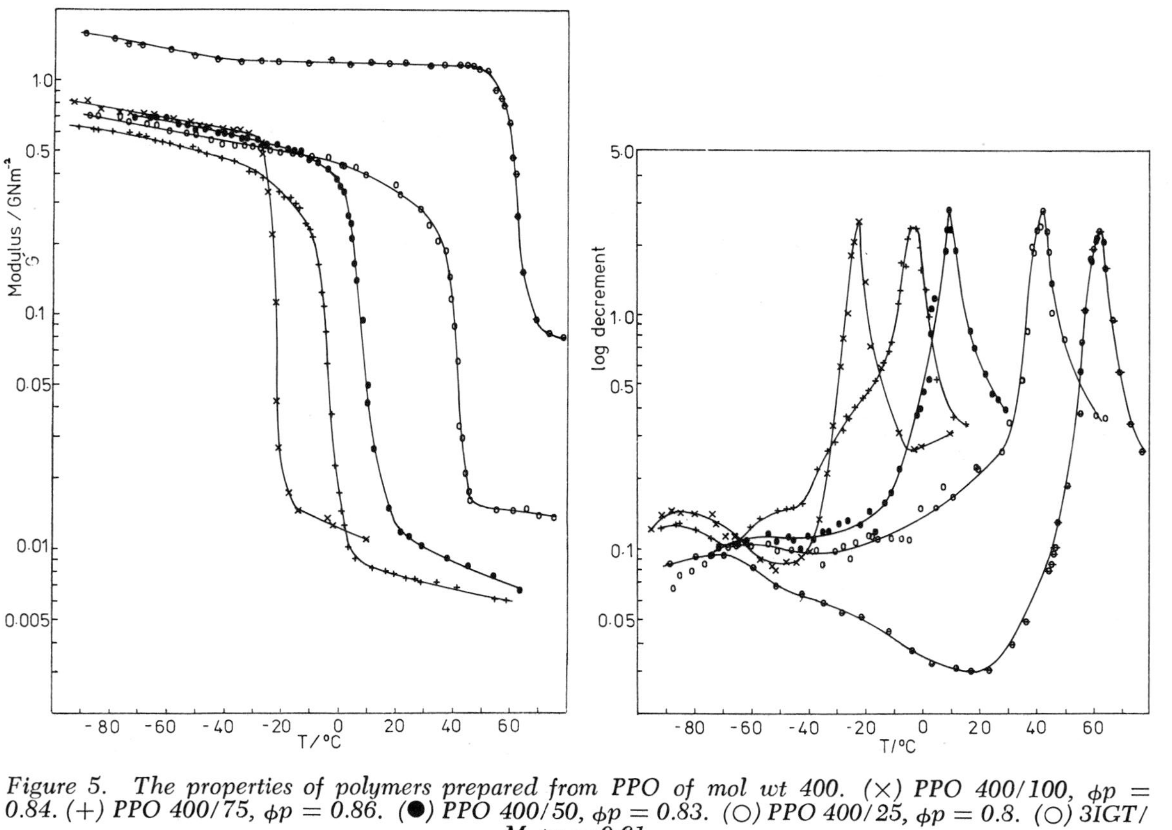

Figure 5. The properties of polymers prepared from PPO of mol wt 400. (×) *PPO 400/100,* $\phi p = 0.84$. (+) *PPO 400/75,* $\phi p = 0.86$. (●) *PPO 400/50,* $\phi p = 0.83$. (○) *PPO 400/25,* $\phi p = 0.8$. (○) *3IGT/M* $\phi p = 0.61$.

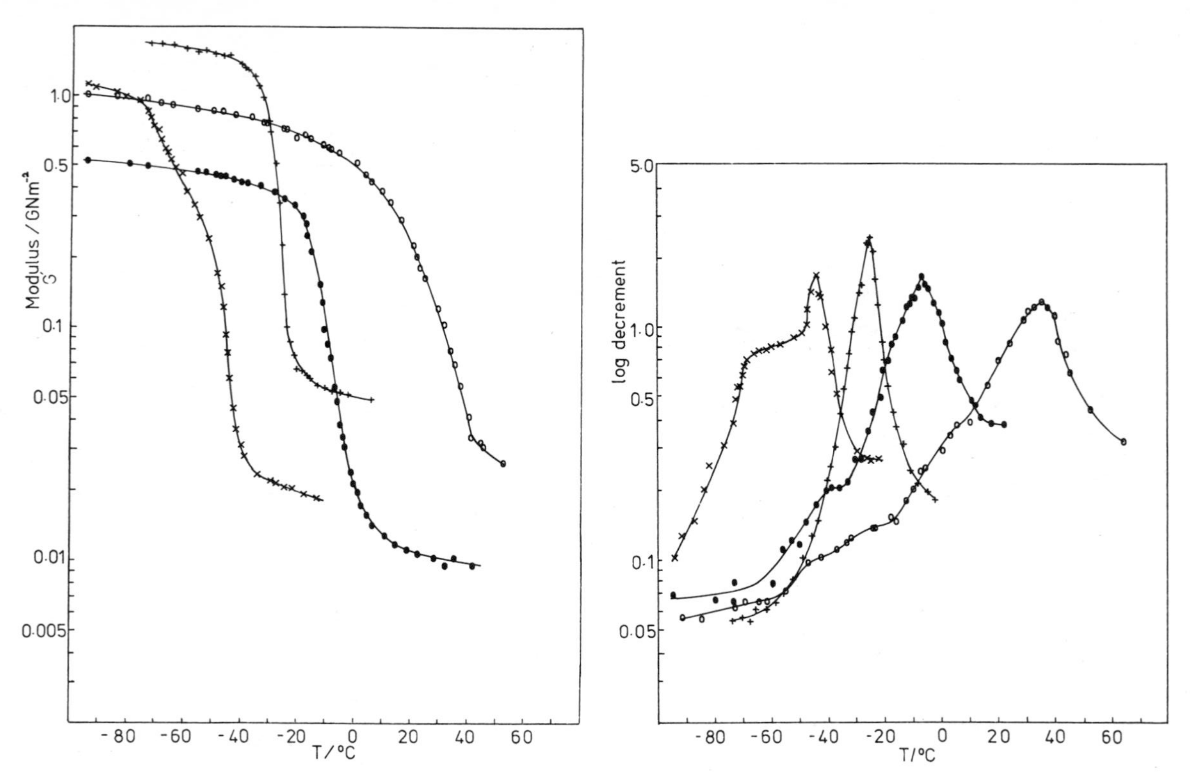

Figure 6. The properties of polymers prepared from PPO of mol wt 1000. (×) *PPO 1000/100,* $\phi p = 0.79$. (+) *PPO 1000/75,* $\phi p = 0.86$. (●) *PPO 1000/50,* $\phi p = 0.85$. (○) *PPO 1000/25,* $\phi p = 0.79$.

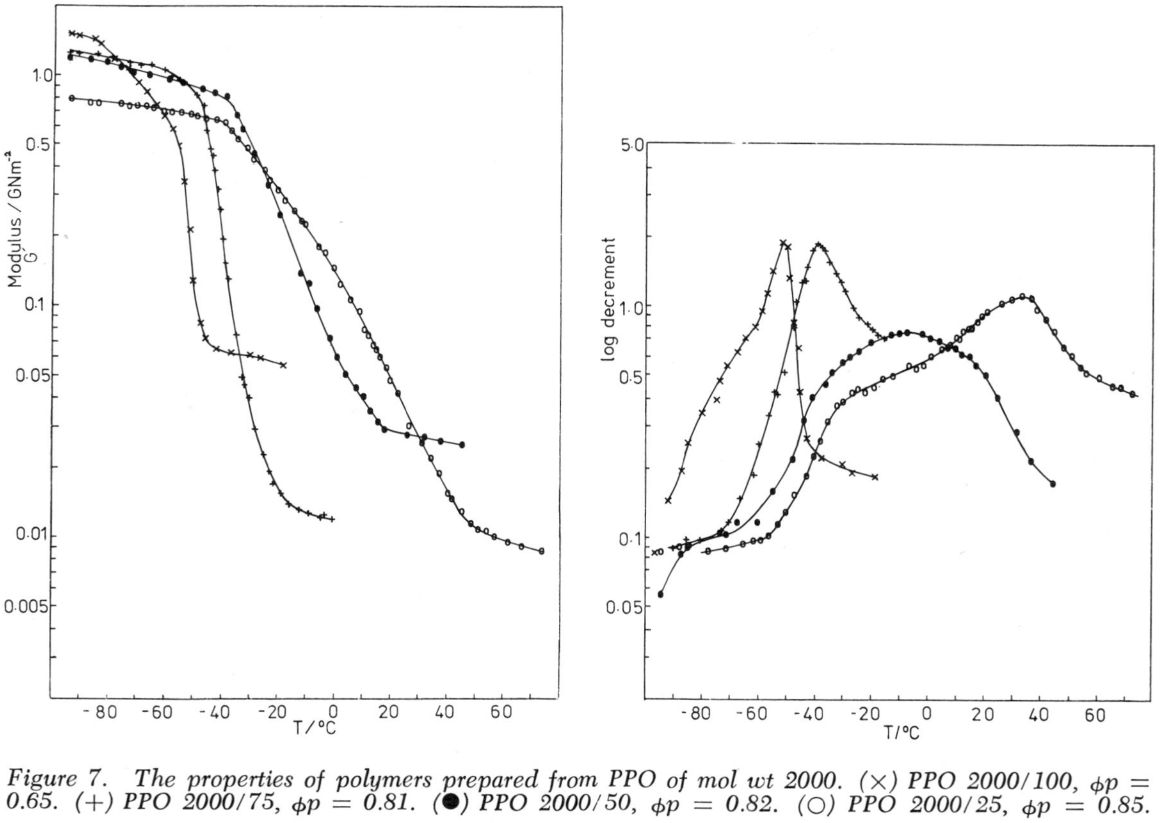

Figure 7. The properties of polymers prepared from PPO of mol wt 2000. (×) *PPO 2000/100,* ϕp = *0.65.* (+) *PPO 2000/75,* ϕp = *0.81.* (●) *PPO 2000/50,* ϕp = *0.82.* (○) *PPO 2000/25,* ϕp = *0.85.*

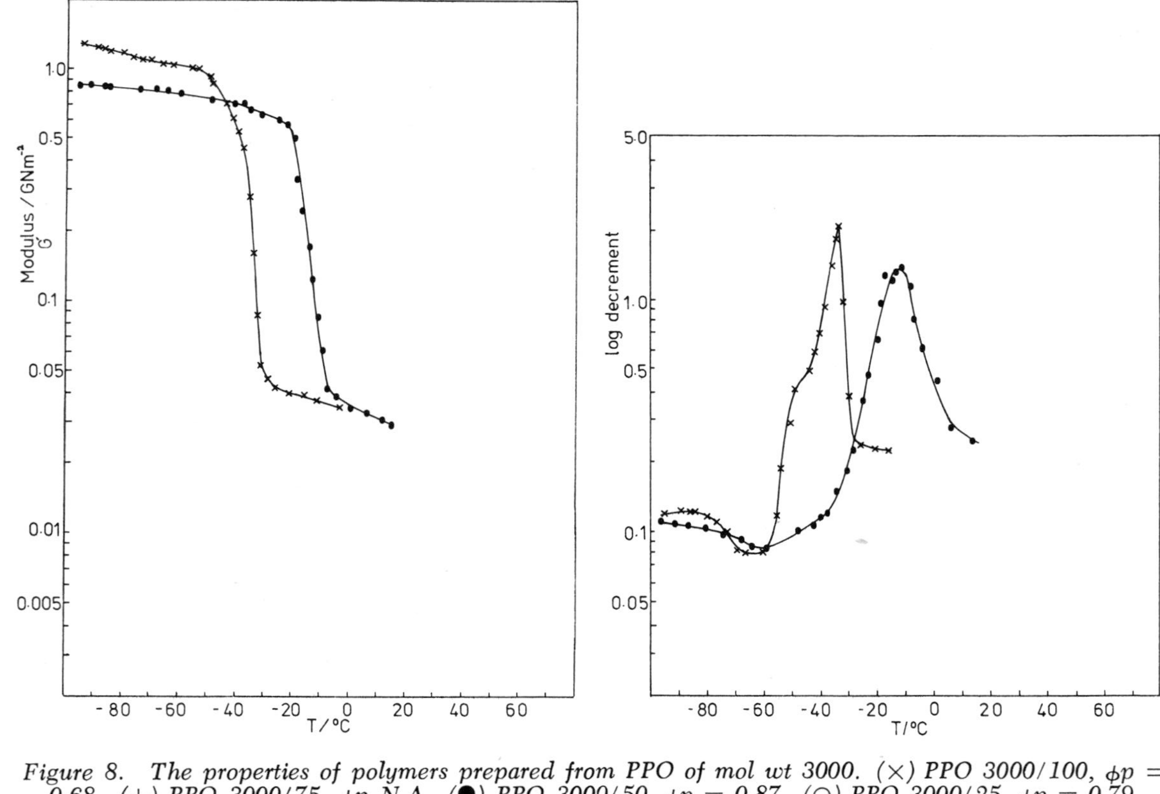

Figure 8. The properties of polymers prepared from PPO of mol wt 3000. (×) *PPO 3000/100, ϕp = 0.68.* (+) *PPO 3000/75, ϕp N.A.* (●) *PPO 3000/50, ϕp = 0.87.* (○) *PPO 3000/25, ϕp = 0.79.*

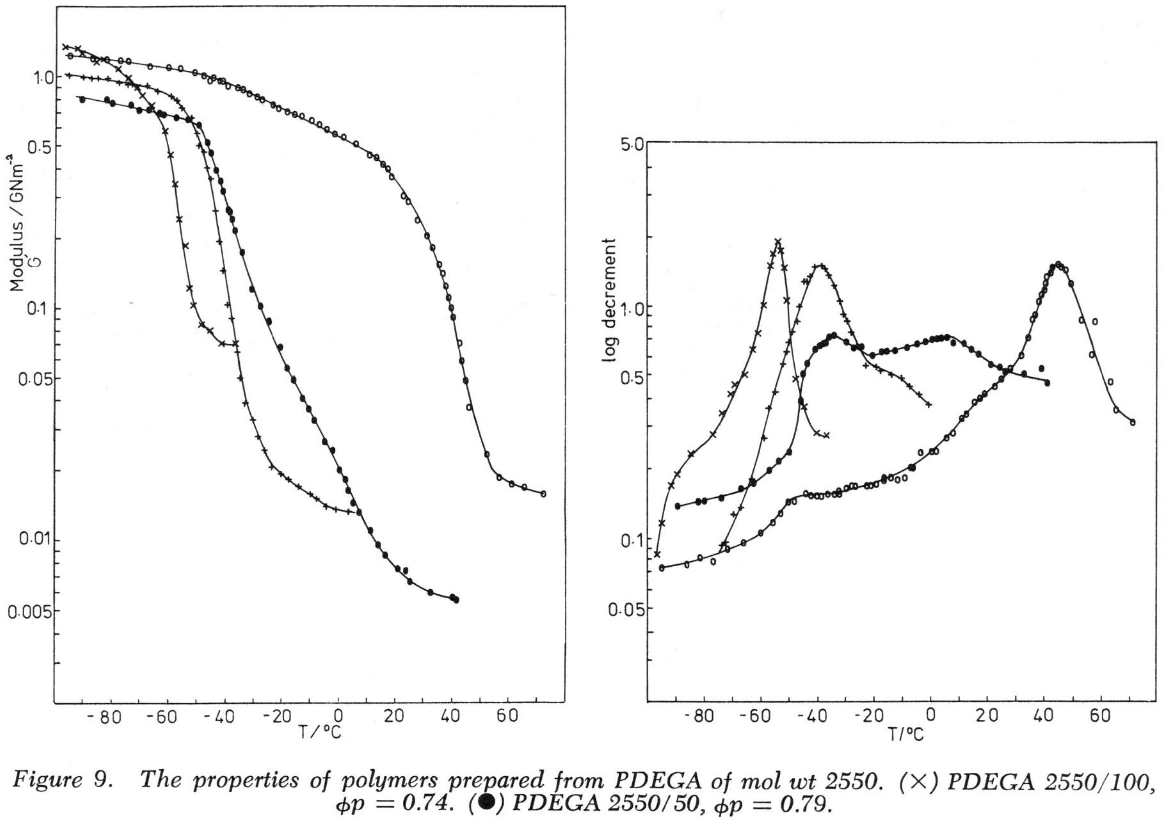

Figure 9. The properties of polymers prepared from PDEGA of mol wt 2550. (×) PDEGA 2550/100, $\phi p = 0.74$. (●) PDEGA 2550/50, $\phi p = 0.79$.

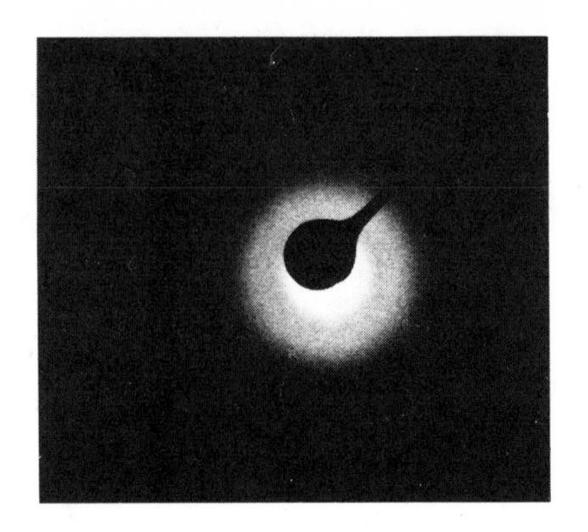

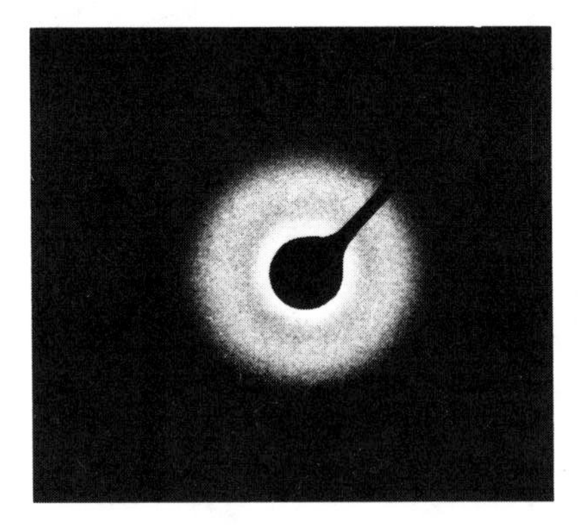

Figure 10. Small angle x-ray pictures of the multiblock polymers. (top) *PPO 3000/25, 24-hr exposure.* (middle) *PPO 3000/50, 24-hr exposure.* (bottom) *PPO 2000/25, 50-hr exposure.*

Discussion

The small angle x-ray results show that regular inhomogeneities were present in samples PPO 3000/50 and 3000/25. The crucial question in any study of this type is what sort of phase separation has occurred in these polymers? The blocks, whole chains with slightly different composition, or low-molecular-weight impurities could have separated out. It is well established that the size of the inhomogeneities produced by phase separation of the blocks in a block copolymer are controlled by the molecular weight of the blocks (*24, 25*) because the joints between the blocks must lie in the interfacial region between the domains (Figure 11).

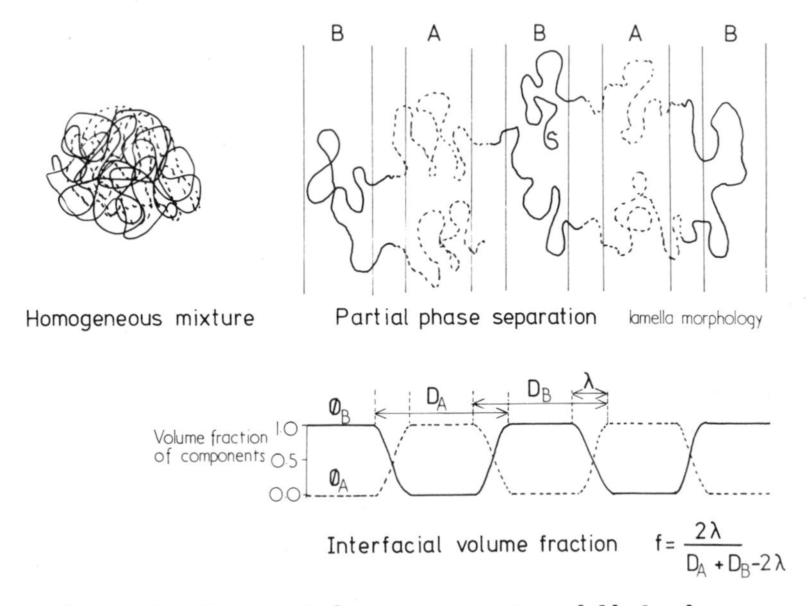

Figure 11. Diagram of phase separation of a multiblock polymer

Table II shows the Bragg spacing and the sum of the unperturbed block dimensions for PPO 3000/50 and 3000/25, $< r_o >$ was estimated from the values for PPO, and poly(oxyethylene oxyterephthaloyl) given in the literature (*26*). As the average block molecular weight increased, the Bragg spacing also increased and is the same order of magnitude as the unperturbed block dimensions. This can be taken as a clear indication that phase separation of the blocks has occurred in PPO 3000/50 and 3000/25.

Table II. Small Angle Scattering Data

Polymer	*PPO (mol wt)*	*3IGT (mol wt)*	*Average (mol wt)*	R[a]	*Bragg Spacing*
PPO 3000/50	3000	3368	3184	10.09 nm	14.4 nm
PPO 3000/25	3000	8852	5926	13.3 nm	19.2 nm

[a] $R = \langle r_o \rangle_{PPO} + \langle r_o \rangle_{3IGT}; \frac{\langle r_o \rangle}{M^{\frac{1}{2}}} = 0.087$ nm.

T_g/*Composition Relation*

If the polymers do not show phase separation they can be considered as random copolymers. The T_g of a random copolymer solely depends on composition if the molecular weight is sufficiently high for T_g to be independent of molecular weight (*27*). It has been established that the molecular weight of 3IGT/C is high enough to make the T_g independent of molecular weight (*16, 17*). In these copolymers the residues of three monomers appear. To define the composition of all of the PPO-based copolymers by one weight fraction, the contributions of the other two components must be added together.

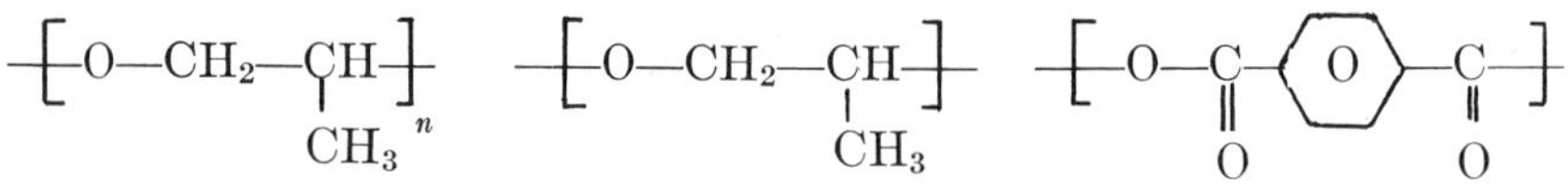

$$\left[\!\!-O-CH_2-\underset{\displaystyle CH_3}{\underset{|}{CH}}-\!\!\right]_n \qquad \left[\!\!-O-CH_2-\underset{\displaystyle CH_3}{\underset{|}{CH}}-\!\!\right] \qquad \left[\!\!-O-\underset{\displaystyle O}{\underset{\|}{C}}-\langle O \rangle-\underset{\displaystyle O}{\underset{\|}{C}}-\!\!\right]$$

PPO residue | propane-1,2-diol residue | oxyterephthaloyl residue

Defining T_g by the maximum in tan δ, a plot of T_g vs. weight-fraction PPO residues (Wf_{PPO}) and T_g vs. weight-fraction oxyterephthaloyl residues (Wf_{TR}) can be compared in Figure 12. The plot against Wf_{TR} shows the majority of the points lying on a curve; those lying off the curve are indexed and correspond to those polymers with unsymmetrical maxima in log decrement. The points lying on the line are those with single, sharp maxima in log decrement and the 50% copolymers that have symmetrically broadened peaks in log decrement. The simple T_g–composition relationship for polymers with a single sharp maxima in log decrement is evidence that they are behaving like homogeneous random copolymers. Similarly, those polymers that do not lie on the line must be phase separated because the regions producing the maxima in log decrement have a different composition to the original polymer. The

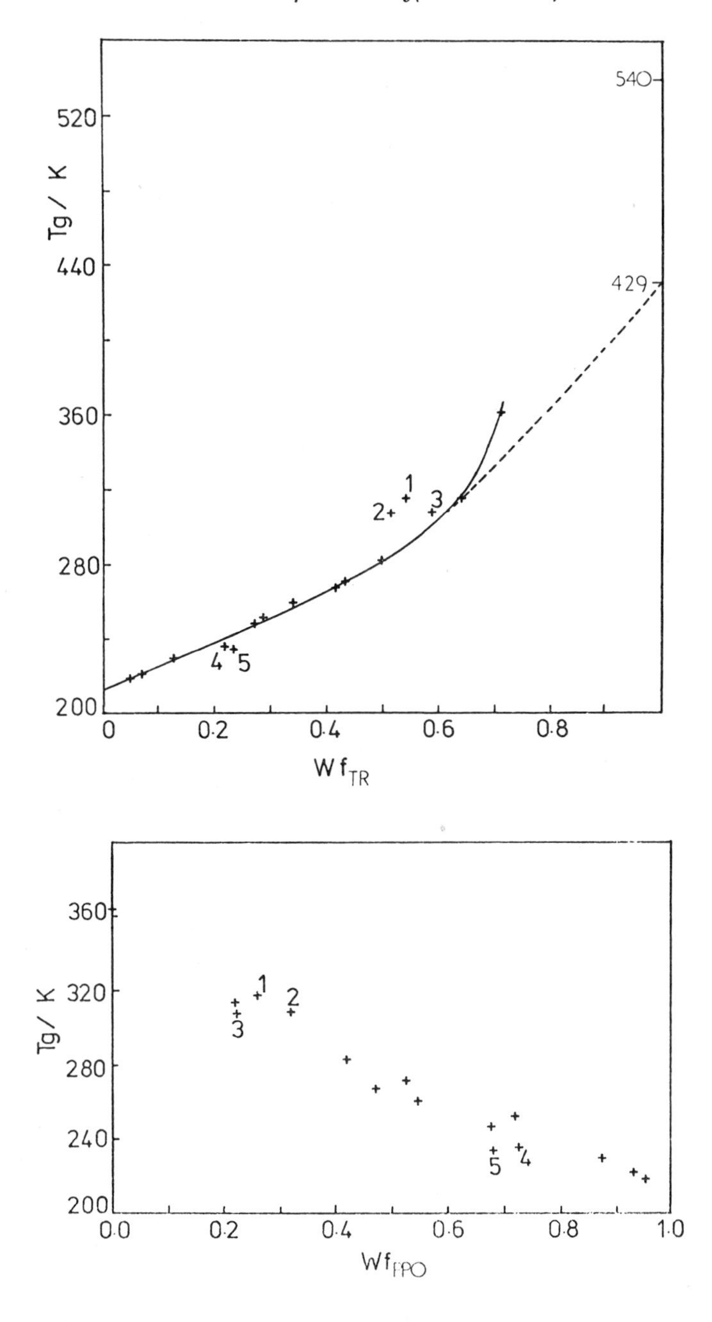

Figure 12. T_g against copolymer composition defined by weight fraction oxyterephthaloyl and PPO residues. (1) PPO 3000/25, (2) PPO 2000/25, (3) PPO 1000/25, (4) PPO 2000/75, (5) PPO 3000/75.

T_g of poly(1,4 phenylene terephthaloyl) has been reported as 540 K (*41*). All of the points could not be fitted by the general T_g/composition formula by Wood (*27*), even when the T_g of poly(oxyterephthaloyl) was allowed to float. If the point for 3IGT was omitted, the rest of the points were fitted by the Fox equation (*40*), shown as a dashed line in Figure 12.

$$\frac{1}{T_g} = \frac{W_1}{T_{g1}} + \frac{W_2}{T_{g2}}$$

The T_g against weight-fraction PPO-residues graph shows a general T_g/composition curve with significantly more scatter than the Wf_{TR} plot (Figure 12). In defining the copolymer by Wf_{PPO}, the T_g is assumed independent of variations in the ratio of terephthaloyl-to-diol residues. In the case of Wf_{TR}, the T_g is assumed independent of variations in the ratio of PPO to diol by weight. Because PPO and propane 1,2-diol have the same repeating unit, their contribution to chain flexibility will be the same and hence Wf_{TR} provides the best measure of the composition of the polymers.

Analysis of the Mechanical Properties

The morphology and composition variation found in the phase-separated materials can be deduced from the shape and position of the damping curves. Two-phase composites are generally found in three basic morphologies: phase 1 dispersed in a matrix of phase 2, phase 2 dispersed in a matrix of phase 1, and a lamella morphology in which both phases are continuous. If phase 1 is a rubber and phase 2 a glass, then the modulus of the composite depends crucially on the morphology. If one of the phases is dispersed, the continuous phase dominates the dispersed phase, leading to an impact-resistant glass in the first case and a filled rubber in the second. In the case of a lamella morphology, both phases contribute equally to the modulus of the composite in a logarithmic law of mixing. Consequently, the size of the tan δ peaks corresponding to the T_g of the phases also will depend on the morphology. The tan δ peak for the continuous phase is larger than that of the dispersed phase, and a lamella morphology will lead to two equally sized peaks. This may be seen in the experimental data on poly(styrene-b-diene) polymers (*28, 45*) and follows from the relationship between tan δ and the differential of the log of the real modulus. The relationship between tan δ and the viscoelastic functions is discussed in Refs. *6, 16, 21,* and *44,* and the properties of composites have been reviewed in Ref. *21*.

The following assumptions are made in analyzing the mechanical property results. In a completely phase-separated system, the T_g of the soft block is the same as the T_g of the PPO "homopolymers," e.g., PPO 3000/100. The T_g of the hard block of a given molecular weight is the same as the T_g of the free homopolymer with same molecular weight. The overall glass–rubber transition is considered to arise from incremental contributions of regions of different composition combined together as in a conventional composite.

PPO 400/75/50/25, 1000/75 (Figures 5 and 6)

These all show a single sharp maxima in log decrement and a simple relationship between T_g and composition, indicating that they are homogeneous materials.

PPO 1000/50, 2000/50 (Figures 6 and 7)

Both samples show a symmetrically broadened peak in log decrement. The width of the peak in PPO 2000/50 indicates that regions with widely varying composition are present and the narrower peak in PPO 1000/50 indicates that less phase separation has occurred in this sample. The fact that the maxima of the log-decrement peaks lie on the T_g/composition curve and these peaks are symmetrically broadened indicates that regions with the same composition as the original polymer are most common and the regions of varying compositions are arranged in a lamella morphology in which they all contribute equally to the overall properties of the composite.

PPO 3000/50 (Figures 4 and 8)

The presence of two peaks in log decrement shows that phase separation has developed sufficiently for regions rich in PPO and 3IGT to be most common. The differences between the temperatures of these two maxima and the T_gs of the blocks if complete phase separation had occurred are shown in Table III. The two peaks in the MEK-cast sample are indicative of a lamella-like morphology. The difference in the size of the peaks in the toluene-cast sample indicates a 3IGT matrix containing PPO-rich dispersed regions.

This dependence of block polymer property on casting solvent also is seen in poly(styrene-b-diene) polymers (*28*). Theoretical work (*23*) has shown that the thermodynamically most stable morphology for a diblock polymer containing 50% of each component is a lamella morphology. For this reason MEK was chosen for the tests on all other polymers.

Table III. T_g Values of the Phase-Separated Regions

Sample	*Maxima in Log. Dec.*	
	High Temp (°C)	*Low Temp (°C)*
PPO 3000/50: MEK Cast	5	−35
PPO 3000/50: Toluene Cast	35	−40
PPO 3000/100	—	−54
3IGT/M mol wt 3368	68	—

PPO 3000/25 (Figure 8)

PPO 3000/25 produces a small angle x-ray diffraction ring but only a single maxima in log decrement with a long, low-temperature tail. The composition of the regions producing the maximum in log decrement are different from the composition of the parent polymer. The fact that a second peak in log decrement is not observed for the PPO-rich regions would suggest that there is a 3IGT-rich matrix that dominates the mechanical properties, with PPO-rich dispersed regions producing the long, low-temperature tail. Block copolymers containing 25% of one component generally form dispersed phases (*23, 29*); also, the low-temperature transition in PPO 3000/50 cast from toluene is only just visible, so it is reasonable that the reduction in PPO weight fraction would lead to the disappearance of this transition.

PPO 2000/25, 1000/25 (Figures 6 and 7)

From the same evidence as above it may be deduced that there is a continuous 3IGT matrix with dispersed PPO-rich regions. The narrowing of the tan δ maxima indicates that the differences in composition get progressively smaller with lower PPO molecular weight, in line with the small angle x-ray results.

PPO 3000/75, 2000/75 (Figures 7 and 8)

The shifting of the maxima in log decrement to low temperatures is the evidence for phase separation. These polymers probably show a long low-temperature tail that has been obscured by the limiting modulus of the braid.

PDEGA 2550/50 (Figure 9)

This sample shows a slightly broadened peak similar to PPO 1000/50. As described in the introduction, the driving force for phase separation comes from the large negative enthalpy of the phase-separated state relative to the homogeneous mixture. The enthalpy (ΔH) can be quantified in terms of the Scott–Hildebrand theory for these multiblocks (*10, 12, 16*):

$$\Delta H \propto M(\delta_1 - \delta_2)^2$$

where M is the average block molecular weight. The difference in solubility parameters for the two copolymers can be calculated from the table of group contributions (*26*).

PPO/3IGT	$\delta_1 - \delta_2 = 2.5$
PDEGA/3IGT	$\delta_1 - \delta_2 = 1.4$

Clearly the difference in solubility parameters is consistent with the difference in block molecular weight required to produce phase separation in these two copolymers.

PPO "Homopolymers" and 3IGT

These polymers represent the homologous series shown in Table IV. The strong secondary relaxation in 3IGT and PPO 400/100 is observed as a shoulder in PPO 1000/100, 2000/100, and 3000/100. The secondary in terephthalate polymers is thought to arise from torsional oscillations of the terephthalate units (*30, 31*); this mechanism is in line with the observation that the relaxation gets weaker for the higher-molecular-weight PPO polymers.

Table IV. Structure of the PPO "Homopolymers"

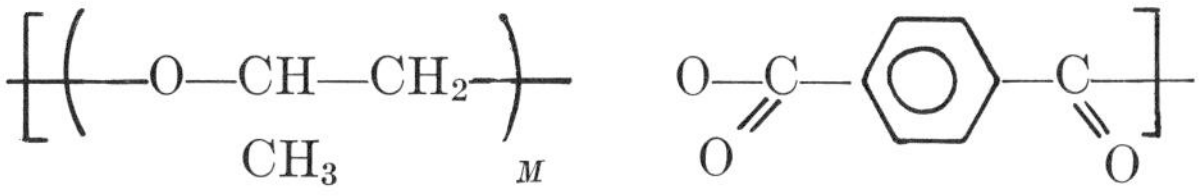

$$\left[\left(-O-\underset{CH_3}{CH}-CH_2-\right)_M O-\overset{O}{\overset{\|}{C}}-C_6H_4-\overset{O}{\overset{\|}{C}}-\right]$$

Polymer	M	Wf_{TR}
3IGT	1	0.718
PPO 400/100	6.8	0.28
PPO 1000/100	17.3	0.131
PPO 2000/100	34.5	0.07
PPO 3000/100	51.7	0.049

Sources of Uncertainty

The major source of uncertainty in these results comes from the influence the glass substrate might have on phase separation; preferential absorption of one of the blocks onto the surface of the glass (*32*) would obviously affect phase separation. The low-molecular-weight peak observed in the GPC curves also could affect the results. In so far as the T_g/composition curve (Figure 12) shows a direct T_g/composition relation with symmetrical displacement of the 25% and 75% copolymers above and below the line, there is no evidence to suggest that these other factors are affecting phase separation. The detailed analysis of torsional pendulum data is also rather uncertain at high damping values (*33*), as is the analysis of multiphase systems by single-frequency isochronal data (*34*). For most of the curves discussed here, the maximum in log decrement is about 1.0, and these effects are likely to be "second order" compared with the effect of the braid and the large differences observed between the samples.

Conclusions

The sequence of curves in Figures 5–11 shows the stages of partial phase separation that occur in an amorphous multiblock polymer with polydisperse hard blocks. By systematically varying the block molecular weights and copolymer composition, the breadth and shape of the damping peaks changed in a systematic and predictable way. Controlled partial phase separation would appear to be a powerful way to tailoring polymer properties for a particular application.

The difference between these polymers and conventional Hytrel-type poly(ether-b-esters) is emphasized by the attempt to polymerize PPO 3000/50 by melt polycondensation, the method normally used for Hytrel (*42*). Complete phase separation occurred, showing that the incompatibility between PPO/3IGT is much greater than poly(oxytetramethylene-b-oxytetramethylene oxyterephthaloyl).

Acknowledgment

One of us (M.P.C.W.) would like to thank the Science Research Council for financial support.

Literature Cited

1. Ungar, E. E., Hatch, D. K., *Prod. Eng.* (1961) **32**, 41.
2. Ball, G. L., Salyer, I. O., *J. Acoust. Soc. Am.* (1966) **39**, 663.
3. Sperling, L. H., Thomas, D. A., *National Technical Information Service Tech. Bull.* (1975) AD/A-003 852.

4. Hiejboeur, J., "Physics of Non-crystalline Solids," p. 231, J. A. Prins, Ed., North Holland, Netherlands, 1965.
5. Manson, J. A., Sperling, L. H., "Polymer Blends and Composites," Plenum, New York, 1976.
6. McCrum, N. G., Read, B. E., Williams, M. L., "Anelastic and Dielectric Effects in Polymeric Solids," Wiley, New York, 1967.
7. Kollinsky, F., Markert, G., "Multicomponent Polymer Systems," N. A. Platzer, Ed., ADV. CHEM. SER. (1971) **99**, 175.
8. Kraus, G., Rollman, K. W., "Multicomponent Polymer Systems," N. A. Platzer, Ed., ADV. CHEM. SER. (1971) **99**, 189.
9. Mitzumachi, J., *Adhesion* (1970) **2**, 292.
10. Meier, D. J., *Polym. Prepr.* (1974) **15**, 171.
11. Helfand, E., *J. Chem. Phys.* (1975) **62**, 99.
12. Krause, S., *Macromolecules* (1970) **3**, 84.
13. Leary, D. F., Williams, M. C., *J. Polym. Sci., Polym. Phys. Ed.* (1974) **12**, 265.
14. Allport, D. C., James, W. H., "Block Copolymers," p. 208, Applied Science, United Kingdom, Division of Elsevier, New York, 1973.
15. Flory, P. J., "Principles of Polymer Chemistry," Cornell University, Ithaca, NY, 1953.
16. Watts, M. P. C., Ph.D. Thesis, University of Manchester, Institute of Science and Technology, 1977.
17. Watts, M. P. C., White, E. F. T., unpublished data.
18. Peebles, L. H., *Macromolecules* (1974) **9**, 58.
19. Goodman, I., Rhys, J. H., "Polyesters," Vol. 1, IUife Books, London, 1965.
20. Gillham, J. K., *A.I.Ch.E. J.* (1974) **20.6**, 1066.
21. Neilsen, L. E., "Mechanical Properties of Polymers and Composites," Vol. I, Dekker, 1974.
22. Marrs, W., Still, R. H., Peters, R. H., *J. Appl. Polym. Sci.*, in press.
23. Meier, D. J., "Block and Graft Copolymers," J. J. Burke, V. Weiss, Eds., Syracuse University, Syracuse, NY, 1973.
24. Kromer, H., Hoffmann, M., Kampf, G., *Ber Bunsenges. Phys. Chem.* (1970) **74**, 859.
25. Helfand, E., Wasserman, Z. R., *Polym. Eng. Sci.* (1977) **17**, 73.
26. "Polymer Handbook," 2nd ed., J. Brandrup, E. H. Immergut, Eds., Wiley-Interscience, 1975.
27. Wood, L. A., *J. Polym. Sci.* (1958) **28**, 319.
28. Miyamoto, S., Kodama, K., Shibayama, K., *J. Polym. Sci., Polym. Phys. Ed.* (1970) **8**, 2095.
29. Folkes, M. J., Keller, A., "Physics of Glassy Polymers," R. N. Howard, Ed., p. 548, Applied Science, 1973.
30. Yip, H. K., Williams, H. L., *J. Appl. Polym. Sci.* (1976) **20**, 1217.
31. Sacher, E., *J. Polym. Sci., Polym. Phys. Ed.* (1968) **6**, 1935.
32. Lipatov, Y. S., *Adv. Polym. Sci.* (1977) **22**, 1.
33. Struick, L. C. E., *Rheol. Acta.* (1967) **6**, 119.
34. Tschoegl, N. W., Cohen, R. E., *175th National ACS Meeting, California, 1978.*
35. Toporowski, G. M., Roovers, J. E. L., *J. Polym. Sci., Polym. Chem. Ed.* (1976) **14**, 2233.
36. Kraus, G., Rollman, K. W., *J. Polym. Sci., Polym. Phys. Ed.* (1976) **14**, 1133.
37. Senich, G. A., Macknight, W. J., ADV. CHEM. SER. (1979) **176**, 97.
38. Matzner, M., Noshay, A., Robeson, L. M., Merriam, N., Barclay, R. McGrath, J. E., *App. Polym. Symp.* (1973) **22**, 143.
39. Peebles, L. H., *Macromolecules* (1974) **7**, 872.
40. Fox, T. G., *Bull. Am. Phys. Soc.* (1956) **1**, 123.

41. Frosina, V., Levita, G., Landis, J., Woodward, A. E., *J. Polym. Sci.* (1977) **15**, 239.
42. Cella, R. J., *J. Polym. Sci., Part C* (1973) **42**, 727.
43. Sperling, L. H., "Recent Advances in Polymer Blocks, Blends and Grafts," p. 152, Plenum, New York, 1974.
44. Schwartzl, F. R., Struick, L. C. E., *Adv. Mol. Relaxation Processes* (1967) **1**, 210.
45. Robinson, R. A., White, E. F. T., "Block Copolymers," S. L. Aggarwal, Ed., p. 123, Plenum, New York, 1970.
46. Stadniki, S. J., Gillham, J. K., Boyer, R. F., *J. Appl. Polym. Sci.* (1976) **20**, 1245.
47. Neilsen, L. E., *Polym. Eng. Sci.* (1977) **17**, 713.

RECEIVED April 14, 1978.

BLOCK COPOLYMERS

9

Properties and Morphology of Amorphous Hydrocarbon Block Copolymers

MITCHEL SHEN

Department of Chemical Engineering, University of California, Berkeley, CA 94720

The morphologies of heterogeneous block copolymers are determined by the block composition and by sample preparation conditions. The resulting microstructures exert a profound influence on the properties of these materials. In this review we shall first discuss the thermodynamic conditions under which homogeneous block copolymers can be formed. These systems are interesting because their underlying chain dynamics can be treated by the accepted molecular models. Next we present some examples of block copolymer morphologies. Statistical theories capable of satisfactorily explaining the observed morphologies are then briefly discussed. Finally the elastic, viscoelastic, and rheological properties of these materials are described. In all instances the dominant effect of the microdomain structure on these properties is demonstrated.

During the past decade, there has been an upsurge of interest in studying the relation between morphology and properties of block copolymers. The interest is generated mainly by the technological importance of these materials, e.g., their ability to form thermoplastic elastomers or impact-resistant plastics. Although there are some block copolymers that are homogeneous, most of them show microphase separation. The type of structure depends on such variables as chemical composition, block configuration, solvent power, etc. Advances in characterization techniques such as electron microscopy and low-angle x-ray scattering now render it possible to investigate their detailed morphologies. In this chapter we shall review the morphological and property studies of both homogeneous

0-8412-0457-8/79/33-176-181$06.00/0

and heterogeneous block copolymers as well as the thermodynamic theories in microphase separation in these materials. The review is not intended to be exhaustive, rather it will focus on the more current works in the field. Further information is available in the recent research monographs (*1, 2, 3, 4, 5*) and review papers cited (*6, 7, 8, 9, 10*).

Homogeneous Block Copolymers

The thermodynamic criterion for the mixing of two or more systems is that the free energy of mixing must be negative. For polymeric systems, the entropy increases accompanying the mixing of long chain molecules are very small. Since the enthalpy of mixing is usually positive, it is therefore not too surprising that phase separations often occur in polymeric mixtures. The thermodynamic basis for microphase separation in block copolymers has been presented by Krause (*11, 12*) and Meier (*13*). In the work of Krause, the enthalpy change on microphase separation is given by the Hildebrand–Van Laar–Scatchard expression (*14*):

$$\Delta H = -kT(V/V_z)v_A v_B \chi_{AB}(1 - 2/z) \qquad (1)$$

where V is the total volume of the mixture, V_z the volume of each lattice site, z is the coordination number, k is Boltzmann constant v_A and v_B are volume fractions of A and B blocks respectively, T is absolute temperature, and χ_{AB} is the Flory–Huggins interaction parameter between As and Bs.

The entropy change accompanying microphase separation is (*12*) for each copolymer molecule. In Equation 2, m is the number of blocks

$$\Delta S/k = (v_A \ln v_A + v_B \ln v_B) - 2\,(m - 1)\,(\Delta S_d/k) + \ln\,(m - 1) \qquad (2)$$

in the block copolymer molecule. The entropy change attributable to the demixing from a homogeneous mixture to a phase-separated system is given by the first term on the right-hand side of Equation 2. The second term accounts for the entropy decrease attributable to the immobilization of the segments linking the A and B blocks (where ΔS_d is disorientation entropy). If the number of blocks in the copolymer is large ($m > 3$), then it is necessary to recognize the fact that after the first block-linking segment has been placed on the interface, the possible number of sites available to the subsequent links is now constrained. The entropy change for this effect is given by the third term. Combining Equations 1 and 2, the free energy change on microphase separation can be written. The critical interaction parameter then can be readily obtained by setting the free energy change to zero:

$$(\chi_{AB})_{cr} = \frac{zV_z}{(z - 2)V_A n_A n_B} [-(v_A \ln v_A + v_B \ln v_B) + 2(m - 1)(\Delta S_d/k) - \ln(m - 1)] \quad (3)$$

where n_A and n_B are the numbers of A and B units in each copolymer molecule. Calculations on the basis of Equation 3 show that the microphase separation becomes more difficult with increasing m. Values of critical interaction parameters computed for a mixture of homopolymers are much lower than those for the block copolymer with identical composition. Thus it is more difficult for microphase separation to take place when polymers are linked together via covalent bonds as block copolymers. Experimentally it has been found that although polyblends of polystyrene and poly(α-methyl styrene) tend to be heterogeneous, the corresponding block copolymers are often homogeneous (*15–21*).

Although there are very few homogeneous block copolymers available they are nevertheless of interest because their viscoelastic behavior can be studied within the existing theoretical framework to elucidate the molecular dynamics of block copolymers. The most accepted model is the molecular theory of polymer viscoelasticity proposed many years ago by Rouse (*22*), Bueche (*23*), and Zimm (*24*). The RBZ model divides the polymer molecule into $N + 1$ submolecules (beads) held together with N springs. The springs are stretched when the polymer coil is disturbed by a shear gradient. The spring constant is given by $3kT/b^2$, where b^2 is the average end-to-end distance of the submolecule. As the beads move through the medium, a viscous drag is exerted on them whose magnitude is given by a friction coefficient f. At equilibrium the viscous and elastic forces are equal to each other. A simplified form of the equation of motion can be written as follows:

$$\dot{\mathbf{x}} = \sigma \mathbf{Z} \mathbf{x}$$

where $\mathbf{x}$ and $\dot{\mathbf{x}}$ are column vectors of bead positions and bead velocities, $\mathbf{Z}$ is the nearest neighbor matrix, and $\sigma = 3kT/b^2 f$.

In the case of block copolymers (*25, 26, 27, 28, 29*), Equation 4 must be modified to take into account the fact that not all of the beads are the same (as is the case for homopolymers). For a triblock copolymer such as poly(styrene-b-α-methyl styrene-b-styrene), the equation of motion is (*26*):

$$\dot{\mathbf{x}} = -\sigma_s \mathbf{D}^{-1} \mathbf{Z} \mathbf{x}$$

where $\sigma_s = 3kT/b_s^2 f_s$, the subscripts s refer to the PS submolecule. The matrix $\mathbf{D}^{-1}$ is the inverse of

$$D = \begin{bmatrix} 1 & & & & & & \\ & 1 \cdot\cdot\cdot & & & & \bigcirc & \\ & & \delta_A & & & & \\ & & & \delta_A \cdot\cdot\cdot & & & \\ & & & & 1 & & \\ & \bigcirc & & & & 1 \cdot\cdot\cdot & \end{bmatrix} \quad (6)$$

where $\delta_A = b_A^2 f_A / b_s^2 f_s$ and subscripts A refer to PαMS submolecules. Thus the elements in the diagonal of this matrix take into account the differences between the PS and PαMS submolecules.

The solution of the equation of motion yields the distribution of viscoelastic relaxation times. For ease of comparison with experimental data, maximum relaxation times for a number of block copolymers with different block configurations were computed (*26*). These are given in Table I. Figure 1 shows the stress–relaxation isotherms determined for two diblock copolymers of styrene and α-methylstyrene of two different molecular weights (*21*). These are shifted into smooth viscoelastic master curves (Figure 2). Their shift factors (Figure 3) are seen to follow the WLF equation closely, indicating the essential homogeneity of the block copolymer samples. Similar experiments were also carried out for a number of triblock copolymers of styrene and α-methylstyrene (*19*). Maximum relaxation times were determined from the master curves by

Table I. Block Configurations and Maximum Retardation Times of Block Copolymers (26)

Polymer	*Block Configuration*	*Log* τ_{max} [a] *10% A*	*Log* τ_{max} [a] *50% A*
I. Diblock copolymer	A_xB_y	−0.42	0.65
II. Triblock copolymers	(a) $A_xB_yA_x$	0.64	1.22
	(b) $B_xA_yB_x$	−0.62	0.65
III. Alternating (segmented) block copolymers	(a) $A_xB_xA_xB_x$ . . .	0.26	0.97
	(b) $A_xB_yA_xB_y$ . . .	—	0.97
IV. Multiblock copolymers	(a) $B_xA_yB_z$	−0.60	0.65
	(b) $B_xA_yB_zA_yB_x$	0.34	1.01

[a] Computed for $\partial_A = 185$.

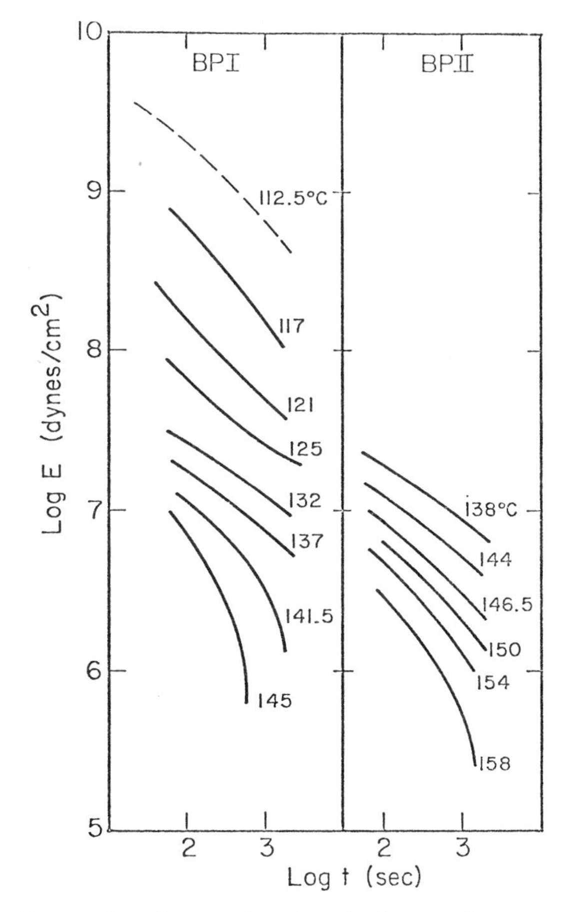

Figure 1. Stress–relaxation isotherms for two samples of poly(styrene-b-α-methylstyrene), BPI = $\overline{M}_w$ = 0.8 × 10^5; BPII = $\overline{M}_w$ = 1.5 × 10^5. Solid curves: tensile data; broken curve: flexural data (21).

Procedure X of Tobolsky and Murakami (*30*). Table II shows that the agreement is satisfactory between the calculated and experimental values (*10*).

Morphology of Heterophase Block Copolymers

Five fundamental domain structures are possible for block copolymers consisting of two types of blocks. Generally lamellar structures will form at compositions with approximately equal proportions of the two components. As the proportion of one component increases at the expense of the other, cylindrical morphologies will result. The matrix phase will

Table II. Maximum Viscoelastic Relaxation Times for Block Copolymers of Styrene and α-Methylstyrene

Sample	*Wt % αMS*	*log $(\tau_m/\tau_m{}^\circ)$* *Expt'l*	*Cal'd*	*Ref.*
SAS	5	0.50	0.50	*19*
	17	0.59	0.55	*19*
	34	0.01	1.10	*19*
	42	0.89	1.50	*19*
	65	1.76	1.90	*19*
ASA	73	2.26	2.30	*19*
AS	50	1.75	1.75	*21*

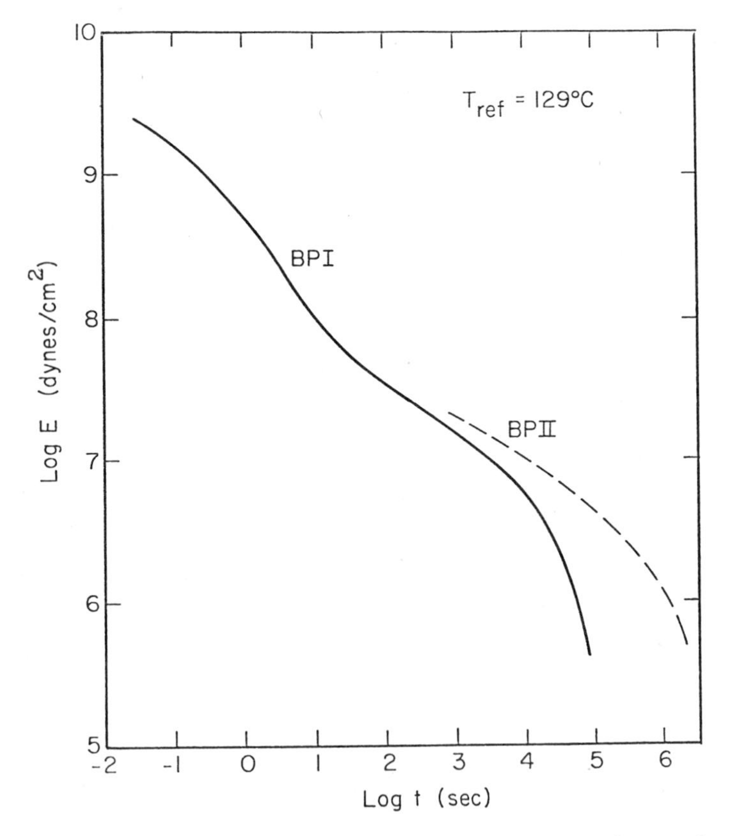

Figure 2. Viscoelastic master curves of poly(styrene-b-α-methylstyrene). Solid curve: sample BPI; broken curve: sample BPII (21).

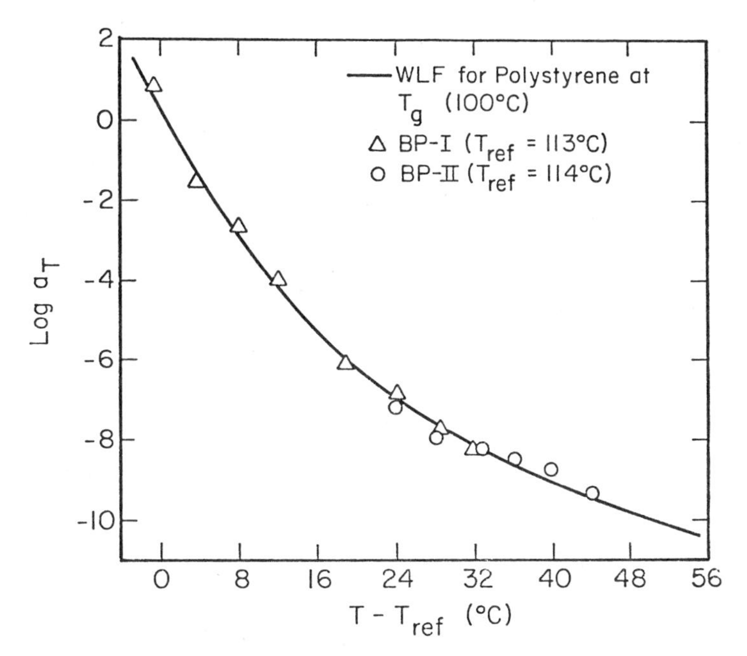

Figure 3. Viscoelastic shift factor data for samples BPI (triangles) and BPII (circles) of poly(styrene-b-α-methylstyrene). The solid curve was calculated from the WLF equation (21).

be composed of the component in greater abundance. As the proportion of one component continues to increase, eventually the morphology of sperical domains of minor component embedded in the matrix of the other component appears. These structures have been observed for diblock copolymers of isoprene and styrene cast from toluene (*31*). Their electron micrographs are shown in Figure 4. The dark regions belong to the polyisoprene (PIP) phase which was selectively stained by OsO_4. The domain structure of the 20/80 styrene/isoprene block copolymer (Figure 4a) shows tiny spheres of polystyrene (PS) blocks dispersed in a matrix of polyisoprene. Electron micrographs of 40/60 and 50/50 compositions (Figure 4b and 4c) appear as alternating stripes which are actually profiles of the three dimensional lamellar structures. For the 60/40 block copolymer, cylindrical domains of the isoprene component in PS matrix can be observed (Figure 4d). The dark dots represent ends of the cylindrical rods.

The progressive changes in morphology with changing compositions also can be achieved by adding homopolymers to the block copolymer (*32, 33, 34, 35*). The added homopolymer is solubilized into the corresponding domains in the block copolymer if the molecular weight of the added homopolymer is equal to or less than that of the corresponding

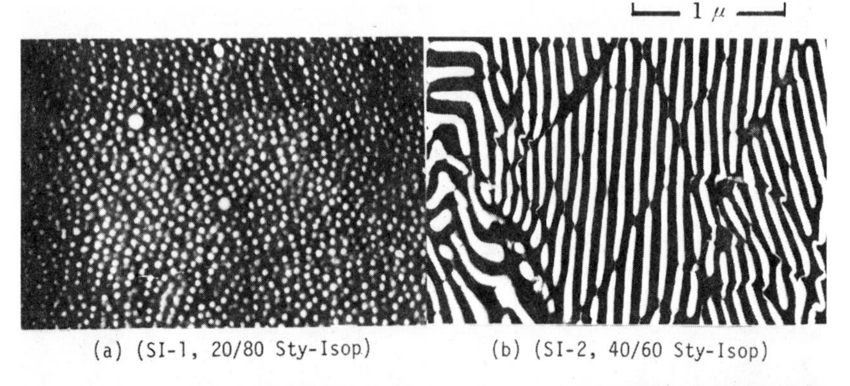

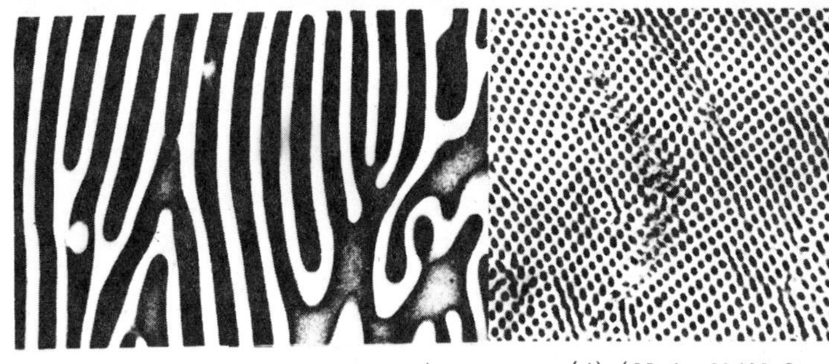

Memoirs of the Faculty of Engineering, Kyoto University

Figure 4. Electron micrographs of diblock copolymers of styrene and isoprene cast from toluene and microtomed normal to the surface as indicated (31).

block in the copolymer. Figure 5a shows the electron photomicrograph of a triblock copolymer of styrene–butadiene–styrene (SBS) cast from a mixed solvent of THF/methyl ethyl ketone. Incorporation of a low molecular weight polystyrene (PS) in the block copolymer enlarged the PS domains (light regions), as seen in Figure 5b. However, if the added PS has a molecular weight that is greater than that in SBS, then separate domains of pure PS appears (Figure 5c).

Under appropriate conditions it is possible to observe long-range order in block copolymers, e.g., if samples are prepared by melt extrusion, thermal annealing, or slow rate of casting (*36–42*). An example for a styrene–butadiene block copolymer containing 68% styrene is shown in Figure 6. These structures are often referred to as "macrolattice." In some instances imperfections in the long-range order may appear as "grain boundaries" normally found in metallic systems. These electron microscopic observations are supported by small angle x-ray scattering (SAXS) and optical light scattering studies (*39, 40, 41, 42*).

Theories of Microdomain Formation

The basic driving force for microdomain formation in block copolymers is the reduction in the positive surface free energy of the system resulting from the increase of the domain size. This domain size increase gives rise to a decrease in the volume fraction of interfacial region in which junction points of the copolymers must be distributed. In addition, configurations of the block chains must also change in order to even-up the density deficiency in the interior of the domains.

A number of statistical thermodynamic theories for the domain formation in block and graft copolymers have been formulated on the basis of this idea. The pioneering work in this area was done by Meier (*43*). In his original work, however, he assumed that the boundary between the two phases is sharp. Leary and Williams (*43, 44*) were the first to recognize that the interphase must be diffuse and has finite thickness. Kawai and co-workers (*31*) treated the problem from the point of view of micelle formation. As the solvent evaporates from a block copolymer solution, a critical micelle concentration is reached. At this point, the domains are formed and are assumed to undergo no further change with continued solvent evaporation. Minimum free energies for an AB-type block copolymer were computed this way.

Helfand (*45, 46*) used a mean field approach to treat the problem of microdomain formation (Figure 7). For a diblock copolymer with a high degree of polymerization, the following free-energy expression can be written (*46*):

$$\frac{G}{NkT} = \left(\frac{2\gamma}{kT}\right)\left(\frac{x_A}{\rho_A} + \frac{x_B}{\rho_B}\right)\left(\frac{1}{d}\right) + \log\left(\frac{d}{2a_J}\right) + 0.141\, d^{5/2}\left\{\frac{(x_A{}^{1/2}/b_A\rho_A)^{5/2} + (x_B{}^{1/2}/b_B\rho_B)^{5/2}}{[(x_A/\rho_A) + (x_B/\rho_B)]^{5/2}}\right\} - \frac{\alpha(x_A/\rho_A)(x_B/\rho_B)}{(x_A/\rho_A) + (x_B/\rho_B)} \qquad (7)$$

The first term on the right-hand side of Equation 7 accounts for the energy of mixing at the interphase, and the entropy loss resulting from the fact that an A chain (or B chain) which has penetrated into the B phase (A phase) must turn back. In this term, γ is the interfacial tension, x is the degree of polymerization, and ρ is the density of pure A or B. In addition, the interfacial term must decrease with increasing domain size, which goes as $1/d$ where d is the domain repeat distance. The second term of this equation is attributable to the necessity of the segment linking A and B blocks being confined to the interphase. It is proportional to the logarithm of the ratio of the volume available to the

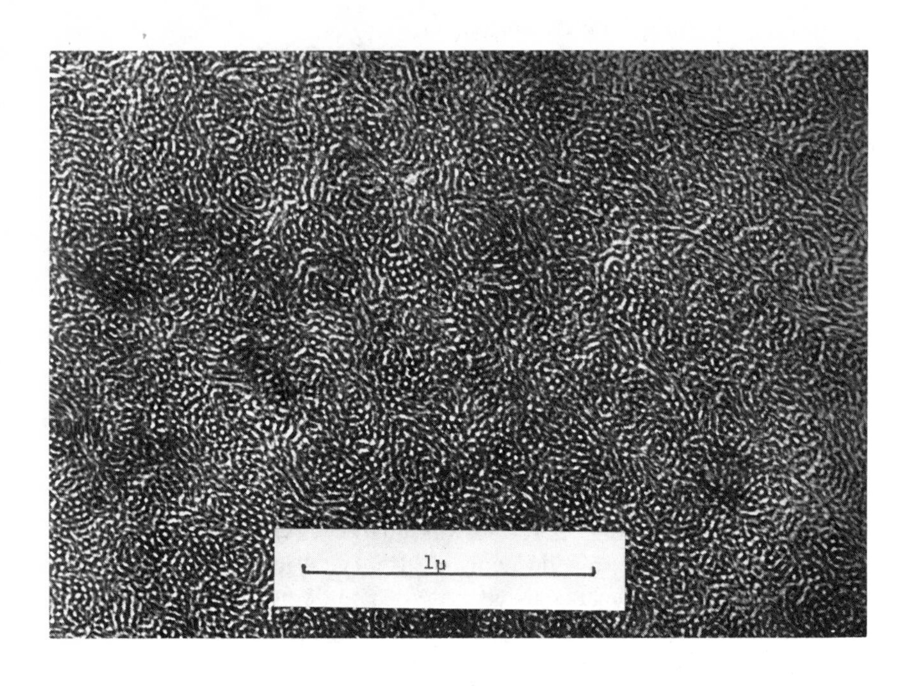
1μ

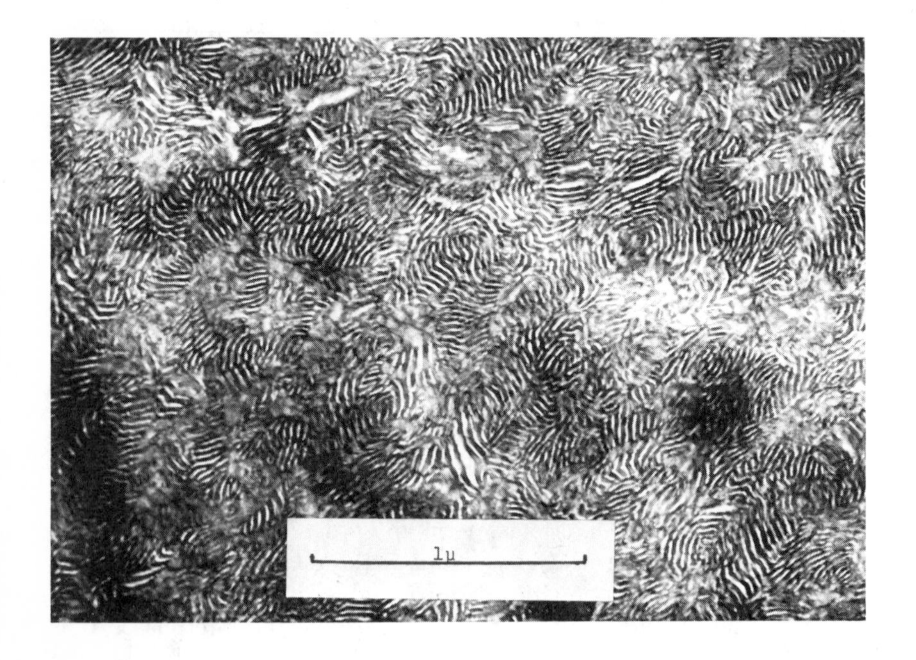
1μ

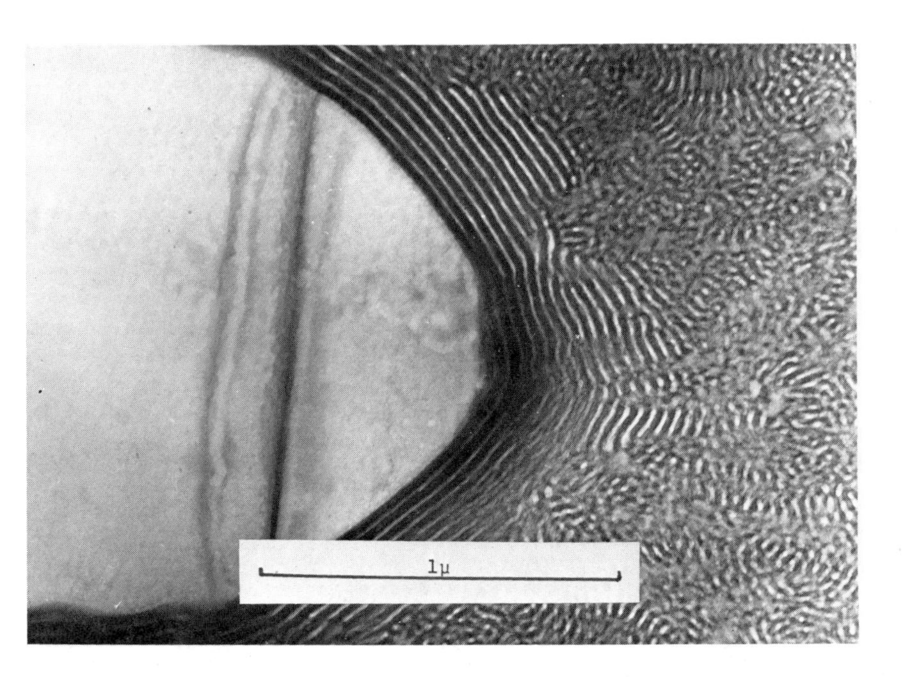

Figure 5. Electron micrographs of triblock copolymers of styrene and butadiene. (a) (top left) *As cast from THF/methyl ethyl ketone; (b)* (bottom left) *cast from the same solvent with 20% polystyrene ($\overline{M}_n = 3{,}000$); (c)* (above) *cast from the same solvent with 20% polystyrene ($\overline{M}_n = 30{,}000$).*

Figure 6. Electron micrograph of diblock copolymer of styrene and butadiene cast from xylene (courtesy of M. Hoffman)

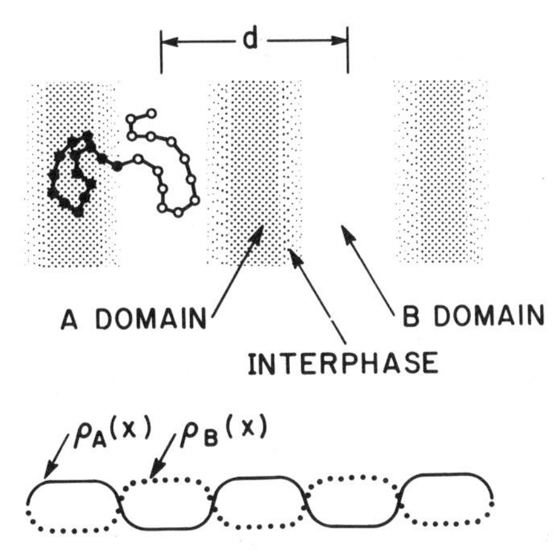

Figure 7. Diagram of a lamellar microdomain structure in block copolymers (46)

link in a mixed homogeneous state to that in the microdomains. The width of the interfacial region is given by a_J and is generally of the order of nanometers. Another consequence of the confinement of the link to the interphase is the density deficiency in the interior of the domain. The system tends to statistically reduce the conformations which lead to the inhomogeneous density and favor the rarer conformations in the center of the domain. The loss of conformational entropy will increase with increasing size of the domain, which is represented by the third term of Equation 7. The symbol b in this term is the statistical length of a monomer unit. The last term in the equation is independent of domain sizes and fixes the standard state of the system as that of a homogeneous mixed state. Here α is a measure of the repulsion between A and B blocks. By minimizing Equation 7, d can be calculated. Table III shows that the computed values of ds are in satisfactory agreement with available experimental data for a number of block copolymers of styrene and butadiene.

In their statistical model for microphase separation of block copolymers, Leary and Williams (*43*) proposed the concept of a separation temperature T_s. It is defined as the temperature at which a first-order transition occurs when the domain structure is at equilibrium with a homogeneous melt, i.e.,

$$\Delta G = \Delta H - T\Delta S = 0 \quad (8)$$

or

$$T_s = (\Delta H/\Delta S)_{\mathrm{demix}} \quad (9)$$

Table III. Microdomain Repeat Distances in Block Copolymers of Styrene and Butadiene (*46*)

Polymer	*Mol Wt* (*kg/mol*)	d_{exp} (*nm*)	d_{calc} (*nm*)
I. Lamellar morphology	32–48	44.5	51
S–B	35.5–54.5	49	55
	71–46	74	63
	48.9–32.4	46	49
B–S–B	19.4–72–19.4	40	38
	24–72–24	44	41
	37.5–72–37.5	48	48
	73–72–73	66	64
S–B–S	14.1–27.9–14.1	27–30	24
	17–68–17	30	38
	14–30–14	26	25
II. Spherical morphology			
S–B	7.2–33	8.6	8.4
	8–40	10.7	9.3
	11–47	10.8	11.1
	12–147	11.2	11.6
	12–163	10.9	11.4
	13–59	12.8	12.4
	15–32	11.2	14.3
	15–83	12.2	13.8
S–B–S	13–75–13	13.5	13.1
	10–71–10	10	10.7
	7–35–7	9.3	8.5
	14–63–14	11.6	13.5
	21–98–21	17.0	18.1
	120–660–120	21	58

Polymer Engineering and Science

Values of the separation temperature for a series of poly(styrene-b-butadiene-b-styrene) were determined by light transmission, calorimetry, and electron microscope observations (*44*). A comparison between these experimental and calculated values of T_s is given in Table IV. Further evidence for the existence of such "structured–unstructured" transitions through rheological measurements will be given in a later section.

Elasticity of Heterophase Block Copolymers

The stress–strain behavior of heterogeneous block copolymers depends on their chemical composition. Those consisting of a soft rubbery component and a hard glassy component may either be rubberlike or plasticlike. In triblock copolymers where the former is the major component, the stress–strain curves would exhibit high elasticity up to nearly 1000% before fracture. The rubberlike elasticity arises from the fact that the plastic domains "anchor" the rubbery network chains as pseudo-crosslinks. In addition, these domains also have the reinforcing effect of fillers (*47*). Leonard (*48*) derived an equation of state for such systems:

$$f = (NRT/v_r^{1/3}L_o)\ (1.0 + 2.5\ \mu_p + 14.1\ \mu_p^2)\ (\lambda - 1/\lambda^2) \qquad (10)$$

where f is the elastic force, R is the ideal gas constant, T is the absolute temperature, L_o is the unstretched length, v_r and v_p are volume fractions of the rubbery and plastic components, respectively, N is the number of rubber chains anchored between $2N$ domains, and λ is the elongation ratio. The theory was derived from entropy considerations of the heterogeneous system although the resulting equation is identical to the classical statistical theory of rubber elasticity for filled systems.

Block copolymers in which the plastic component is sufficiently abundant to form continuous regions can be regarded as microcomposite materials. The dispersed domains in these materials are microscopic rather than macroscopic in dimensions. A number of existing theories for the elasticity of composites has been successfully applied to calculate the elastic moduli of these materials. Takayanagi (*49*) and Kawai (*50*) and their co-workers were among the first to treat the elastic moduli as composites. They chose an equivalent model to represent composites, using the degree of mixing (λ) of the dispersoids and the composition (ϕ) of the dispersoids and matrix as independent variables. Perfect material contact between the phases is assumed. When the equivalent model is stretched, the elastic force can be borne by the matrix alone or by both the matrix and the dispersed phases. The modulus of the equivalent model can be calculated by either the Series Model or the Parallel Model. For the Series Model, the modulus of the composite is:

$$M = \lambda \left[\frac{\phi}{M_d} + \frac{(1 - \phi)}{M_m} \right]^{-1} + (1 - \lambda)\ M_m, \qquad (11)$$

and for the Parallel Model:

$$M = \left[\frac{\phi'}{\lambda' M_d + (1 - \lambda')M_m} + \frac{(1 - \phi')}{M_m} \right]^{-1}, \qquad (12)$$

where subscripts d and m refer to dispersed and matrix phases respectively, v_s are the volume fractions of the two phases, and $\lambda\phi = v_d$. The

Table IV. Separation Temperatures in Triblock

Polymer	*Molecular Weight × 10⁻³* $M_s - M_B - M_s$	$\overline{M}_n$	*Volume Fraction Styrene* (ϕ_s)
Kraton 1101	12.5–75–12.5	100	0.245
Kraton 1102	5.4–32.3–5.4	43	0.245
	9.4–56.2–9.4	75	0.245
TR–41–1647	7–36–6	49	0.25
TR–41–1648	16–78–16	112	0.275
TR–41–1649	14–30–14	58	0.463

unprimed λ and ϕ refer to the Series Model and the primed ones to the Parallel Model. The two models are in fact equivalent, if $\lambda' = 1 - v_d - \phi$ (*51, 52*). Equations 11 and 12 have been used by a number of authors (*51, 52, 53*) to compare with experimentally determined elastic moduli of heterophase block copolymers. However, the Series–Parallel Model is only valid for soft dispersoids in hard matrix in concentration ranges where geometry of the dispersed phase is not important. For the inverse case of hard dispersoids in soft matrix, the moduli data cannot be adequately predicted by the model. Halpin (*54*) and Nielsen (*55, 56*) proposed a more general equation that covers the complete composition range. But as the composition of the block copolymer changes, a phase inversion can occur at a certain point. For such a situation, the use of some empirical mixing rules is necessary. Recently, Faucher (*57*) pointed out that by using the "polyaggregate" model of Kerner (*58*) it is not necessary to postulate the existence of the matrix phase. In fact, the model implies the equivalence of the two phases. Since neither one can be regarded as the matrix for the other, the difficulty of treating the phase inversion is circumvented. The resulting equations are lengthy, but the predictions appear to agree well with literature data.

For plastic-like heterophase block copolymers, the stress–strain behavior is strongly dependent on morphology. Kawai and co-workers (*59*) found that for a 50/50 diblock copolymer of styrene/isoprene cast from a mixed solvent system of toluene and methyl ethyl ketone, the stress–strain curve shows regions of yielding and drawing. Transmission electron micrographs show that there is extensive elongation of the plastic domains in the region of drawing. These authors hypothesized that such

Copolymers of Styrene and Butadiene (*44*)

T_s *(K)*		
Model	*Observed*	*Method*
590	583–593	light transmission
	> 343	calorimetry
265	373–423	light transmission
430	> 340	calorimetry
298	< 310	calorimetry
735	> 600	light transmission
	> 347	calorimetry
515	< 530	electron microscope
	> 500	calorimetry
	573–598	light transmission

Journal of Polymer Science, Polymer Physics Edition

morphological changes can be attributable to heat transformed from the strain energy, thereby causing the flow to take place upon stretching.

Under appropriate conditions of sample preparation, the phenomena of "strain-induced plastic–rubber transition" can be observed. For block copolymers exhibiting yielding and drawing regions in the first stress–strain cycle, there is usually considerable strain-softening in the second and subsequent deformation (*60, 61, 62, 63*). The drawing process occurs when the narrowing of the cross-sectional area of the sample suddenly appears at one point in the sample and subsequently propagates until the entire sample is transformed. Such phenomena are similar to that in conventional plastics except that in this instance the necked regions are not plastic but rubbery. After the necking process has propagated throughout, the sample which was initially a plastic has now become a rubber. The electron micrographs show that there is extensive disruption of the continuous polystyrene domains in the stretched sample. If the sample is annealed at elevated temperature, then the sample returns to the plastic state (*63*). These morphological changes also can be observed by small-angle x-ray scattering (SAXS) in Figure 8. The unstretched sample of a poly(styrene-b-butadiene-b-styrene) blended with 20% polystyrene shows a rather sharp peak but becomes broadened upon stretching. The scattering curve for the annealed sample, however, is more similar to that of the unstretched sample, indicating a partial restoration of the original morphology (*64*).

The mechanical properties of a macrolattice of SBS has been investigated (*65*). The sample consists of a hexagonal array of polystyrene cylinders embedded in the polybutadiene matrix. The stress–strain curves

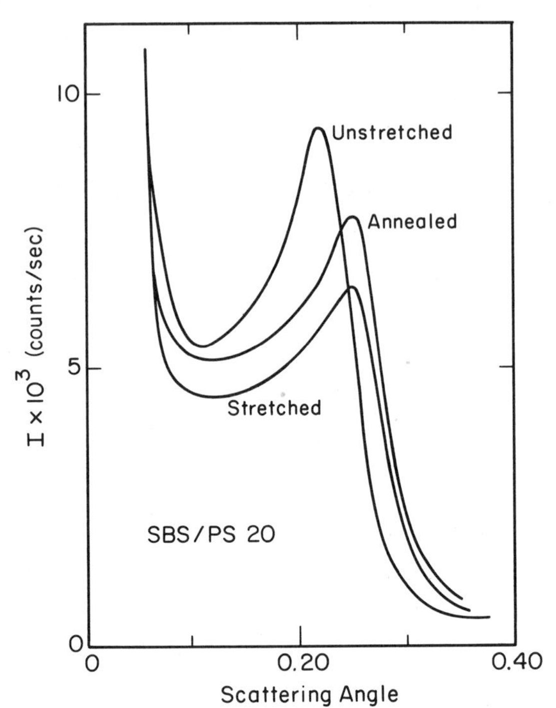

Figure 8. Small-angle x-ray scattering (SAXS) data for poly(styrene-b-butadiene-b-styrene) blended with 20% polystyrene and cast from THF/methyl ethyl ketone (after Ref. 64)

of the macrolattice show a decisive anisotropy. The moduli data were found to be in excellent agreement with the Takayanagi–Kawai model if the longitudinal sample is represented by parallel coupling and the transverse sample by series coupling.

Viscoelasticity of Heterophase Block Copolymers

In an earlier section, we have shown that the viscoelastic behavior of homogeneous block copolymers can be treated by the modified Rouse–Bueche–Zimm model. In addition, the Time–Temperature Superposition Principle has also been found to be valid for these systems. However, if the block copolymer shows microphase separation, these conclusions no longer apply. The basic tenet of the Time–Temperature Superposition Principle is valid only if all of the relaxation mechanisms are affected by temperature in the same manner. Materials obeying this Principle are said to be thermorheologically simple. In other words, relaxation times at one temperature are related to the corresponding relaxation times at a reference temperature by a constant ratio (the shift factor). For heterogeneous systems, the constituent polymers exist in separate phases and must undergo relaxation processes individually. Such heterogeneous block copolymers therefore do not satisfy the said stipulation and should be considered thermorheologically complex (*66, 67, 68, 69*). Their master curves are in fact different in shape at different temperatures because the relaxation times of the two different phases are affected by temperature differently. However, the experimentally accessible range (which Fesko and Tschoegl (*66*) call "the experimental window") is small. Within this window the neighboring isotherms appear to be superposable by simple horizontal shifting along the logarithmic time axis, but the result of such shifting would give rise to an erroneous master curve. A useful way to represent the viscoelasticity of heterogeneous system is the contour plot, an example for which is shown in Figure 9. Such a plot shows simul-

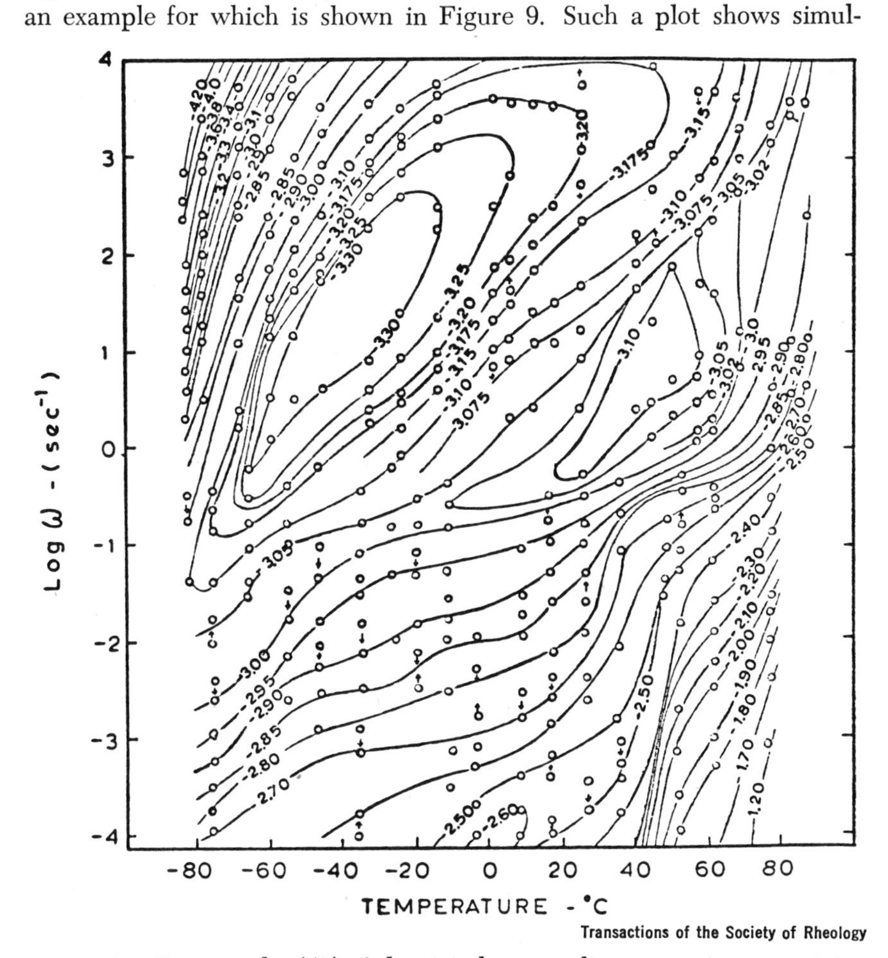

Transactions of the Society of Rheology

Figure 9. Contour plot (70) of dynamic loss compliance as a function of frequency and temperature for poly(styrene-b-butadiene-styrene).

taneously how a given viscoelastic parameter, in this case the dynamic loss compliance, depends on both the frequency (or time) and temperature (*70*).

The effect of morphology on the viscoelasticity of block copolymers has been investigated (*71, 72*). The most important factor appears to be the connectivity of domains. If the sample was cast from a solvent, which results in extensive interconnections among the hard domains (for instance the glassy PS domains in SBS), then the modulus in the region (above the T_g of PS) will be relatively high. On the other hand, if the hard domains are dispersed in a soft matrix (the rubbery PB domains), then the moduli in the same region will be lower for the same sample (*71*). In addition, the ratio of storage moduli (E'/G') in tensile and shear modes was found to be nearly three for the PB-continuous SBS, which is as expected for elastomers. However, the ratio for the same sample which was cast from solvents that render them PS continuous is now greater by an order of magnitude (*72*). The anomalously high value is attributed to the anisotropic PS-domain connectivity in the form of long fibrils or lamellae.

Rheology of Heterogeneous Block Copolymer Melts

Because the existence of domain structure in heterogeneous block copolymers persists even in the molten state, thier rheological behavior is rather unique when compared with homogeneous polymer melts.

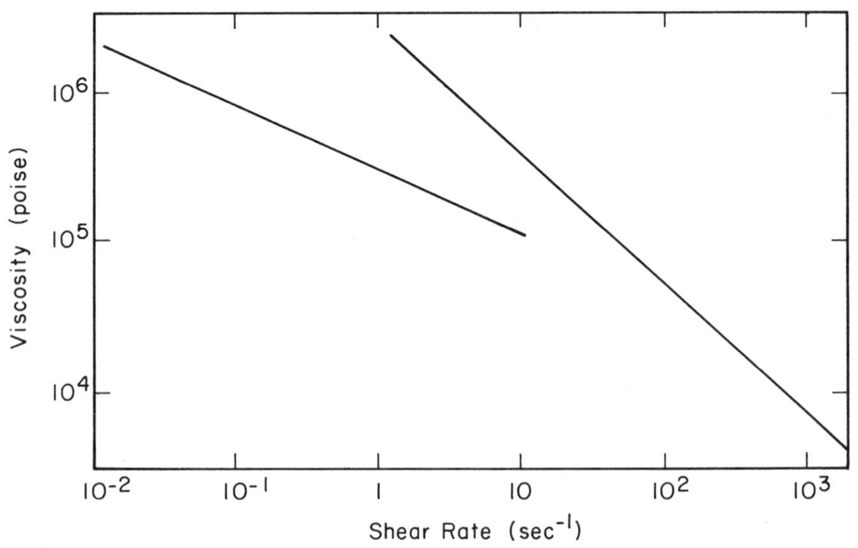

Figure 10. Shear viscosity as a function steady-state shear rate for poly(styrene-b-butadiene-b-styrene) at 150°C (after Ref. 47)

Holden et al. (*47*) first noted the peculiar characteristics in the steady shear behavior of the SBS block copolymer melts. For a certain composition of styrene and butadiene, no limiting Newtonian viscosity was found at low shear rates. For some of the others, there exist two distinct viscosity vs. shear rate relationships (Figure 10). Arnold and Meier (*73*) carried out the experiments in oscillatory shear and found the same

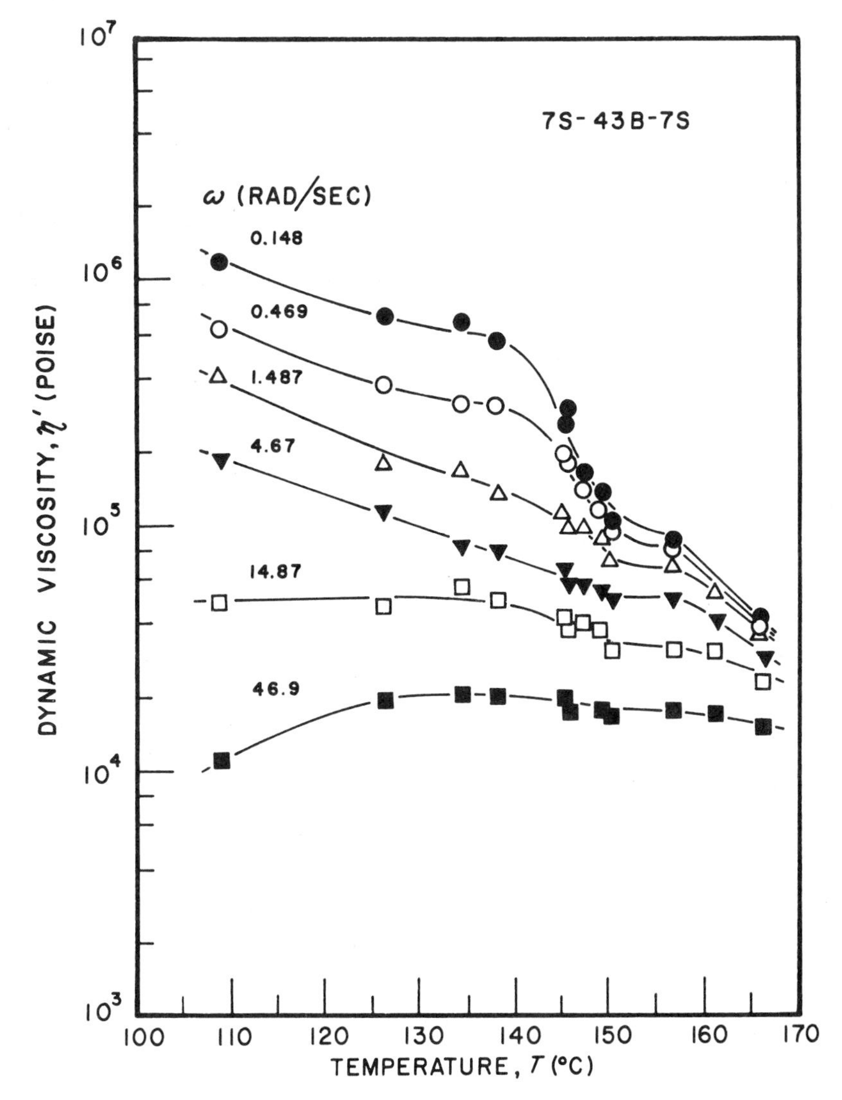

Figure 11. Dynamic shear viscosity as a function of temperature for poly(styrene-b-butadiene-b-styrene) at various angular frequencies (77)

anomaly. In addition, these authors found that the viscosities obtained were much higher than that of either homopolymer of the same molecular weight as the block copolymer. The absence of a Newtonian viscosity was explained in terms of a fluid domain structure in the melt that was progressively disrupted, causing the viscosity to decrease markedly with increasing shear rate. The high viscosity is attributed to the additional work needed to overcome the thermodynamic resistance to the mixing process for the different block species to flow past each other.

Kraus et al. (*74, 75*) studied the steady flow and oscillatory flow behavior of linear triblocks of S–B–S and B–S–B and radial block copolymers of the type $(B–S–)_3$, $(S–B–)_3$, and $(S–B–)_4$. For block copolymers of the same molecular weight and composition, those with end blocks of PS always have higher viscosities. However, when compared with corresponding linear block copolymers, the viscosities of the radial block copolymers are generally lower.

More recently, it has been demonstrated that many of the unusual rheological behavior of block copolymers will disappear when the measurements were carried out at temperatures higher than the separation temperature proposed by Leary and Williams (*43*). Figure 11 shows that for bulk SBS block copolymers with a composition of 7–43–7 ($x10^3$), the transition occurs around 145°C (especially clear at low frequencies) (*76, 77*). These data are consistent with those of Pico and Williams on plasticized block copolymers (*78*).

Acknowledgment

This work was supported by the Office of Naval Research. We wish to thank M. Hoffman, R. E. Cohen, E. Helfand, and C. I. Chung for the use of their figures.

Literature Cited

1. Allport, D. C., Janes, W. H., Eds., "Block Copolymers," Wiley, New York, 1973.
2. Burke, J. J., Weiss, V., Eds., "Block and Graft Copolymers," Syracuse University, Syracuse, 1973.
3. Sperling, L. H., Ed., "Recent Advances in Polymer Blends, Grafts and Blocks," Plenum, New York, 1973.
4. Platzer, N. A., Ed., "Copolymers, Polyblends and Composites," ADV. CHEM. SER. (1975) **142.**
5. Noshay, A., McGrath, J. E., "Block Copolymers: Overview and Critical Survey," Academic, New York, 1976.
6. Estes, G. M., Cooper, S. L., Tobolsky, A. V., *J. Macromol. Sci.* (1970) **C4,** 313.
7. Krause, S., *J. Macromol. Sci.* (1972) **C7,** 251.
8. Bever, M., Shen, M., *Mater. Sci. Eng.* (1974) **15,** 145.
9. Aggarwal, S. L., *Polymer* (1976) **17,** 938.

10. Shen, M., Kawai, H., *Am. Inst. Chem. Eng. J.* (1978) **24,** 1.
11. Krause, S., *J. Polym. Sci., Polym. Phys. Ed.* (1969) **7,** 249.
12. Krause, S., *Macromolecules* (1970) **3,** 84.
13. Meier, J., *J. Polym. Sci., Polym. Symp.* (1969) **C26,** 81.
14. Hildebrand, J. H., Prausnitz, J. M., Scott, R. L., "Regular and Related Solutions," van Nostrand, New York, 1970.
15. Baer, M., *J. Polym. Sci., Part A* (1964) **2,** 417.
16. Dunn, D. J., Krause, S., *J. Polym. Sci., Polym. Lett. Ed.* (1974) **12,** 591.
17. Robeson, L. M., Matzner, M., Fetters, L. J., McGrath, J. E., "Recent Advances in Polymer Blends, Grafts and Blocks," L. H. Sperling, Ed., p. 281, Plenum, New York, 1973.
18. Shen, M., Hansen, D. R., *J. Polym. Sci., Polym. Symp.* (1974) **46,** 55.
19. Hansen, D. R., Shen, M., *Macromolecules* (1975) **8,** 903.
20. Krause, S., Dunn, D. J., Seyed-Mozzaffari, A., Biswas, A. M., *Macromolecules* (1977) **10,** 786.
21. Soong, D., Shen, M., *Macromolecules* (1977) **10,** 357.
22. Rouse, P. E., *J. Chem. Phys.* (1953) **21,** 1272.
23. Bueche, F., *J. Chem. Phys.* (1954) **22,** 603.
24. Zimm, B., *J. Chem. Phys.* (1956) **24,** 269.
25. DeWames, R. E., Hall, W. F., Shen, M., *J. Chem. Phys.* (1967) **46,** 2782.
26. Hansen, D. R., Shen, M., *Macromolecules* (1975) **8,** 343.
27. Stockmayer, W. H., Kennedy, J. W., *Macromolecules* (1975) **8,** 351.
28. Hall, W. F., Kennedy, J. W., *Macromolecules* (1975) **8,** 349.
29. Wang, F. W., *Macromolecules* (1975) **8,** 364.
30. Tobolsky, A. V., Murakami, K., *J. Polym. Sci.* (1959) **40,** 443.
31. Kawai, H., Soen, T., Inoue, T., Ono, T., Uchida, T., *Mem. Fac. Eng. Kyoto Univ.* (1971) **33,** 383.
32. Molau, G. E., Wittbrodt, W. M., *Macromolecules* (1968) **1,** 260.
33. Hashimoto, T., Nagatoshi, K., Todo, A., Hasegawa, H., Kawai, H., *Macromolecules* (1974) **7,** 364.
34. Toy, L., Niinomi, M., Shen, M., *J. Macromol. Sci., Phys.* (1975) **11,** 281.
35. Niinomi, M., Akovali, G., Shen, M., *J. Macromol. Sci., Phys.* (1977) **13,** 133.
36. Fischer, E., *J. Macromol. Sci., Chem.* (1968) **A2,** 1285.
37. Kampf, G., Hoffman, M., Kromer, H., *J. Macromol. Sci., Phys.* (1972) **B6,** 167.
38. Dlugosz, J., Keller, A., Pedemonte, E., *Kolloid Z. Z. Polym.* (1970) **242,** 1125.
39. McIntyre, D., Campos-Lopez, E., *Macromolecules* (1970) **3,** 322.
40. Price, C., *Polymer* (1972) **13,** 20.
41. Pedemonte, E., Turturro, A., Bianchi, U., Devetta, P., *Polymer* (1973) **14,** 145.
42. Folkes, M. J., Keller, A., Haward, R. N., Eds., "Physics of Glassy Polymers," Wiley, New York, 1973.
43. Leary, D. F., Williams, M. C., *J. Polym. Sci., Polym. Phys. Ed.* (1973) **11,** 345.
44. Leary, D. F., Williams, M. C., *J. Polymer Sci., Polym. Phys. Ed.* (1974) **12,** 265.
45. Helfand, E., *Acc. Chem. Res.* (1975) **8,** 295.
46. Helfand, E., Wasserman, Z., *Polym. Eng. Sci.* (1977) **17,** 582.
47. Holden, G., Bishop, E. T., Legge, N. R., *J. Polym. Sci., Polym. Symp.* (1969) **C26,** 37.
48. Leonard, W. J., Jr., *J. Polym. Sci., Polym. Symp.* (1976) **C54,** 237.
49. Takayanagi, M., Uemura, S., Minami, S., *J. Polym. Sci., Polym. Symp.* (1964) **C5,** 113.
50. Fujino, H., Ogawa, I., Kawai, H., *J. Appl. Polym. Sci.* (1964) **8,** 2147.
51. Dickie, R. A., *J. Appl. Polym. Sci.* (1973) **17,** 45.

52. Kaplan, D., Tschoegl, N. W., *Polym. Eng. Sci.* (1974) **14**, 43.
53. Kraus, G., Rollman, K. W., Gruver, J. T., *Macromolcules* (1970) **3**, 92.
54. Haplin, J. C., *J. Compos. Mater.* (1969) **3**, 732.
55. Nielsen, L. E., *J. Appl. Phys.* (1970) **41**, 4626.
56. Nielsen, L. E., *Rheol. Acta* (1974) **13**, 86.
57. Faucher, J. A., *J. Polym. Sci., Polym. Phys. Ed.* (1974) **12**, 2153.
58. Kerner, E. H., *Proc. Phys. Soc., London* (1956) **69B**, 808.
59. Inoue, T., Ishihara, H., Kawai, H., Ito, Y., Kato, K., "Mechanical Behavior of Materials," Vol. 3, p. 149, Society of Materials Science—Japan, Tokyo, 1972.
60. Kraus, G., Childers, C. W., *Rubber Chem. Technol.* (1967) **40**, 1183.
61. Smith, T. L., Dickie, R. A., *J. Polym. Sci., Polym. Symp.* (1969) **C26**, 163.
62. Aggarwal, S. L., Livigni, R. A., Marker, L. F., Dudek, T., "Block and Graft Copolymers," J. J. Burke, V. Weiss, Eds., p. 157, Syracuse University, Syracuse, 1973.
63. Akovali, G., Diamant, J., Shen, M., *J. Macromol. Sci., Phys.* (1977) **B13**, 117.
64. Hong, S. D., Shen, M., Russell, T., Stein, R. S., "Polymer Alloys," D. Keemper, K. C. Frisch, Eds., Plenum, New York, 1977.
65. Folkes, M. J., Keller, A., *Polymer* (1971) **12**, 222.
66. Fesko, D. G., Tschoegl, N. W., *J. Polym. Sci., Polym. Symp.* (1971) **C35**, 51.
67. Kaniskin, V. A., Kaya, A., Ling, A., Shen, M., *J. Appl. Polym. Sci.* (1973) **17**, 2695.
68. Shen, M., Kaniskin, V. A., Biliyar, K., Boyd, R. H., *J. Polym. Sci., Polym. Phys. Ed.* (1973) **11**, 2261.
69. Kaya, A., Choi, G., Shen, M., "Deformation and Fracture of High Polymers," H. H. Kausch, J. A. Hassell, R. E. Jaffe, Eds., p. 27, Plenum, New York, 1974.
70. Cohen, R. E., Tschoegl, N. W., *Trans. Soc. Rheol.* (1976) **20**, 153.
71. Shen, M., Cirlin, E. H., Kaelble, D. H., "Colloidal and Morphological Behavior of Block and Graft Copolymers," G. E. Molau, Ed., p. 307, Plenum, New York, 1971.
72. Kraus, G., Rollman, K. W., Gardner, J. O., *J. Polym. Sci., Polym. Phys. Ed.* (1972) **10**, 2061.
73. Arnold, K. R., Meier, D. J., *J. Appl. Polym. Sci.* (1970) **14**, 427.
74. Kraus, G., Gruver, J. T., *J. Appl. Polym. Sci.* (1967) **11**, 2121.
75. Kraus, G., Nayler, F. E., Rollman, K. W., *J. Polym. Sci., Polym. Phys. Ed.* (1971) **9**, 1839.
76. Chung, C. I., Gale, J. C., *J. Polym. Sci., Polym. Phys. Ed.* (1976) **14**, 1149.
77. Chung, C. I., Lin, I. L., *J. Polymer Sci., Polym. Phys. Ed.* (1978) **16**, 545.
78. Pico, E., Williams M. C., *J. Polym. Sci., Polym. Phys. Ed.* (1977) **15**, 573.

RECEIVED April 14, 1978.

10

Block Size and Glass-Transition Temperatures of the Polystyrene Phase in Different Block Copolymers Containing Styrene Blocks as the Hard Phase

S. KRAUSE and M. ISKANDAR

Department of Chemistry, Rensselaer Polytechnic Institute, Troy NY 12181

Glass-transition temperatures of the styrene microphases in diblock copolymers of styrene and dimethyl siloxane have been obtained as a function of the molecular weights of the styrene blocks using DSC and DTA techniques. The magnitudes of the glass-transition temperatures of the styrene microphases in these systems were, within experimental error, exactly the same as the glass-transition temperatures taken from the literature of the styrene microphases in styrene–isoprene diblock and triblock copolymers and in styrene–ethylene oxide diblock and triblock copolymers in which the styrene blocks had comparable molecular weights. It appears, therefore, that the chemical nature of the other blocks which are attached to the styrene blocks has no effect on the glass-transition temperature of the styrene microphase; some implications of this observation are discussed.

Often, when T_g data have been obtained on block copolymer samples containing small, low-molecular-weight blocks, a systematic variation of T_g with block length has been observed. If we consider only block copolymers in which microphase separation has been demonstrated by the observation of two different glass-transition temperatures, we find that the higher T_g, to be referred to henceforth as the glass-transition

0-8412-0457-8/79/33-176-205$05.00/0

temperature of the glassy phase, varies quite differently with changes in molecular weight of the relevant blocks than does the lower T_g, to be referred to henceforth as the glass-transition temperature of the rubbery phase.

In this chapter, we shall deal mainly with the T_g of the glassy phase, which, in almost all cases that have been investigated, has been polystyrene. In one of the earliest investigations of this type, Kraus, Childers, and Gruver (*1*) noticed that the temperature of the loss peak attributed to the styrene blocks in dynamic mechanical measurements of styrene–butadiene block copolymers decreased from 104° to 60°C when the styrene block length decreased by a factor of about four (exact molecular weights of the samples were not known). In a more recent work, Kraus and Rollmann (*2*) studied triblock copolymers of styrene with isoprene over a copolymer-molecular-weight range from 2×10^4 to about 10^5, again using dynamic mechanical techniques. The temperature of the upper (polystyrene) loss maximum varied from 120° to 20°C as the molecular weight of the styrene blocks decreased from 2×10^4 to about 5000. Differential scanning calorimetry (DSC) was used by Robinson and White (*3*) to study glass-transition temperatures in triblock copolymers of styrene and isoprene and by O'Malley et al. (*4*) to study the melting transition of the ethylene oxide and the glass transition of the styrene phase in diblock and triblock copolymers of styrene and ethylene oxide. The T_g of the polystyrene microphases in both DSC studies decreased with decreasing polystyrene block lengths as usual.

Additional T_g data on very low molecular weight styrene–isoprene diblock and triblock copolymers were obtained by Toporowski and Roovers (*5*), also using DSC techniques. Only the T_g of the polystyrene microphases was given in those cases in which the T_g of both phases was obtained, i.e., where microphase separation definitely occurred, and this T_g also decreased with decreasing size of the styrene blocks; however, it increased slightly as the size of the isoprene blocks was increased while holding constant the sizes of the styrene blocks. This is in sharp contrast to the data of Kraus and Rollmann (*2*), which indicated that the T_g of the polystyrene microphase depended only on the molecular weight of the styrene blocks, whether the other phase was polybutadiene or polyisoprene. It is of interest to note that the data of Kraus and Rollmann (*2*) were obtained on block copolymers of much higher molecular weight than were those of Toporowski and Roovers (*5*); the latter authors, as a matter of fact, studied block copolymers of molecular weight only slightly greater than that at which microphase separation no longer occurs. They studied, in addition to the microphase-separated block copolymer samples, many lower-molecular-weight samples in which microphase separation did not occur.

A number of explanations exist for the decrease in glass-transition temperature of the glassy microphase with decreasing block length of the glassy blocks. Childers and Kraus (*6*) speculated that this T_g lowering might possibly be attributed to some mixing of the block segments in styrene–butadiene block copolymers, that is, they postulated increased compatibility of the blocks as the block lengths decreased. Many investigators have postulated that mixing of the chemically different block segments occurs in an interlayer with finite volume between microphases; the volume of this mixed interlayer would presumably depend on the molecular weights of the blocks and on their mutual compatibility. Kraus and Rollmann (*2*) assumed an interlayer in their styrene–butadiene and styrene–isoprene block copolymers that was disproportionately rich in styrene, and they presented calculations which showed that such an interpretation was consistent with their data while an interpretation that relied solely on the free volume provided by polymer chain ends, whose concentration increases with decreasing molecular weight, predicted a variation of T_g with molecular weight that was much smaller than that which was observed. The presence of a more uniformly mixed interlayer had been postulated earlier by Fesko and Tschoegl (*7*) to account for the dynamic mechanical properties of some styrene–butadiene–styrene triblock copolymers.

The observed T_gs of the polystyrene microphases in all of the work quoted above (*1, 2, 3, 4, 5*) were lower than those expected for polystyrene homopolymers having the same molecular weights as the blocks in the block copolymers. We present a detailed comparison of such data below.

Even if there is no mixed interlayer with finite volume between microphases, it is obvious that contacts between the chemically different blocks occur at the surfaces between microphases and that the number of these contacts must increase as the size of the microphases decreases, that is, as the surface area between microphases increases. Bares and Pegoraro (*8*) analyzed the observed glass-transition temperatures in ethylene–propylene copolymers with vinyl–chloride grafts in terms of such surface mixing. This type of mixing provides a noticeable decrease in T_g only when the microphases are extremely small; Bares and Pegoraro (*8*) calculated that effects would only be observed when spherical microphases were less than 25 monomer units in diameter or when cylindrical microphases were less than 10 monomer units in diameter. In a later publication, Bares (*9*) devised a less restrictive argument for calculating the T_g of the glassy microphases as the molecular weight of the blocks comprising these microphases decreased and as the surface-to-volume ratio of these microphases simultaneously increased. He essentially used the Fox–Flory (*10*) equation for T_g lowering caused by low molecular weight in homopolymers and in random copolymers and added another term which provided for an additional lowering of T_g as the surface:volume ratio of a particular microphase increased. Since the surface:volume ratio of microphases with different morphology is not the same, the T_g lowering would be expected to vary with morphology. Bares (*9*) found qualitative agreement between his predictions and some of the data of Robinson and White (*3*).

Couchman and Karasz (*11*) recently have made some calculations indicating that spherical microphases should exhibit increased glass-transition temperatures because of an increased pressure inside such microphases attributed to the surface tension between microphases. Since there is some doubt about the existence of a surface of tension in the Gibbs' sense (*12*) between chemically linked microphases, we shall simply note that these calculations are the only ones in existence that indicate a possible reason for an increase in the T_g of a glassy microphase and, in addition, that these calculations also postulate differences in T_g with differences in morphology. For example, this surface-tension-dependent effect would not be expected in samples with lamellar morphology, no matter how small the width of each lamella.

To summarize the observations made in the preceding paragraphs, we can say that there is no doubt about the fact that the T_g of the glassy microphases in phase-separated block copolymers decreases when the lengths of the glassy blocks decrease, and furthermore, that the T_gs of such microphases are generally lower than are those of the corresponding homopolymers of comparable molecular weight. We have discussed mostly block copolymers in which styrene comprises the glassy blocks; however, similar data have been obtained for the glass-transition temperatures of the polycarbonate blocks in alternating multiblock copolymers of dimethylsiloxane and bisphenol-A carbonate by Kambour (*13*). Explanations that have been advanced for this general T_g lowering include: (1) an explanation similar to the molecular weight effect in homopolymers, (2) mixing of segments from chemically different blocks by surface contact between microphases, and (3) mixing of segments from chemically different blocks in interlayers of finite volume.

The decrease in homopolymer T_g with decreasing molecular weight has generally been attributed to an increase of free volume in low-molecular-weight bulk polymers caused by the increased concentration of chain ends. However, end blocks in block copolymer molecules have only a single-chain end while center blocks have no free-chain ends. One would therefore expect that the T_g of a microphase comprised of end blocks would be lower than that of a microphase comprised of center blocks of comparable molecular weight.

The volume of a mixed interlayer between microphases should depend on compatibility between the blocks as measured by the difference between their solubility parameters or by their interaction parameter. Surface contacts between microphases, on the other hand, should be affected only by microphase size and morphology.

To disentangle the various possible effects on the glass-transition temperatures of glassy microphases, we have begun studies on a block copolymer system in which the blocks are expected to be extremely incompatible so that the formation of a mixed interlayer with finite volume is much less probable than it is in a system like styrene–isoprene. The slight but significant compatibility of styrene blocks with isoprene blocks is shown directly in the data of Toporowski and Roovers (*5*), who were able to observe complete mixing between styrene and isoprene blocks in their block copolymers when molecular weights were very low. This compatibility makes the presence of a mixed interlayer between microphases reasonable. Some experimental studies of possible dimensions of this interlayer using small angle x-ray scattering (SAXS) are beginning to appear in the literature (*14*), although older studies using SAXS indicated a sharp boundary between microsphases (*15, 16*).

The system with which we have begun our investigations is the styrene–dimethylsiloxane system. The dimethylsiloxane blocks should be considerably less compatible with polystyrene blocks than either polybutadiene or polyisoprene since the solubility parameter of dimethylsiloxane is much farther from that of polystyrene than are the solubility parameters of polybutadienes or of polyisoprenes (*17*), no matter what their microstructure. Furthermore, even hexamers of polystyrene and of polydimethylsiloxane are immiscible at room temperature and have an upper critical-solution temperature above 35°C (*18*). In addition, the microphases in this system can be observed without staining and with no ambiguity about the identity of the phases in the transmission electron microscope (TEM); silicon has a much higher atomic number than carbon or oxygen, making the polydimethylsiloxane microphases the dark phases in TEM (*19, 20*).

Experimental

Polymers. The polystyrene standards used in this work were Arro Laboratories' "Monodisperse." One group of styrene–dimethylsiloxane diblock copolymers were those prepared and characterized by Zilliox, Roovers, and Bywater (*21*), kindly sent to us by J. E. L. Roovers. These polymers have been given the designation R in Table II below; molecular weights of the blocks were calculated from the experimentally determined weight-average molecular weights and compositions of the copolymers (*21*) except for sample R13, for which only the number-average molecular weight was available. Gel permeation chromatograms were obtained in toluene solution on a Waters Associate's Anaprep GPC run in the analytical mode using five columns ranging from 2×10^3 to 7×10^5 Å in exclusion limit and were compared with chromatograms supplied to us by Roovers. The chromatograms were virtually identical.

A second group of diblock copolymers was kindly supplied to us by J. W. Dean of the Silicone Products Department of the General Electric Co. These samples are designated D in Table II; molecular weights of the blocks were calculated from weight-average molecular weights obtained using GPC in our laboratory and compositions of the copolymers as determined by Dean (*22*). Weight-average molecular weights obtained from GPC in our laboratory were calculated as follows: the set of five columns described above were calibrated using Arro Laboratories' "Monodisperse" polystyrenes, and then we assumed that the block copolymers followed the same calibration curve as polystyrene. The validity of this procedure is presently under examination in our laboratory.

Thermal Analysis. Differential thermal analysis (DTA) data were obtained using a Fisher Thermalyzer Model 300 QDTA with a heating rate of 20°C/min and static air in the sample holder, while differential scanning calorimetry (DSC) data were obtained using a Perkin–Elmer DSC-1B with a heating rate of 10°C/min and a steady flow of nitrogen. The cooling rate was also controlled at 10°C/min between runs on the same sample in the DSC experiments; cooling rates varied in the DTA runs. The temperature scales on both instruments were calibrated using the melting transitions of *p*-nitrotoluene and naphthalene below 100°C and those of lead, tin, indium, and benzoic acid above 100°C. On both instruments, the average error of the temperature scale for all these transitions was ±1.0% (°C).

All glass-transition temperatures recorded in Tables I and II are in the middle of the transition range of the heat capacities. We are aware that some workers (*23*) feel that the temperature at which the heat capacity just begins to change during heating is somewhat more reproducible, though probably less meaningful than the middle of the transition range. Recorded temperatures are the averages of many runs for each sample; if the T_g was not completely reproducible, the range of values obtained is shown in the tables. There was one sample, R14, for which two T_g-like transitions sometimes appeared on first heating the

Table I. Glass-Transition Temperatures of Standard Polystyrene Samples Obtained in This Laboratory

Mol Wt	T_g (°C)	
($\times 10^{-4}$)	*DTA*	*DSC*
0.20	63	55
1.00	95 ± 1	95 ± 2
2.00	100.5 ± 0.5	99 ± 2
5.00	101.5 ± 0.5	104
9.72	101 ± 1	106
180	104 ± 2	105

Table II. Glass-Transition Temperatures of Styrene–Dimethylsiloxane Diblock Copolymers Obtained in This Laboratory

Sample	$\overline{M}_w$ *(Styrene Block)*	*Wt % Styrene*	$\overline{M}_w/\overline{M}_n$ *(Whole Sample)*	T_g (°C)	
				DTA	*DSC*
R13	9,100	53.5	1.06	77	79 ± 4
R14	19,300	56.8	1.10	75 ± 2[a]	72 ± 4[a]
				98 ± 3[b]	102 ± 2[b]
D2	20,300	89.7	1.26	—	96
D1	23,900	91.9	1.30	—	95 ± 1
R15	47,100	47.1	1.14	103 ± 3	104 ± 2
R8	121,000	59.6	1.16	101.5 ± 0.5	108 ± 1
R5	136,000	73.5	1.19	102 ± 1	107 ± 1
R25	327,000	66.5	1.15	98.5	104
R26	362,000	58.3	1.13	103 ± 2	109

[a] The lower T_g appeared on the first heating of the sample only, sometimes in conjunction with the higher T_g.

[b] The higher T_g sometimes appeared in conjunction with the lower T_g on the first heating of the sample; when a transition appeared on subsequent heating of the sample, it was always the higher T_g.

sample; both of these are shown even though the lower transition disappeared on subsequent heating. When a large endotherm or, in one case, an exotherm, appeared in the glass-transition range of a sample on first heating, no attempt was made to deduce a T_g from it.

Transmission Electron Microscopy. Films of all samples designated R were obtained by evaporation of toluene from solutions of the block copolymers and were observed without staining using a Hitachi Hu-125 or a JEOL JEM 100 S electron microscope. Methods of preparing the films have been described previously (*24*). So far, we have obtained evidence for microphase separation in only the four highest molecular-weight samples by TEM. We have not obtained continuous films of the lower molecular-weight samples; we plan to examine sections of these samples later. Because of the very small compatibility of styrene and polydimethylsiloxane, however, we expect phase separation in all of these samples.

Results and Discussion

Table I shows the T_g values obtained in this work on standard polystyrene samples; in most cases, the values obtained from DTA data are slightly higher than those obtained by DSC. Since these data were to be used for detailed comparisons with data obtained by others as well as by ourselves on styrene block copolymers, we first compared our data on anionically polymerized standard polystyrenes with those obtained on other anionically polymerized polystyrenes by other workers also using DTA and DSC techniques. This comparison is shown graphically in Figure 1. In Figure 1, the solid curve is drawn through a combination of our data with that obtained by Wall et al. (*25*) using a Dupont 900

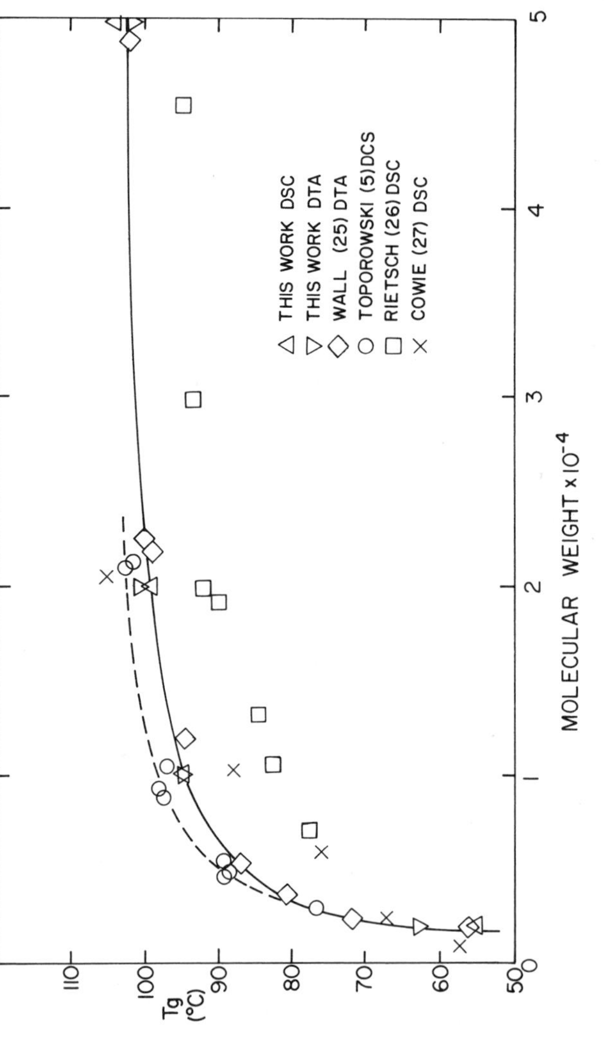

Figure 1. Glass-transition temperatures of anionically polymerized polystyrenes as determined by DTA or DSC. The solid curve is drawn through a combination of our data and that of Wall et al. (25); the dashed line is drawn through the data of Toporowski and Roovers (5).

Differential Thermal Analyzer at a heating rate of 20°C/min. This curve will be considered our standard curve for polystyrene homopolymer and is redrawn on Figure 2. On Figure 1, the dashed curve is drawn through the data of Toporowski and Roovers (*5*); only their lowest molecular-weight sample falls on our standard curve. The discrepancy between the two curves may possibly be caused by some polydispersity in Toporowski and Roovers' samples; they reported only $\overline{M}_n$ values for their homopolymers. The data of Rietsch et al. (*26*), which were obtained using a Perkin–Elmer DSC-2 at various heating rates and extrapolating the calculated T_gs to very low heating rate, fall well below our standard curve; this is to be expected because of the extrapolation to low heating rate. The data of Cowie (*27*), obtained using a DuPont 900 DSC module at a heating rate of 20°C/min, cannot be explained simply by noting that Cowie reported the initial change of heat capacity on heating as his glass-transition temperature.

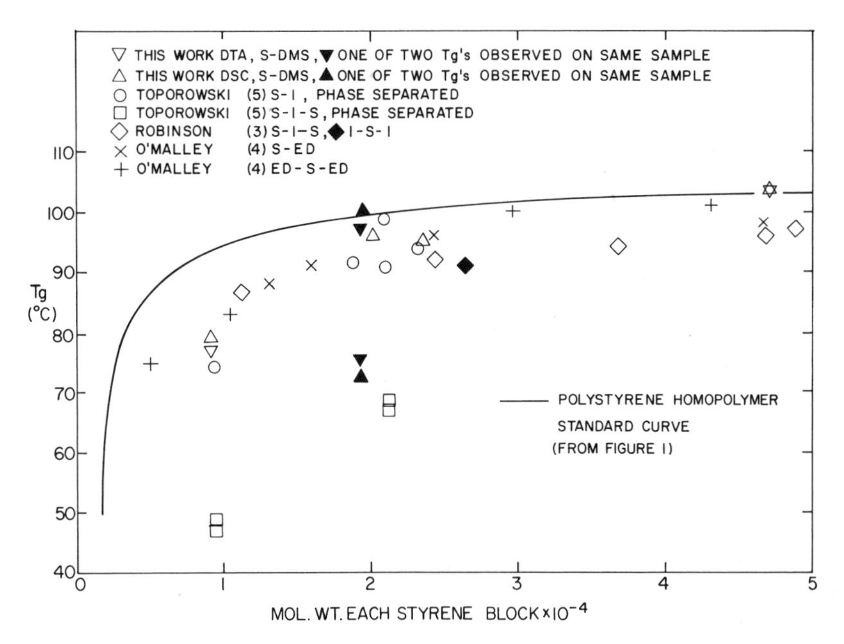

Figure 2. Glass-transition temperatures of the glassy microphases of various block copolymers containing styrene blocks vs. the molecular weights of the styrene blocks as determined by DTA or DSC.

S–DMS refers to styrene–dimethylsiloxane diblock copolymers; S–I refers to styrene–isoprene diblock copolymers; S–I–S and I–S–I refer to styrene–isoprene–styrene and to isoprene–styrene–isoprene triblock copolymers, respectively; and S–EO and EO–S–EO refer to styrene–ethylene oxide diblock copolymers and to ethylene oxide–styrene–ethylene oxide triblock copolymers, respectively.

Table II shows the T_g values obtained for the glassy phase of the styrene–dimethylsiloxane block copolymers used in this work; in the case of the block copolymers, neither DSC nor DTA gave a consistently higher value of T_g. The only peculiar sample in the table is sample R14, which exhibited two T_gs on both instruments at least occasionally during the first heating cycle. Although the presence of two T_gs implies some kind of phase separation, possibly in addition to microphase separation, the GPC of this sample shows no double peak, shoulder, or other peculiarity which might explain its peculiar phase behavior. This sample is an anomaly unless one wishes to dismiss all first heating data.

Figure 2 shows our polystyrene standard curve, the data from Table II, and data obtained for the T_g of the glassy microphase in various styrene–rubber block copolymers by other investigators using DTA and DSC techniques. It is remarkable that the data for all but a few samples fall in a narrow band of the order of 10°C below the standard polystyrene homopolymer curve. The lowest molecular-weight styrene blocks have T_gs that are somewhat farther than 10°C below the standard curve. It should be noted that some of the block copolymers for which data appear on Figure 2 have styrene as one of the blocks of a diblock copolymer, some have styrene as the center block in a triblock copolymer, and others have styrene as the two end blocks in a triblock copolymer. For those samples in which styrene appears as the two end blocks in a triblock copolymer, the molecular weight of only one of the styrene blocks has been used in the figure. One can note that, in the case of these samples, the data would no longer be in a band with most of the other data if the combined molecular weight of the two styrene blocks had been used in the figure. One also should note that there are no systematic differences between the data for styrene–isoprene, styrene–ethylene oxide, and styrene–dimethylsiloxane block copolymers. The molecular weight of each styrene block appears to be the major variable that controls its glass-transition temperature, not its compatibility with its comonomer and not its position in the block copolymer molecule. It is not obvious from Figure 2, though it may be deduced from Table II and from a perusal of the percentage composition data for all the block copolymers shown in the figure, that the glass-transition temperatures of the styrene microphases do not depend, in any systematic way, on the percentage compositions of the block copolymers which compose them; this implies that the morphology of the block copolymers, which at least at equilibrium is controlled by percentage composition, has no major effect on the glass-transition temperature of the glassy microphase.

The "maverick" points on Figure 2 can be rationalized. One set of them belongs to our anomalous sample, R14, that exhibited two T_gs. If one desires to dismiss all "first-heating" data, the lower two points will disappear. The other four points belong to some of Toporowski and Roovers' styrene–isoprene–styrene block copolymers in which the molecular weights of the isoprene blocks were not very much above the molecular weight at which microphase separation occurs in block copolymers with styrene-block molecular weights equal to those in the block copolymers for which data are shown in Figure 2. This implies that these particular block copolymers of Toporowski and Roovers (5) are ones in which segment mixing of the chemically different blocks is extremely likely. We would guess that the low T_gs of the styrene microphases in these four block-copolymer samples are connected with the compositions of these microphases; considerable isoprene is probably admixed within these microphases.

We plan to continue studies on the glass-transition temperatures of the glassy microphases in block copolymers and, in the future, we also shall investigate the glass-transition temperatures of the rubbery microphases.

Conclusion

The glass-transition temperature of the glassy microphase in a glass–rubber block copolymer depends almost entirely on the molecular weights of the glassy blocks and not on the chemical nature of the rubbery blocks, the position of the glassy block or blocks within the block copolymer molecule, or the percentage composition of the block copolymer. The chemical nature and/or molecular weight of the rubbery block influences the glass-transition temperature of the glassy microphase only when the molecular weight and composition of the block copolymer is near that at which microphase separation no longer takes place.

Acknowledgment

Many thanks to J. E. L. Roovers and to J. W. Dean for providing us with the styrene–dimethylsiloxane block copolymers; thanks to the National Science Foundation, Contract No. DMR76-19488, for helping us with this work; and thanks to the National Institutes of Health for providing one of us (S.K.) with a Research Career Award.

Literature Cited

1. Kraus, G., Childers, C. W., Gruver, J. T., *J. Appl. Polym. Sci.* (1967) **11,** 1581.
2. Kraus, G., Rollmann, K. W., *J. Polym. Sci., Polym. Phys. Ed.* (1976) **14,** 1133.
3. Robinson, R. A., White, E. F. T., "Block Polymers," S. L. Aggarwal, Ed., p. 123, Plenum, New York, 1970.

4. O'Malley, J. J., Crystal, R. G., Erhardt, P. F., "Block Polymers," S. L. Aggarwal, Ed., p. 163, Plenum, New York, 1970.
5. Toporowski, P. M., Roovers, J. E. L., *J. Polym. Sci., Polym. Chem. Ed.* (1976) **14**, 2233.
6. Childers, C. W., Kraus, G., *Rubber Chem. Technol.* (1967) **40**, 1183.
7. Fesko, D. G., Tschoegl, N. W., *Int. J. Polym. Mater.* (1974) **3**, 51.
8. Bares, J., Pegoraro, M., *J. Polym. Sci., Polym. Phys. Ed.* (1971) **9**, 1287.
9. Bares, J., *Macromolecules* (1975) **8**, 244.
10. Fox, T. G, Flory, P. ., *J. Appl. Phys.* (1950) **21**, 581.
11. Couchman, P. R., Karasz, F. E., *J. Polym. Sci., Polym. Phys. Ed.* (1977) **15**, 1037.
12. Gibbs, J. W., "The Scientific Papers, Vol. I, Thermodynamics," Dover, New York, 1961.
13. Kambour, R. P., "Block Polymers," S. L. Aggarwal, Ed., p. 263, Plenum, New York, 1970.
14. Hashimoto, T., Todo, A., Itoi, H., Kawai, H., *Macromolecules* (1977) **10**, 377.
15. Kim, H., *Macromolecules* (1972) **5**, 594.
16. Skoulios, A. E., "Block and Graft Copolymers," J. J. Burke, V. Weiss, Eds., p. 121, Syracuse University, 1973.
17. Burrell, H., "Polymer Handbook," J. Brandrup, E. H. Immergut, Eds., 2nd ed., IV–337, Wiley, New York, 1975.
18. Okazawa, T., *Macromolecules* (1975) **8**, 371.
19. Saam, J. C., Fearon, F. W. G., *Polym. Prepr., Am. Chem. Soc., Div. Polym. Chem.* (1970) **11**, 455.
20. Morton, M., Kesten, Y., Fetters, L. J., *Polym. Prepr., Am. Chem. Soc., Div. Polym. Chem.* (1974) **15**(2), 175.
21. Zilliox, J. G., Roovers, J. E. L., Bywater, S., *Macromolecules* (1975) **8**, 573.
22. Dean, J. W., Private communication.
23. Griffiths, M. D., Maisey, L. J., *Polymer* (1976) **17**, 869.
24. Krause, S., Iskandar, M., "Polymer Alloys, Blends, Blocks, Grafts, and IPNS's," D. Klempner, K. C. Frisch, Eds., p. 231, Plenum, New York, 1977.
25. Wall, L. A., Roestamsjah, Aldridge, M. H., *J. Res. Natl. Bur. Stand., Sect. A* (1974) **78A**, 447.
26. Rietsch, F., Daveloose, D., Froelich, D., *Polymer* (1976) **17**, 859.
27. Cowie, J. M. G., *Eur. Polym. J.* (1975) **11**, 297.

RECEIVED April 14, 1978.

11

Dilatometry and Relaxations in Triblock Polymer Systems

J. B. ENNS[1], C. E. ROGERS, and ROBERT SIMHA

Department of Macromolecular Science, Case Western Reserve University, Cleveland, OH 44106

Dilatometry has been used for the investigation of relaxation processes in styrene–butadiene (SBS) triblock systems of varying total molecular weight and relative lengths of center and end blocks, and in S–B blends. Over the temperature range from about −100°C to the glass temperature of polystyrene, several characteristic temperature regions are noted in the thermal expansivity, one a relaxation at $T/T_g \simeq 0.87$*, the other a minimum at* T/T_g *of 0.92–0.95. A significant transition is seen at* T/T_g *equal to 0.78–0.83, corresponding to the range of the* β *relaxation reported for polystyrene. The sensitivity of the results to annealing effects indicates that the magnitude of the change in* α *is greater for systems with an initially more perfected interphase structure. An increase in temperature thus allows those samples to gain the largest configurational freedom corresponding to a loosening of structure.*

Dilatometry has long been recognized as a valuable technique for obtaining information on transition properties of polymers. Its use had been generally limited to the glass-transition region and the melting process which display a change in slope and discontinuity, respectively, in the volume–temperature curves. In recent years it has been demonstrated (*1, 2, 3, 4, 5*) that relaxation processes of low intensity in amorphous and semicrystalline polymers can be investigated with good reso-

[1] Current address: Polymer Materials Program, Princeton University, Princeton, NJ 08540.

0-8412-0457-8/79/33-176-217$05.00/0

lution by using the first and second derivatives of the specific volume V as a function of temperature T, namely: $\alpha = (1/V)(dV/dT)$ and $d\alpha/dT$, where α is the thermal volume expansivity. In a plot of α vs. T, second-order type transitions appear as steps in the curve, with $d\alpha/dT$ exhibiting maxima.

In the present study, the effects of composition, molecular weight, and heat treatment on the relaxation behavior of styrene–butadiene–styrene (SBS) block polymers are investigated. There is evidence (e.g., *6, 7, 8*) that these types of multicomponent multiphase systems exhibit unusual phenomena in their dynamic mechanical behavior and in other physical properties. These are apparently related to the presence of the so-called interphase mixing region between the elastomeric and glassy domains. Similar evidence has been obtained by gas diffusion and sorption studies on the copolymer samples used in this investigation (*9*).

Experimental

The SBS copolymer materials were kindly supplied by the Small Development Company, Torrance, California. The characteristics of the additive-free samples are given in Table I.

The test specimens were prepared by casting from a 20% by weight solution in purified toluene at room temperature. Solvent evaporation in a controlled solvent atmosphere was carried out over a 24-hour period. The specimens were then vacuum dried for periods of time ranging from 36 hr (for 75°C) to seven days (for 40°C). This procedure yielded transparent, defect free, uniform thickness membranes with thicknesses between 0.012 and 0.018 in. Samples were prepared at the following final drying temperature: TR-41-2443 at 55°C, TR-41-2444 at 55° and 75°C, and TR-41-2445 at 40° and 55°C.

The samples were placed into a sample chamber, immersed in mercury, and the change in volume as a function of temperature obtained

Table I. Characterization of Experimental Block Copolymers

Quantity	*TR-41-2443*	*TR-41-2444*	*TR-41-2445*
Styrene content (weight fraction)	0.277	0.443	0.254
Molecular weights (in thousands)			
$\overline{M}_{S1}$	16	14	7
$\overline{M}_{B}$	85	33	43
$\overline{M}_{S2}$	17	12	7
$\overline{M}_{total}$	118	59	57
Polybutadiene microstructure (%)			
cis-1,4	40	45	40
trans-1,4	49	45	50
1,2	11	10	10

by observing the height of the mercury in the column attached to the sample chamber. The apparatus, technique, and numerical procedures for the determination of first and second volume derivations have been described elsewhere (*2, 10*). The procedures enable the volume to be measured with a maximum error of 1.3×10^{-4} cm^3/g. The consequent maximum error in the determination of α is 0.014%. The maximum error in the determination of the temperature variation of α is 0.14%.

Results and Discussion

Dilatometric measurements were performed on each of the SBS block copolymers, as well as the polybutadiene homopolymer and a solution blend of polystyrene and polybutadiene. Figure 1 shows a plot of the thermal expansion coefficient and the specific volume vs. temperature for the TR-41-2444 sample dried at 75°C. In the range 20°–30°C below the glass-transition temperature of the polystyrene component, there is a minimum in the α vs. T curve. At 20° to 30°C lower there is an apparent transition, as indicated by a step increase in α and a change in slope V_s. These two features appear at approximately the same location on a normalized temperature scale of T/T_g for other systems (Figures 1, 2, 3, 4, and 5). The transition occurs at 0.87, and the minimum at 0.92–0.95. For the TR-41-2445 copolymers (Figures 3 and 4) these two features are suppressed by varying amounts. In the sample dried at 40°C no transitions or minima were observed between $T/T_g = 0.85$ and $T/T_g = 0.95$. In the 55°C dried material a transition occurred at $T/T_g = 0.90$, but no minimum was found. In all four samples there is a significant transition which occurs at $T/T_g = 0.78$–0.83. This range of temperature corresponds to that reported for the β transition in polystyrene (*11*).

Another feature is the dependence of the polystyrene glass temperature on both the copolymer type and the drying temperature. This is a combination of effects attributable to both molecular weight and annealing effects. The T_g of polystyrene depends on molecular weight according to the equation $T_g = T_g^{\infty} - K/M$, where K is a constant equal to 10^5 for polystyrene, M is the molecular weight, and $T_g^{\infty} = 96.5°C$. This would predict a polystyrene T_g in the TR-41-2444 material at 89°C and in the TR-41-2445 copolymer at 82°C. Calculated values correspond reasonably well with those determined experimentally for the samples dried at higher temperatures, i.e., 88° and 76.5°C for TR-41-2444 and TR-41-2445, respectively. When the samples were dried at lower temperatures, the T_gs shifted to even lower values: 83°C for TR-41-2444 dried at 55°C and 58.5°C for TR-41-2445 dried at 40°C.

The effect of molecular weight is affirmed by the results for TR-41-2443 (Figure 5). This copolymer is similar to TR-41-2445 except that the polystyrene molecular weight is about 2.5 greater than that of TR-41-2445. The T_g in this case is 93.5°C, and since it was prepared in the

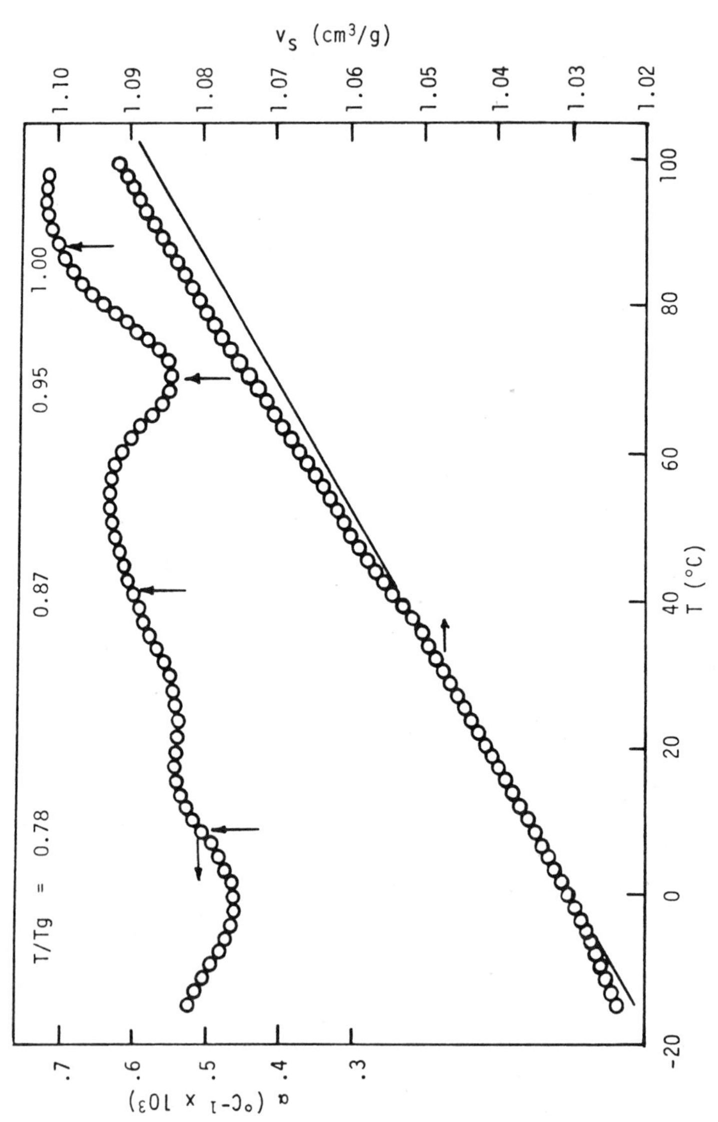

Figure 1. Thermal expansion coefficients and specific volumes for TR-41-2444 dried at 75°C. Arrows indicate transitions.

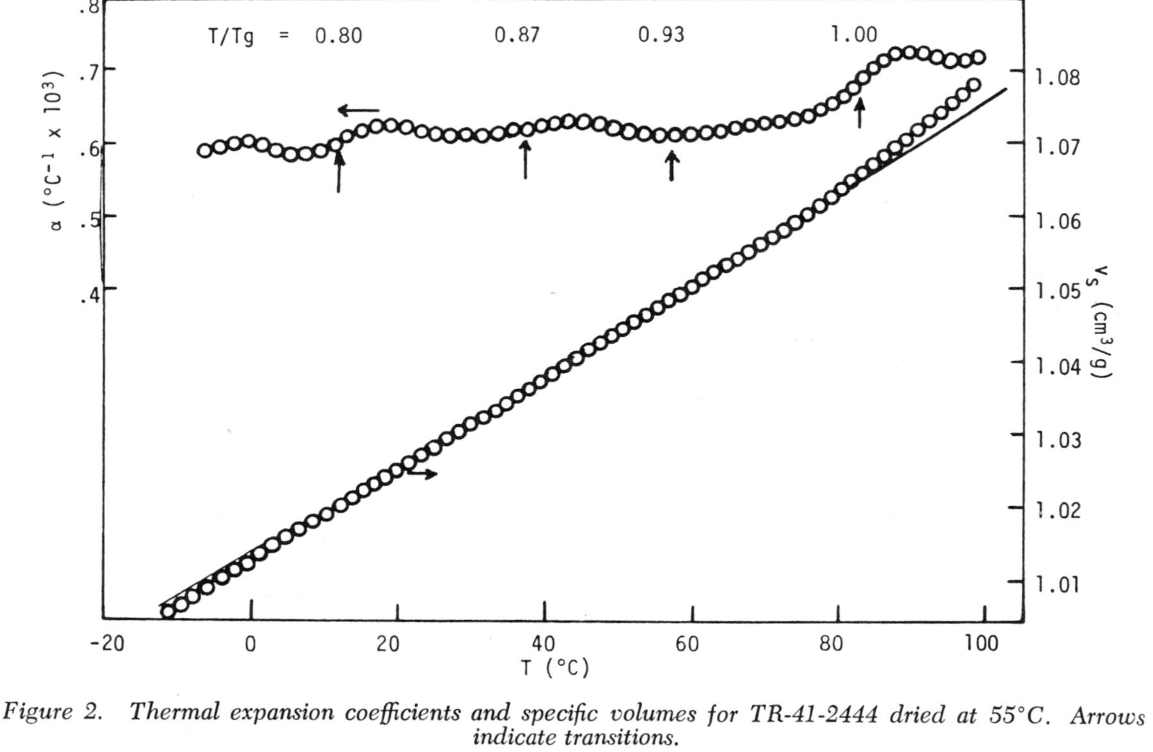

Figure 2. Thermal expansion coefficients and specific volumes for TR-41-2444 dried at 55°C. Arrows indicate transitions.

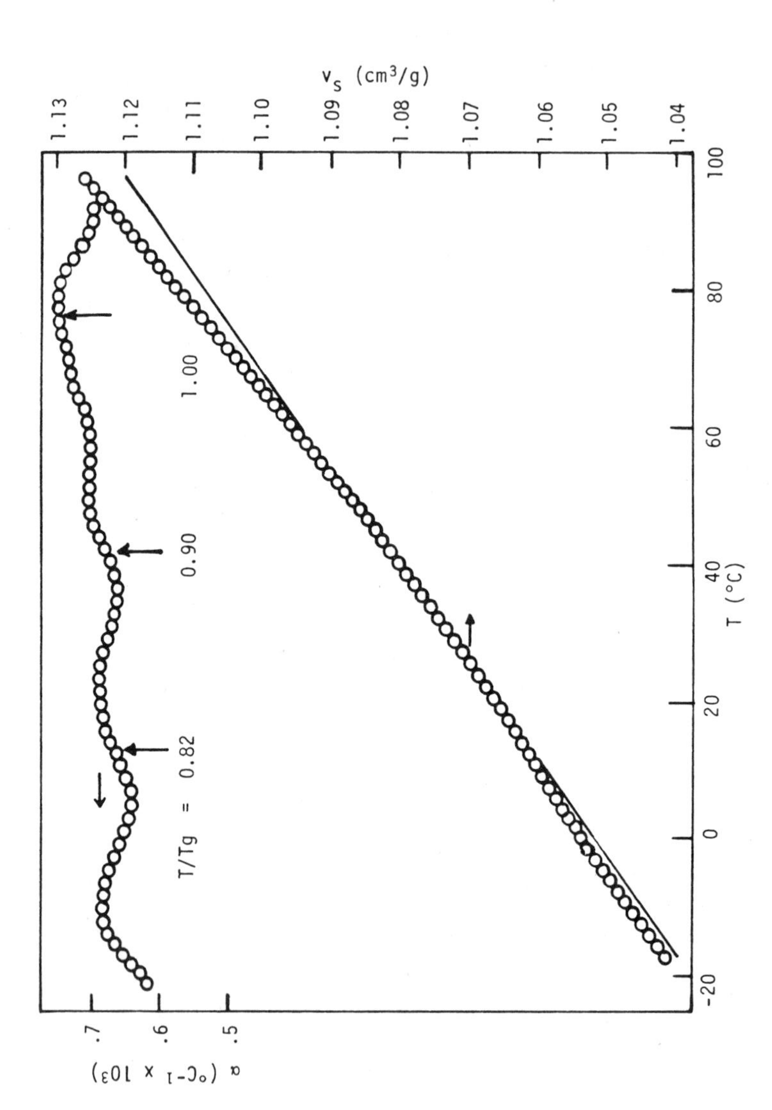

Figure 3. Thermal expansion coefficients and specific volumes for TR-41-2445 dried at 55°C. Arrows indicate transitions.

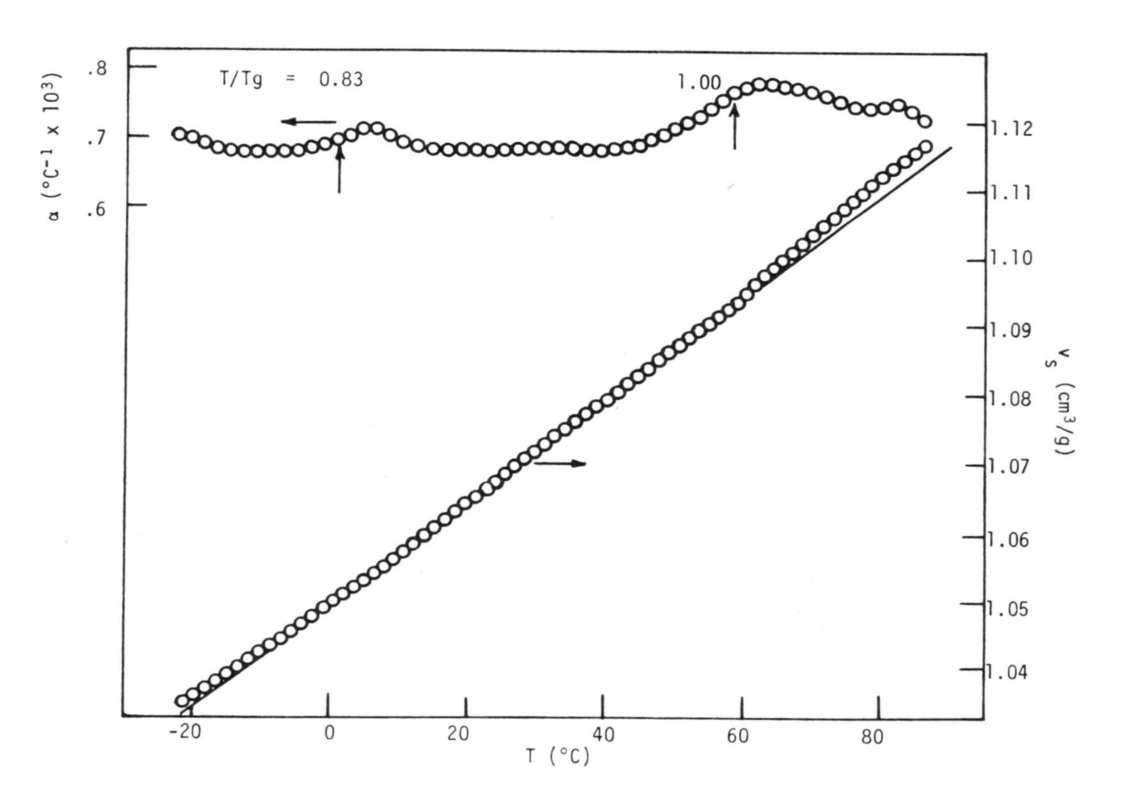

Figure 4. Thermal expansion coefficients and specific volumes for TR-41-2445 dried at 40°C. Arrows indicate transitions.

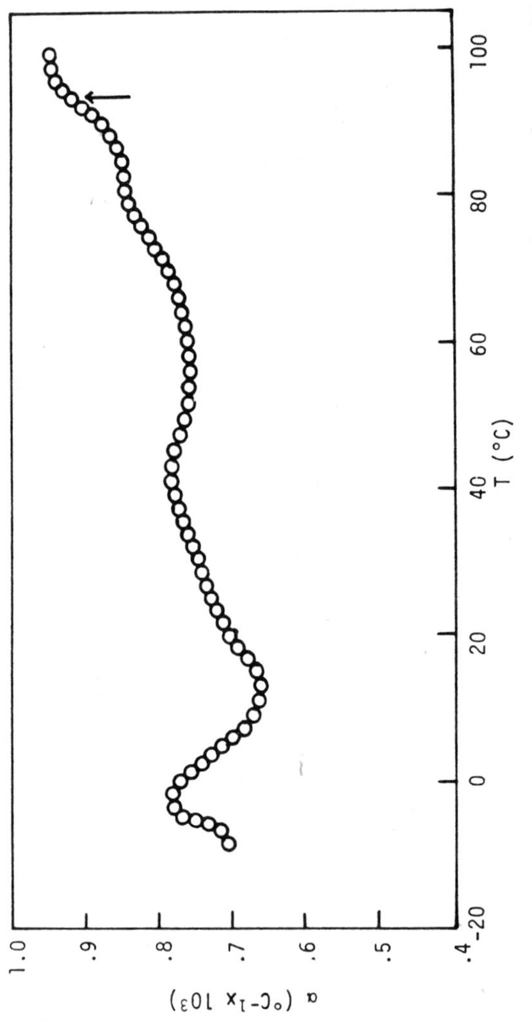

Figure 5. Thermal expansion coefficients for TR-41-2443. Arrow indicates polystyrene T_g.

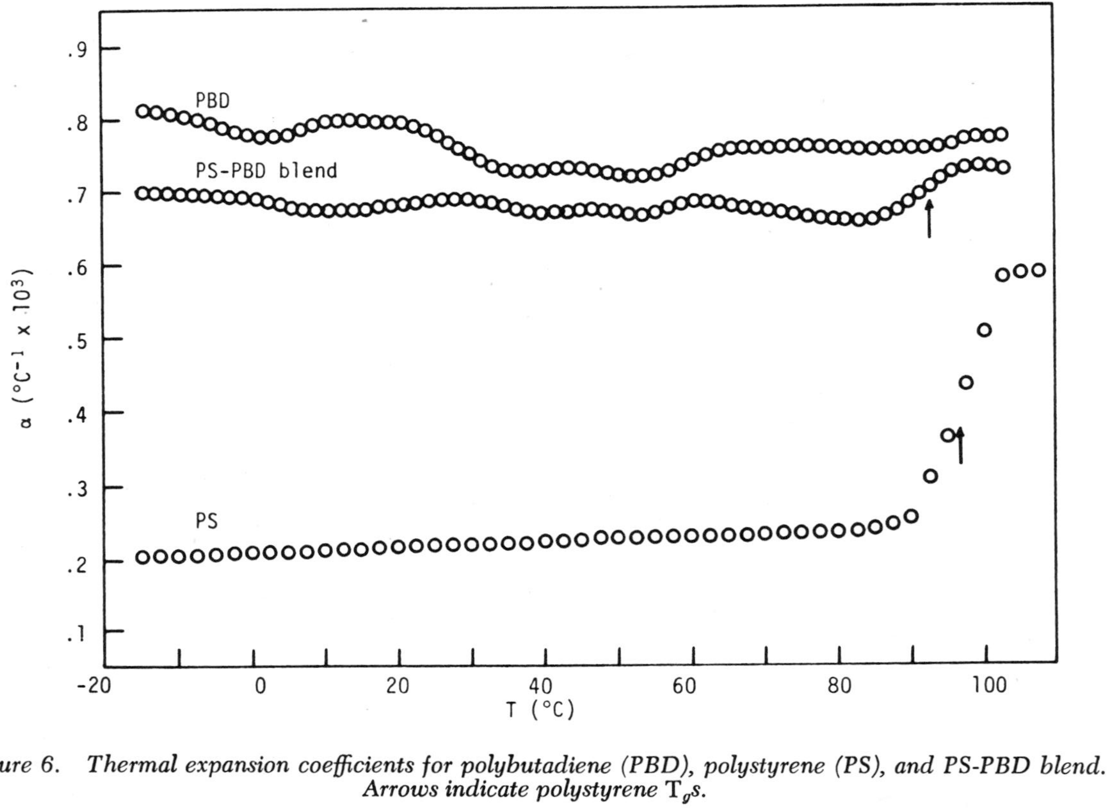

Figure 6. Thermal expansion coefficients for polybutadiene (PBD), polystyrene (PS), and PS-PBD blend. Arrows indicate polystyrene T_g*s.*

same manner as the TR-41-2443 dried at 55°C, the T_g effect must be attributable to the difference in molecular weight. Figure 5 also shows the minimum in α and the transition found for the other block copolymers.

The specific volume and expansion coefficient of the solution-blended material are shown in Figure 6, along with data for pure polybutadiene and pure polystyrene. None of the three polymers has any distinguishing features below the polystyrene T_g, illustrating that the observed transition and minimum are the results of the unique structural morphology of the block copolymers. It should be noted that the substantial difference in the thermal expansion coefficients of polybutadiene and polystyrene can be expected to be an important factor affecting the structure and properties of block copolymer samples prepared under various conditions.

From the pure component data it is possible to calculate the expected α behavior as a function of temperature for a blend of the two polymers using the equation $\alpha_b = \phi_s\alpha_s + \phi_r\alpha_r$, where the subscripts b, s, and r refer to the blend, polystyrene, and rubber, respectively, and the ϕs represent the volume fractions of the two components in the blend. The calculated curves (Figure 7) are reasonably smooth and exhibit only the polystyrene T_g. The calculated curve for TR-41-2445 is in good agreement with that found experimentally for the solution-blended material. The only significant difference is that below the polystyrene T_g the calculated values of α are about 0.5×10^{-4} deg^{-1} lower than the experimentally determined data points. This may be attributable to the density differences in the samples, particularly for the blended material where density variations and void formation can occur at the interfaces between the polymer phases.

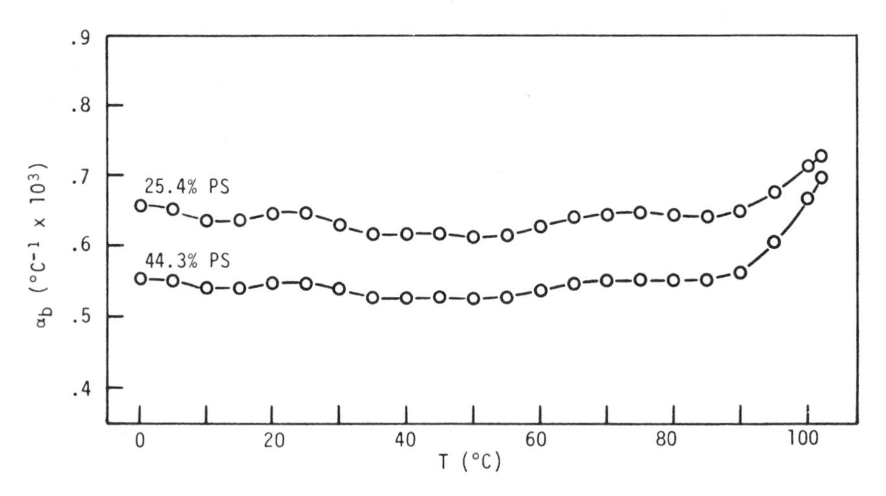

Figure 7. Calculated thermal expansion coefficients for PS-PBD blends

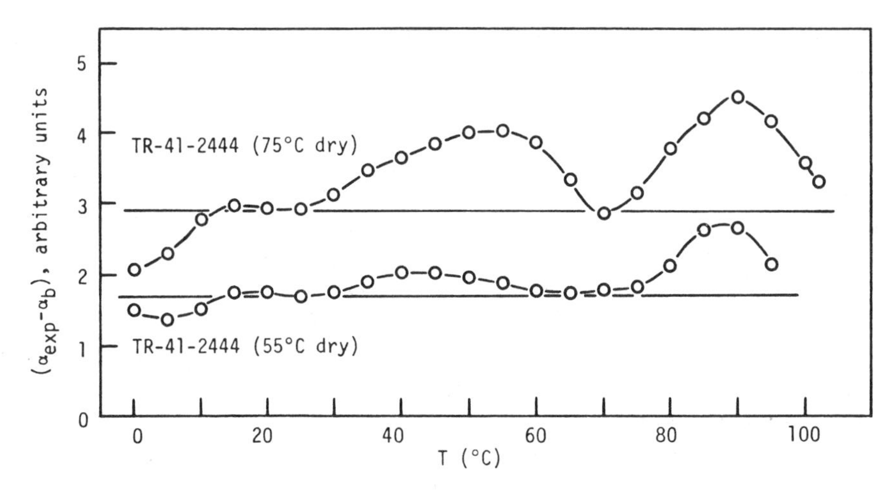

Figure 8. Excess thermal expansion coefficients for TR-41-2444 samples

If the data calculated for the blend are then subtracted, point by point, from the experimental results of the block copolymers, an excess α curve can be derived. The calculated data are used rather than the experimental results since the two are comparable and because it is desired to avoid the possible interfacial effects of the blended material.

An excess α curve would illustrate clearly the differences between the blended materials and the block copolymers; i.e., demonstrating what behavior, if any, is unique to the copolymers. Figures 8 and 9 show the results of this calculation for TR-41-2444 and TR-41-2445. The excess α values at higher temperatures are attributable to the lowering of the

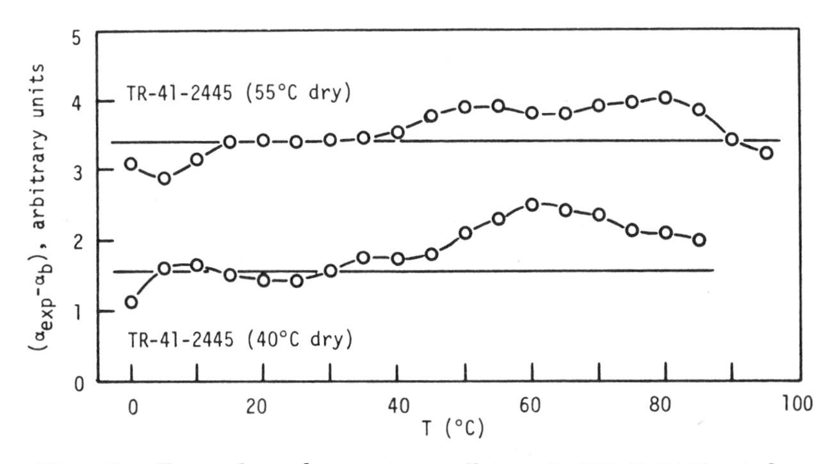

Figure 9. Excess thermal expansion coefficients for TR-41-2445 samples

polystyrene T_g relative to that of the higher molecular weight polystyrene used to calculate the α_b curve.

The existence of regions of excess α suggests that there is a loosening of the membrane's polymer network, which implies an increase in specific volume. Since the α and excess α plots demonstrated that the region between $T/T_g = 0.83$ and 0.87 was reasonably constant, this portion of the specific volume vs. temperature curves could be fitted quite well with straight lines. When these lines are extrapolated to higher temperatures, an excess specific volume term can be arrived at by subtracting from the measured values of V_s the extrapolated values of the lines.

The temperature position of the minimum in the α curve always falls within 5°C of the final drying temperature of the sample, with the corresponding step increase occurring 15° to 20°C lower. This, combined with the fact that the observed glass-transition temperature of the polystyrene phase is also a function of the membrane drying temperature, demonstrates that the observed property changes depend strongly on the preparation conditions. Drying the sample at an elevated temperature would have the effect of annealing the polymer, with a resultant perfection of the phase-separated domain structure. Electron micrograph evidence indicates that the domain boundaries of the TR-41-2445 sample dried at 46°C are more diffuse than the boundaries of the TR-41-2444 sample dried at 56°C. A diffuse boundary layer, or interphase, would have the effect of diluting the polystyrene with polybutadiene segments, resulting in the observed T_g decrease beyond that attributable to molecular weight effects alone.

Conclusions

We observe that the thermal expansion coefficient and the specific volume increase are greater for the materials that had an initially more perfected interphase structure, i.e., a lower interphase configurational entropy. As the temperature is raised, the chains in the interphase become more free to move and the sample with the tightest initial structure (lowest entropy) will be able to gain the most configurational freedom, as reflected by the largest increase in specific volume and thermal expansivity. The current observations and interpretation are consistent with the results obtained from gas sorption and diffusion. Annealing at higher temperatures indicated more perfected interphase structures in terms of those experiments as well.

Acknowledgment

This work was supported by the National Science Foundation under Grants GH-36124 and DMR 75-15-401.

Literature Cited

1. Schell, W. J., Simha, R., Aklonis, J. J., *J. Macromol. Sci. Chem.* (1969) **3,** 1297; with references to earlier work.
2. Roe, J. M., Simha, R., *Int. J. Polym. Mater.* (1974) **3**, 193.
3. Lee, S., Simha, R., *Macromolecules* (1974) **7,** 909.
4. Enns, J. B., Simha, R., *J. Macromol. Sci. Phys.* (1977) **13,** 11.
5. Ibid., 25.
6. Beecher, J. F., Marker, L., Bradford, R. D., Aggarwal, S. L., *J. Polym. Sci.* (1969) **C26,** 117.
7. Jamieson, R. T., Kaniskin, V. A., Ouano, A. C., Shen, M., "Advances in Polymer Science and Engineering," K. D. Pae, D. R. Morrow, Y. Chen, Eds., p. 163, Plenum, New York, 1972.
8. Uchida, T., Soen, T., Inoue, T., Kawai, H., *J. Polym. Sci.,* A2 (1972) **10,** 101.
9. Ostler, M. I., Ph.D. Thesis, Case Western Reserve University (1975).
10. Wilson, P. S., Simha, R., *Macromolecules* (1973) **6,** 902.
11. Boyer, R. F., *Plast. Polym.* (1973) **41,** 15, 71.

RECEIVED June 5, 1978.

12

Molecular Theories of the Interdomain Contribution to the Deformation of Multiple Domain Polymeric Systems

RICHARD J. GAYLORD

Department of Metallurgy and Mining Engineering and Materials Research Laboratory, University of Illinois at Urbana–Champaign, Urbana, IL 61801

A number of theories of the contribution of interdomain polymeric material to the stress–strain, modulus, and swelling behavior of block copolymers and semicrystalline polymers are examined. The conceptual foundation and the mathematical details of each theory are summarized. A critique is then made of each theory in terms of the validity of the theoretical model, the mathematical development of the theory, and the ability of the theory to explain experimental findings.

A number of polymeric systems exhibit domain formation. This results in some polymeric material being confined in regions between the domains. The deformation properties of these systems depend on the types of polymer chains lying between the domains, as well as on the shape and spatial arrangement of the domains. Several theories have been proposed to date for the contribution of the interdomain material to different deformation properties in semicrystalline polymers and block copolymers. We will present and analyze these theories herein.

Semicrystalline Polymers

Modulus. Jackson et al. (*1*) calculate the contribution of amorphous material to the shear modulus of a semicrystalline polymer by assuming that only tie chains (chains whose ends are attached to different crystallites) contribute to the modulus and that these chains follow Gaussian statistics. They assume that the chains deform affinely. The predicted modulus values are lower than the observed values. The

0-8412-0457-8/79/33-176-231$05.00/0

authors note three possible effects which were neglected in their treatment: (a) the crystallites act as rigid inclusions within an elastic medium. This effect depends on the volume fraction of crystals, their shape, and their dispersion in the surrounding medium; (b) the tie molecules are highly extended and therefore do not follow Gaussian statistics; (c) the crystals introduce interfaces which are impenetrable to the amorphous chains and thereby limit their configurations.

Nielson and Stockton (*2*) attempt to account for the effect of the crystallites as rigid fillers by multiplying the shear modulus result of Jackson et al. by the "filler effect" correction term which had been derived by Guth and Smallwood for the Young's modulus. The predicted shear modulus values are still too low. The authors explain this by the fact that: (a) at low crystallinities, amorphous chain entanglements may be significant; (b) at high crystallinities, the crystallites may impinge on each other and form a continuous crystal phase; and (c) at high crystallinities, the tie chains may be very short and follow non-Gaussian behavior. The authors also point out that while experimentally the modulus of crystalline polymers decreases with temperature, Gaussian elasticity theory predicts the opposite effect. However, if the degree of crystallinity decreases with temperature then one can predict a negative temperature coefficient of the modulus, using Gaussian statistics.

Krigbaum et al. (*3*) account for the fact that tie molecules may be in a highly extended state even in the absence of an external macroscopic strain, by using inverse Langevin chain statistics to calculate the Young's modulus. It is assumed that in the undeformed state, the crystallites are randomly oriented (in the previous two theories, the arrangement of crystallites is unspecified, although it is presumably random). An additional assumption is that while the overall deformation of the semicrystalline polymer is affine, the crystallites themselves do not deform. Therefore, the deformation of the tie molecules is greater than affine. Crystal shear and reorientation under deformation are both neglected. The expression which they obtain for the Young's modulus contains the degree of crystallinity and the total number of segments in the semicrystalline chain as variables. It predicts that the Young's modulus should increase with crystallinity and that, for constant crystallinity, the modulus is proportional to temperature. Their theory is further developed by using a relation between the degree of crystallinity and temperature which they had previously derived for the crystallization of folded-chain crystallites in an isotropic, undeformed sample (*4*). The final expression predicts a decrease in the Young's modulus with increasing temperature.

Lohse and Gaylord (*5*) have examined the role of crystallite impenetrability on the Young's modulus. The crystallites are assumed to form pairs of parallel lamellae. Two macroscopic morphologies are considered:

the stacked lamellae structure, in which all the lamellae pairs are parallel to each other, and the spherulitic structure, in which the lamellae pairs are distributed in a spherically symmetric manner along the radii of the spherulite. It is assumed, as in the Krigbaum et al. model, that the overall deformation of the material is affine, while the lamellae are undeformable. Crystal shear and reorientation are incorporated into the model. To consider impenetrability effects, a lamellar surface, although not infinite, is approximated by an infinite plane. The configurational statistics of a chain confined by infinite, parallel, impenetrable walls (6; 7) are then used. It is predicted that cilia (chains with one end attached to a crystal surface), loops (chains with both ends attached to the same crystal surface), and unattached chains all contribute to the Young's modulus as a result of domain impenetrability effects and that their contribution decreases with decreasing chain contour length. Tie molecules, when they are very large, contribute to the Young's modulus in the same manner as the other types of chains, but when the contour length becomes very small, the modulus begins to rise with a further decrease in chain contour length. The modulus is always greater in a stacked lamellar structure than in a spherulite for each type of amorphous chain. The authors also calculate the Young's modulus dependence on temperature, at constant crystallinity, with the use of the Rotational Isomeric State scheme. It is predicted that the Young's modulus of cilia, loops, and unattached chains should always decrease with increasing temperature. The tie molecule shows the same behavior at low temperatures but at high temperatures follows Gaussian behavior as its Young's modulus begins to rise with further temperature increases. The prediction of a decrease in modulus with increasing temperature at constant crystallinity agrees with experiment. No attempt is made by the authors to relate the degree of crystallinity to the temperature.

Block Copolymers

Stress–Strain Relation for Uniaxial Extension. Leonard (8) has calculated the stress–strain relation for an interdomain tie molecule in a spherical domain morphology. He first writes the total entropy of deformation as the sum of the entropy of deformation which one gets from Gaussian elasticity theory and the entropy of domain formation (in both terms, the extension ratio refers to the interdomain strain and not to the macroscopic strain). This sum is then multiplied by the Guth and Smallwood "filler effect" correction term. The stress is calculated by differentiating the entropy expression with respect to the length of the deformed interdomain region. The ratio of the length of the undeformed interdomain region to the initial overall sample length is set equal to the volume fraction of interdomain material, raised to the one-third power. Leonard's calculation contains a great many flaws: it is incorrect to use the entropy of domain formation in the stress calculation because domain formation occurs prior to deformation and does not change thereafter; the stress calculation should be performed by differentiating with respect to the macroscopic sample dimension rather than the interdomain dimension; the expression relating the interdomain and macroscopic extension ratios is incorrect since it fails to predict that the former quantity becomes unity as the latter quantity goes to one; and, the relation given for the ratio of the initial overall sample length to the initial interdomain length fails to change when the number and size of the domains are varied while the total volume fraction of interdomain material is kept constant.

Meier (*9*) has modeled the spherical domain morphology by a simple cubic lattice in which domains are arranged on the lattice sites. The tie molecules run between nearest-neighbor domains and are assumed to be confined by pairs of infinite, parallel walls. The extension ratio for the interdomain region is set equal to the macroscopic extension ratio divided by the volume fraction of the interdomain material. The ratio of the initial interdomain dimension to the domain dimension is set equal to the ratio of the volume fractions of the interdomain and domain material. Using this three-chain model, Meier calculates the stress–strain relation by differentiating his entropy expression with respect to the interdomain extension ratio. The Meier calculation has some difficulties: the interdomain deformation fails to vanish in the absence of an applied macroscopic deformation; the relation between the ratio of the domain dimension to the initial interdomain dimension and the ratio of volume fractions is incorrect; and the differentiation should be carried out with respect to the macroscopic extension ratio.

Gaylord and Lohse (*10*) have calculated the stress–strain relation for cilia and tie molecules in a spherical domain morphology using the same type of three-chain model as Meier. It is assumed that the overall sample deformation is affine while the domains are undeformable. It is predicted that the stress increases rapidly with increasing strain for both types of chains. The rate of stress rise is greatly accelerated as the ratio of the domain thickness to the initial interdomain separation increases. The results indicate that it is not correct to use the stress–strain equation obtained by Gaussian elasticity theory, even if it is multiplied by a "filler effect" correction term. No connection is made between the initial dimensions and the volume fractions of the domain and interdomain material in this theory.

Partial Molar Elastic Free Energy of Swelling. Leonard (*8*) calculated the partial molar elastic free energy of swelling for an interdomain tie molecule in a spherical domain morphology. He included the entropy of domain formation and the entropy term from Gaussian elasticity theory. The cube of the extension ratio for isotopic swelling is taken equal to the inverse of the volume fraction of polymer in the interdomain region. Leonard stated that his expression reduces to the Flory–Rehner equation when the domain size goes to zero. The use of the entropy of domain formation by Leonard is incorrect. Additionally, Leonard's final equation for the partial molar elastic free energy of swelling is incorrectly written. The actual expression never reduces to the Flory–Rehner equation, but is positive over the entire range of interdomain polymer volume fraction and goes to zero as the volume fraction becomes zero.

Meier (*11*) considers the swelling of a tie molecule in block copolymers with lamellar, cylindrical, and spherical domain morphologies. The statistics used for the lamellar domain morphology is that of a chain confined between a pair of infinite, parallel impenetrable walls. The cylindrical and spherical domain structures are modeled by a chain confined between infinite, concentric cylinders and concentric spheres, respectively. The inverse of the volume fraction of polymer in the interdomain region is taken equal to the isotropic swelling ratio, raised to the first, second, and third power in the lamellar, cylindrical, and spherical domain morphologies, respectively. The results indicate that the behavior of the partial molar elastic free energy of swelling as a function of interdomain polymer volume fraction is quite different in the different morphologies. An objection can be raised about Meier's cylindrical and spherical domain models. Meier takes one particular domain and then constructs a confining shell around that domain, which passes through nearest-neighbor domains. The tie chain is then confined between the domain and the surrounding shell. However, a tie molecule is attached to two different domains, around each of which one can construct a confining shell. An interdomain tie molecule should therefore be confined to the volume defined by the intersection of these two shells if Meier's approach is to be consistent.

Gaylord and Lohse (*10*) have examined the swelling behavior of tie molecules, loops, cilia, and unattached chains in different domain morphologies. Each chain is confined between a pair of infinite, parallel impenetrable walls although the domains are not assumed to be infinite. There are one, two, and three orthogonal pairs of parallel walls in the lamellar, cylindrical, and spherical domain strucutres, respectively. The relation between the swelling extension ratio and the interdomain polymer fraction for the different domain morphologies is the same as that used by Meier. The results indicate that cilia, loops, and unattached chains all favor swelling over the entire range of dilution. If the initial interdomain separation is not too large relative to the interdomain chain contour length, a tie molecule also will favor swelling at low degrees of swelling but will oppose swelling at high levels of dilution. When the tie chain is short or the initial interdomain separation is large, the curves are similar to those obtained by Meier, and swelling is always opposed.

Young's Modulus. Gaylord and Lohse (*10*) have examined the Young's modulus of the various types of interdomain chains in the different domain morphologies, using the model described in the previous section. The results indicate that in any given morphology, the Young's modulus behavior of the loop, cilium, and unattached chain all arise from the impenetrability of the domains and decreases with decreasing chain length. The tie molecule shows this same behavior at long chain contour lengths but at sufficiently small chain length behaves in a Gaussian elastic manner. It is predicted that the Young's modulus is greatest for a lamellar domain structure stretched normal to the lamellar plane. The cylindrical domain structure stretched normal to the cylindrical axis has a lower modulus, and the modulus of the spherical domain structure is even lower. The modulus is lowest for cylinders stretched along the cylindrical axis and lamellae stretched along the lamellar plane. It also is predicted that the modulus increases rapidly with increasing volume fraction of domain material and that, at low temperatures, the modulus decreases with increasing temperature. These last two predictions are in agreement with experiment.

Acknowledgment

The author is appreciative of many enlightening discussions on various aspects of this work with David J. Lohse. This work was supported, in part, by the U.S. Energy Research and Development Administration under contract ERDA-EY-76-C-02-1198.

Literature Cited

1. Jackson, J. B., Flory, P. J., Chaing, R., Richardson, M. J., *Polymer* (1963) **4**, 237.
2. Nielson, L. E., Stockton, F. D., *J. Polym. Sci., Part A* (1963) **1**, 1995.
3. Krigbaum, W. R., Roe, R. J., Smith, K. J., Jr., *Polymer* (1964) **5**, 533.
4. Roe, R. J., Smith, K. J., Jr., Krigbaum, W. R., *J. Chem. Phys.* (1961) **35**, 1306.
5. Lohse, D. J., Gaylord, R. J., *Polym. Eng. Sci.* (1978) **18**, 512.
6. Gaylord, R. J., Lohse, D. J., *J. Chem. Phys.* (1976) **65**, 2779.
7. Lohse, D. J., Gaylord, R. J., *J. Chem. Phys.* (1977) **66**, 3843.
8. Leonard, W. J., Jr., *J. Polym. Sci., Polym. Symp.* (1976) **54**, 237.
9. Meier, D. J., *Polym. Prepr., Am. Chem. Soc., Div. Polym. Chem.* (1973) **14**(1), 280.
10. Gaylord, R. J., Lohse, D. J., *Polym. Eng. Sci.* (1978) **18**, 359.
11. Meier, D. J., *J. Appl. Polym. Symp.* (1974) **24**, 67.

Received April 14, 1978.

13

Viscoelastic Properties of Homopolymers and Diblock Copolymers of Polybutadiene and Polyisoprene

R. E. COHEN and A. R. RAMOS[1]

Department of Chemical Engineering, Massachusetts Institute of Technology, Cambridge, MA 02139

Three diblock copolymers of cis-*1,4 polyisoprene (IR) and 1,4-polybutadiene (BR) have been studied in dynamic mechanical experiments, transmission electron microscopy, and thermomechanical analysis. The block copolymers had molar ratios of 1/2, 1/1, and 2/1 for the isoprene and butadiene blocks. Homopolymers of polybutadiene and polyisoprene with various diene microstructures also were examined using similar experimental methods. Results indicate that in all three copolymers, the polybutadiene and polyisoprene blocks are essentially compatible whereas blends of homopolymers of similar molecular weights and microstructures were incompatible.*

Considerable research efforts have been directed towards gaining an understanding of heterogeneous block copolymer systems. This is attributable, in a large part, to the wide range of such materials now available, the unusual and often unique physical properties which may be obtained, and the wide range of morphologies which can form during the process of microphase separation. A corresponding level of interest in homogeneous block copolymers has not emerged, owing to similar considerations, i.e., such systems are relatively rare, properties are not dramatically different from those of random copolymers or homopolymers, and no distinct morphological features are available for analysis. Never-

[1] Current address: Centre de Recherches sur Les Macromolecules, 6 Rue Boussingault, Strasbourg, France.

0-8412-0457-8/79/33-176-237$05.00/0

theless, homogeneous block copolymers are interesting materials in their own right because they can provide guidance in the study of polymer compatibility and because they are of potential interest as morphology regulators or "homogenizing agents" in polymer blends.

In the rare case that two homopolymers form a homogeneous molecular mixture in bulk, it also would be expected that homogeneity would be observed in high-molecular-weight block copolymers made from sequences of the two homopolymers. A more interesting case, however, is that of a block copolymer system which does not exhibit phase separation even though a blend of the corresponding homopolymers is heterogeneous under the same conditions. A qualitative explanation of this latter case lies in considerations of the covalent link between each successive pair of block segments. If the constituent blocks are not strongly incompatible, the constraint provided by this covalent linkage may be sufficient to force the overall block copolymer material into a compatible or semicompatible state. The semicompatible state (*1*) may be viewed as a case in which interfacial mixing takes place to a sufficient degree so that regions of either pure component are never achieved. Experimental determination of the composition fluctuations in this semicompatible state will be hindered by the gentleness of the fluctuations themselves and by the small-size range over which they are constrained to occur in block copolymers.

Few examples of the "homogeneous diblock–incompatible homopolymer" behavior have been reported. One that has received considerable attention is the system polystyrene–poly-α-methylstyrene (*2*). Block copolymers of styrene and α-methylstyrene exhibit a single loss peak in dynamic experiments (*2, 3*) and have been shown to be thermorheologically simple (*4*); hence they are considered to be homogeneous. Mechanical properties data on these copolymers also has been used to validate interesting extensions of the molecular theories of polymer viscoelasticity (*2, 3, 4*).

Polycarbonate–polysulfone block copolymers with rather short segments, i.e., molecular weight of segments in the order of 5000, also have been found to be homogeneous while the corresponding homopolymers are incompatible at the same molecular weight level (*5*). It is not clear, however, whether copolymers with longer blocks will also be homogeneous since the solubility parameter differential ($\sim$ 1.4 J/cm^3) is larger than that of the polystyrene–poly-α-methylstyrene system ($\sim$ 0.4 J/cm^3). Angelo et al. (*6*) observed compatibility of the polybutadiene and polyisoprene block segments (both with less than 10% 1,4 units) in styrene–butadiene–isoprene and butadiene–styrene–isoprene triblocks. Blends of styrene–butadiene and styrene–isoprene diblocks also showed some mixing of the two diene segments, although mixing was not as complete as that obtained through terpolymerization, where some indication of the effect of segment position on the blending also was noted (*6*). Finally, styrene–butadiene block copolymers have exhibited semicompatibility under appropriate conditions. Low segmental molecular weights (*7*) and/or higher temperatures (*8*) promote mixing of the dissimilar segments in agreement with theoretical predictions (*9, 10*).

The present work considers in detail a new example of the homogeneous block–heterogeneous homopolymer system, for which preliminary results have been reported (*1, 11, 12, 13*). The two components, high *cis*-1,4-polyisoprene (IR) and medium *cis*-1,4-polybutadiene (BR) are both elastomeric materials of considerable practical importance. These two homopolymers have been found to form heterogeneous blends (*14*) although their solubility parameters are very nearly identical, i.e., 17.0–17.2 J/cm^3 for the polybutadiene and 16.4–16.8 J/cm^3 for polyisoprene (*15*). The system under consideration here may be considered to be the diene analog of the styrene–α-methyl styrene system; in each case the repeat units of the two blocks differ only by a single methyl group. An additional structural difference arises in the polymers of interest here, however, owing to the capability of the isoprene and butadiene monomers to polymerize in four different ways—*cis*-1,4; *trans*-1,4; 1,2; and 3,4 addition—with the last two being equivalent for the case of butadiene.

Materials and Experimental Methods

Three polybutadiene/polyisoprene diblocks were provided to us by Paul Rempp of CRM, Strasbourg. Details of the molecular weights and the expected diene microstructures were discussed earlier (*1, 13*) and are summarized in Table I. Homopolymers of corresponding microstructures and molecular weights (*1, 13*) are also described in Table I. The molecular-weight distributions of the homopolymers are considerably broader than those of the block polymers, owing to a mastication procedure (*13*) carried out on these samples to lower their molecular weights. To insure that the microstructures of the homopolymers were well matched to the constituent blocks in the copolymers, 60-MHz NMR spectra were obtained on a Varian T-60 apparatus for the diblocks, the individual homopolymers, and three equivalent homopolymer blends (*13*).

Samples for the viscoelastic experiments were prepared by a conventional slow-solvent-evaporation technique (*1*) followed by vacuum drying. For ease in handling in certain experiments, some samples were lightly cured using a 30-MRad dose of electrons; other experiments were carried out on uncured materials. Transmission electron microscopy (Phillips Model 200) was used to investigate possible morphological features in the block polymers and blends. Details of the various staining techniques used are presented elsewhere (*1, 11, 12, 13*).

Viscoelastic experiments were conducted on a Rheovibron DDV-II-C apparatus and on a Rheometrics Mechanical Spectrometer. The Rheovibron studies were carried out in tension on cured samples at 3.5, 11,

Table I. Characterization of Polymers

Polymer[e]	*Isoprene Wt %*	*Molecular Weight*[a]	*Percent*[b] *Cis-1,4, trans-1,4, vinyl*
Diblock 2143 (1/1)	56	250,000	B: 45,45,10 I: 90,—,—
Diblock 2144 (2/1)	39	264,000	B: 45,45,10 I: 90,—,—
Diblock 2148 (1/2)	71	270,000	B: 45,45,10 I: 90,—,—
Polyisoprene[c]	100	133,000	92,—,—
Polybutadiene[d]	0	120,000	45,45,10

[a] Diblocks via light scattering; homopolymers via intrinsic viscosity (after mastication (*1*)).
[b] Expected microstructure based on anionic polymerization conditions used in the synthesis (*see* Figure 1 for verification of matched microstructures in the homopolymers and block segments).
[c] Cariflex 309, Shell Nederland Chemie, unmasticated molecular weight 1.5×10^6.
[d] Solprene 233, Phillips Petroleum, unmasticated molecular weight 160,000.
[e] The numbers in parenthesis after each copolymer designation indicate the ratio of moles of butadiene repeat units/moles of isoprene repeat units.

35, and 110 Hz at various temperatures between − 135° and 40°C. A nitrogen blanket was used at all temperatures to protect the specimens from degradation. The end-butt method of Voet (*16*) was used for gripping the specimens, and the appropriate correction factors (*17, 18*) were applied to account for required changes in the specimen geometry during the experiment. The Rheometrics apparatus was used in the eccentric rotating disk mode (*8*) on uncured samples at frequencies between 0.1 and 100 rad/sec and over a temperature range generally extending from − 40° to 40°C; in a few cases the upper limit of the temperature range was extended to 175°C. To obtain reliable information in the sample temperature, it was necessary to shield (*13*) the upper test fixture from the nitrogen gas stream used to blanket the specimen. Also, all measured forces and displacements were corrected to account for the instrument compliance. The appropriate correction equations are presented elsewhere (*13*). The compliance values for our apparatus are $K_x = 3.2 \times 10^{-6}$ m/N and $K_y = 2.7 \times 10^{-6}$ m/N; these are close to the values reported by Gouinlock and Porter (*8*) for a similar instrument.

Thermal analysis on a DuPont 900 thermomechanical analyzer provided additional information on transition temperatures. A blunt-end expansion probe was used with heating rates in the range of 5°–10°C/min.

Results

Because diene microstructure is known to have a marked influence on physical properties, we felt that proper interpretation of our viscoelastic studies would require detailed information of this type. The information provided by the suppliers of our samples (Table I) was

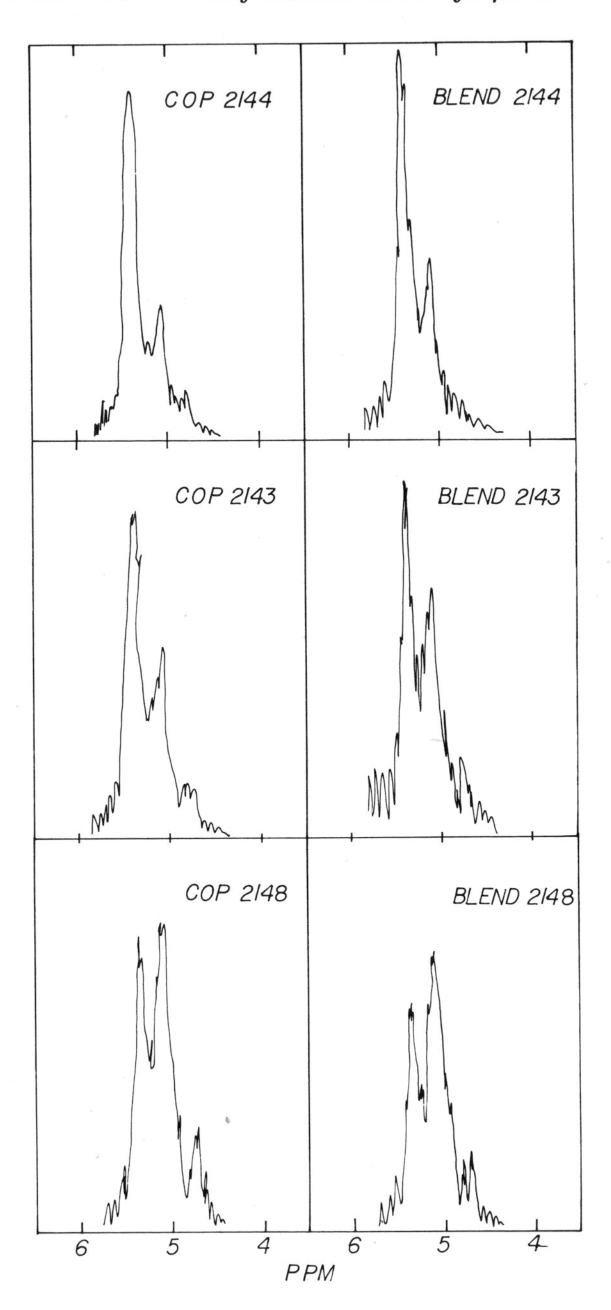

Figure 1. NMR spectra of diblock copolymers and corresponding homopolymer blends. Spectra obtained in carbon tetrachloride solution with tetramethylsilane internal reference.

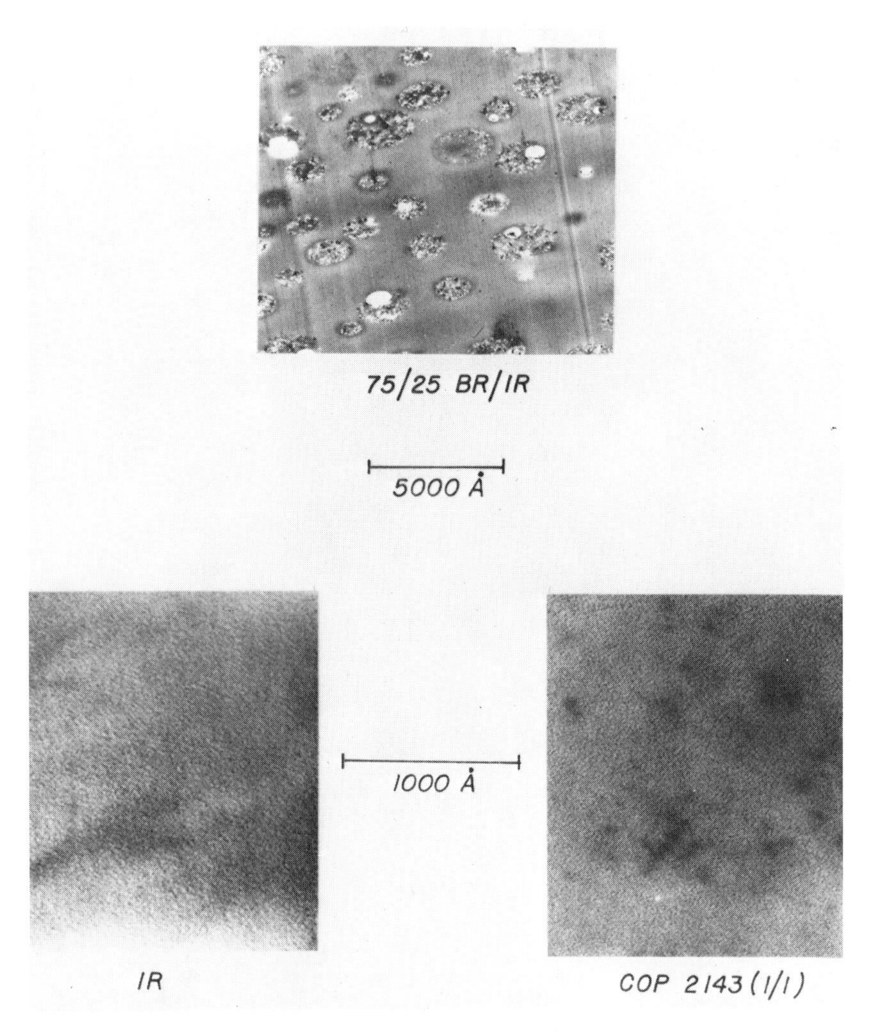

Figure 2. Transmission electron micrographs of (a) a blend of polybutadiene (25 wt %) and polyisoprene (75 wt %); (b) polyisoprene homopolymer; (c) diblock copolymer 2143. Magnifications as indicated.

based on the method of anionic polymerization used in the sample preparation; to support the data presented in Table I, various NMR experiments were performed. Results obtained on the polybutadiene homopolymer showed that less than 10% vinyl addition was present, as expected, but it was not possible to determine the cis/trans ratio for the 1,4 units with our NMR apparatus. Likewise, it was only possible to demonstrate that polyisoprene homopolymer contained more than 92%, 1,4 addition. The NMR spectra of the various diblocks could not be interpreted quantitatively, owing to the complicated overlap of peaks from the various isoprene and butadiene repeat units; however, it was clear that all three samples contained predominantly 1,4 linkages. To demonstrate that the microstructures of the constituent blocks of our copolymers were well matched with the homopolymers used in our investigation, three homopolymer blends also were studied in NMR. Blend 2144 contained 39-wt % polyisoprene, identical to the polyisoprene content of copolymer 2144; similarly the compositions of blends 2143 and 2148 corresponded to the polyisoprene/polybutadiene ratios of copolymers 2143 and 2148, respectively. The close matching of the NMR spectra for each pair (Figure 1) confirms that the diblock segments and corresponding homopolymer have the same microstructure. Figure 1, along with the information obtained on the homopolymers alone, also provides support of the quantitative microstructural information listed in Table I.

Figure 2 presents some transmission electron micrographs obtained using the ebonite method (*13, 19*) for obtaining contrast. Figure 2a clearly shows the incompatibility of the IR and BR homopolymers and also provides a typical example of the degree to which the staining technique can differentiate between polyisoprene and polybutadiene phases when present. A high-magnification micrograph of a pure IR sample, also treated by the ebonite method, is shown in Figure 2b. No morphological features are evident in this micrograph except for the "salt and pepper" texture characteristic of nearly all materials at very high resolution on our microscope. Micrographs of all three diblocks at low and medium magnification showed no evidence of phase separation. At high magnifications the micrographs of the diblocks were very similar to those obtained on the homopolymers except for a few diffuse darker regions randomly scattered around the field of view; Figure 1c is a typical example for copolymer 2143 (1/1), containing 56-wt % polyisoprene.

Glass-transition temperatures of the three diblocks and the two homopolymers are plotted against isoprene content in Figure 3. The values plotted in Figure 3 were determined by TMA at a heating rate of 5°C/min. The points fall near a straight line which can be described by a simplified version of the Gordon–Taylor equation (*20*):

$$T_g = W_I T_{gI} + (1 - W_I) T_{gB} \quad (1)$$

where T_{gI} and T_{gB} represent the glass transitions of the two homopolymers and W_I is the weight fraction of isoprene in the material.

Curves of tan δ vs. temperature for the three diblocks and the two homopolymers are presented in Figure 4. The two arrows along the abscissa indicate the positions of the peaks in the tan δ curves for the two homopolymers; the peak locations for the three copolymers fall between these two temperatures in a systematic fashion. Also, we note the absence of any peak broadening in the curves for the block copolymers, indicating that the single peak for each of these materials is not attributable to the merging of two poorly resolved peaks (*1*). A plot of the temperature corresponding to the maximum in tan δ as a function of composition is shown in Figure 5. As in Figure 3, all points fall near a straight line which, again, is well described by Equation 1; in this case the limiting values of T_{gB} and T_{gI} are taken as the tan δ-peak locations for the two homopolymers. The fact that values of T_g obtained in the dynamic mode on the Rheovibron are all slightly greater than those obtained in the quasistatic TMA experiment is qualitatively consistent

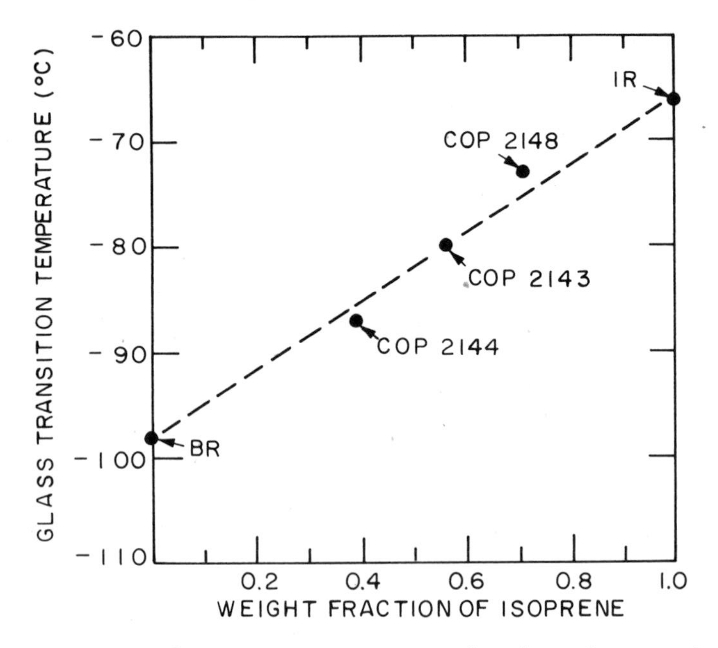

Figure 3. Glass-transition temperature plotted as a function of isoprene content for BR–IR homopolymers and diblocks; TMA, 5°C/min

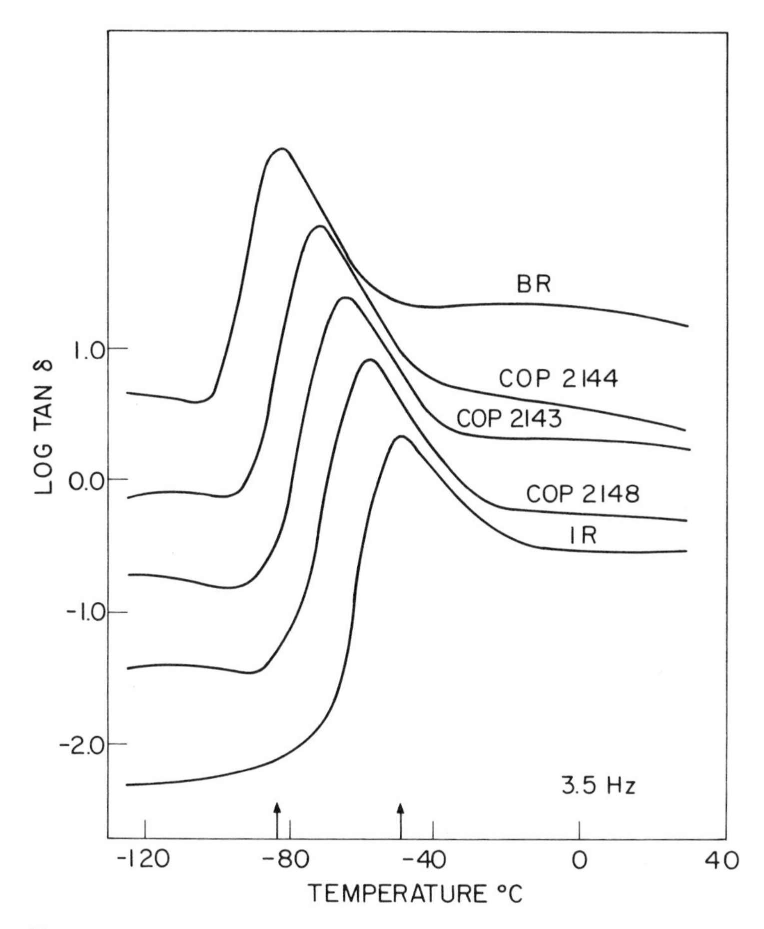

Figure 4. Tan δ vs. temperature curves at 3.5 Hz for the diblocks and homopolymers. The scale on the ordinate corresponds to the bottom curve only, the other curves have been shifted upwards for clarity. The two arrows along the temperature scale indicate the polybutadiene (−82°C) and polyisoprene (−49°C) transitions.

with well-known rate effects which influence all measurements of transition temperatures in polymers.

Isothermal measurements of the dynamic mechanical behavior as a function of frequency were carried out on the five materials listed in Table I. Numerous isotherms were obtained in order to describe the behavior in the rubbery plateau and in the terminal zone of the viscoelastic response curves. An example of such data is shown in Figure 6 where the storage shear modulus for copolymer 2148 (1/2) is plotted against frequency at 10 different temperatures.

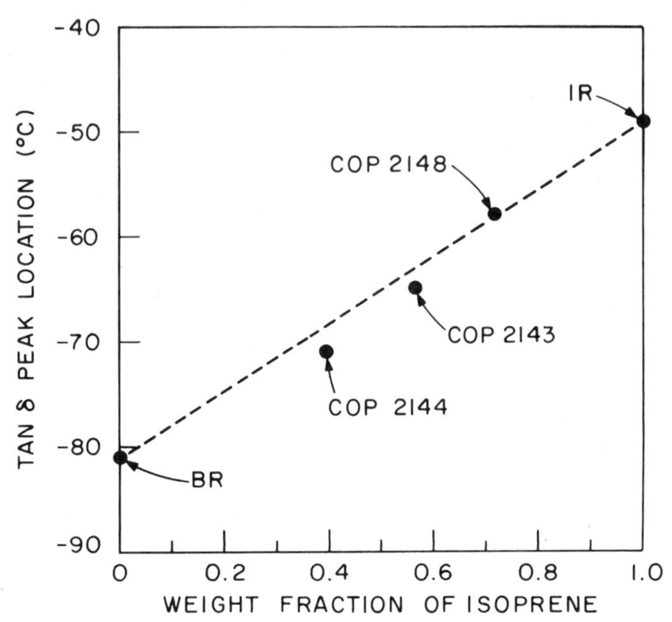

Figure 5. Location of tan δ peak as a function of composition for BR–IR homopolymers and diblocks; Rheovibron, 3.5 Hz

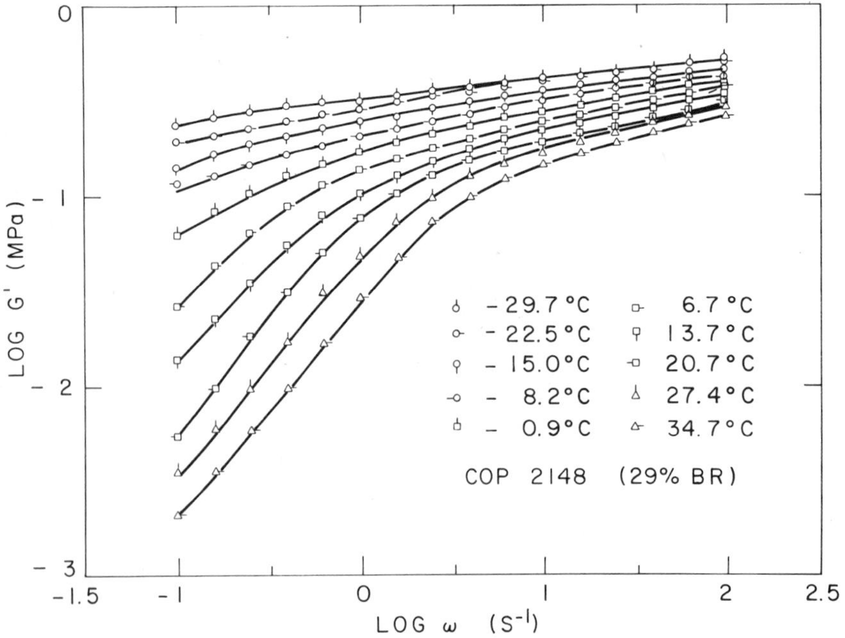

Figure 6. Storage-modulus isotherms for copolymer 2148(1/2)

Discussion

The method used to provide contrast in transmission electron microscopy was successful in demonstrating the presence of a two-phase structure in homopolymer blends of BR and IR (Figure 2a). The opposite situation, i.e., a clear absence of any phase separation in the block copolymers, also is demonstrated, but much less convincingly by the comparison of Figures 2b and 2c. It is necessary to consider the evidence from all of the mechanical and thermal analysis experiments, along with the evidence from microscopy.

Close examination of Figure 2c and others of similar or higher magnification does not reveal any hint of the regular morphology, with characteristic domain sizes in the range of hundreds of angstroms, which is expected for diblock copolymers of this type. We did frequently observe, however, diffuse darker regions about 200–400 Å in size scattered randomly around the field of view in micrographs of all the diblocks. These regions were not observed in any of the micrographs of stained homopolymers and therefore cannot be attributed directly to artifacts introduced by the staining technique. Instead, we believe that the three block copolymers described in Table I are "semicompatible," i.e., they contain significant fluctuations in composition throughout the bulk material; these gentle fluctuations do not appear to be part of any well-organized morphology and, based on all the data obtained here, it is unlikely that the fluctuations ever lead to regions which consist of either pure BR or pure IR.

The Gordon–Taylor Equation, which in simplified form was successful in fitting data (Figures 3, 4, and 5) on transition temperatures for the BR/IR block copolymers, is based on considerations of segmental interaction energies; the equation carries the underlying assumption that the interactions between segments of the dissimilar polymers can be expressed as the arithmetic mean of the interaction energies for the two homopolymers. The close agreement with our experimental results suggests a large degree of mixing at the segmental level in the three block copolymers of interest here.

Detailed analysis of the isothermal dynamic mechanical data obtained as a function of frequency on the Rheometrics apparatus lends strong support to the tentative conclusions outlined above. It is important to note that heterophase (*21*) polymer systems are now known to be thermorheologically complex (*22, 23, 24, 25*), resulting in the inapplicability of traditional time–temperature superposition (*26*) to isothermal sets of viscoelastic data; limitations on the time or frequency range of the data may lead to the appearance of successful superposition in some ranges of temperature (*25*), but the approximate shift factors (*26*) thus obtained show clearly the transfer from the dominance of the viscoelastic response

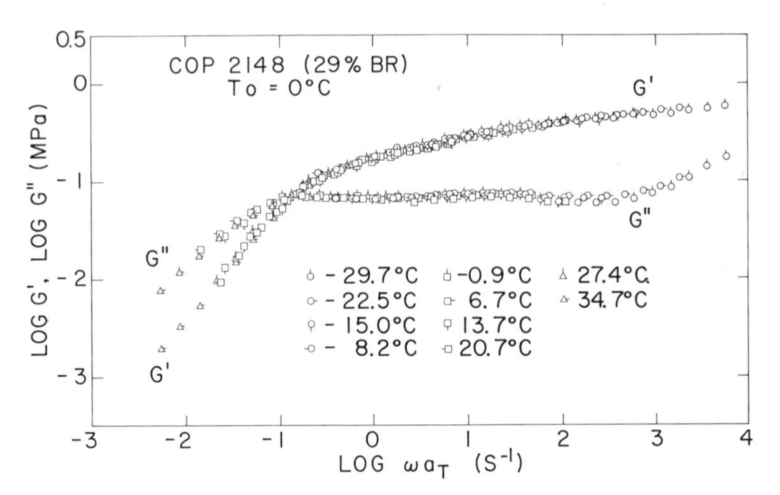

Figure 7. Storage- and loss-modulus master curves for copolymer 2148(1/2) reduced to 0°C according to the WLF constants shown in Table II

of one phase to that of a second phase in the material. On the other hand, polymeric materials which are essentially homogeneous at the segmental level (homopolymers, random copolymers, and homogeneous blocks and blends) are thermorheologically simple (*25*) materials to which concepts of time–temperature superposition can be readily applied.

In the present case, all of our dynamic mechanical data could be reduced successfully into master curves using conventional shifting procedures. As an example, Figure 7 shows storage and loss-modulus master curves and demonstrates the good superposition obtained. In all cases, the shifting was not carried out empirically in order to obtain the best possible superposition; instead the appropriate shift factors were calculated from the WLF equation (*26*):

$$\log a_T = \frac{-C_1(T - T_o)}{C_2 + T - T_o} \tag{2}$$

using C_1 and C_2 values which were weighted averages of the values reported for the two homopolymers (*26*). The values of these WLF constants are listed in Table II.

Figure 8 presents the storage and loss-modulus master curves obtained on all five samples of interest. The dashed lines indicate extensions of the master curves, using appropriately reduced data from the Rheovibron experiments in tension. Storage-modulus data in the rubbery plateau region vary systematically with composition, i.e., the modulus increases with the isoprene content of the materials. The interpretation of the behavior in the flow region (low reduced frequency) is not as straightforeward; in particular, both of the homopolymers exhibit very much lower slopes than any of the block copolymers. This is probably a consequence of the relatively broad molecular-weight distributions of these materials, introduced during the milling procedure (*1, 13*) used to reduce their molecular weights. Similar overall trends are seen in the loss-modulus master curves.

All of the evidence presented above supports the conclusion that the diblock copolymers are essentially homogeneous. On the other hand, the corresponding homopolymers have been shown to be incompatible in essentially all proportions under similar conditions of sample preparation. Thus, if at room temperature the BR and IR can be made compatible by the addition of a single chemical bond, i.e., the one linking the two segments in the diblock copolymer, it is not unreasonable to expect that an upper critical-solution temperature for the homopolymers might exist not far above room temperature. The direct determination of this temperature by visual methods (*27*) was not feasible in the present case because of the nearly equal indices of refraction of BR and IR. As an alternative, the dynamic shear properties of a 50/50 blend of IR and BR were determined in the Mechanical Spectrometer from 30° to 200°C.

Isochronal plots of the storage modulus G' vs. temperature for this blend are shown in Figure 9. A clear discontinuity, which is more pronounced at lower frequencies, is observed for all curves around 80°C. This transition also was accompanied by a noticeable decrease in the slope of each of the isochrones at the higher temperatures. A similar, although more dramatic, drop in modulus was observed (*8*) near 142°C for a styrene–butadiene–styrene triblock copolymer of rather low molecular weight (7000–43,000–7000). It was concluded that above this temperature the polystyrene blocks are compatible with the continuous polybutadiene phase. By analogy, one may then conclude that above 80°C the dispersed IR domains in the BR–IR blend under consideration are disrupted, resulting in an essentially homogeneous blend.

Table II. WLF Equation Parameters at 0°C

	C_1^0	C_2^0
IR[a]	7.10	127
COP 2148[b] (1/2)	6.26	137
COP 2143[b] (1/1)	5.82	142
COP 2144[b] (2/1)	5.32	148
BR[a]	4.19	162

[a] Adjusted to 0°C from Ref. *26*.
[b] Weighed averages based on composition (wt %) of the diblock.

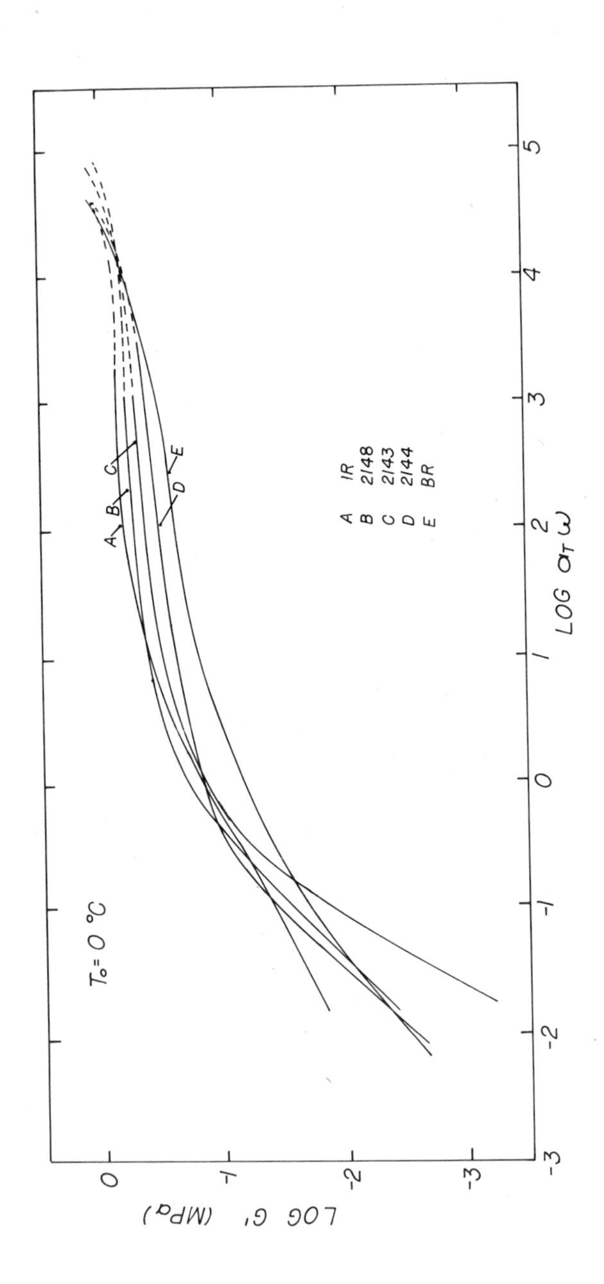
T_o = 0 °C
A IR
B 2148
C 2143
D 2144
E BR
LOG G' (MPa)
LOG $a_T\omega$

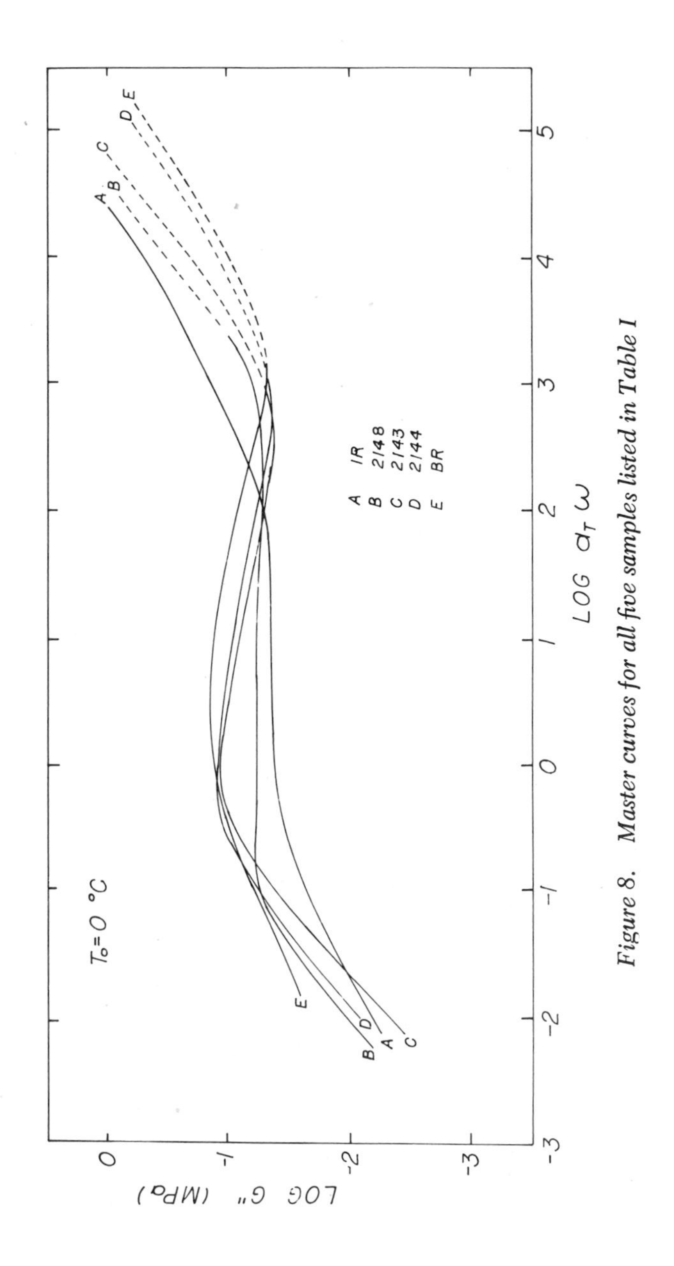

Figure 8. *Master curves for all five samples listed in Table I*

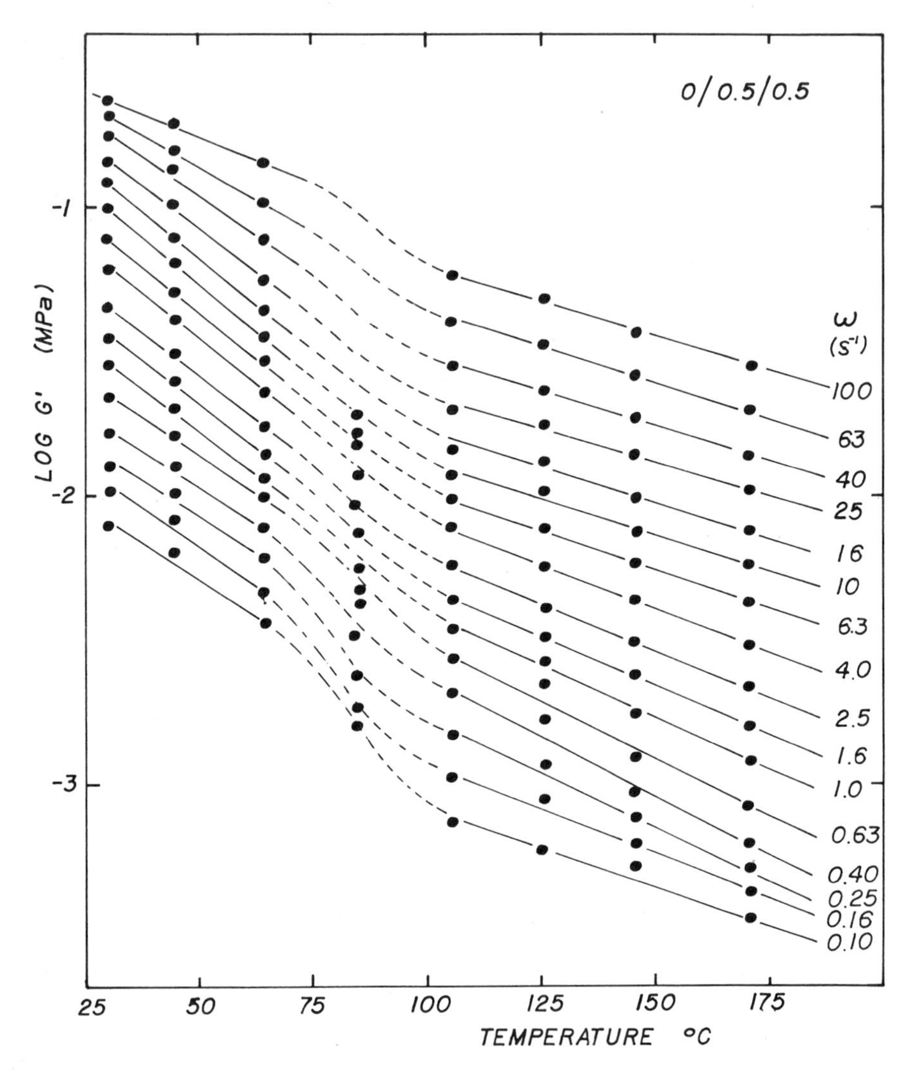

Figure 9. Storage-modulus isochrones for a 50/50 (wt %) blend of BR and IR

As a final point regarding compatibility in BR/IR systems, we consider the effect of diene microstructure. In all of the experiments described above, microstructure was not a variable (Table I), with predominantly 1,4-addition appearing in both polymers. In this case it is possible that the incompatibility of the homopolymers might arise from the presence of the methyl group in the isoprene repeat unit and/or from the partial trans-1,4 structure of the polybutadiene. On the other hand, it is interesting to consider whether changes in microstructure (e.g., higher vinyl content in the polybutadiene) would make the homopolymers themselves compatible. To answer questions of this type, we studied the compatibility of the polyisoprene (IR) of Table I in blends with several polybutadienes supplied to us by the Phillips Petroleum Co. The microstructures and molecular weights of these polymers are shown in Table III, along with the corresponding data for IR and BR taken from Table I.

Binary blends (50/50 wt%) of different combinations of the polymers of Table III were prepared from benzene solution as described above. Either electron microscopy, measurements of dynamic mechanical properties, or both were used to ascertain the heterogeneity (or homogeneity) of each blend.

Blends of IR and medium-cis BR have already been discussed in detail and are clearly heterogeneous systems. The blend of IR and high-trans BR resulted in a completely opaque, white material which was obviously heterogeneous. A blend of IR and high-cis BR produced a transparent film which, nonetheless, appeared heterogeneous in the electron microscope (*see* Figure 10a). Using the same phase-contrasting techniques, a micrograph of a 50/50 blend of high-vinyl BR and medium-cis BR exhibited two phases, as shown in Figure 10b. On the other hand, micrographs of the blend of IR and high-vinyl BR did not show any evidence of heterogeneity. The loss tangent curve of the same blend exhibited a single peak at a position which was between the loss peak postions of the homopolymers (− 49°C for IR, − 33°C for high-vinyl BR, − 42°C for the 50/50 blend). This latter system would then offer one possibility for examining the case in which the homopolymers are compatible as well as the corresponding diblock copolymer.

Table III. Microstructure and Molecular Weights of Butadiene and Isoprene Polymers

	Trans *(%)*	*Cis* *(%)*	*Vinyl* *(%)*	$\overline{M}_w$
High-cis BR	4.2	92.0	3.8	416,000
Medium-cis BR[a]	45.0	45.0	10.0	120,000
High-vinyl BR	15.9	19.7	64.4	468,000
High-trans BR	90.7	—	2.6	156,000
High-cis IR[a]	—	92.0	—	133,000

[a] Homopolymers used throughout this work.

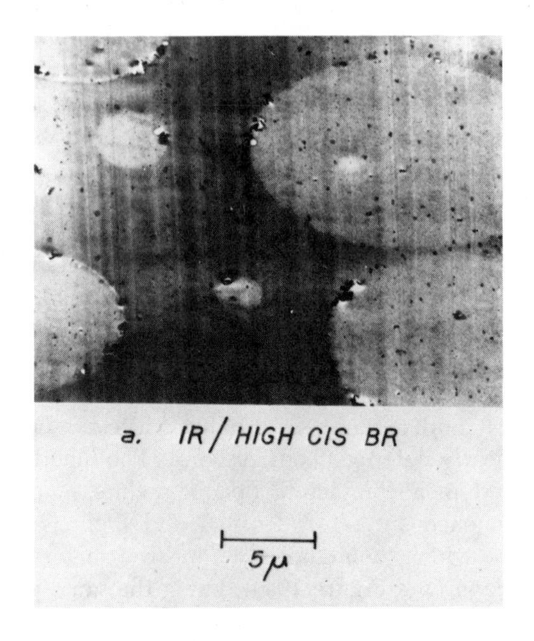

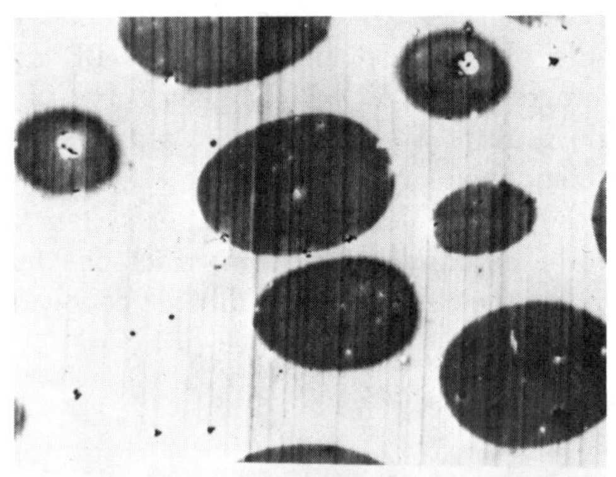

Figure 10. Electron micrographs of (a) a 50/50 blend of IR and high-cis polybutadiene and (b) a 50/50 blend of medium-cis polybutadiene and high-vinyl polybutadiene. See Table III for details of diene microstructures.

Acknowledgment

The authors are grateful to Paul Rempp of the Centre de Recherches sur les Macromolecules, Strasbourg, France, for generously supplying the diblock copolymers used in this study. Homopolymers were supplied by Shell Chemical and Phillips Petroleum. Support for this research was provided by the Office of Naval Research and by the Petroleum Research Fund, administered by the American Chemical Society. The Rheometrics Mechanical Spectrometer used in this study was obtained through the support of the RIAS Program of the National Science Foundation.

Literature Cited

1. Ramos, A. R., Cohen, R. E., *Polym. Eng. Sci.* (1977) **17,** 639.
2. Shen, M., Hansen, D. R., *J. Polym. Sci. (Part C)* (1974) **46,** 55.
3. Hansen, D. R., Shen, M., *Macromolecules* (1975) **8,** 344, 903.
4. Soong, D., Shen, M., *Macromolecules* (1977) **10,** 357.
5. McGrath, J. E., Ward, T. C., Shchori, E., Wnuk, A., *Polym. Prepr., Am. Chem. Soc., Div. Polym. Chem.* (1977) **18**(1), 346.
6. Angelo, R. J., Ikeda, R. M., Wallach, M. L., *Polymer* (1965) **6,** 141.
7. Kraus, G., Rollman, K. W., *J. Polymer Sci., Polym. Phys. Ed.* (1976) **14,** 133.
8. Gouinlock, E. V., Porter, R. S., *Polym. Eng. Sci.* (1977) **17,** 535.
9. Leary, D. F., Williams, M. C., *J. Polymer Sci., Polym. Phys. Ed.* (1973) **11,** 345.
10. Leary, D. F., Wililams, M. C., *J. Polymer Sci., Polym. Phys. Ed.* (1974) **12,** 265.
11. Ramos, A. R., Cohen, R. E., *Polym. Prepr., Am. Chem. Soc., Div. Polym. Chem.* (1977) **18**(1), 335.
12. Ramos, A. R., Cohen, R. E., *Polym. Prepr., Am. Chem. Soc., Div. Polym. Chem.* (1978) **19**(1), 87.
13. Ramos, A. R., Sc.D. Thesis, Massachusetts Institute of Technology (1977).
14. Walters, H. H., Keite, D. N., *Rubber Chem. Technol.* (1965) **38,** 62.
15. Brandrup, J., Immergut, E. H., Eds., "Polymer Handbook," Wiley, New York, 1966.
16. Voet, A., Morawski, J., *Rubber Chem. Technol.* (1974) **47,** 758.
17. Massa, D. J., *J. Appl. Phys.* (1973) **44,** 2595.
18. Ramos, A. R., Bates, F. S., Cohen, R. E., *J. Polymer Sci., Polym. Phys. Ed.* (1978) **16,** 753.
19. Smith, R. W., Andries, J. C., *Rubber Chem. Technol.* (1974) **47,** 64.
20. Gordon, M., Taylor, J. S., *J. Appl. Chem.* (1952) **2,** 493.
21. Tschoegl, N. W., Cohen, R. E., *Polym. Prepr., Am. Chem. Soc., Div. Polym. Chem.* (1978) **19**(1), 49.
22. Fesko, D. G., Tschoegl, N. W., *J. Polym. Sci. (Part C)* (1971) **35,** 51.
23. Cohen, R. E., Tschoegl, N. W., *Trans. Soc. Rheol.* (1976) **20,** 153.
24. Cohen, R. E., Sawada, Y., *Trans. Soc. Rheol.* (1977) **21,** 157.
25. Lim, C. K., Cohen, R. E., Tschoegl, N. W., ADV. CHEM. SER. (1971) **99,** 397.
26. Ferry, J. D., "Viscoelastic Properties of Polymers," 2nd ed., Wiley, New York, 1970.
27. Bernstein, R. E., Cruz, C. A., Paul, D. R., Barlow, J. W., *Macromolecules* (1977) **10,** 681.

RECEIVED April 14, 1978.

14

Strain-Induced Plastic-to-Rubber Transition of a SBS Block Copolymer and Its Blend with PS

T. HASHIMOTO, M. FUJIMURA, K. SAIJO, and H. KAWAI

Department of Polymer Chemistry, Faculty of Engineering,
Kyoto University, Kyoto 606, Japan

J. DIAMANT and M. SHEN

Department of Chemical Engineering, University of California,
Berkeley, CA 94720

The strain softening phenomenon, a strain-induced plastic-to-rubber transition, of poly(styrene-b-butadiene-b-styrene) and its blends with homopolystyrene, was investigated by electron microscopy and small angle x-ray scattering. The specimens have morphology of randomly oriented alternating lamellar domains of the two components, and the transition is believed to occur as a result of structural change from alternating lamellar domains to fragmented PS domains dispersed in a PB matrix. After the strain has been removed, the specimens will spontaneously "heal" themselves, i.e., an inverse transition from rubber-like to plastic-like behavior. The healing effect is investigated by the same techniques and is interpreted in terms of interfacial domain-boundary relaxations activated with the fragmentation of the PS lamellar domains.

It has been shown that some block copolymers and their blends with corresponding homopolymers exhibit the so-called strain softening effect (*1, 2, 3, 4, 5*). When the plastic specimen is stretched beyond the yield point, it becomes rubbery and exhibits high-elasticity with large recoverable deformation attributable to the break-up of their original rigid

0-8412-0457-8/79/33-176-257$05.00/0

structure (*1, 6, 7*). Moreover, specimens after stretching exhibit a healing effect in that properties of the original undeformed specimens are recovered upon removal of the applied stress (*1, 4, 5*). The effect has been attributed to the reformation of the original microdomain structure. In this chapter, the structural changes in SBS block copolymer and its blends with homopolystyrene accompanying the strain-induced plastic-to-rubber transition and the healing process are investigated by means of electron microscopy and small angle x-ray scattering (SAXS). The healing effect is rationalized in light of the domain-boundary relaxation mechanism.

Experimental

Research grade poly(styrene-b-butadiene-b-styrene), designated as TR-41-1647, TR-41-1648, and TR-41-1649, were received from Shell Development Co. These block copolymers contain 26.8, 29.3, and 48.2 wt% polystyrene (PS), respectively. The average molecular weights, determined by intrinsic viscosity measurements in toluene at 30°C, were found to be 7-36-6, 16-78-16, and 14-30-14 in units of thousands. The microstructure of polybutadiene (PB) blocks was found to contain about 40 mol% in *cis*-1,4, 50% in *trans*-1,4, and 10% in 1,2 units.

Film samples were prepared by spin-casting (*8*) from 10% solutions of the polymers in a mixed solvent of THF and methyl ethyl ketone (90/10 by volume). Residual solvent was removed by heating in vacuo at 60°C until constant weight was reached. For electron microscope observations, samples were stained by osmium tetroxide and then cut normal to the film surface by a LKB ultramicrotome to a thickness of about 300 Å. Unstrained samples and samples that have been highly stretched and then returned to the unstrained state were stained in aqueous solution of OsO_4 for 24 hours at room temperature and subsequently cut into ribbon shape, embedded in epoxy resin, trimmed, and stained again in the aqueous solution of OsO_4 for 5–12 hours at 50°C and then sectioned. To observe the morphology of samples in the stretched state, the specimens were first elongated to 85 and 500%, fixed into metal frames, and subsequently stained by OsO_4 vapor for 48 hours at room temperature under the stretched state. The stained specimens were then released from metal frames and treated in the manner described above for restaining and ultrathin sectioning. The fixation of the specimens by OsO_4 is more effective near the surfaces, therefore a partial contraction of the specimens occurs upon release from stretched state leading to a reduction of effective bulk strain to 64 and 200%, respectively. For the specimens to investigate the healing effect, more quantitatively, the embedding process of the stained specimen into epoxy resin was not performed in order to avoid heat generation during the process.

The SAXS patterns were obtained with nickel-filtered CuKα radiation at 40 KV and 100 mA using a rotaflex RU-100PL generator (Rigaku-Denki) and with a point focusing system so arranged that distances of the first and second pinholes, specimen, and photographic film from the focal spot are 128, 378, 438, and 738 mm, respectively. Sizes of the first and second pinholes are 0.5 and 0.2 mm in diameter. The SAXS intensity distributions were detected by a scintillation counter with a pulse-height analyzer. The same generator and the same power as in the photographic experiment were used as an incident x-ray source. The size of focal spot was 0.5×5 mm^2 on target and 0.07×5 mm^2 in projection. The incident x-ray source was collimated by using the following arrangement: the first, second, and third slits, the specimen, and the counter and scattering slits were placed at 128, 378, 418, 443, 703, and 743 mm from the focal spot, respectively. The sizes of the first, second, counter, and scattering slits were 0.1×10, 0.1×10, 0.1×15, and 0.05×15 mm^2. The intensity distribution was measured by a conventional low-angle x-ray goniometer (No. 2202, Rigaku–Denki), using a step-scanning device with a step interval of 0.6 min, each for a fixed time of 100 sec. The measured scattered intensity distributions were corrected for collimation error by using the weighting function calculated from the Hendricks–Schmidt equations (*9*) and by using Schmidt's method of desmearing (*10*), the detailed procedure of which has been described elsewhere (*11*).

Results and Discussion

Mechanical Properties. Figure 1 shows the cyclic tensile stress–strain behavior of the three block copolymer film specimens at 25°C at a constant rate of tensile strain of 50%/min. The tensile stress is expressed in terms of true stress, i.e., tensile force divided by actual cross-sectional area of the elongated specimen, and the tensile strain is expressed by extension ratio. The number of small arrows attached to the cyclic deformation curves indicate the first, second, and third cycles. As can be seen in the figure, the yielding phenomenon becomes more pronounced with increasing styrene content (from TR-41-1647 to TR-41-1648 to TR-41-1649). The strain-softening effect resulting from the strain-induced plastic-to-rubber transition is particularly clear for the TR-41-1649 specimen. Presumably the randomly oriented styrene lamellae have been elastically deformed up to the yielding point, beyond which the lamellae disintegrate into fragments and are dispersed in the butadiene matrix to result in strain softening.

Figure 2 shows again the tensile stress–strain behavior of TR-41-1649 and its PS blend measured in first and second stretches. The experiment was conducted at room temperature at a strain rate of 50%/min. For clarity only the stretching half-cycles are shown. Here the tensile stress is expressed in terms of nominal stress, i.e., tensile force divided by original cross-sectional area of the specimen, and the tensile strain is expressed as percent elongation. As can be seen in the first stretching half-cycle in the figure, after the yielding has taken place there is a plateau region followed by a rapid increase of the stress. These regions are more clearly discernable than those shown in Figure 1 where true stress was used.

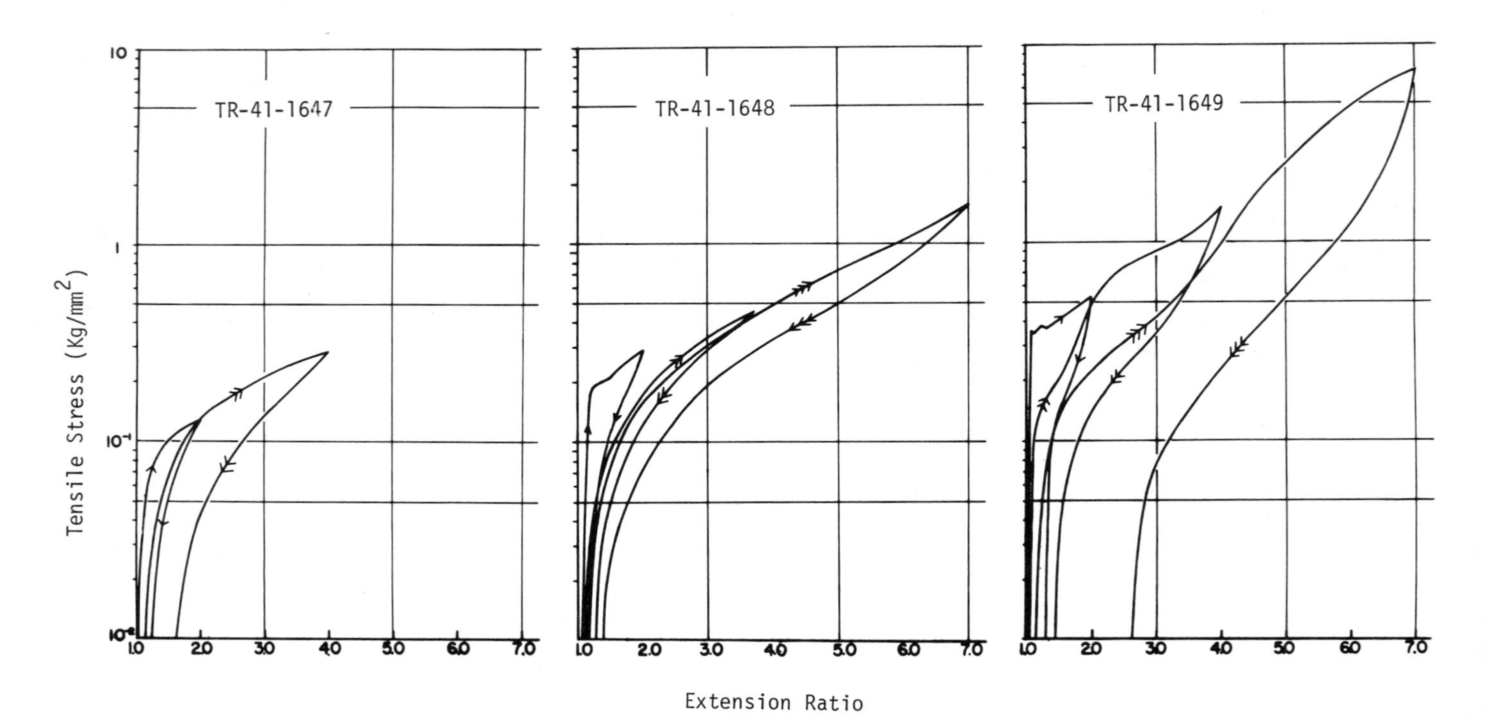

Figure 1. Cyclic tensile stress–strain behavior of the spin-cast specimens, TR-41-1647, TR-41-1648, and TR-41-1649, at 25°C at a constant rate of tensile train, 50%/min. The tensile stress is expressed in terms of true stress. The specimens were air dried without heating.

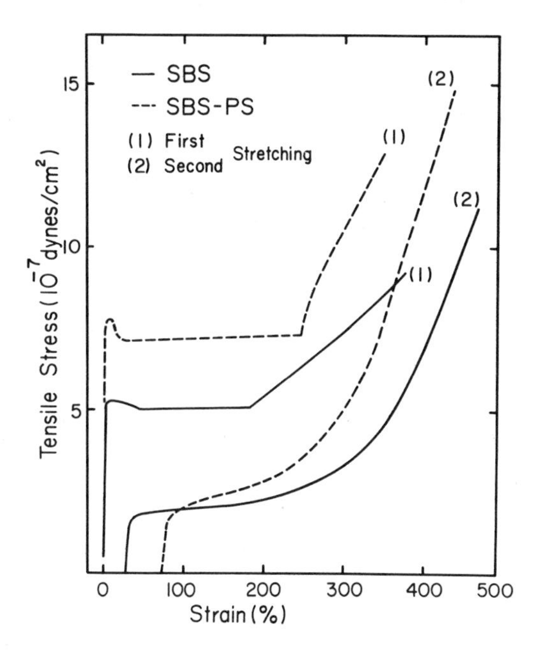

Rubber Chemistry and Technology

Figure 2. Cyclic tensile stress–strain behavior of TR-41-1649 and its blend with polystyrene ($\bar{M}$: 20,400) at room temperature at a constant rate of tensile strain, 50%/min. Curves (1) and (2) refer to the first and second stretching half-cycles, respectively, and the tensile stress is expressed in terms of nominal stress (29).

The block copolymer exhibits a yield point at around 5% strain, after which the stress decreases until about 20% strain. The stress then remains constant with further elongation up to about 180% strain. This is typical plastic-like behavior. When the applied stress exceeds the yield stress, necking suddenly appears at a localized region in the specimen, which subsequently grows continuously until the whole specimen is covered. The necking process gives rise to the plateau region observed in the stress–strain curve. Upon further stretching, the stress rises rapidly and fracture soon follows. If the deformation process is reversed, there is now substantial strain recovery. Thus the initially plastic-like specimen becomes rubber-like at the completion of the necking process. The stress–strain behavior during the second cyclic deformation is also rubber-like and does not show any yielding and necking phenomena except for a small but rapid increase in stress at the beginning of the second stretching half-cycle.

The blended specimen exhibits similar strain-induced plastic-to-rubber transition or the strain softening, as clearly seen in Figure 2.

Because of the higher overall PS content of the specimen, it shows a higher value of Young's modulus (initial modulus), yield stress, stress at the plateau region, and the tangent modulus after completion of the necking. The observed residual strain after removal of the applied stress is also higher than the SBS specimen. It is interesting to note that the difference in stress–strain behavior of the block copolymer and the blend becomes less pronounced for the second stretch half-cycle.

Electron Microscopy. Figure 3 shows electron micrographs of ultrathin sections of film specimens of the three kinds of block copolymers. As can be seen in the figure, TR-41-1647 and TR-41-1648 specimens have a heterogeneous structure in which the polystyrene domains are dispersed within a polybutadiene matrix and are connected to each other to form a swirl-like structure. On the other hand, TR-41-1649 specimen is seen to consist of alternating lamellar domains of the two components. Changes of the domain structure with fractional compositions of styrene and butadiene components are consistent with predictions of the current theories of micro-phase separation (*12, 13, 14, 15*) for block copolymers cast from such a nearly nonselective solvent as the mixture of THF and methylethylketon (90/10 in volume ratio).

Figure 4 shows the electron micrographs of ultrathin sections of film specimens of the TR-41-1649 block copolymer (as a reference) and of two blends of the block copolymer with 20% homopolystyrenes having average molecular weights of 4,800 and 20,400, respectively. The two blend systems also have alternating lamellar domain structures similar to

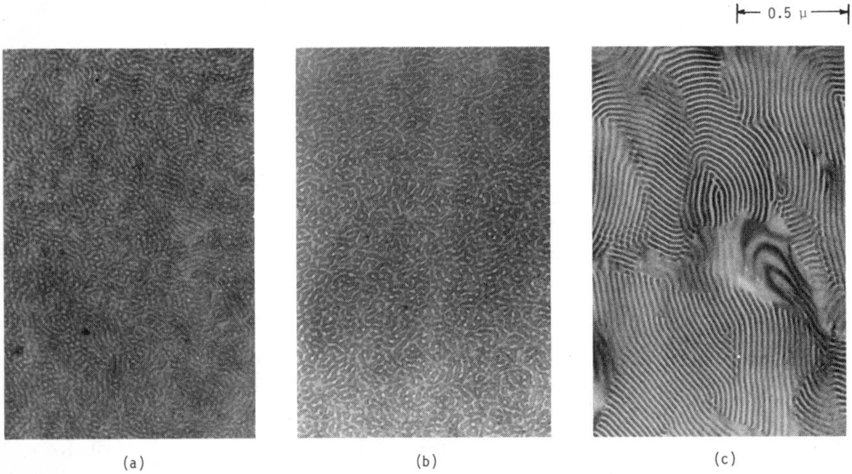

0.5 μ

(a) (b) (c)

Figure 3. Electron micrographs of ultrathin sections of film specimens spin-cast from 10% solutions of a series of tri-block copolymers; (a) TR-41-1647, (b) TR-41-1648, and (c) TR-41-1649, all in a mixture of THF and methylethylketone in volume ratio of 90/10 and stained by OsO_4

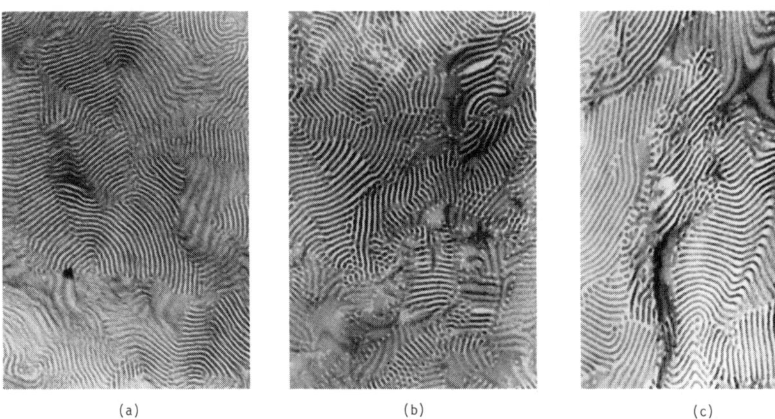
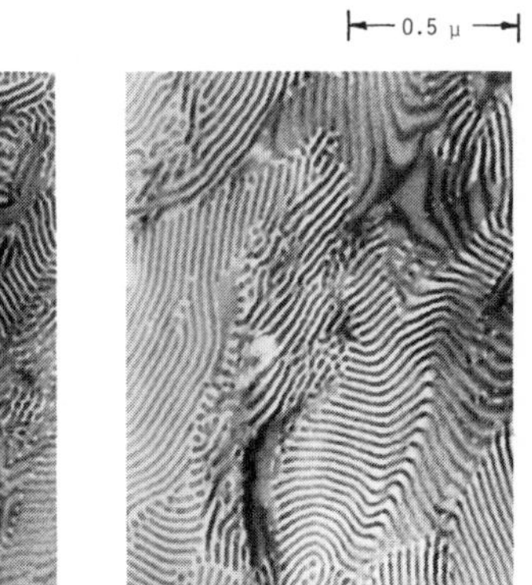

Figure 4. Electron micrographs of ultrathin sections of film specimens spin-cast from 10% solutions of: (a) TR-1649 alone, (b) mixture of TR-41-1649 with homopolystyrene ($\overline{M}$: 4,800) in weight ratio of 80/20, and (c) mixture of TR-41-1649 with homopolystyrene ($\overline{M}$: 20,400) in weight ratio of 80/20, all in a mixture of THF and methylethylketone in volume ratio of 90/10 and stained by OsO_4 (29)

that of the pure block copolymer. Since the molecular weights of the added homopolystyrenes are either lower than or comparable with those of the styrene segments in the block copolymer, they are well-solubilized (*16*) into the styrene domains of the original block copolymer specimen shown in Figure 4a. Figures 4b and 4c show that the solubilization has resulted in thickened PS domains.

Hereafter, the film specimens of the TR-41-1649 block copolymer and the blend of the block copolymer with the high molecular weight polystyrene are redesignated as "SBS" and "SBS-PS" specimens, respectively, and are used mainly for assuring the above postulate on the structural changes associated with the strain-softening and healing processes by means of electron microscopic and small angle x-ray scattering investigations.

The micrographs indicate that the original unstretched specimens of the SBS and SBS-PS have randomly oriented alternating lamellar microdomains of styrene and butadiene components. The blended homopolystyrene in the SBS-PS specimen is solubilized into the polystyrene lamellae, resulting in thickening of the PS lamellae. Stretching the SBS specimen by 85% elongation produces irregular deformation of the lamellar microdomains accompanied by shearing, kinking, disruption, and orientation of the lamellar microdomains (Figure 5a). These defor-

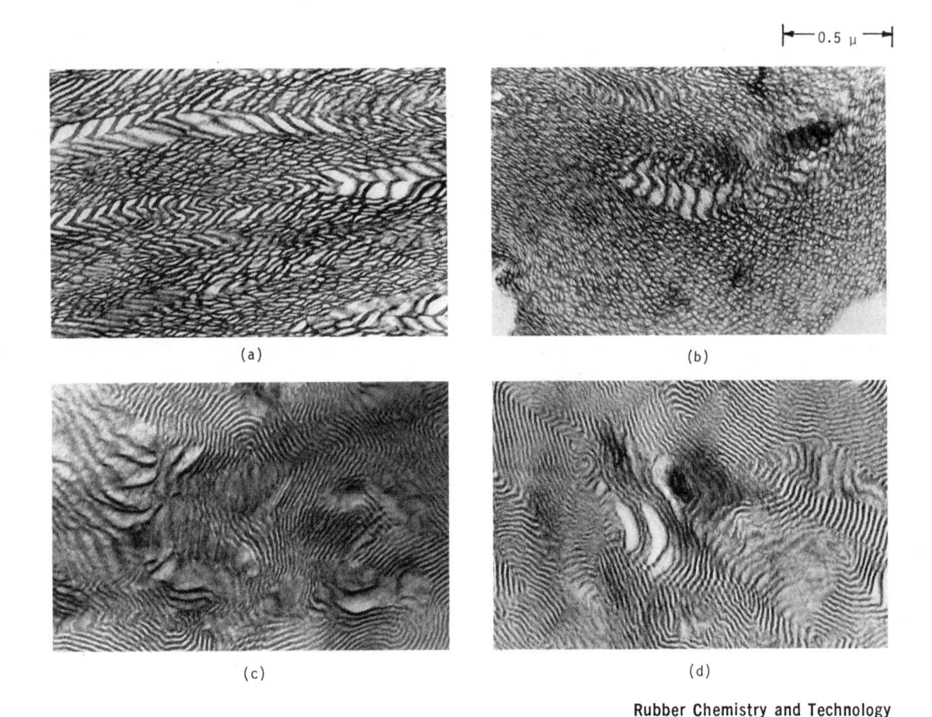

Rubber Chemistry and Technology

Figure 5. Electron micrographs of ultrathin section of SBS film specimens; (a) stretched to 80%, (b) stretched to 500%, (c) released from 600% and left unstretched at room temperature for several days, and (d) released from 600% elongation and heat treated at 100°C for 2 hr. The sectioning was made parallel to the stretch direction and normal to the film surfaces. The stretch direction is horizontal (29).

mation processes are responsible for the yielding and necking of the specimen and are dominant until the necking is completed.

Upon further stretching, fragmentation of the lamellar microdomains prevails. Finally, fragmented polystyrene domains are dispersed in the matrix of polybutadiene, as can be seen in Figure 5b. These fragmentation processes must have been responsible for the plastic-to-rubber transition. The polystyrene fragments act as surface-active filler particles for polybutadiene, and the elasticity of the specimen is essentially entropic in origin. Quantitative aspects of the deformation processes can best be analyzed by the small angle x-ray scattering (SAXS) experiments.

It is also observed in Figure 5c that the fragmented polystyrene domains are transformed into the original lamellar domains after "resting" in the unstretched state for several days at room temperature. The heat generated during the embedding and restaining processes, however, may have contributed to the reformation process also. The reformed structure is more perfect when it is annealed at elevated temperature, as seen in Figure 5d, and less perfect for the SBS/PS. The reappearance of lamellar morphology, however, is obvious in both types of healing processes.

Small Angle X-Ray Scattering. Figure 6 shows the logarithm of the desmeared relative intensity as a function of scattering angle (2θ in minutes). The unstretched SBS specimen (bottom curve) shows the first-order and the third-order scattering maxima at $2\theta = 19.7$ and 60.0 minutes, respectively. The second-order scattering maximum which is supposed to appear at $2\theta \simeq 40$ minutes is not clearly observed because volume fractions of the two lamellae are nearly equal. This results in decreased intensity of the second-order scattering maximum (extinction rule). Each scattering maximum corresponds to the first- and higher-order scattering maxima of a single lamellar spacing of 269 Å. The homopolystyrene which was blended into the SBS is solubilized, as shown in Figure 4, so that the lamellar spacing of the unstretched SBS-PS specimen (upper curve) is increased to 319 Å, as indicated in the figure. Because of the blending, volume fraction of the polystyrene lamellae is now greater than 50%. The intensity of the second-order maximum increases and thereby becomes clearly discernible.

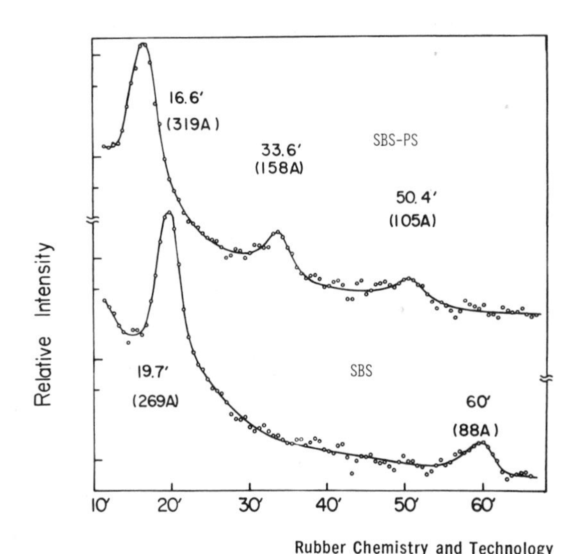

Rubber Chemistry and Technology

Figure 6. Logarithm of relative desmeared intensity distributions of SAXS from the SBS-PS (upper curve) and SBS (bottom curve) specimens plotted against scattering angles 2θ in minutes (29)

Figure 7 shows the change in SAXS patterns upon stretching the SBS specimen in the first stretch half-cycle. The figure also includes representations of the SAXS patterns with only the first-order scattering maximum. The numbers in the figure indicate percent elongations of the specimens. Upon stretching the meridional SAXS maxima tend to shift toward smaller angles. Its intensity decreases until 30% elongation where it disappears. This indicates that the lamellae, originally oriented with their boundaries normal to the stretch direction, increase their spacing with elongation. However, the spacing also becomes irregular because of such deformation processes as shearing, kinking, and destruction of the lamellae. The relation between the extension ratio in bulk to that of the lamellar spacing ($d_{||}/d_0$), estimated from the change of the first- and the third-order scattering maxima, is plotted in Figure 8a for the initial stage of stretching (less than 30% elongation). On the other hand, the equatorial SAXS maxima remain nearly invariant under the same condi-

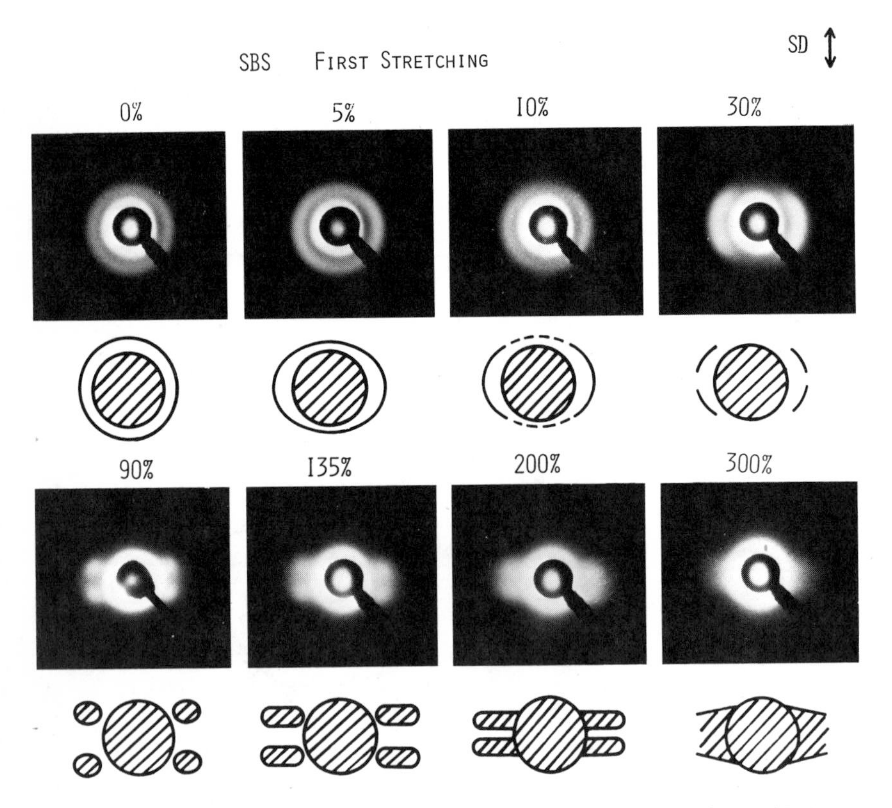

Rubber Chemistry and Technology

Figure 7. Change in SAXS patterns of the SBS specimen during the course of the first stretch half-cycle. The stretch direction is vertical (29).

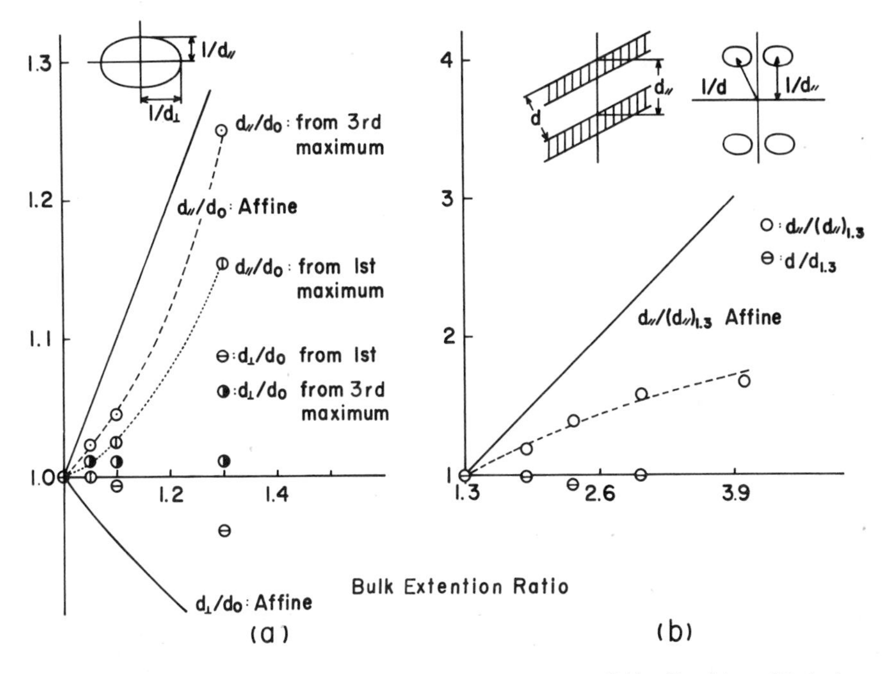

Rubber Chemistry and Technology

Figure 8. The relation between the extension ratio in bulk and that of the spacing (a) at the initial stage of stretching (less than 30% elongation) and (b) at large elongations (29)

tions, as seen in the same figure for the $d_{\perp}/d_0$ data, which indicates that the spacings do not vary for those lamellae which were originally oriented parallel to the stretch direction. This accounts for the volume dilatation of the specimen.

At higher elongations, the equatorial scattering maxima disappear because of the deformation processes discussed above. The scattering patterns in the plateau region of the stress–strain curve after the yield point are characteristic of lamellar surfaces inclined to the stretch direction, as schematically shown in Figure 8b. With increasing bulk extension ratio, the scattering lobes tend to be elongated parallel to the equator, i.e., the lateral breadth of the lobe increases, indicating that the lamellae are fragmented so that their lateral continuity decreases. Moreover, the spacing parallel to the stretch direction ($d_{||}$) is seen to increase with extension ratio as the distance between scattering lobes in the stretch direction decreases with increasing bulk extension ratio. This is quantitatively demonstrated in the plot of $d_{||}/(d_{||})_{1.3}$ vs. bulk extension ratio in Figure 8b, where $d_{||}$ is the spacing parallel to the stretch direction and $(d_{||})_{1.3}$ is that at the bulk extension ratio of 1.3 at which the four-point pattern begins to develop. The spacing d in the direction perpendicular to the lamellar surfaces was measured also as a function of the bulk extension ratio, as indicated in Figure 8b. Since $d = d_{||}\cos\alpha$, where α is the angle between the stretch direction and the lamellar normal, the fact that the spacing d hardly changes with the bulk extension ratio, in contrast to $d_{||}$, indicates that the lamellar surfaces tend to orient parallel to the stretch direction so that the angle α increases.

Upon further stretching, scattering lobes disappear. The scattering pattern becomes diffuse and is more or less independent of the azimuthal angle. The implication is that fragmented polystyrene domains are randomly dispersed in the rubber matrix. This is consistent with the conclusion obtained by a qualitative examination of the electron micrographs in Figure 5. The polystyrene domains now act as filler particles, on whose surfaces the polybutadiene chains (the middle block segments of the SBS) must be anchored, and the specimen exhibits rubber-like elasticity.

Figure 9 shows a representation of the change of the SAXS patterns for the SBS (*left*) and SBS-PS (*right*) specimens. The general trend in the change of SAXS patterns for the SBS-PS specimen is identical to that for the SBS specimen. For the SBS-PS specimen, however, the structural regularity tends to be more easily destroyed, resulting in more extensive fragmentation than the SBS specimen.

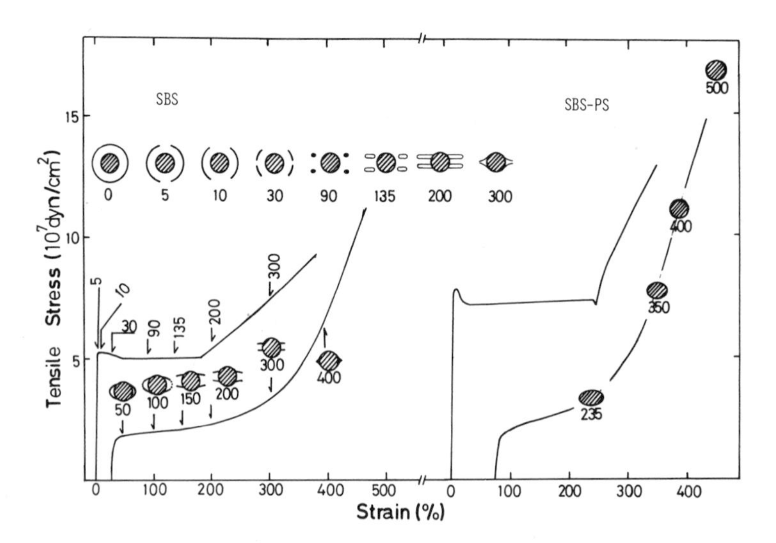

Figure 9. Representation of changes in SAXS patterns for the SBS (left) *and SBS-PS* (right) *specimens during the first and second stretch half-cycles* (29)

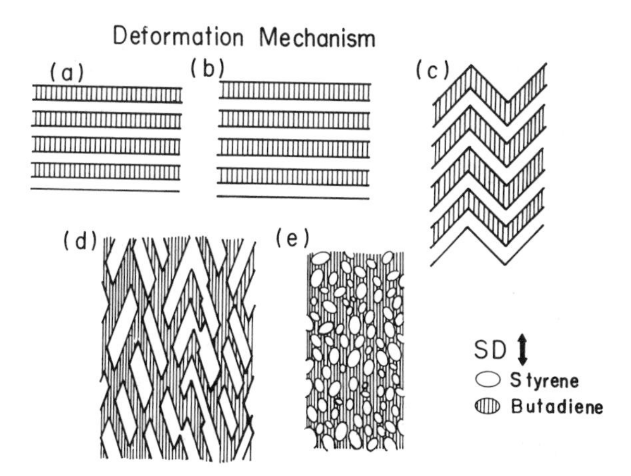

Rubber Chemistry and Technology

Figure 10. Representation of the deformation processes involved in the strain-induced plastic-to-rubber transition (29)

Figure 10 summarizes schematically the structural changes occuring in the strain-induced plastic-to-rubber transition. The change in structure from (a) to (b) illustrates the initial stage of deformation of the microdomain up to the yield point, as observed in the expansion of the lamellar spacing for the lamellae oriented perpendicular to the stretch direction. The changes in structure from (b) to (d) illustrate the expected changes in the yielding and necking processes in which the irregular deformation of the lamellar microdomains occurs. The deformation involves kinking, shearing, destruction, and orientation of the lamellae. Finally, fragmented polystyrene domains are randomly dispersed in the rubber matrix, as illustrated in (e), and act as surface-active filler particles for the polybutadiene chains in the matrix.

To further examine the effect of healing on morphology, we show in Figures 11a and 11b the SAXS patterns of the SBS and SBS-PS specimens released from 355% elongation and rested at room temperature for a few days, and in Figures 11c and 11d the patterns of the SBS and SBS-PS specimens released from 390% and 500% elongations, respectively, then both annealed at 59°C for 5 hours. In all figures the stretch direction is vertical. As can be seen in Figures 11a and 11b, no significant change in the SAXS patterns is observed upon healing, except for a faint reappearance of the first-order maximum at equatorial zone for the SBS specimen. Therefore, the structure does not become sufficiently regular to give the scattering maxima up to the third order. As expected the reformation process is slow at room temperature. On the other hand, Figure 11c shows that the annealing at 59°C substantially restores the

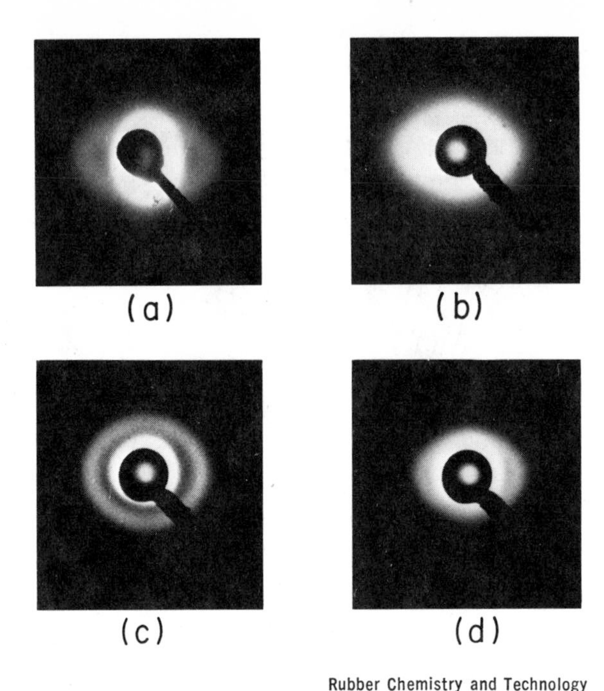

Rubber Chemistry and Technology

Figure 11. SAXS patterns of the SBS and SBS-PS specimens taken during the healing process at room temperature and at an elevated temperature; (a) SBS specimen released from 355% elongation and left in unstretched state at room temperature for a few days, (b) SBS-PS specimens released from 355% elongation and left in unstretched state at room temperature for a few days, (c) SBS specimen released from 390% elongation and annealed at 59°C for 5 hr, and (d) SBS-PS specimen released from 500% elongation and annealed at 59°C for 5 hr (29).

original lamellar domain structure, especially for the SBS specimen, though the lamellar spacing parallel to the stretch direction is still slightly more expanded than that normal to it. It is of interest to note that structural reformation does occur at a temperature as low as 59°C, which is considerably lower than the glass-transition temperature of polystyrene.

Figures 12 and 13 show electron micrographs of ultrathin sections of the SBS and SBS-PS specimens, respectively. Both display the healing effect after releasing from 500% elongation (the stretch direction is horizontal). Here (a) is after healing at room temperature for a few days and (b), (c), and (d) show the healing effect followed by annealing at 60°, 88°, and 116°C, respectively, all for 5 hours. Although the initial stretching of the specimens up to 500% elongation is larger than that for the SAXS test in Figure 11, the structural reformation is generally much more pronounced for the SBS specimen, and its general trend is consistent with that expected from SAXS patterns shown in Figure 11. For example, in Figure 12a the reformed domain structure is still largely oriented in the stretch direction with less regular lamellar spacing parallel to the stretch direction than that perpendicular to it, giving rise to the SAXS pattern shown in Figure 11a. The electron micrograph of Figure 5c more closely resembles that of Figure 12b than Figure 12a, suggesting that heat generated during the embedding and restaining processes has a nonnegligible effect on the observed structure.

Figures 14a and 14b show the changes of equatorial SAXS intensity distribution around the first-order maximum for various durations of healing at room temperature and at an elevated temperature of 60°C, respectively. The SAXS intensity distribution from unstretched (original) and stretched (500% elongation) specimens also are included for comparison. These figures demonstrate the recovery of the diffuse SAXS intensity distribution of the stretched specimen to relatively sharp first-order maximum of the original specimen. At room temperature no significant recovery can be seen up to 4 hours, while at 60°C almost complete recovery, except for a slight shift of the maximum to higher scattering

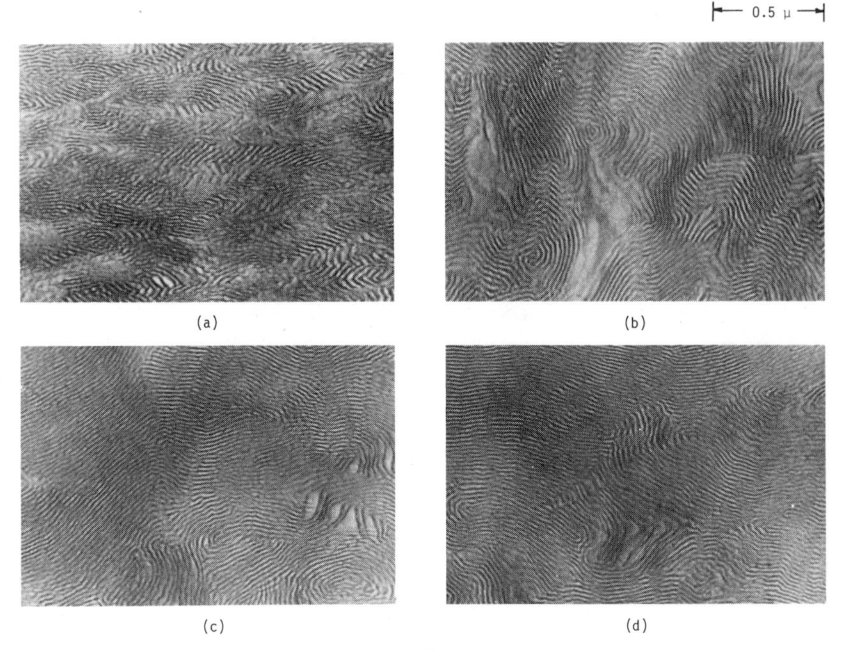

Figure 12. Electron micrographs of ultrathin sections of the SBS specimens released from 500% elongation and (a) left in unstretched state at room temperature for a few days, and (b), (c), and (d) annealed at 60°, 88°, and 116°C, respectively, all for 5 hr. Sectioning was made parallel to the stretch direction and normal to the film surfaces. The stretch direction is horizontal.

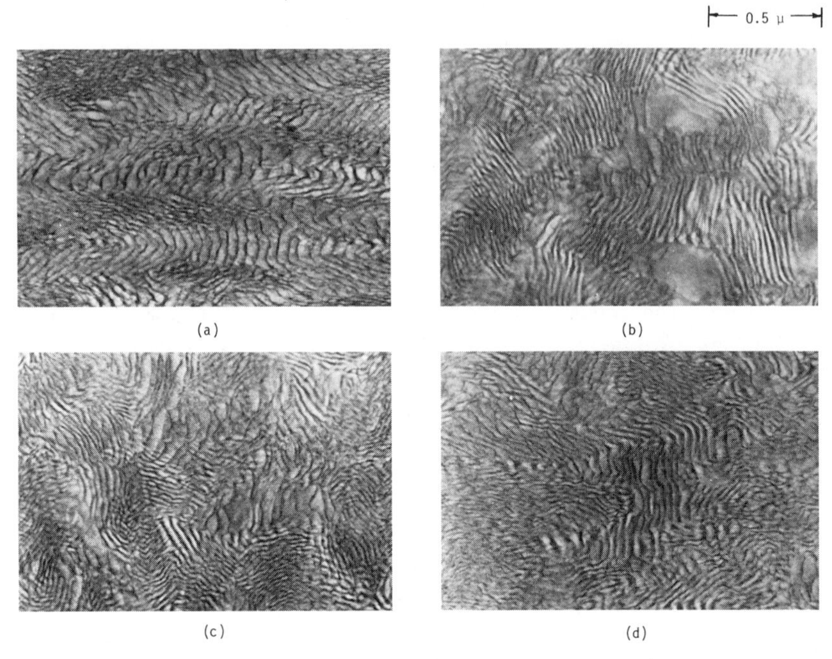

Figure 13. Electron micrographs of ultrathin sections of the SBS-PS specimens released from 500% elongation and (a) left in unstretched state at room temperature for a few days, and (b), (c), and (d) annealed at 60°, 88°, and 116°C, respectively, all for 5 hr. The sectioning and stretching directions are the same as those in Figure 12.

angles, can be obtained in a duration as short as 10 minutes. When the duration is longer than 10 minutes, the first-order maximum becomes more intensive and closer in peak position to that of the original specimen, suggesting not only the recovery of the domain structure in lamellar spacing but also the increased regularity of lamellar structure, i.e., the annealing effect.

Dynamic Mechanical Properties. Figure 15 shows the temperature dispersion of isochronal complex, dynamic tensile modulus functions at a fixed frequency of 10 Hz for the SBS-PS specimen in unstretched and stretched (330% elongation) states. The two temperature dispersions around − 100° and 90°C in the unstretched state can be assigned to the primary glass-transitions of the polybutadiene and polystyrene domains. In the stretched state, however, these loss peaks are broadened and shifted to around − 80° and 80°C, respectively. In addition, new dispersion, as emphasized by a rapid decrease in $E'(\omega_0)$, appears at around 40°C. The shift of the primary dispersion of polybutadiene matrix toward higher temperature can be explained in terms of decrease of the free volume because of internal stress arisen within the matrix. On the other

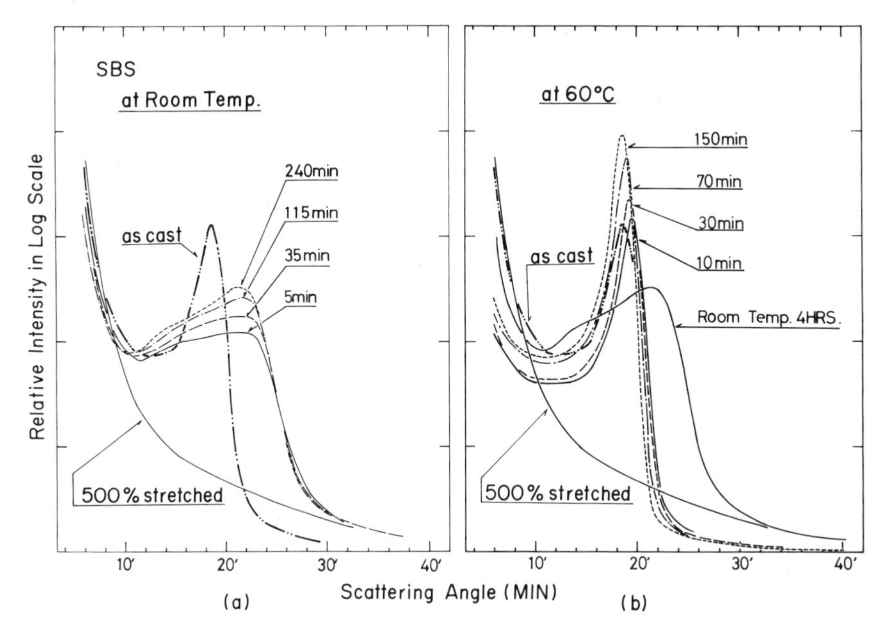

Figure 14. Healing effect of the SBS specimen released from 500% elongation, investigated from the change of SAXS intensity distribution around the first-order scattering maximum with duration of staying (a) at room temperature and (b) at 60°C

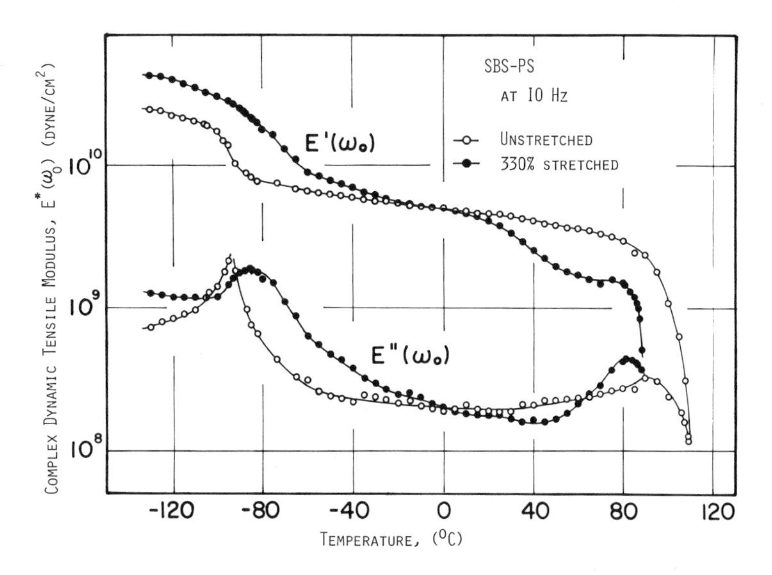

Figure 15. Temperature dispersion of isochronal complex, dynamic tensile modulus function at a fixed frequency of 10 Hz, observed for the SBS-PS specimen at unstrethed and stretched (330% elongation) states

hand, the opposite shift of the primary dispersion of polystyrene domain and the appearance of the additional dispersion can be attributed to the fragmentation of the polystyrene lamellar domains.

The existence of an additional relaxation mechanism has been noted by several authors for the heterogeneous system of block and graft copolymers and been assigned to a type of grain-boundary phenomena (*17*), i.e., interfacial domain-boundary relaxation (*18–24*), though the existence of the additional relaxation mechanism has not always been found as a discrete loss peak but as a monotonous and rather rapid decrease in $E'(\omega_0)$. The existence of the interfacial domain boundary also has been investigated theoretically (*13, 14, 15*) and experimentally from SAXS intensity distribution in terms of the domain-boundary thickness (*11, 25, 26, 27*). Fragmentation of the polystyrene lamellae attributable to stretching must greatly increase the specific surface areas of the styrene phase and therefore the volume fraction of the interfacial domain-boundary region (*27*). This results in the opposite shift of the primary dispersion of polystyrene domain, as suggested by Braes (*28*), as well as the appearance of the additional dispersion.

The fact that the structural reformation of the original lamellar domains from the fragmented ones can occur even at a temperature as low as 60°C and a duration as short as 10 minutes, as demonstrated in Figure 14b, can be explained in terms of thermodynamic driving force, i.e., the fragmented system being associated with an excess free energy relative to the original lamellar system owing to: (A) orientation of the polybutadiene chains (giving rise to decreased entropy), and (B) enormous increase of the specific surface areas in the fragmented system (giving rise to increased interfacial energy). The structural reformation is associated with the process of the enthalpy and entropy relaxations. In the interfacial region, the volume fraction of which is enormously increased in the fragmented system, the polystyrene segments must be intermixed with polybutadiene segments so that the polystyrene chains can gain the amount of mobility required for the structural reformation even under conditions as mild as annealing at 60°C for 10 minutes.

Acknowledgment

The authors are indebted to the Shell Development Co. for supplying the tri-block copolymers, TR-41-1647, TR-41-1648 and TR-41-1649. H. Kawai and M. Shen also wish to thank the Japan Society for Promotion of Science, the Dreyfus Foundation, and the Office of Naval Research for support which has enabled them to cooperate in their research projects on heterophase systems, including this chapter. Finally, thanks are owed to the Editor of Rubber Chemistry & Technology for permission to reproduce Figures 2 and 6–11 from the original paper (*29*).

Literature Cited

1. Beecher, J. F., Marker, L., Bradford, R. S., Aggarwal, S. L., *J. Polym. Sci., Polym. Symp.* (1969) **26**, 117.
2. Henderson, J. F., Grundy, K. F., Fischer, E., *J. Polym. Sci., Polym. Symp.* (1968) **16**, 3121.
3. Fischer, E., Henderson, J. F., *J. Polym. Sci., Polym. Symp.* (1969) **26**, 149.
4. Akovali, G., Niinomi, M., Diamant, J., Shen, M., *Polym. Prepr., Am. Chem. Soc., Div. Polym. Chem.* (1976) **17**, 560.
5. Hong, S. D., Shen, M., Russell, T., Stein, R. S., "Polymer Alloys," D. Klempner, K. C. Frisch, Eds., Plenum, New York, 1977.
6. Hendus, H., Illers, K. H., Ropte, E., *Kolloid Z. Z. Polym.* (1967) **216**, 110.
7. Fischer, E., *J. Macromol. Sci., Chem.* (1968) **A2**, 1285.
8. Toy, L., Niinomi, M., Shen, M., *J. Macromol. Sci. Phys.* (1975) **B11**, 281.
9. Hendricks, R. W., Schmidt, P. W., *Acta Phys. Austriaca* (1973) **37**, 20.
10. Schmidt, P. W., *Acta Crystallogr.* (1965) **19**, 938.
11. Todo, A., Hashimoto, T., Kawai, H., *J. Appl. Crystallogr.* (1978) **11**, 40.
12. Inoue, T., Soen, T., Hashimoto, T., Kawai, H., *J. Polym. Sci., Polym. Phys. Ed.* (1969) **7**, 1283.
13. Leary, D. J., Williams, M. C., *J. Polym. Sci., Polym. Lett. Ed.* (1970) **8**, 335.
14. Meier, D. J., *Polym. Prepr., Am. Chem. Soc., Div. Polym. Chem.* (1974) **15**, 171.
15. Helfand, E., *Acc. Chem. Res.* (1975) **8**, 295.
16. Inoue, T., Soen, T., Hashimoto, T., Kawai, H., *Macromolecules* (1970) **3**, 87.
17. Saito, N., Okano, K., Iwayanagi, S., Hideshima, T., "Solid State Physics," H. Ehrenreich, F. Seitz, D. Turnbull, Eds., Vol. 14, p. 458, Academic, New York, 1963.
18. Aggarwal, S. L., Livigni, R. A., Marker, L. F., Dudek, T. J., "Block and Graft Copolymers," J. J. Burke, V. Weiss, Eds., Syracuse University Press, New York, 1973.
19. Shen, M., Kaelble, D., *J. Polym. Sci., Polym. Lett. Ed.* (1970) **8**, 149.
20. Fesko, D. G., Tschoegl, N. W., *Int. J. Polym. Mater.* (1974) **3**, 51.
21. Soen, T., Ono, T., Yamashita, K., Kawai, H., *Kolloid Z. Z. Polym.* (1972) **250**, 459.
22. Soen, T., Shimomura, M., Uchida, T., Kawai, H., *Coll. Polym. Sci.* (1974) **252**, 933.
23. Kraus, G., Rollmann, K. W., *J. Polym. Sci., Polym. Phys. Ed.* (1976) **14**, 1133.
24. Akovali, G., Diamant, J., Shen, M., *J. Macromol. Sci., Phys.* (1977) **B13**, 117.
25. Hashimoto, T., Nagatoshi, K., Todo, A., Hasegawa, H., Kawai, H., *Macromolecules* (1974) **7**, 364.
26. Hashimoto, T., Todo, A., Itoi, H., Kawai, H., *Macromolecules* (1977) **10**, 377.
27. Todo, A., Uno, H., Miyoshi, K., Hashimoto, T., Kawai, H., *Polym. Eng. Sci.* (1977) **17**, 587.
28. Bares, J., *Macromolecules* (1975) **8**, 244.
29. Fujimura, M., Hashimoto, T., Kawai, H., *Rubber Chem. Tech.* (1978) **51**, 215.

RECEIVED April 14, 1978.

15

Morphology and Dynamic Viscoelastic Behavior of Blends of Styrene–Butadiene Block Copolymers

GERARD KRAUS, L. M. FODOR, and K. W. ROLLMANN

Phillips Petroleum Co., Bartlesville, OK 74004

Different block length distributions in styrene–butadiene block copolymers of the linear SBS or $(SB)_x$ *"star" type and their mixtures can cause wide changes in domain morphology at constant overall monomer composition (75 wt % styrene). Block polymers of substantially uniform block length had the expected spherical, polystyrene-continuous morphology. Broadening the styrene block length distribution by blending polymers of different block lengths led to appearance of cylindrical and lamellar structures and, ultimately, to complex polybutadiene-continuous morphologies. Polymers and blends were characterized by electron microscopy and by their viscoelastic behavior. Correlations were established between morphology on one hand and anisotropy in the storage modulus and the height and position of the polybutadiene tan δ maximum on the other.*

When a block copolymer is blended with the homopolymer of one of the monomers of which it is composed, the homopolymer will enter the block polymer domain structure only when its molecular weight does not greatly exceed that of the block sequences of like composition (*1, 2*). When it does so, the homopolymer forms its own, usually much larger, domains which may absorb some of the like-block sequences in their surface regions.

In the present study we examine the situation where both constituents of the blend are block copolymers of the same two monomers, but where the block lengths may vary widely between constituents. As a

0-8412-0457-8/79/33-176-277$05.00/0

constraint on the enormous number of such blends possible, the total composition of the blend is held fixed. The monomers chosen are styrene and butadiene at an overall blend composition of 75% styrene (by weight).

Experimental

Block polymers were prepared by organolithium-initiated polymerization in cyclohexane solution by using the sequential monomer addition technique (3). Polymers were both of the linear-SBS and "radial"-branched $(SB)_x$ type. Blends were prepared in cyclohexane solution, either before or after coupling the initially linear SBLi precursor. Coupling agents investigated were ethyl acetate (for linear coupling), epoxidized soybean oil (ESO), and $SiCl_4$.

Block molecular weights were calculated from monomer charges and initiator levels and corrected for "scavenger level," i.e., the amount of RLi destroyed by system impurities. These nominal block lengths were, in general, in good agreement with gel-permeation chromatographic molecular weights.

Block length polydispersity indices for blends were calculated on the assumption that the block polymers, as prepared, were composed of monodisperse blocks. This is ,of course, an approximation justified only by the narrowness of the molecular weight distribution in polymerizations of the present type. The block heterogeneity indices given here should, therefore, be regarded as relative measures of breadth of distribution.

Polymers and blends were worked up by evaporating the cyclohexane solvent and massing the polymer on a 140°C roll mill. Films were then prepared by compression molding (5 min at 200°C) or, in one set of experiments, by extrusion through a slit die. Dynamic viscoelastic meas-

Table I. Blend Compositions

Composition	*Component*	*Wt Fraction*	*Precoupling Block Length* $\overline{M}_B/1000$	$\overline{M}_S/1000$	*Styrene (%)*	$(\overline{M}_w/\overline{M}_n)_s$
A	—	1.00	19	56	75	1
B	1	0.64	30	120	80	
	2	0.36	7.6	13.4	63.8	
	Blend	1.00	—	—	74.2	2.4
C	1	0.64	13	137	91.3	
	2	0.36	11	10	47.6	
	Blend	1.00	—	—	75.7	3
D	1	0.64	6	144	96.3	
	2	0.36	13.5	7.5	35.9	
	Blend	1.00	—	—	74.5	3.5
E	1	0.64	—	150	100	
	2	0.36	14.9	6.1	29.1	
	Blend	1.00	—	—	74.5	3.7

urements were made with a Rheovibron Model DDV-II viscoelastometer in the tensile mode at 35 Hz. Ultrathin sections of polymer films were prepared by cryomicrotomy, stained with OsO_4 vapor (*4*), and examined under a Philips EM-300 electron microscope.

Results

Table I describes five compositions, each of 75% styrene content, prepared by coupling diblock SBLi molecules of varying block length with a polyfunctional epoxide. The data are arranged in order of increas-

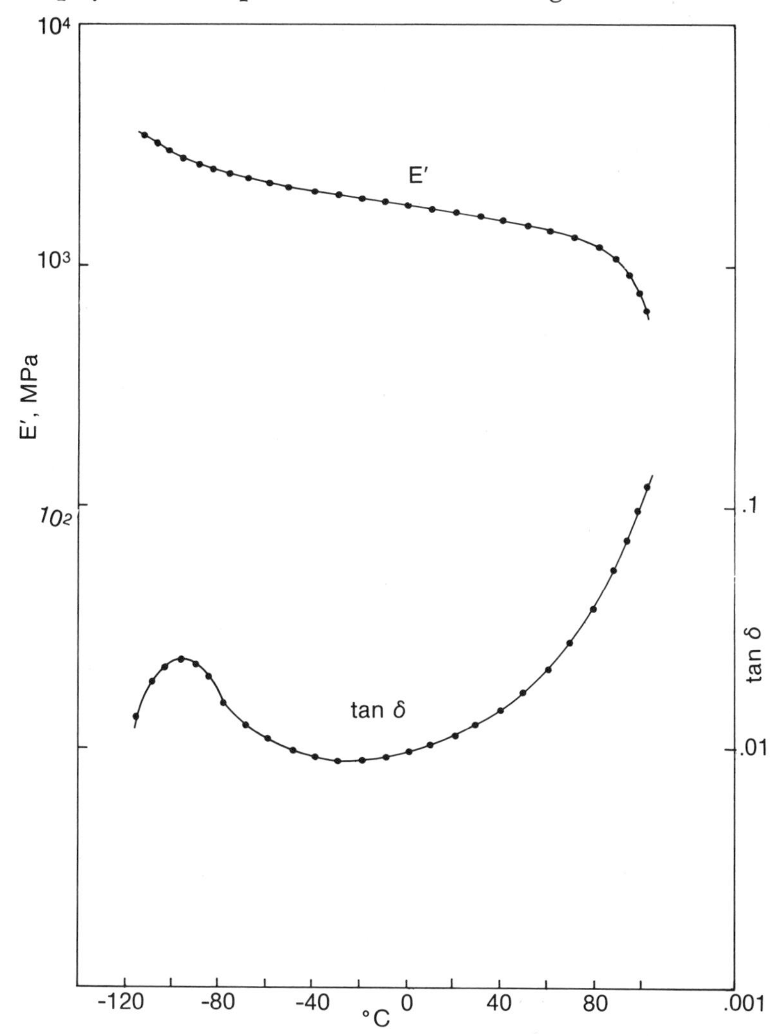

Figure 1. Storage modulus and loss tangent (35 Hz) for block polymer with uniform polystyrene blocks (composition A). Compression molded.

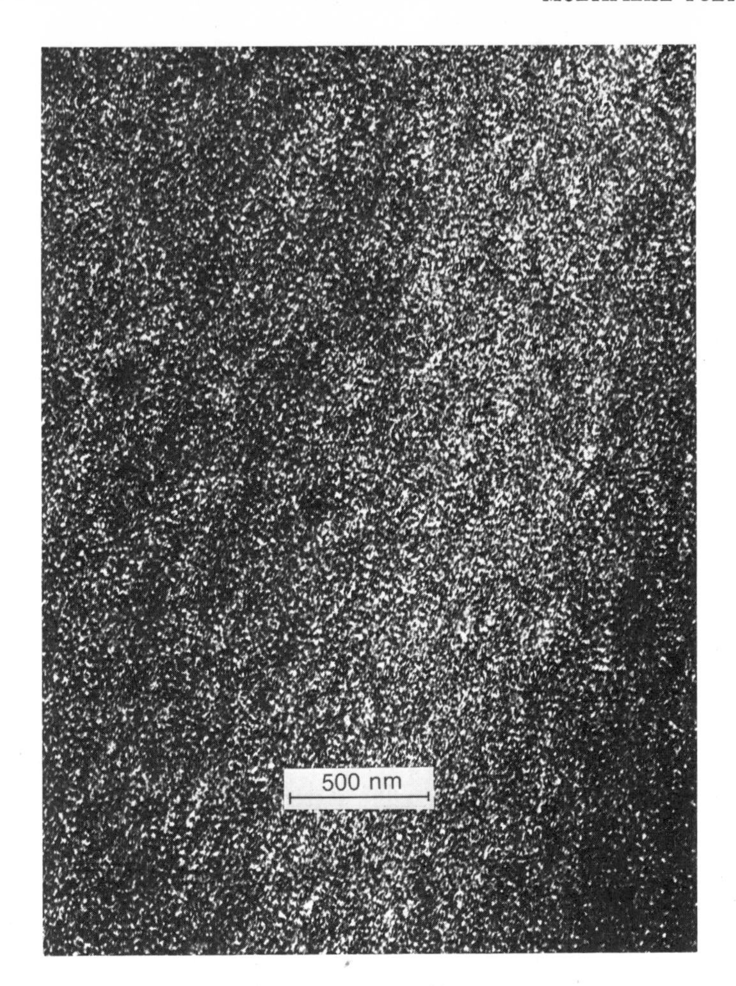

Figure 2. Electron micrograph of composition A

ing polydispersity of (nominal) polystyrene (PS) block lengths. While the polybutadiene (PB) blocks also vary in length, their polydispersity is considerably less.

Figure 1 shows storage modulus and loss tangent vs. temperature plots for composition A, which is not a blend. As shown by Figure 2, the morphology is spheres of PB in a continuum of PS. The loss tangent clearly shows the PB glass transition at $-90°C$ and the ascending branch of the PS maximum near 100°C. The results are exactly as expected from the spherical morphology, except that the temperature of the PB tan δ maximum lies several degrees lower than that of polybutadiene of the appropriate microstructure (ca. 50% trans, 40% cis, and 10% vinyl) for which $T(\tan \delta_{max}) \cong -80°C$.

Figure 3 shows the same kind of data for blend C, in which PB block lengths are similar, but the PB blocks differ greatly in length; Figure 4 shows an electron micrograph of this composition. The morphology of this blend is clearly lamellar, with considerable orientation in one direction. (The uneven spacings of light and dark bands result from lamellae sectioned at various angles.) The direction of orientation is that of mold flow. The dynamic data are entirely consistent with this

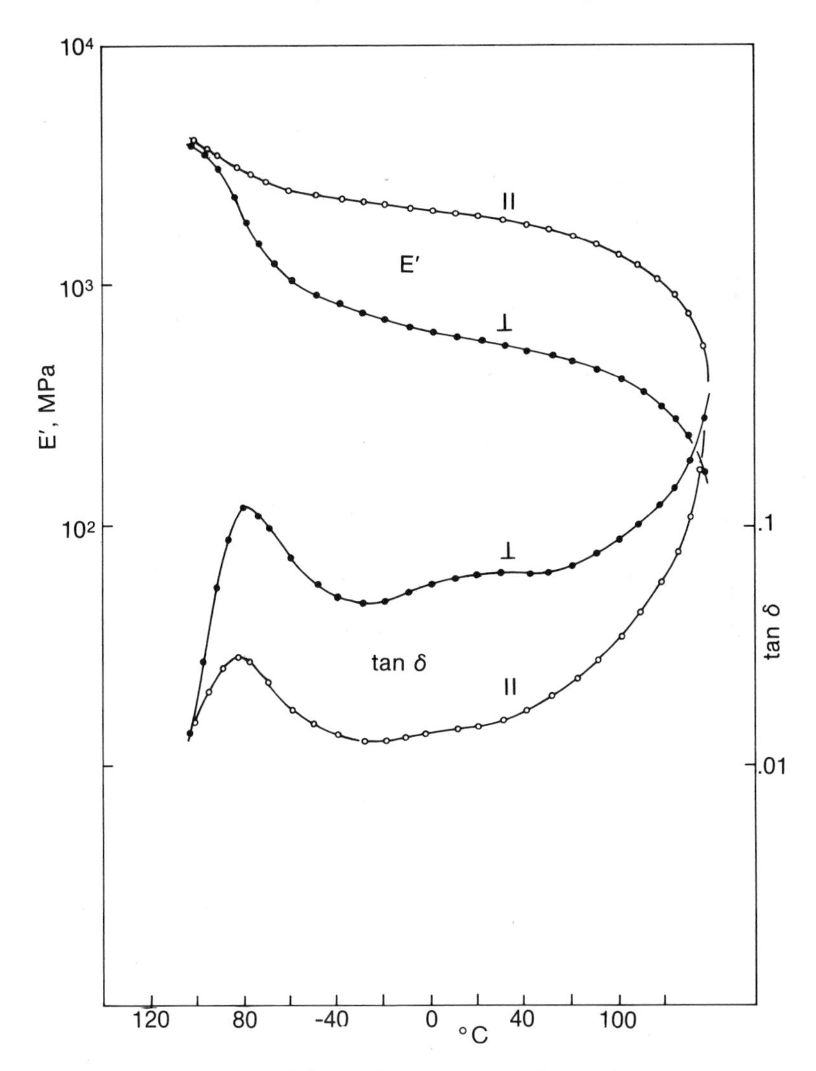

Figure 3. Storage modulus and loss tangent (35 Hz) for composition with bimodal polystyrene block length distribution (composition C). Compression molded: (||) parallel to direction of mold flow; (⊥) normal to direction of mold flow.

Figure 4. Electron micrograph of composition C

morphology. In the direction of mold flow, E' and tan δ are not too greatly different from Figure 1 but normal to the flow direction, the resin is much softer (smaller E'), and the PB tan δ peak is strongly accentuated.

The above results are for compression-molded samples. A closer investigation of these resins in extruded film is summarized in Figures 5 and 6. Note the relative isotropy in mechanical properties characteristic of the spherical morphology for the single polymer and the strong anisotropy for the blend. Note also that for the lamellar blend $T(\tan \delta_{max})$ is consistently $-80°$ to $-81°C$, the normal value for polybutadiene independent of orientation. The reason for the depression of $T(\tan \delta_{max})$ in the single polymer is evidently the constraint the polybutadiene domains find themselves under as the result of differences in thermal contraction of the phases as they cool from T_g(polystyrene). The smaller coefficient of expansion of glassy polystyrene causes the cavities accommodating the polybutadiene inclusions to shrink less than the free contraction of polybutadiene, placing the latter phase in a state of hydrostatic tension and lowering T_g. In the lamellar morphology there is no such constraint; the polybutadiene lamellae merely thin out and the

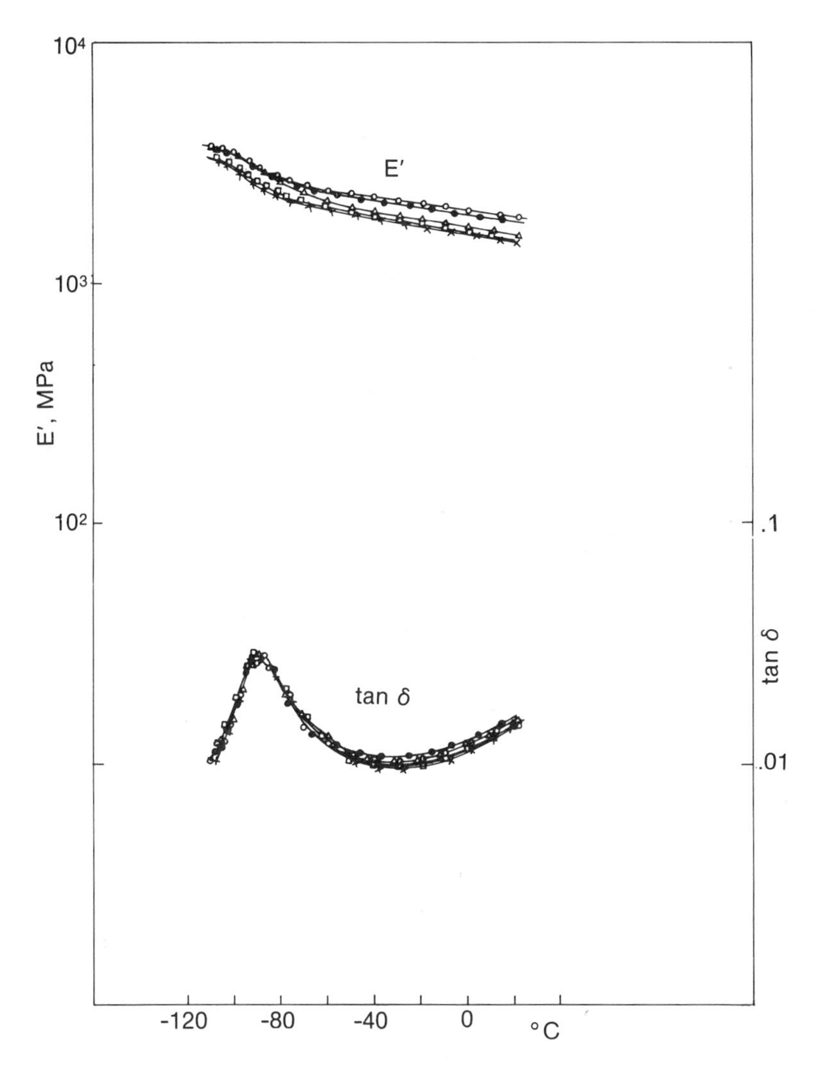

Figure 5. Effect of orientation on E′ *and tan δ in extruded film (35 Hz) of composition A. Direction of measurement with respect to extrusion direction:* (○) *0°,* (●) *22.5°,* (△) *45°,* (□) *67.5°,* (×) *90°.*

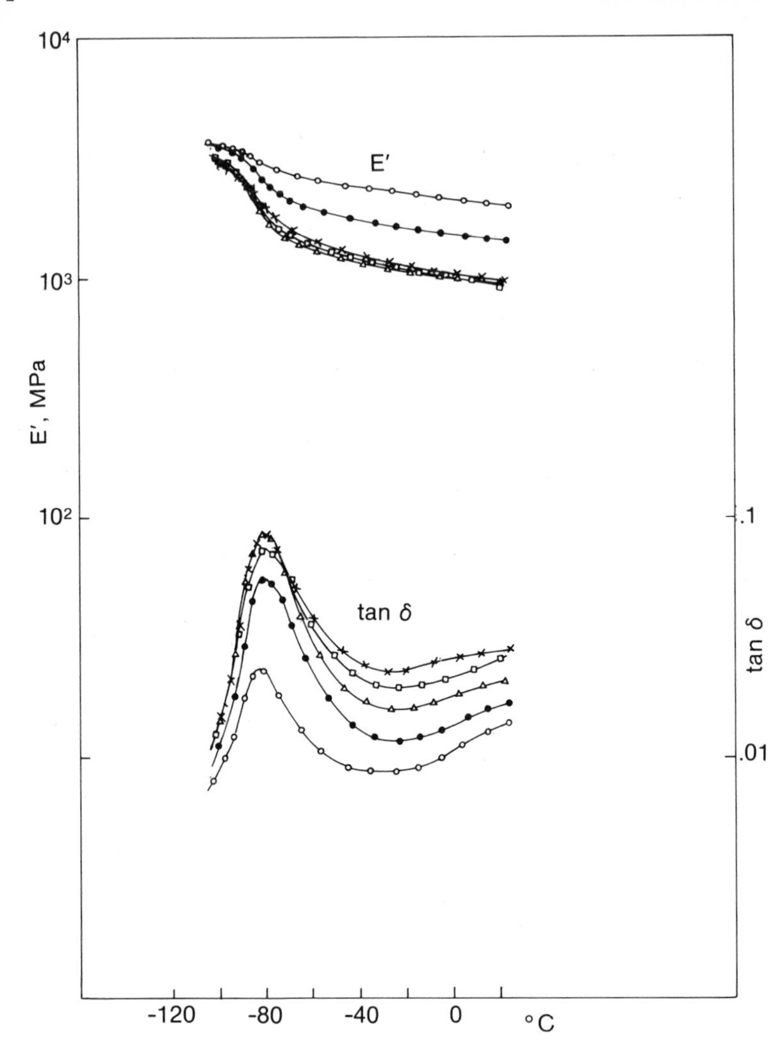

Figure 6. Effect of orientation on E′ *and tan* δ *in extruded film (35 Hz) of composition C. Notation as in Figure 5.*

normal T_g is observed. Comparison of Figures 3 and 6 indicates the degree of orientation to be greater in the compression-molded sample.

Returning now to composition B, in which the PS block distribution is less severe than in C, we note that evidently both rod-like and lamellar morphologies are about equally probable. The morphologies shown in Figures 7 and 8 were obtained on presumably identically prepared samples; they appear to be the result of small adventitious variations in molding technique and/or thermal history. The data of Table II, which is a summary of the principal morphology-related features of the dynamic viscoelastic data, clearly confirm the different morphologies. The rod-like PB domains of Figure 7 cause only modest anisotropy since PS remains continuous in both directions of orientation. $T(\tan \delta)$ is again depressed, as cylindrical PB domains cannot contract freely under the constraint of the glassy continuum.

Electron micrographs of compositions D and E are shown in Figures 9 and 10. It is evident that in E polybutadiene is the continuous phase (with some rubber in the polystyrene domains) while D represents a transition from lamellar to polybutadiene-continuous morphology. Again the dynamic mechanical data (Table II) are consistent with these obser-

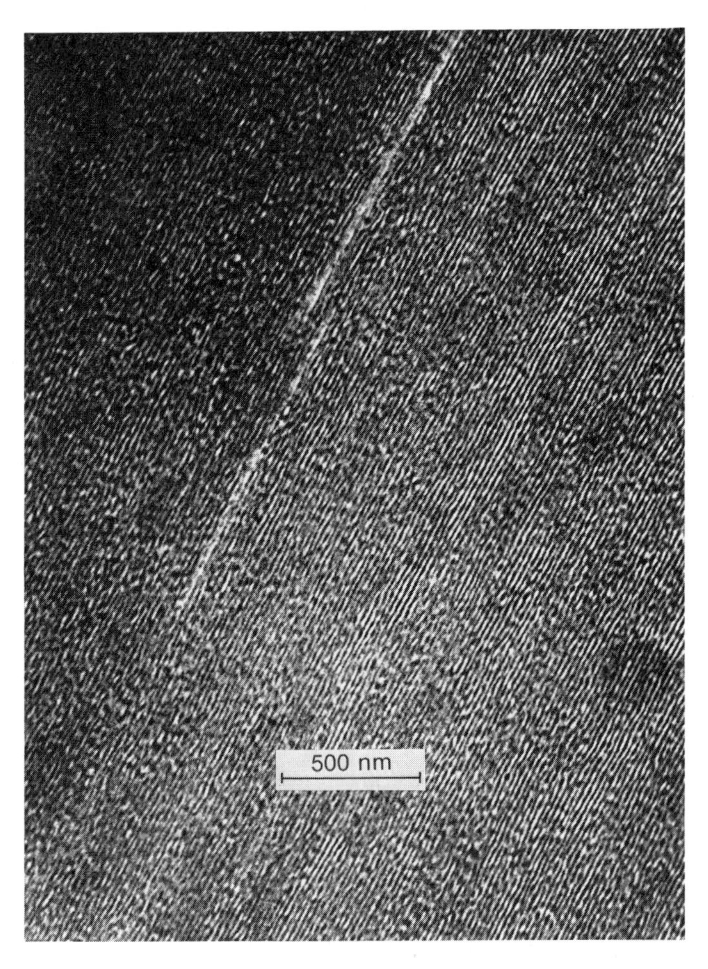

Figure 7. Composition B in rod-like form

Table II. Morphology and

Resin	$(\overline{M}_w/\overline{M}_n)_s$	*Continuous Phase*	*Discrete Phase*
A	1.0	PS	PB (spheres)
B [c]	2.4	PS	PB (rods)
		alternating lamellae	
C	3	alternating lamellae	
D	3.5	PB [d]	PS (complex
E	3.7	PB	PS (ellipsoids)

[a] Compression-molded samples.
[b] E'_{20} = storage modulus at 20°C.

Figure 8. Composition B in lamellar form

Dynamic Viscoelastic Properties[a]

Parallel to Flow[b]			Normal to Flow[b]		
$\tan \delta_{max}$	$T(\tan \delta_{max})$ (°C)	E'_{20} (MPa)	$\tan \delta_{max}$	$T(\tan \delta_{max})$ (°C)	E'_{20} (MPa)
0.021	−94	1760	0.026	−94	1680
0.024	−90	1550	0.044	−90	1300
0.061	−84	1260	0.137	−83	520
0.030	−80	1850	0.120	−80	530
0.181	−76	330	0.244	−76	170
0.296	−79	140	0.300	−78	130

[c] This composition has been observed in two distinct morphologies; *see* text.
[d] Predominantly.

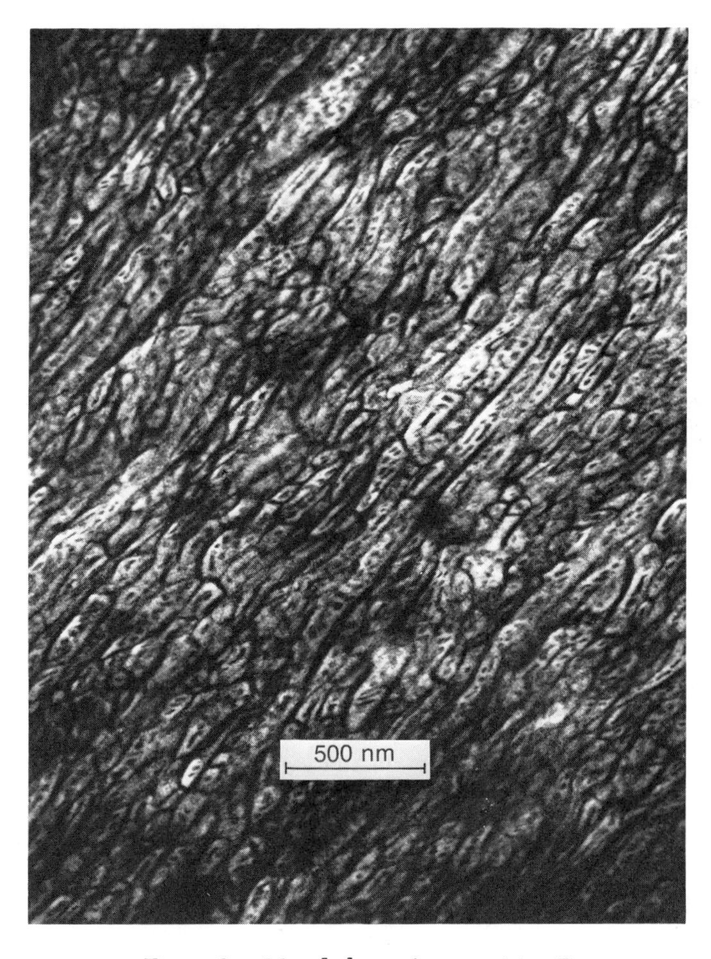

Figure 9. Morphology of composition D

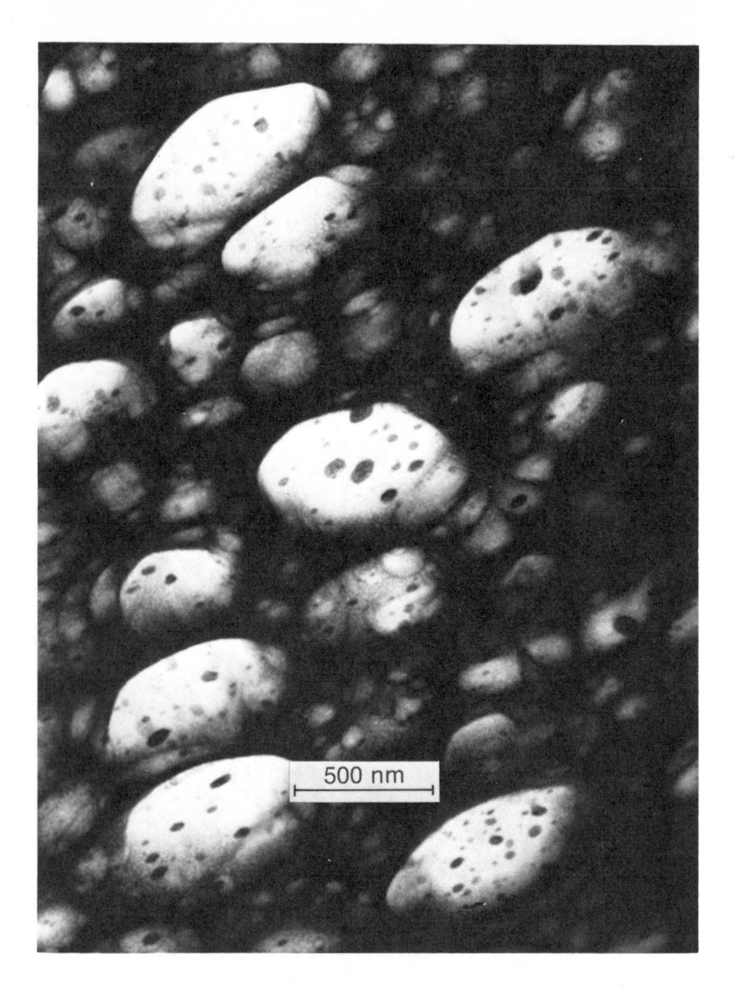

Figure 10. Morphology of composition E

vations, showing much smaller storage moduli and large tan δ maxima near − 80°C for the PB domains. Mechanical anisotropy is absent in E.

The above results show clearly that, in the system at hand, the broader the distribution of PS block length, the greater the tendency of the minor PB phase to become continuous. It is also obvious that the dynamic mechanical data tell a great deal about the morphology. The height of the PB loss maximum increases as the rubber becomes increasingly load bearing while at the same time E' (between the transitions) decreases. Mechanical anisotropy resulting from orientation is most pronounced for the lamellar structure. Finally, the position of the polybutadiene tan δ peak is depressed for those morphologies in which polystyrene is the continuous phase.

Some seventy blends were examined by dynamic viscoelastic measurements only. They differed in molecular weight of the constituents, linearity of the constituent block polymer molecules, (SBS vs. $[SB]_x$), type and stoichiometry of coupling, order of coupling (before and after blending), composition of the fractions, and blend ratio—always, however, subject to the constraint of 75% styrene content. Although differences in viscoelastic behavior were observed, the most decisive variable by far was block-length heterogeneity. Figure 11 shows a plot of the height of the tan δ maximum vs. $(\overline{M}_w/\overline{M}_n)_s$. One can easily spot the ranges of block heterogeneity in which different morphologies are to be expected. This pattern is confirmed by Figure 12 in which the height of tan δ is plotted against its position. There are several reasons, aside from experimental error, for the variability in properties at equal $(\overline{M}_w/\overline{M}_n)_s$. One is that $\overline{M}_w/\overline{M}_n$ is only one of many possible, nonequivalent ways of expressing block length heterogeneity and is not necessarily the most relevant one to the present situation. Also, as in Table I, in the expanded study the PB blocks do vary in size, even if much less than the PS blocks. Lack of morphological uniqueness, as in blend B, complicates the picture in the overlap region near $\overline{M}_w/\overline{M}_n = 2.5$. Finally, there is a tendency for abnormally high $T(\tan \delta_{max})$ in the compositions with the broadest PS block distribution. Attainment of such distributions requires use of substantial amounts of polymer with PS blocks of less than 10,000

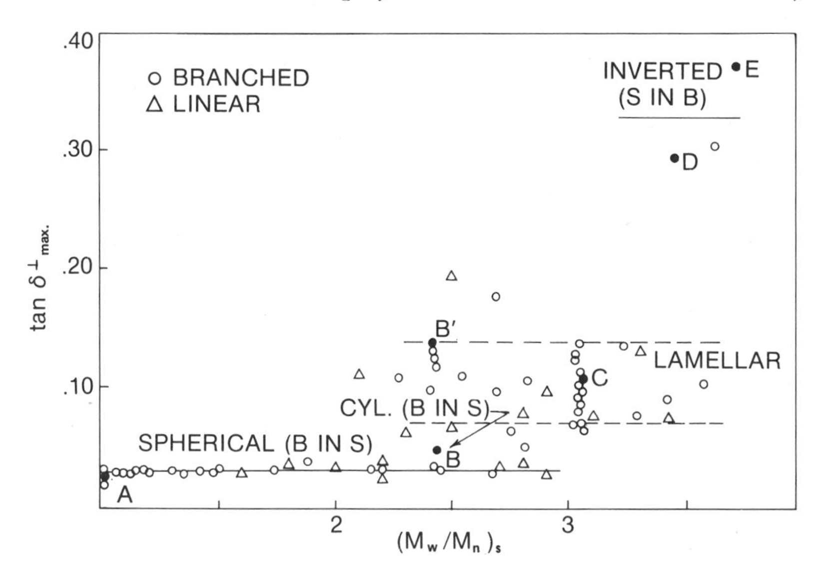

Figure 11. Maximum low temperature loss tangent (35 Hz) measured normal to mold flow vs. styrene block length heterogeneity. Circles—branched polymers, triangles—linear polymers, solid symbols—electron micrographs displayed.

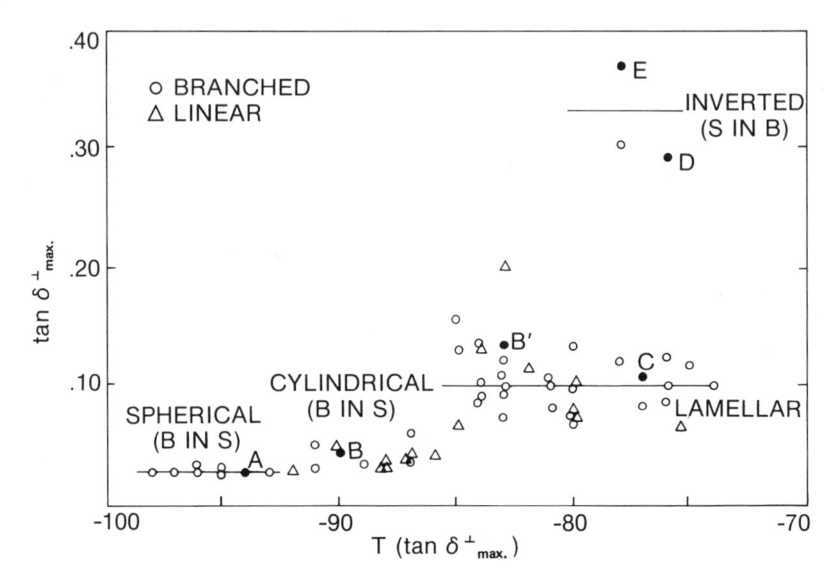

Figure 12. Height and position of low temperature loss maximum. Notation as in Figure 11.

molecular weight (5000 was the shortest block used in this work). In this range of block molecular weights, interphase effects begin to have an effect on the position of the PB loss maximum (*5*).

Extreme differences in PB block length curiously appear to extend the range in PS block heterogeneity in which polystyrene-continuous morphologies are possible. For example, Figure 13 shows a straight blend of linear SBS polymers in which both kinds of blocks vary tenfold in length (composition F):

	Wt Fraction	*S/B/S*	*Styrene (%)*	$(\overline{M}_w/\overline{M}_n)_s$
Component 1	0.60	150000/100000/150000	75	(1)
Component 2	0.40	15000/10000/15000	75	(1)
Blend	1.00	—	75	2.9

In spite of $(\overline{M}_w/\overline{M}_n)_s = 2.9$, the morphology appears to be basically spherical, albeit with considerable connectivity of polybutadiene domains. Moreover, for this blend tan $\delta_{max} = 0.028$, $T(\tan \delta_{max}) = -88°C$, with virtually no anisotropy in storage modulus, consistent with spherical or short rod-shaped polybutadiene domains.

Figure 13. Morphology of composition F

Discussion

The observation that broad, bimodal styrene, block length distributions tend to favor continuity of the polybutadiene phase is not confined to 75% styrene content. Thus, a limited study at 50% styrene showed that polybutadiene-continuous compositions could be prepared by broad blending in place of the normal alternating lamellar structures characteristic of this composition.

Since simple blending of the finished block polymers and coupling blends of SBLi di-block polymers did not produce markedly different results, it seems clear that the block length distribution per se is more important in governing morphology than the disposition of these blocks over individual molecules.

Extensive use is made in this work of the effects of orientation on mechanical properties in block polymers with cylindrical and lamellar structures. These effects are, in general, known from earlier studies (*6, 7*); they add convincing evidence to the morphological assignments made.

It should be clear that the conclusions of this work are limited to block polymers isolated from the polymerization solvent (cyclohexane) by evaporation and subsequently processed by conventional thermal mixing and shaping techniques. Obviously, other morphologies could be realized in many instances by casting films from solvents of varying quality for the two block sequences.

Conclusions

Different block length distributions in SBS and $(SB)_x$ block polymers and their mixtures can cause wide changes in domain morphology at constant overall monomer composition, which lead to characteristically different linear viscoelastic properties.

Acknowledgment

The authors are indebted to J. O. Gardner for the electron micrographs displayed in this report.

Literature Cited

1. Inoue, T., Soen, T., Hashimoto, T., Kawai, H., *Macromolecules* (1970) **3**, 87.
2. Niinomi, M., Akovali, G., Shen, M., *J. Macromol. Sci.* (1977) **B13**, 133.
3. Zelinski, R. P., Childers, C. W., *Rubber Chem. Technol.* (1968) **41**, 161.
4. Kato, K., *Polym. Eng. Sci.* (1967) **7**, 38.
5. Kraus, Gerard, Rollmann, K. W., *J. Polym. Sci., Polym. Phys. Ed.* (1976) **14**, 1133.
6. Charrier, Jean-Michel, Ranchoux, Robert J. P., *Polym. Eng. Sci.* (1971) **11**, 381.
7. Folkes, M. J., Keller, A., *Polymer* (1971) **12**, 222.

RECEIVED April 14, 1978.

16

Poly(Arylene Ether Sulfone)-Poly(Aryl Carbonate) Block Copolymers

T. C. WARD, A. J. WNUK, E. SHCHORI, R. VISWANATHAN, and J. E. MC GRATH

Department of Chemistry, Virginia Polytechnic Institute and State University, Blacksburg, VA 24061

The influence of block molecular weight and slightly varying chemical compositions on microphase separation in poly-(arylene ether sulfone)–poly(aryl carbonate) block copolymers was examined. Compositional variety was achieved by replacing the isopropylidene unit in the bisphenol-A derived aryl sulfone blocks by either a thiol or a sulfonyl group. Synthetic methods and characterization techniques were used that allowed careful control of average block size in these $[AB]_n$ copolymer systems, all of which were overall 50% by weight of each type segment. Although physical blends of the homopolymers are incompatible, DSC and mechanical measurements indicated that either one- or two-phase films could be molded from the block copolymers. Kinetic effects regulating the formation of two phases from the melt were observed.

Multiphase polymeric materials very often are based on either graft copolymers or block copolymers. It is interesting to note that graft copolymers have been largely associated with impact thermoplastic technology whereas block copolymers have to date found most of their practical applications in the form of thermoplastic elastomers (*1–8*). We have been interested in studying ductile glassy–glassy and glassy–crystalline block copolymers as models for homogeneous and multiphase engineering materials. Very little quantitative experimental information is available concerning the important parameters governing the development of the microphases in such systems. Qualitatively, block molecular weight and segment interaction parameters most likely are important. Secondly, almost nothing is known with respect to the effect of microphase development on mechanical properties such as ductility, impact strength, environmental stress cracking, and physical aging.

0-8412-0457-8/79/33-176-293$05.00/0

We recently have reported our initial studies on step-growth block copolymers containing segments of poly(aryl ethers) and poly(aryl carbonates) (*9, 10*). The multiblock $[A\text{–}B]_n$ block copolymers were prepared by phosgenation in methylene chloride/pyridine solution either by what was termed an "in situ" or by a "coupled oligomer" technique (*10*). The choice of polycarbonates and poly(aryl ethers) for initial studies was based on the several considerations. Copolymerization is feasible since the end groups in the two oligomers can be identical, as shown in Structures **1** and **2**. Considerable information is available in the

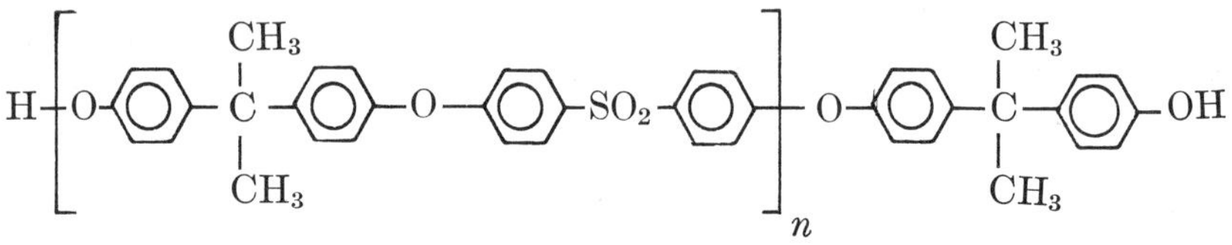

$$H\left[O-C_6H_4-C(CH_3)_2-C_6H_4-O-C_6H_4-SO_2-C_6H_4\right]_n O-C_6H_4-C(CH_3)_2-C_6H_4-OH$$

hydroxyl terminated
bisphenol-A polysulfone

1

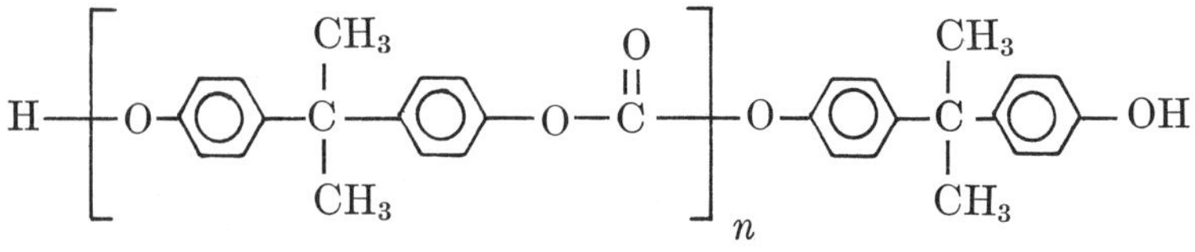

$$H\left[O-C_6H_4-C(CH_3)_2-C_6H_4-O-\overset{O}{\overset{\|}{C}}\right]_n O-C_6H_4-C(CH_3)_2-C_6H_4-OH$$

hydroxyl terminated
bisphenol-A polycarbonate

2

literature on both homopolymers (*11, 12, 13, 14*). Both polymers are amorphous as prepared, which allows characterization of the oligomers and copolymers by solution methods (*10*). However, the polycarbonate segments subsequently can be crystallized by certain solvents (*15, 16*). Thus the same copolymer can, in principle, be studied with either a glassy–glassy or glass–crystalline morphology. Lastly, wide structural variations are possible within the general classes of poly(aryl ethers) and poly(aryl carbonates) simply by changing the chemical nature of the bisphenol. These variations, in turn, can be used to model the effect of the "differential solubility parameter" (*8*) (or other more quantitative expressions) on the development of a multiphase system at constant block molecular weights.

Physical blends of Structures **1** and **2** were not miscible (*9, 10*), even at very low molecular weights (*10*). Our initial investigation suggested that the block copolymer derived from coupling Structures **1** and **2** with phosgene were homogeneous at somewhat higher block molecular weights than one would have expected (*10*). Evaluation of these copolymers also suggested that it would be desirable to develop additional synthesis routes which did not require quantitative removal of residual pyridine and pyridine hydrochloride. Either residue can promote degradation of the copolymer during studies of its mechanical behavior. One possibility appeared to be an interfacial technique (*17, 18, 25, 26*) which uses phase transfer catalysis (*19*). Both oligomers are soluble in methylene chloride and are reasonably compatible at low concentrations (*10*). More detailed end-groups analyses techniques for the oligomers based on UV–visible spectroscopy (*20, 21*) and potentiometric titrations (*22, 23*) also were developed which have greatly improved the assessment of segment number-average molecular weights.

Experimental

Synthesis. High purity bisphenol A (Bis-A) and 4,4′-dichlorosulfone (DCDPS) were obtained from Union Carbide. 4,4′-Thiodiphenol (Bis-T) and 4,4′-sulfonyldiphenyl (Bis-S) were supplied by Crown Zellerbach. 4,4′-Difluorodiphenylsulfone was either obtained from Aldrich or by reaction of DCDPS with anhydrous KF.

The poly(arylene ether sulfones) were prepared as previously described (*10, 14, 24*). Toluene/DMAC was the usual solvent system used. The more reactive 4,4′-difluorodiphenyl sulfone was used in reactions with the relatively acidic 4,4′-sulfonyldiphenol.

The polycarbonate oligomers were prepared by solution or interfacial techniques (*10, 17, 18*). Methylene chloride and tetraethyl ammonium chloride served as the solvent and phase transfer catalyst, respectively. The block copolymerizations were performed essentially under interfacial reaction conditions. In the case of copolymerizations using the Bis-S polysulfone oligomers, it was necessary to use tetrachloroethane as the organic solvent.

The oligomer molecular weights were characterized by both UV-visible spectra (*20, 21*) and/or potentiometric titrations (*22, 23*). Details of the measurements are provided in these papers. The block copolymers also were characterized by intrinsic viscosity and in some cases by membrane osmometry and gel permeation chromatography. Additional characterization studies are continuing and will be reported later. A typical synthesis of a 5000–5000 polysulfone-S–polycarbonate-A copolymer via "interfacial" polymerization is described below.

Bis-S-polysulfone oligomer (5.0 g, $\overline{M}_n \sim 5300$ by titration) and Bis-A-polycarbonate oligomer (5.0 g, $\overline{M}_n \sim 5000$ by UV) were dissolved in 300 mL of tetrachloroethane in a hood. The solution was somewhat hazy even at these concentrations. (By contrast, Bis-A-polysulfone oligomers and Bis-A-polycarbonate oligomer yield clear solutions at comparable molecular weights and concentrations.) Separately, 0.2 g of sodium hydroxide and 2.0 g of tetraethyl ammonium chloride were dissolved in 120 mL of distilled water. The oligomer solutions and the aqueous solution were combined in a Waring blender. The blender was fitted with a phosgene inlet. A combination pH electrode connected to a digital pH meter (Orion 601) was used to monitor pH during the polymerization. The two layers were rapidly mixed and phosgene addition was started. A reaction time of 30 min was used, although high molecular weight could be achieved at considerably shorter times. The pH was maintained at about 8.8 via addition of 20% sodium hydroxide solution from a buret. After the phosgene flow was stopped, the reaction product was placed in a separation funnel for 30 min. An organic phase and a "foamy" aqueous layer separated. The organic phase was precipitated in excess isopropyl alcohol separately from the foam layer. Each precipitate was collected by vacuum filtration, washed with three 200-mL volumes of isopropyl alcohol, three 200-mL volumes of distilled H_2O, and then one 200-mL volume of isopropyl alcohol. The two precipitates were dried 24 hr at 120°C under an aspirator vacuum ($\sim$ 30 Torr).

7.59 g were collected from the organic phase and 2.33 g from the "foamy" layer. DSC studies of each precipitate were identical. Two T_gs were observed at 166° and 229°C. Total weight was 9.92 g of copolymer (99% yield). Cast films from tetrachloroethane solutions were transparent and ductile. A compression-molded (280°C) film was clear but brown.

Dynamic Mechanical Measurements. A Rheovibron DDV-II was used to measure mechanical properties at frequencies of 3.5 and 110 Hz. Heating rates were approximately 1°C/min. Films of 5-mil thickness were compression molded at 260°C from dried powders for these studies. No annealing treatments were applied to the films. Complex, storage, and loss moduli as well as tan δ were calculated over a $-$ 160° to 240°C range. For clarity, all of the data are not shown in the figures.

Thermal Analysis Measurements. DSC scans at 40°C/min were obtained on a Perkin–Elmer DSC-2. Dried copolymer powder was used. In situ annealing treatments were applied as discussed below by programming the desired temperature profiles on the instrument.

Materials

The oligomeric poly(arylene ether sulfone) structures on the following page were studied. It was initially expected that the series **1**, **3**, and **4** would allow one to observe a gradation in polarity attributable to the diphenyl linking agent. However, solubility parameter calculations show that **1** and **3** are in fact rather similar (eg. $\delta \cong 10.3$). It was possible to accurately assess the end group concentration, and hence the number-average molecular weight by potentiometric titration of the oligomers by tetraalkyl ammonium hydroxide in dry DMAC. The more acidic phenol end groups of **4** show larger breaks.

The block copolymers of this study were, of course, $[\text{AB}]_n$ in type, as pointed out above. Overall composition was adjusted to a 50–50 wt %

$$H{-}\left[O{-}C_6H_4{-}C(CH_3)_2{-}C_6H_4{-}O{-}C_6H_4{-}SO_2{-}C_6H_4\right]_n{-}O{-}C_6H_4{-}C(CH_3)_2{-}C_6H_4{-}OH$$

1

$$H{-}\left[O{-}C_6H_4{-}S{-}C_6H_4{-}O{-}C_6H_4{-}SO_2{-}C_6H_4\right]_n{-}O{-}C_6H_4{-}S{-}C_6H_4{-}OH$$

3

$$H{-}\left[O{-}C_6H_4{-}SO_2{-}C_6H_4{-}O{-}C_6H_4{-}SO_2{-}C_6H_4\right]_n{-}O{-}C_6H_4{-}SO_2{-}C_6H_4{-}OH$$

4

ratio of polysulfone to polycarbonate. For ease of discussion, each of the samples has been given a characteristic code to indicate both compositions and block lengths. Since each of the polycarbonate blocks was derived from the bisphenol-A monomer, its occurrence in the copolymer can be symbolized simply by AC. However, three chemically different polysulfone blocks were coupled with the polycarbonate and are discussed above as the Bis-A, Bis-T, and Bis-S oligomers. Accordingly, these sulfone units are coded AS, TS, and SS, respectively. Notation of the block number-average molecular weight in thousands of daltons completes the code, with the polysulfone notation always preceeding that of the polycarbonate. As an example, a block copolymer consisting of 26,000 g/mol polysulfone segments originating from **1** above when coupled with a 22,000 g/mol bisphenol-A polycarbonate would be designated AS/AC–26/22.

We are reporting the thermal analysis and dynamic mechanical data for a total of four copolymers ranging in average block molecular weight from 5000 g/mol up to blocks in which the polycarbonates and polysulfones were each in the 20,000 g/mol range. The information in Table I shows that four copolymers of the AS/AC type have been investigated, two of the SS/AC, and one of the TS/AC type. Before proceeding directly to the results of the measurements, a few comments concerning the homopolymer oligomers are appropriate.

Background. Homopolymers of the carbonate and of the sulfone type in this work have been investigated extensively with respect to

Table I. Glassy–Glassy Block Copolymer Properties

Sample Code	Polysulfone/ Polycarbonate Mol Wt (g/mol)	Glass Transitions (°C)		
		T_g	$(T_g)_L$	$(T_g)_H$
AS/AC–5/5[a]	5000/5000	175	—	—
AS/AC–10/10	10000/10000	175	—	—
AS/AC–16/17	16000/17000	—	165	185
AS/AC–26/22	26000/22000	—	162	190
SS/AC–5/5	5000/5000	—	167	229
SS/AC–10/10	10000/10000	—	161	231
TS/AC–10/10	10000/10000	160	—	—

[a] From Ref. *1*.

thermal and mechanical behavior (*1, 27, 28*). Table II lists some notable features of the homopolymers appropriate for our oligomers. Interesting combinations of glass-transition temperatures and specific interactions are seen to be possible by appropriate choices among the constituent ingredients. We also estimate, in response to current interest in equation-of-state approaches to polymer–polymer compatibility, that the sulfone and carbonate moieties have thermal-expansion mismatches of about 16% in each of the three series of materials.

The low-temperature (β) transition in polycarbonates and polysulfones has received much attention because of its postulated role in a mechanism of impact-strength enhancement (*27, 31, 32, 33*). However, this low-temperature relaxation appears to be relatively insensitive to annealing, in contrast to the polymer's impact properties, an observation leading to the conclusion that the β-process contributes to, but does not exclusively determine, toughness in bisphenol-A based polymers (*33*). The great breadth of the β-peak has been discussed by a number of workers as arising from the almost superposition of two different loss peaks having different activation energies (*32, 34*).

Table II. Homopolymer Properties

Polymer Repeat Unit	T_g(°C)[a]	Solubility Parameter[b]
Bishenol-A–carbonate	145	9.6
Bisphenol-A–sulfone	190	10.3
Bisphenol-S–sulfone	231	12.6
Bisphenol-T–sulfone	180	10.3

[a] At 3.5 Hz.
[b] Calculated (*26*), however, *see* Ref. *27*.

Results and Discussion

Both solution and interfacial methods of polymerization produced block copolymers having essentially identical mechanical and thermal properties. Below, the dynamic mechanical results are presented, followed by thermal analyses for each of the compositions AS/AC, SS/AC, and TS/AC.

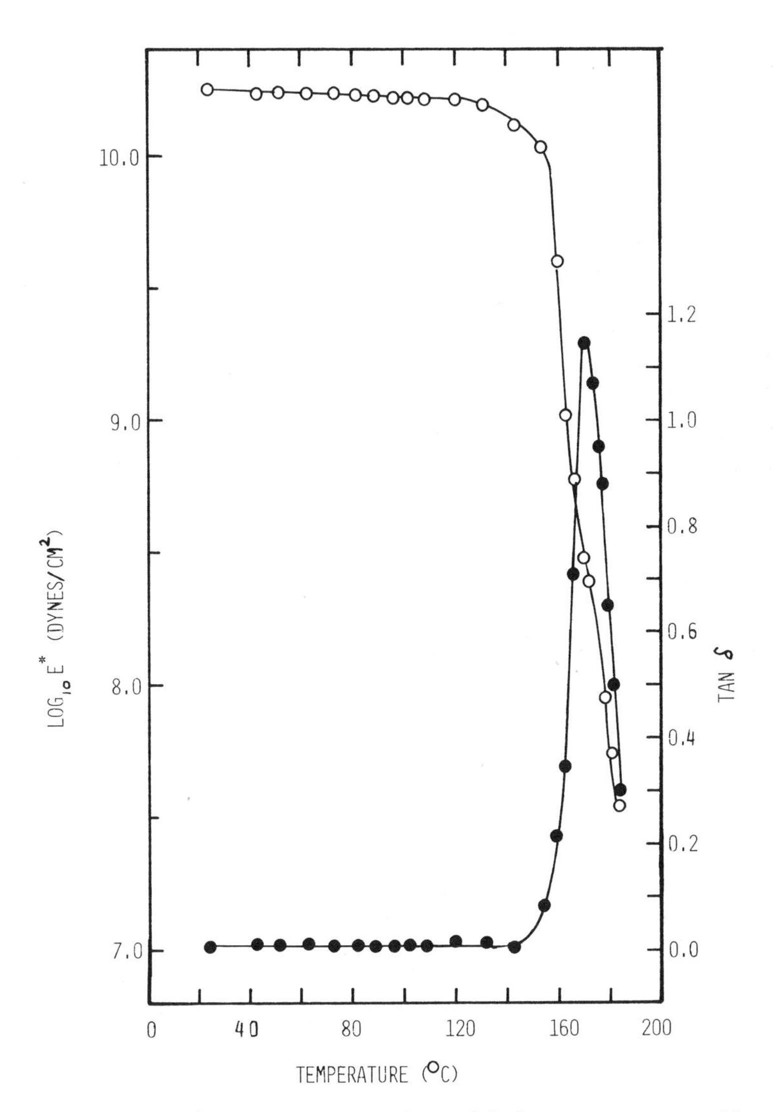

Figure 1. High-temperature mechanical behavior of interfacially prepared bis-A-polysulfone/bis-A-polycarbonate (10,000/10,000) block copolymer

AS/AC Copolymers. It is apparent on comparison of Figures 1 and 2 that a critical average block length for the AS/AC system must lie somewhere between 10,000 and 22,000 g/mol. Figure 1 shows mechanical response typical of a one-phase, homogeneous material having a T_g intermediate between that of the respective homopolymers. We note that AS/AC–5/5 previously was observed to have virtually identical behavior (9). By contrast, two loss-modulus peaks are seen in Figure 2, indicative of a two-phase solid. A closer approach to the molecular weight necessary for phase separation is illustrated in Figure 3, where the twin peaks in the loss modulus appear for copolymer AS/AC–16/17. Parenthetically, the relative magnitudes of these E'' α-loss peaks emphasized

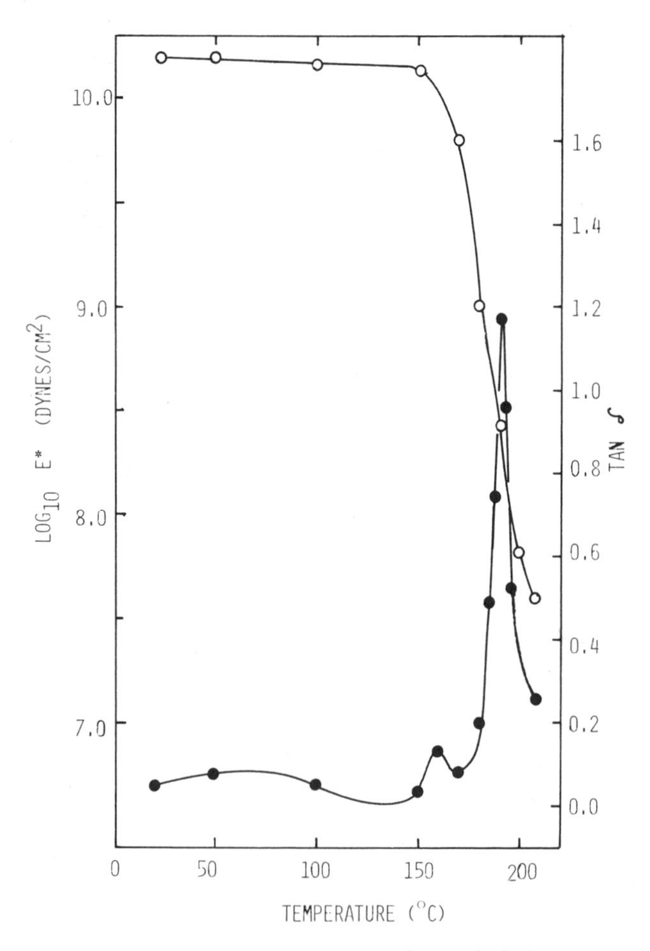

Figure 2. High-temperature mechanical behavior of solution-prepared bis-A-polysulfone/bis-A-polycarbonate (26,000/22,000) block copolymer.

the low-temperature (polycarbonate) T_g; a plot of tan δ vs. temperature (not shown) indicated a much larger area under the high-temperature T_g in contrast.

Also appearing in Figures 2 and 3 is a small intermediate peak in the 0°–100°C range, which is similar to one previously reported (*33, 35*) for the homopolymers and postulated to arise from the relaxation of frozen stresses. Its absence in Figure 1 (and in data reported below on all of the one-phase systems) suggests that this intermediate peak appears because of nonequilibrium strain states originating during the formation of microphases. Future studies will feature this observation.

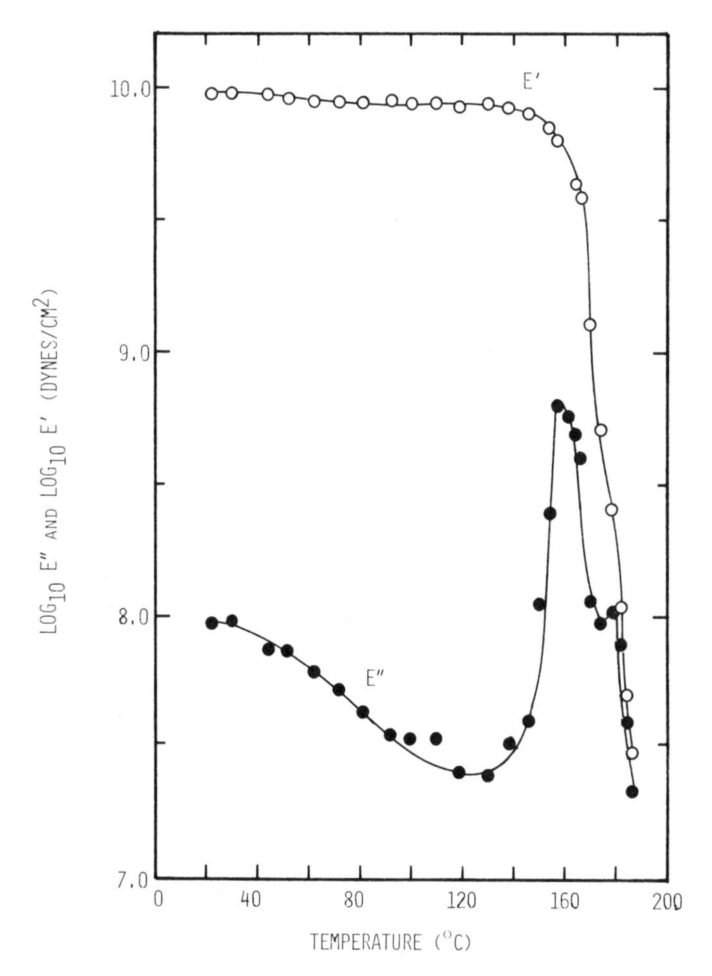

Figure 3. High-temperature mechanical behavior of bis-A-polysulfone/bis-A-polycarbonate (16,000/17,000) block copolymer. Compression molded at 260°C.

Low-temperature mechanical properties of the entire AS/AC series, regardless of molecular weight, are well represented by the data in Figure 4, actually determined for the 10,000–10,000 g/mol copolymer. Very little sample-to-sample variation appeared; all plots revealed a large, broad β-loss-modulus maximum occurring at $-110°C$. The typical asymmetry of this peak also is apparent from this figure. Varying test frequencies indicated an activation energy of approximately 10 kcal/mol associated with the peak maximum. The presence of either one or two phases was not reflected by changes in the β transition; indeed this peak, especially when represented by E'', was practically superimposable with that obtained for homopolymers of bisphenol-A polycarbonate and bisphenol-A polysulfone.

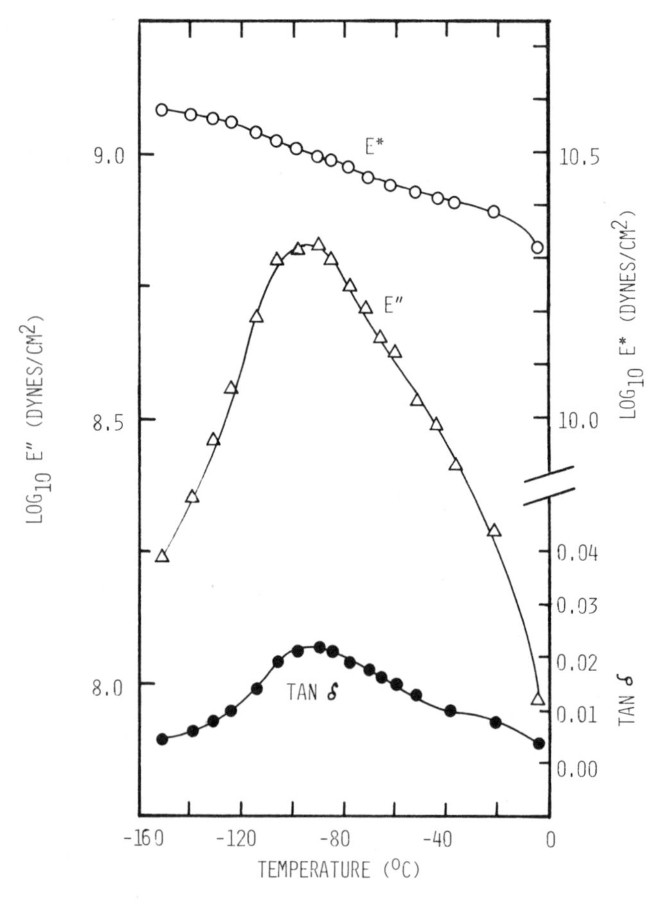

Figure 4. Low-temperature dynamic mechanical response of a typical block copolymer, interfacially prepared bis-A-polysulfone/bis-A-polycarbonate (10,000/10,000)

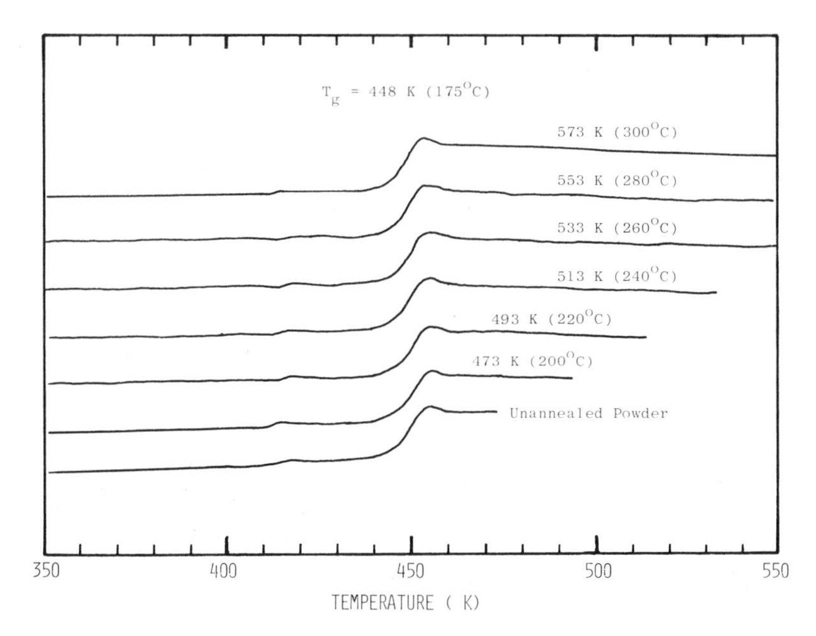

Figure 5. DSC thermograms of bis-A-polysulfone/bis-A-polycarbonate (10,000/10,000) block copolymer after annealing at indicated temperatures for 15 min. Heating rate, 40 K/min. Range, 5 mcal/sec.

General features observed in the mechanical testing were confirmed by DSC. Figure 5 shows results for copolymer AS/AC–10/10. A single glass transition is readily apparent at 175°C. In the vicinity of 147°C, a small apparent T_g appeared in some, but not all, samples. Probably this was a result of some residual homopolymer. Recognizing that in these systems kinetic control of domain formation was a possibility, various annealing experiments were conducted. We observe in Figure 5 that repeated annealing of the AS/AC–10/10 at successively higher temperatures up to 300°C and for 15 minutes in each case followed by a thermal quenching produced no phase separation. In contrast, the typical thermal behavior of AS/AC-series two-phase solids is represented in Figure 6. In the bottom DSC trace (A), the virgin powder was scanned; only one T_g at 175°C was observed. Just beyond this temperature one notes an exotherm at 207°C, undoubtedly a consequence of crystallization in these carbonate domains of the copolymer. In curve B a one-minute annealing at 220°C followed by quenching produced a sample with latent two-phase behavior. More patient thermal-annealing treatments at 220°C led to samples where the two glass transitions were readily detected, as illustrated by curves C and D of Figure 6.

To further explore the time dependence of formation of microphases, DSC studies were conducted in which annealing at different temperatures

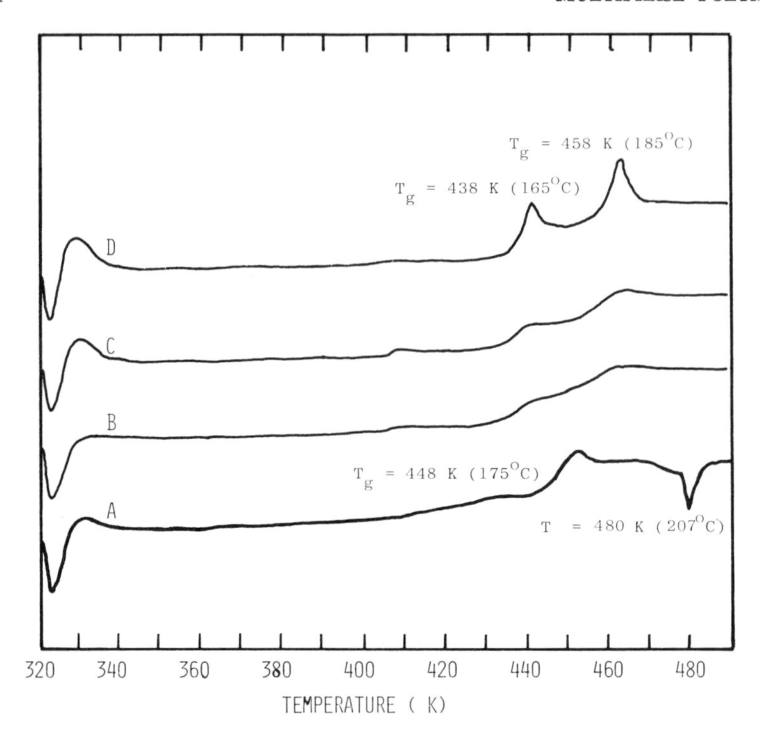

Figure 6. Kinetic effects appearing in DSC thermograms for bis-A-polycarbonate (16,000/17,000) block copolymer. Heating rate, 40 K/min. Range, 5 mcal/sec. (A) Dried polymer powder; no thermal pretreatment. (B) Annealed 1 min at 493 K and quenched to 320 K. (C) Annealed 30 min at 493 K and quenched to 320 K. (D) Annealed 30 min at 493 K and quenched to 320 K. (D) Annealed 30 min at 493 K and cooled to 320 K at 1.25 K/min.

was used. Results are shown in Figure 7 where the virgin-precipitated and dried powder's thermogram is shown as the bottom curve for reference. Then considering the uppermost trace first and moving downward, two T_gs clearly resulting from 15 minutes annealing at 300°C. Further lowering of annealing temperature resulted in two less-well-defined transitions whose locations tend toward an intermediate temperature. The point of convergence to a single T_g suggests here an analogy to the lower critical solution temperature (LCST) often noticed in polymer blends and solutions. We were prevented from actually determining this proposed singularity as experimental temperature was lowered because of onset of the T_g of the polysulfone phase at about 185°C, immobilizing further chain migration. Not shown are data on annealing AS/AC–16/17 at 190°C, in which isotherms of three hours were required for the samples to develop the double baseline shift in the DSC scans. Clearly, the kinetics have intervened here to mask our observation of the proposed LCST. Current experiments are underway to determine the domain formation kinetics and, if possible, the heat of domain formation.

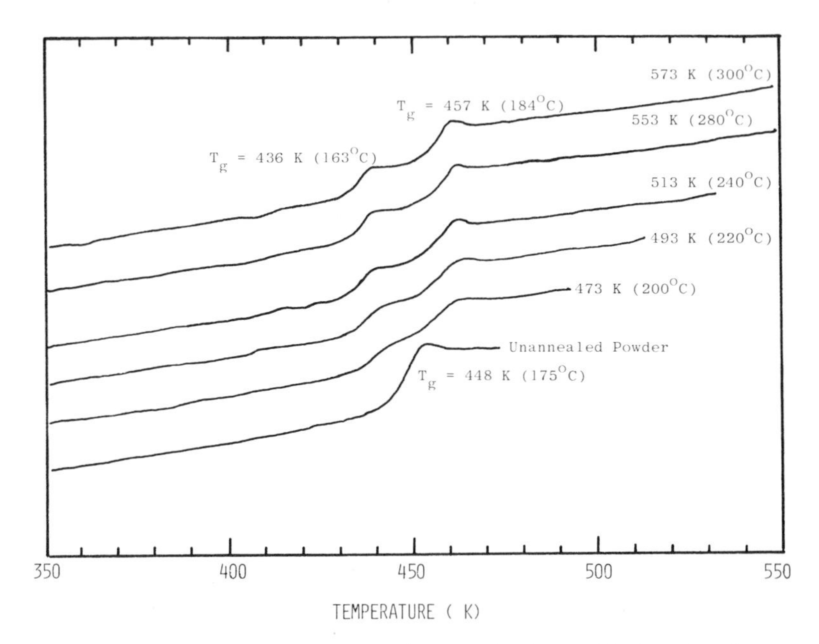

Figure 7. DSC thermograms of bis-A-polysulfone/bis-A-polycarbonate (16,000/17,000) block copolymer after annealing at indicated temperatures for 15 min. Heating rate, 40 K/min. Range, 5 mcal/sec.

In Table I the single-phase block copolymer glass transitions (T_g) or, where appropriate, the low and high observed T_gs in the case of two phases, $(T_g)_L$ and $(T_g)_H$, are presented. Bisphenol-A-polycarbonate homopolymer's T_g is approximately 145°C, some 20°C lower than $(T_g)_L$. However, the $(T_g)_H$ falls quite close to that of homopolysulfone based on bisphenol-A, a point to which we shall later return.

SS/AC Copolymers. This series of block copolymers also was investigated in view of determining a critical block length for microdomain formation. In addition, by now examining the same block lengths as in the AS/AC samples above, the opportunity exists to estimate specific interaction influence on phase separation attributable to the large solubility parameter mismatch between blocks). Both 5000/5000 and 10,000/10,000 average block-length materials were studied mechanically and thermally. Figures 8 and 9 present data only for the longer block-length specimen, but these are representative of the behavior on both samples. Both types of tests indicate two-phase materials. Broad intermediate peaks were observed as before. Again, the polycarbonate moiety T_g is elevated (about 15°C) while that of the polysulfone is essentially unperturbed. Exact values are shown in Table II. While no quantitative conclusions can be drawn, at least the ranges of domination of the two effects, block length and chemical composition, are better defined by comparison of the SS/AC with the AS/AC polymers. Roughly, a one-

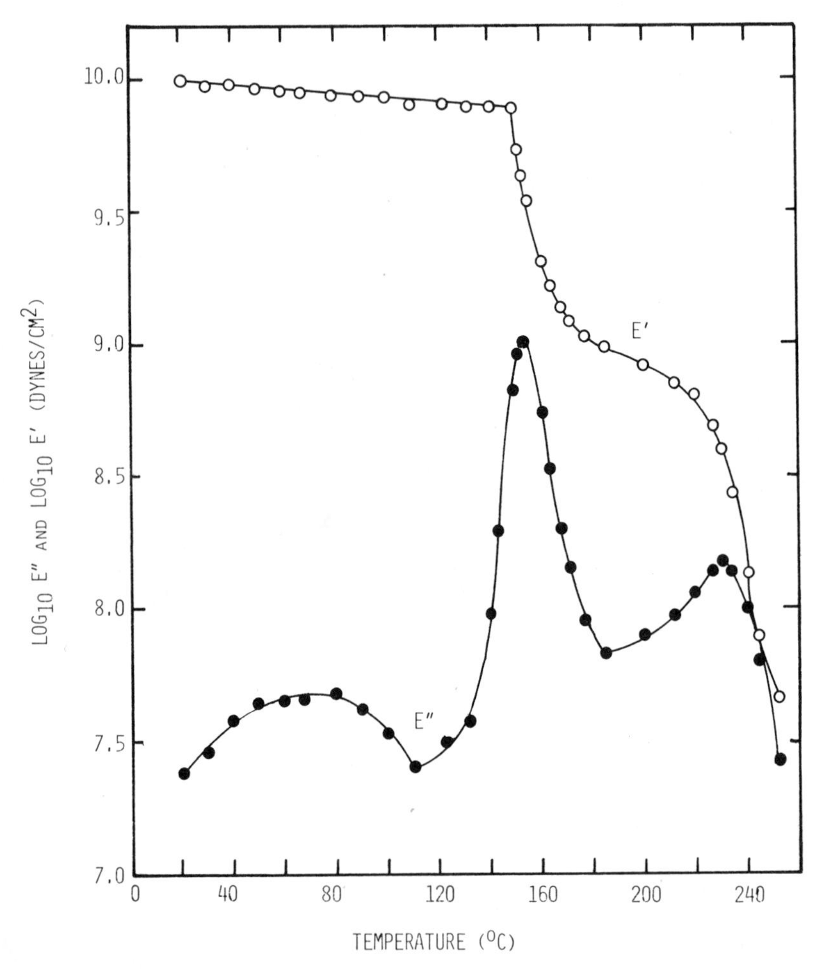

Figure 8. Dynamic mechanical spectrum at higher temperatures for bis-S-polysulfone/bis-A-polycarbonate (10,000/10,000) block copolymer

third increase in molecular weight was required of the AS/AC–10/10 sample to achieve the phase segregation demonstrated by blocks of one-half the length but differing by two solubility parameter units instead of 0.7 unit. Low-temperature mechanical spectra for the SS/AC samples were equivalent to those described above.

TS/AC Copolymers. As a final composition variation within the 50/50 overall weight-percent framework, tests were run on copolymer based on the bisphenol-T-polysulfone oligomer. Recall that this oligomer has a solubility parameter equal to that of the bisphenol-A polysulfone. Only the 10,000–10,000 g/mol block-size material have been studied. The results appear in Figures 10 and 11. Clearly, a single transition is indicated, regardless of annealing procedures. This T_g lies at 159°C, intermediate between the 140° and 180°C homopolymer T_gs, as one would expect from a 50/50 copolymer (*see* Table II). Supporting earlier data, Figure 10 reveals there were no intermediate relaxations for this homogeneous solid.

Conclusions

Two-phase, yet optically transparent, glass–glassy block copolymers can be produced from polycarbonate and polysulfone oligomers either by increasing average block weight beyond 16,000 g/mol or by increasing

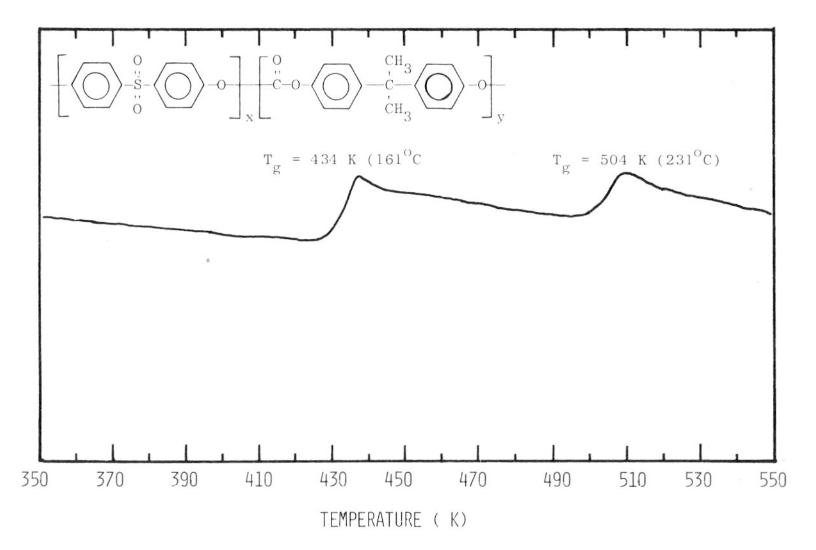

Figure 9. DSC thermogram for bis-S-polysulfone/bis-A-polycarbonate (10,000/10,000) block copolymer. Heating rate, 40 K/min. Range, 5 mcal/sec.

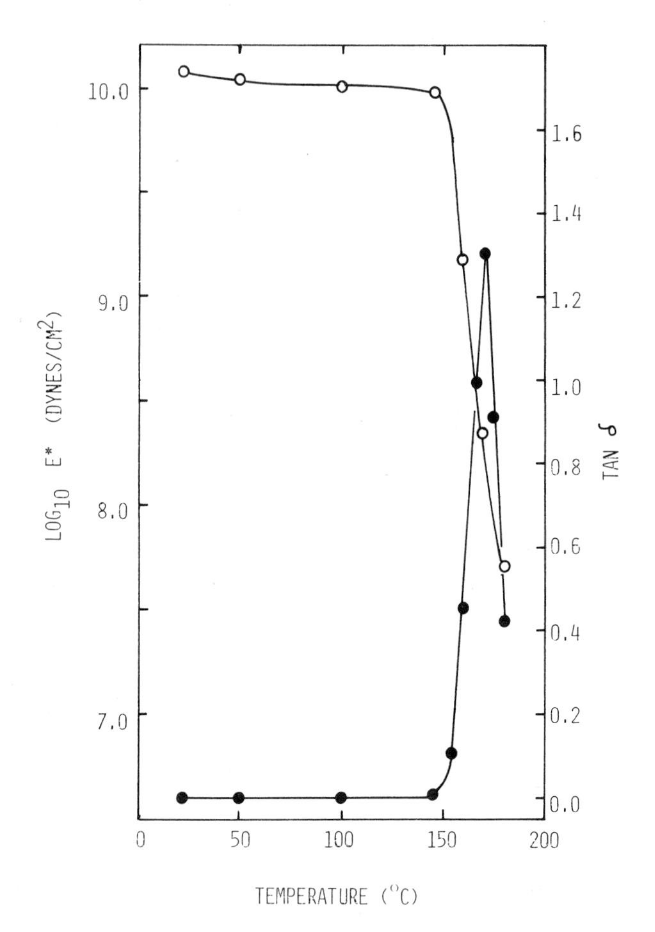

Figure 10. Dynamic mechanical spectrum at higher temperatures for bis-T-polysulfone/bis-A-polycarbonate (10,000/10,000) block copolymer

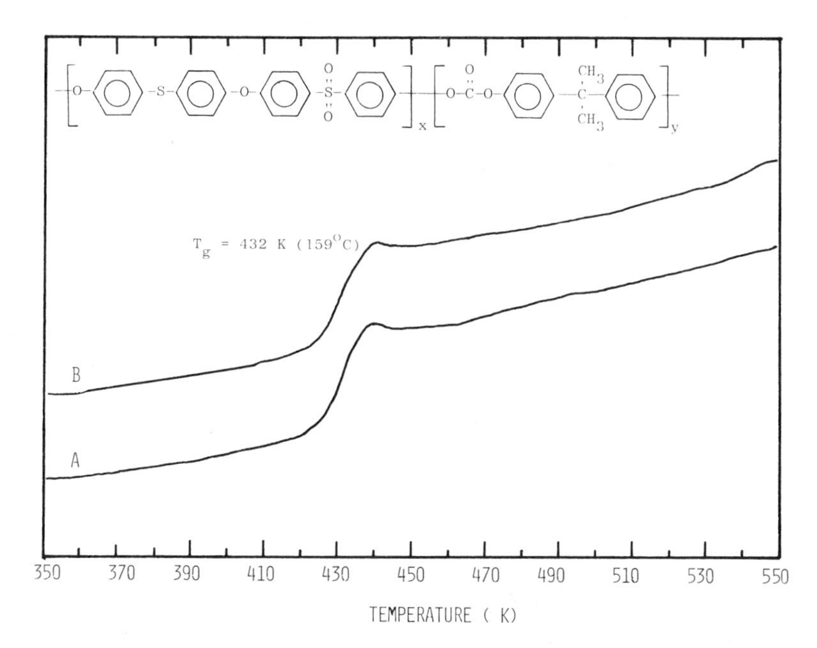

Figure 11. DSC thermogram for bis-T-polysulfone/bis-A-polycarbonate (10,000/10,000) block copolymer. Heating rate, 40 K/min. Range, 5 mcal/sec. (A) After annealing at 573 K (300°C) for 30 min. (B) Dried polymer powder, no thermal pretreatment.

solubility parameter differences by minor compositional variations at about 5000 g/mol block weight. When two phases do result in these materials, the phase having the lower T_g must be heated to a temperature considerably above that of the corresponding homopolymer before showing a glass transition. This is probably attributable to the surrounding glassy matrix in which the lower T_g phase is immobilized; however, it is possible that phase mixing of the two components is responsible. Consistent with the first point of view is the observation that the phase having the higher T_g in the copolymer experiences a glass point, $(T_g)_H$, which is virtually the same as that of the homopolymer of the same composition. Here, there is no rigid matrix but rather a rubbery environment around the microphase as $(T_g)_H$ is approached from below.

Low-temperature mechanical properties in the series of block copolymers having either one or two phases appear essentially unaltered from those of the homopolymers. Intermediate relaxations may appear in these systems when two phases are present.

Acknowledgment

The authors would like to thank the Army Research Office for support of this research under grant DAAG-76-G-0312 and the National Science Foundation Polymer Program for partial support under Grant DMR 76-11963.

Literature Cited

1. Noshay, A., McGrath, J. E., "Block Copolymers: Overview and Critical Survey," Academic, New York, 1977.
2. Matzner, M., Robeson, L. M., Noshay, A., McGrath, J. E., "Block and Graft Copolymers," *Encycl. Polym. Sci. Technol.* (1977) Supplement 2, 129.
3. Allport, D. C., James, W. H., Eds., "Block Copolymers," Halstead, New York, 1972.
4. Ceresa, R. J., Ed., "Block and Graft Copolymerization," Wiley, New York, 1973.
5. Matzner, M., Noshay, A., Schober, D. L., McGrath, J. E., *Ind. Chim. Belge* (1973) **38**, 1104.
6. "Sagamore Conference on Block and Graft Copolymers," Sept. 5–8, 1972. J. J. Burke, V. Weiss, Eds., Syracuse University Press, 1973.
7. Matzner, M., Noshay, A., Robeson, L. M., Merriam, C. N., Barclay, R., Jr., McGrath, J. E., *Appl. Polym. Symp.* (1973) **22**, 143.
8. Matzner, M., Noshay, A., McGrath, J. E., *Trans. Soc. Rheol.* (1977) **21**(2), 272.
9. McGrath, J. E., Robeson, L. M., Matzner, M., Barclay, R., Jr., *Midwest Regional ACS Meeting, 8th, Akron, Ohio, 1976, J. Polym. Sci.*, in press.
10. McGrath, J. E., Ward, T. C., Shchori, E., Wnuk, A. J., *Polym. Eng. Sci.* (1977) **17**(8), 647.
11. Ryan, J. T., *Polym. Eng. Sci.* (1978) **18**(4), 264.
12. Cornes, P. L., Smith, K., Haward, R. N., *J. Polym. Sci., Polym. Lett. Ed.* (1977) **15**, 955.
13. Haward, R. N., Ed., "The Physics of Glassy Polymers," Halstead, London, 1973.
14. Robeson, L. M., Farnham, A. G., McGrath, J. E., *Appl. Polym. Symp.* (1975) **26**, 373.
15. Rebenfeld, L., Makarewicz, P. J., Weigmann, H. D., Wilkes, G. L., *J. Macromol. Sci., Rev. Macromol. Chem.* (1976) **C15**(2), 279.
17. Morgan, P. W., "Condensation Polymers by Interfacial and Solution Methods," Interscience, New York, 1965.
18. Morgan, P. W., *Macromolecules* (1970) **3**, 536.
19. Dehmlow, E. V., "Advances in Phase-Transfer Catalysis," *Angew. Chem., Int. Ed. Engl.* (1977) **16**(8), 493.
20. Shchori, E., McGrath, J. E., *Polym. Prepr., Am. Chem. Soc., Div. Polym. Chem.* (1978) **19**(1), 494–505.
21. Shchori, E., McGrath, J. E., "Polymer Analysis," T. K. Wu, J. Mitchell, Eds., Wiley, in press.
22. Wnuk, A. J., Davidson, T. F., McGrath, J. E., *Polym. Prepr., Am. Chem. Soc., Div. Polym. Chem.* (1978) **19**(1), 506.
23. Wnuk, A. J., Davidson, T. F., McGrath, J. E., "Polymer Analysis," T. K. Wu, J. Mitchell, Eds., Wiley, in press.
24. Johnson, R. N., Farnham, A. G., Clendinning, R. A., Hale, W. F., Merriam, C. N., *J. Polym. Sci., Polym. Chem. Ed.* (1967) **5**, 2399.

25. Millich, F., Carraher, C. E., Jr., Eds., "Interfacial Synthesis, Volume I, Fundamentals," Dekker, New York, 1977.
26. Millich, F., Carraher, C. E., Jr., "Interfacial Synthesis, Volume II, Polymer Applications and Technology," Dekker, New York, 1977.
27. Kurz, J. E., Woodbrey, J. C., Ohta, M., *J. Polym. Sci., Polym. Phys. Ed.* (1970) **8**, 1169.
28. McCrum, N. G., Read, B. E., Williams, G., "Anelastic and Dielectric Effects in Polymeric Solids," Wiley, New York, 1967.
29. Van Krevelen, D. W., "Properties of Polymers," Elsevier, Amsterdam, 1976.
30. Shaw, M. T., *J. Appl. Polym. Sci.* (1974) **18**, 449.
31. Nielsen, L. E., "Mechanical Properties of Polymers," Reinhold, London, 1962.
32. Locati, G., Tobolsky, A. V., *Adv. Mol. Relaxation Processes* (1970) **1**, 375.
33. Watts, D. C., Perry, E. P., *Polymer* (1978) **19**, 248.
34. Aoki, Y., Brittain, J. O., *J. Polym. Sci., Polym. Phys. Ed.* (1976) **14**, 1297.
35. Illers, K. H., Breuer, H., *Kolloid-Z.* (1961) **176**, 110.

RECEIVED June 5, 1978.

BLENDS

17

A Brief Review of Polymer Blend Technology

D. R. PAUL and J. W. BARLOW

Department of Chemical Engineering, The University of Texas at Austin, Austin, TX 78712

Thermodynamically miscible, single amorphous-phase multicomponent polymer blends with good physical properties are rapidly being discovered as researchers become aware that miscible polymer pairs require chemical structures which form strong specific interactions. Thermodynamically immiscible, multiphase blends continue to be developed as a variety of methods, including use of interfacial agents, morphology control, and interpenetrating network formation, are used to improve physical properties by enhancing interphase stress transfer. Recent commercial applications of both types of blends are given and their economic and property advantages are discussed.

The concept of appropriately combining together two or more different polymers to obtain a new material system with the desirable features of its constituents is not new. Over the years, numerous systems based on the chemical combination of different monomers through random, block, and graft copolymerization methods have been developed with this goal in mind. For similar reasons the coating and rubber industries have long blended together different low molecular weight polymers; and particularly over the last decade, the interest in polymer blend systems as a way to meet new market applications with minimum development cost has increased rapidly.

It is the purpose of this chapter to briefly examine the present state of the art of polymer blends from both a technical and commercial standpoint. Because of the large variety of blend systems currently being investigated and used commercially, it will not be possible to present an in-depth discussion of specific results. Instead, the first two sections of this chapter will concentrate on the key concepts and fundamental

0-8412-0457-8/79/33-176-315$05.25/0

considerations associated with polymer blends in order to provide some technical perspective for the specific technical discussions in later chapters. The last section of this chapter discusses several specific examples of commercial blend systems and serves to emphasize the property advantages inherent in properly formulated polymer blend systems.

Miscible and Immiscible Polymer Blends

Over the last decade, the poor economics of new polymer and copolymer production and the need for new materials whose performance/cost ratios can be closely matched to specific applications have forced polymer researchers to seriously consider purely physical polymer blend systems. This approach has been comparatively slow to develop, however, because most physical blends of different high molecular weight polymers prove to be immiscible. That is, when mixed together, the blend components are likely to separate into phases containing predominantly their own kind. This characteristic, combined with the often low physical attraction forces across the immiscible phase boundaries, usually causes immiscible blend systems to have poorer mechanical properties than could be achieved by the copolymerization route. Despite this difficulty a number of physical blend systems have been commercialized, and some of these are discussed in a later section. Also, the level of technical activity in the physical blend area remains high, as indicated by the number of reviews published recently (*1–10*).

The basic issue confronting the designer of polymer blend systems is how to guarantee good stress transfer between the components of the multicomponent system. Only in this way can the component's physical properties be efficiently used to give blends with the desired properties. One approach is to find blend systems that form miscible amorphous phases. In polyblends of this type, the various components have the thermodynamic potential for being mixed at the molecular level and the interactions between unlike components are quite strong. Since these systems form only one miscible amorphous phase, interphase stress transfer is not an issue and the physical properties of miscible blends approach and frequently exceed those expected for a random copolymer comprised of the same chemical constituents.

Although the number of miscible blends is rapidly being increased, immiscibility is generally the result when unlike polymers are mixed. Consequently, a great deal of research has been and is being done on ways to improve the mechanical properties of immiscible blends. A widely practiced approach at the present time is to connect the minor dispersed phase to the major continuous phase through a covalent bond. This approach can take several forms. The oldest and most basic is to chemically link one phase to the other as is done in the manufacture of rubber-modified impact grades of polystyrene (*11, 12, 13*). Alternately, block or graft copolymers of the form A–B can be used as compatibilizers (*see* chapter 12 of Ref. *10* by D. R. Paul) or interfacial agents to improve adhesion between immiscible A-rich and B-rich phases. The physical affinity of the A portion of the copolymer for the A phase and the B portion for the B phase serves to locate the copolymer at the interface and to connect physically the two phases through the covalent bonds in the copolymer backbone. The net result of this improved adhesion in a variety of systems is a great improvement in the ultimate mechanical properties, elongation, and tensile strength, and a finer dispersion of the minor component (*14–19*).

The tendency of immiscible blocks to segregate into domains or phases comprised of like or miscible species is incidentally used to good advantage to form an important class of thermoplastic rubbers. These materials are comprised of A–B–A tri-block sequences in which the B blocks are elastomeric and the A blocks are either glassy or crystalline at room temperature and immiscible with the B blocks. The "hard" domains comprised of A blocks anchor the rubber blocks at their ends to form thermally reversible equivalents of crosslinks. Styrene–butadiene–styrene block copolymers as well as the segmented polyurethanes and related polyesters form commercial examples of this class. A variety of mechanical properties can be obtained by blending these materials with others. Provided the thermoplastic elastomer is retained as the continuous phase, the blend will show rubbery behavior in proportion to the amount of rubber in the blend. Should it become discontinuous, the blend properties become controlled by the major continuous phase and the degree of miscibility between the hard blocks and the third component (*see* chapter 20 of Ref. *10* by E. N. Kresge).

Another method to improve the mechanical properties of immiscible blends is to co-mingle the immiscible phases in such a way that each phase remains continuously connected throughout the bulk of the blend. The idea behind this approach is to minimize the importance of the interphase adhesion by developing co-continuous or interpenetrating phase morphologies capable of direct stress transfer. By analogy, with the improved mechanical behavior of continuous fiber-reinforced plastics relative to systems containing discontinuous reinforcement (*20*), immiscible blends with interpenetrating phases show improved mechanical properties relative to the usual dispersed phase-continuous phase mixtures (*see* chapters 1 and 16 of Ref. *10*). Although in its infancy, phase morphology control without chemical reaction to form interpenetrating phases would seem to be especially appropriate for preparing immiscible thermoplastic blends with improved properties. By judicious choice of binary component concentrations and viscosities it is possible to achieve co-continuous phases in the melt because the phase in higher concentration tends to be continuous when both phases have similar viscosities, whereas at similar concentrations the phase with the lower viscosity tends to be continuous. Kresge has demonstrated co-continuous phase formation in blends of polyethylene or polypropylene with ethylene–propylene copolymers when the copolymers are present at high concentrations and have higher viscosities by a factor 5–10. He further suggests that both polyolefins, when separately blended with styrene–butadiene–styrene block copolymers, form interpenetrating phases.

A chemical method of combining polymeric materials is to form what are called interpenetrating networks (IPN). The conventional method of synthesizing an IPN involves swelling a cross-linked primary polymer network with a solution of monomer and cross-linking agent, followed by in situ polymerization of the second network (*22, 23*). Ideally what results is an extended primary network intermingled with but not reacted to a co-continuous secondary network to give a quasi-single-phase material. When properly formed, IPNs have been shown by Frisch et al. (*5, 24, 25, 26*) to have many of the attributes of miscible physical mixtures, including single glass transition temperatures, maximum tensile strengths at intermediate compositions, and increased densities relative to the usual phase-separated mixtures. The ideal phase topology is difficult to obtain, however, and some phase segregation of the second network often occurs (*27*). A great deal of work is presently being done to elucidate the chemical and physical parameters necessary for single-phase formation in these systems and to determine the relationships between these parameters and the resulting physical properties.

Thermodynamic Considerations

Leaving aside for the moment the relative advantages of immiscible vs. miscible blend systems, it is clear from the brief review above that the blend properties are strongly dependent on their phase structures and on the adhesion between phases. The presence and composition of phases as well as the surface energy of interaction between phases are, in principle, functions of the thermodynamics of interaction between the polymer components of the blend. Consequently, there is a need to be able to predict this interaction.

The theory for predicting polymer–polymer miscibility is not well developed at this time, but a brief and incomplete survey of experimental observations on a variety of systems suggests that many miscible systems exist (*28–51*). The existence of miscible-phase behavior depends on the chemical structures of the blend components (*21, 44, 46*), including copolymer compositions where used as a blend component (*30, 37, 38, 42–44, 47*), the molecular weights of the components (*48, 49*), the tacticity of the components (*50, 51*), the observation temperature (*38, 40, 44, 52*), and the method of blend fabrication (*47, 53, 54*). In this section, we will attempt to bring some of these observations together with a qualitatively useful, if quantitatively incomplete, thermodynamic overview.

Basic thermodynamics suggests that spontaneous solution to give a miscible mixture occurs whenever the free energy of mixing is negative, as given by

$$\Delta G_{\mathrm{mix}} = \Delta H_{\mathrm{mix}} - T\Delta S_{\mathrm{mix}} \tag{1}$$

where ΔH_{mix} is the enthalpy of mixing and ΔS_{mix} is the entropy of mixing. According to Scott (*55*) who applied the Flory–Huggins equation (*56*) to mixtures of dissimilar polymers, the enthalpy and entropy of mixing of polymers 1 and 2 are given by

$$\Delta H_{\mathrm{mix}} = BV\Phi_1\Phi_2 \tag{2}$$

$$\Delta S_{\mathrm{mix}} = \frac{-RV}{V_{\mathrm{r}}}\left(\frac{\Phi_1\ln\Phi_1}{X_1} + \frac{\Phi_2\ln\Phi_2}{X_2}\right) \tag{3}$$

where V_{r} is the reference volume per monomer repeat unit, V is the volume of the system, X_i is the degree of polymerization of species i, Φ_i is the volume fraction of i in the binary, and B is the interaction energy density characteristic of the polymer–polymer segmental interactions in the blend.

It is clear from these equations that the entropy of mixing is a function of the molecular sizes being mixed, decreasing rapidly toward zero as the degrees of polymerization of the components approach the values typically found in commercial polymers. The enthalpy of mixing, on the other hand, primarily depends on the energy change associated with changes in nearest neighbor contacts during mixing (*57*) and is much less dependent on molecular lengths. The net result of these considerations is that the free energy of mixing is primarily influenced by the sign and magnitude of ΔH_{mix} for high molecular weight mixtures.

Since ΔH_{mix} could be predicted for weakly interacting materials via

$$\Delta H_{\mathrm{mix}} = V(\delta_1 - \delta_2)^2\,\Phi_1\Phi_2 \tag{4}$$

to be positive and dependent on the difference in pure component solubility parameters, δ_i (*57*), early polymer blend work was concentrated on closely matching component solubility parameters to achieve miscibility. Despite considerable work (*58*), this approach has not been particularly successful, partly because of the lack of sufficiently accurate solubility parameters and partly because of the theoretical requirement that the parameters be matched to within $0.1(\text{cal/cm}^3)^{1/2}$ or less. As a result, most early workers concluded that formation of miscible polymer blends was a highly unlikely event.

By rough count, more than 50 miscible systems have been reported in the literature, and from intensive studies of these systems a more optimistic view of the potential for miscibility is beginning to develop. Most of these miscible polymer pairs have chemical structures that are capable of forming strong specific interactions such as donor–acceptor complexes and hydrogen bonds. While they have not all been completely studied, those that have show behavior which can only result from the presence of exothermic or negative heats of mixing.

For example, Olabisi (*59*) recently proposed the possibility of charge-transfer interactions between poly(vinyl chloride) and the ester oxygen groups on poly(ϵ-caprolactone) to explain the negative interaction parameter estimated from solvent probes of the miscible blends in the inverse gas chromatography technique and from spectroscopic measurements. Matzner et al. (*60*) also have proposed that the specific interaction between the basic disubstituted amide on copolymers of ethylene and *N,N*-dimethyl acrylamide and the α hydrogen on PVC is responsible for the observed miscibility. Kwei et al. (*52*) found that the interaction parameter for polystyrene–poly(vinyl methyl ether) blends was likewise negative in the range 35–65% polystyrene. Nishi and Wang (*61*) showed that the heat of mixing parameter in classical theory of melting-point-depression miscible was negative when that theory was applied to miscible blends of poly(vinylidene fluoride), PVF_2, and poly(methyl methacrylate), and, following that lead, workers at The University of Texas have demonstrated that the interaction parameter, B, is similarly negative for miscible PVF_2 blends with poly(ethyl methacrylate), poly(methyl acrylate), poly(ethyl acrylate), and poly(vinyl methyl ketone) (*46, 62, 63, 64, 65*). What is particularly encouraging about this latter work is that it is possible to relate the magnitude of the interaction parameter to the availability of the carbonyl for interaction with the PVF_2 segment. This rank order also agrees well with that estimated from the literature values of solution dipole moments of the amorphous polymers, a result that suggests that strong dipolar interactions can be responsible for the observed miscibility.

Whatever the reason, it is clear that systems with negative heats of mixing have a good chance of being miscible and that the simple solubility parameter approach, embodied in Equation 4, cannot be used to describe the solution thermodynamics of these systems. Blanks and Prausnitz (*66*) suggest a scheme for characterizing polar interactions between polymers and solvents which could prove useful when extended to polymer–polymer systems. Even their approach, however, is not intended to estimate strong specific interactions. Considering the void that presently exists in our ability to predict negative heats of mixing, the best approach for estimating the potential for miscibility of polymer pairs may be simply to measure the heats of mixing of low molecular weight analogs. Should the analogs show negative heats of mixing, the polymers could be expected to have a high probability of being miscible. Initial work in this direction at The University of Texas appears promising, and if successful, it should be possible to use the large body of calorimetric data on low molecular weight mixtures that currently exists as a qualitative guide for formulating miscible polymer blends.

A complete quantitative description of the thermodynamics of polymer–polymer solutions also might need to include the effects of polymer tacticity. As demonstrated recently by Schurer et al. (*50*), changing the stereo configuration of poly(methyl methacrylate) from isotactic to syndiotactic causes it to become miscible with PVC. These results suggest the importance of the spatial articulation of interacting segments in the polymer.

One of the more interesting aspects of miscible polymer blend studies is the finding that many miscible systems show cloud points on heating, which signal the existence of a Lower Critical Solution Temperature, LCST (*38, 40, 46, 62–65*). The observation of an LCST establishes a liquid–liquid phase diagram which has unambiguous thermodynamic significance. In addition, comparison of the temperature locations of LCST behavior for various blends has the potential for providing a quantitative scale for ranking the specific interactions associated with miscibility. With care, a study of LCST behavior could provide a basis for rigorous thermodynamic analysis.

It is particularly interesting that many of the systems which show LCST behavior also show evidence of negative heats of mixing below the LCST. In this regard, polymer blends appear to be following the same behavior seen for low molecular weight binaries. Rowlinson (*67*), for example, observes that low molecular weight binaries which show LCST behavior invariably have negative heats of mixing, negative volumes of mixing, and positive excess heat capacities as a result of the strong solute–solvent bonds.

Further, data of Nishi and Kwei (*68*) show that the LCST for the polystyrene–poly(vinyl methyl ether) system is constant to within 10°C when the polystyrene $\overline{M}_w$ lies between 50,000 and 1,000,000. This result again suggests that entropic contributions to the phase transition are of secondary importance when the component molecular weights are high and that the phase instability at LCST is governed by enthalpic considerations. Using this argument, Paul et al. (*46*) were able to demonstrate that the cloud point temperature was directly proportional to the magnitude of the negative heat of mixing observed from melting point depression data of the PVF_2 for the series of blends with PVF_2 noted above.

Obviously, much remains to be done, both experimentally and theoretically, before we will be able to predict polymer–polymer miscibility with any degree of certainty. The situation is far from discouraging, however, inasmuch as an experimental pattern is developing which strongly suggests the use of specific interactions to obtain miscible systems. Continuation of this work and the theoretical reformulations which will surely follow ultimately should provide the tools required for formulation of new and improved polymer blends.

Commercial Polymer Blends

Previous sections have dealt with some of the fundamental issues of the technology of polymer blends, and it should be quite clear that there are many important questions which remain unanswered. Despite this lack of fundamental guidance, there has been a strong effort to develop commercially attractive products from polymer blends, and a considerable number of these products are on the market today. In this section, we will give a brief overview of the status of this commercial practice; however, it will be useful to first give some rationale for this commercial interest in the concept of polymer blending.

Why Blend? All new materials attract interest on the basis of their property–processing–cost performance. With regard to properties, polymer blends can be expected to exhibit any of the following three possibilities for a given property.

Perhaps the most commonly expected property vs. composition relationship is the concept of additivity. By this we mean that when polymer A is mixed with polymer B, we can expect the blend to have a property which is a weighted average of that property for pure A and pure B. In a general sense, we need not be concerned here with the exact form of the weighting function involved. Such averaging of properties has many potential benefits for formulating a new material. For example, polymer A may have a modulus which is too high for a particular application, whereas polymer B may have a modulus which is too low for this application. Thus, if blends of A and B have moduli that are intermediate to those of pure A and pure B, then the possibility exists for formulating a blend which meets the precise modulus requirements of the particular application. Since unoriented glassy polymers or wholly crystalline polymers have moduli of the order of 10^{10} dyn/cm^2, whereas rubbery amorphous polymers have moduli of the order of 10^7 dyn/cm^2, it is clear that intermediate values of modulus can only be had through two-phase composites of these extreme possibilities. Semicrystalline polymers are a good example of such two-phase composites which allow intermediate values of modulus that depend on the degree of crystallinity and morphology. Similarly, blends of rigid and soft materials might pose another way of achieving intermediate modulus levels, just as block and graft copolymers of similarly constructed phases would also allow. Generally, blends represent a less expensive route than block or graft copolymers because of the less involved chemistry. Modulus is a property that can be expected to obey some additivity relationship for blends where the weighting functions of composition will be sensitive to phase morphology.

In many cases, important properties like strength and toughness do not follow an additive relationship, and in fact, these properties are less than predictions based on additivity. Often it is found that these properties exhibit a minimum when plotted vs. blend proportion such that many mid-range blends have strengths or toughnesses that are less than either pure component (*69*). This situation usually arises from a poor degree of interfacial adhesion between components that provides a multiplicity of defects for early failure. This situation is the real dilemma that prohibits a more general use of blending two polymers. Such behavior is often cause for terming two polymers as incompatible. Often the term compatibility is used synonomously with the thermodynamic concept of miscibility; however, we prefer to make a distinction between these two terms as explained earlier.

A very intriguing possibility, although less frequently observed, is when blends of polymer A and B show synergism with respect to some property. Here, we refer to the situation where some property, such as tensile strength, for the blend is larger than that for either pure A or pure B. Clearly, such a maxima in the plot of property vs. composition is not predicted by any additivity relationship. A few examples of this kind of behavior are known and offer unique possibilities when they exist. Two examples of synergism of tensile strength will be mentioned later.

In the above discussion, we were concerned with how a single property depends upon blend composition. However, the success of a new material depends on more than one such factor. In this context, we may identify two general situations which provide an impetus for blending polymer A and B for commercial applications. The first of these is an advantageous combination of properties. For example, polymer A may have a very desirable high thermal resistance but its processing characteristics may be very poor. On the other hand, polymer B may have very good processability but poor thermal resistance. Therefore, it can be of interest to attempt to combine the high thermal resistance of polymer A with the good processability of polymer B although there will generally be a compromise involved in this attempt to combine the desirable characteristics of each polymer into one material. Specifically, the blend can be expected to have poorer thermal resistance than polymer A and not to process as easily as polymer B; but, nevertheless, the blend may have adequate combinations of the desirable attributes for certain applications that neither pure A nor pure B alone could meet. Clearly, the exact relationship between properties, in this example—thermal resistance and processability, will be critical to the success of the blend.

A second general reason for interest in blending for commercial products is cost dilution. In this situation, polymer A may have excellent properties, and, in fact, greater levels of some properties than are needed for certain applications. However, the price of polymer A can prohibit its use in some applications. Dilution of polymer A with a cheaper polymer B can reduce the properties to a level still acceptable for the particular application but can bring the price of the blend to within a range where it can be competitive in this market. Thus in these situations, blending can be an attractive means to engineer a material so that the user does not have to pay for more than he needs. This is one of the stronger driving forces for developing blend products.

Examples of Commercial Blends. In this subsection we will review some of the commercial activity in polymer blends. We find it interesting and informative to categorize examples into specific areas that relate to both technical issues associated with these mixtures, such as miscibility or crystallinity, and the intended commercial applications, such as rubbers or fibers. Other schemes of classification could be used, and the present one is not intended to be exhaustive. Likewise, there is no intent to mention all of the commercially interesting polymer blends, but rather, the present purpose is to illustrate some of the possibilities. Information about the examples used here was obtained from product literature supplied by the companies who sell these blends and from various literature references that have attempted to review commercial developments in polymer blends (*70–76*).

MISCIBLE BLENDS. *Both Components Amorphous.* Certainly one of the most commercially important and publicized examples of a miscible polymer blend system is that based on polystyrene and poly(phenylene oxide), which is sold under the trade name Noryl by General Electric. Many fundamental studies of this system have been published, many of which were devoted to proving that these two components are miscible in a thermodynamic sense (*see* chapter 5 of Ref. *10* by MacKnight, Karasz, and Fried). Commercial interest in this system involves both the concepts of property combination and cost dilution. Poly(phenylene oxide), PPO, is well known for its high thermal stability but it is quite difficult to process. On the other hand, polystyrene is easily processed but has poorer thermal resistance. PPO when mixed with polystyrene yields a blend which can be processed more easily but has a reduced thermal resistance compared with that of pure PPO. The degree of reduction in thermal resistance depends on the amount of polystyrene introduced and thus permits various grades of different thermal resistance to be made. As a further benefit, addition of the cheaper polystyrene results in a dilution of the cost of the more expensive poly(phenylene oxide). From a technical point of view, it is interesting to note that this system exhibits a synergism of tensile strength (*77*). Blends are stronger than either pure polystyrene or pure poly(phenylene oxide), and it is expected that future fundamental work will be devoted to elucidating the mechanism for this behavior.

The scientific literature contains many references to other miscible blends of wholly amorphous components. One example is the system based on poly(methyl methacrylate), PMMA, and certain styrene–acrylonitrile copolymers, SAN, for which there may be some commercial interest and possibilities (*71, 73*).

One Component Crystalline. Here, we are concerned with miscible mixtures of two polymers of which one component can partially crystallize as a pure phase. Many examples of this are known (*46, 62–65, 78, 79–81*, and *see* chapter 22 of Ref. *10* by Koleske). The extent to which crystallization occurs will depend on many circumstances of processing and thermal history. The remaining amorphous phase will consist of a homogeneous mixture of the two components. Over the last several years, it has been shown that poly(vinylidene fluoride), PVF_2, is miscible with a variety of oxygen-containing polymers typically of the acrylic type. Pennwalt has commercialized various blends based on PVF_2 and PMMA. These blends, or alloys, have chemical and solvent resistance that is reduced from that of pure PVF_2 but that is greatly improved over that of acrylates, especially towards polar solvents. These blends are expected to find application where the high cost and performance of the PVF_2 is not needed but where PMMA is deficient. These blends offer combinations of transparency, toughness, weatherability, and self-extinguishing behavior. Rexham also has introduced blends of PVF_2 and an unspecified acrylic polymer that are trade-named Flourex. Some of these materials have found applications in the automotive area as decorative striping and labeling where clarity, chemical resistance, and adhesion to acrylic paint are requirements. The PVF_2 in these blends may or may not crystallize, depending on its proportion in the blend and thermal history.

There is a large body of patent literature and a growing amount of scientific literature on blends of polycarbonate with various crystallizable polyesters. The latter would include poly(ethylene terephthalate), poly-(butylene terephthalate), polycaprolactone, and certain copolyesters derived from mixtures of terephthalic acid and isophthalic acid co-reacted with 1,4-cyclohexanedimethanol (*79, 80, 81, 82*). As shown recently, some of these mixtures form miscible blends although the polyester possesses the possibility of crystallizing. The number of patents on such systems indicates a degree of commercial interest.

One Component Acts as a Plasticizer. Considerable quantities of poly(vinyl chloride), PVC, and nitrocellulose are used in a flexible form created by the addition of plasticizers. Conventionally, plasticizers are relatively low molecular weight liquids which are miscible with the polymer in question and, thus, cause a lowering of the glass transition temperature to make a rubbery-like product. It is commonly believed that this rubbery state is stabilized by physical cross-links resulting from a small amount of residual crystallinity. The low molecular weight nature of these plasticizers creates certain problems associated with their lack of permanence. This is evidenced by the undesirable loss of plasticizer which eventually may stiffen the polymer and create adverse side effects to the environment. For example, a frequent problem of vinyl seat coverings in automobiles is the evaporation of plasticizer at high temperatures and subsequent condensation on cooler glass surfaces to produce an obnoxious oily film. Similarly, low molecular weight plasticizers can be leached from the polymer when contacted by liquids such as water with resulting pollution and contamination. This has led to an interest in replacing the low molecular weight plasticizers with polymeric components that will serve the same function but will not be leached or will not evaporate from the mixture. A number of polymeric plasticizers, primarily for PVC and to a lesser extent nitrocellulose, have been identified and commercialized. Some of these include certain copolymers of ethylene and vinyl acetate; certain terpolymers of ethylene, vinyl acetate, and SO_2; certain terpolymers of ethylene, vinyl acetate, and carbon monoxide; certain copolymers of acrylonitrile and butadiene (nitrile rubber); and polycaprolactone. Hammer (*see* chapter 17 of Ref. *10*) has reviewed the use of polymeric plasticizers and has pointed out the advantages and disadvantages of using this type of polymer blend to replace conventionally plasticized polymer products.

IMMISCIBLE BLENDS. *Rubber.* Elastomer/elastomer blends are used extensively for commercial applications, particularly in the construction of automobile tires. There is an extensive patent and technological literature on this subject. A recent review (*see* chapter 19 of Ref. *19* by McDonel, Baranwal, and Andries) summarizes a great deal of this information. There seems to be no question that most elastomers are incompatible with one another and that the reasons for blending are primarily to achieve combinations of properties not possessed by any single component. For example, ethylene–propylene-based elastomers are frequently blended with diene-based elastomers to achieve a measure of oxidation or ozone resistance while retaining cost and property advantages of the latter. Other advantages are noted but these seem to be poorly understood presently. There are at least two fundamental technological problems associated with producing useful blends of elastomers. The first is to create the desired phase morphology through mixing and processing steps while the second is to develop a chemical method for proper covulcanization of the two phases. In a poorly designed system, the curatives can largely migrate into one phase, resulting in its being overcured while the other phase is undercured. Balanced vulcanization of the two phases can be achieved by proper attention to solubilities of the curative components in the two phases and the relative rates of vulcanization of the two elastomers. In addition, it is desirable to achieve a certain level of adhesion between phases which can partly stem from cross links formed across the interface.

Fibers. Textile technology has long used the concept of blending various fiber types, such as polyester and cotton, into fabrics. However, for many years there have been products based on polymer–polymer blends within a single fiber (*see* chapter 16 of Ref. *10* by Paul). Such mixtures can be highly structured composites like side-by-side bicomponent structures which possess self-crimping capability of interest in developing "bulk" and "stretch." Alternately, the two polymers can be co-formed into a sheath–core arrangement in which one polymer forms the exterior surface while another is imbedded in the interior and thus offers such possibilities as adhesion or dyeing of the external fiber while retaining cost or mechanical advantages of the interior polymer. Less-structured mixtures are also possible and are commercially used. For example, one component containing ionic dye sites can be dispersed within a matrix of a base polymer which does not contain these sites. The state of dispersion can range all the way from complete miscibility to rather finite domain sizes. Several commercial products use a similar concept in which a highly conductive polymer is mixed with a conventional fiber-forming polymer to yield carpets that do not exhibit the undesirable characteristic of high static-charge buildup in low-humidity environments. The blend concept is viewed as a means of solving existing problems and of creating new products without the necessity of synthesizing entirely new chemical structures.

"Thermoplastic Elastomers." A new generation of materials was initiated by the introduction of phase-separated block copolymers which

offered a modulus similar to elastomeric materials but with thermoplastic-like behavior. This was advantageous in that it eliminated the necessity for introducing cross links chemically. The effective cross links in such systems are provided by physical factors associated with the phase separation of "hard segments" dispersed in a matrix of "soft segments" that are covalently bonded one to another. Various chemical approaches have been commercialized, such as anionic polymerization, to produce styrene–diene multiblock copolymers plus segmented polyurethanes and related polyester formulations based on condensation syntheses. While these materials possess very desirable characteristics for which there is a market, they also require rather sophisticated and expensive chemistry. Certain types of polymer blends have been introduced in recent years which offer similar behavior at lower cost (*see* chapter 20 of Ref. *10* by Kresge). Most of these are based on mixtures of crystalline polyolefins, such as low density polyethylene and high density polyethylene, with various types of ethylene–propylene-based elastomers. In some cases the components can be partially cross-linked during mixing but to a degree that does not prohibit subsequent flow and processing, whereas others are based on a unique concept of an interpenetrating network of phases that provides unusual characteristics. Such materials are the subject of very active research and commercial development. Some commercially available products that are based, in one form or another, on the above-mentioned ideas are TPR (by Uni-Royal), TELCAR (by B. F. Goodrich), REN–FLEX (by Ciba–Geigy), and VISTAFLEX (by Exxon). Some interesting work along these lines is described in the current volume (*83*) in which an example of synergism in strength is provided. In this case, it has been shown that some ethylene–propylene-based elastomers, when blended with low density polyethylene, yield tensile strength vs. composition plots that are just slightly below that expected by simple additivity, whereas other such elastomers can be blended with low density polyethylene to give blends that are as much as 25% stronger than the strongest component, i.e., synergism. The difference which causes these two patterns of behavior evidently has to do with the ratio of ethylene to propylene in the elastomer used for blending. It is suggested that elastomers containing high proportions of ethylene, and thus long sequences of ethylene capable of crystallizing, are responsible for this behavior. It is interesting to compare this example of synergism with that shown by the polystyrene–PPO system described earlier since the mechanisms involved in these two cases clearly must be different. It will be of interest to elucidate more clearly by further research all of the possible causes for synergism of this kind because of the many practical implications.

In addition to the polyolefin blends designed for "thermoplastic elastomer" applications, a great deal of interest also has centered on other kinds of blends of polyolefins as has been reviewed recently (*see* chapter 21 of Ref. *10* by Plochocki). In a recent paper (*84*), we showed that blends involving polypropylene–high density polyethylene–low density polyethylene in various proportions and combinations exhibit additivity of tensile strength; however, there are serious losses in ductility in some cases such that the blends are less ductile than either pure component. It is interesting to note, however, that these losses in ductility can largely be restored by addition of rather small amounts of an amorphous ethylene–propylene rubber (*84*).

Plastics. Plastic–Elastomer Blends (Impact Modifications). Many of the commodity thermoplastics lack toughness to a degree that excludes them from many applications. However, it has been found that this deficiency can be eliminated by properly blending these glassy polymers with small amounts of suitable rubbery polymers. Recent reviews have extensively described the technology of rubber toughening of brittle plastics (*77;* also *see* Ref. *10,* chapter 13 by Newman and chapter 14 by Bucknall). The mechanism of this phenomenon is rather complex and is currently believed to involve craze initiation within the brittle plastic caused by the inclusion of the rubbery particles plus certain elements of crack arresting. It is well known that the size of the included rubbery particles and other aspects of blend morphology are very influential parameters in the efficiency of impact modification. Rubber toughening of plastics generally involves some synergism of properties.

It is now recognized that adhesion at the rubber–glass interface is a necessary condition for this phenomenon to occur. In many systems, adhesion does not occur naturally and must be promoted by the presence of an interfacial agent. Typically, impact modification of polystyrene is accomplished by producing a reactor-generated blend with a diene rubber. In this case, rubber is dissolved in the styrene monomer which is subsequently polymerized. Proper control of the attending phase inversion is known to play a substantial role in generating a desired blend morphology. However, it has been shown that chain-transfer processes accompany polymerization and result in a certain amount of graft-polymer formation simultaneous with polystyrene homopolymer. This graft material is believed to play an interfacial role in the blend to provide the necessary adhesion between phase domains. A recent review (*see* chapter 12 of Ref. *10* by Paul) has considered the fundamental aspects of interfacial agents in polymer–polymer blends. PVC also is impact modified by blending with rubbery particles; however, in this technology the two polymers are made separately and are post-reactor blended. However, pre-formed rubber particles for blending with PVC usually contain some grafted material at the surface that provides adhesion with PVC and thus can be regarded to contain an interfacial agent as well. The diverse family of materials known as ABS also use these concepts. The base polymer is typically a glassy copolymer of styrene and acrylonitrile while the dispersed rubbery particles are generally composed of copolymers of butadiene and acrylonitrile. Once again, some graft material generally is involved to provide the interfacial function described above. In principle, either reactor blending or post-reactor blending may be used.

On the other hand, it is not always necessary that an interfacial agent be present. Polypropylene is available in impact-modified grades which are made by simply blending polypropylene with suitable olefin-based elastomers. Most often the elastomer is a suitably chosen ethylene–propylene-based rubber. Evidently, the required adhesion develops naturally in these systems without the need for an interfacial agent. However, proper control of phase morphology during mixing is essential.

Plastic–Plastic Blends. In recent years, a number of alloys of two or more plastics which are believed to be largely immiscible have achieved commercial prominence or are being considered for commercialization. The rationale for these products is different than simply impact modification and generally involves the concepts of property combinations and cost dilution. We will consider a number of examples of such systems.

It was indicated above that ABS itself is a polymer blend and actually constitutes a family of materials that differ widely as a result of the many chemical and physical variations that are possible. Interestingly, ABS has been blended with other plastic materials to achieve several new products. For example ABS has been blended with PVC, thermoplastic polyurethanes (PU), and polycarbonate. Table I shows a comparison of some of the properties of blends of ABS with these other plastics (*71, 73*).

Table I. Properties

Property[b]	*ABS/PU (Rigid)*
Notched Izod impact (ft-lb/in.)	8.0
Yield strength (psi)	4,450
Heat deflection temperature at 264 psi (°F)	2.2
Hardness, Rockwell R	82
Tensile modulus, 10^5 (psi)	182
Specific gravity	1.04

[a] Adapted from Refs. *71* and *73*.

Several commercial varieties of ABS/PVC blends for injection molding or extrusion are commercially available under trade names such as Abson (from Abtec), Cycovin (from Borg–Warner), and Polyman (from Schulman). Geoffroy (*76*) has studied the mechanical properties of ABS/PVC blends and has found that for certain ABS grades the blends show a synergism in Izod impact strength such that some blends have considerably greater impact strengths than either pure ABS or pure PVC. The modulus of these blends appears to follow a simple additivity form. The heat distortion temperature of PVC and ABS are quite similar, and this property is relatively unchanged by blending. Addition of PVC to ABS results in a product with a certain degree of flame retardance. Further, ABS/PVC blends have been found to process more easily than either pure PVC or impact-modified PVC. These blends provide good rigidity, toughness, and heat and chemical resistance for their cost. A further advantage is that various modulus grades of ABS may be used to produce a broad family of blend products. Some investigations have considered substituting chlorinated polyethylene for the PVC.

Blend products based on ABS and polycarbonate for extrusion and injection molding are available under the trade names Cycloloy (from Borg–Warner) and Bayblend (from Mobay). Geoffroy has shown that these blends exhibit mechanical properties which are more or less intermediate to those of the pure components. The same applies for the heat-distortion temperature. These blends have higher impact strength, tensile strength, modulus, and heat-distortion temperature than pure ABS; however, these gains are accompanied by an increased cost by virtue of adding the more expensive polycarbonate. These blends often compete with the so-called engineering thermoplastics primarily because they have properties similar to polycarbonate at a lower cost as a result of the presence of the less expensive ABS. These blends exhibit good processability. However, they lack the advantage of transparency that provides polycarbonate with many of its important markets. These blends did not exhibit the critical thickness problem shown by pure polycarbonate that

of ABS Alloys[a]

Standard ABS	*ABS/PVC (Rigid)*	*ABS/PVC (Flexible)*	*ABS/PC (Rigid)*
6.5	12.5	15.0	10.3
6,000	5,450	3,000	8,200
3.4	3.2	1.0	3.7
103	102	50	118
210	147	—	246
1.04	1.21	1.13	1.14

[b] All properties at room temperature.

prohibits it from developing its high impact strength in thick sections. More will be said later about this "critical thickness" problem of polycarbonate.

Another commercially important blend family is derived from PVC and acrylics (primarily PMMA). Some common trade names for these products for extrusion, injection molding, and most importantly, thermoforming, include DKE-450 (from du Pont), Kydek (from Rohm & Haas), and Polydene (from Schulman). Geoffroy (*76*) has shown that the mechanical and thermal properties of this system follow a more or less additive-type behavior. These blends get flame retardance, toughness, and chemical resistance from PVC and rigidity and formability from the acrylic component. They are most noteworthy because they permit deep draw and exhibit high melt strength needed for thermoforming. ABS blends with a wide spectrum of polysulfones have been described in the patent literature (*85*), and some of these are believed to be the basis for recent products.

The "critical thickness" problem of polycarbonate was mentioned earlier. Polycarbonate is noted for its very high impact strength which, combined with its clarity and high heat distortion temperature, make it a very versatile but expensive engineering thermoplastic. However, it is well known that this high impact strength is only available in relatively thin sections. Beyond a critical thickness of about ⅛ in., the impact strength of polycarbonate is dramatically reduced because of a complex change in mechanism of load bearing (*86, 87*). It has been observed that this critical thickness shifts to higher values and can, in effect, be eliminated by addition of small amounts of relatively soft polymers such as polyethylene. However, this solution to the problem is accompanied by a loss in transparency and a small reduction in heat distortion temperature. However, this approach still has some commercial value.

Recent literature (*72, 74*) has indicated a considerable interest in blending ionomers (such as du Pont's Surlyn) with a variety of polymers. It is believed that an ionomer–nylon blend is the basis for a new engineering thermoplastic.

Over the past 20 years a considerable chemical research effort was devoted to developing new high temperature polymers. It is interesting to note that some of these materials are now finding new applications and solutions to old problems of processing through blending. A commercial series of products trade-named Tribolon XT has been announced which are based on an aromatic polymide (Upjohn's 2080) with Phillips' poly(phenylene sulfide), trade-named Ryton. A recent publication (*88*) describes some of the unique characteristics of this new family of materials.

Literature Cited

1. H. Keskkula, Ed., "Polymer Modification of Rubbers and Plastics," *Appl. Polym. Symp.* (1968) **7.**
2. P. F. Bruins, Ed., "Polyblends and Composites," *Appl. Polym. Symp.* (1970) **15.**
3. N. A. J. Platzer, Ed., "Multicomponent Polymer Systems," ADV. CHEM. SER. (1971) **99.**
4. G. E. Molau, Ed., "Colloidal and Morphological Behavior of Block and Graft Copolymers," Plenum, New York, 1971.
5. L. H. Sperling, Ed., "Recent Advances in Polymer Blends, Grafts, and Blocks," Plenum, New York, 1974.
6. N. A. J. Platzer, Ed., "Copolymers, Polyblends, and Composites," ADV. CHEM. SER. (1975) **142.**
7. D. Klempner, K. C. Frisch, Eds., "Polymer Alloys: Blends, Blocks, Grafts, and Interpenetrating Networks," Plenum, New York, 1977.
8. Manson, J. A., Sperling, L. H., "Polymer Blends and Composites," Plenum, New York, 1976.
9. Bucknall, C. B., "Toughened Plastics," Applied Science, London, 1977.
10. Paul, D. R., Newman, S., Eds., "Polymer Blends," Vols. I and II, Academic, New York, 1978.
11. Amos, J. L., *Polym. Eng. Sci.* (1974) **14,** 1.
12. Rosen, S. L., *Polym. Eng. Sci.* (1967) **7,** 115.
13. Lundstedt, O. W., Bevilacqua, E. M., *J. Polym. Sci.* (1957) **24,** 297.
14. Locke, C. E., Paul, D. R., *J. Appl. Polym. Sci.* (1973) **17,** 2597, 2791.
15. Barentsen, W. M., Heikens, D., Piet, P., *Polymer* (1974) **15,** 119.
16. Paul, D. R., Locke, C. E., Vinson, C. E., *Polym. Eng. Sci.* (1973) **13,** 202.
17. Locke, C. E., Paul, D. R., *Polym. Eng. Sci.* (1973) **13,** 308.
18. Ide, F., Hasegawa, A., *J. Appl. Polym. Sci.* (1974) **18,** 963.
19. Molau, G. E., "Block Polymers," S. L. Aggarwal, Ed., p. 79, Plenum, New York, 1970.
20. Holister, G. S., Thomas, C., "Fibre Reinforced Materials," Elsevier, New York, 1966.
21. Work, J. L., *Polym. Eng. Sci.* (1973) **13,** 46.
22. Manson, J. A., Sperling, L. H., "Polymer Blends and Composites," Plenum, New York, 1976.
23. Sperling, L. H., *J. Polym. Sci., Macromol. Rev.* (1977) **12,** 141.
24. Frisch, H. L., Frisch, K. C., Klempner, D., *Polym. Eng. Sci.* (1974) **14,** 646.
25. Frisch, K. C., Klempner, D., Migdal, S., Frisch, H. L., Ghiradella, H., *Polym. Eng. Sci.* (1974) **14,** 76.
26. Kim, S. C., Klempner, D., Frisch, K. C., Frisch, H. L., *Macromolecules* (1976) **9,** 263.
27. Huelek, V., Thomas, D. A., Sperling, L. H., *Macromolecules* (1972) **5,** 340.
28. Koleske, J. V., Lundberg, R. D., *J. Polym. Sci.* (1969) **7A-2,** 795.
29. Zakrzewski, G. A., *Polymer* (1973) **14,** 347.
30. Hammer, C. F., *Macromolecules* (1971) **4,** 69.
31. Brode, G. L., Koleske, J. V., *J. Macromol. Sci., Chem.* (1972) **A6,** 1109.
32. Kargin, K. A., *J. Polym. Sci.* (1963) **40,** 1601.
33. Hughes, L. J., Brown, G. L., *J. Appl. Polym. Sci.* (1961) **5,** 580.
34. Schultz, A. R., Gendron, B. M., *Am. Chem. Soc., Div. Polym. Chem., Prepr.* (1973) **14,** 571.
35. Shultz, A. R., Gendron, B. M., *J. Polym. Sci.* (1973) **43C,** 89.
36. Shaw, M. T., *J. Appl. Polym. Sci.* (1974) **18,** 449.
37. Kruse, W. A., Kirste, R. G., Haas, J., Schmitt, B. J., Stein, D. J., *Makromol. Chem.* (1976) **177,** 1145.
38. McMaster, L. P., *Macromolecules* (1973) **6,** 760.

39. Kenney, J. F., *J. Polym. Chem. Ed.* (1976) **14**, 123.
40. Bernstein, R. E., Cruz, C. A., Paul, D. R., Barlow, J. W., *Macromolecules* (1977) **10**, 681.
41. Paul, D. R., Barlow, J. W., Cruz, C. A., Mohn, R. N., Nassar, T. R., Wahrmund, D. C., *Am. Chem. Soc., Div. Org. Coat. Plast. Chem., Prepr.* (1977) **37**(1), 130.
42. Shur, Y. J., Randby, B. G., *J. Appl. Polym. Sci.* (1975) **19**, 2143.
43. Shur, Y. J., Randby, B. G., *J. Appl. Polym. Sci.* (1975) **19**, 1337.
44. Alexandrovich, P., Karasz, F. E., MacKnight, W. J., *Polymer* (1977) **18**, 1022.
45. Hammel, R., MacKnight, W. J., Karasz, F. E., *J. Appl. Phys.* (1975) **46**, 4199.
46. Paul, D. R., Barlow, J. W., Bernstein, R. E., Wahrmund, D. C., "Polymer Blends Containing Poly(Vinylidene Fluoride), Part IV," unpublished data.
47. Shultz, A. R., Beach, B. M., *Macromolecules* (1974) **7**, 902.
48. Noshay, A., Robeson, L. M., *J. Polym. Sci., Polym. Chem. Ed.* (1974) **12**, 689.
49. Massa, D. J., *Am. Chem. Soc., Div. Polym. Chem., Prepr.* (1978) **19**(1), 157.
50. Schurer, J. W., deBoer, A., Challa, G., *Polymer* (1975) **16**, 201.
51. Lo, J. H. G. M., Kransen, G., Tan, Y. Y., Challa, G., *J. Polym. Sci., Polym. Lett. Ed.* (1975) **13**, 725.
52. Kwei, T. K., Nishi, T., Roberts, R. F., *Macromolecules* (1974) **7**, 667.
53. Hughes, L. J., Britt, G. E., *J. Appl. Polym. Sci.* (1961) **5**, 337.
54. Ichihara, S., Komatsu, A., Hata, T., *Polym. J.* (1971) **2**, 640.
55. Scott, R. L., *J. Chem. Phys.* (1949) **17**, 279.
56. Flory, P. J., "Principles of Polymer Chemistry," Chap. XII, pp. 495–540, Cornell, Ithaca, NY, 1962.
57. Hildebrand, J. H., Scott, R. L., "The Solubility of Nonelectrolytes," 3rd ed., Chap. —, pp. 346–396, Reinhold, New York, 1950.
58. Krause, S., *J. Macromol. Sci., Rev. Macromol. Chem.* (1972) **C7**(2), 251.
59. Olabisi, O., *Macromolecules* (1975) **9**, 316.
60. Matzner, M., Robeson, L. M., Wise, E. W., McGrath, J. E., *Am. Chem. Soc., Div. Org. Coat. Plast. Chem., Prepr.* (1977) **37**(1), 123.
61. Nishi, T., Wang, T. T., *Macromolecules* (1975) **8**, 909.
62. Imken, R. L., Paul, D. R., Barlow, J. W., *Polym. Eng. Sci.* (1976) **16**, 593.
63. Wahrmund, D. C., Bernstein, R. E., Barlow, J. W., Paul, D. R., "Polymer Blends Containing Poly(vinylidene Fluoride)," Part I, *Polym. Eng. Sci.* (1978) **18**(**9**), 677.
64. Bernstein, R. E., Paul, D. R., Barlow, J. W., "Polymer Blends Containing Poly(vinylidene Fluoride)," Part II, *Polym. Eng. Sci.* (1978) **18**(**9**), 683.
65. Bernstein, R. E., Wahrmund, D. C., Barlow, J. W., Paul, D. R., "Polymer Blends Containing Poly(vinylidene Fluoride)," Part III, unpublished data.
66. Blanks, R. F., Prausnitz, J. M., *Ind. Eng. Chem. Fundam.* (1964) **3**(1), 1.
67. Rowlinson, J. S., "Liquids and Liquid Mixtures," pp. 159–190, Academic, New York, 1967.
68. Nishi, T., Kwei, T. K., *Polymer* (1975) **16**, 285.
69. Paul, D. R., Vinson, C. E., Locke, C. E., *Polym. Eng. Sci.* (1972) **12**, 157.
70. Hauck, J. E., *Mater. Eng.* (1967) **66**(1), 60.
71. Jalbert, R. L., *Mod. Plast. Encycl.* (1975) **52**, 107.
72. *Mod. Plast.* (1976) **August**, 42.
73. Jalbert, R. L., Smejkal, J. P., *Mod. Plast. Encycl.* (1976) **53**, 108.
74. Forger, G. R., *Mater. Eng.* (1977) **July**, 44.
75. *Mod. Plast.* (1977) **November**, 42.
76. Geoffroy, R. R., M.S. thesis, Lowell Technological Institute (1975).

77. Yee, A. F., *Polym. Eng. Sci.* (1977) **17,** 213.
78. Paul, D. R., Altamirano, J. O., "Copolymers, Polyblends, and Composites," ADV. CHEM. SER. (1975) **142,** 371.
79. Wahrmund, D. C., Paul, D. R., Barlow, J. W., *J. Appl. Polym. Sci.,* in press.
80. Nassar, T. R., Paul, D. R., Barlow, J. W., *J. Appl. Polym. Sci.,* in press.
81. Mohn, R. N., Paul, D. R., Barlow, J. W., Cruz, C. A., *J. Appl. Polym. Sci.,* in press.
82. Cruz, C. A., Paul, D. R., Barlow, J. W., *J. Appl. Polym. Sci.,* in press.
83. Lindsay, G. A., Singleton, C. J., Carman, C. J., Smith, R. W., ADV .CHEM. SER. (1979) **176,** 367.
84. Robertson, R. E., Paul, D. R., *J. Appl. Polym. Sci.* (1973) **17,** 2579.
85. Ingulli, A. F., Alter, H. C., Uniroyal, Inc., U. S. Patent **3,636,140** (1972).
86. Kazanjian, A. R., *Polym.–Plast. Technol. Eng.* (1973) **2,** 123.
87. Yee, A. F., *J. Mater. Sci.* (1977) **12,** 757.
88. Alvarez, R. T., Driscoll, S. B., Nahill, T. E., *Soc. Plast. Eng., Tech. Pap.* (1977) **35,** 308.

RECEIVED April 14, 1978.

18

Morphology, Viscoelastic Properties, and Stress–Strain Behavior of Blends of Polycarbonate of Bisphenol-A (PC) and Atactic Polystyrene (PST)

G. GROENINCKX, S. CHANDRA[1], H. BERGHMANS, and G. SMETS

Katholieke Universiteit Leuven, Department of Chemistry, Laboratory of Macromolecular and Organic Chemistry, Celestijnenlaan 200 F, 3030 Heverlee, Belgium

The modulus-temperature behavior of polycarbonate–polystyrene blends has been investigated and correlated with sample composition and morphology. Stress–relaxation experiments were carried out as a function of time at different temperatures. The time–temperature superposition principle was not applicable in the temperature range between the glass transitions of the two polymers; shift factors were found to be a function of time in addition to temperature. Valid master curves were calculated by means of equivalent mechanical models. In the temperature range between the two glass transitions, the tensile stress–strain behavior of the blends with polycarbonate as the continuous phase is characteristic of a toughened system. Two toughening mechanisms, crazing and shear-band formation, have been observed.

Blends of two incompatible polymers with different glass-transition temperatures have properties which differ from the pure components. Their viscoelastic behavior is complex and deviation from the simple time–temperature superposition is generally observed (*1*). The high

[1] Current address: H. B. Technological Institute, Kanpur, India.

0-8412-0457-8/79/33-176-337$07.50/0

strain properties, i.e., yielding, deformation, and fracture, depend on the morphology and the concentration of the constituent phases (*2, 3, 4, 5, 6*). In this chapter, the results of an investigation of the influence of the two-phase structure of polycarbonate–polystyrene blends on mechanical behavior are presented. An attempt was made to relate their low and high strain properties to sample composition and morphology.

Experimental

Materials. Polycarbonate of Bisphenol-A(PC) was a commercial product (Lexan, General Electric Co., U.S.A.), with a viscosimetric average molecular weight $M_v = 40{,}000$. High molecular-weight atactic polystyrene ($M_v = 280{,}000$) was obtained from Schuchardt, München.

Preparation of the Samples. Blends of different composition were prepared by freeze drying dioxane solutions. Sheets were compression molded at 249°C and cut into samples of desired dimensions. Rectangular samples were used for the stress–relaxation measurements and dumbbell-shaped samples were used for the tensile stress–strain experiments. The compositions by weight of the PC–PST blends studied are as follows: 95/5, 90/10, 80/20, 75/25, 50/50, and 25/75.

Measurements. The morphology of the blends was studied by optical microscopy (Leitz Dialux Pol), transmission electron microscopy (Jeol 100 U), and scanning electron microscopy (Cambridge MK II). Ultramicrotome sections were made with an LKB Ultratome III. Samples for scanning electron microscopy were obtained by fracturing sheets at low temperature. The fracture surfaces were etched with a 30% potassium hydroxide solution to hydrolyse the polycarbonate phase. Stress–relaxation and tensile stress–strain experiments were performed with an Instron testing machine equipped with a thermostatic chamber. Relaxation measurements were carried out in flexion ($E > 10^8$ dyn/cm^2) or in traction ($E < 10^8$ dyn/cm^2). Prior to each experiment, the samples were annealed to obtain volumetric equilibrium.

Morphology

The morphology of PC–PST blends depends on the concentration of the components. To facilitate their observation by electron microscopy, the contrast between both phases was increased by crystallizing the polycarbonate. This also allows exact identification of the two phases by optical microscopy with polarized light; crystallized PC is a highly birefringent phase and PST appears as a black phase. When one of the components is in excess, a morphology of dispersed particles in a continuous matrix is observed. The dispersed particles have a large size distribution which becomes more pronounced as the concentration is increased. This is illustrated in Figures 1 and 2 for PST in PC and vice versa.

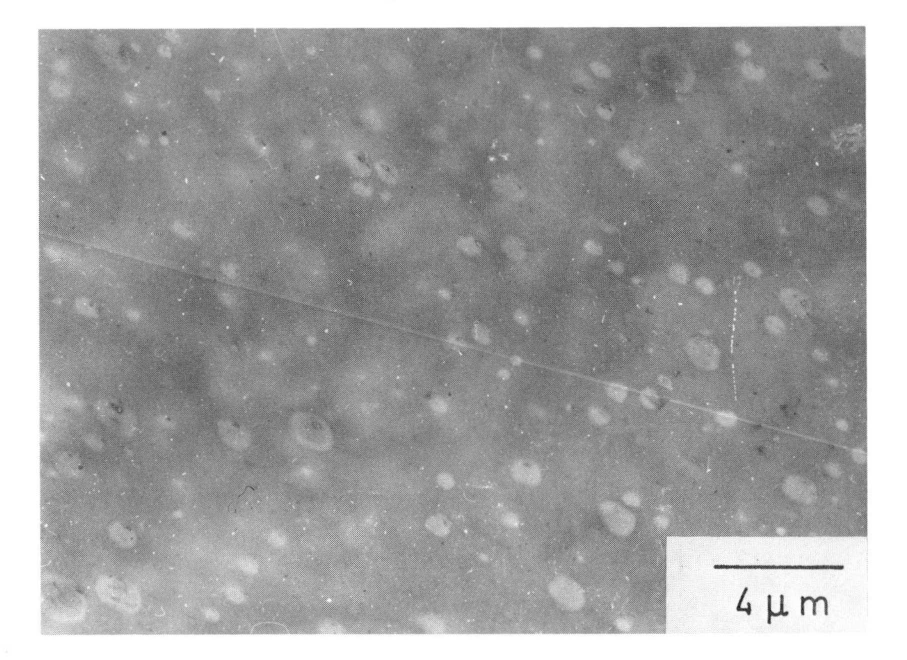

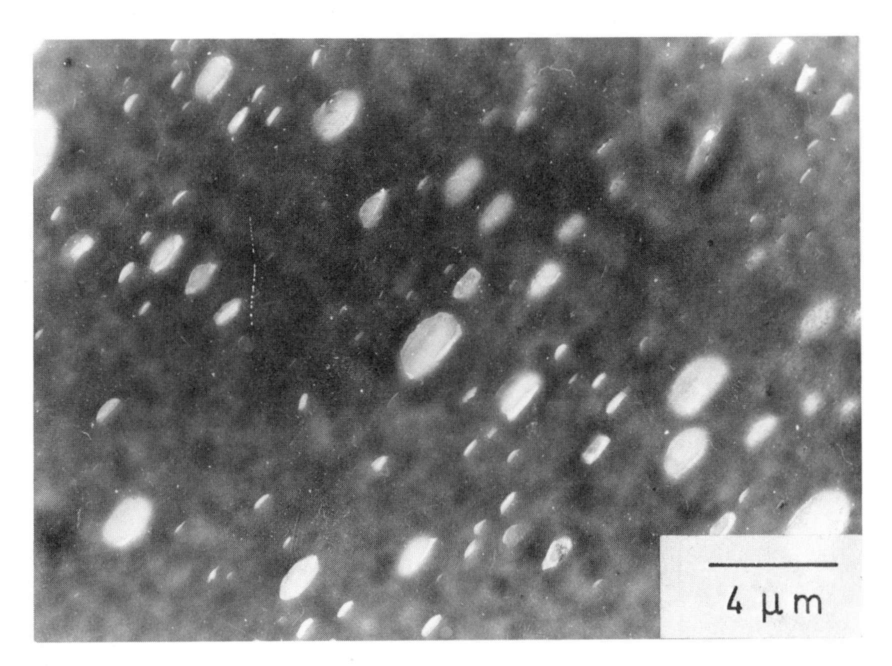

Figures 1a, 1b. (a) (top) *Transmission electron micrograph of the 95/5 PC–PST blend. (b)* (bottom) *Transmission electron micrograph of the 90/10 PC–PST blend.*

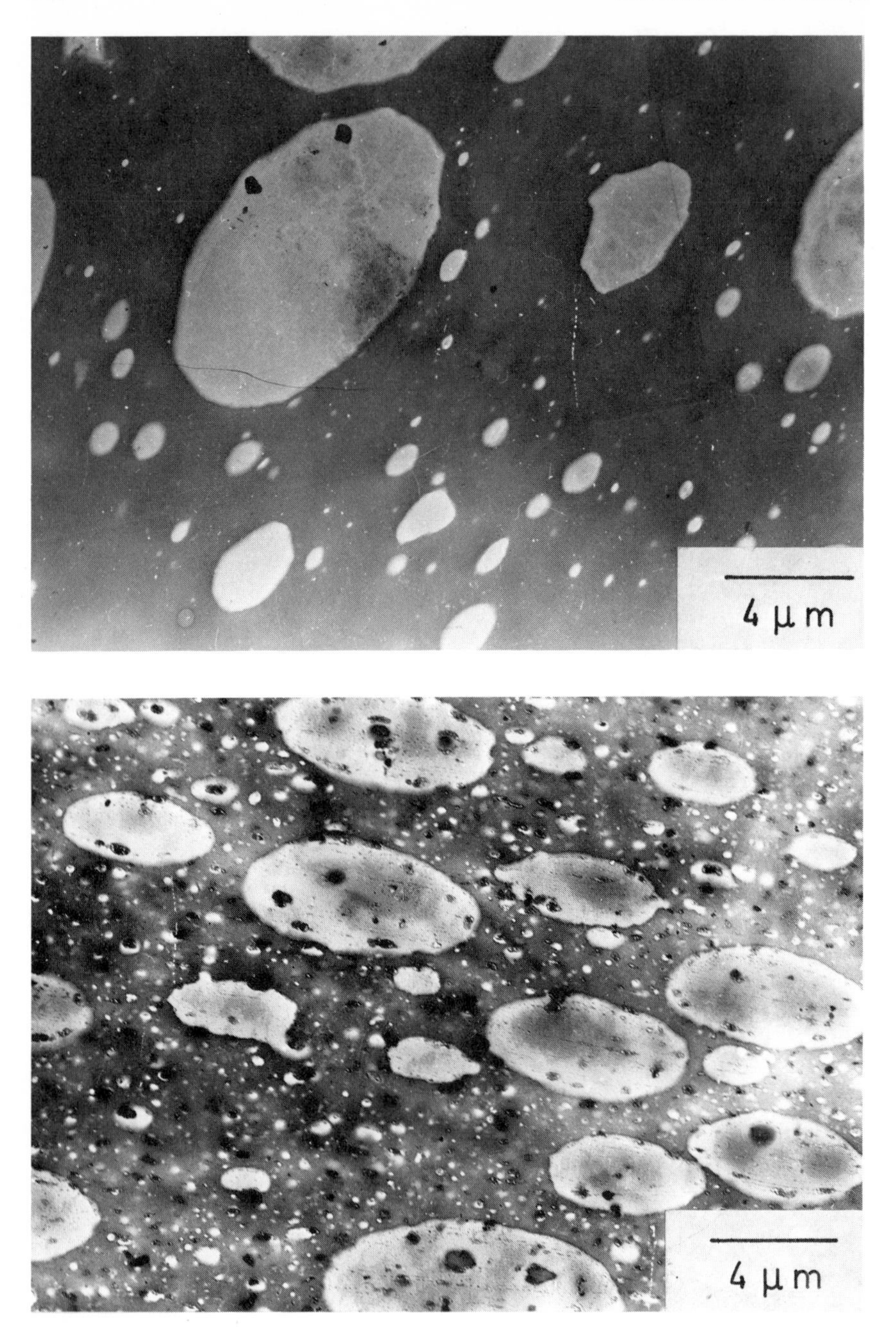

Figures 1c, 1d. (c) (top) *Transmission electron micrograph of the 80/20 PC–PST blend. (d)* (bottom) *Transmission electron micrograph of the 75/25 PC–PST blend.*

Figure 1e. Scanning electron micrograph of the etched fracture surface of the 75/25 PC–PST blend

Figures 1a, 1b, 1c, and 1d show transmission electron micrographs of the blends with 5, 10, 20, and 25 wt% PST, respectively. PC forms the continuous phase and PST is dispersed as spherical particles without any aggregation. The observed oval shape of the dispersed particles results from the deformation of the specimens during ultramicrotome sectioning. Their normal spherical shape is evident from scanning electron micrographs of the etched fracture surface of these blends (Figure 1e).

Figures 2a, 2b, 2c, and 2d represent the morphology of the 25/75 (PC–PST) blend. PST forms the continuous phase and the dispersed spherical PC particles show some aggregation (Figure 2a). From the transmission electron micrographs (Figures 2b and 2c) it can be seen that small PST particles are present in the dispersed PC phase. Examination of the etched fracture surface by scanning electron microscopy reveals depressions of completely removed PC particles as well as spheres with dark periphery because of partial hydrolysis of PC (Figure 2d).

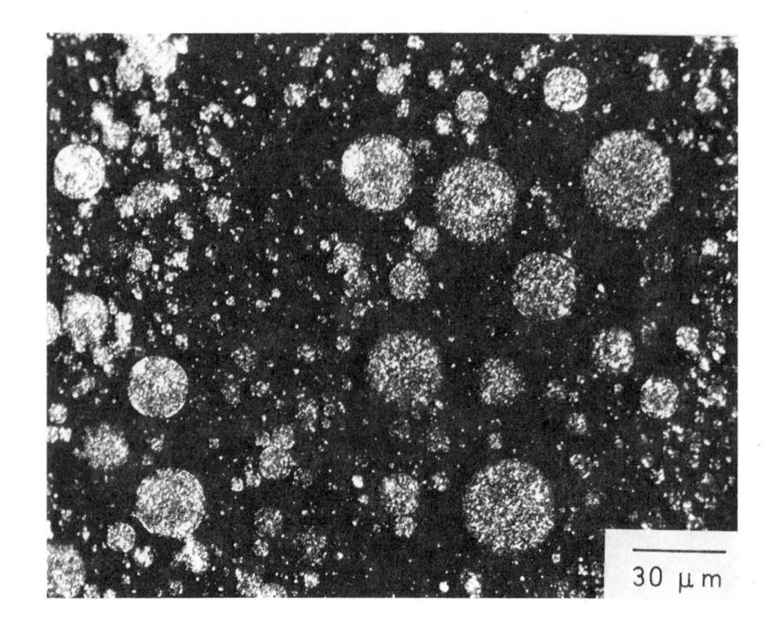

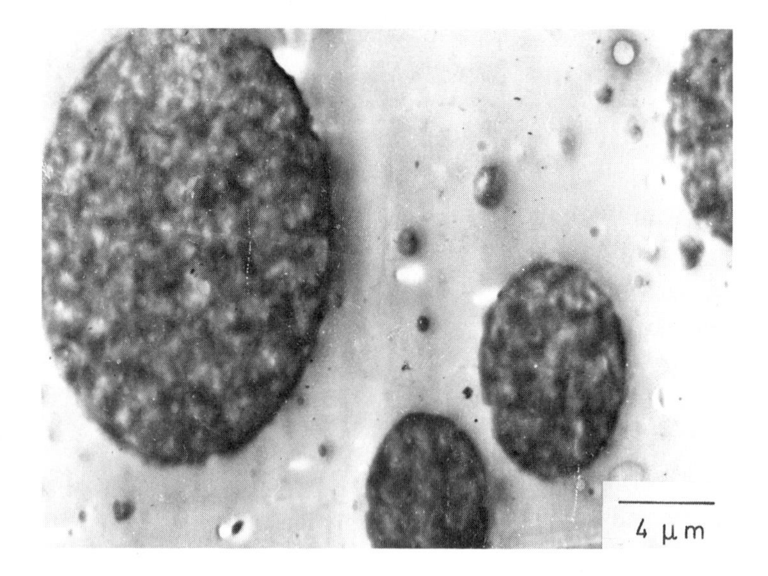

Figures 2a, 2b.—(a) (top) *Optical micrograph of the microtomed section of the 25/75 PC–PST blend. (b)* (bottom) *Transmission electron micrograph of the 25/75 PC–PST blend.*

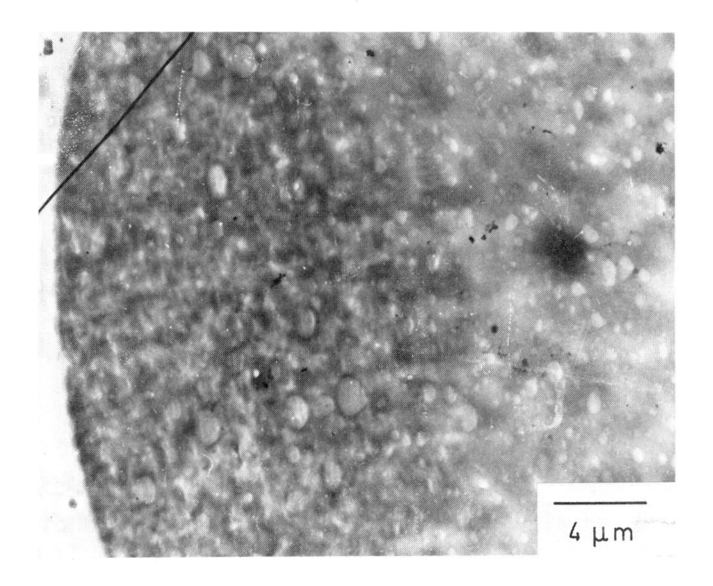

Figures 2c, 2d. (c) (top) *Transmission electron micrograph of the 25/75 PC–PST blend showing only the PC phase with sperical inclusions of PST. (d)* (bottom) *Scanning electron micrograph of the etched fracture surface of the 25/75 PC–PST blend.*

In the 50/50 blend, large continuous phases of both components are present. Moreover, inside the phases inclusions of the other component are observed (Figure 3a). The scanning electron micrograph of the PST phase, taken at higher magnification, shows distinct trapped spherical PC particles (Figure 3b). The etched fracture surface shows large hydrolyzed PC phases, alternating with PST domains (Figure 3c).

Modulus-Temperature Behavior

The temperature dependence of the relaxation modulus at 500 seconds of polycarbonate (*7*), polystyrene (*8*), and their blends (75/25, 50/50, and 25/75) was obtained from stress–relaxation experiments (Figure 4, full lines). In the modulus-temperature curves of the blends, two transition regions are generally observed in the vicinity of the glass–rubber transitions of the pure components. The inflection temperatures T_i in these transition domains are reported in Table I; they are almost independent on composition. The presence of these two well-separated transitions is a confirmation of the two-phase structure of the blends, deduced from microscopic observations.

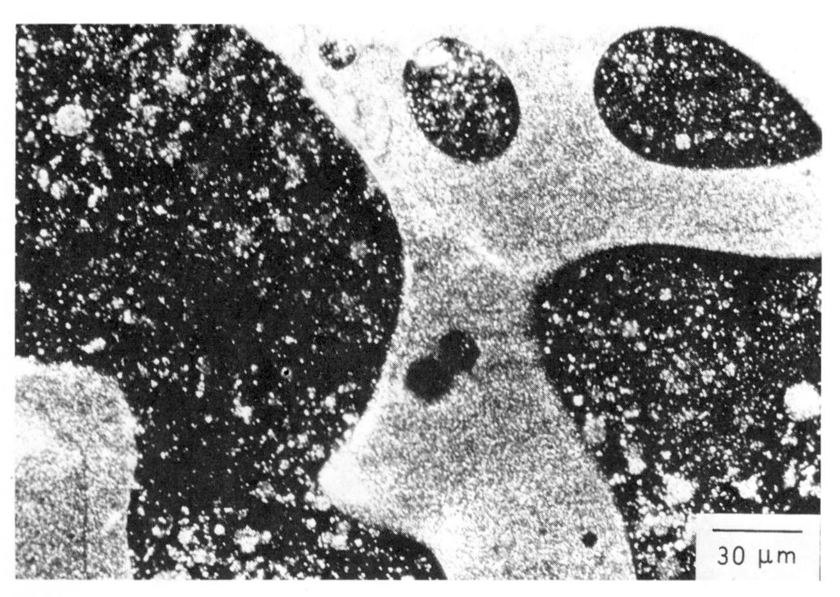

Figures 3a, 3b, 3c. (a) (above) *Optical micrograph of the microtomed section of a 50/50 PC–PST blend. (b)* (top right) *Scanning electron micrograph of the PST phase in the 50/50 PC–PST blend. (c)* (bottom right) *Scanning electron micrograph of the etched fracture surface of the 50/50 PC–PST blend showing both continuous phases.*

5 µm

50 µm

The change of the relaxation modulus with composition gives interesting information about the phase distribution in the blends. In the glassy state (below 95°C), the relaxation modulus is high and independent of composition. Above the glass transition of the PST phase, the mechanical behavior is affected by the composition. The 25/75 PC–PST blend behaves almost like pure PST, shifted slightly to higher temperatures. The dispersed PC phase is then present as a hard filler and has only a secondary influence on the rubbery modulus. With the

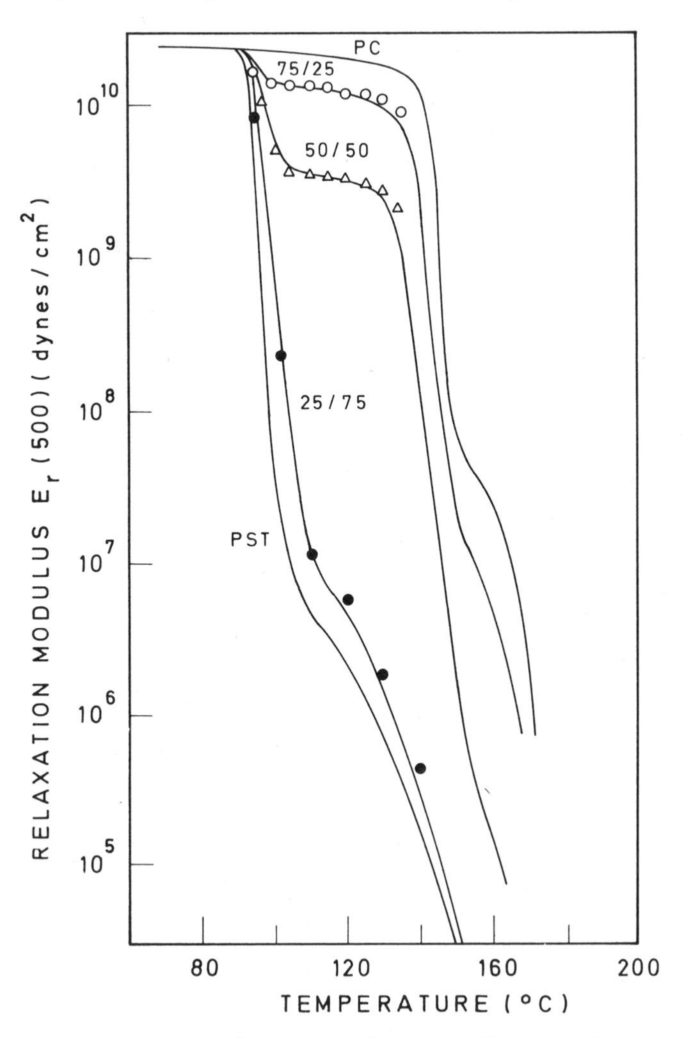

Figure 4. Stress–relaxation modulus at 500 seconds as a function of temperature (experimental data: solid lines; calculated values: (○) 75/25 PC–PST, (△) 50/50 PC–PST, (●) 25/75 PC–PST)

Table I. Characteristic Parameters Deduced from the Modulus-Temperature Curves

Sample	*Inflection Temperature*[a] T_i^A (°C)	T_i^B (°C)	*500-sec Relaxation Modulus at 120°C* $E_r(500)$ (dyn/cm²)
PC	—	147	2.10^{10}
75/25	100	145	$1,25.10^{10}$
50/50	101	145	$3,30.10^{9}$
25/75	103	—	5.10^{6}
PST	100	—	$2,5.10^{6}$

[a] A and B correspond to PST and PC phases respectively.

50/50 blend, a real plateau with high modulus is obtained between the glass–rubber transitions of the components. With further increase of the PC content, phase inversion occurs; PC then forms the continuous matrix, and the modulus value of the blend in the plateau region approaches that of pure PC.

The change of the relaxation modulus with composition at 120°C is given in Figure 5. The sudden increase of the relaxation modulus between 0.3–0.4 volume fraction of PC is attributable to the increasing continuity of the PC phase. As a consequence, the mechanical response of the blend becomes more and more dominated by the rigid PC phase, as this phase increases in continuity throughout the whole polymer mass.

Modulus-Time Behavior

Experimental Relaxation Isotherms. To characterize the viscoelastic properties of the PC–PST blends completely, stress–strain–relaxation measurements were carried out over a period of 10^4 seconds at different temperatures. The relaxation isotherms of the blends are represented in Figures 6a, 6b, and 6c; on each curve the deformation given to the sample is indicated. In Figure 6a for the 75/25 blend and Figure 6b for the 50/50 blend, the two homopolymer transitions are very well observed; this is typical for the two-phase structure of these blends and confirms previous observations. The relaxation isotherms of the 25–75 blend (Figure 6c) resemble those of pure PST (*8*), indicating that the relaxation behavior of this blend is completely dominated by the PST phase over the entire temperature and time scale. The glass transition of the dispersed PC phase is masked by the flow region of PST.

Horizontal shift distances at different points between the relaxation isotherms were measured. For PC (*7*), PST (*8*), and the 25/75 blend, the shift factors are independent of time, and complete master curves can be obtained by simple translation of the relaxation isotherms along

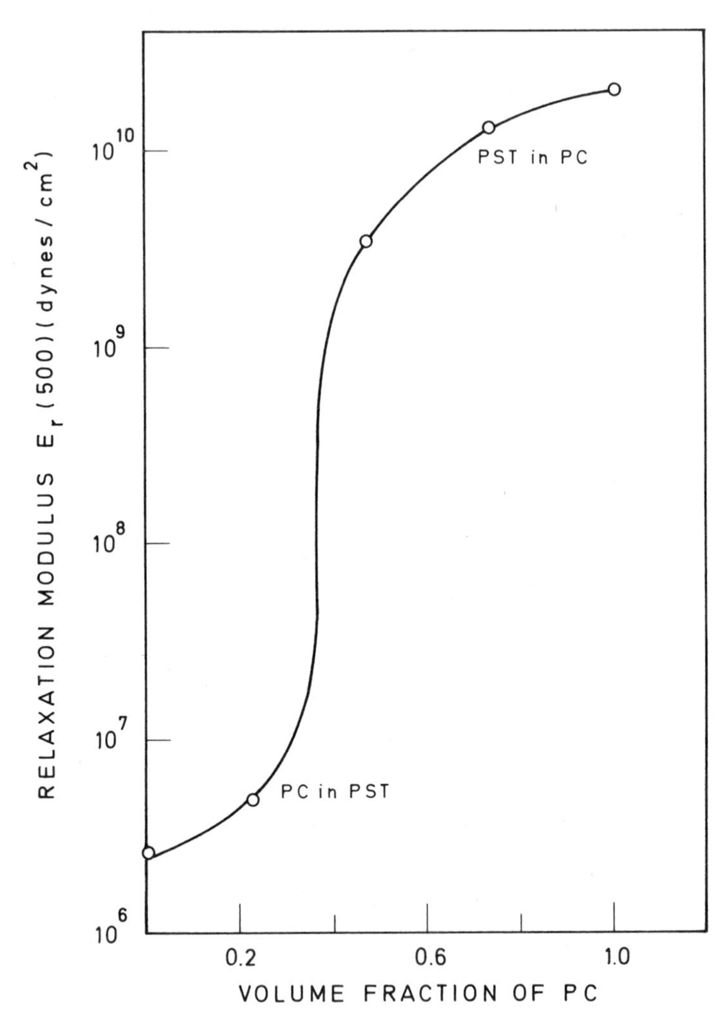

Figure 5. Change of the 500-second relaxation modulus with composition at 120°C

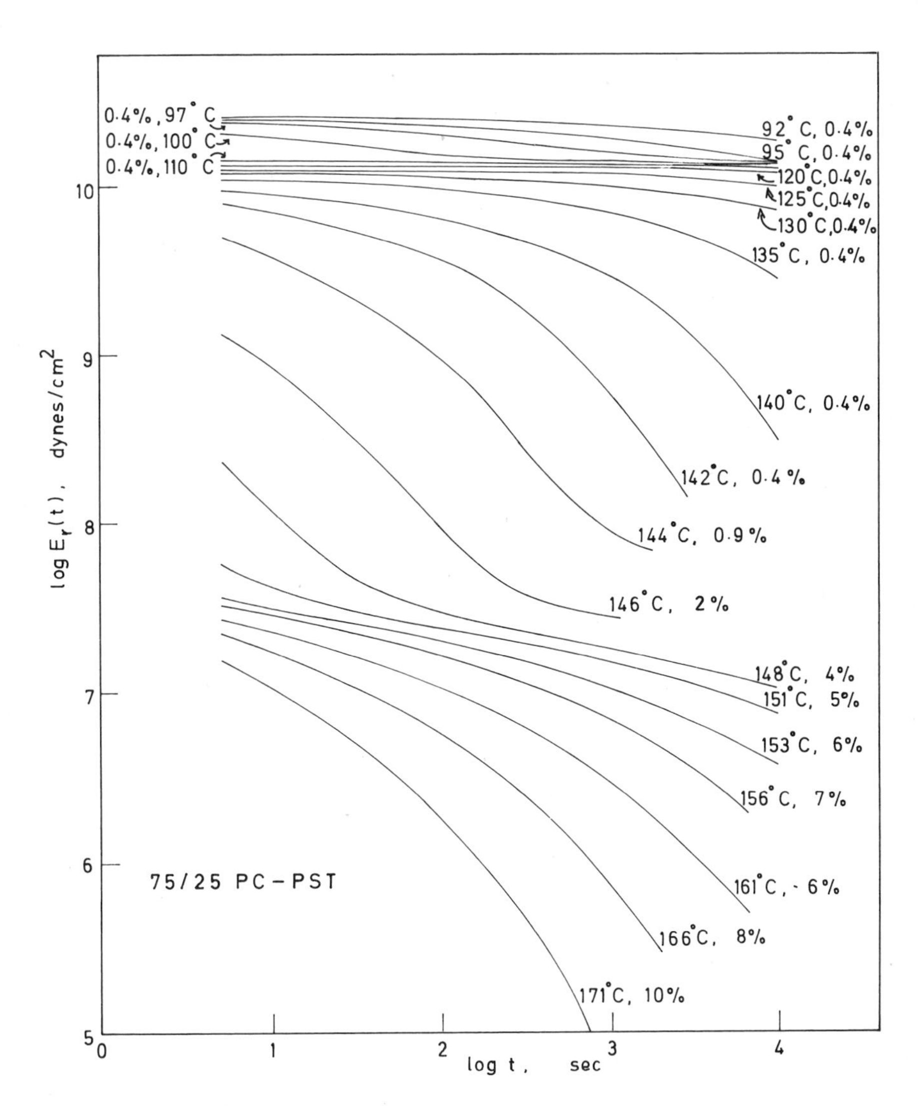

Figure 6a. Stress–relaxation isotherms for the 75/25 PC–PST blend

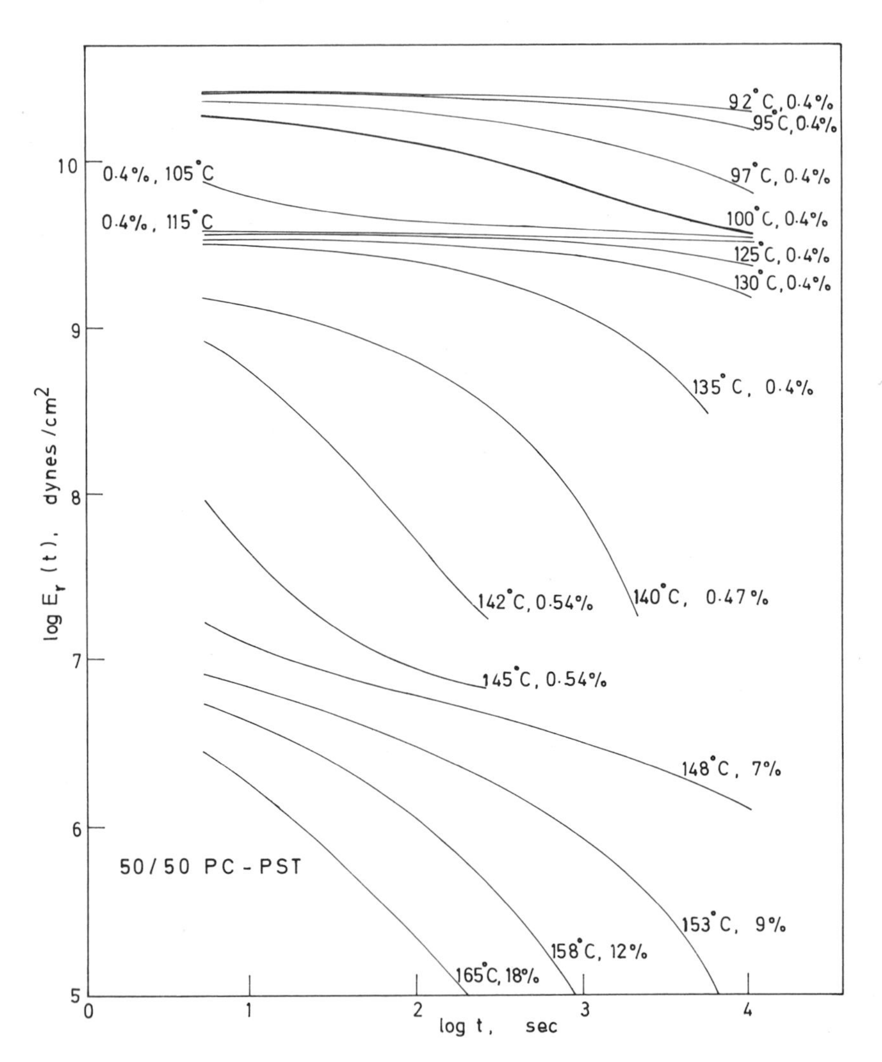

Figure 6b. Stress–relaxation isotherms for the 50/50 PC–PST blend

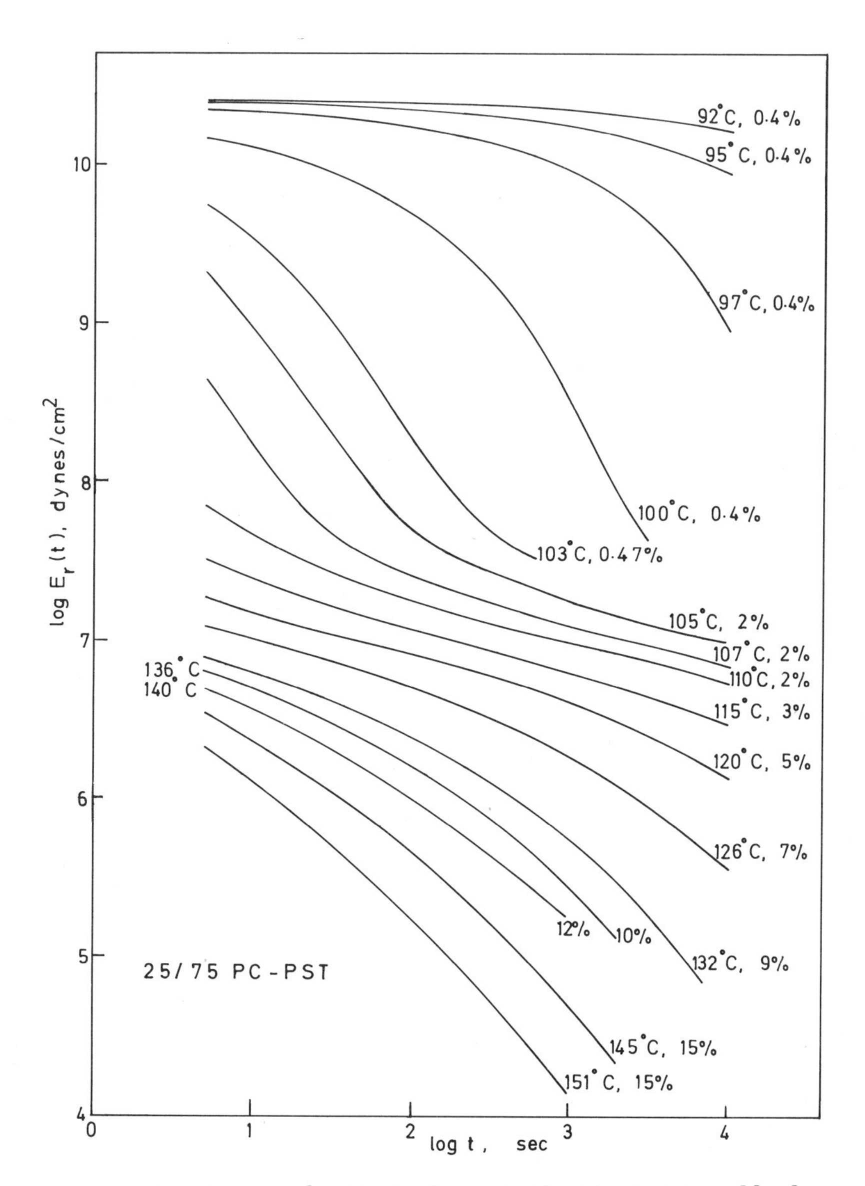

Figure 6c. Stress–relaxation isotherms for the 25/75 PC–PST blend

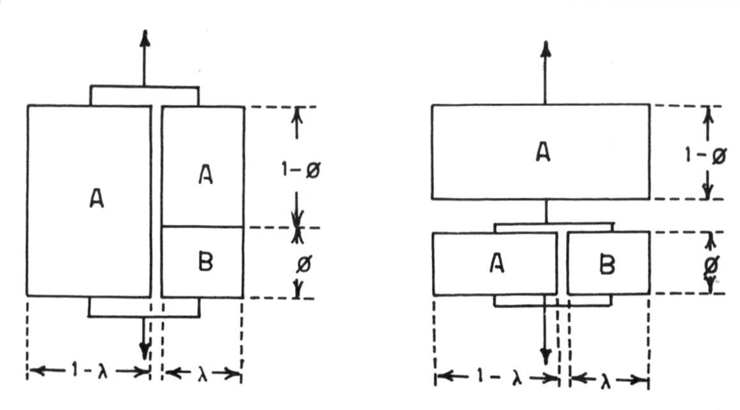

Figure 7. Mechanical models. (a) Parallel model; (b) series model.

the logarithmic time scale. The shift factors of the 75/25 and 50/50 blends between 100° and 140°C not only depend on temperature but also on time. Consequently, for these blends no master curves can be constructed by simple horizontal shifting of the isotherms. This thermorheologically complex behavior is generally observed for two-phase polymer systems (*1*). It results from the difference in time dependence of the relaxation mechanisms of the two phases. Complete master curves of PC, PST and their blends at 140°C are represented in Figure 8. For the 75/25 and 50/50 blends, the part of the curves between 10^{-8} and 10 seconds has been calculated from the equivalent mechanical model; at higher relaxation times, the curves were obtained by horizontal shifting. In the case of the 25/75 blend, although a valid master curve could be obtained by a horizontal shift of the experimental relaxation isotherms, a master curve also has been calculated from the model to show its validity (Figure 9).

Calculation of Master Curves from Mechanical Models. The only way to obtain valid master curves for the thermorheologically complex systems (75/25 and 50/50 blends) is to calculate the moduli of the blends as a function of time, using an appropriate mechanical model. This method requires knowledge of the time and temperature dependence of the mechanical properties of the constituent phases.

In this work the mechanical model proposed by Takayanagi (*9*) will be used. Two variants were proposed, both assuming that the two phases are connected partly in parallel and partly in series (Figures 7a and 7b). Kaplan and Tschoegl (*10*) have shown that the two variants of the Takayanagi model are equivalent. The series model (Figure 7b) will be used for our calculations. The modulus is given by:

$$E(\mathrm{t,T}) = \left[\frac{1-\phi}{E_{\mathrm{A}}} + \frac{\phi}{(1-\lambda)E_{\mathrm{A}} + \lambda E_{\mathrm{B}}}\right]^{-1} \quad (1)$$

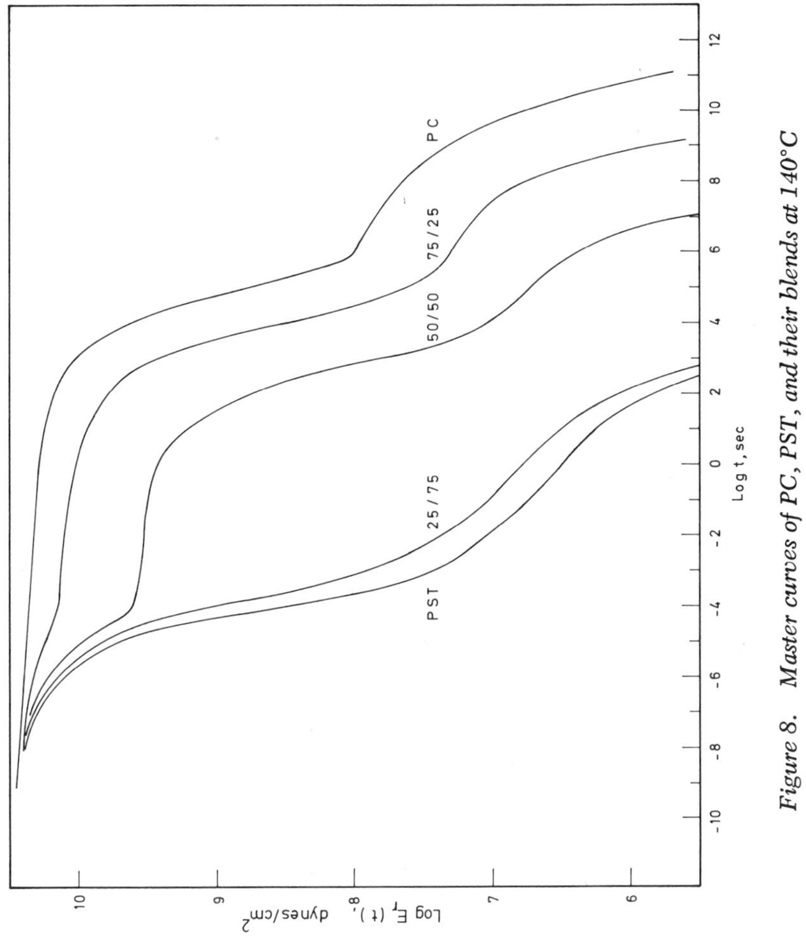

Figure 8. Master curves of PC, PST, and their blends at 140°C

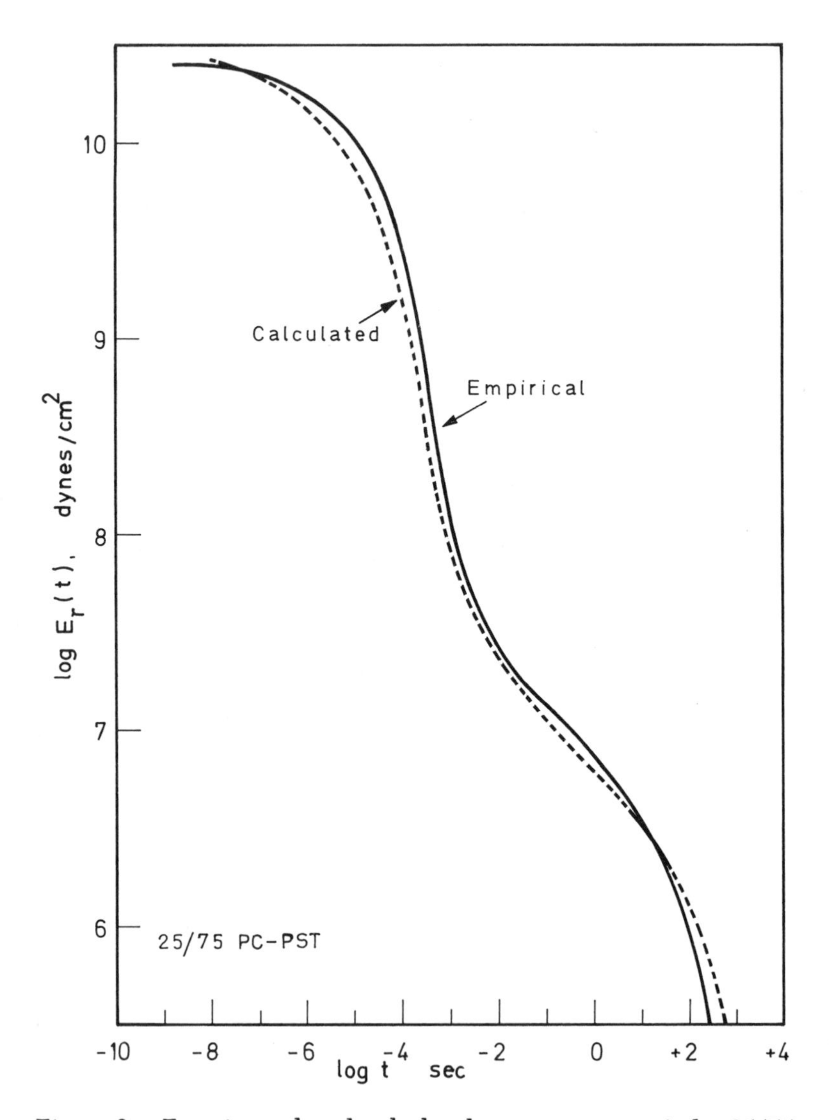

Figure 9. Experimental and calculated master curves of the 25/75 PC–PST blend at 140°C

with $\lambda\varnothing = v_B$. E_A and E_B are the moduli of phase A and phase B (pure components), λ and $\varnothing$ are model parameters relating to the degree of coupling between both phases, and v_B is the volume fraction of the dispersed phase B.

To apply Equation 1, the model parameters λ and $\varnothing$ have to be determined. They are derived from the calculated modulus-temperature curves which best fit the experimental data of Figure 4. To perform these calculations, one of the components has to be taken as the continuous phase. For the 75/25 and 50/50 blends, PC was taken as the continuous phase while for the 25/75 blend, PST was taken as this phase. This choice is based on the morphological study and the mechanical behavior reported earlier. The λ and $\varnothing$ values used to fit the data are reported in Table II. A fairly good agreement is found in the temperature range between 95° and 140°C.

Table II. Takayanagi Model Parameters of PC–PST Blends

Sample	*Volume Fraction of Dispersed Phase*	λ	ϕ
75/25	0.28 (PST)	0.57	0.49
50/50	0.53 (PST)	0.90	0.58
25/75	0.23 (PC)	0.40	0.58

Deduction of Shift Factors. Time–temperature shift factors for the blends were obtained by shifting the experimental relaxation isotherms to the calculated master curves (*10*). The temperature and time dependence of the shift factors of the 75/25 and 50/50 blends are represented in Figures 10a and 10b at $t = 10$ sec and $t = 1000$ sec for a reference temperature of 140°C. The empirically determined shift factors of the pure components are given in these figures by dotted lines; their temperature dependence is of the WLF type.

The calculated shift factors for the 75/25 and 50/50 blends in the low temperature region (below 100°C) are close to the empirical shift factors for the pure PST phase. Above 140°C, a WLF-type behavior is found but with important deviations from PC. In between, the shift factors are time and temperature dependent. For the 25/75 blend (Figure 10c), no time dependence of log a_T is found because time–temperature superposition is valid over the whole temperature domain. The relaxation behavior of this blend is completely dominated by the PST phase. The good agreement between the calculated and empirical values of the shift factors confirms again the validity of the mechanical model.

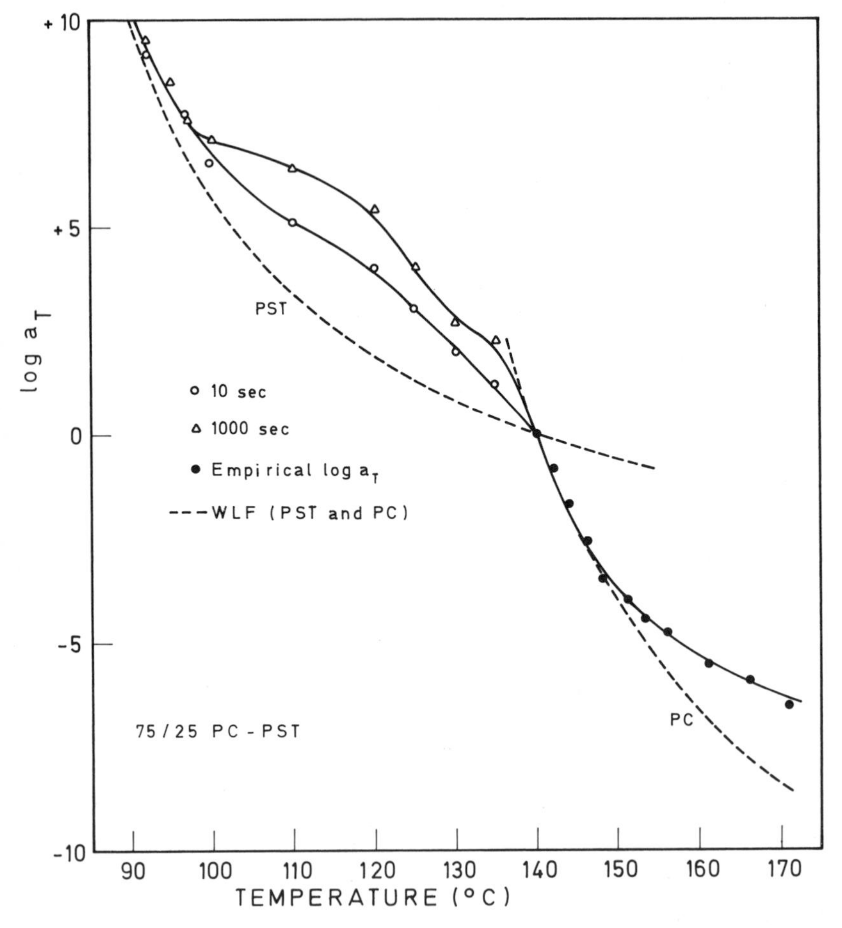

Figure 10a. Shift factors, log a_T, as a function of temperature for the 75/25 blend at 10 and 1000 sec; reference temperature: 140°C

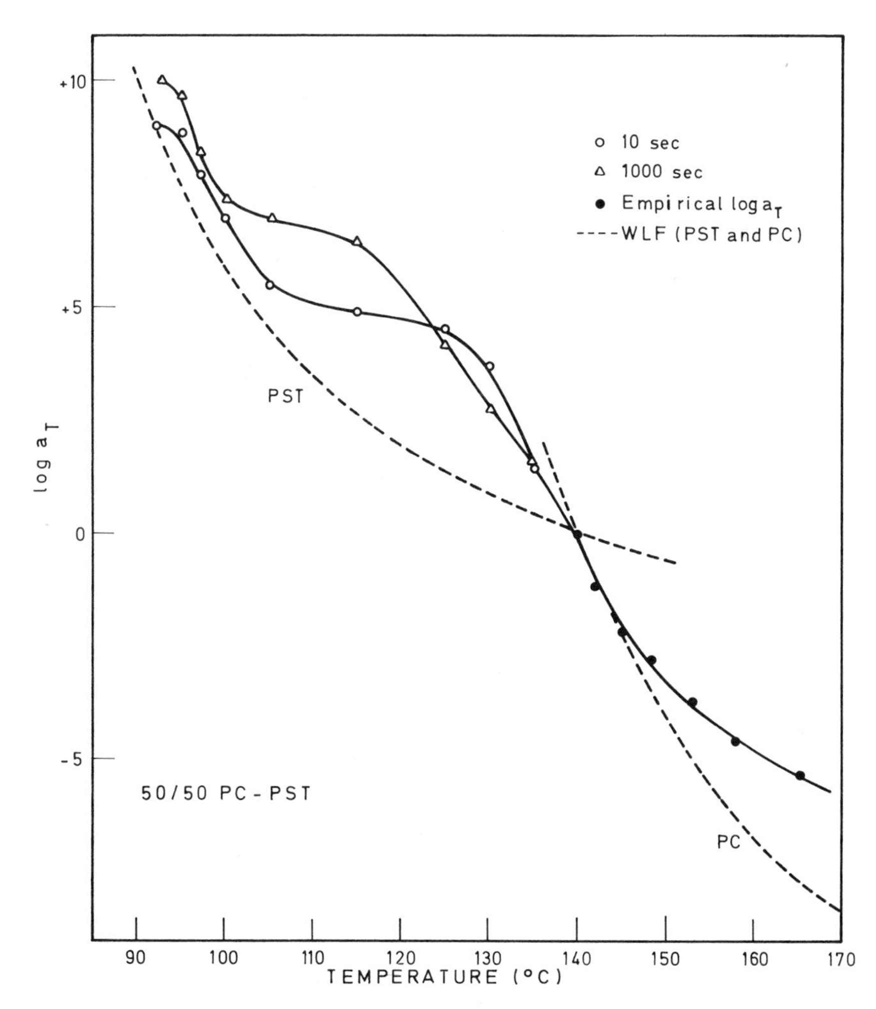

Figure 10b. Shift factors, log a_T, as a function of temperature for the 50/50 blend at 10 and 1000 sec; reference temperature: 140°C

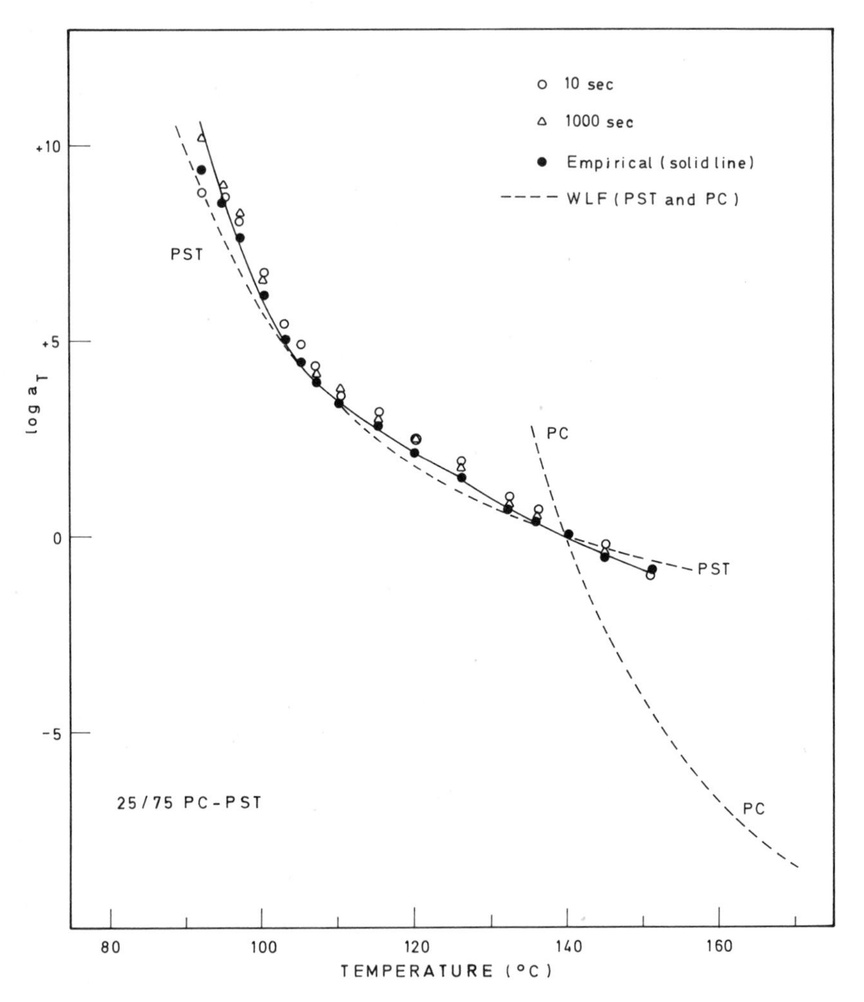

Figure 10c. Shift factors, log a_T, as a function of temperature for the 25/75 blend at 10 and 1000 sec; reference temperature: 140°C

Stress–Strain Properties

Stress–Strain Curves. The tensile stress–strain behavior of the blends in which PC is the continuous phase (blends with 5, 10, 20, and 25 wt% PST) also has been investigated. Some preliminary results regarding the influence of composition, strain rate, and temperature on the yield and fracture behavior of these blends will be reported.

It has been observed that these blends behave as toughened systems in the temperature range between the glass transitions of the two polymers. In this temperature domain, the PC matrix is in the glassy state and the dispersed PST phase is in the rubbery state. Below the glass transition of PST, the blends are very brittle, although pure PC exhibits ductile behavior. Figure 11 shows the effect of the PST content on the stress–strain curves at 120°C (strain rate $\dot{\epsilon} = 10\%/\text{min}$). Pure PC and the different blends exhibit a yield point and thus plastic deformation. As can be seen from this figure, the yield stress σ_y decreases and the elongation at rupture ϵ_r increases with increasing PST content. The blend with 25 wt% PST (75/25 blend) shows very pronounced ductile behavior; the elongation at rupture ϵ_r of this specimen is remarkably high ($\epsilon_r = 95\%$). The toughness, defined as the amount of energy stored in a sample during deformation prior to failure, can be deduced from the area under the stress–strain curve. For the blends studied, it can be observed that the toughness rises markedly with increasing PST content.

The influence of the strain rate on the tensile behavior has been investigated for the 75/25 blend at 120°C and is represented in Figure 12. The yield stress σ_y increases with increasing strain rate $\dot{\epsilon}$, but the fracture strain $\dot{\epsilon}_r$ is almost independent of the applied strain rate within the range considered. A plot of σ_y/T vs. log $\dot{\epsilon}$ at 120°C shows a linear relationship which means that the simplified Eyring equation (*11, 12, 13, 14*) describing the yielding process can be applied to our yield data. To obtain the characteristic Eyring parameters, the temperature dependence of the yield stress also has to be determined. These experiments still have to be carried out and the results will be reported elsewhere.

Mechanisms of Plastic Deformation. The mechanisms of plastic deformation of the 95/5, 90/10, 80/20, and 75/25 blends will now be examined. In glassy polymers, two different toughening mechanisms have been identified: crazing and shear deformation (*3, 4, 5*). Crazing is a tensile yielding process accompanied by molecular orientation in the stretching direction and extensive formation of microvoids. Crazes occur in a direction perpendicular to the tensile-stress axis and cause an increase in the sample volume. Shear yielding involves elongation by shearing under a certain angle with regard to the tensile-stress direction, with no

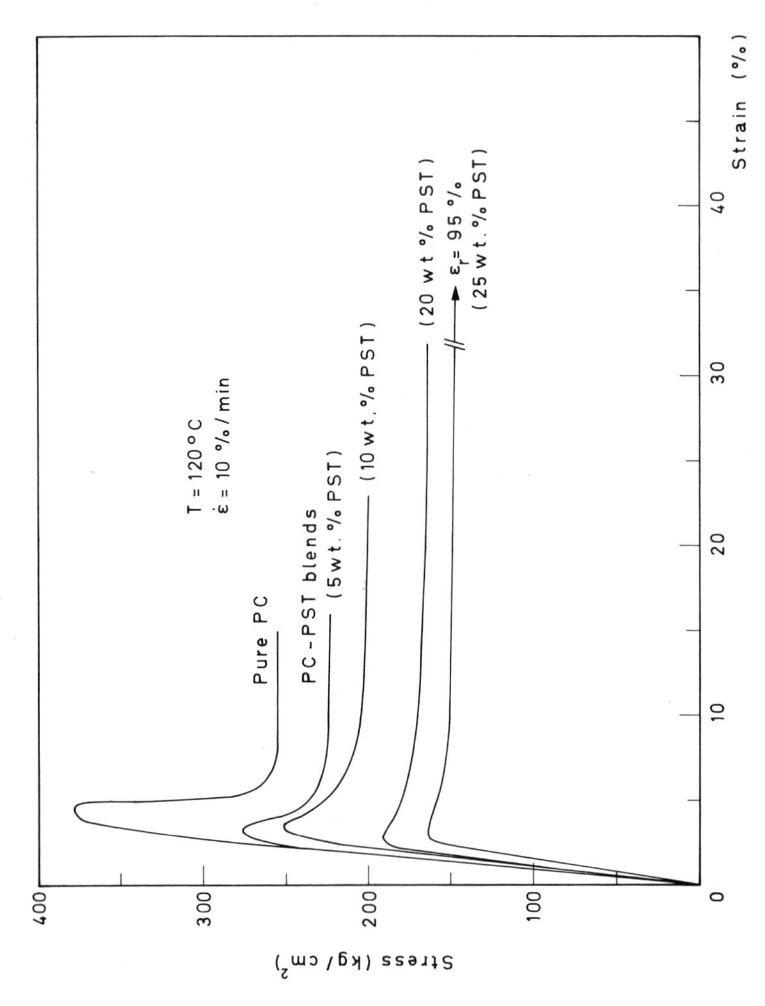

Figure 11. Stress–strain curves of PC–PST blends at $\dot{\varepsilon} = 10\%/min$ and 120°C. Effect of PST content.

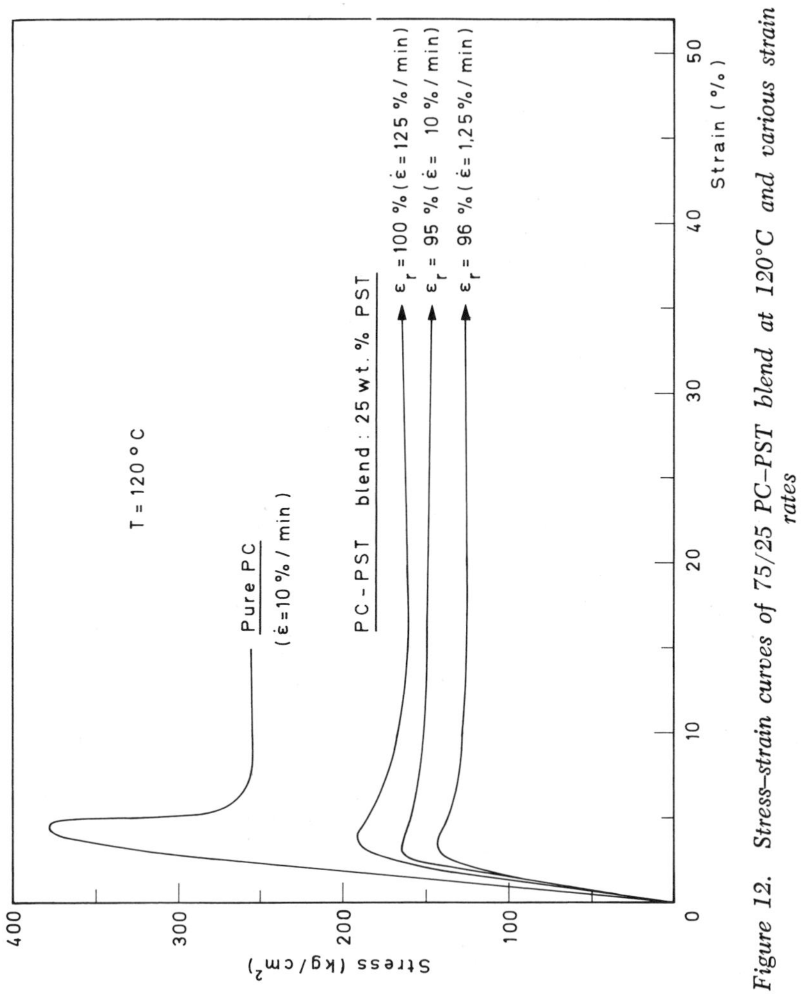

Figure 12. Stress–strain curves of 75/25 PC–PST blend at 120°C and various strain rates

change in sample density. It is well known that the impact strength and the toughness of glassy polymers can be improved by the incorporation of finely dispersed elastomer particles. When such a composite material is stretched, the stress concentrations around the rubber particles cause the glassy matrix to yield locally by allowing crazing and shear mechanisms to operate (*3, 4, 5*).

In the case of the blends studied, when the applied stress exceeds the yield point, the samples first show homogeneous whitening throughout their whole length, indicating that crazing occurs, and then undergo a localized deformation in the form of a neck. For the 75/25 blend, the

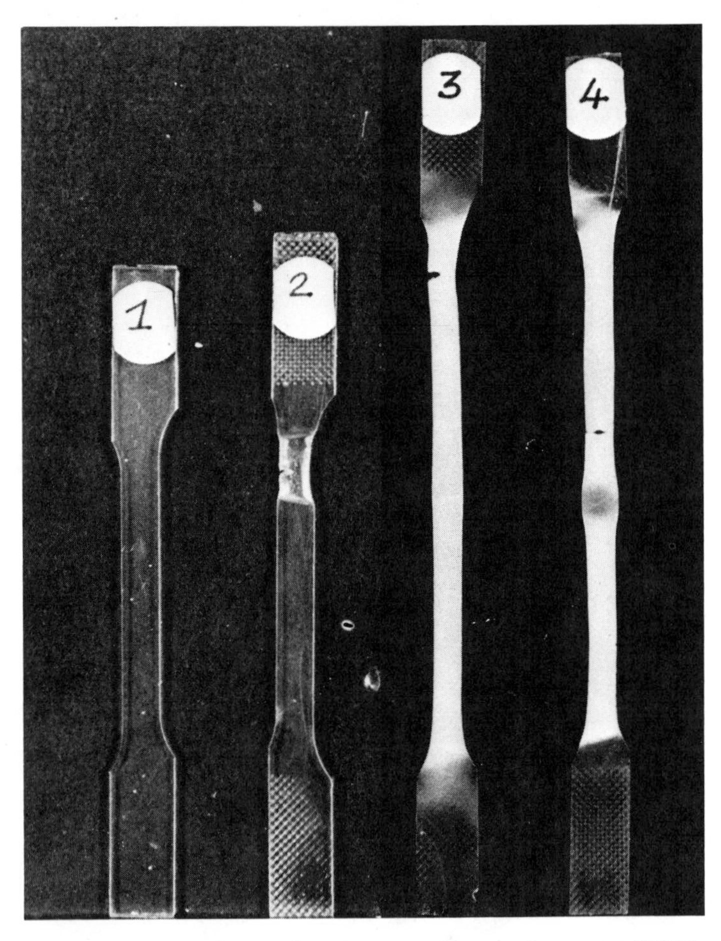

Figure 13. Samples used in stress–strain experiments at 120°C. (1) undeformed sample of 75/25 PC–PST blend; (2) sample of pure PC ($\dot{\varepsilon}$ = 10%/min); (3) sample of 75/25 PC–PST blend ($\dot{\varepsilon}$ = 1,25%/min); (4) sample of 75/25 PC–PST blend ($\dot{\varepsilon}$ = 1,25%/min).

Figure 14. Optical micrograph showing crazes and dispersed PST particles in sample 3 of Figure 13. Stretching direction is perpendicular to the crazes.

initiated neck subsequently grows until the whole sample is covered and then fracture occurs. For this sample, a very high fracture strain was obtained. For the 95/5, 90/10, and 80/20 blends, the neck only propagates over a certain length of the samples and early fracture results. Consequently, for these blends, the elongation at break is much less than that of the 75/25 blend.

The observation that the different blends exhibit stress whitening, necking, and cold drawing during a tensile test indicates that crazing and shear processes contribute to the tensile strain. This is illustrated in Figure 13 for samples of the 75/25 blend. The whitening observed in the samples during the initial stage of plastic deformation points out to the formation of numerous crazes. This is evident from the optical micrograph in Figure 14, where many small crazes are observed; they are induced by the dispersed PST particles. These particles correspond to the largest particles of the bimodal size distribution shown in Figure 1d. Shear bands appear very clearly in the necked zone of the samples (Figure 15).

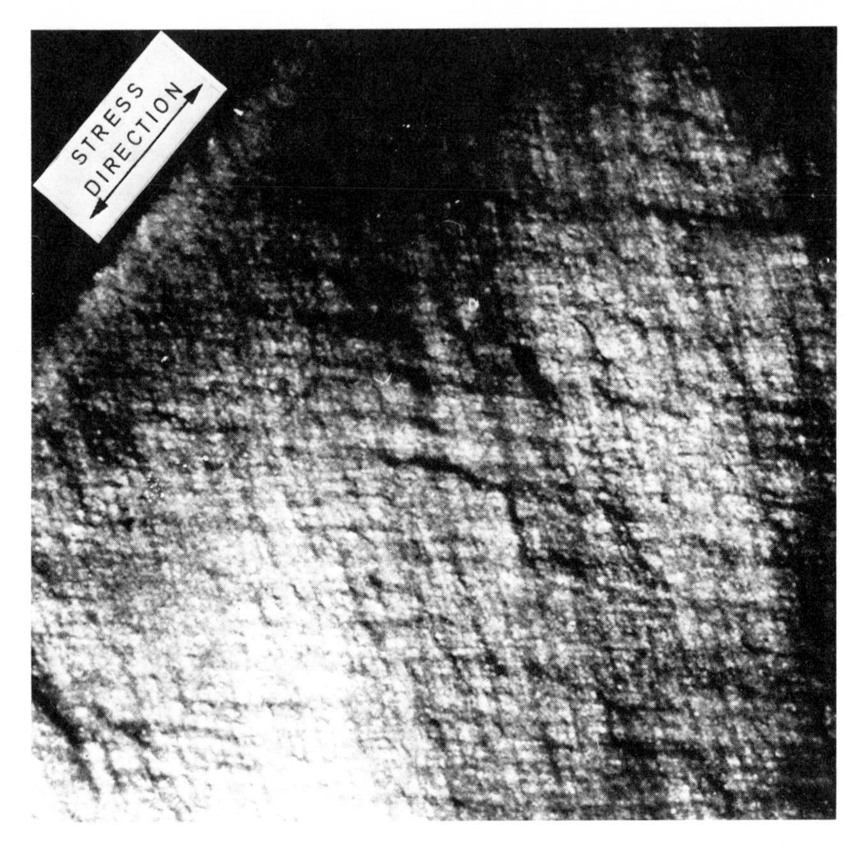

Figure 15. Photomicrograph of the surface of sample 3 in Figure 13 showing shear bands

Consequently, the plastic deformation of the blends, in the temperature range between the glass transitions of the two phases, occurs partly by crazing and partly by shear band formation. Measurements of longitudinal and lateral strains can be used to evaluate the contribution of crazing and shear yielding to the extension of the samples (*15, 16*). This analysis is based on the assumption that shear yielding occurs at constant volume, so that any volume increase has to be attributed to crazing. For the 75/25 blend, these measurements show that about 85% of the total deformation arises from shear processes while crazing accounts for about 15%.

The whitening developed in the 95/5 and 90/10 blends with the smallest particles (Figure 1a and 1b) appears to be less than that in the 80/20 and 75/25 blends in which a bimodal distribution of particle sizes was observed (Figures 1c and 1d). These observations indicate that crazing is more pronounced in the blends with the largest particles.

From this it is concluded that shear deformation is enhanced by the presence of relatively small particles whereas crazing is favored by larger particles.

Similar observations were made by Sultan and McGarry (*17*) in rubber-modified epoxy resins and by Groeninckx (*18*) in rubber-toughened gelatin. Among the different blends studied, the 75/25 blend exhibits the highest toughness. The relatively high concentration of PST particles (28 vol%) and their bimodal size distribution, which allow both shear and craze deformations to occur, give excellent tensile properties to this blend. As a conclusion, the improved mechanical properties exhibited by the PC–PST blends studied between the glass transitions of the pure components and the combination of crazing and shear yielding make them very attractive for a further fundamental study of the kinetics of plastic deformation mechanisms.

Acknowledgment

The authors are indebted to N. Overbergh for electron microscope investigations, A. Van Dormael for the preparation of the ultramicrotome sections and R. De Wil for the photographic work. They are also indebted to the Ministry of Scientific Programming and to the National Fonds voor Wetenschappelijk Onderzoek (N.F.W.O.) for equipment and financial support. We also wish to thank the Belgian Ministry of National Education for a fellowship (S.C.) and to thank the Ministry of Education of India.

Literature Cited

1. Fesko, D. G., Tschoegl, N. W., *J. Polym. Sci., Part C* (1971) **35**, 51.
2. Sperling, L. H., Ed., "Recent Advances in Polymer Blends, Grafts, and Blocks," *Polym. Sci. Technol.* (1974) **4.**
3. Kambour, R. P., *J. Polym. Sci., Macromol. Rev.* (1973) **7**, 1.
4. Bucknall, C. B., "Toughened Plastics," p. 359, Applied Science, London, 1977.
5. Haward, R. N., "The Physics of Glassy Polymers," p. 620, Applied Science, London, 1973.
6. Henning Kausch, H., Hassell, J. A., Jaiffee, R. I., "Deformation and Fracture of High Polymers," p. 644, Plenum, New York, 1973.
7. Mercier, J. P., Groeninckx, G., *Rheol. Acta* (1969) **8**, 510.
8. Van Cutsem, R., Thesis License, Université Catholique de Louvain, Belgium (1971).
9. Takayanagi, M., Uemura, S., Minami, S., *J. Polym. Sci., Part C* (1964) **5**, 113.
10. Kaplan, D., Tschoegl, N. W., "Recent Advances in Polymer Blends, Grafts and Blocks," *Polym. Sci. Technol.* (1974) **4**, 415.
11. Eyring, H., *J .Chem. Phys.* (1936) **4**, 283.
12. Ree, T., Eyring, H., "Rheology," F. R. Eirich, Ed., Vol. II, Chap. III, Academic, New York, 1958.

13. Bauwens-Crowet, C., Bauwens, J. C., Homès, G., *J. Polym. Sci., Part A-2* (1969) **7**, 735.
14. Bauwens, J. C., Bauwens-Crowet, C., Homès, G., *J. Polym. Sci., Part A-2* (1969) **7**, 1745.
15. Bucknall, C. B., Clayton, D., Keast, Wendy E., *J. Mater. Sci.* (1972) **7**, 1443.
16. Bucknall, C. B., Drinkwater, I. C., *J. Mater. Sci.* (1973) **8**, 1800.
17. Sultan, J. N., McGarry, F. J., *Polym. Eng. Sci.* (1973) **13**, 29.
18. Groeninckx, G., in press.

RECEIVED June 5, 1978.

19

Morphology of Low Density Polyethylene/EPDM Blends Having Tensile Strength Synergism

GEOFFREY A. LINDSAY, CHLOE J. SINGLETON,
CHARLES J. CARMAN, and RONALD W. SMITH

B. F. Goodrich Co., Chemical and Corporate Divisions, 9921 Brecksville Road, Brecksville, OH 44141

Blends of low density polyethylene (LDPE) with certain crystalline ethylene–propylene–diene monomer (EPDM) rubbers have tensile strengths greater than either component. Differential calorimetry scans of these blends and their components have been compared with those of low strength blends of LDPE with amorphous EPDM rubber. Upon cooling the high strength blend from the melt, LDPE crystallites appear to nucleate the crystallization of some high ethylene segments of the EPDM rubber. Nodules, believed to be crystallites, were observed on ion-etched surfaces by scanning electron microscopy. High strength blends had a 50% larger nodule size than other blends, which may also result from high ethylene segments of EPDM crystallized on the surfaces of LDPE crystallites.

Batiuk, Herman, and Healy discovered that blends of certain ethylene–propylene–diene monomer (EPDM) rubbers with polyethylene have surprisingly high tensile strengths (*1*). Several patents have been published on these and related blends (*2, 3, 4*). These blends are not cross-linked in the normal sense, but can be melted and molded repeatedly, retaining their high strength upon cooling.

We have begun to study the morphology of these blends in an attempt to learn what causes the tensile strength synergism. This chapter gives our initial results and interpretations.

0-8412-0457-8/79/33-176-367$05.00/0

Table I. Physical Properties of Polymers

Polymer type	*EPDM*	*EPDM*	*LDPE*
Trade name	Epcar 845	Epcar 847	Epolene C-14
Obtained from	B. F. Goodrich	B. F. Goodrich	Eastman
Crystallinity (%)[a]	0	6	32
Density, g/cm^3	.87	.88	.918
Wt % ethylene	56	71	100
Wt % ENB[b]	4	4	0
Mooney viscosity[c]	56	55	—
Melt index, g/ 10 min[d]	—	—	1.6

[a] From x-ray diffraction scans.
[b] Ethylidene norbornene.
[c] Four minute warm up/four minute run at 120°C.
[d] ASTM D1238 2.16 kg/190°C.

Experimental

The polymers used and some of their physical properties are listed in Table I. Polymers were mixed and blended on a two-roll mill at 450 K. Samples were compression molded at 450 K for 7 min and cooled in the press with tap water for 5 min. ASTM D412 6.35-mm (¼ in.) dumbbells were cut parallel to the mill grain from sheets having 1.9-mm (75 mils) thickness. Instron tensile tests were carried out at least 48 hr after molding. Pull rate was 50.8 cm/min (20 in./min).

A Perkin–Elmer DSC-2 differential scanning calorimeter equipped with an Automatic Scanning Zero was used. DSC conditions were 20 K/min on a 20.8-mW sensitivity with a 24-mg aluminum reference.

A JEOL JSM 50A scanning electron microscope was used. Polymer surfaces were etched with a Hummer II sputtering device using argon ions and coated with a 20-nm thick sputtered film of gold/palladium alloy.

Results and Discussion

The structure of the EPDM rubber is very important to the strength of these blends. Those EPDM rubbers which have crystallinity melting above room temperature give blends with the highest tensile strengths. EPDM rubbers having crystallinity levels (by x-ray diffraction measurements) in the range 1–20% are of greatest interest. Above 20%, the EPDM behaves more like a plastic than a rubber. Blends of amorphous EPDM rubber with LDPE do not have tensile strength synergism.

Commercial EPDM rubbers have monomer sequence distributions ranging between random and alternating in the first-order Markovian sense. From our previous experience they must contain at least about 68 wt % ethylene to have runs of ethylene long enough to crystallize above room temperature at rest.

A low level of crystallinity (1–3%) is sometimes difficult to detect by x-ray diffraction techniques. Differential scanning calorimetry (DSC) does detect these first-order thermal transitions. However, crystallinity is difficult to quantify by DSC because of uncertainties in locating the baseline. This uncertainty is a result of an overlap of both the EPDM T_g and LDPE melting endotherm with the two ends of the EPDM melting endotherm. ^{13}C NMR has proved to be a good method of characterizing the monomer sequence distribution of ethylene–propylene copolymers (*3, 4, 5*). Those EPDM rubbers with an appropriate fraction of long ethylene runs also give blends with polyethylene having unusually high tensile strengths. We plan to make this the subject of a future paper.

Figure 1 shows tensile strengths for the amorphous and crystalline EPDM blended with various levels of LDPE. The strength of the crystalline EPDM blend goes through a maximum between 20–40 wt % LDPE, where it is greater than either component. The strength of the amorphous EPDM blend is much lower.

Tensile properties of these two EPDM rubbers are fairly representative of what can be obtained using commercial polymers having a 125°C Mooney viscosity near 55. Tensile strengths as high as 2900 psi have

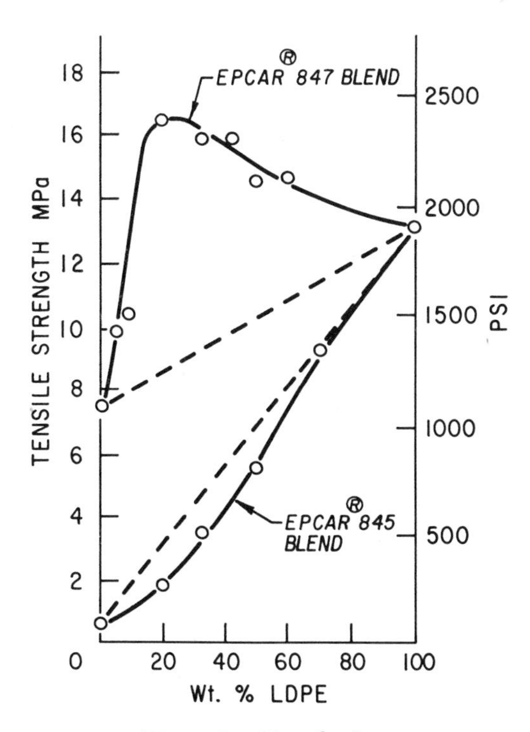

Figure 1. Tensile data

been obtained with blends of LDPE and an EPDM having a slightly higher ethylene content. Lower tensile strengths have been reported for blends of LDPE and EP copolymer having only 43 wt % ethylene (*6*). Tensile strength also decreases as the molecular weight of the EP(D)M decreases.

Figure 2 shows tensile-yield strengths for blends of the crystalline EPDM with various levels of LDPE. The curve increases monotonically as expected if no phase inversion occurs. Since amorphous LDPE has a glass-transition temperature near that of EPDM (*7*) and since the LPDE has only 27% crystallinity, one should not expect a rubber-to-rigid phase transition.

Our wide-angle x-ray diffraction (WAXD) measurements have shown that stretching pure Epcar 847 induces considerable orientation and crystallinity—just as is the case with natural rubber. The blend of Epcar 847 with 50 phr LDPE also undergoes stress-induced crystallization. However, quantitative or qualitative differences in amounts of

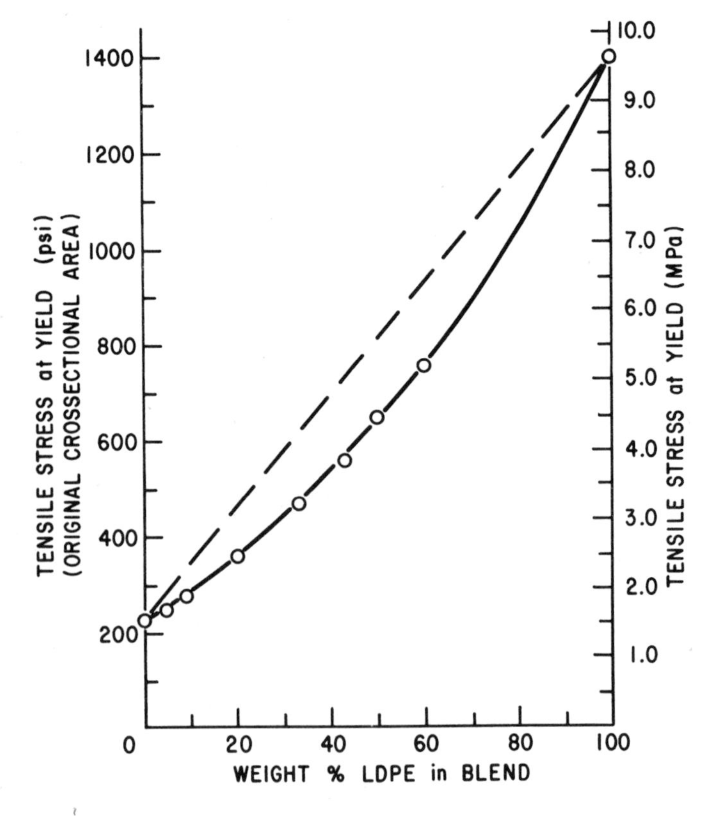

Figure 2. Tensile yield strength vs. composition for blends of Epcar 847 and LDPE

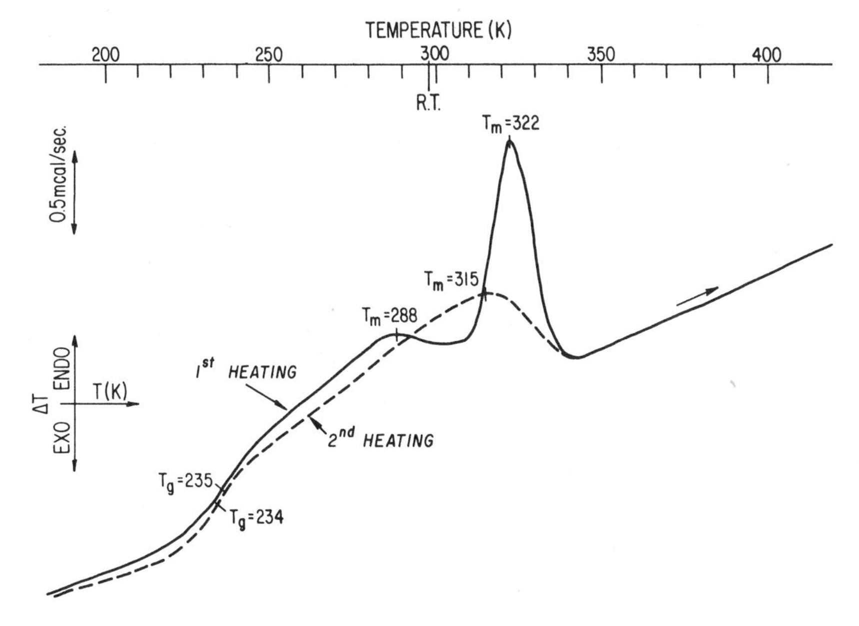

Figure 3. DSC scans of Epcar 847

Figure 3. DSC SCANS of EPCAR 847.

stress-induced orientation between the pure crystalline rubber and its blend with LDPE are impossible to measure. The amorphous Epcar 845 and its blend with LDPE show no stress-induced orientation or crystallization by WAXD measurements. The crystallinity as measured by WAXD agrees with that calculated from a weighted average of the crystallinity of the individual components.

In our differential scanning calorimetry (DSC) measurements, all samples were quenched in liquid nitrogen from room temperature to 170 K, heated at 20 K/min to 470 K, cooled at 20 K/min to 170 K, and reheated to 470 K. Figure 3 shows the first and second heating scans of the crystalline EPDM rubber. The first DSC heating scans give good representations of the morphology present in tensile dumbbell specimens which were annealed at least two days at room temperature. The location and shape of the endotherm changes considerably in the second heating scan. Shih and Cluff also have shown how sensitive the endothermic response of crystalline ethylene copolymers can be to the thermal history of the sample (*8*).

Figure 4 shows the first and second DSC heating scans of the blend of crystalline EPDM with 50 phr LDPE. The LDPE endotherm changes very little between the first and second heating scan. The rubber's endotherms in the blend change similar to those of the pure rubber as shown in Figure 3.

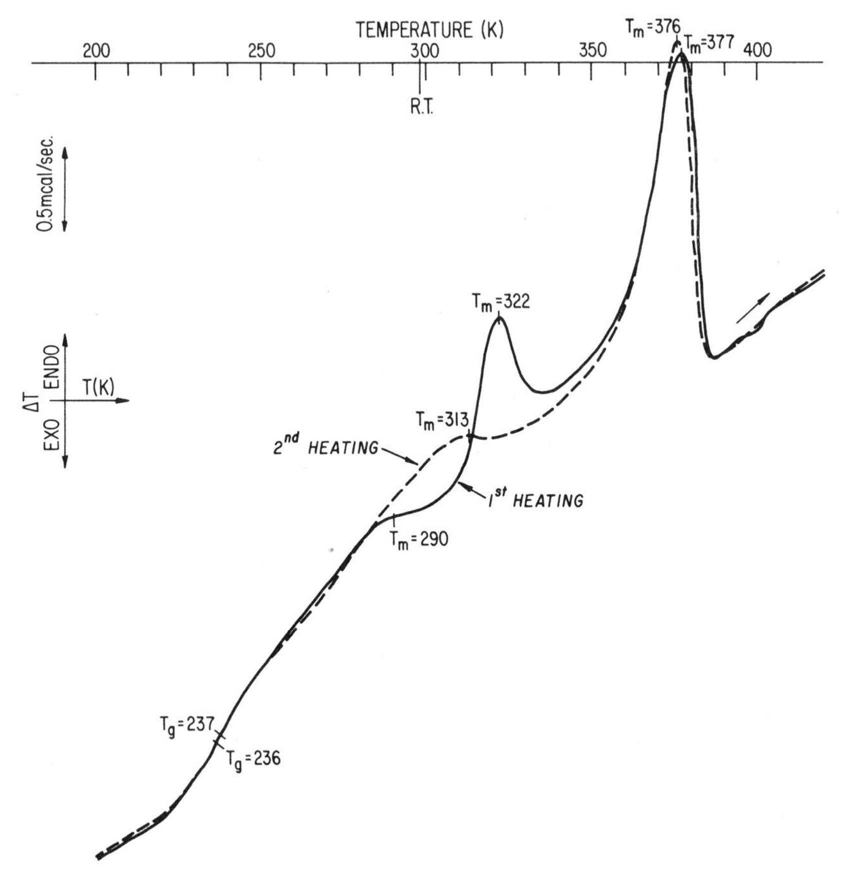

Figure 4. DSC scans of a blend of Epcar 847 with 50 phr LDPE

Figure 5 compares the second DSC heating scans of pure LDPE, a blend of amorphous EPDM with 50 phr LDPE, and a blend of crystalline EPDM with 50 phr LDPE. The LDPE endotherm is decreased three or four kelvins in the blends compared with pure LDPE. This could be caused by: (a) slightly smaller or less perfect crystallites in the blend and/or (b) better heat transfer to the LPDE crystallites in the blend attributable to intimate contact with molten EPDM chains .

Figures 6 and 7 show DSC cooling scans of the pure rubbers, LDPE, and blends of rubber with 50 phr LDPE. The crystallization temperature (T_{cr}) of LDPE (largest exothermic peak) is decreased six kelvins in the presence of the amorphous EPDM and is decreased 12 kelvins in the presence of the crystalline EPDM. This T_{cr} decrease could be attributable to partial solubility of EPDM rubber and LDPE. The soluble EPDM chains would have to be expelled from the LDPE crystal-

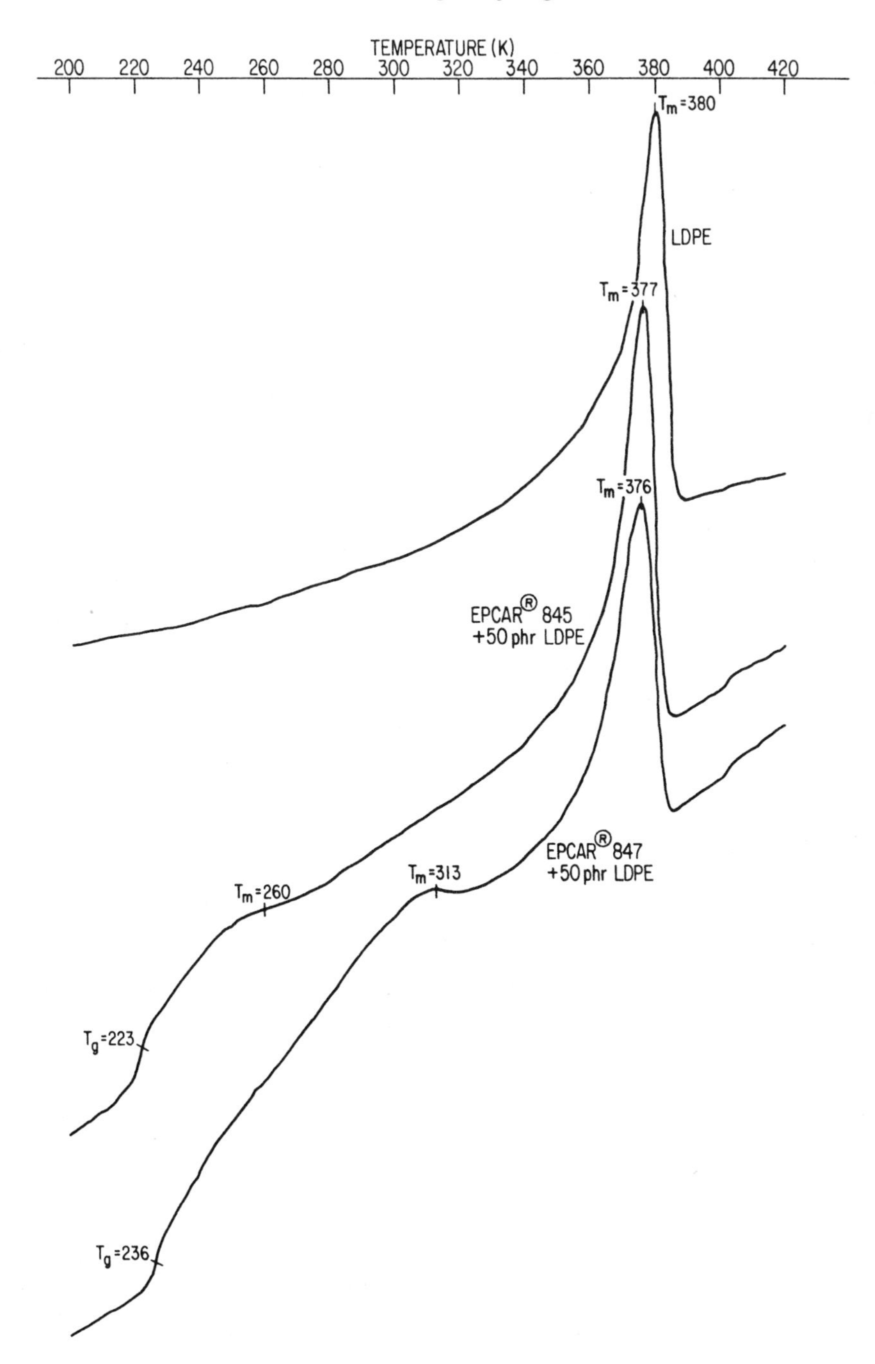

Figure 5. DSC heating scans immediately following cooling from the melt

lization zones. This would retard the LDPE crystallization rate and lower T_{cr}. Because the crystalline EPDM has a higher ethylene content than the amorphous EPDM, it should be more soluble with LDPE.

The DSC cooling scan of LDPE shows a small exotherm at 325 K (*see* Figure 6). This could be caused by less perfect crystals or paracrystalline LDPE regions. Perhaps it represents LDPE segments containing the branch points (*9*) which must lie outside the normal chain-folded lamellae. Depending on crystallization conditions, some segments are normally trapped as loose loops or cilia (*10*). In the amorphous rubber/LDPE blend, this small exotherm at 325 K has not been shifted. However, in the crystalline rubber/LDPE blend (*see* Figure 7),

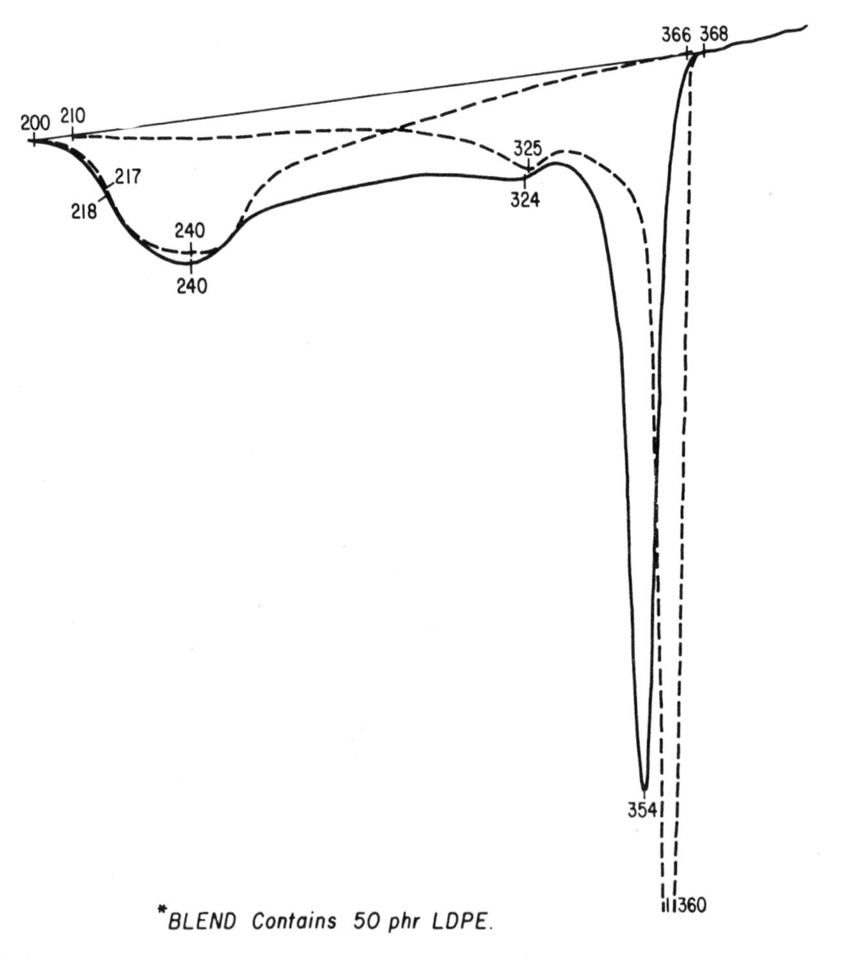

Figure 6. DSC cooling scans of an Epcar 845/LDPE blend (——) and its individual components (– – –). Blend contains 50 phr LDPE.

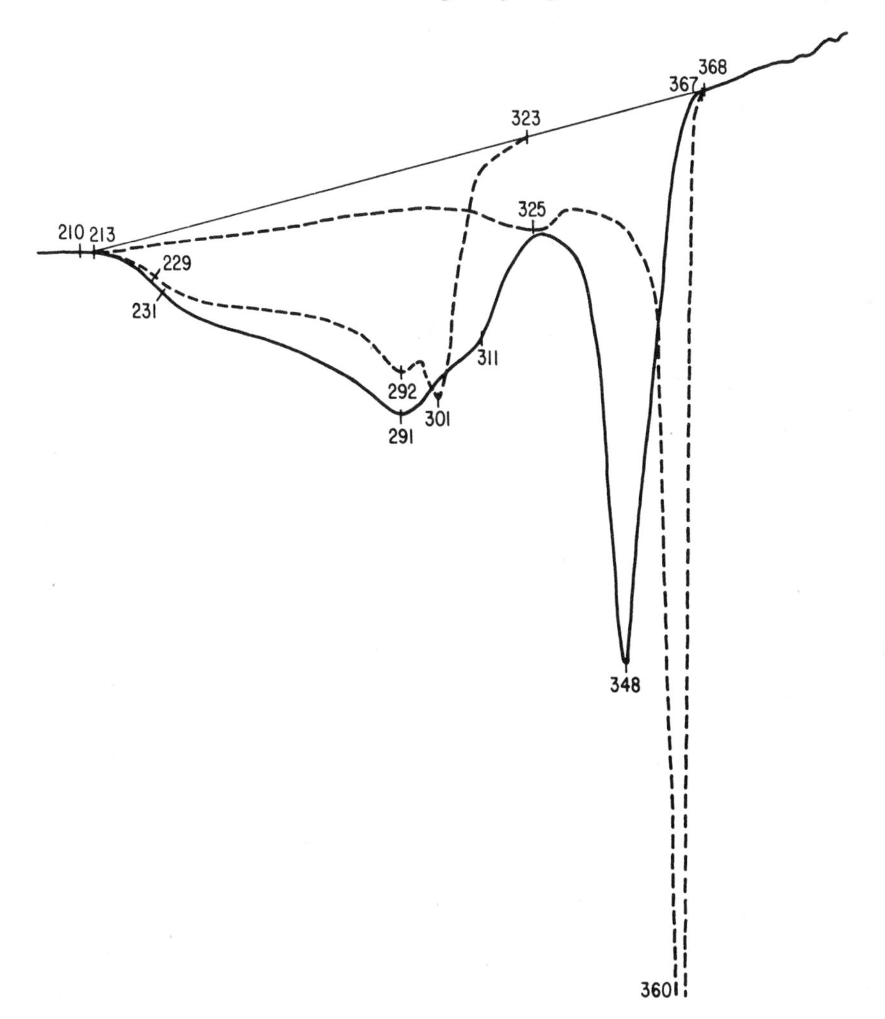

Figure 7. DSC cooling scans of an Epcar 847/LDPE blend (——) and its individual components (– – –). Blend contains 50 phr LDPE.

this small exotherm is absent or shifted and overlapped. This may indicate an interaction of this LDPE structure and high ethylene EPDM segments.

In Figure 7 one can see that the temperature of a portion of the EPDM exotherm is increased in the blend compared with the pure EPDM. A possible explanation is that LDPE crystallites are nucleating the crystallization of high-ethylene-EPDM segments. In this case, EPDM crystallites would form layers on LDPE crystallites.

Figure 8 shows scanning electron photomicrographs of ion-etched surfaces of the three pure polymers. Presumably, argon-ion bombard-

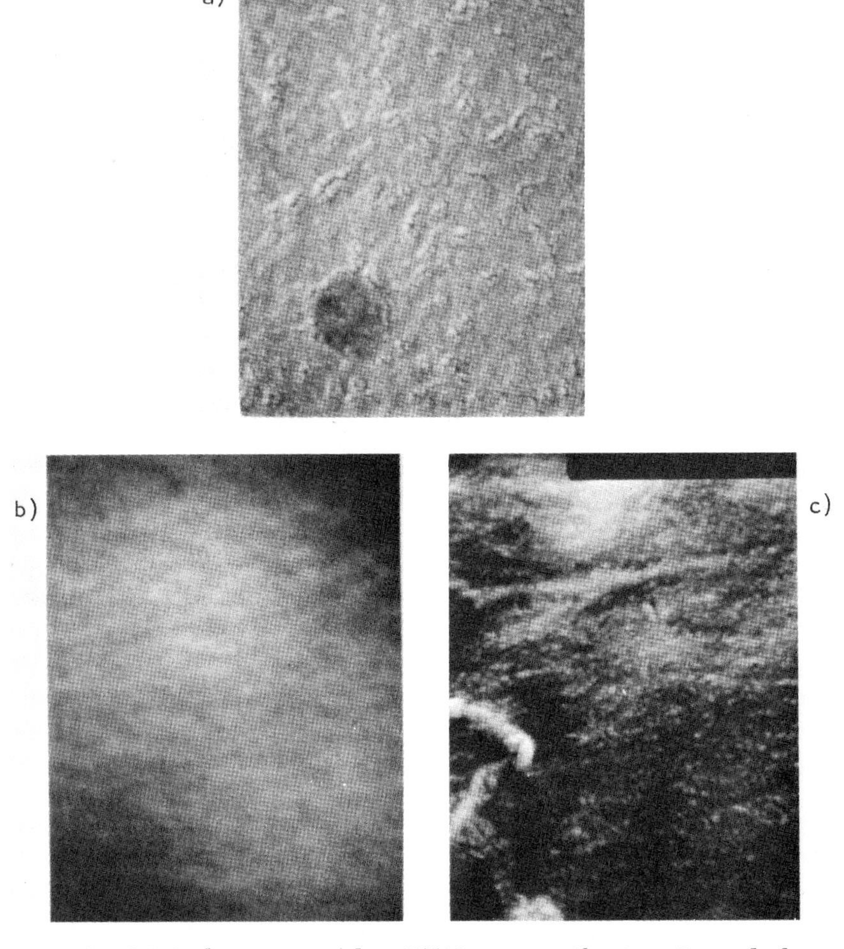

Figure 8. SEM photomicrographs. 25500 × magnification. Ion-etched surfaces. Polymers and % crystallinity as follows: (a) (top) *Epcar 847 (EPDM), 6; (b)* (bottom left) *Epcar 845 (EPDM), 0; (c)* (bottom right) *Epolene C-14-(LDPE), 32.*

ment selectively etches a soft amorphous polymer to a greater extent than a hard crystalline polymer. Hence, the nodules seen in these photomicrographs are probably crystallite residues. Pure amorphous Epcar 845 has no nodules visible by SEM as expected. Diameters of nodules in pure LDPE and Epcar 847 are about 60 nm on the average.

Figure 9 shows SEM photomicrographs of ion-etched surfaces of blends of LDPE/amorphous Epcar 845 and of LDPE/crystalline Epcar 847. Nodule diameters of the LDPE/amorphous rubber blend average

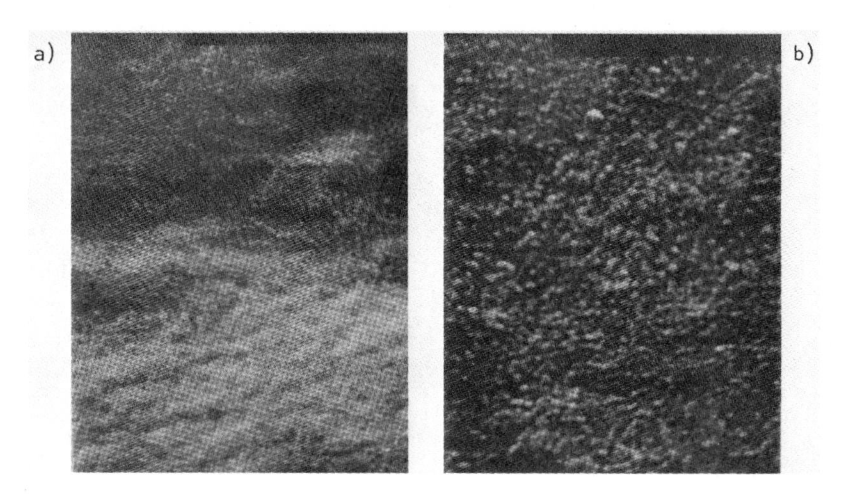

Figure 9. SEM photomicrographs. 24000 ×. Ion-etched surfaces. (a) (left) *Epcar 845 + 50 phr LDPE (amorphous EPDM). (b)* (right) *Epcar 847 + 50 phr LDPE (crystalline EPDM).*

about 60 nm—the same as for pure LDPE. Nodule diameters of the LDPE/crystalline rubber blend average about 90 nm. This larger size may be attributable to EPDM crystallized on surfaces of LDPE crystallites. This is in line with DSC data which indicates LDPE crystallites nucleate the crystallization of high ethylene EPDM.

One might expect the nodule diameter of pure LDPE to be the same as that in the amorphous rubber/LDPE blend. This could result if the same proportion of LDPE nucleated the crystals and if no amorphous EPDM lay inside the LDPE crystallites. However, the concentration of crystallites would be lower in the blend. It is impossible for us to measure the concentration of crystallites in this blend. The resolution is inadequate and the etching depth is not accurately known. We will have to look at blends containing less LDPE to see if the crystallite concentration decreases. No spherulites are seen in these blends by polarized optical microscopy. However, these nodules are too small for optical resolution, and may indeed be spherulites or aggregates of lamellae.

Figure 10 compares a transmission electron (TEM) photomicrograph, (a), with SEM photomicrographs, (b) and (c), of the same blend of crystalline EPDM and 50 phr LDPE. A thin section of the blend was stained with osmium tetroxide and examined with our Phillips EM100B TEM. In the TEM photomicrograph, 10a, the light spots are probably LDPE crystallites, and the gray regions are probably EPDM rubber (black spots are probably osmium residue). Figure 10b is a view of a solvent-etched surface which has been gently swabbed with xylene, a good solvent for amorphous EPDM. Figure 10c is the ion-etched surface.

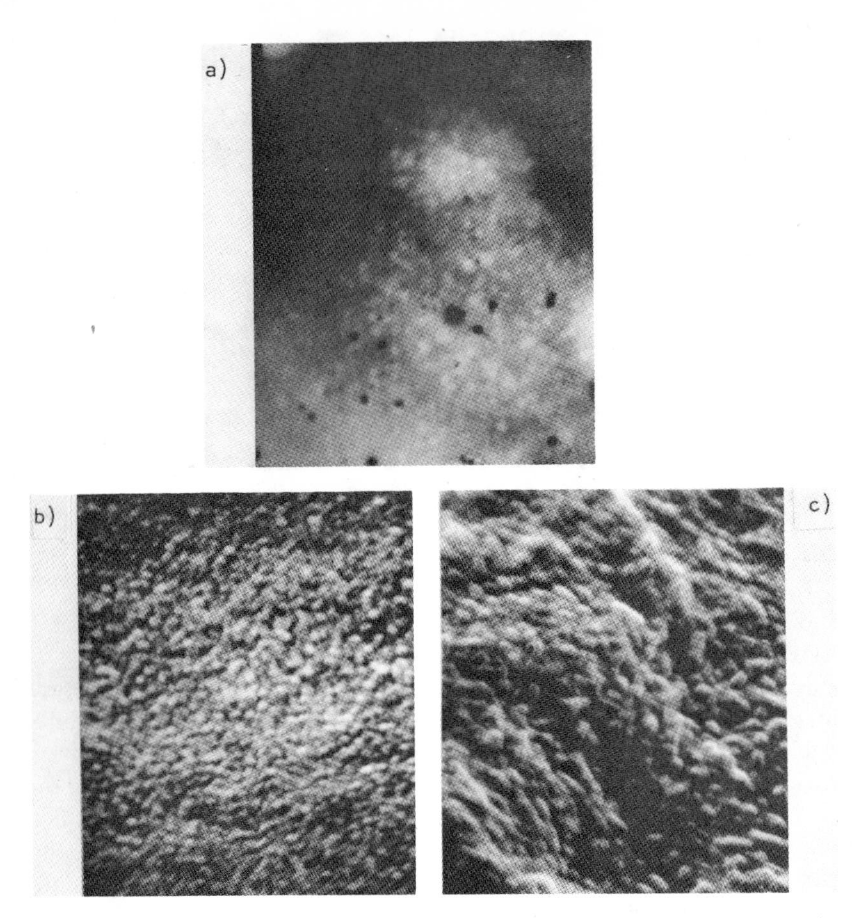

Figure 10. Epcar 847 + 50 phr LDPE. (a) (top) *TEM photomicrograph; 10560 ×; OsO_4-stained thin section. (b)* (bottom left) *SEM photomicrographs; 24000 ×; ion-etched surface; (c)* (bottom right) *xylene-etched surface.*

Crystallite diameters estimated from Figure 10a, 10b, and 10c, are 140, 120, and 90 nm, respectively. Hence, these three different sample preparation techniques give crystallite-diameter-size estimates of the same order of magnitude.

Summary and Conclusions

Blends of LDPE with certain partially crystalline EPDM rubbers exhibit tensile strengths greater than that of either pure component. Blends of LDPE with amorphous EPDM have tensile strengths less than expected from a weighted average of the pure components' tensile strengths.

DSC measurements show that the crystallization temperature of LDPE is decreased 6 and 12 kelvins in blends with EPDM rubber having 56 and 72 wt % ethylene respectively. We believe this indicates partial miscibility between melts of LDPE and high ethylene EPDM rubber. DSC cooling scans also give evidence that upon cooling the high ethylene EPDM/LDPE blend from the melt, the LDPE crystallites nucleate the crystallization of high ethylene EPDM segments.

SEM observations of ion-etched surfaces of these blends add more evidence that high ethylene EPDM rubber crystallizes on LDPE lamellae in the surface regions of LDPE crystallites. Hence, the morphology of these high tensile blends appears to consist of EPDM reinforced with many tiny LDPE spherulites—much like carbon-black-reinforced SBR. A fraction of high ethylene EPDM chains crystallizes in surface regions of neighboring LDPE spherulites, tying them together in a virtually crosslinked network. The crystalline EPDM rubber undergoes further crystallization upon stretching, just like natural rubber. The blend of crystalline EPDM with LDPE also stress crystallizes, and its special morphology must help distribute the tensile load more evenly.

Acknowledgment

The authors would like to thank the B. F. Goodrich Co. for granting permission to publish our findings.

Literature Cited

1. Batiuk, M., Herman, R. M., Healy, J. C., U.S. Patent **3,919,358** (1975), assigned to BFGoodrich.
2. Batiuk, M., Herman, R. M., Healy, J. C., U.S. Patent **3,941,859** (1976), assigned to BFGoodrich.
3. Carman, C. J., Batiuk, M., Herman, R. M., U.S. Patent **4,046,840** (1977), assigned to BFGoodrich.
4. Stricharczuk, P. T., U.S. Patent **4,036,912** (1977), assigned to BFGoodrich.
5. Carman, C. J., Harrington, R. A., Wilkes, C. E., *Macromolecules* (1977) **10,** 536.
6. Robertson, R. E., Paul, D. R., *J. Appl. Polym. Sci.* (1973) **17,** 2579.
7. Maurer, J. J., *Rubber Chem. Technol.* (1965) **38,** 979.
8. Shih, C. K., Cluff, E. F., *J. Appl. Polym. Sci.* (1977) **21,** 2885.
9. Bovey, F. A., Schilling, F. C., McCrakin, F. L., Wagner, H. L., *Macromolecules* (1976) **9,** 76.
10. Boyer, R. F., *J. Macromol. Sci., Phys.* (1973) **B8**(3–4), 503.

RECEIVED April 14, 1978.

20

Fracture Surface Morphology and Phase Relationships of Polystyrene/Poly(methyl Methacrylate) Systems

Low-Molecular-Weight Poly(methyl Methacrylate) in Polystyrene

RAYMOND D. PARENT[1] and EDWARD V. THOMPSON

Department of Chemical Engineering, University of Maine at Orono, Orono, ME 04473

Samples of low-molecular-weight poly(methyl methacrylate) (PMMA) in polystyrene (PS) of varying number-average molecular weight (3600–49,000) and composition (5–30 wt %) of PMMA were prepared, and the surface morphology and phase relationships studied by scanning electron microscopy. Three distinct types of phase relationships were observed: (1) a single phase consisting of PMMA dissolved in PS; (2) PMMA dispersed in PS; and (3) PS dispersed in PMMA. Values of the size of the dispersed particles are reported and are compared with the results of a prior study of the low-molecular-weight PS/PMMA system. Finally, the size of the dispersed particles and the various types of phase relationships are discussed in terms of the ternary poly(methyl methacrylate)/polystyrene/styrene phase diagram.

Considerable interest has developed in recent years concerning the morphology, phase relationships, and mechanical properties of two-phase, polyblend-like polymer systems; that is, systems in which one polymer component appears as a discontinuous, dispersed phase in a

[1] Present address: S. D. Warren and Co., Westbrook, ME 04092.

0-8412-0457-8/79/33-176-381$07.75/0

continuous phase of the other polymer. Probably the bulk of this effort has centered around high-impact plastics based on a rubbery, dispersed phase such as polybutadiene contained in a glassy, continuous phase such as polystyrene, although many other systems have been examined for various features. In spite of this interest, however, very little systematic data have appeared, in particular data concerned with the influence of molecular weight and concentration of the dispersed phase on the properties of the final system. In this and several other reports we will discuss results that we have obtained for the polystyrene/poly(methyl methacrylate) system with the aim of helping to elucidate the dependence of morphology and phase behavior, and in later reports, mechanical properties, on the molecular weight and weight percent of the dispersed phase. First, however, before proceeding to a description of the experimental procedures and a discussion of the results, it seems appropriate to briefly review prior work that bears on the present report.

Studies of two-phase, polymer/polymer systems have been reported from at least the turn of the century (*1*); interest, if only empirical, in polymer/polymer solubility or compatibility (or incompatibility, the far more usual situation) developed in the 1940s (*2*) and 1950s (*3*). The development of the technology and subsequent manufacture of systems based on two incompatible polymers, so-called high-impact plastics, followed; microscopic observations of the structures of the dispersed phase were probably first made in a study of high-impact polystyrene by Claver and Merz (*4*). Shortly thereafter, Traylor reported studies involving the use of phase contrast microscopy to investigate two-phase systems (*5*); more recently, transmission electron microscopy has been applied to the study of the dispersed phase in high-impact materials (*6, 7, 8, 9*).

It has been suggested that both the size of the dispersed particles (*10, 11*) and the amount of the dispersed phase (*12*) are of critical importance in determining the mechanical properties and, in particular, the impact strength of high-impact plastics. These observations are based on the belief that the dispersed particles dissipate energy of impact by the formation of countless, small internal crazes (*13, 14*). Baer studied the effect of variations of polybutadiene molecular weight on the size of dispersed particles and found that at molecular weights above 60,000 the size increases linearly with molecular weight when plotted on a log–log plot (*15*). Further, Moore demonstrated that the concentration of the original polymer in the initial polymer/monomer mixture is an important factor in determining the size of the dispersed particles (*16*). Unfortunately, these data are incomplete in that the systematic variation of both molecular weight and concentration of the original polymer was not investigated. In a recent report we described the results of a study of the polystyrene/poly(methyl methacrylate) system in which the molecular weight and weight percent of the initial polystyrene was varied from 2100 to 49,000 and 5 to 40%, respectively (*17, 18*). These results will be reviewed in subsequent sections and compared with the results to be presented in this chapter.

The phase relationships of two-phase polymer systems also have been of considerable interest in recent years. In an important series of papers, Molau and co-workers (*19–24*) studied systems, which were denoted POO emulsions (polymeric oil-in-oil), prepared by dissolving a given polymer in monomer and then polymerizing the monomer. During polymerizations of this type the composition of the respective phases reverses, and a phase inversion process was proposed to explain this. A similar process has been suggested as the mechanism by which polybutadiene forms the dispersed phase in the manufacture of high-impact polystyrenes (*22, 25*). Recently, Kruse has pointed out that this phase-inversion point may correspond to that point on a ternary phase diagram at which the reaction line bisects a tie line (*26*), and we have advanced a similar point of view in our earlier reports (*17, 18, 27*).

Because of the very limited data available correlating both molecular weight and concentration of the initial polymer to the size and distribution of the dispersed phase resulting after polymerization, we undertook the study of the polystyrene/poly(methyl methacrylate) system to help elucidate these relationships. In particular, we were especially interested in the morphology and phase relationships after polymerization and the nature of the dispersed and continuous phases. The results for the case in which polystyrene is the initial polymer present have been reported (*18*), and in this chapter we will discuss the situation when poly(methyl methacrylate) is the initial polymer. The results of a morphological study using scanning electron microscopy and the particle size and distribution will be described, and the various phase relationships will be discussed in terms of the ternary poly(methyl methacrylate)/polystyrene/styrene phase diagram.

Experimental Procedures

The polymers chosen for the initial stages of our studies were poly(methyl methacrylate) and polystyrene. There were several reasons for this choice, the most important being: (1) well-characterized, low-molecular-weight monodisperse samples of poly(methyl methacrylate) and polystyrene are readily available, or at least relatively easy to synthesize; and (2) the poly(methyl methacrylate)/polystyrene system is especially amenable to study by scanning electron microscopy, as we will discuss in more detail below. In what now follows we describe the various synthesis and experimental techniques that were used during our studies.

Styrene (hereafter denoted S; Eastman Kodak) and methyl methacrylate (hereafter denoted MMA; Eastman Kodak) were dried for two days with calcium chloride and then for an additional few hours with calcium hydride, passed through a column packed with silica gel to remove the inhibitor, distilled under vacuum, and stored for future use at − 30°C. Poly(methyl methacrylate) (hereafter denoted PMMA) samples of number-average molecular weight $\overline{M}_n$ = 3600; 5700; 8400; 19,900, and 49,000 were prepared in our laboratories through the use of free radical polymerization and standard fractionation techniques. The weight-average to number-average molecular weight ratios of these samples were in the 1.1 to 1.3 range, and thus the PMMA samples were reasonably monodisperse. Polystyrene (hereafter denoted PS) samples for which $\overline{M}_n$ = 2100 (Pressure Chemical, Pittsburgh, PA) and 3100; 9600; 19,650; and 49,000 (Waters Associates, Milford, MA) were used as received. The weight-average to number-average molecular weights of these samples ranged from 1.02 to 1.10, and they could be considered to be reasonably monodisperse. The initiator, 2,2′-azobis(isobutyronitrile) (AIBN; Eastman Kodak), was recrystallized from toluene and stored until needed.

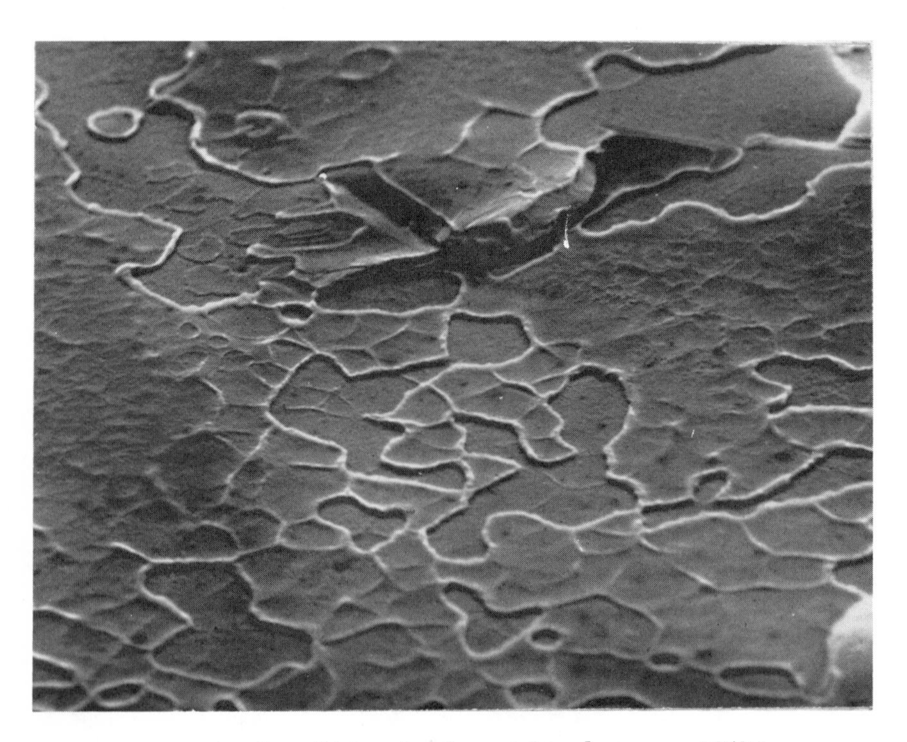

Figure 1. Pure, high-molecular-weight polystyrene. 1150×.

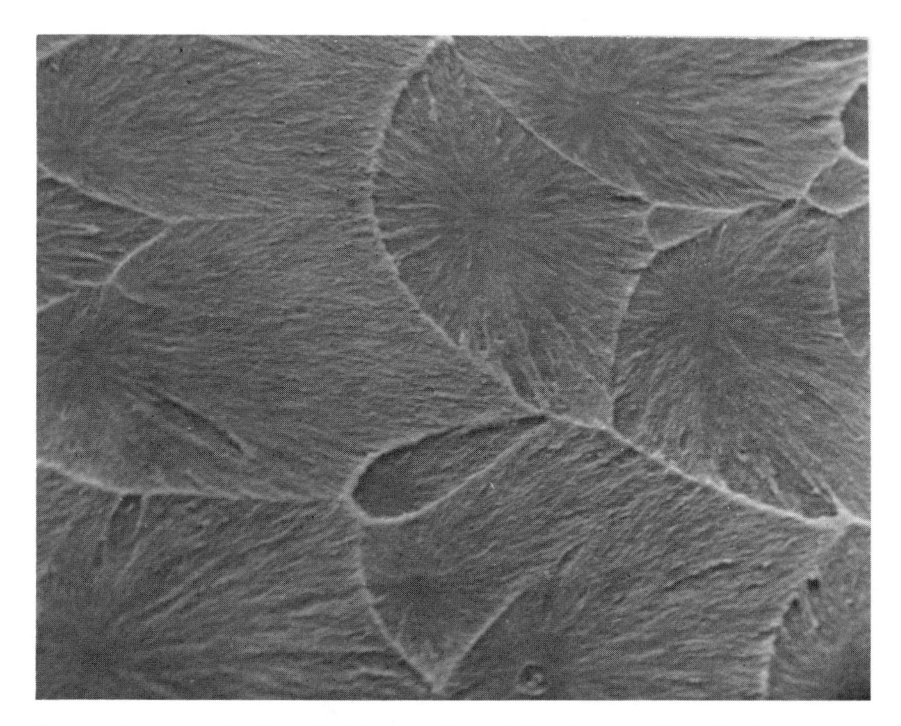

Figure 2. Pure, high-molecular-weight poly(methyl methacrylate). 1200×.

Two-phase polymer samples were prepared by first weighing a given amount of low-molecular-weight PMMA (or PS) into a 5-mm i.d. borosilicate glass tube sealed at one end and then introducing S (or MMA) monomer precatalyzed with ¼ wt % AIBN. The tube was then reweighed and the open end sealed. The polymer was dissolved in the monomer by physical agitation, and until complete solution of the polymer took place the tubes were kept at temperatures well below 50°C to prevent any polymerization of the monomer. After the polymer was completely dissolved, the monomer was polymerized at 50°C for two days, and then postcured for about 12 hr at 80°C and 2 hr at 100°C in an oven to polymerize any remaining traces of monomer.

Several aspects of the characterization of the samples obtained by the procedure described above deserve comment, and these are summarized as follows.

(1) No residual monomer could be detected by gas–liquid chromatographic techniques, and hence is at most no greater than about 0.15 % by weight.

(2) The molecular weight of PS or PMMA polymerized under the conditions given above is very high, being over one million. We compared the intrinsic viscosities of several samples of low PMMA or PS content with that of pure PS or PMMA polymerized under identical conditions and since the viscosities were similar, concluded that the

molecular weight of PS or PMMA in our samples was also very high, at least when they form the continuous phase.

(3) When PMMA is the dissolved polymer and S the polymerizing monomer, little if any grafting would be expected. However, in the reverse situation where MMA is polymerizing in the presence of dissolved PS, the α-hydrogen presents an opportunity for grafting. In an effort to check on the possible extent of this, we determined the weight percent of PS after polymerization by gel permeation chromatography and compared it with the known, initial concentration. Since the ratio of PS to MMA before and to PMMA after polymerization appeared to be virtually unaltered, we concluded that, at most, relatively little grafting took place.

The morphological details of the final polymer samples were examined using a Stereoscan S4 scanning electron microscope. The samples were brittle fractured and mounted and examined using routine SEM procedures. Typically, a specimen was coated with a 100-Å-thick layer of gold metal to provide proper surface conduction; and the microscope was operated with #1 aperture opening, 20-kV acceleration potential, 45° tilt angle, 180–220 μA beam current, and 2.3–2.5 A filament current. Magnifications ranged from 700 to more than 10,000X.

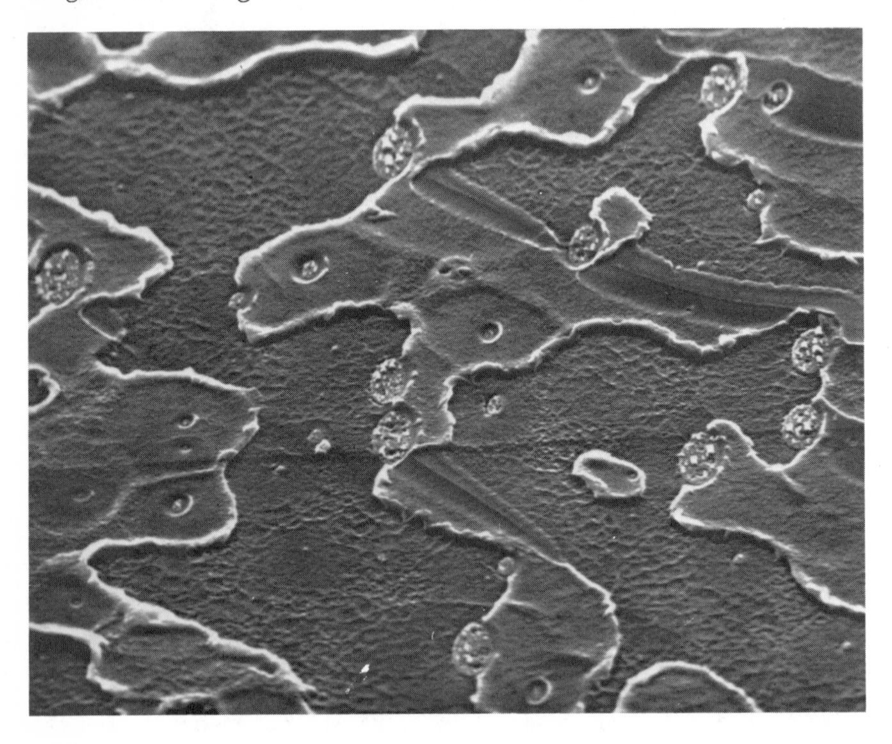

Figure 3. PMMA/PS system with PMMA ($\overline{M}_n = 8400$; W = 5.01 wt %) the dispersed phase and PS the continuous phase. 1000×.

Figure 4. PS/PMMA system with PS ($\overline{M}_n$ = 19,650; W = 4.96 wt %) the dispersed phase and PMMA the continuous phase. 960×.

Clearly, successful use of SEM microscopy for this study depends on the ability to distinguish PS regions from PMMA regions. Pure PS and pure PMMA display characteristically different fracture–surface features, the latter displaying rather regular, recurrent, and intersecting parabolic structures while the former is characterized by much more irregular surface patterns (*28, 29, 30, 31*). These features are shown in Figures 1 and 2; and, furthermore, these distinctive features are retained when dispersed particles are present as seen in Figures 3 and 4. Based on the overall surface morphology, therefore, it is always possible to identify the continuous phase unambiguously. However, it is not always possible from the visual appearance of the surface to specify the nature of a dispersed phase, or more particularly, multiple emulsions within a dispersed phase, but it happens that a second criterion is available for the system under study.

It is well known that PMMA depolymerizes under a variety of influences, including electron irradiation (*32, 33*). Indeed, if a PMMA sample is overexposed to the electron beam of the scanning electron

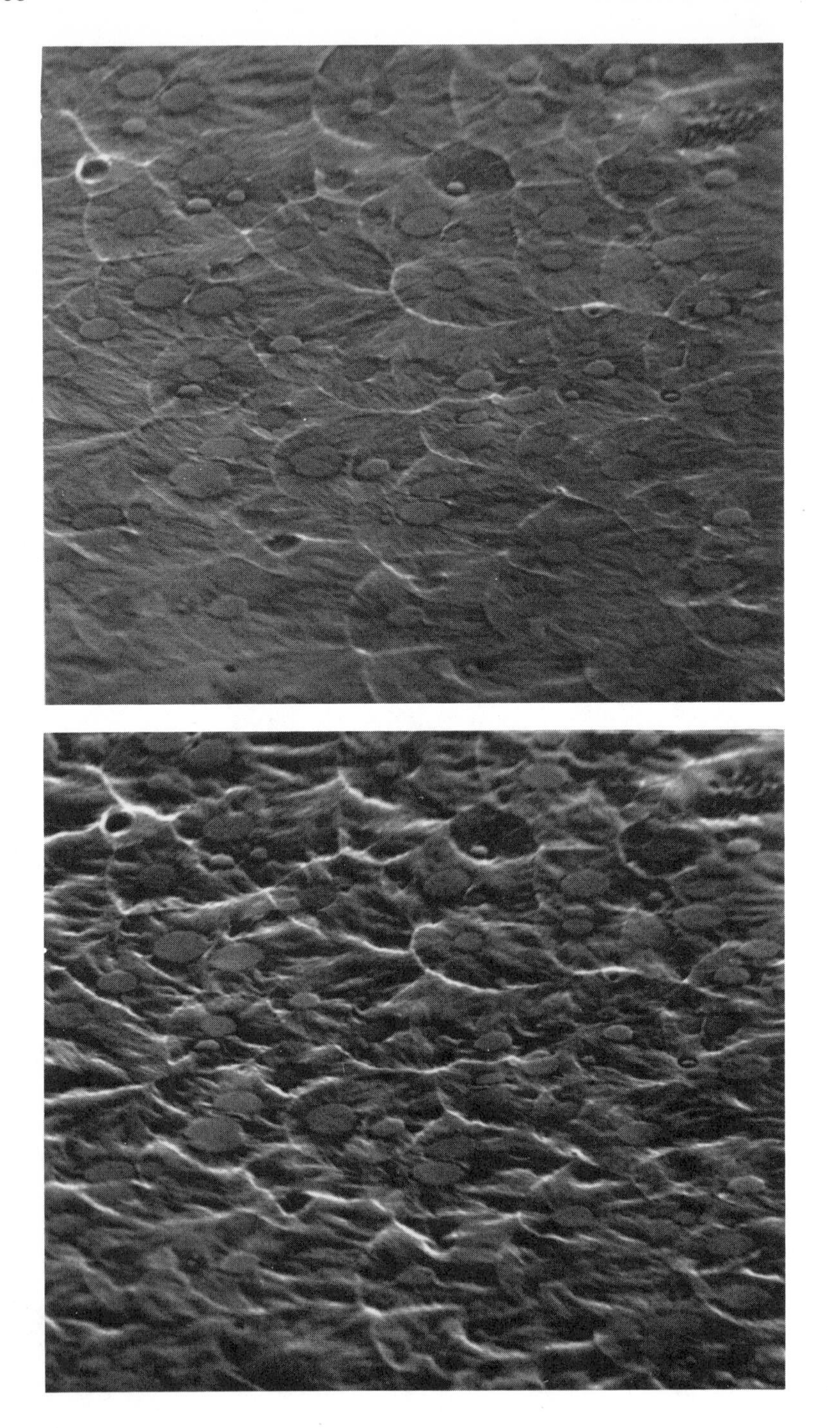

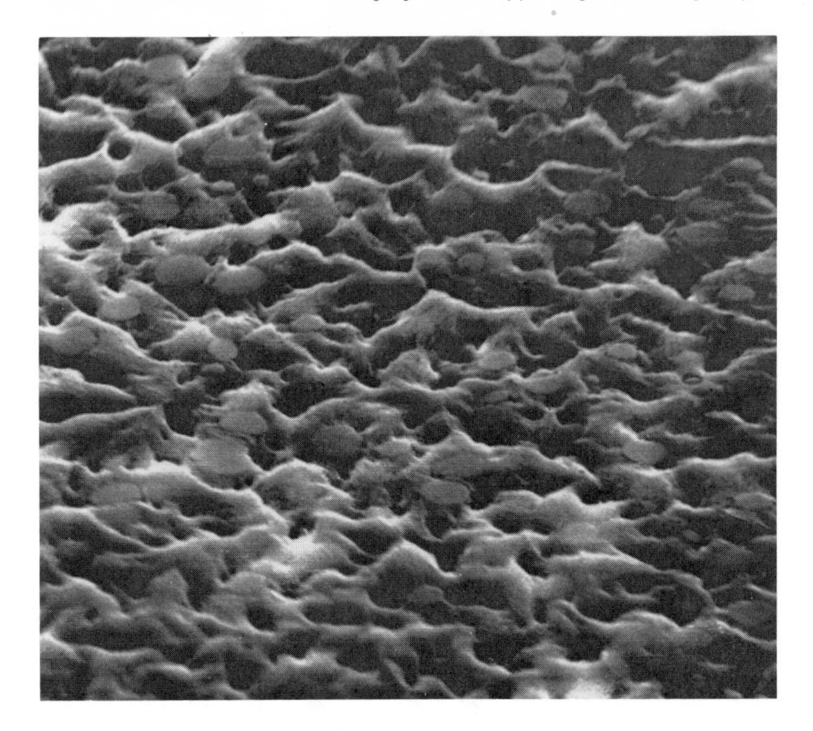

Figure 5. PS/PMMA system with PS ($\overline{M}_n = 9600$; W = 5.04 wt %) the dispersed phase and PMMA the continuous phase: (a) prior to any depolymerization; (b) some depolymerization; (c) substantial depolymerization. 2000×.

microscope, this phenomenon soon manifests itself with attendant bubbling and loss of surface features. Because PS is not similarly affected at all by the electron beam, this feature proves to be a very sensitive identifying criterion regardless of whether PMMA is the continuous or dispersed phase or even present as multiple emulsions in the dispersed phase. This effect is shown in Figures 5a, 5b, and 5c where PMMA is the continuous phase and in Figures 6a, 6b, and 6c where PMMA is the dispersed phase. In both sets, a continuous erosion of the surface features of the PMMA phase is observed, these changes taking place rather rapidly at high magnifications. Parenthetically, as it might be surmised, the depolymerization of the PMMA phase, especially at magnifications above 2000X, was also somewhat troublesome, and required the development of special techniques emphasizing speed to get high quality photomicrographs.

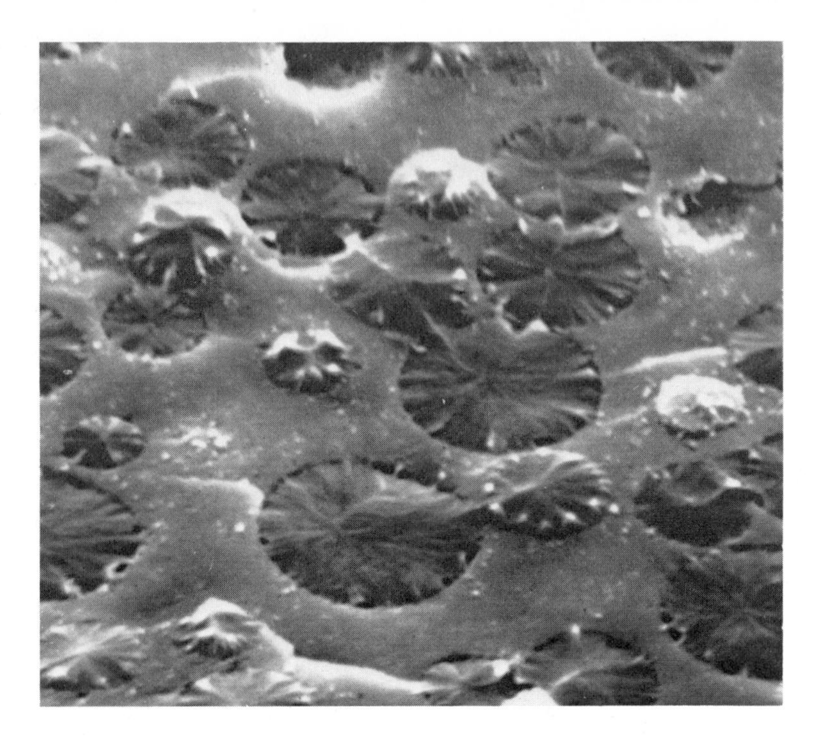

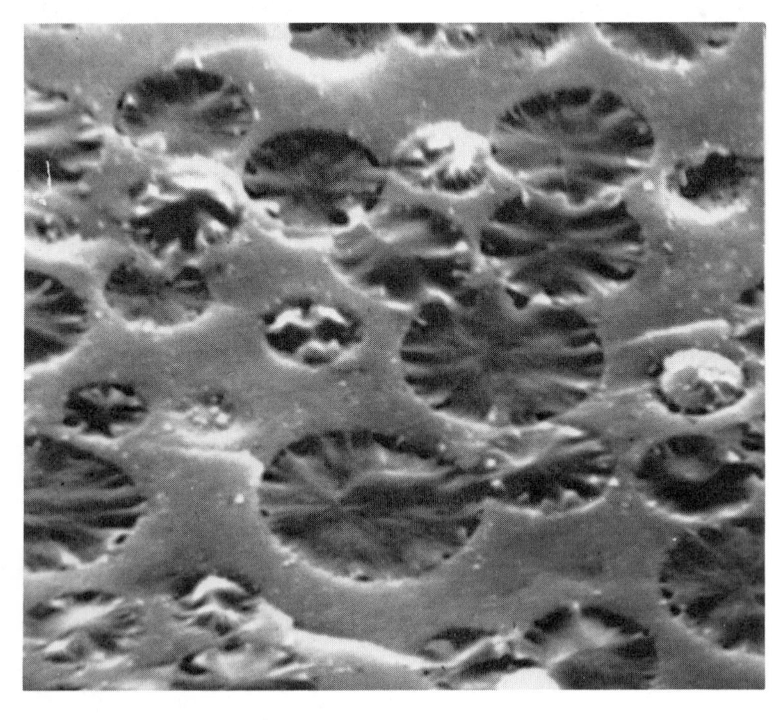

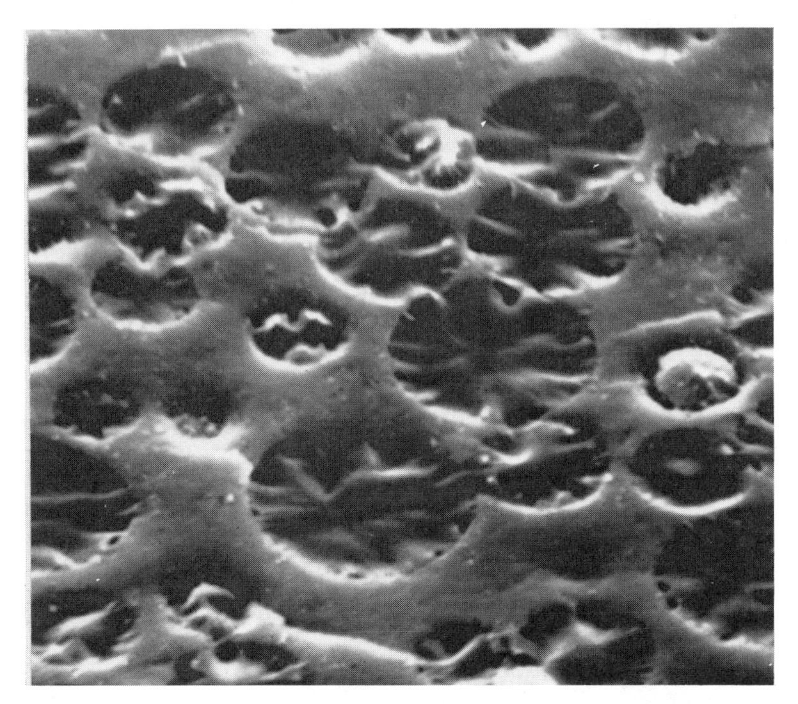

Figure 6. PS/PMMA system with PMMA the dispersed phase and PS ($\overline{M}_n$ = 19,650; W = 30.02 wt %) the continuous phase: (a) prior to any depolymerization; (b) some depolymerization; (c) substantial depolymerization. 3500×.

Experimental Results

In Table I we list the specimens that were prepared in conjunction with this study, giving the number-average molecular weight ($\overline{M}_n$) and weight percent (*W*) of the PMMA present in the initial PMMA/S mixture and an indication of the opacity of the final system.

For the PMMA/PS system summarized in Table I, three distinct types of fracture morphologies and phase relationships are observed, these being:

Type (1) A single phase which has the morphological features of pure PS and in which no dispersed phase can be detected at magnifications as high as 15,000X. Figure 7 depicts an example of this type.

Type (2) PMMA is the dispersed phase and PS is the continuous phase. Almost all of the samples were of this type, and Figures 8, 9, 10, 11, and 12 illustrate the group, showing the phase behavior and particle sizes for the 10-wt % PMMA series.

Type (3) PS is the dispersed phase and PMMA is the continuous phase. This type was observed in only three samples, one of which is illustrated in Figure 13.

Table I. Summary of PMMA/PS Specimens[a]

Wt % PMMA	$\overline{M}_n$ *PMMA*	*Opacity*[b]	*Wt % PMMA*	$\overline{M}_n$ *PMMA*	*Opacity*[b]
4.99	3600	cl.	4.96	19,900	op.
10.05	3600	trans.	9.76	19,900	op.
20.26	3600	trans.	14.87	19,900	op.
29.92	3600	op.	20.19	19,900	op.
			25.17	19,900	op.
4.94	5700	trans.			
10.01	5700	op.	4.93	49,000	op.
20.13	5700	op.	9.98	49,000	op.
			15.13	49,000	op.
5.01	8400	op.	20.25	49,000	op.
9.65	8400	op.			
20.42	8400	op.			
25.03	8400	op.			

[a] PMMA the polymer present in the initial PMMA/S mixture.
[b] cl. = clear; trans. = translucent; op. = opaque.

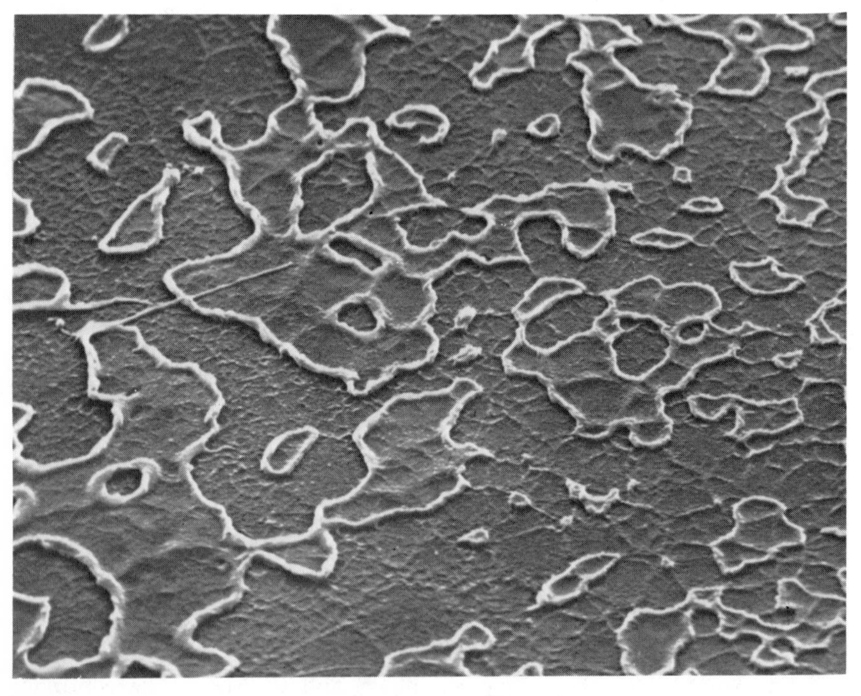

Figure 7. PMMA/PS system with PMMA ($\overline{M}_n$ = 3600; W = 4.99 wt %) in PS. 920×.

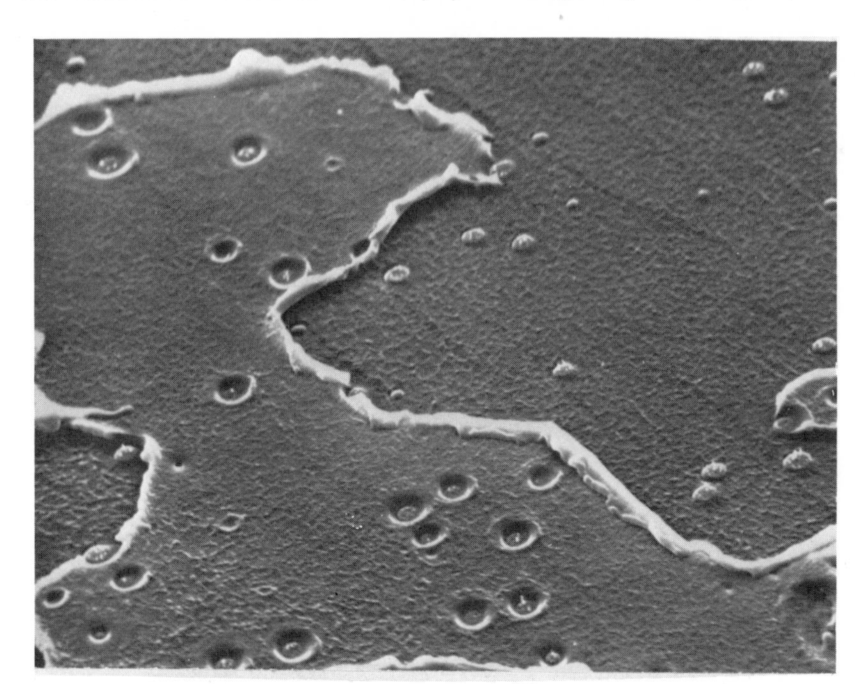

Figure 8. PMMA/PS system with PMMA ($\overline{M}_n = 3600$; W = 10.05 wt %) the dispersed phase and PS the continuous phase. 1530×.

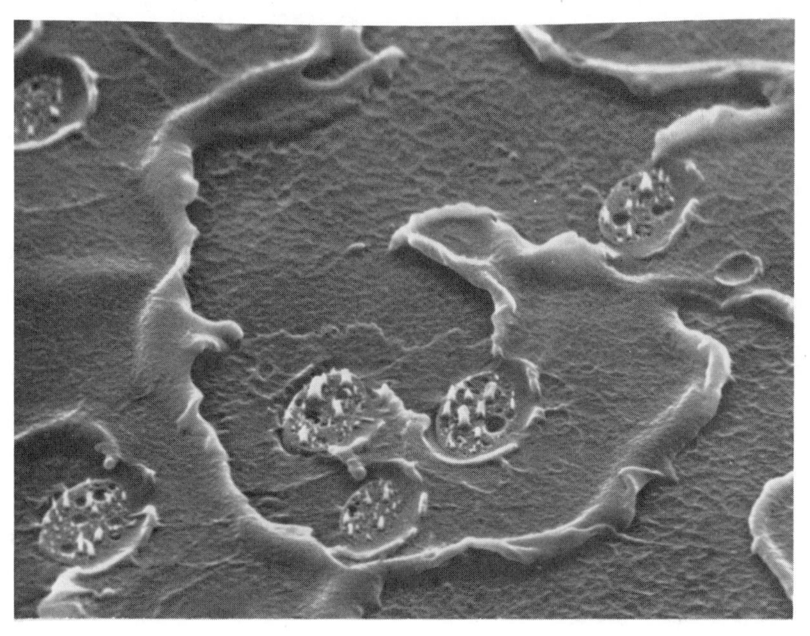

Figure 9. PMMA/PS system with PMMA ($\overline{M}_n = 5700$; W = 10.01 wt %) the dispersed phase and PS the continuous phase. 1980×.

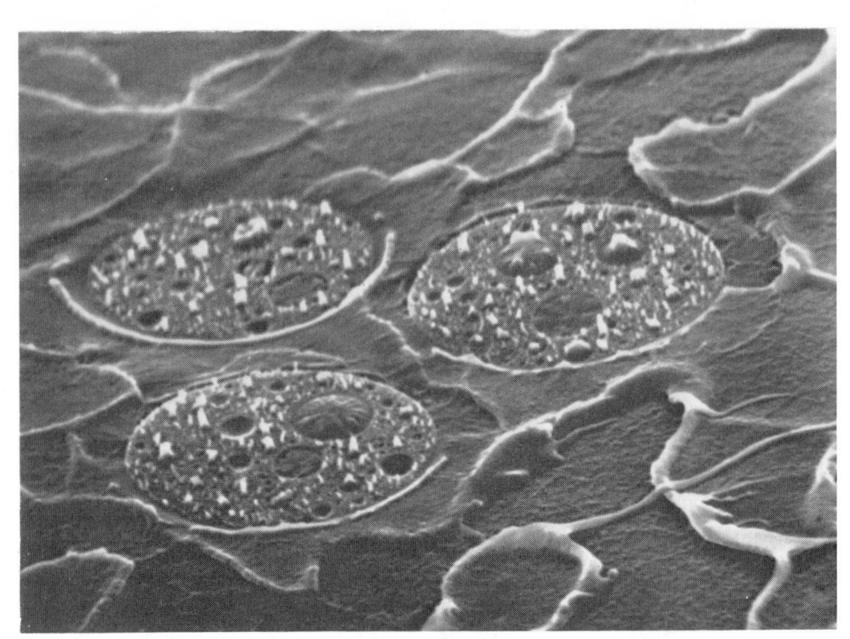

Figure 10. PMMA/PS system with PMMA ($\overline{M}_n = 8400$; W = 9.65 wt %) the dispersed phase and PS the continuous phase. 1620×.

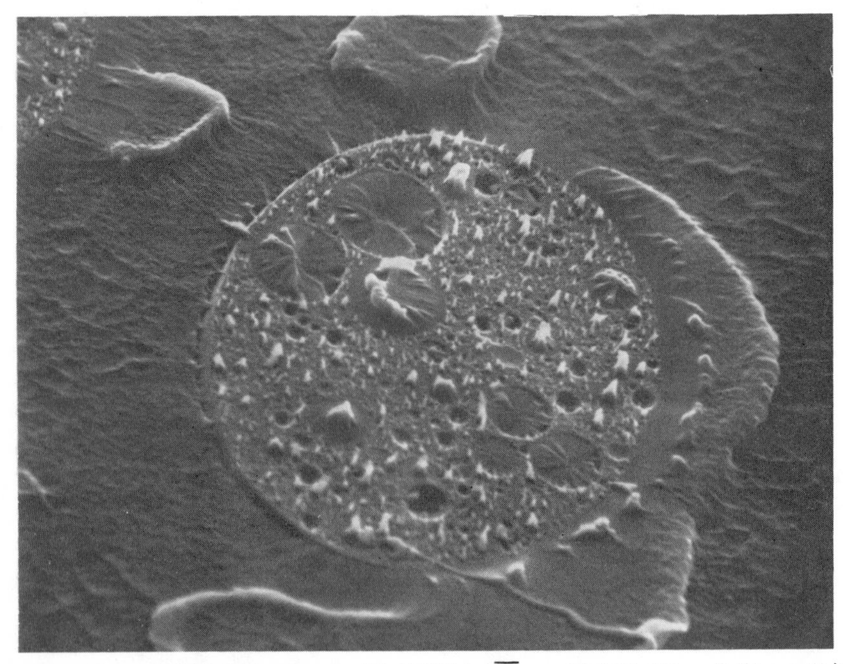

Figure 11. PMMA/PS system with PMMA ($\overline{M}_n = 19{,}900$; W = 9.76 wt %) the dispersed phase and PS the continuous phase. 1840×.

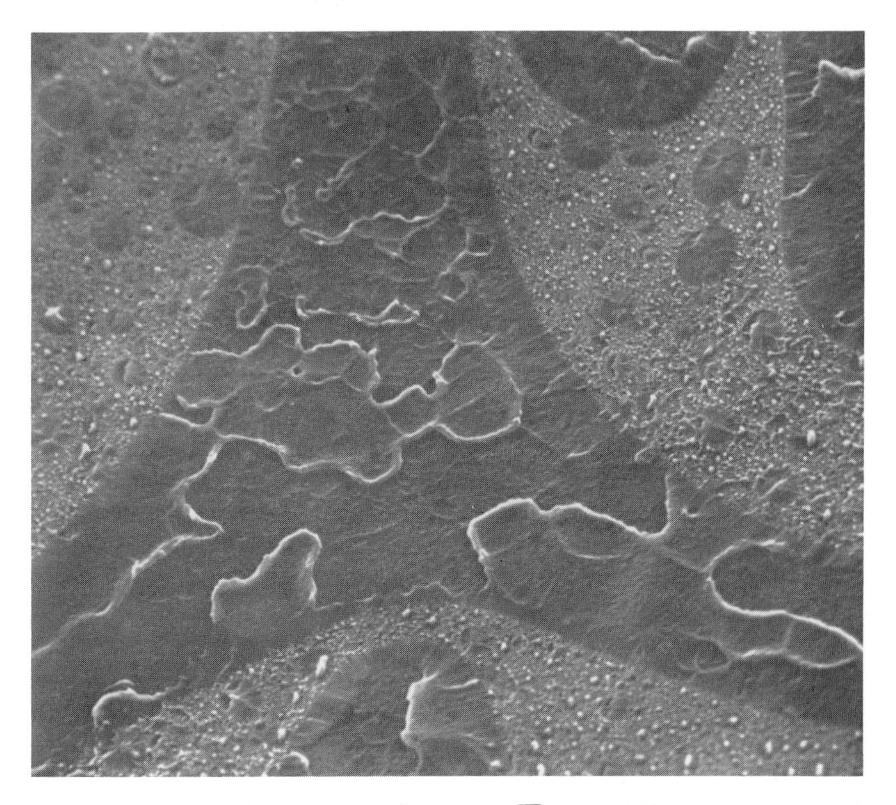

Figure 12. PMMA/PS system with PMMA ($\overline{M}_n$ = 49,000; W = 9.98 wt %) the dispersed phase and PS the continuous phase. 1080×.

In Table II we list the values of the size of the dispersed particles ($\overline{S}$) in microns, defined perhaps somewhat arbitrarily as the average of the geometrical mean diameter of the particles in a given representative area. Only one sample appeared to be a homogeneous single phase (type (1)), that being $W = 5.0\%$ and $\overline{M}_n = 3600$, for which $\overline{S} = 0$. This sample was, of course, perfectly transparent; all other samples were opaque, although for $\overline{S} < 4.0$ the samples are only slightly opaque, or translucent, and are so designated in Table I. In Table II, unstarred values refer to type (2) where the initial polymer, PMMA, forms the dispersed phase; starred values refer to type (3) where PS forms the dispersed phase. Some care must be taken that underestimates of $\overline{S}$ are not obtained (*16, 34*); the standard deviation of our estimates was usually within 5% and never more than 10%. The relationship between the three types of phase behavior described above and their dependence on W and $\overline{M}_n$ is summarized schematically in Figure 14.

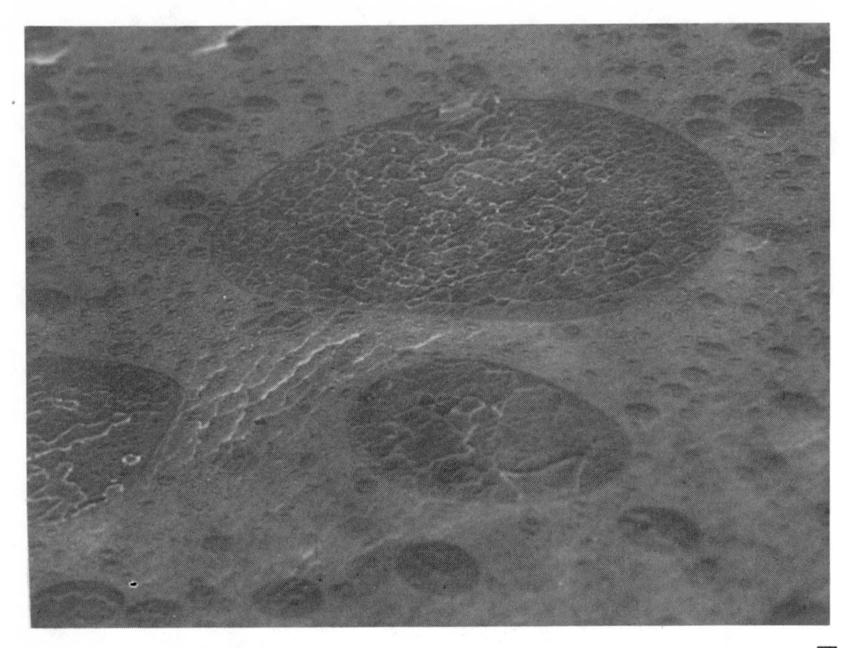

Figure 13. PMMA/PS system with PS the dispersed phase and PMMA ($\overline{M}_n$ = 49,000; W = 20.25 wt %) the continuous phase. 504×.

WEIGHT PERCENT PMMA

NUMBER-AVERAGE MOLECULAR WEIGHT PMMA

3600 5700 8400 19,900 49,000

5.0
10.0
15.0
20.0
25.0
30.0

ONE PHASE

PMMA DISPERSED PHASE; PS CONTINUOUS PHASE

PS DISPERSED PHASE; PMMA CONTINUOUS PHASE

Figure 14. Regions of three types of phase relationships for the low-molecular-weight PMMA/PS system for various values of number-average molecular weight ($\overline{M}_n$) of PMMA and weight percent (W) of PMMA

Table II. Dispersed Particle Size ($\overline{S}$)[a] in Microns as a Function of Weight Percent PMMA (W) and Number-Average Molecular Weight PMMA ($\overline{M}_n$) for the PMMA/PS System

Wt % PMMA	$\overline{M}_n$ *PMMA*				
	3600	*5700*	*8400*	*19,900*	*49,000*
5.0	0	2.6	4.7	6.6	10.8
10.0	1.2	4.5	12.5	21.6	$>$ 130
15.0	—	—	—	55	~ 80*
20.0	3.8	9.1	93	129	$<$ 60*
25.0	—	—	$>$ 150	8.2*	—
30.0	5.2	—	—	—	—

[a] Unstarred values of $\overline{S}$ refer to type (2), PMMA dispersed in PS; and starred (*) values to type (3), PS dispersed in PMMA.

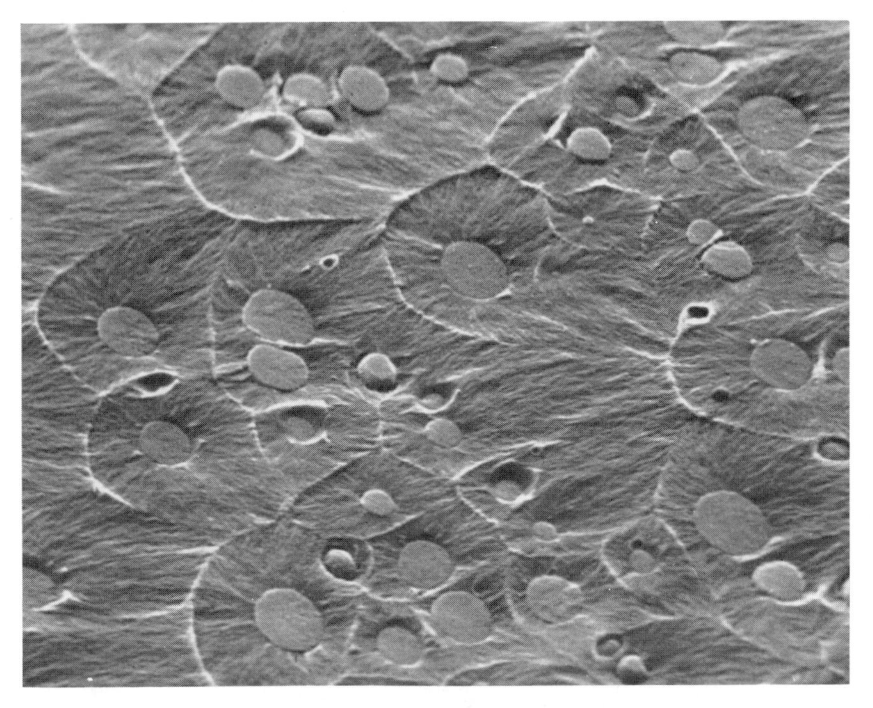

Figure 15. PS/PMMA system with PS ($\overline{M}_n$ = 9600; W = 5.04 wt %) the dispersed phase and PMMA the continuous phase. 1900×.

It will be of interest to compare the results described above with those obtained when low-molecular-weight PS is the initial polymer and MMA the monomer. Although these data do not, strictly speaking, belong to the present experimental results, we will nevertheless include a summary of them here. For this situation we observed four types of phase behavior.

Type (1′) A single phase which has the morphological features of pure PMMA with no indication of a second, dispersed phase. Samples of this type showed SEM photomicrographs identical to pure PMMA as shown in Figure 2.

Type (2′) PS is the dispersed phase and PMMA is the continuous phase. Figure 15 is an example of this type, although there is considerable variation in the shape of the dispersed PS particles, these tending to be spherical at low W and/or $\overline{M}_n$ and becoming more ellipsoidal or irregular at higher values of W and $\overline{M}_n$.

Type (3′) PMMA is the dispersed phase and PS is the continuous phase. Figure 16 is a typical example of this type.

Type (4′) Regions coexisting in which both PMMA and PS form the continuous phases, with PS and PMMA forming the dispersed phases,

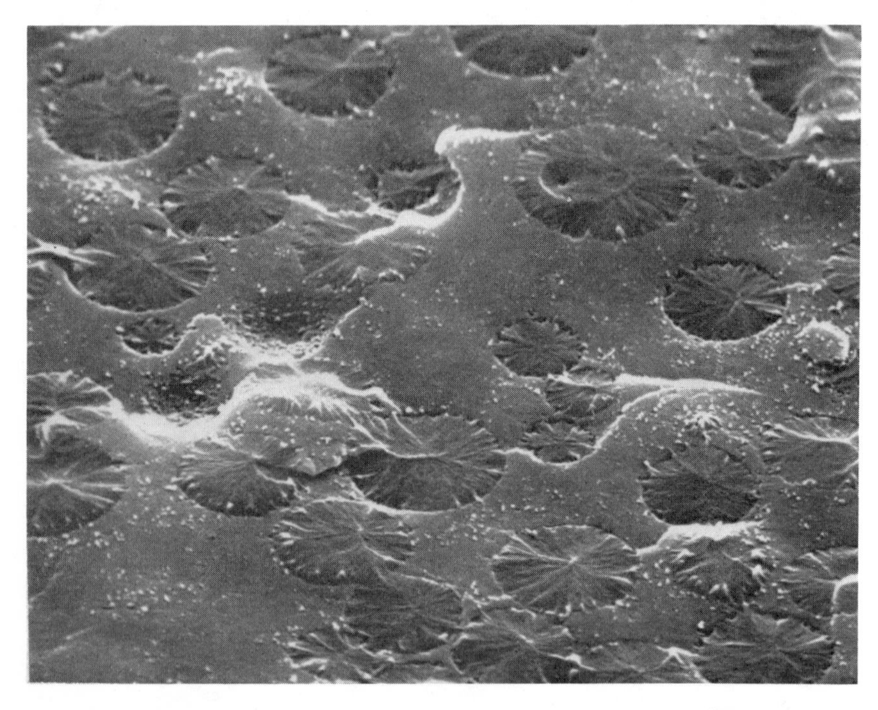

Figure 16. PS/PMMA with PMMA the dispersed phase and PS ($\overline{M}_n = 19{,}650$; W = 30.02 wt %) the continuous phase. 1780×.

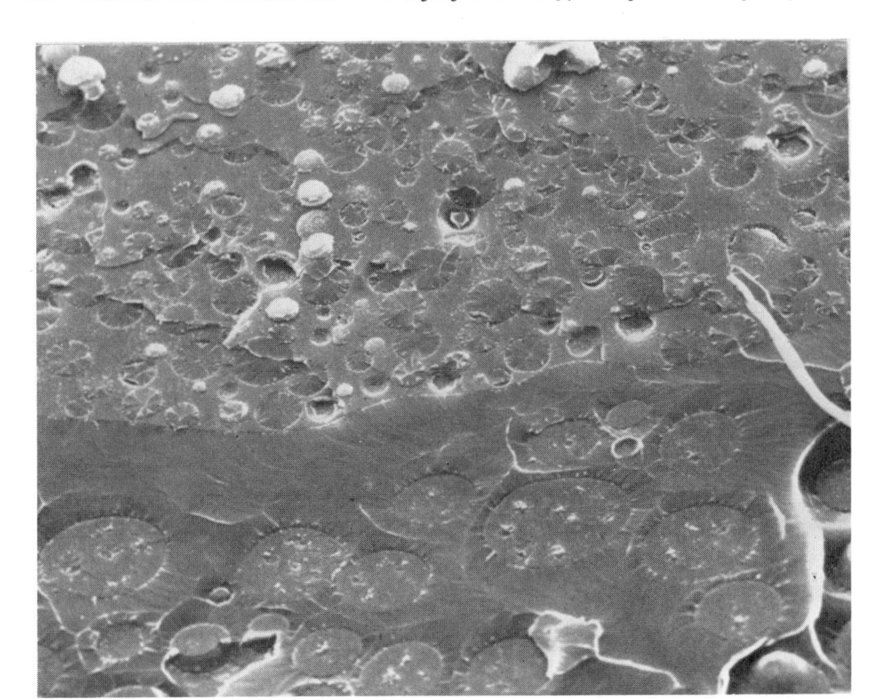

Figure 17. PS/PMMA with PMMA the dispersed phase in PS (upper part) and PS the dispersed phase in PMMA (lower part). ($\overline{M}_n = 9600$; W = 20.31 wt %.) 855×.

WEIGHT PERCENT PS

NUMBER-AVERAGE MOLECULAR WEIGHT PS

2100 3100 9600 19,650 49,000

5.0
10.0
15.0
20.0
25.0
30.0
35.0
40.0

ONE PHASE

PS DISPERSED PHASE; PMMA CONTINUOUS PHASE

BOTH PMMA AND PS FORM THE CONTINUOUS PHASES

PMMA DISPERSED PHASE; PS CONTINUOUS PHASE

Figure 18. Regions of four types of phase relationships for the low-molecular-weight PS/PMMA system for various values of number-average molecular weight ($\overline{M}_n$) of PS and weight percent (W) of PS

respectively. At least 11 samples of this type were studied, and Figure 17 represents a typical example of the morphology observed.

The quantitative data associated with these samples have been reported earlier (*18*), and for comparison with Figure 14 the dependence of the various regions of phase behavior on $\overline{M}_n$ and W of the initial PS is summarized in Figure 18.

Discussion

In the previous section we have described the three types of phase behavior observed in the low-molecular-weight PMMA/PS system and reviewed the four types observed in the low-molecular-weight PS/PMMA system. These various phase relationships have been studied in terms of their dependence on the molecular weight ($\overline{M}_n$) and weight percent (W) of the initial polymer present. Further, we have presented quantitative data concerning the sizes of the dispersed particles, again correlated to variations in $\overline{M}_n$ and W. In this section we will discuss the results in terms of the poly(methyl methacrylate)/polystyrene/styrene and polystyrene/poly(methyl methacrylate)/methyl methacrylate ternary phase diagrams, whichever is appropriate.

It is important to note that discussions involving phase diagrams imply, strictly speaking, the attainment of equilibrium. This certainly will not be the case for a polymerizing system at a high degree of conversion which is the situation during the later stages of the polymerization process that we used, that is, relatively high S-to-PS conversion. However, several interesting insights result from the analysis of our data in terms of the phase diagrams and the properties of the polymerizing system, and for this reason we feel justified in making use of the equilibrium phase diagrams in the discussion which follows.

Phase Relationships. The first systematic investigation of the two-phase behavior of polymer/polymer/solvent systems was probably made by Dobry and Boyer–Kawenoki (*2*) for a variety of polymer pairs, and more recently this work was extended by Kern and Slocombe (*3*) and Paxton (*35*) to a number of other systems including several vinyl polymers. Typically, the three-component phase behavior is as shown in Figure 19 for the polystyrene/polybutadiene/benzene system (*2*), where a one-phase (polystyrene/polybutadiene/benzene) region is separated by a phase boundary from a two-phase (polystyrene-rich/benzene and polybutadiene-rich/benzene) mixture. As with any three-component system of this type, a critical point exists somewhere near the maximum of the phase boundary, and appropriate tie lines give the compositions and amounts of the respective phases in the two-phase region.

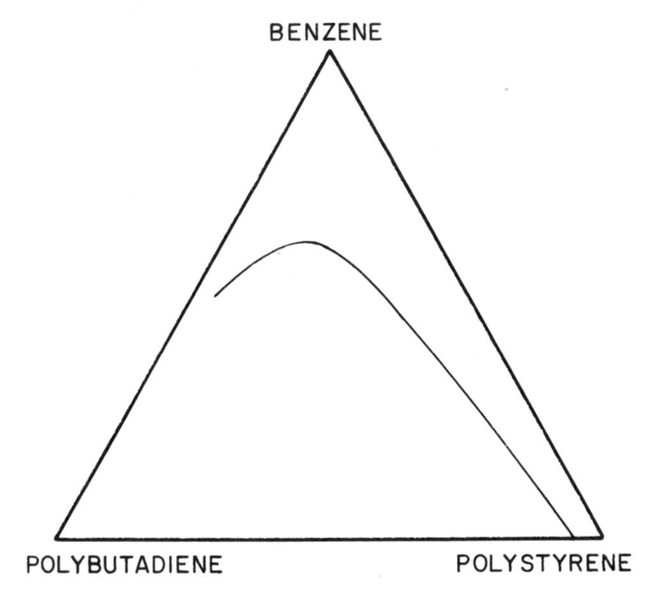

Figure 19. Ternary phase diagram for the polystyrene/polybutadiene/benzene system. (After Dobry and Boyer-Kawenoki (2).)

The details of the poly(methyl methacrylate)/polystyrene/styrene (hereafter denoted PMMA/PS/S) system have not, to our knowledge, been worked out, particularly as a function of molecular weight of PMMA. However, based on the evidence provided by other systems, we assume that the general form is as shown in Figure 20, where we might make the following points:

(1) Two different molecular weights of PMMA are indicated in Figure 20. Clearly, the one-phase region will be more limited for the higher-molecular-weight PMMA, as the mutual miscibility of two polymers in a common solvent is highly molecular-weight dependent. Further, we expect relatively little effect of molecular weight of PMMA on the solubility of PS in the vicinity of the PMMA apex since the molecular weight of PS is assumed to be very high; however, the reverse is true in the vicinity of the PS apex where even minor changes in the molecular weight of PMMA will have pronounced effects on its solubility in this region.

(2) The critical point is not symmetrically located on the phase boundary but skewed towards the lower-molecular-weight polymer, that is, the PMMA apex (for a discussion of this print *see* Ref. *36*).

(3) Although probably similar in general form, the PMMA/PS/S phase diagram discussed here is not a mirror-image of the low-molecular-weight PS/PMMA/MMA phase diagram which we have described in an earlier report (*18*). There are at least two indications of this, namely as follows.

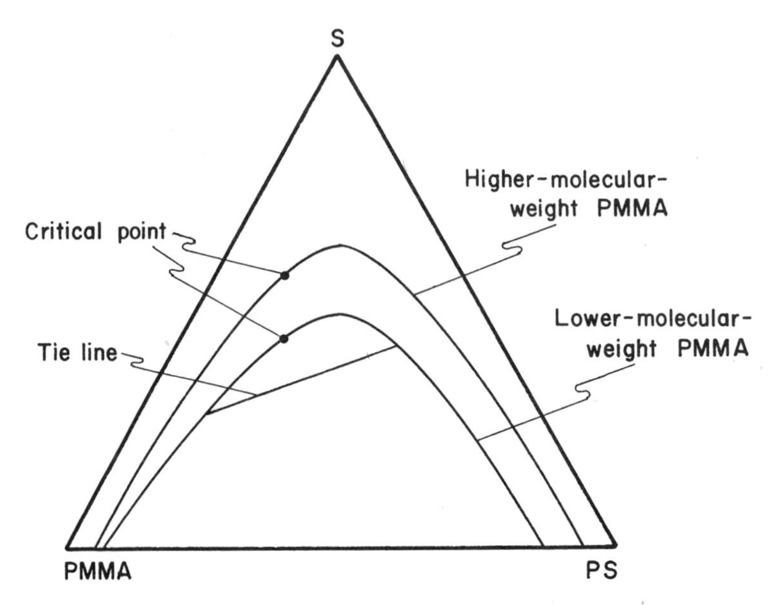

Figure 20. Assumed ternary phase diagram for the poly(methyl methacrylate)/polystyrene/styrene system for two molecular weights of poly(methyl methacrylate)

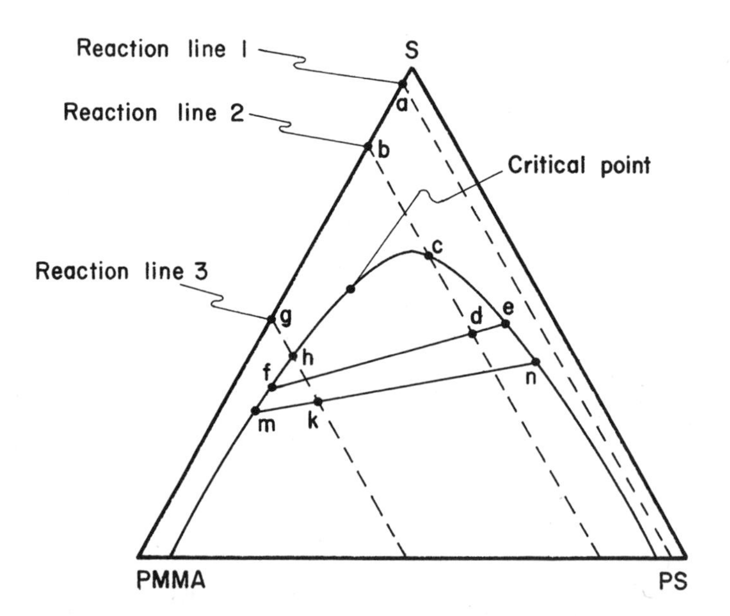

Figure 21. PMMA/PS/S phase diagram with three reaction lines indicating: reaction line 1, one phase system (type (1)); reaction line 2, PMMA dispersed in PS (type (2)); and reaction line 3, PS dispersed in PMMA (type (3))

(a) The one-phase region appears to be somewhat more restricted for the low-molecular-weight PMMA/PS/S system.

(b) We found no evidence in the PMMA/PS/S system of transitional behavior from type (2) (PMMA dispersed) to type (3) (PMMA continuous) systems. This behavior, denoted type (4′) in the section above, was very common in the PS/PMMA/MMA system.

Turning our attention now to the three types of phase relationships resulting from the PMMA/PS/S system (types (1), (2), and (3) as defined in the section above), we refer to Figure 21 and summarize our analysis in terms of the three reaction lines (hereafter denoted RL) as follows.

1. Reaction line 1. The initial composition of the PMMA/S mixture is given by point a: since RL1 never intersects the phase boundary, a one-phase solution of PMMA in PS results. This corresponds to type (1).

2. Reaction line 2. The initial composition is given by b, and initial phase separation occurs to the right of the critical point at point c, the intersection of RL2 with the phase boundary. Further polymerization results in a two-phase system of overall composition d, with the predominant phase rich in PS (point e) and the lean phase rich in PMMA (point f) and the relative amounts of the PS-rich to PMMA-rich phases being fd/de. The final system consists of a PMMA-rich phase dispersed in a continuous, PS-rich phase, and corresponds to type (2).

3. Reaction line 3. The initial composition is given by g, and initial phase separation occurs to the left of the critical point at point h. Further polymerization results in a two-phase system of overall composition k, with the predominant phase rich in PMMA (point m) and the lean phase rich in PS (point n) and the relative amounts of the PMMA-rich to PS-rich phases being kn/mk. The final system consists of a PS-rich phase dispersed in a continuous, PMMA-rich phase and corresponds to type (3).

We can now further investigate the effect of the molecular weight of PMMA on the phase relationships of the final system by referring to Figure 22. For example, it is seen that RL1 does not intersect the lower-molecular-weight phase boundary but does intersect the higher-molecular-weight phase boundary to the right of the critical point. Polymerization along RL1, therefore, results in a type (1) system in the former and a type (2) system in the latter case. On the other hand, along RL2, again for a constant weight percent PMMA, intersection of the lower-molecular-weight phase boundary is to the right of the critical point and of the high-molecular-weight boundary to the left of the critical point; the former leads to type(2) and the latter to type (3) systems.

Examination of Table II or Figure 14 indicates that the lowest values of molecular weight ($\overline{M}_n$) and weight percent (W) favor the formation of a one-phase, type (1) system; higher values of $\overline{M}_n$ and/or W favor

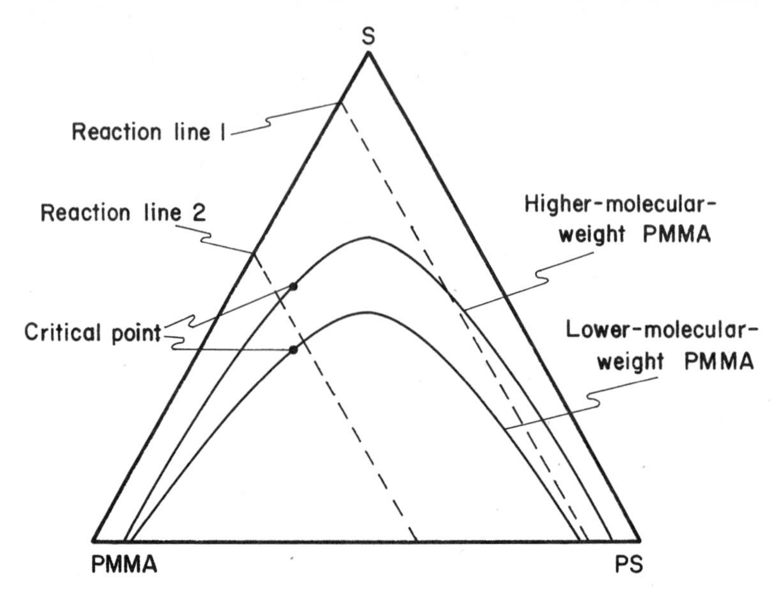

Figure 22. PMMA/PS/S phase diagram with two reaction lines indicating possible final phase relationships

type (2) systems; and the highest values of $\overline{M}_n$ and/or W, type (3) systems. These trends seem to be in complete accordance with the explanation offered in terms of the PMMA/PS/S phase diagram and the polymerizing systems as depicted in Figures 21 and 22.

Finally, referring to Figure 18 we note that the dependence of the phase relationships denoted as types (1′), (2′), and (3′) for the PS/PMMA/MMA system where low-molecular-weight PS is the initial polymer are similar to those discussed above, and we have offered a similar explanation for this behavior (*17, 18*). The major departure is the appearance of a rather extended transitional region, denoted type (4′), between type (2′) where PS is the sole dispersed phase to type (3′) where PMMA is the sole dispersed phase. This region is characterized by intermediate values of $\overline{M}_n$ and/or W and displays areas coexisting in which both PS and PMMA form the continuous phases, with PMMA and PS forming the dispersed phases, respectively. We discussed type (4′) systems in terms of reaction lines which intersect the phase boundary very close to the critical point, thereby giving rise to tie-line ratios near unity. Under these circumstances, neither phase can become predominant in the sense that it becomes the sole continuous phase. As we have pointed out, we have not observed any type (4) samples for the PMMA/PS/S system, and, if they exist, this behavior must be limited to a very narrow $\overline{M}_n$–W band. We suspect that the tie lines for the phase diagram of this system are so steep, that is, essentially parallel to the PMMA-S edge of the phase diagram, as to preclude tie-line ratios approaching unity except in the very immediate vicinity of the critical point.

Dispersed Particle Size. We will continue now to consider the variations of the dispersed particle size, $\overline{S}$, with W and $\overline{M}_n$, and their relationship to the PMMA/PS/S phase diagram. The experimental values of $\overline{S}$ reported in Table II are shown graphically in Figure 23 where, in all

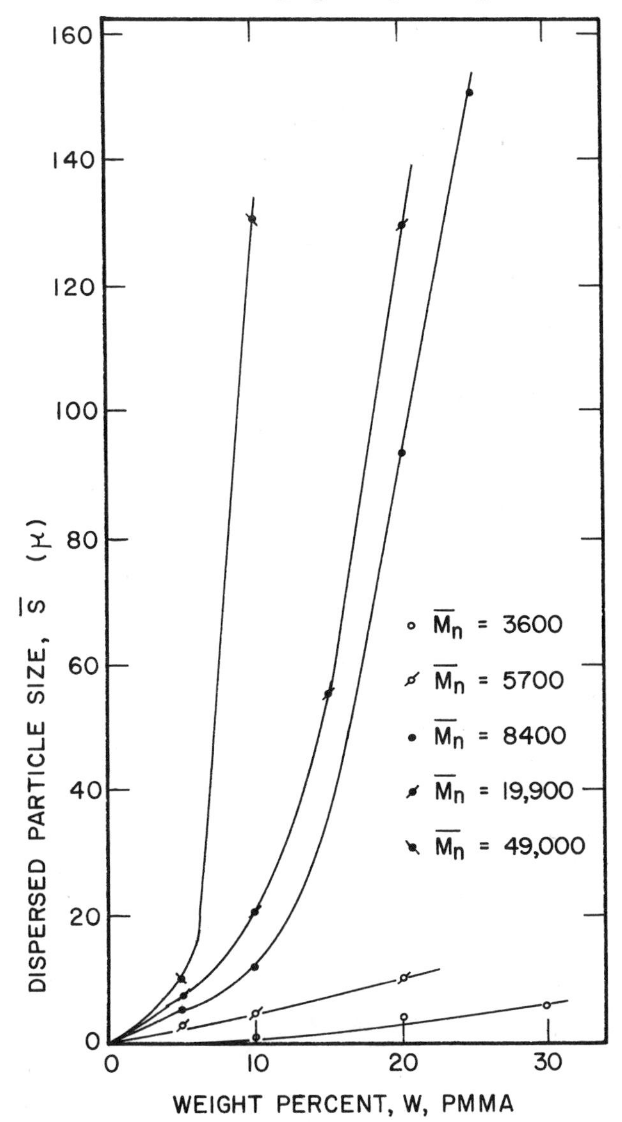

Figure 23. Dispersed particle size vs. wt % PMMA for various values of number-average molecular weight PMMA: PMMA is the dispersed phase

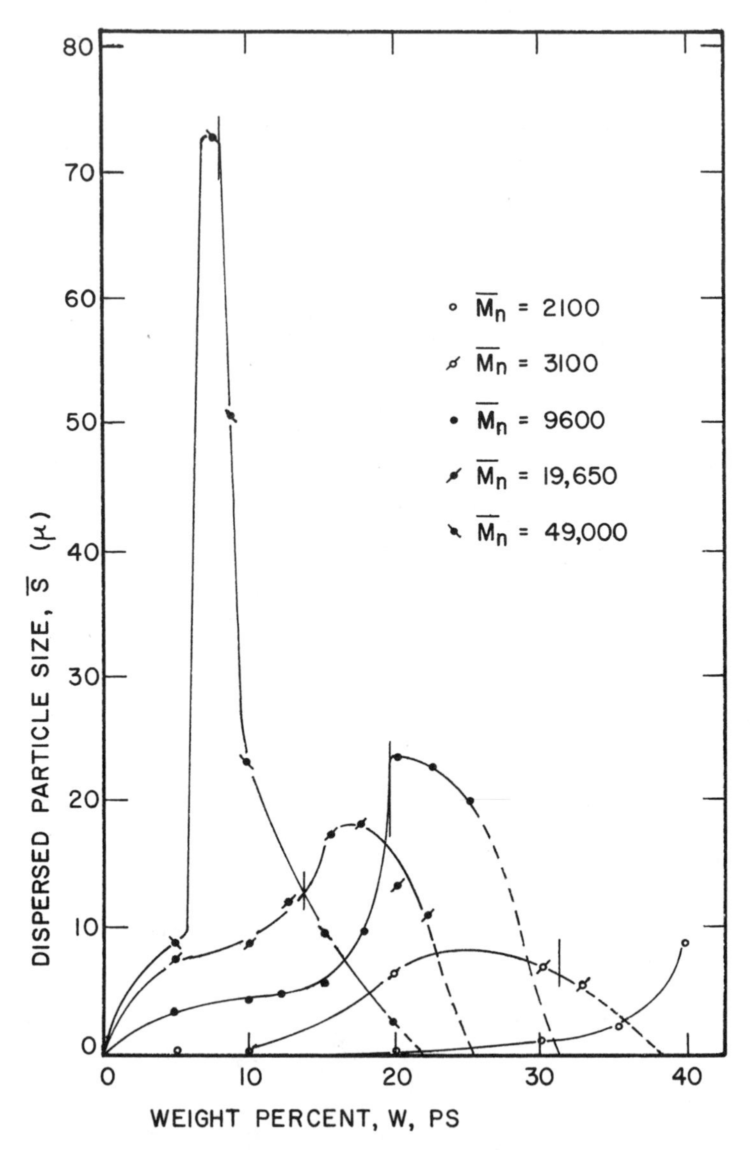

Figure 24. Dispersed particle size vs. wt % PS for various values of number-average molecular weight PS: PS is the dispersed phase; the region in which both PS and PMMA coexist as the continuous phases is to the right of the single bar

cases, PMMA is the dispersed phase. For comparison, the results from our prior study of the low-molecular-weight PS/PMMA/MMA system are shown in Figures 24 and 25; where PS and PMMA are the dispersed phases, respectively, and the regions where both PS and PMMA coexist as the continuous phases are to the right of the single bar in Figure 24 and to the left of the double bar in Figure 25.

The variations of $\overline{S}$ with W and $\overline{M}_n$ can be understood in terms of the phase diagram, Figure 20, and of certain assumptions concerning the viscosity of the polymerizing solution at the point of phase separation. In particular, we assume that at the point of phase separation and beyond, the dispersed particles in a less viscous medium have a higher coalescence rate and therefore larger sizes. Considering first the type (2) system, we compare RL1 and RL2 in Figure 26 for two values of W at a constant value of $\overline{M}_n$. We note that phase separation occurs at a much lower value of overall monomer conversion along RL2 (point b) than along

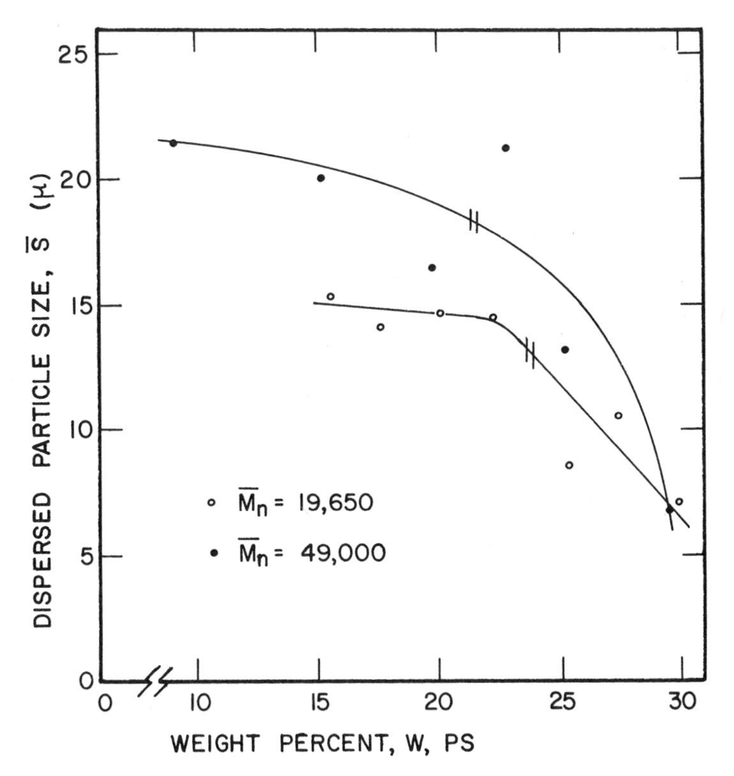

Figure 25. Dispersed particle size vs. wt % PS for various values of number-average molecular weight PS: PMMA is the dispersed phase; the region in which both PS and PMMA coexist as the continuous phases is to the left of the double bar.

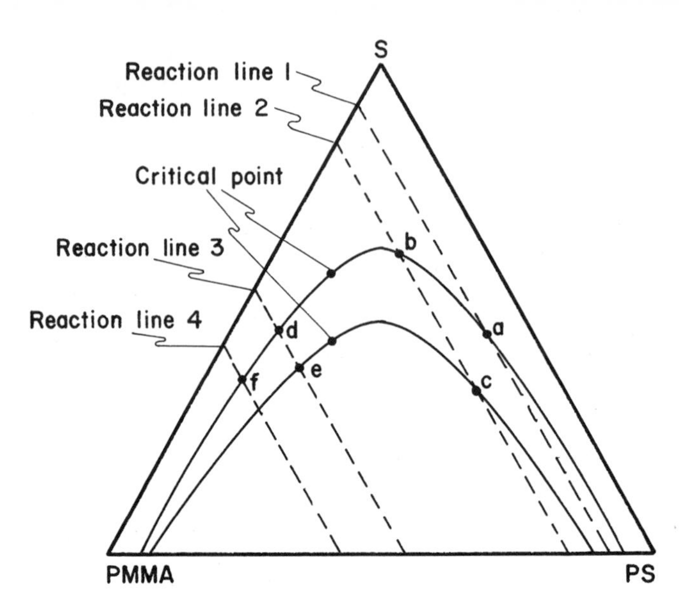

Figure 26. PMMA/PS/S phase diagram indicating various relationships between molecular weight and composition of PMMA and the dispersed particle size

RL1 (point a). Therefore, the viscosity at phase separation along RL2 is lower than that along RL1 (assuming that this effect, attributable to lower monomer conversion, is not offset by the increase in viscosity because of the somewhat higher initial concentration of PMMA). Similarly, comparing the effect of two different values of $\overline{M}_n$ for a constant value of W, we observe that along RL2, for example, phase separation occurs at a lower overall monomer conversion for the higher value of $\overline{M}_n$ (point b) compared with the lower value (point c), and thus results in a lower viscosity (assuming this effect is not offset by the increase in initial viscosity of the system attributable to a higher molecular of PMMA). In both cases the higher value of either W or $\overline{M}_n$ favors lower viscosity at phase separation and hence increased particle coalescence and increased particle size. These conclusions are in accord with the experimental observations seen in Figure 23.

Turning to the situation where PS forms the dispersed phase (type (3)), we have relatively sparse data for the PMMA/PS/S system; however, these data do suggest trends analogous to those displayed by the PS/PMMA/MMA system and illustrated in Figure 25. Again referring to Figure 26, for a constant value of W the behavior is exactly analogous to that discussed above, where now phase separation is depicted as occurring along RL3 at points d and e. The behavior with respect to variations in W, however, is not analogous. We note that in this region the phase boundary is almost parallel to the PMMA–S edge of the triangular diagram, and thus phase separation will occur at roughly the same overall conversion along both RL3 and RL4. However, since the initial viscosity of the mixture polymerizing along RL4 will be greater because of its higher initial concentration of PMMA, we expect its viscosity also to be greater at the point of phase separation as compared with the mixture reacting along RL3. This being the case, the mixture reacting along RL4 will ultimately produce smaller dispersed particles. For the type (3) region we conclude that increased $\overline{M}_n$ results in increased particle size while increased W results in decreased particle size; both of these conclusions are in accord with the experimental results shown in Figure 25.

Finally, we note that type (4′) behavior also can be discussed in terms of the appropriate ternary diagram and we have done so in an earlier report (*18*). However, since we did not observe this type of phase relationship in our present study, we will not pursue this point here.

Multiple Emulsions. Multiple emulsions, or subinclusions, were noted in virtually all of the samples containing PMMA as the dispersed phase. They are, for example, clearly visible in Figures 9, 10, 11, and 12, with Figures 10 and 11 providing striking examples; even in Figure 8 where the dispersed particles are very small, multiple emulsions can be seen at higher magnification. This phenomenon can also be understood

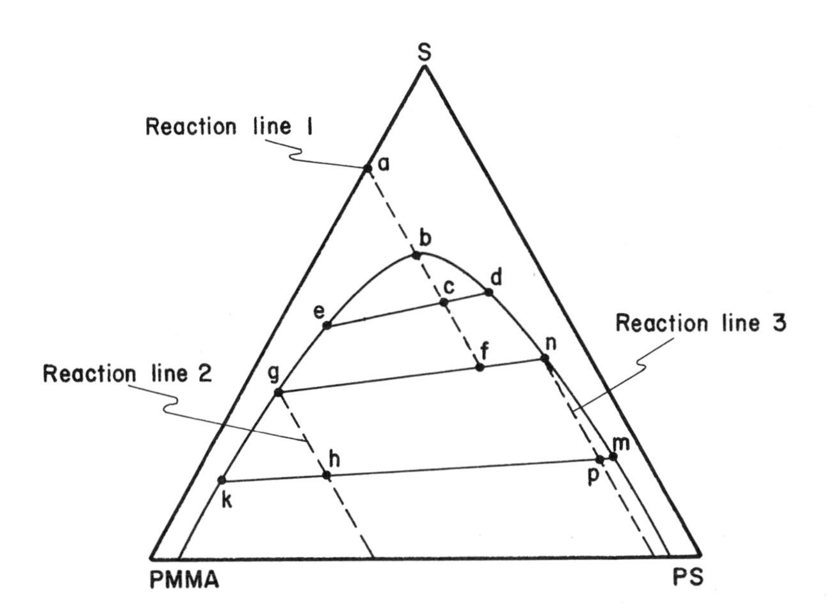

Figure 27. PMMA/PS/S phase diagram reaction lines leading to multiple emulsions

in terms of the ternary phase diagram and the viscosity of the polymerizing system, and we conclude with this discussion.

Suppose a PMMA/S mixture of initial concentration indicated by point a in Figure 27 polymerizes along reaction line 1. Phase separation occurs at point b, and soon afterwards the system reaches the overall composition c with a PS-rich phase of composition d and a PMMA-rich phase of composition e in the relative amounts ce/cd. As the system continues to polymerize the viscosity increases, and eventually it seems reasonable to suppose that the viscosity becomes so high that the two phases become essentially isolated from each other. If this point is represented by the overall composition f, we then have two isolated phases of compositions g and n. Further polymerization of the dispersed PMMA-rich phase of composition g will follow along RL2; and at point h, for example, two phases will coexist, a PMMA-rich phase of composition k and a PS-rich phase, the multiple emulsions, of composition m. Similarly, further polymerization of the original continuous phase (point n) will follow along RL3; at point p we would expect a PS-rich continuous phase of composition m and a new, PMMA-rich dispersed phase of composition. Indeed, we have studied several samples in which this behavior is observed although they are not among the samples discussed here and will be mentioned in a future report.

Literature Cited

1. Beijerinck, M. W., *Kolloid-Z* (1910) **7**, 16.
2. Dobry, A., Boyer-Kawenoki, F., *J. Polym. Sci.* (1947) **2**, 90.
3. Kern, R. J., Slocombe, R. J., *J. Polym. Sci.* (1955) **15**, 183.
4. Claver, G. C., Jr., Merz, E. H., *Off. Dig. Fed. Paint Varn. Prod. Clubs* (1956) **28**, 858.
5. Traylor, P. A., *Anal. Chem.* (1961) **33**, 1629.
6. Mann, J., Bird, R. J., Rooney, G., *Makromol. Chem.* (1966) **90**, 207.
7. Keskkula, H., Traylor, P. A., *J. Appl. Polym. Sci.* (1967) **11**, 2361.
8. Kato, K., *Jpn. Plast.* (1968) **2**(2), 6.
9. Matsuo, M., *Jpn. Plast.* (1968) **2**(3), 6.
10. Haward, R. N., Brough, I., *Polymer* (1969) **10**, 724.
11. Keskkula, H., *Appl. Polym. Symp.* (1970) **15**, 51.
12. Williams, R. J., Hudson, R. W. A., *Polymer* (1967) **8**, 643.
13. Bucknall, C. B., Smith, R. R., *Polymer* (1965) **6**, 437.
14. Matsuo, M., *Polym. Eng. Sci.* (1969) **9**, 206.
15. Baer, M., *J. Appl. Polym. Sci.* (1972) **16**, 1109.
16. Moore, J. D., *Polymer* (1971) **12**, 478.
17. Parent, R. R., Thompson, E. V., *Polym. Prepr.* (1977) **18**(2), 507.
18. Parent, R. R., Thompson, E. V., *J. Polym. Sci.* (in press).
19. Molau, G. E., *J. Polym. Sci., Part A* (1965) **3**, 1267.
20. Ibid., 4235.
21. Molau, G. E., *J. Polym. Sci., Part B* (1965) **3**, 1007.
22. Molau, G. E., Keskkula, H., *J. Polym. Sci., Part A-1* (1966) **4**, 1595.
23. Molau, G. E., Wittbrodt, W. M., Meyer, V. E., *J. Appl. Polym. Sci.* (1969) **13**, 2735.

24. Molau, G. E., *Kolloid Z. Z. Polym.* (1970) **238**, 493.
25. Bender, B. W., *J. Appl. Polym. Sci.* (1965) **9**, 2887.
26. Kruse, R. L., "Copolymers, Polyblends, and Composites," N. A. J. Platzer, Ed., ADV. CHEM. SER. (1965) **142**, 141–147.
27. Parent, R. R., Thompson, E. V., *Polym. Prepr.* (1978) **19**(1), 180.
28. Berry, J. P., *J. Polym. Sci., Part C* (1963) **3**, 91.
29. Haward, R. N., Mann, J., *Proc. R. Soc. London, Ser. A* (1964) **282**, 120.
30. Doyle, M. J., Maranci, A., Orowan, E., Stork, F. R. S., Stork, S. T., *Proc. R. Soc. London, Ser. A* (1972) **329**, 137.
31. Beahan, P., Bevis, M., Hull, D., *Polymer* (1973) **14**, 96.
32. Chapiro, A., "Radiation Chemistry of Polymeric Systems," pp. 509–512, Interscience, New York, 1962.
33. Thompson, E. V., *J. Polym. Sci., Part B* (1965) **3**, 675.
34. Schwartz, D. M., *J. Microsc. Oxford* (1972) **96**(1), 25.
35. Paxton, T. R., *J. Appl. Polym. Sci.* (1963) **7**, 1499.
36. Scott, R. L., *J. Chem. Phys.* (1949) **17**, 279.

RECEIVED April 14, 1978. This paper drawn from the thesis of Raymond R. Parent submitted in partial fulfillment of the Master of Science degree requirements of the University of Maine at Orono.

21

Physical Properties of Blends of Poly(vinyl Chloride) and a Terpolymer of Ethylene

E. W. ANDERSON, H. E. BAIR, G. E. JOHNSON, T. K. KWEI, F. J. PADDEN, JR., and DENISE WILLIAMS[1]

Bell Laboratories, Murray Hill, NJ 07974

The compatibility of blends of poly(vinyl chloride) (PVC) and a terpolymer (TP) of ethylene, vinyl acetate, and carbon monoxide was investigated by dynamic mechanical, dielectric, and calorimetric studies. Each technique showed a single glass transition and that transition temperature, as defined by the initial rise in E'' *at 110 Hz,* ϵ'' *at 100 Hz, and* C_p *at 20°C/min, agreed to within 5°C. PVC acted as a polymeric diluent which lowered the crystallization temperature,* T_c*, of the terpolymer such that* T_c *decreased with increasing PVC content while* T_g *increased. In this manner, terpolymer crystallization is inhibited in blends whose value of* $(T_c - T_g)$ *was negative. Thus, all blends which contained 60% or more PVC showed little or no crystallinity unless solvent was added.*

By blending with any one of a multitude of additives, PVC can be transformed into a broad spectrum of resins ranging from highly plasticized to impact resistant. The use of polymeric plasticizers has attracted a great deal of attention because they provide superior permanence in physical properties over their low molecular weight counterparts. Recently a terpolymer of ethylene, vinyl acetate, and carbon monoxide was reported to be miscible with PVC (*1, 2*). The system is of interest because blends of PVC and ethylene-vinyl acetate copolymers range from incompatible to miscible, depending on the content of vinyl acetate in the copolymer (*3, 4, 5*). We have therefore undertaken x-ray,

[1] Present address: University of Massachusetts, Amherst, MA 01002.

0-8412-0457-8/79/33-176-413$05.00/0

morphological, calorimetric, dielectric, and dynamic mechanical studies to elucidate the compatibility and physical properties of blends of PVC with the terpolymer.

A second objective of this study concerns the crystallization of the terpolymer in the mixture. While the terpolymer acts as a plasticizer for PVC, the latter also can be viewed as a polymeric diluent which influences the crystallization of the former. The melting point of the terpolymer, about 77°C, is slightly lower than the glass transition temperature of PVC, and it is conceivable that a mixture containing a minor quantity of the terpolymer can have a glass temperature approaching or exceeding the crystallization temperature of the terpolymer in the mixture. One would then expect the crystallization process to be retarded or inhibited by the highly viscous matrix. If, however, the T_g of the mixture is lowered by the addition of a solvent, then crystallization still may be able to proceed. We have conducted preliminary experiments to explore this possibility.

Experimental

Materials. PVC, Geon 103 EP from B. F. Goodrich Co., and a terpolymer, Elvaloy 741 from E. I. du Pont de Nemours and Co., were used. Blends of the two polymers were mixed on a two-roll mill at 160°C and then pressed into films on a heated press at 150°C. The compositions of the mixtures are reported as weight percentages in our text.

X-Ray and Light Microscopy. X-ray diffraction patterns of powdered specimens of Elvaloy 742 and its blends with PVC were recorded on photographic films in a Guinier–DeWolff focusing camera. Light microscopy was performed on a Reichert Zetopan polarizing microscope with a Mettler FP21 hot stage.

Dynamic Modulus. Measurement of dynamic viscoelasticity was made by the use of a direct-reading dynamic viscoelastometer from the Toyo Measuring Instrument Co., at a frequency of 110 Hz.

Tensile Strength. The tensile strength and elongation of the two polymers and their blends were measured at room temperature with a table model, Instron tensile tester at a strain rate of 111 percent per minute.

Calorimetry. Glass and melting transitions and apparent heats of fusion (ΔQ_f) were analyzed by a differential scanning calorimeter, Perkin–Elmer DSC-2. Measurements were made in a helium atmosphere at a heating or cooling rate of 40°C/min in a manner which has been reported elsewhere (*6*). Where accurate absolute values of heat capacity (rather than merely the change in C_p) are desired, a Scanning Auto Zero accessory was used to produce an essentially flat baseline from −100° to 150°C. Heat-capacity calculations were performed by a Perkin–Elmer programmable calculator system which includes a Tektronix Model 31 calculator. In these measurements, the heating rate was 20°C/min. The heat capacities of a standard Al_2O_3 sample in the temperature range of −30° to 120°C were found to be within 1% of the values reported in the literature (*7*).

Dielectric Properties. Dielectric measurements were conducted at 10^2, 10^3, 10^4, and 10^5 Hz. The data were obtained by combining a Princeton Applied Research 124 lock-in amplifier and a General Radio 1615A capacitance bridge. The bridge was connected to a Balsbaugh LD3 research cell inside a test chamber. Temperature was monitored with a thermocouple taped to the cell. After the chamber was equilibrated at −160°C for 1 hr, measurements were made at approximately 10°-intervals with a 15-min waiting period between each temperature change.

Results

X-Ray Studies. Lines in the x-ray, powder diffraction pattern of the terpolymer were similar to those found in polyethylene (PE). Since the terpolymer contains a majority of ethylene segments, we assumed that it crystallized in an orthorhombic unit cell as polyethylene does. The lattice parameters for the terpolymer were calculated to be a = 7.76 Å, b = 5.06 Å, and c = 2.56 Å. These values are to be compared with the dimensions of PE unit cell: a = 7.40 Å, b = 4.93 Å, and c = 2.54 Å (*8*). Thus x-ray data suggest that the terpolymer crystallizes with an expanded PE unit cell.

The low intensity of the x-ray lines found in the diffraction pattern of the terpolymer compared with the high intensity observed for PE signified that the level of crystallinity in the former was significantly below that of PE. In spite of this, an amorphous sample could not be prepared by quenching. The presence of crystalline regions was indicated in many mixtures, but crystallinity decreased markedly as the weight fraction of the terpolymer decreased.

Light Microscopy. Although films of the terpolymer were birefringent, no spherulites could be detected regardless of the manner in which specimens were prepared. Samples cast from a 2% solution in tetrahydrofuran (THF) and annealed at 60°C overnight had optical melting points which ranged from 63° to 75°C.

Dynamic Viscoelasticity. The dynamic modulus and loss tangent curves are shown in Figures 1, 2, and 3. The E'' peak temperature of the main transition in PVC occurs at 90°C; the secondary transition, which has been ascribed to local mode motion of the PVC main chain, extends from −120° to 40°C, with a maximum located at about −30°C. These results are in good agreement with earlier investigations (*9*). The terpolymer has a sharp glass transition near −22°C; the γ relaxation in the neighborhood of −135°C occurs at a lower temperature and is broader than the corresponding relaxation in linear or branched polyethylene (*9*).

The blends show three regions of mechanical relaxation. The low temperature dispersion near −135°C, characteristic of the terpolymer, remains prominent. For each mixture, a single loss maximum above 0°C is accompanied by an abrupt decrease in E' and is undoubtedly associated

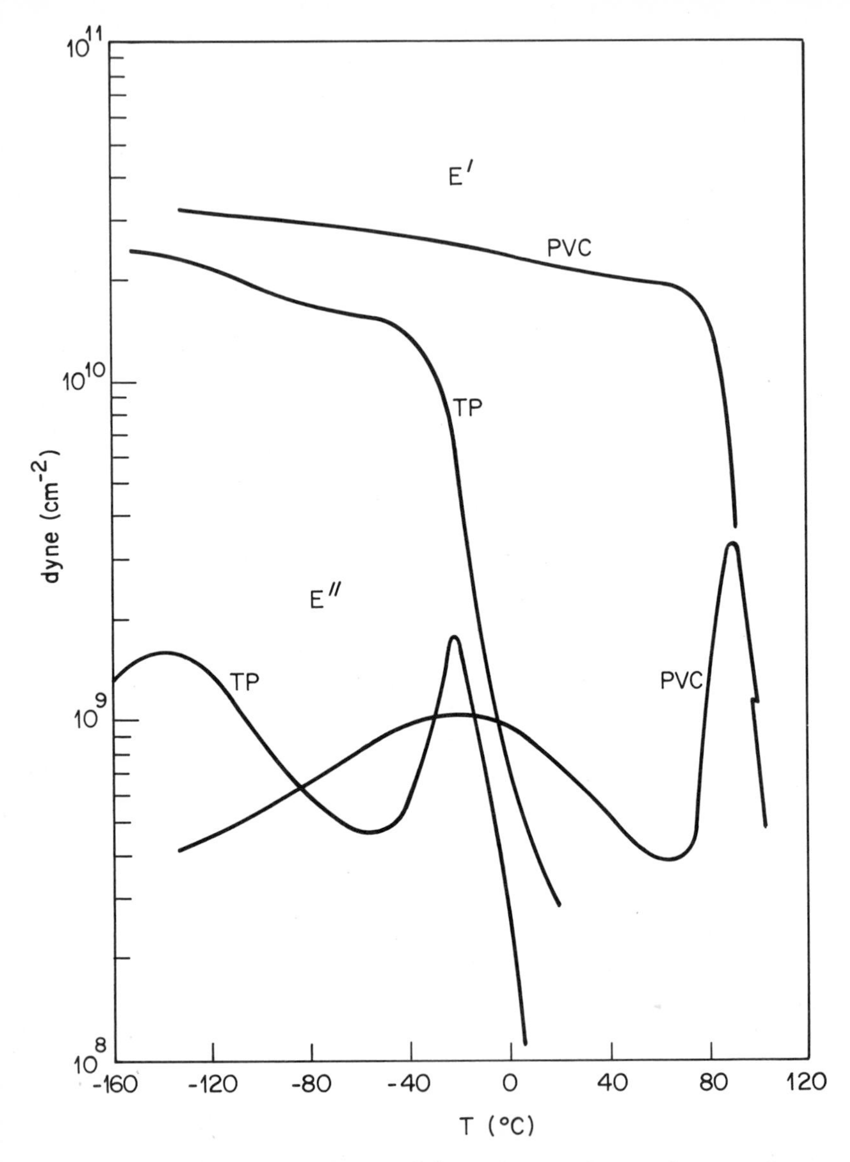

Figure 1. The storage and loss modulus curves for the terpolymer and PVC

with the glass transition of the mixture. In the region between −20° and −80°C, where the local mode motion of PVC occurs, the loss tangent curves are almost flat. This observation suggests that the local environment of PVC segments has been influenced by the terpolymer as a result of extensive mixing. The flat region also appears to shift to a lower temperature and covers a narrower temperature span as the amount of terpolymer increases. Similar results were reported in blends of PVC with a copolyester of poly(tetramethylene ether) glycol terephthalate and tetramethylene terephthalate (*10*), poly(ϵ-caprolactone (*11*), ethylene–ethyl acrylate–carbon monoxide terpolymer (*12*), ethylene-*N*,*N'*-dimethylacrylamide copolymer (*13*), and ethylene–vinyl acetate copolymer (*4*).

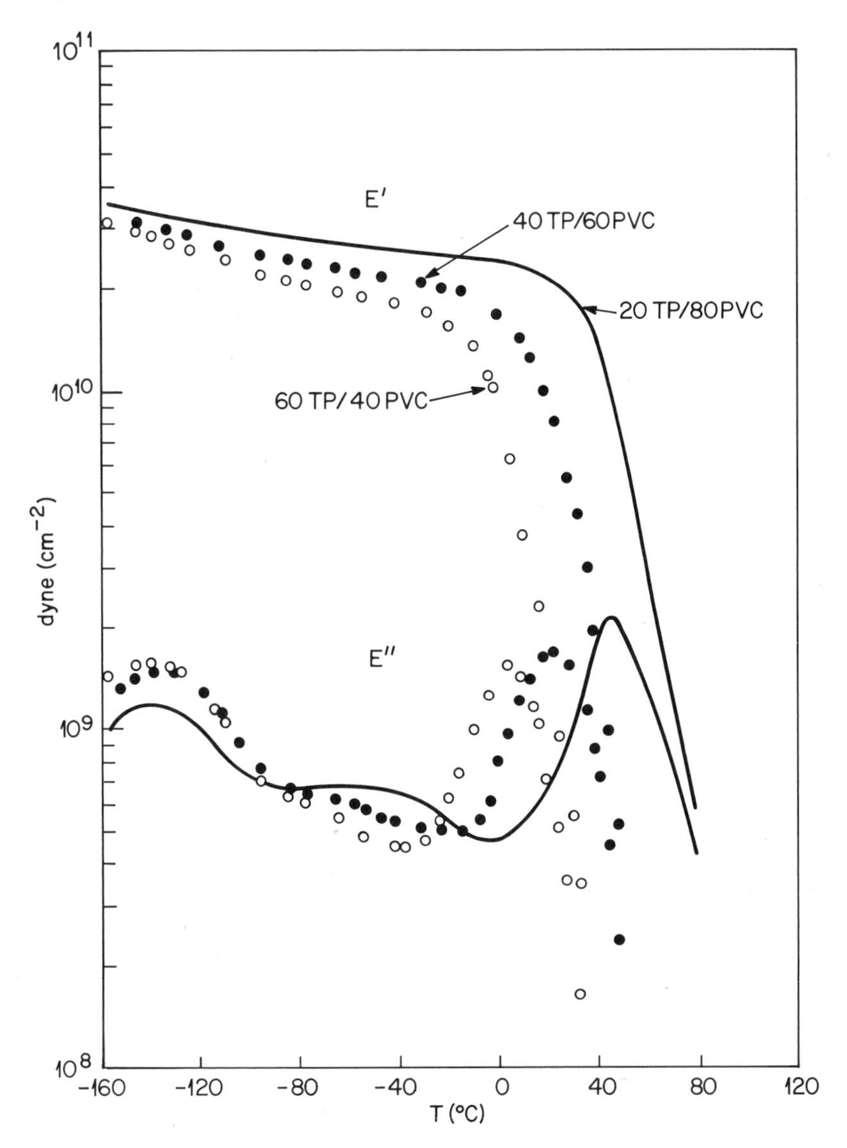

Figure 2. Temperature dependence of dynamic–mechanical properties for 60 TP/40 PVC (○), 40 TP/60 PVC (●), and 20 TP/80 PVC (—)

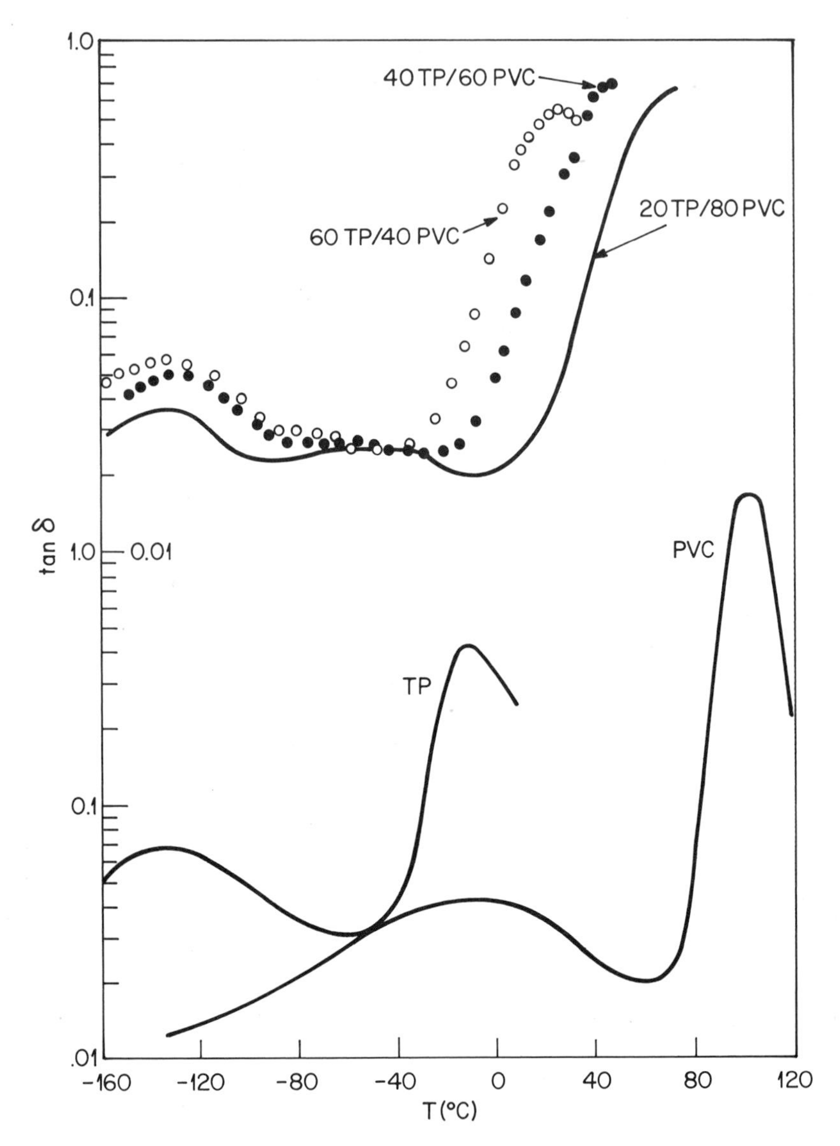

Figure 3. Mechanical loss tangent curves of TP, PVC, 60 TP/40 PVC, 40 TP/60 PVC, and 20 TP/80 PVC

Dielectric Measurements. The dielectric loss (ϵ'') curves at different frequencies for samples containing 100, 80, 40, and 0% PVC, respectively, are shown in Figures 4, 5, 6, and 7. Figure 8 is a composite of the dielectric loss data at 1 kHz for each sample. The general characteristics of α and β relaxation peaks of the component polymers and their mixtures parallel the results of dynamic mechanical measurements. For each

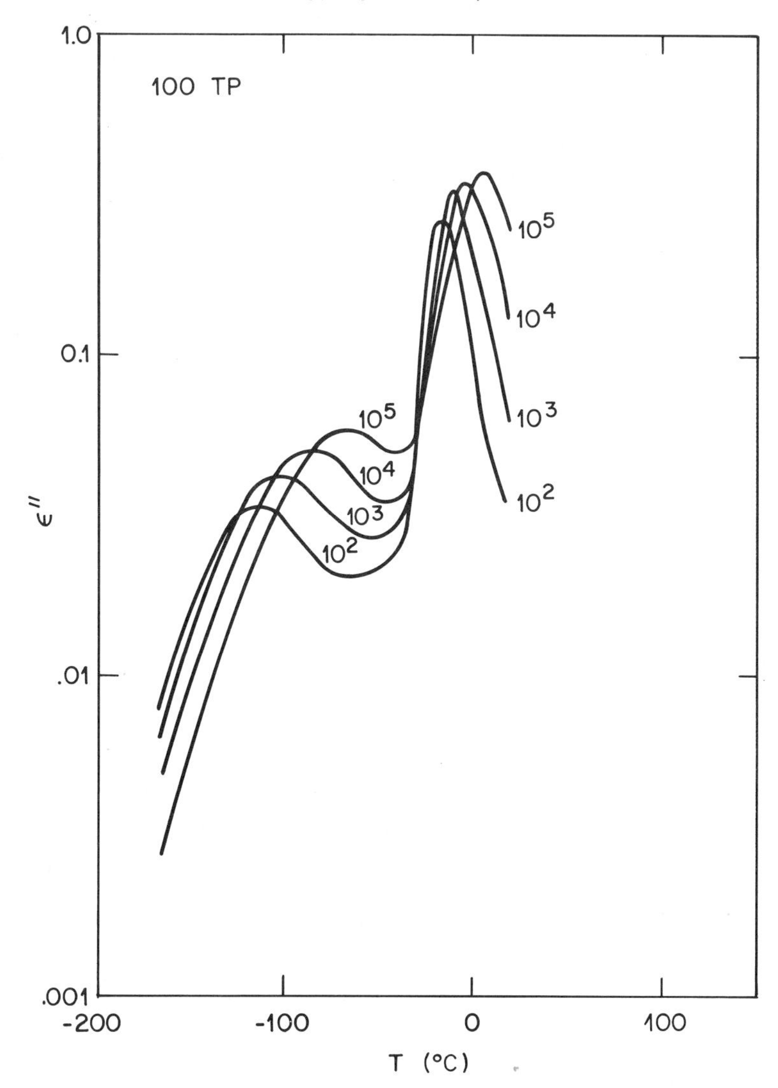

Figure 4. Temperature dependence of the dielectric loss behavior from 10^2 to 10^5 Hz of the terpolymer (100 TP)

mixture, a single α peak was detected. It is noticed, however, that the α transition of the 80% PVC sample is broader than that of the 40% PVC sample and that small shoulders between − 20° and − 80°C are observed for the former at 10^2 and 10^3 Hz. These results are again in harmony with the viscoelastic data.

Calorimetry. Prior to C_p measurements, each sample was heated to 140°C and then cooled at a rate of 40°C/min to − 90°C unless otherwise stated. Measured C_p values are listed in Table I.

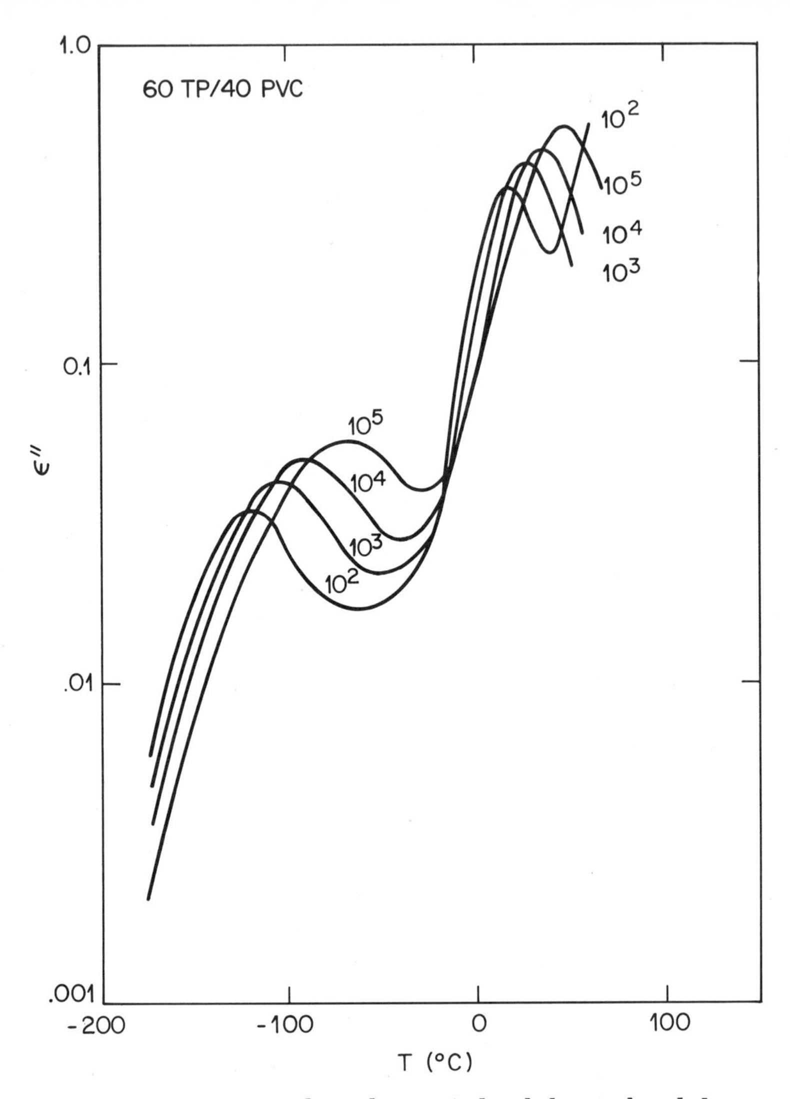

Figure 5. Temperature dependence of the dielectric loss behavior from 10^2 to 10^5 Hz of the 60 TP/40 PVC

The C_p of PVC (lower curve in Figure 9) increases linearly with temperature from − 60° to 80°C. Between 85° and 93°C, the glass transition is manifested by a discontinuous increase in C_p and corresponds to the α transition observed in the dielectric and dynamic mechanical experiments. Above the glass-transition temperature, C_p increases smoothly to 120°C where the experiment was terminated.

The C_p curve of the terpolymer (upper curve Figure 9) is more complex. It rises smoothly from − 120° to − 40°C but above − 34°C increases abruptly, and a broad endothermic peak, typical of fusion of polymer crystals, occurs between 50° and 77°C. We attribute the initial discontinuity in C_p at − 34°C to the onset of glass transition and the subsequent increase in C_p to melting. The point of inception of melting of terpolymer crystals can be estimated by extrapolating the C_p curve

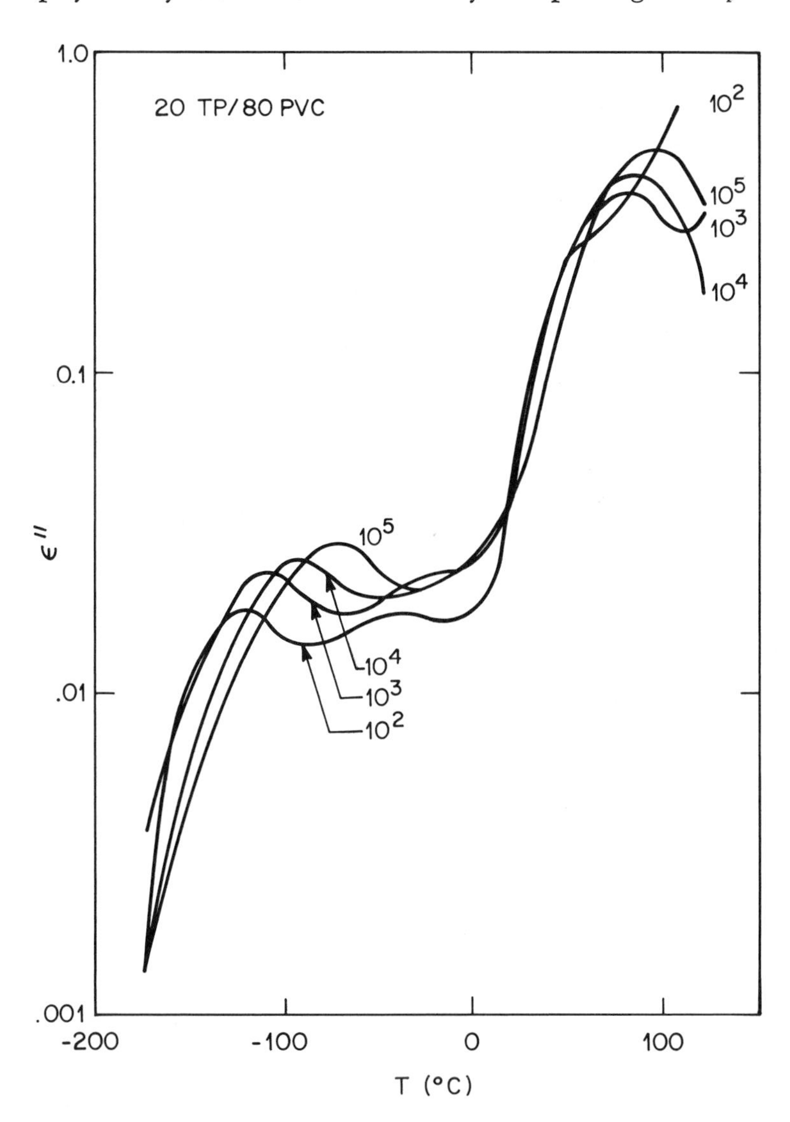

Figure 6. Temperature dependence of the dielectric loss behavior from 10^2 to 10^5 Hz of the 20 TP/80 PVC

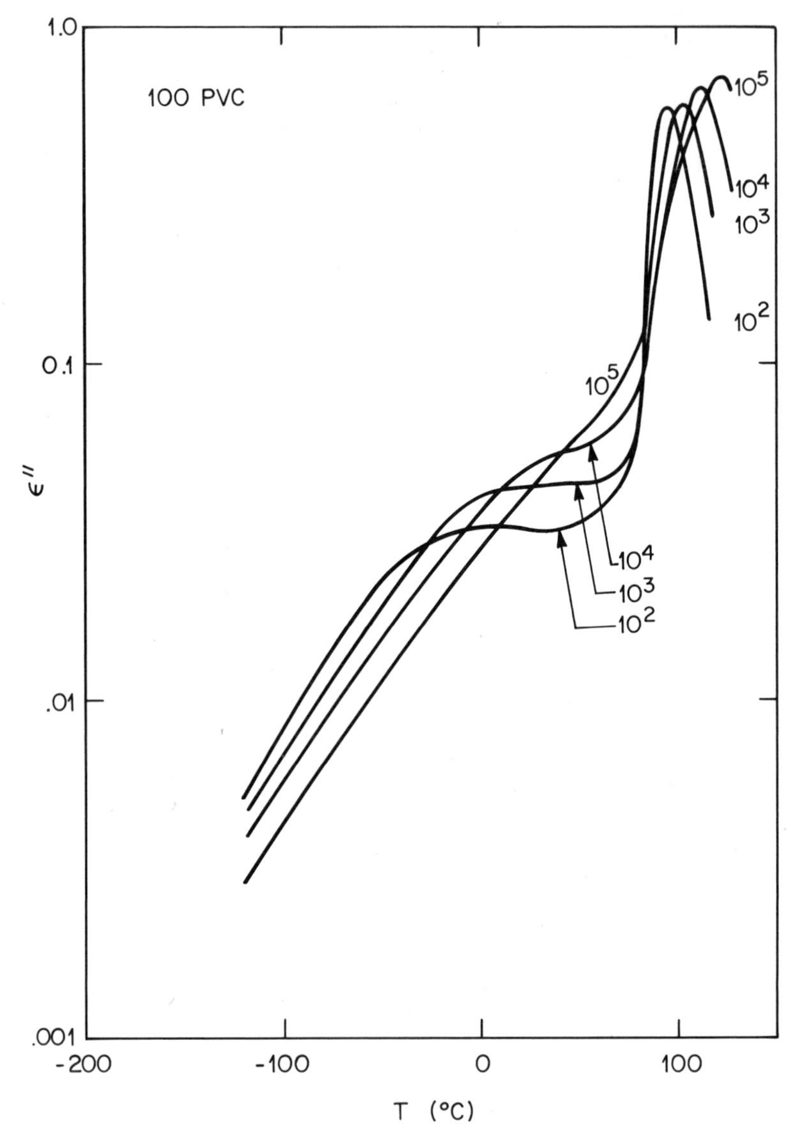

Figure 7. Temperature dependence of the dielectric loss behavior from 10^2 to 10^5 Hz of the PVC (100 PVC)

of the liquid between 80° and 150°C to lower temperatures until it intersects the C_p curve at about − 15°C. The broad temperature range of fusion, from − 15° to 77°C, most likely reflects the presence of lamellae of varying thickness or defects. Accepting the above interpretation, we estimate the apparent heat of fusion, ΔQ_f, from the area of the C_p curve between the inception of melting at $-15°C$ and its termination at 77°C. The value of ΔQ_f is 8.3 cal/g. (As a point of reference, the heat of fusion per crystallized unit is about 68 cal/g for linear polyethylene (*14, 15*).)

Slow cooling from the melt at 0.31°C/min results in only minor changes in the C_p curve. Annealing of a quenched sample near its T_g, on the other hand, produces noticeable differences in the thermal scan.

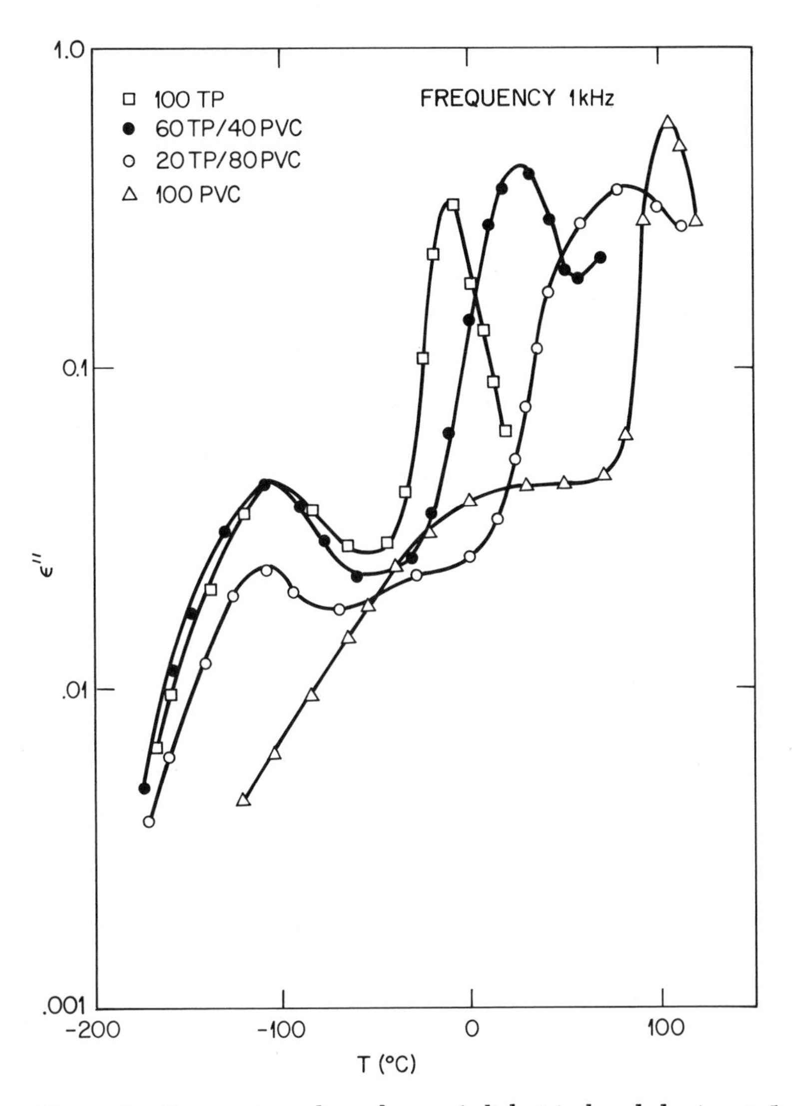

Figure 8. Temperature dependence of dielectric loss behavior at 1 kHz for PVC (△), 20 TP/80 PVC (○), 60 TP/40 PVC (●), and TP (×)

A melted sample of the terpolymer was cooled in the calorimeter at 320°/min to − 40°C and was allowed to remain at − 40°C for 16 hr after which the temperature was lowered to − 150°C. Upon reheating, the T_g of the annealed sample increased to − 31°C; moreover, an additional adsorption of thermal energy was superimposed upon the normal increase in C_p during the glass transition. The additional increase in enthalpy, 0.5 cal/g, is the result of enthalpy relaxation occurring during

Table I. Heat Capacity of TP and PVC and Two Blends[a]

	C_p (cal g^{-1} K^{-1})			
	100 TP	*60 TP/40 PVC*	*40 TP/60 PVC*	*100 PVC*
T(K)		C_p		
215.0	0.278	—	0.212	0.170
220.8	0.288	—	0.216	0.173
225.7	0.296	—	0.223	0.176
230.6	0.306	0.261	0.228	0.180
234.6	0.316	0.266	0.235	0.182
240.5	0.338	0.274	0.239	0.184
245.9	0.354	0.284	0.247	0.186
248.9	0.372	0.289	0.250	0.187
254.9	0.409	0.298	0.255	0.191
260.8	0.438	0.309	0.262	0.195
263.8	0.453	0.316	0.267	0.196
269.7	0.481	0.335	0.275	0.200
275.6	0.508	0.358	0.285	0.204
281.6	0.532	0.371	0.300	0.208
284.5	0.542	0.369	0.306	0.209
290.5	0.562	0.361	0.319	0.213
293.4	0.572	0.362	0.325	0.215
299.4	0.600	0.377	0.334	0.219
305.3	0.621	0.406	0.342	0.222
311.3	0.630	0.441	0.354	0.226
317.2	0.643	0.476	0.368	0.230
323.1	0.655	0.506	0.385	0.234
329.1	0.691	0.527	0.401	0.238
335.0	0.737	0.534	0.409	0.242
341.0	0.763	0.513	0.406	0.247
346.9	0.701	0.469	0.398	0.251
352.8	0.538	0.445	0.397	0.257
358.8	0.513	0.445	0.401	0.264
364.7	0.514	0.448	0.404	0.282
370.6	0.517	0.451	0.409	0.319
376.6	0.519	0.454	0.412	0.341
379.6	0.520	0.456	0.414	0.345
385.5	0.522	0.459	0.417	0.350
391.4	0.524	0.462	0.420	0.356

[a] All samples were quenched from 400 to 200 K at a cooling rate of −160 K min^{-1}.

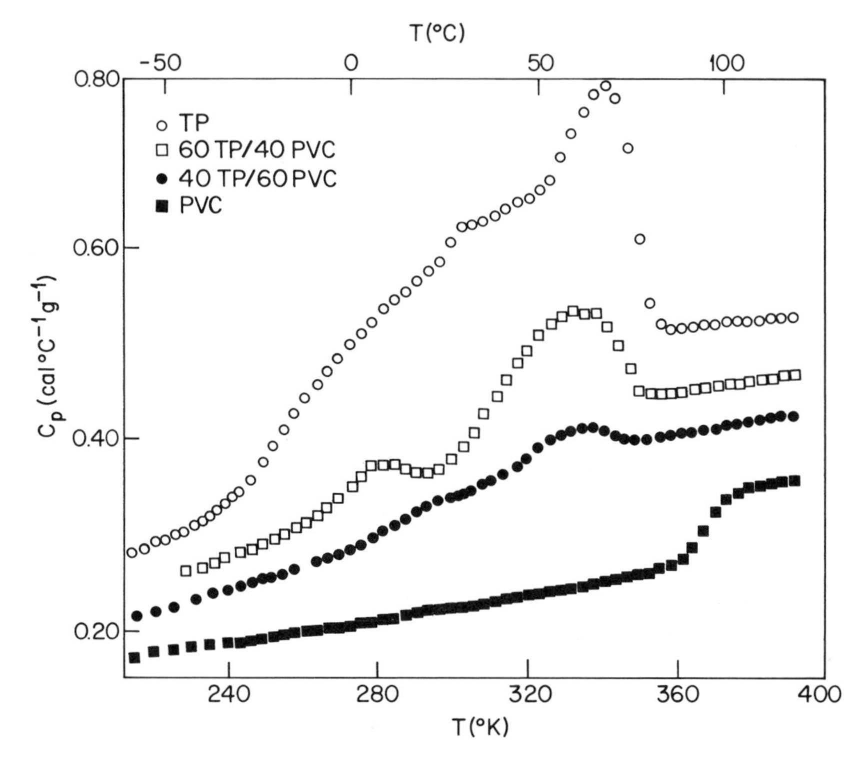

Figure 9. Specific heat as a function of temperature for TP, 40 TP/60 PVC, 60 TP/40 PVC, and PVC

annealing at − 40°C. We have observed similar reduction in excess enthalpy after annealing poly(butylene terephthalate), a semicrystalline polymer, below its T_g (*16*). Furthermore, the annealing process caused a small increase in the amount of crystals which melted between 21° and 51°C.

Annealing at temperatures near the melting point of the terpolymer also resulted in additional features in the C_p curve. In Figure 10, the melting curve for a terpolymer sample with a complex thermal history is depicted. After melting, this sample was placed in a 60°C oven for 24 hr, then cooled slowly to 40°C, and finally stored at 23°C. Although its T_g was unchanged at − 34°C, the sample exhibited three melting peaks at 43°, 56°, and 72°C. The end of the fusion process occurred at a higher temperature, 92°C, than previously observed. Annealing at 20°, 40°, and 60°C for 30 min each produced crystals which melted at 30°, 48°, 67°C, respectively. Apparently each increase in annealing temperature produced lamellae of increased thickness and correspondingly higher melting points.

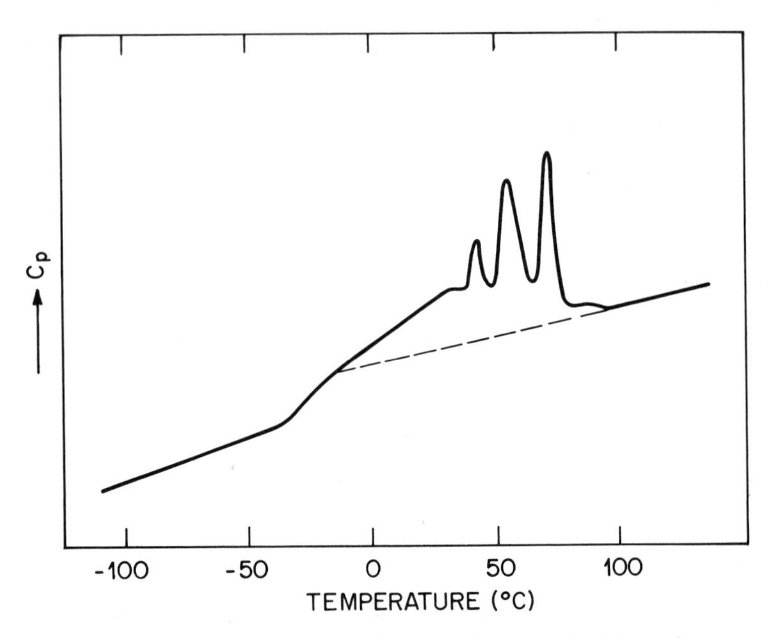

Figure 10. C_p behavior of TP after annealing at 60°, 40°, and 23°C

Similar heat treatment of the blends also yielded multiple melting peaks (Table II). Because the specific heats of blends are influenced by the thermal histories of the samples, only two mixtures containing 40 and 60% terpolymer, respectively, were selected for quantitative C_p measurements. Each sample experienced the same thermal history as the two component polymers, namely, cooling at 40°C/min from 140° to − 90°C. The C_p curves of the two mixtures are also shown in Figure 9. In the liquid state, the C_p of the blend was found to be the weight average of

Table II. Thermal Properties of Elvaloy

Elvaloy 741 (*Wt %*)	T_g [a] (°C)	T_c [a] (°C)	T_m [a] (°C)
100	−34	43	77
80	−29	34	76
60	−19	27	73
40	−6	15	71
20	11	—	—
10	41	—	—
0	85	—	—

[a] Cooled from melt at 40°C/m.
[b] Annealed at 60°C for 24 hr, then cooled slowly to 40°C and finally stored at 23°C for several weeks.

the heat capacities of PVC and the terpolymer. For example, the calculated C_p of a blend of 40% terpolymer is 0.528 cal°C^{-1}g^{-1} at 118°C while the experimental value is 0.524 cal°c^{-1}g^{-1}. Both the melting point and the apparent heat of fusion of the terpolymer decrease as the amount of PVC increases. We made no attempt for theoretical analysis in this regard.

Only a single T_g was detected for each blend and it increased in temperature as the amount of PVC was increased (Figure 11). The value of T_g, however, did not conform to the volume-fraction average of component polymers as had been found in poly(vinylidene fluoride) and poly(methyl methacrylate) mixtures (*17*). The reasons are twofold; first, the composition of the amorphous portion of the terpolymer is different from that of the crystalline region. Therefore the T_g of the terpolymer is not representative of the overall composition. Secondly, blends containing more than 40% terpolymer still retain some crystallinity even in the quenched state. As crystallinity decreases continuously in the blend, the composition of the amorphous fraction also changes through the incorporation of more methylene units which are normally crystallizable in the pure polymer. Consequently, the glass-transition temperatures of the mixtures do not obey simple relationships.

At this point, a comparison of the calorimetric results with dielectric and viscoelastic data, with regard to glass transitions, is in order. In both viscoelastic and dielectric measurements, the α-relaxation processes encompass both the glass transition of the mixture and the melting of terpolymer crystals, if present. Therefore, the peak position is not necessarily a good measure of the glass-transition temperature. On the other hand, the calorimetric results are relatively free of such ambiguities. For the purpose of correlating data between these three different measurements, we have compiled the temperatures of initial rise of E'' at 110 Hz, ϵ'' at 100 Hz, and C_p at 20°C/min for these samples and plotted the

741 and Its Blends with PVC

T_m[b] (°C)	T_m[b] (°C)	ΔQ_f[a] (cal/g)	ΔQ_f[b] (cal/g)	ΔQ_f[c] (cal/g)
92, 74, 66, 49	79	8.3	11.8	13.8
79, 69, 52	78	7.3	8.0	9.2
78, 67, 52	—	2.8 (ϕ)	4.9	—
67, 62, 53	74	0.6	2.0	3.0
62, 54	74	—	—	1.6
—	73	—	—	0.8
—	—	—	—	—

[c] Solution crystallized.

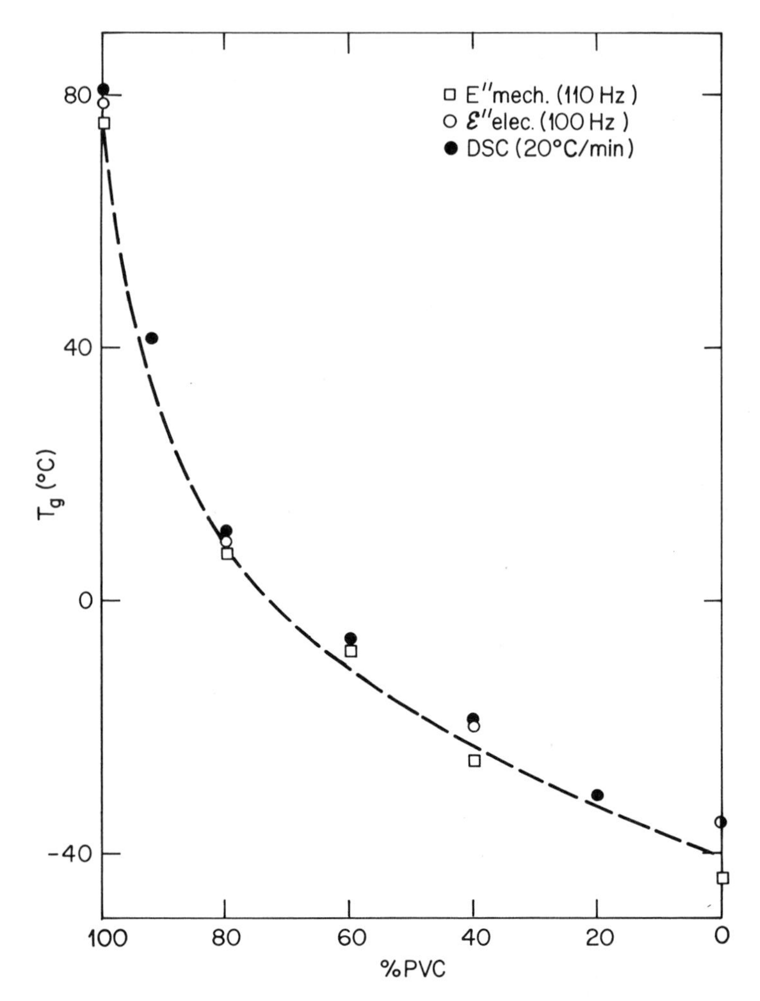

Figure 11. T_g as a function PVC concentration as defined from initial rise in E'', ε', and C_p by dynamic mechanical (□), dielectric (○), and calorimetric (●) measurements, respectively

values against composition in Figure 11. The agreement among the three methods is within 5°C. Although such a comparison appears obvious, we do not recall similar efforts in the literature.

Crystallization of the Terpolymer in the Mixture. Crystallization of the terpolymer occurred in samples containing 60% or more of the terpolymer when cooled from the melt at a rate of 40°C/min. The specimen containing 40% terpolymer at first did not exhibit a measurable exotherm because of crystallization, although a small amount of crystals was detected in the sample upon subsequent reheating. Additional cooling studies conducted at 40°, 20°, 10°, and 5°C/min on samples containing 40% TP have revealed that the onset of crystallization, T_c, occurred at about 15°C, which was just before the sample began to undergo vitrification. Mixtures containing 20 and 10% terpolymer, respectively, did not crystallize upon cooling from their melts.

When T_c (determined at a cooling rate of 40°C/min) and T_g were plotted vs. weight percent PVC, it was noticed that the former decreased with increasing PVC content while the latter increased (Figure 12). The differences between T_c and T_g were 77°, 63°, 46°, and 21°C for samples containing 100, 80, 60, and 40% terpolymer, respectively. If the T_c data is extrapolated linearly to other compositions, the value of

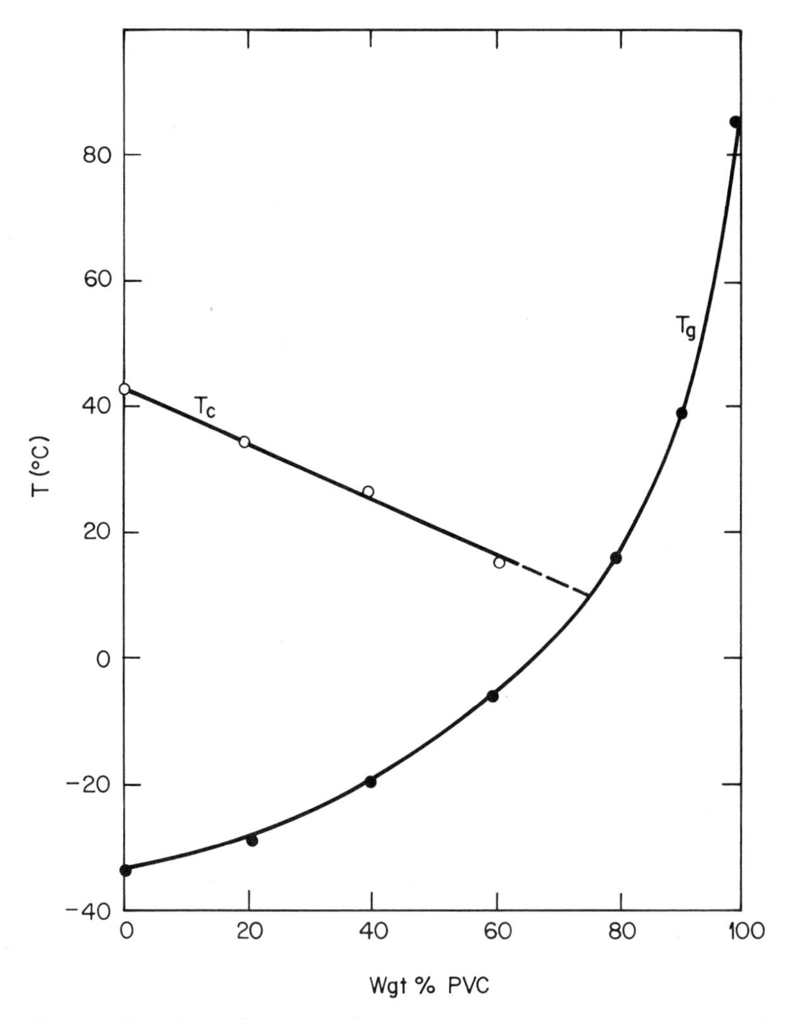

Figure 12. Crystallization and glass temperatures as a function of PVC concentration

($T_c - T_g$) would have been 0°C for a 25% terpolymer sample. The fact that the extrapolated T_c has fallen below T_g in the 20% TP sample probably accounts for the lack of crystallization in this sample.

When the quenched samples were thermally treated at 60°C for 24 hr, they underwent further crystallization as indicated by the higher values of heat of fusion obtained in subsequent thermal scans (Table II). In addition to the annealing study, a second approach also was attempted. It was based on the thought that crystallization might be facilitated by lowering the T_g of the mixtures with a suitable solvent. Three samples were selected for this investigation, containing 40, 20, and 10% terpolymer, respectively. The crystallization temperature was chosen to be 0°C because T_c for the 10% terpolymer sample in the bulk was estimated to be below room temperature. For the purpose of comparison, films containing 80 and 100% terpolymer also were cast from solution.

Our rudimentary experiments were conducted as follows: each sample was dissolved in THF to form a 7% solution. An aliquot of the solution was poured into a Petri dish which was surrounded by an ice-water mixture. The content in the dish was placed under a well-ventilated hood and the solvent was allowed to evaporate. Within 15 minutes, the solution began to turn opaque, indicative of crystallization. At the end of about one hour, the mixture was completely opaque. It still contained more than 30% solvent but could be lifted from the dish as a free film. The opacity of the film was retained even after heating at 100°C overnight.

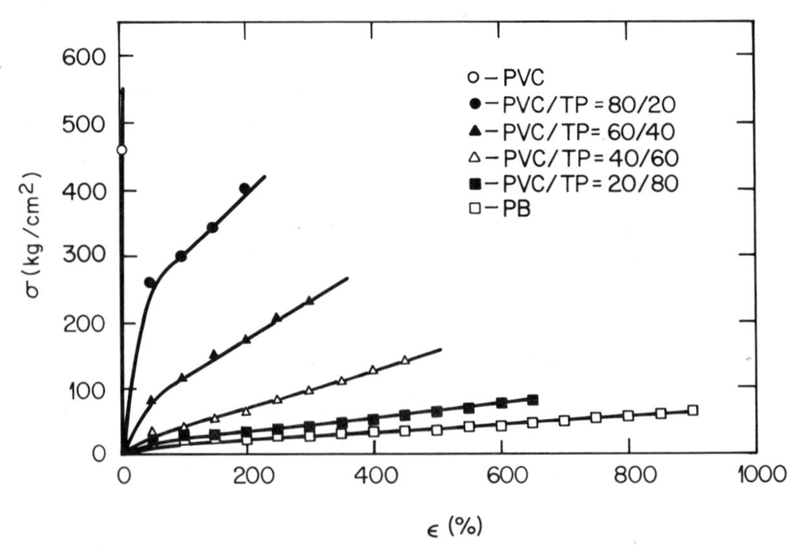

Figure 13. Stress–strain curves of PVC, TP and their blends at a strain rate of 111%/min

The calorimetric results of the solution-crystallized films are given in Table II. The apparent heat of fusion is enhanced for all films containing 40% or more of TP relative to the bulk crystallized samples. In addition, the 10 and 20% TP blends were found to be partially crystalline with T_ms only lowered by 5° or 6°C.

Tensile Properties. Figure 13 shows the stress–strain behavior at room temperature of the two component polymers and their mixtures. The tensile properties range from rigid response for PVC to elastomeric, ductile behavior for the terpolymer. Note that the 80% PVC sample exhibits a high Young's modulus followed by extensive yielding at high stress before breaking.

Conclusion

The compatibility of blends of PVC and the terpolymer was investigated by dynamic mechanical, dielectric, and calorimetric studies. Not only did each technique show a single glass transition for each mixture, but also the temperature of the transition, as defined by the initial rise in E'' at 110 Hz, ϵ'' at 100 Hz, and C_p at 20°/min, agreed to within 5°C. T_g was found to increase with increasing concentration of PVC.

Although the terpolymer plasticized PVC, the latter was found to act as a polymeric diluent which lowered the crystallization temperature, T_c, of the former. In this maner, the crystallization of the terpolymer could be retarded or inhibited in blends whose value of $(T_c - T_g)$ was near or below 0°C. Mixtures containing 60% or more PVC were found to be in this category and showed little or no sign of bulk crystallization. However mixtures containing 80 and 90% PVC which were completely amorphous in the bulk, crystallized when solvent was added. Though we believe crystallization occurred because of a lowering of T_g, separation of this from the possibility of crystallization from solution would require further studies.

Acknowledgment

The authors wish to acknowledge the valuable advice and assistance rendered by P. C. Warren and J. E. Adams in the course of this study.

Literature Cited

1. Tordella, J. P., Hyde, T. J., Gordon, B. S., Hammer, C. F., *Mod. Plast.* (1976) **54**(1), 64.
2. Hammer, C. F., *Am. Chem. Soc., Div. Org. Coat. Plast. Prepr.* (1977) **37,** 234.
3. Jyo, Y., Nozaki, C., Matsuo, M., *Macromolecules* (1971) **4,** 517.
4. Hammer, C. F., *Macromolecules* (1971) **4,** 69.

5. Elmquist, C., *Eur. Polym. J.* (1977) **13**, 95.
6. Bair, H. E., *Polym. Eng. Sci.* (1974) **14**, 202.
7. Ginnigs, D. C., Furukawa, G. T., *J. Am. Chem. Soc.* (1953) **75**, 522.
8. Kavesh, S., Schultz, J. M., *J. Polym. Sci. Part A2* (1970) **8**, 243.
9. Takayanagi, M., Harima, H., Iwata, Y., *Mem. Fac. Eng., Kyushu Imp. Univ.* (1963) **23**, 1.
10. Nishi, T., Kwei, T. K., Wang, T. T., *J. Appl. Phys.* (1975) **46**, 4157.
11. Koleske, J. V., Lundberg, R. D., *J. Polym. Sci. Part A2* (1969) **7**, 795.
12. Robeson, L. M., McGrath, J. E., *Polym. Eng. Sci.* (1977) **17**, 300.
13. McGrath, J. E., Ward, T. C., Wnuk, A., *Polym. Prepr., Am. Chem. Soc., Div. Polym. Chem.* (1977) **18**, 346.
14. Mandelkerm, L., *Rubber Chem. Technol.* (1959) **32**, 1392.
15. Roe, R. J., Bair, H. E., *Macromolecules* (1970) **3**, 454.
16. Bair, H. E., Bebbington, G. H., Kelleher, P. G., *J. Polym. Sci., Polym. Phys. Ed.* (1976) **14**, 2113.
17. Kwei, T. K., Frisch, H. L., Radigan, W., Vogel, S., *Macromolecules* (1977) **10**, 157.

RECEIVED June 5, 1978.

22

Physical Properties of Blends of Polystyrene with Poly(methyl Methacrylate) and Styrene/(methyl Methacrylate) Copolymers

DENNIS J. MASSA

Research Laboratories, Eastman Kodak Co., Rochester, NY 14650

The effect of polystyrene (PS) molecular weight (M) on compatibility and physical properties of blends of PS with poly(methyl methacrylate) (PMMA) and PS with styrene/(methyl methacrylate), PMMA, and PS with styrene/(methyl methacrylate) copolymers has been investigated. The polystyrene M ranged from 600 to 110,000, and the styrene fraction in the copolymers varied from 0 to 0.8. The conclusions of this work are as follows: (i) PS is compatible at low M with PMMA over a range of PS concentrations, up to nearly 40% for M = 600; (ii) increasing the styrene content in the styrene/(methyl methacrylate) copolymers increases compatibility with PS; (iii) molecular weight is important in all cases; compatibility increases with decreasing M; and (iv) the Flory–Huggins–Scott solubility parameter approach is consistent with qualitative trends in the results.

The miscibility, or compatibility, of polymers in the bulk state is of immense importance to their use in mixtures and blends. The phase morphology and resulting physical properties of polymer blends are critically dependent upon the equilibrium thermodynamic interactions among polymers as well as upon the nonequilibrium effects that the thermal and physical histories can impress on the blend during preparation.

Although the prediction of polymer–polymer miscibility based on chemical structure and/or equilibrium thermodynamic properties has been a goal of polymer chemists for decades (*1, 2*), the complexity of

0-8412-0457-8/79/33-176-433$05.00/0

the problem has limited successful efforts to relatively few cases. Instead, compatible polymer pairs are often discovered empirically and only thereafter studied intensively. Nonetheless, despite difficulties with quantitative predictions, the way in which molecular variables affect polymer–polymer miscibility is understood well enough to allow qualitative judgements of their influence on polymer blend properties.

For example, the Flory–Huggins–Scott (*3–8*) expression for the free energy of mixing, shown below in the form given by Krause (*1*), where

$$G_{mix} = \frac{RTV}{V_r}\left[\frac{V_A}{X_A}\ln V_A + \frac{V_B}{X_B}\ln V_B + \chi_{AB}V_AV_B\right],$$

V is the total volume of the mixture; V_r is the reference volume, the volume of polymer repeat unit; V_A and V_B are the volume fractions of polymers A and B; X_A and X_B are the degrees of polymerization of A and B; X_{AB} is related to the enthalpy of interaction of the polymer repeat units, each of molar volume V_r, indicates that whereas at high degrees of polymerization (X_A and X_B) the value of the interaction parameter X_{AB} would have to be nearly zero or negative to result in a favorable (negative) free energy of mixing (*9*), even fairly dissimilar polymers might be compatible at low molecular weights (small X_A and X_B). The relatively few studies of compatibility of low-molecular-weight polymers reported in the literature, an example being the work of Allen et al. (*10*), confirm the latter prediction. A recent treatment of molecular weight effects in polymer compatibility is contained in the paper of Casper and Morbitzer (*11*).

The extent of incompatibility of high-molecular-weight polymers has been elegantly shown by Yuen and Kinsinger (*12*) in their work on light scattering from blends of polystyrene, PS, and poly(methyl methacrylate), PMMA. For these high-molecular-weight blends, incompatibility was detected when the PS concentration exceeded 0.008 wt%. The effect of molecular weight, M, was not investigated by Yuen and Kinsinger.

In an effort to define more clearly the effects of molecular weight and composition on polymer compatibility, a study was undertaken on the effect of PS molecular weight on compatibility and physical properties of blends of PS with PMMA and PS with styrene(S)/methyl methacrylate(MMA) copolymers. The range in M investigated for PS was 600–110,000, and the S/MMA ratio was varied from 0 to 4.0. The results of this study are reported below.

Experimental

The polymers studied are listed in Table I along with their sources and characterizations. Bulk samples were prepared in two steps. First, a thin film (2–10 mils) was obtained by casting onto glass from a solution

Table I. Materials and Characterization

Polymer	T_g (°C)[a]	η_{inh}[b]	$\overline{M}_w/\overline{M}_n$	$\overline{X}_n$ (*estimated*)	δ_{Calc}[k]
PMMA[c]	115	0.34	1.77[e]	400[f]	9.26
S-MMA-10 (9.8)[d]		2.19		4500[g]	9.24
S-MMA-25 (26.6)[d]	109	1.10		1800[g]	9.21
S-MMA-60 (62.6)[d]	106	1.18		2000[g]	9.15
S-MMA-80 (80.0)[d]		1.19		2000[g]	9.13
PS-600[h]	−15		< 1.10[i]	5.8[j]	
PS-2100	60		< 1.10	20.2[j]	
PS-4000	78		< 1.10	38.5[j]	
PS-10,000	96		< 1.06	96.2[j]	
PS-20,400			< 1.06	196[j]	
PS-37,000			< 1.06	356[j]	
PS-110,000	108		< 1.06	1058[j]	9.10

[a] T_g taken as midpoint in baseline transition at a heating rate of 10°C/min.
[b] Inherent viscosity, $(\ln\eta_{rel})/c$, where $c = 0.25$ g/dL, in benzene.
[c] Rohm and poly(methyl methacrylate). Plexiglas type V-811.
[d] Prepared by radical suspension bead polymerization by J. L. Tucker. Dissolved in benzene, precipitated, and freeze dried from benzene. Numbers in parentheses are the percentages of styrene determined by NMR.
[e] Polystyrene-equivalent ratio as determined by GPC.
[f] Estimated from the inherent viscosity and the Mark–Houwink parameters $K = 5.2 \times 10^{-5}$ and $a = 0.76$ (*23*).
[g] Estimated from the η_{inh}, using $K = 5.75 \times 10^{-5}$ and $a = 0.77$, and assuming $M_w/M_n = 2$. This degree of precision was sufficient for calculation of miscibility because in most cases the molecular weights were high compared with those of PS.
[h] Pressure Chemical Co. The PS numbers indicate the molecular weights.
[i] The polydispersities are those given by the manufacturer.
[j] Calculated assuming monodispersity.
[k] Solubility parameters calculated using the tables of Hoy, as given by Krause (*1*).

of the two polymers in methylene chloride, evaporating to dryness, heating for 2 hr in vacuum at about 150°C, and cooling. Following this, the films were broken up into small pieces and compression molded in vacuum at 200°C for 2 hr to form a disc 1 in. in diameter and ¼ in. thick.

Our method of sample preparation was chosen for the following reasons. The thick moldings allowed us to check for optical clarity without concern for the apparent clarity that can result from thin films. Also, it was expected that the effect of the solvent used to cast the films would be eliminated by heating the thin film above T_g to remove the solvent and subsequently molding at 200°C. The efficacy of this technique could be checked for the PS-600/PMMA samples by diffusing PS-600 into PMMA at 180°C in the absence of solvent. No differences were observed.

Two criteria for compatibility were used: optical clarity and a single glass-transition temperature, T_g. Optical clarity indicates either that compatibility is present or that the domains involved in the phase separation are small compared with optical wavelengths. (The refractive indices must differ, of course. In these studies the refractive index between blend components varied from 0.02 for PS/S-MMA-80 blends to

0.10 for PS/PMMA.) Cloudiness or opacity indicates unequivocally that phase separation has occurred. The thermal measurements reveal whether a single T_g or two T_gs are present, the latter a confirmation of incompatibility. For blends of high-molecular-weight PS ($T_g = 108°C$) with PMMA ($T_g = 115°C$) and S/MMA copolymers ($105° \leqslant T_g \leqslant 115°C$), the glass-transition temperatures are too close to be resolved by thermal measurements. Only for PS-600 ($T_g = -15°C$) and PS-2100 ($T_g = 60°C$) are the PS glass-transition temperatures sufficiently lower than those of PMMA and S/MMA to be resolved. Because most of the blends exhibiting compatibility are at low molecular weight and because where both optical and thermal criteria were used there was agreement between the two methods, the optical criterion was considered a useful one at higher molecular weights where the thermal criterion could not be used.

Dynamic mechanical properties were measured on some PS/PMMA blends, using a Rheovibron model DDV-II system previously described (*13*). These measurements are a sensitive indicator of phase separation and can be used to obtain semiquantitative information about phase morphology as well (*14*).

Results and Discussion

Experimental Results. Regions of compatibility at room temperature were determined as described in the section above for PS in PMMA and the S/MMA copolymers as a function of PS molecular weight and weight-fraction PS. Blends were prepared at weight-fractions separated by intervals of 5 wt% so that the line between compatible and incompatible compositions for each series was determined to no better than ± 2.5 wt% in most cases. The results are shown in Figure 1 and are discussed below.

PS/PMMA. The PS-600 and PS-2100 show compatibility (optically clear and a single T_g) up to about 38 and 10 wt%, respectively (*see* Figure 1 and preceding paragraph). The thermal data of Figure 2 show the onset of phase separation by the appearance in the 42-wt% blend of a second T_g at $-10°C$, or five degrees higher than the T_g of pure PS-600. Assuming that the slightly elevated T_g is caused by the dissolution of a small amount of PMMA in the phase-separated PS-600, one can calculate that approximately 5 wt% PMMA would produce the observed T_g (from $T_g = 263$ K for the separated phase vs. 258 K for pure PS-600 and the approximate relation $w_1/T_{g1} + w_2/T_{g2} = 1/T_g$, where w_1 and w_2 are the weight-fractions of each blend component). Similarly, the higher T_g phase has nearly 40 wt% PS-600. There are other mechanisms by which the T_g of the separated phase could be elevated over that of pure PS-600, owing to the interaction between the rubbery PS-600 phase and the rigid, glassy phase surrounding it. The preferred mechanism is not established by this work.

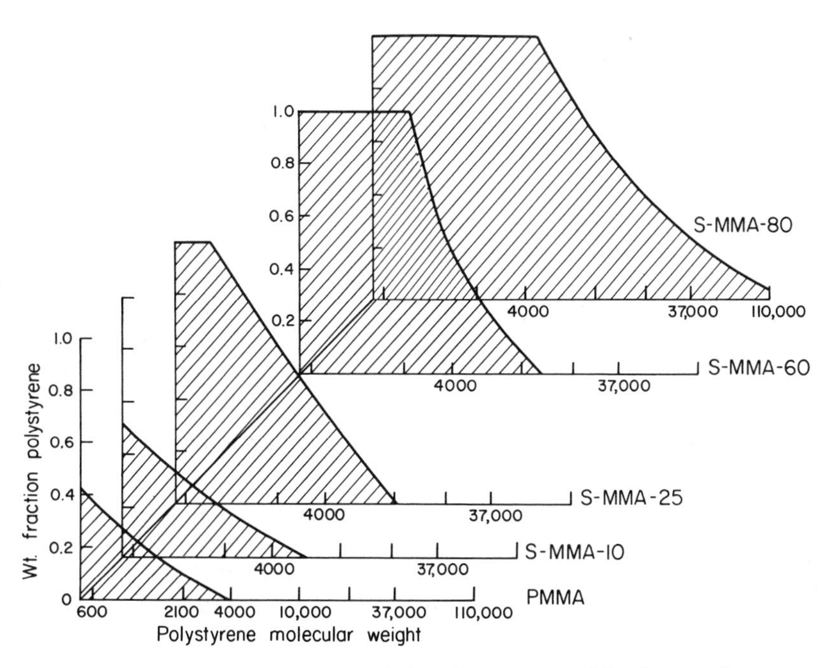

Figure 1. Room-temperature miscibility diagrams for blends of polystyrene with poly(methyl methacrylate) and styrene/(methyl methacrylate) copolymers. Shaded area is compatible region.

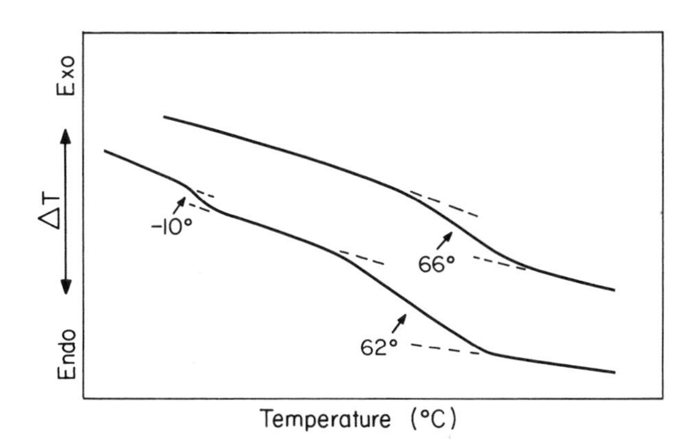

Figure 2. Differential scanning calorimetry curves (10°C/min) for blends of polystyrene, PS-600 with PMMA: (upper) *34 wt% PS-600;* (lower) *42 wt% PS-600.*

The PS-2100/PMMA thermal data of Figure 3 also show two glass-transition temperatures for the 25% blend, indicating incompatibility. The dynamic mechanical results for blends of PS-600 in PMMA are consistent with the thermal and optical results, as Figure 4 shows. The PMMA secondary loss shoulder is diminished somewhat as the concentration of PS-600 is increased, in a manner similar to that observed in other polymer-diluent systems in which the diluent is monomeric (*15*) until phase separation occurs. Thereafter, the phase-separated PS-600 shows up as a characteristically narrow peak at about − 10°C.

Similar results have been reported recently by Parent and Thompson (*16*) for blends prepared by polymerizing MMA in the presence of PS. In their work, in which the M of PS was varied from 2100 to 49,000, a single phase was observed under electron microscopy up to 20 wt% PS-2100, a concentration somewhat higher than observed in this work but in general agreement with our results. Trends in miscibility with M were in agreement with this work as well.

PS/S-MMA-10. The inclusion of 10-wt% styrene in the host polymer had the predictable effect of increasing the compatibility with PS slightly, as shown in Figure 1. The increase would have been greater were it not for the much higher degree of polymerization of the S-MMA-10 vs. the PMMA, as Table I shows.

PS/S-MMA-25. Putting 25-wt% styrene into the host polymer makes it compatible in all proportions with PS-600 and compatible to substantial weight fractions with the PS-2100 and PS-4000 (Figure 1).

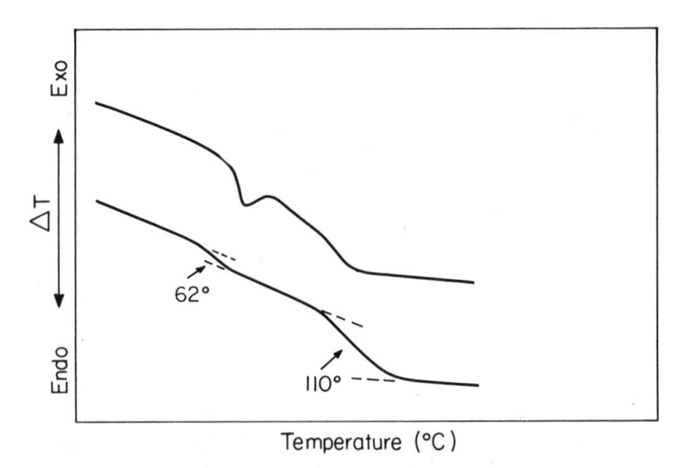

Figure 3. Differential scanning calorimetry curves (10°C/min) for blend of 25 wt% PS-2100 in PMMA: (upper) after one year at room temperature; (lower) after quenching from approximately 150°C.

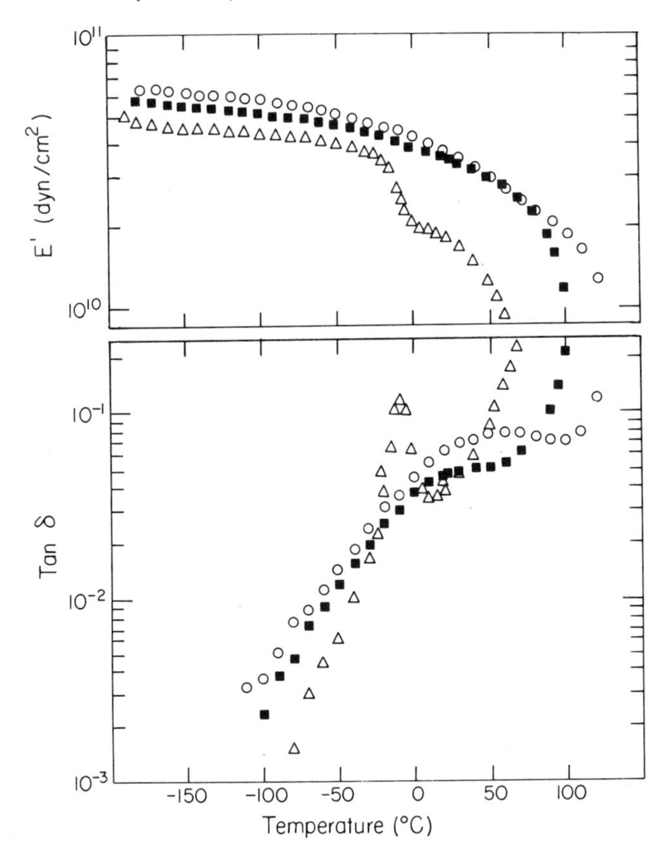

Figure 4. Dynamic mechanical properties of PMMA and blends of PS-600 with PMMA at 11 Hz. (○) PMMA; (■) 20 wt% PS-600 in PMMA; (△) 50 wt% PS-600 in PMMA.

PS/S-MMA-60. At this value of copolymer composition, one observes further increases in compatibility, with PS-600 and PS-2100 compatible in all proportions and even PS-10,000 showing slight miscibility. It is worth emphasizing here that for PS M = 4000 and higher, the compatibility criterion is optical only and therefore is not unequivocal.

PS/S-MMA-80. At the 80-wt% styrene level, even the high-molecular-weight PS samples begin to show compatibility—e.g., PS-37,000 up to approximately 20 wt% and PS-110,000 at a few percent. The lower-M PS samples, 600, 2100, and 4000, are compatible in all proportions (Figure 1).

Comparison with Theory. In view of the qualitative agreement of the data with the predictions of the Flory–Huggins–Scott treatment, i.e., increasing compatibility at lower molecular weights and/or increasing polymer similarity, it is of interest to examine the quantitative agreement between theory and experiment. This, it is recalled (*1, 11*), can be done by calculating χ_{AB} from the solubility parameters using the van Laar–Hildebrand expression given below and comparing it with $(\chi_{AB})_{cr}$

$$\chi_{AB} = \frac{V_r}{RT} (\delta_A - \delta_B)^2$$

calculated from the degrees of polymerization of the blend components, using the expression given below, taken from Krause (*1*).

$$(\chi_{AB})_{rc} = \frac{1}{2} \left[\frac{1}{X_A^{1/2}} + \frac{1}{X_B^{1/2}} \right]^2$$

In these expressions, δ_A and δ_B are the solubility parameters of polymers A and B, defined for simple liquids as the square root of the energy of vaporization per unit volume (*17, 18*), and X_A and X_B are the degrees of polymerization of the blend components. When $X_{AB} < (X_{AB})_{cr'}$ compatibility is predicted in all proportions.

When such calculations were performed using the solubility parameters listed in Table I that were calculated from the tables of Hoy as given by Krause (*1*), the predicted compatibility greatly exceeded the observed compatibility in all cases. For example, for the PS/PMMA blends, for which χ_{AB} is calculated to be 0.004, compatibility in all proportions is predicted up to molecular weights of 20,000, for which $(\chi_{AB})_{cr}$ is 0.007. For the copolymers, compatibility in all proportions was predicted at all molecular weights of PS studied when the fraction of styrene in the copolymers exceeded 0.25. These calculations are, of course, sensitive to the values of solubility parameters chosen for the homopolymers. If, for example, a value of $(\delta_{PMMA} - \delta_{PS})$ equal to 1.0 were used in place of the value of 0.16 obtained from Table I, the fit between theory and experiment is much improved. The limited scope of the present study and the compositional and molecular-weight heterogeneities of the PMMA and S-MMA copolymers cannot support an extensive analysis of the differences between theory and experiment, which must await more exhaustive experimental and theoretical studies of this system. In view of the simplifying assumptions of the Flory–Huggins–Scott treatment and the nonidealities of the system studied here, the qualitative and semiquantitative agreement between theory and experiment is reasonable and provides a framework for considering at least qualitatively the effects of polymer composition and molecular weight.

For all but the more simple trends in polymer–polymer miscibility, however, it seems essential to consider a more complex treatment of the phenomena than the Flory–Huggins–Scott approach. At the least, one should consider the temperature dependence of the free energy of mixing (*11*) since for blends such as those studied here it is the equilibrium miscibility of the polymers in the temperature region just above the glass-transition temperature of the mixture that determines whether miscibility is observed at room temperature in the (nonequilibrium) glassy state. Recent theories on the thermodynamics of polymer mixing predict that lower critical solution temperature (LCST) behavior, i.e., phase separation upon heating, is likely to occur in miscible polymer blends, as summarized by Paul and collaborators (*19*). The number of blends known to exhibit LCST behavior is increasing; for example, the work of Alexandrovich, Karasz, and MacKnight demonstrates LCST behavior in various blends of poly(2,6-dimethyl phenylene oxide) with halogenated analogs of polystyrene (*20*). Differences in thermal expansion coefficient and/or molecular size, which one would expect in the lower-M PS samples, may favor such an occurrence (*21, 22*). It would be of interest to probe the blends studied here and other polymer–oligomer and oligomer–oligomer blends for LCST phenomena.

Summary

The conclusions of this work are as follows: (i) PS is compatible at low M with PMMA over a range of PS concentrations and molecular weights, up to nearly 40% for M = 600; (ii) increasing the styrene ratio in S-MMA copolymers increases compatibility with PS; (iii) molecular weight is important in all cases, compatibility increasing with decreasing M; (iv) the Flory–Huggins–Scott solubility parameter approach, while not quantitatively predictive, is consistent with qualitative trends in the results.

Acknowledgment

The author wishes to acknowledge the experimental assistance of Carol Dona and Peter Rusanowsky.

Literature Cited

1. Krause, S., *J. Macromol. Sci., Rev. Macromol. Chem.* (1972) **C7**(2), 251–314.
2. Flory, P. J., "Principles of Polymer Chemistry," Chapters 12 and 13, Cornell, Ithaca, NY, 1953.
3. Huggins, M. L., *J. Chem. Phys.* (1941) **9**, 440.
4. Huggins, M. L., *Ann. N.Y. Acad. Sci.* (1942) **43**, 1.
5. Flory, P. J., *J. Chem. Phys.* (1941) **9**, 660.
6. Flory, P. J., *J. Chem. Phys.* (1942) **10**, 51.
7. Scott, R. L., *J. Chem. Phys.* (1949) **17**, 279.
8. Tompa, H., "Polymer Solutions," Butterworth, London, 1956.

9. Flory, P. J., "Principles of Polymer Chemistry," p. 555, Cornell, Ithaca, NY, 1953.
10. Allen, G., Gee, G., Nicholson, J. P., *Polymer* (1961) **2,** 8.
11. Casper, R., Morbitzer, L., *Angew. Makromol. Chem.* (1977) **58/59,** 1–35.
12. Yuen, H. K., Kinsinger, J. B., *Macromolecules* (1974) **7**(3), 329.
13. Massa, D. J., Flick, J. R., Petrie, S. E. B., *Am. Chem. Soc., Div. Org. Coat. Plast., Prepr.* (1975) **35**(2), 371.
14. Dickie, R. A., *J. Appl. Polym. Sci.* (1973) **17,** 45.
15. Petrie, S. E. B., Moore, R. S., Flick, J. R., *J. Appl. Phys.* (1972) **43**(11), 4318.
16. Parent, R. R., Thompson, E. V., *Polym. Prepr., Am. Chem. Soc., Div. Polym. Chem.* (1977) **18**(2), 507.
17. Hildebrand, J. H., Scott, R. H., "The Solubility of Nonelectroytes," 3rd ed., Reinhold, New York, 1950.
18. Hildebrand, J. H., Prausnitz, J. M., Scott, R. L., "Regular and Related Solutions," Van Nostrand, New York, 1970.
19. Bernstein, R. E., Cruz, C. A., Paul, D. R., Barlow, J. W., *Macromolecules* (1977) **10**(3), 681.
20. Alexandrovich, P., Karasz, F. E., MacKnight, W. J., *Polymer* (1977) **18,** 1022.
21. McMaster, L. P., *Macromolecules* (1973) **6,** 760.
22. Sanchez, I. C., Lacombe, R. H., *J. Phys. Chem.* (1976) **80,** 2568.
23. Brandrup, J., Immergut, E. H., "Polymer Handbook," 2nd ed., Wiley, New York, 1975.

RECEIVED April 14, 1978.

23

Stress-Strain Behavior of PMMA/ClEA Gradient Polymers

C. F. JASSO,[1] S. D. HONG,[2] and M. SHEN

Department of Chemical Engineering, University of California, Berkeley, CA 94720

Gradient polymers were prepared by diffusing the 2-chloroethyl acrylate (ClEA) monomer into crosslinked poly(methyl methacrylate). The resulting profile of the diffusion gradient is then fixed by polymerizing the monomer in situ by photochemical initiation. If the diffusion of the monomer is permitted to proceed to swelling equilibrium, then interpenetrating networks (IPN) are formed. The stress–strain curves of the gradient polymers are plasticlike in that they exhibit yield points and enhanced fracture strains. Those of interpenetrating networks at comparable chemical composition, on the other hand, show essentially rubbery behavior. Data may be interpreted in terms of the stress transfer mechanism or surface stabilization mechanism.

Gradient polymers are multicomponent polymers whose structure or composition is not homogeneous throughout the material, but varies as a function of position (*1*). In other words, these polymers have gradients in their structures or compositions. In a previous paper (*2*), we have shown that it is possible to produce such materials by diffusing a guest monomer into a host polymer for a period of time sufficient for the establishment of a diffusion gradient profile. This profile is then "fixed" by polymerization of the monomer in situ. The mechanical behavior of such gradient polymers was found to be quite different from the inter-

[1] Current address: Av. Mexico 2286-1, Guadalajara, Jalisco, Mexico.
[2] Current address: Jet Propulsion Laboratory, Pasadena, CA 91103.

0-8412-0457-8/79/33-176-443$05.00/0

penetrating networks (IPNs) of comparable composition. Of course the latter material allows the monomer to reach a swelling equilibrium in the host polymer and therefore has no gradient structure.

In our preliminary report (2), we have chosen poly(methyl methacrylate) or PMMA as a host polymer and methyl acrylate as the guest monomer. They were both crosslinked by a divinyl acrylic monomer. However, because of the similarity in the constitutions of these two components, it was not possible to establish the gradient profile through chemical analysis. In this work, we have selected a halogenated acrylic monomer as the second component to be diffused into PMMA. By analyzing the halogen content, it was possible to determine the profiles of the gradient polymers. Stress–strain measurements of the samples were then carried out on these unique materials.

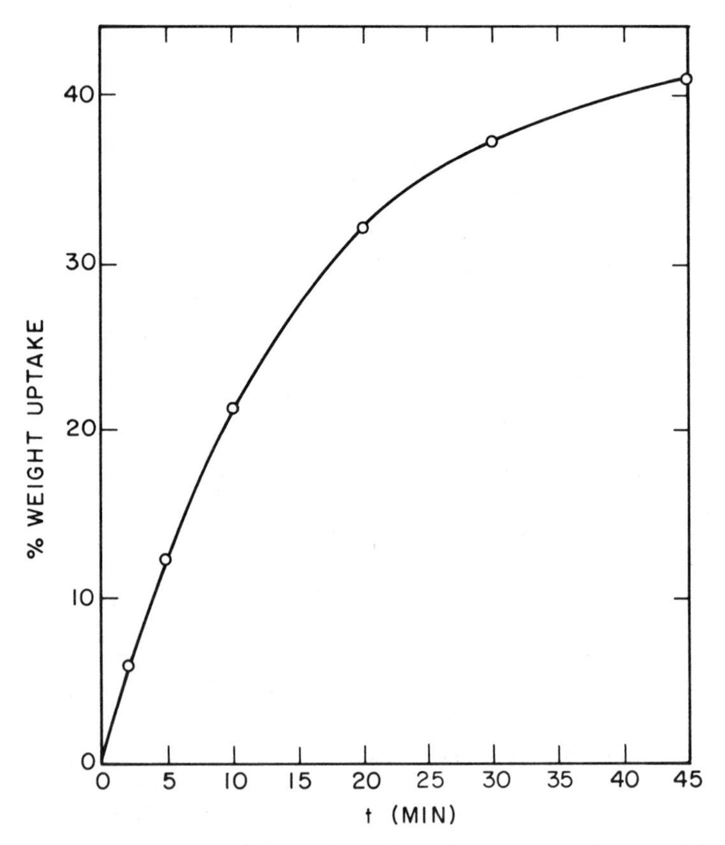

Figure 1. Uptake of 2-chloroethyl acrylate (percent by weight) by crosslinked poly(methyl methacrylate) at 60°C as a function of time of immersion

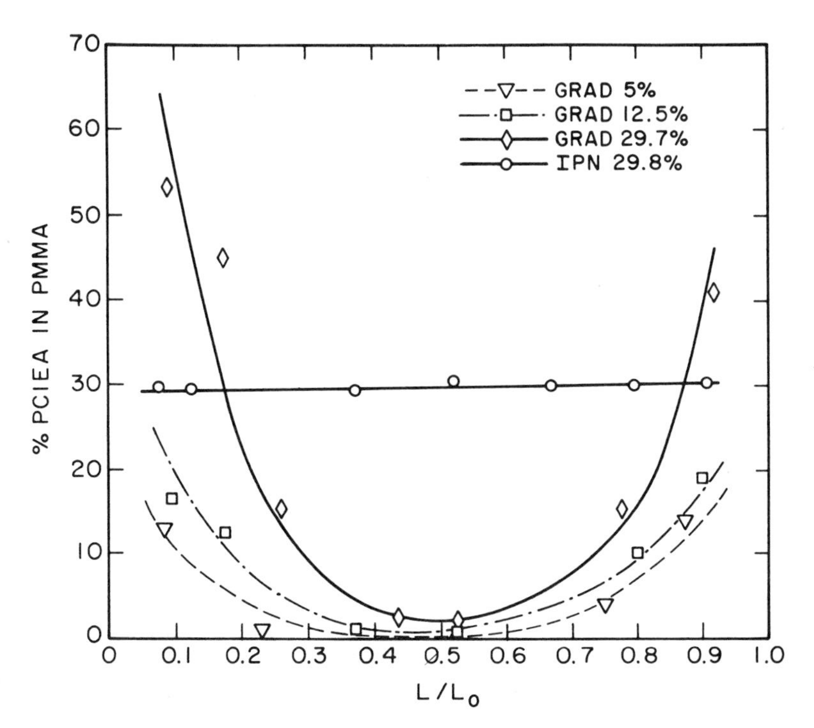

Figure 2. Concentration profiles of poly(2-chloroethyl acrylate) in poly(methyl methacrylate) along the thickness (L_0) dimension of the samples. (GRAD) Gradient polymer, (IPN) interpenetrating networks.

Experimental

Monomers of methyl methacrylate and 2-chloroethyl acrylate (ClEA) were purchased from the Polysciences, Inc. They were distilled and subsequently mixed with 1.3% by weight of the crosslinking agent (ethylene glycol dimethacrylate obtained from the J. T. Baker Co.) and 1.9% by weight of the photosensitizer (benzoin isobutyl ether supplied by the Stauffer Chemical Co.). PMMA was first prepared by photopolymerization in front of UV light for 48 hr. Samples were stored in a vacuum oven at 60°C until a constant weight was achieved to remove remaining monomer. The cross-linked PMMA samples were then immersed in the bath of ClEA monomer for various periods of time by monitoring the weight uptake (Figure 1). When the desired amount of monomer has been imbibed into the host polymer, the latter was removed, surface dried, and then immediately polymerized by UV radiation. IPNs were prepared in a similar manner except that after immersion in the monomer bath the sample was stored in a sealed polymerization cell at 60°C for several days prior to polymerization.

The profiles of these gradient polymers were determined by machining off the samples layer by layer. Shavings from each layer were then analyzed for chlorine content by combustion methods (*3*). The results of these analyses are shown in Figure 2, recalculated in terms of mole percent of ClEA in PMMA.

An Instron Universal Testing Machine Model TM-SM was used to carry out the stress–strain measurements. An environmental chamber, equipped with a Missimers PITC temperature controller, was used in providing constant temperatures to ± 0.5°C. Samples were of the approximate dimensions of 5.0 × 0.5 × 0.1 cm^3. To prevent slippage from the Instron clamps, special aluminum tabs were glued to the ends of the samples.

Results and Discussion

Stress–strain curves for a series of gradient polymers of 2-chloroethyl acrylate in poly(methyl methacrylate), which are designated as PMMA/Grad PClEA, are shown in Figure 3. Quantities in the parentheses indicate the mole percent of the acrylate. Also included are the stress–strain curves of pure PMMA and an interpenetrating network (PMMA/IPN PClEA). All of these measurements were carried out at 80°C and at strain rates of 2–3%/sec. First we note that the stress–strain curve of PMMA has a high initial slope (high elastic modulus) and fractures at

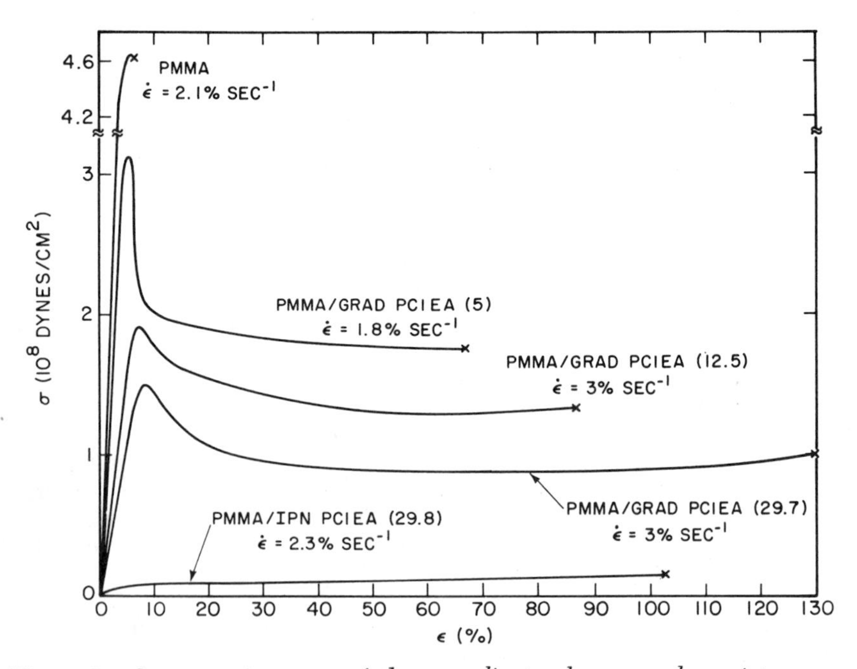

Figure 3. Stress–strain curves of three gradient polymers and one interpenetrating network of poly(methyl methacrylate) with 2-chloroethyl acrylate at comparable strain rates of 2–3%/sec and same temperature of 80°C. The numerals in parentheses indicate concentrations (mole percent) of chloroethyl acrylate in poly(methyl methacrylate).

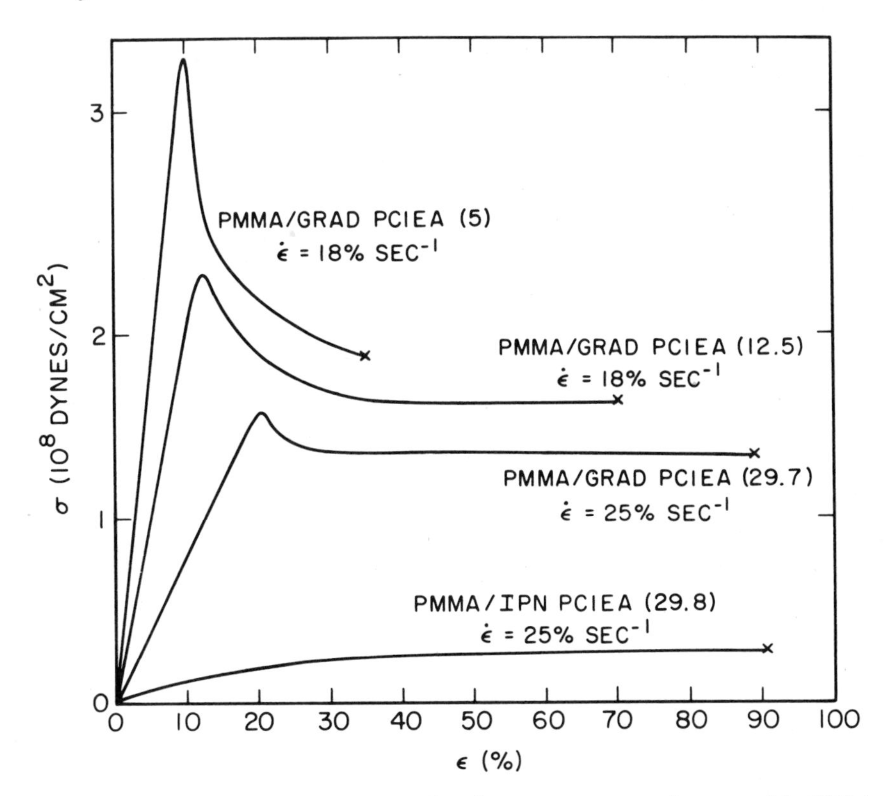

Figure 4. Same as Figure 3 except that the strain rates are between 18–25%/sec and that the PMMA data are not included

about 6% strain. Upon introduction of a 5% PClEA gradient, there is a slight decrease of the initial slope but a dramatic increase in fracture strain (~ 65%). In addition, a pronounced yield region is observed around 5% strain. Further increases in PClEA contents to 12.5 and 29.7% show a continuing decrease in the elastic moduli but even higher fracture strains were attained (85 and 130%, respectively). Both of the latter gradient polymers, however, still exhibit the yielding behavior and are plasticlike in mechanical properties. In contrast, the IPN at 29.8% PClEA content is rubbery and fractures at slightly greater than 100% strain. Apparently the presence of the gradient structure enables the samples to retain the plasticlike properties without sacrificing the enhanced ability to withstand deformation.

The effects of increasing strain rates on PMMA/Grad PClEA and PMMA/IPN PClEA are shown in Figure 4. Increased strain rates are seen to increase the yield stresses as well as stress levels in the plateau regions of the gradient polymers but to decrease the fracture strains. For the interpenetrating networks of comparable composition, there is no observable yielding, but the stress levels are increased by higher strain rates. However, there also appear to be decreases in the fracture strains. As expected, the effects of decreasing strain rates are just the opposite (Figures 5 and 6). Again IPNs behave essentially as rubbers, while the gradient polymers exhibit plasticlike properties.

The effect of temperature on the stress–strain properties of PMMA, its gradient polymers with various compositions, and an IPN are shown in Figures 7, 8, and 9. These experiments were performed at various strain rates at 60°C. Comparison with the 80°C data shows that the main effects of temperature are to increase the stress levels in the plateau regions at lower temperatures without significant differences in other aspects.

In our previous paper (2), we proposed a possible mechanism to interpret the stress–strain behavior of gradient polymers. We perceived the gradient polymer as consisting of infinite number of layers of varying compositions. Upon deformation, the macroscopic strain is the same for the entire sample. Because of the fact that the moduli of the various

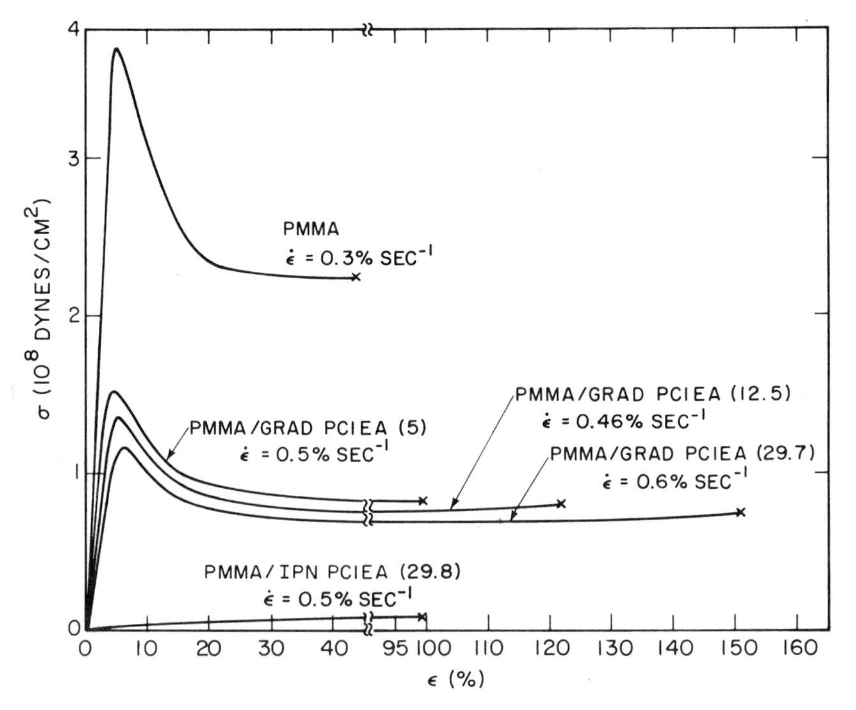

Figure 5. Same as Figure 3 except that the strain rates are between 3–6 × 10^{-1}%/sec

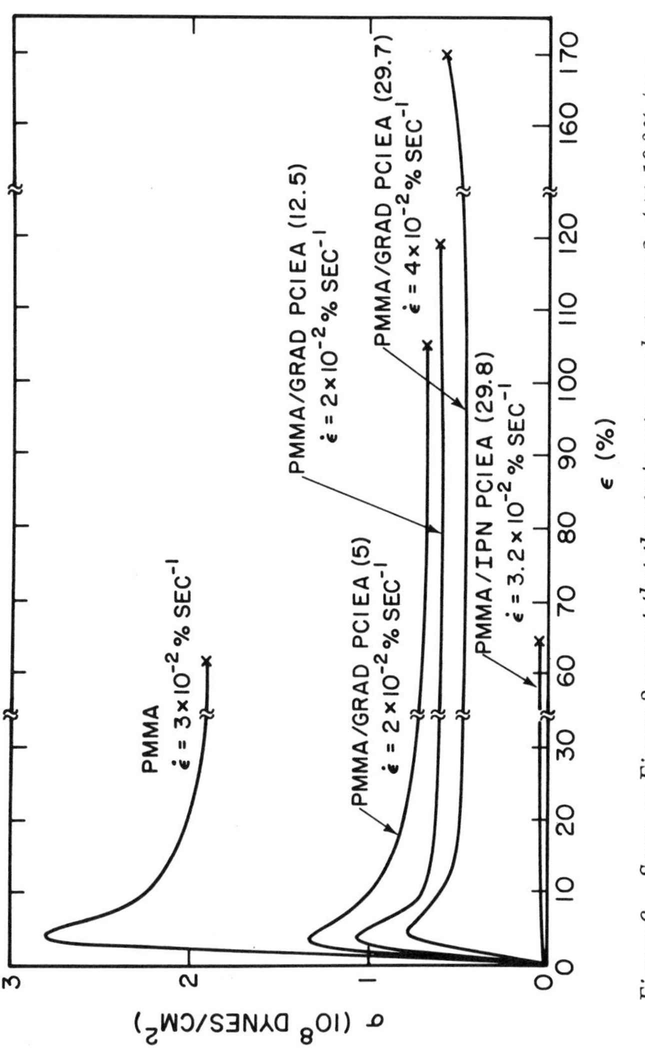

Figure 6. Same as Figure 3 except that the strain rates are between 3–4 × 10^{-2}%/sec

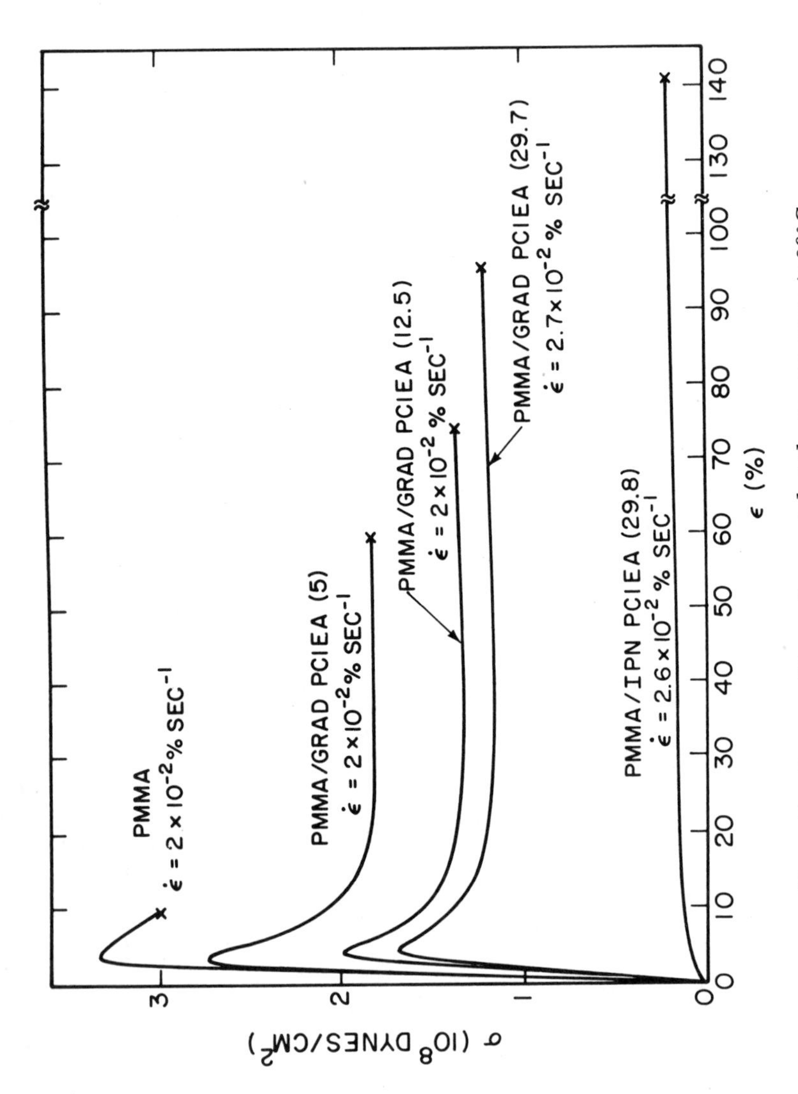

Figure 7. Same as Figure 6 except that the temperature is 60°C

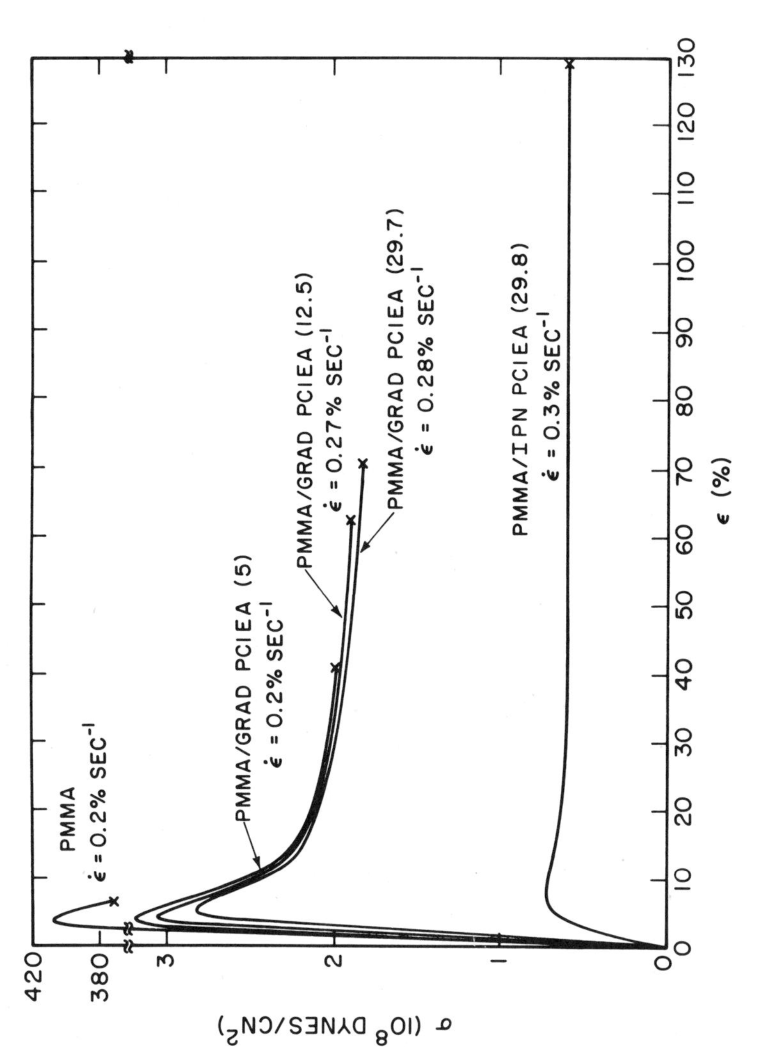

Figure 8. Same as Figure 5 except that the temperature is 60°C

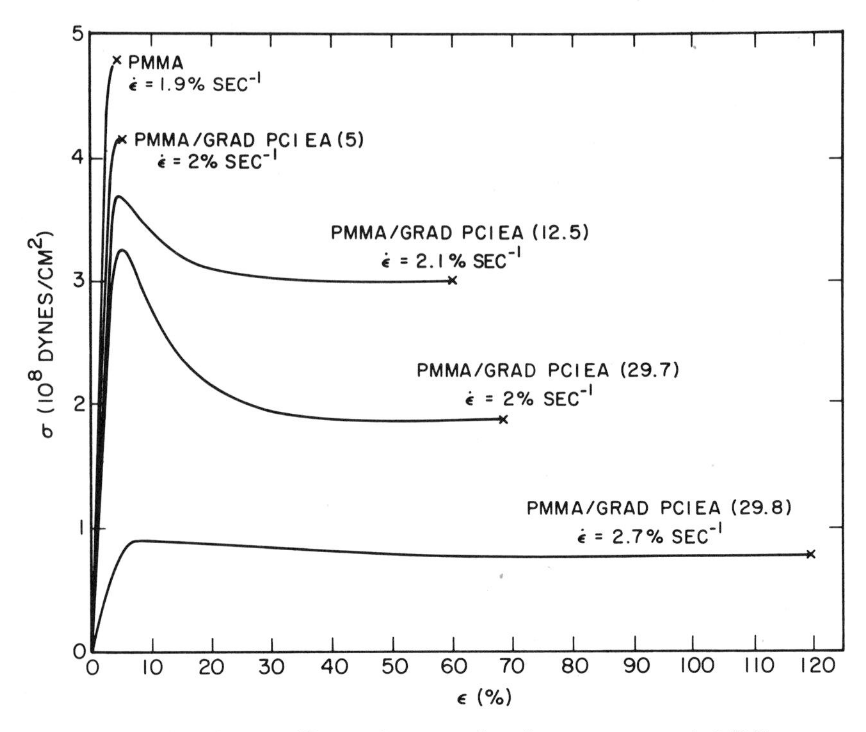

Figure 9. Same as Figure 3 except that the temperature is 60°C

layers are different owing to their differences in composition, those layers with higher moduli must sustain greater stresses (for the same strain). According to Eyring's stress-biased activated rate theory of yielding (*4, 5*), the barrier height for a molecular segment to jump in the forward direction is reduced by the applied stress. As a consequence, the higher modulus layers in the gradient polymer should show greater tendency to yield because of the greater stress biases they have than those with lower moduli. This mechanism will be inoperative for interpenetrating networks because of their uniform composition.

An alternative mechanism for the observed high fracture strain may be the reduction in surface imperfections for the gradient polymers. The surfaces of these materials must be more resistent to fracture because of the relatively high loading of rubbery phases. Thus, under stress they are likely to craze or crack, which would have initiated fracture for the sample as a whole. The validity of either mechanism, however, must await verification.

In conclusion, we find that gradient polymers produced by diffusion polymerization generally show enhanced fracture strain, while retaining their plasticlike properties. This behavior appears to be a consequence of the gradient structure, rather than being attributable to the compositions alone. Interpenetrating networks with comparable composition do not possess the unique mechanical behavior of the gradient polymers.

Acknowledgment

This work was supported by the Office of Naval Research.

Literature Cited

1. Shen, M., Bever, M. B., *J. Mat. Sci.* (1972) **7**, 741.
2. Akovali, G., Biliyar, K., Shen, M., *J. Appl. Polym. Sci.* (1976) **20**, 2419.
3. Sundberg, O. E., Royer, G. L., *Anal. Chem.* (1946) **18**, 719.
4. Eyring, H., *J. Chem. Phys.* (1936) **4**, 283.
5. Matz, D. J., Guldemond, W. G., Cooper, S. L., *J. Polym. Sci., Polym. Phys. Ed.* (1972) **10**, 1917.

RECEIVED April 14, 1978.

24

Crystallization Studies of Blends of Polyethylene Terephthalate and Polybutylene Terephthalate

ANTONIO ESCALA and RICHARD S. STEIN[1]

Polymer Research Institute, Department of Chemistry and Materials Research Laboratory, University of Massachusetts, Amherst, MA 01002

The compatibility of blends of polybutylene terephthalate and polyethylene terephthalate has been studied. A single glass transition for the blend is observed which varies with composition, suggesting that the components are compatible in the amorphous phase. Studies of the effects of time and temperature in the melt show that this is not a consequence of trans-esterification. X-ray, DSC, and IR studies demonstrate that crystallization results in separate crystals of the two components rather than cocrystallization. An IR technique, calibrated by density, was developed for measuring degrees of crystallinity of each component in the presence of the other. Crystallization rates are primarily affected by the degree of supercooling of each component in the blend and by the influence of blending on the glass-transition temperature.

After having studied in our laboratory, polymer blends of amorphous polymers poly-ε-caprolactone and poly(vinyl chloride) (*1*, *2*) (PCL/PVC), blends with a crystalline component PCL/PVC (*3*, *4*), poly(2,6-dimethyl phenylene oxide) (PPO) with isotactic polystyrene (i-PS) (*5*) and atactic polystyrene (a-PS) with i-PS (*6*), we have now become involved in the study of a blend in which both polymers crystallize. The system chosen is the poly(1,4-butylene terephthalate)/poly(ethylene terephthalate) (PBT/PET) blend. The crystallization behavior of PBT has been studied extensively in our laboratory (*7*, *8*); this polymer has a

[1] Author to whom correspondence should be sent.

0-8412-0457-8/79/33-176-455$08.25/0

very high crystallization rate which usually leads to a skin-core morphology during the processing of the polymer. PET is also a crystallizable polymer which has more moderate crystallization rates (*9, 10, 11*). The purpose of the present study is to observe how the crystallization behavior of each component is affected by the presence of the second one.

The evidence from wide angle x-ray scattering (WAXS), differential scanning calorimetry (DSC), and IR spectroscopy (IR) shows that both polymers crystallize separately according to their own unit cell structure. The WAXS diffraction lines of each component are present in the blends; no new bands appear (*12, 13, 14*). By DSC one observes the melting peaks corresponding to each polymer (Figure 1), and IR shows the typical characteristic crystalline bands of the pure polymers in the blends. The IR spectra of the blend can essentially be accounted for as the sum of the spectra of the components.

Studies performed on the amorphous blends by DSC show the presence of only one glass-transition temperature in the blends which is intermediate between those of PBT and PET, changing in value with composition (Figure 2). This is the first evidence we have of the compatibility of both polymers in the amorphous phase.

The overall crystallization behavior was studied by depolarized light intensity (DLI) and DSC (*15, 16, 17*). We observed a decrease in overall crystallization rate in the PBT-rich blends when the PET percentage

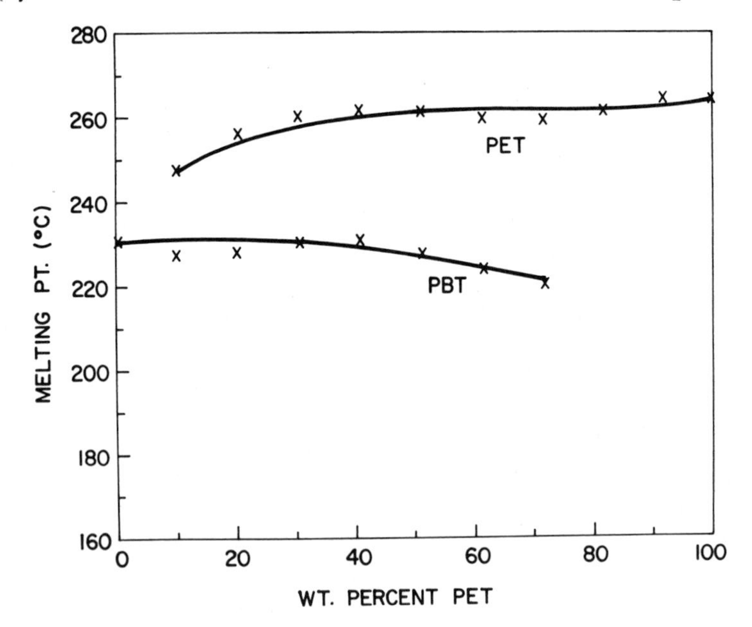

Figure 1. PET and PBT melting points vs. PET percent as determined by DSC. PBT/PET crystallized at 185°C.

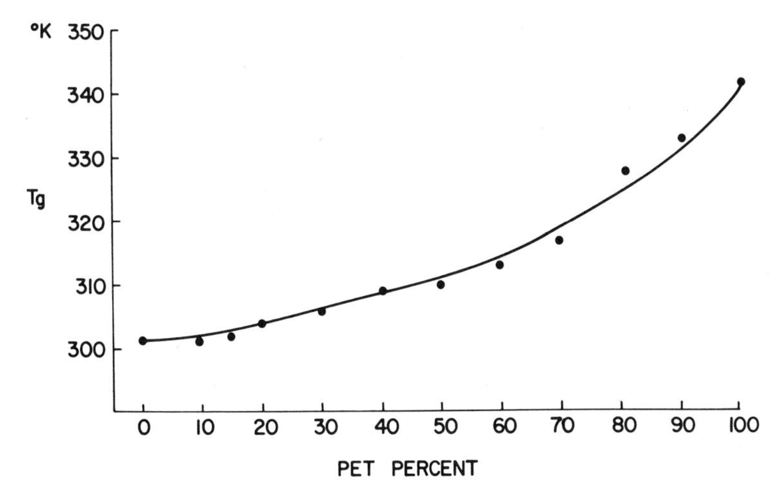

Figure 2. PBT/PET blends glass-transition temperatures. T_g *vs. PET percent as determined by DSC.*

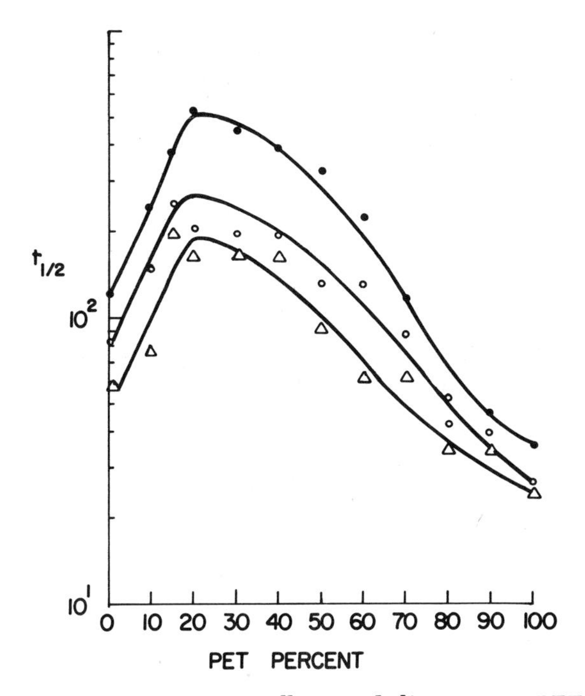

Figure 3. DLI-crystallization half times vs. PET percent at 200° (△), 202° (○), and 205°C (●)

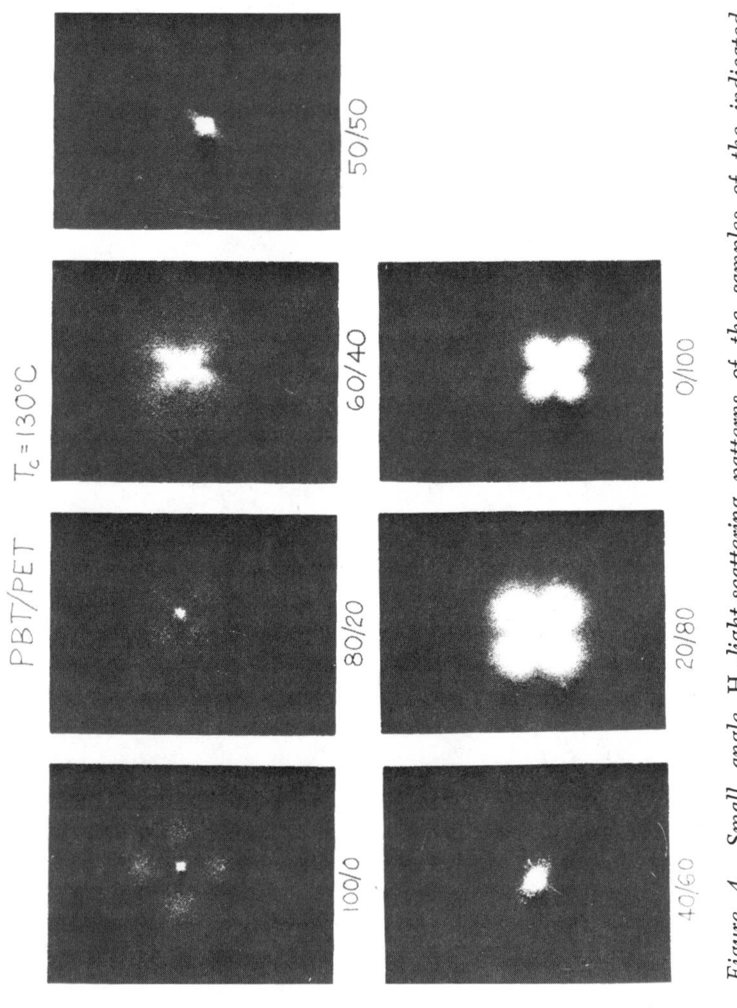

Figure 4. Small angle H_v-light-scattering patterns of the samples of the indicated composition of (PBT/PET) crystallized at 130°C

increases, similarly a decrease in the PET-rich blends with increasing PBT presence is registered, occurring with a minimum in the 85/15 (PBT/PET) blend (Figure 3).

The morphology was studied by small angle light scattering (SALS) at 130°C, showing typical spherulitic morphology for PET and PBT with the maximum in directions 45° to the polaroids for PBT, corresponding to the fact that the extinction crosses of PBT when viewed under a polarizing microscopy are 45° to the directions of polarization. Introduction of small percentages of the second component in each case causes the scattering patterns to become bigger, more diffuse, and less azimuthally dependent, indicating smaller and more disordered spherulites (Figures 4, 5, and 6). Finally with higher concentrations of the

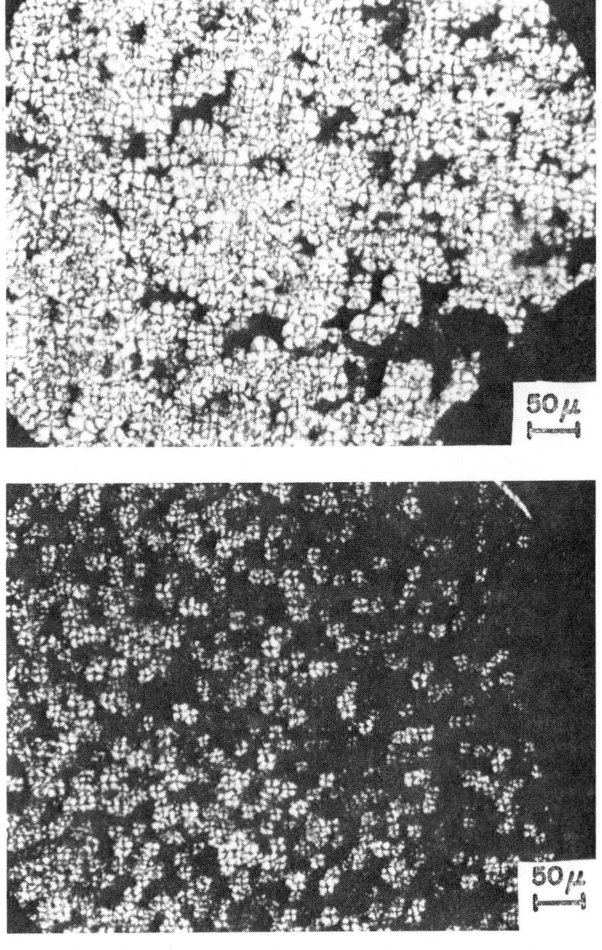

Figure 5. Photomicrographs of the 0/100 (top) *and 10/90* (bottom) *(PBT/PET) blends crystallized at 200°C*

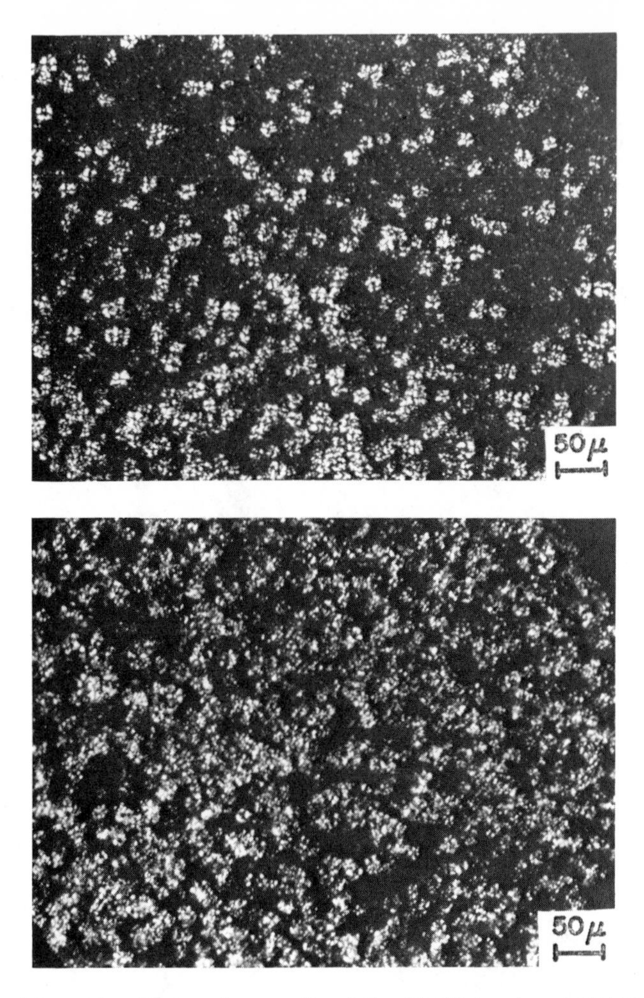

Figure 6. Photomicrographs of the 20/80 (top) *and 40/60* (bottom) *(PBT/PET) blends crystallized at 200°C*

second component the spherulitic morphology is lost, and the scattering patterns show no maximum but an intensity continuously decreasing from the center, indicating rod-like structures. After looking at the overall crystallization behavior, the interpretation of which is very difficult, we have tried to follow the crystallization of each component separately in the blends.

The study of the intensity of WAXS diffraction peaks for this purpose is not useful because of the closeness of the diffraction lines of both polymers because of their similar unit cell structures. An overlapping of the peaks occurs and the patterns cannot be efficiently resolved into the contributions of each component. A similar difficulty is encountered in DSC where the crystallization exotherms overlap each other, and although the peaks are distinctly seen, one cannot separate the overall crystallization exotherms into the contributions of each component. IR has provided a very useful way of following the crystallization of each component separately. The change in intensity of crystalline band of each component, which was not affected by the crystallization of the second one, was followed as crystallization proceeded. One is able, in this way, to follow the crystallization of each component in the blends under different crystallization conditions. By correlating these studies with the simultaneously obtained density measurement, which also provide information on the overall crystallization behavior, we were able to determine the ultimate crystallinity obtained by each component in the blends and follow the changes with blend composition.

Experimental

Samples of the 100/0; 80/20; 60/40; 50/50; 40/60; 20/80; 10/90; 0/100 (PBT/PET) blends were provided by the General Electric Co. by coextruding at 520°F, in a Sterling 1¾ in. extruder, 60 rpm, PBT of a viscosity average molecular weight 25,600 and melt viscosity at 250°C of 2740 poises, and PET of a viscosity average molecular weight of 36,800 and a melt viscosity at 270°S of 1250 poises.

The samples were desiccated in a vacuum oven at 110°C for 24 hr. They were then compression molded on a Pasadena press at 280°C and 30,000 psi for 1 min. They were then immediately quenched in the ice-water mixture and melted again in the press without pressure for another minute. They were then immediately quenched in the ice-water mixture. A set of these amorphous samples was used in the determination of the T_g by DSC. Other samples were crystallized in a fluidized bed at 150°, 130°, 110°, and 90°C for 15, 30, 60, 120, 180, 240, 300, 600, 900, and 1,800 seconds, following which they were quenched again.

Another set of samples of the same blends, after the initial pressing and quenching, were remelted between hot plates at 280°C for 1 min and then transferred immediately to the crystallization bath at 200°C. After the same crystallization times they were quenched in the ice-water mixture.

The morphology studies were performed on two sets of samples, the first set (after the initial melting) was crystallized at 130°C for 10 min in an oil bath. The morphology of these samples was studied by light scattering. The samples were melted again at 280°C for 1 min and then crystallized in the Mettler FP-2 at 200°C for 10 min. The morphology of these samples was studied under the polarizing microscope.

The density studies were carried out in two different density columns. The first one had a density range 1.33–1.40 gm/cm^3 and was made by establishing a gradient using equal volumes of two mixtures of carbon tetrachloride and heptane of the compositions of 70% CCl_4/30% C_7H_{16} and 90% CCl_4/10% C_7H_{16}. The second column of a density range 1.26–

1.40 gm/cm^3 was made by using the two mixtures of 63% CCl_4/37% C_7H_{16} and 82% CCl_4/18% C_7H_{16}. Both carbon tetrachloride and heptane provide excellent wettability of the samples and do not change their crystallinity by solvent-induced crystallization or appear to swell the samples. Samples of PBT and PET were kept in the columns for one week without the occurrence of any change in density. The IR measurements were performed in these crystallized samples using a Perkin–Elmer IR spectrophotometer, Model 283.

Depolarized Light Intensity

The study of the kinetics of crystallization was initiated through the measurement of the depolarized light intensity (DLI), transmitted between crossed polaroids. We basically followed the method initiated by Magill (*17*) in which a thin polymer sample is sandwiched between cover slides, melted, and then rapidly transferred to a Mettler FP-2 hot stage at the crystallization temperature (200°, 202°, and 205°C). We then, through a photocell adapted to the occular, measured the intensity of the light transmitted between crossed polaroids in the light microscope as a function of time. This method, although empirical, describes accurately the primary initial crystallization, that is, until the sample volume becomes filled with structured units. However, the intensity transmitted between crossed polaroids levels off when this occurs, and no further secondary intraspherulitic crystallization is seen. Therefore crystallization half times lower than those obtained by other methods like DSC or density are achieved. This effect can be corrected for by simultaneously measuring the light transmitted with parallel polaroids, which when added to the previous-obtained intensity transmitted between crossed polaroids, gives the turbidity of the sample. These measurements represent more accurately the crystallization process, as they do not only include the initial formation of crystalline structures, but also the further increase of anisotropy as the secondary crystallization proceeds. Therefore, the crystallization half times obtained from the turbidity measurements are more in coincidence with those obtained by other methods.

As we can see in Figure 3, the crystallization half times increase with low amounts of PET up to the 85/15 blend, and from there they go on decreasing with increasing amounts of PET. It is important to notice (in consideration of the melting points given in Figure 1) that our crystallization temperature provides a low undercooling for PBT which explains its slow crystallization, while undercooling for PET is much larger, giving a faster crystallization rate. The DLI response may actually depend upon the size and morphology of the transformed phase as well as its amount. It is apparent from microscopic observations that the morphology of the crystallized blend depends upon its composition, ranging from spherulitic to a random collection of crystals with no superstructure. Thus, the relationship between the amount of transformed phase and the DLI response may depend on composition of the blend and may vary during the course of crystallization.

As we can see, the information provided by DLI on the kinetics of crystallization is useful but very hard to interpret, especially in this case where we have two polymers crystallizing separately. We only obtain an experimental idea of the overall crystallization behavior, but we cannot separate the crystallization of each component. This information was only provided by the IR measurements.

IR Studies

The crystallization of each component in the blends can be followed by IR. This is done by isolating a band that is sensitive to the crystallization of only one of the components.

The crystallization kinetics of PET have been followed by IR by Cobbs and Burton (*18*). It was observed that the 972-cm^{-1} band is sensitive to the crystallization while the 795 cm^{-1} remains unchanged. In the PBT/PET blends the 972-cm^{-1} band is affected by the presence of PBT so that it cannot be used. Koenig and Boerio (*19, 20*) have studied the PET spectra by Fourier transform IR and assigned the 848-cm^{-1} band to the rocking mode of the trans conformation of the ethylene glycol segments in the crystalline regions. Since we did not detect any change of intensity of this band with PBT crystallization, it was used to follow the crystallization of PET.

PBT recently has been studied by Koenig who looked at the -CH_2- rocking region. He identified a high-energy band at 917 cm^{-1}, which showed a marked increase in intensity with crystallization (*21*). To follow the crystallization of PBT we simultaneously monitored the changes in intensity of the 917- and 810-cm^{-1} bands, but the final results are plotted in terms of the 917-cm^{-1} band, which shows a more sensitive change during the crystallization. As a reference band, the 632-cm^{-1} band was selected which is assigned to the C-C-C bending mode in the benzene ring.

The intensity of the crystalline bands was monitored simultaneously during the crystallization. To correct for changes in density or thickness in the different samples, the intensities were normalized by the reference band. These normalized intensities were plotted vs. log time for each of the blends at the different crystallization temperatures. The curves obtained are sigmoidal in nature and they level off when the final crystallinity is achieved. A typical curve for the normalized intensity of the 848-cm^{-1} band vs. log time is plotted in Figure 7 for PET.

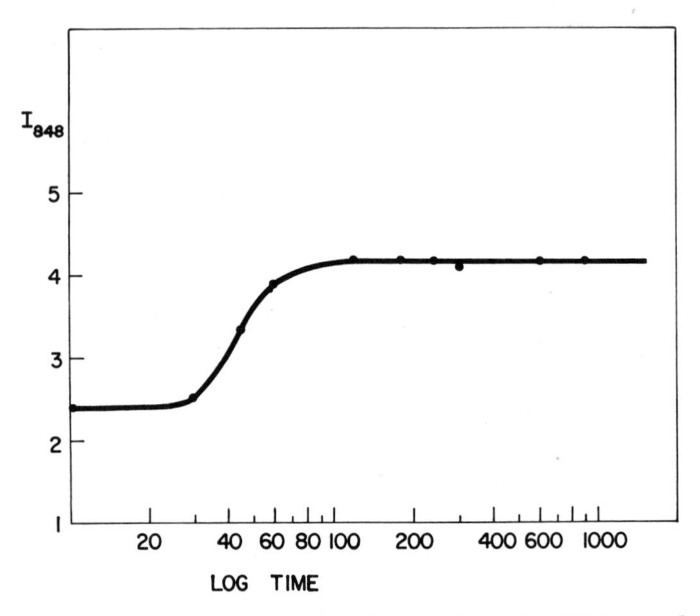

Figure 7. IR intensity of the 848 cm^{-1} band for PET crystallized at 200°C vs. log time

Because of the limitations of signal-to-noise ratio, we can follow the crystallization of PBT only for blends having at least 50% PBT. Similarly, PET crystallization can be followed only in blends having at least 50% PET.

The intensity for the PET and PBT crystalline bands in the blends must be corrected to take into account the contribution to these bands of the other component and the dilution effect also caused by it. The corrected intensity reflects the change in crystallinity of the PET and PBT phases individually, based upon the weight of that phase. In other words, the corrected intensity would be the observed intensity if the sample had only one component, and it would have crystallized at the same rate and manner as it did in the blend.

The correction is done according to Equations 1 and 2, where

$$I_{917\text{-corr}} = \frac{I_{917\text{-exp}} - M_{\text{PET}}\, I_{917\text{-PET}}}{M_{\text{PBT}}} \quad (1)$$

$$I_{848\text{-corr}} = \frac{I_{848\text{-exp}} - M_{\text{PBT}}\, I_{848\text{-PBT}}}{M_{\text{PET}}} \quad (2)$$

$I_{917\text{-corr}}$ and $I_{848\text{-corr}}$ are the corrected intensities for the 917- and 848-cm^{-1} bands. $I_{848\text{-exp}}$ and $I_{917\text{-exp}}$ are the experimentally measured values in the blends. $I_{917\text{-PET}}$ is the value for the intensity of the 917-cm^{-1} band for

PET and $I_{848\text{-PBT}}$ is the value of the intensity of the 848-cm^{-1} band for PBT, neither of which change with crystallization. M_{PBT} and M_{PET} are the monomer mole fractions for PBT and PET in the blends.

For the pure polymers, since the values of the IR crystalline bands and the density are known, a linear correlation can be established between the intensity of the crystalline band for each crystallizing polymer and its degree of crystallinity obtained from the density measurements. By assuming that the same relationship exists in the blends between the IR intensity and the degree of crystallinity, the partial degrees of crystallinity of each component in the blends are obtained.

Density Studies

Considering the semicrystalline polymers as a two-phase system with a sharp delineation between the crystalline and the amorphous material, we can use the specific volumes to calculate the weight-fraction degree of crystallinity.

$$X_c = \frac{\overline{V}_a - V}{\overline{V}_a - \overline{V}_c} \tag{3}$$

or expressing it in terms of the densities:

$$X_c = \frac{\rho_c}{\rho}\,\frac{\rho - \rho_a}{\rho_c - \rho_a} \tag{4}$$

where ρ_c is the density of the crystalline materials and ρ_a that of the amorphous.

There is a dispute in the literature over the values of the crystalline density of PET; for our calculation we have chosen the value proposed by Bunn (*22*) and Tadokoro (*23*) of 1.455 g/cm^3, which traditionally has been used more than the value 1.515 g/cm^3 proposed by Alter and Bonart (*24*), which would give lower degrees of crystallinity for a given density. A similar situation occurs with the PBT crystalline density. Boye (*25*) proposes 1.39 g/cm^3 and this value has been used by Misra and Stein (*7*) and by Slagowski (*26*). Tadokoro (*23*) proposes the value 1.404 g/cm^3 and a similar value, 1.406 g/cm^3, is obtained by Mencik (*12*). In our calculations we have used the value obtained by Alter and Bonart (*24*) of 1.433 g/cm^3 because their WAXS pattern is in very good agreement with the one obtained by us.

The amorphous density for PET is 1.33 g/cm^3 and for PBT is 1.28 g/cm^3. It is apparent that only two decimal digits are significant, which must be kept in mind when interpreting the results.

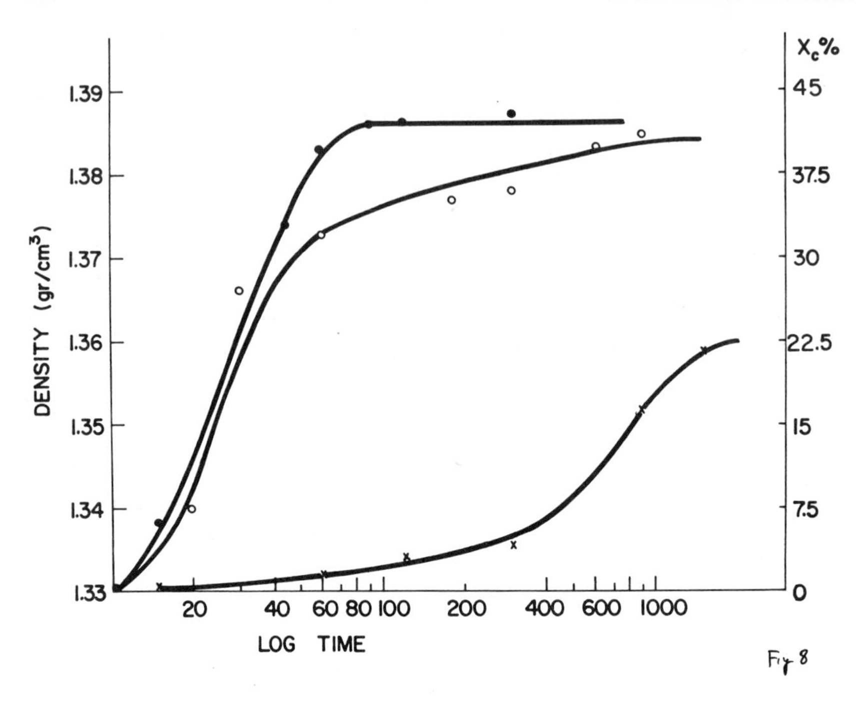

Figure 8. Density and degree of crystallinity vs. log time for PET crystallized at 200° (●), 150° (○), and 110°C (×)

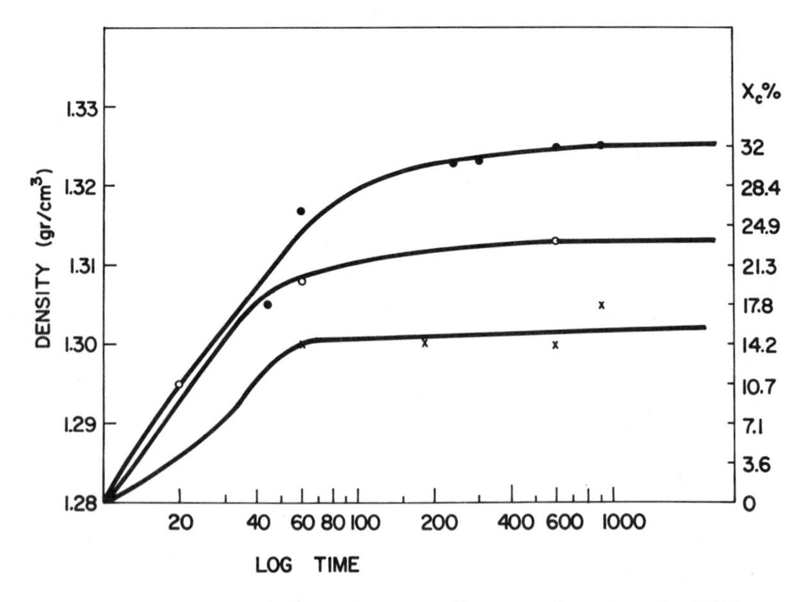

Figure 9. Density and degree of crystallinity vs. log time for PBT crystallized at 200° (●), 150° (○), and 110°C (×)

With the above densities the crystallization of PBT and PET can be easily followed. Figures 8 and 9 show the change in density and degree of crystallinity of PET and PBT with log time for the samples crystallized at 200°, 150°, and 110°C. These density results are used to establish the correlation with IR data.

By plotting the change of the degree of crystallinity at different temperatures vs. log time, as it is done in Figures 8 and 9, it is possible to observe that the curves would not superimpose by means of a horizontal shift which would correct for the rate difference but that the ultimate degree of crystallinity is a function of crystallization temperature. Mandelkern (*27*) has shown that when the crystallization temperatures are close to T_m and close to each other, the curves can be superimposed by a horizontal shift, but still the final crystallinities depend on the crystallization temperature. With much bigger differences in crystallization temperatures, the superposition is lost, especially in those samples annealed from the glass where the crystallization occurs very fast but the ultimate crystallinity depends on the crystallization temperature. This same superposition is observed by Hoffman (*28*) when looking at wide ranges of crystallization temperatures.

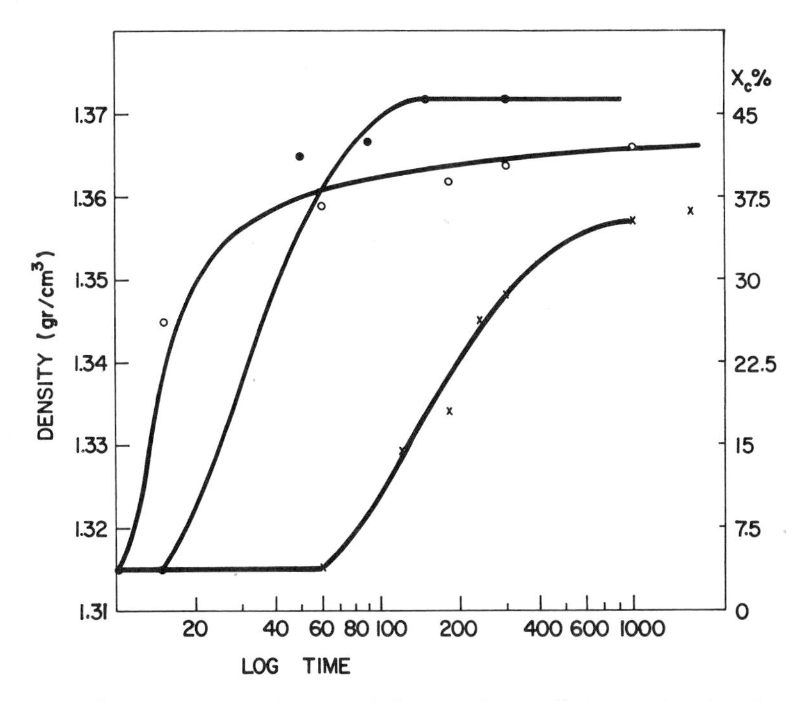

Figure 10. Density and apparent degree of crystallinity vs. log time for the 20/80 PBT/PET blend crystallized at 200° (●), 150° (○), and 110°C (×)

In the melt-crystallized samples, we can observe how at 200°C the ultimate crystallinity of PET is higher than that of PBT. The behavior upon annealing is different for PET and PBT. PET reaches high degrees of crystallinity, the samples become very turbid, and there is microscopic evidence of the formation of small spherulites with their extinction patterns in the directions of the polaroids. Upon annealing, PBT reaches much lower degrees of crystallinity, the samples do not become turbid, and when viewed under the polarizing microscope one observes only slight birefringence with no structures resolvable.

The crystallization of the blends cannot be followed adequately with density measurements since both components crystallize separately; the density measurements indicate only the change of the overall density with time and its interpretation is difficult.

We have plotted in Figures 10, 11, and 12 for the 20/80, 40/60, and 50/50 blends the change of density with log time at the three crystallization temperatures, 200°, 150°, and 110°C. Their interpretation in terms of the apparent degree of crystallinity (obtained through use of analysis), which is also plotted in the same graphs, is difficult but is done in these PET-rich blends to obtain further information about the PET-crystallization behavior and final crystallinities in the presence of PBT.

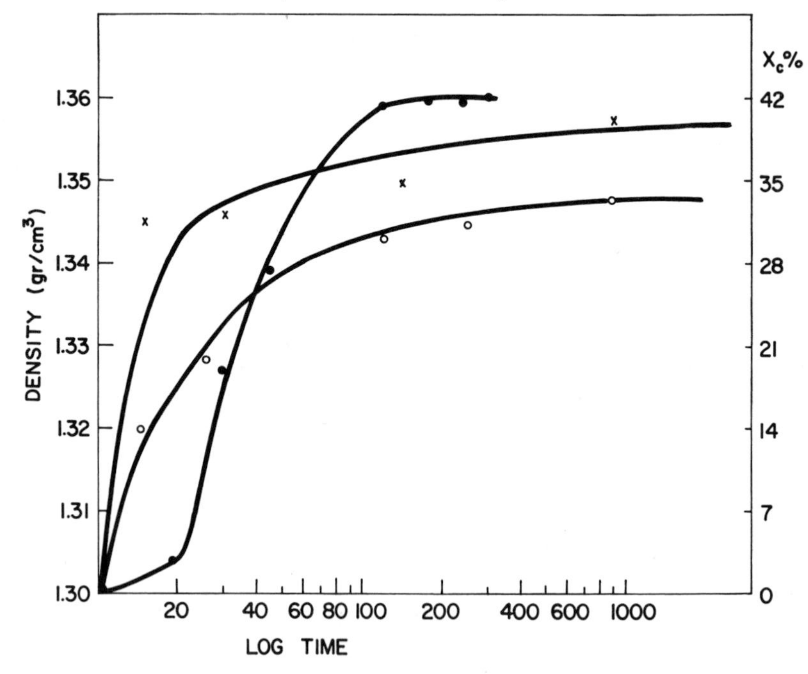

Figure 11. Density and apparent degree of crystallinity vs. log time for the 40/60 PBT/PET blend crystallized at 200° (●), 150° (×), and 110°C (○)

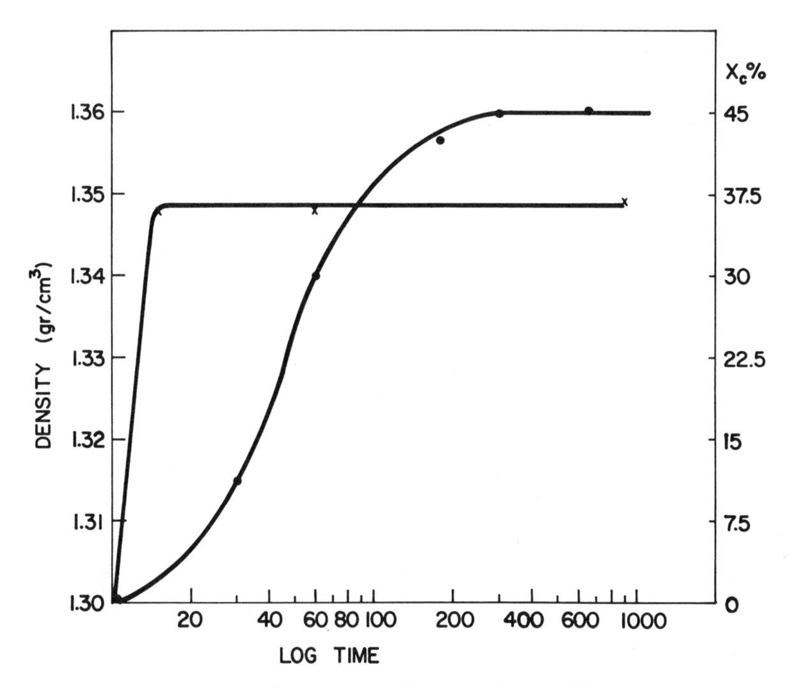

Figure 12. Density and apparent degree of crystallinity vs. log time for the 50/50 PBT/PET blend crystallized at 200° (●) and 150°C (×)

Interpretation of Results

From the IR measurements we have obtained the crystallization half-times of each component in the blends at the different temperatures; they are plotted in Figures 13, 14, 15, 16, and 17. They can be transformed into the crystallization rate vs. temperature of crystallization for the different blends, which are plotted in Figures 18, 19, 20, and 21. From the density–IR correlation we obtained the ultimate degrees of crystallinity of each component in the blends, and their change with blend composition is plotted in Figures 22 and 23. Since the crystallization behavior varies with temperature of crystallization, we will approach its interpretation looking at the behavior at each crystallization temperature separately.

Crystallization at 200°C. While normally PBT is known to have a faster crystallization rate than PET, at 200°C its undercooling is only 24°C while that of PET is 65°C. Therefore, the crystallization rate at 200°C is higher for PET than for PBT, which was already observed by DLI and now can be verified by IR and density. While the crystallization half-time of PBT in the blends remains quite constant, that of PET increases with PBT content, showing a decrease in the crystallization rate

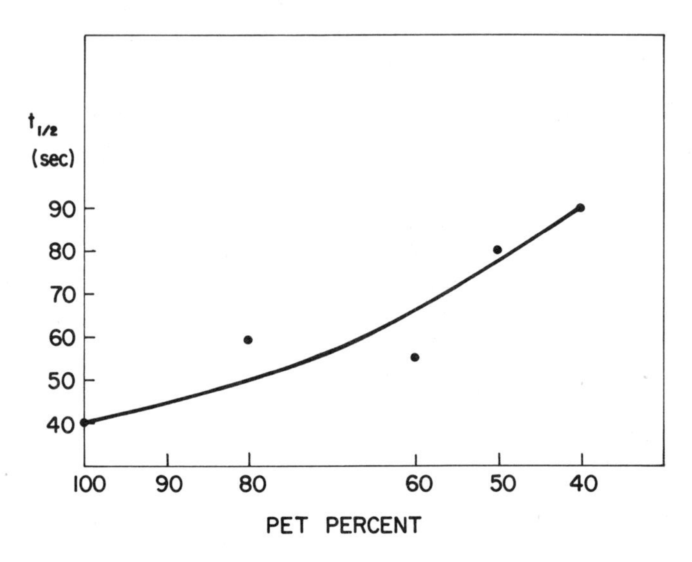

Figure 13. PET-crystallization half times vs. PET percent at 200°C

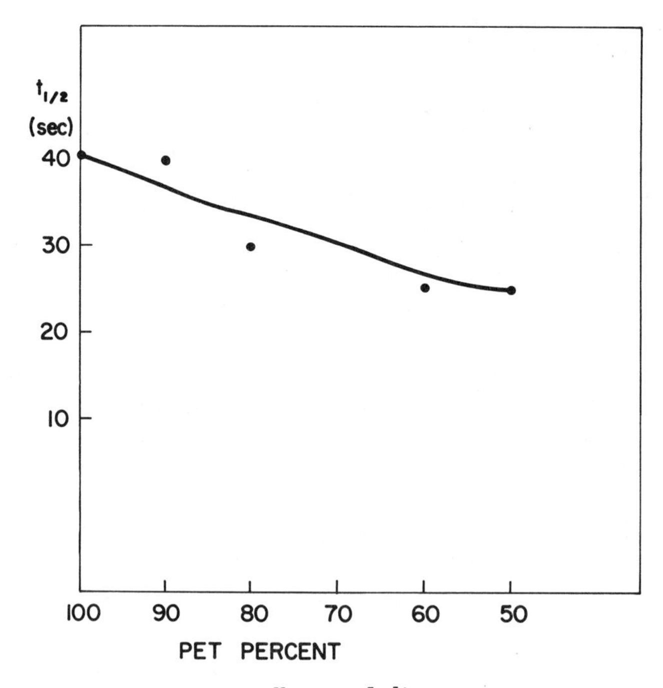

Figure 14. PET-crystallization half times vs. PET percent at 150°C

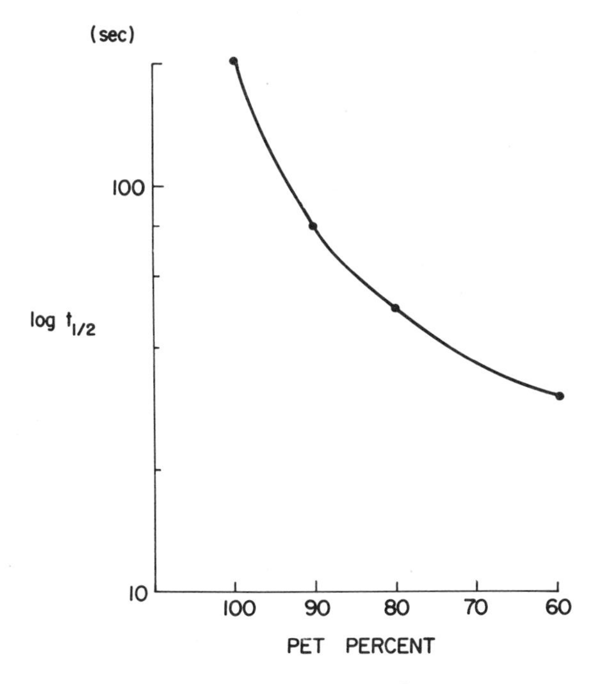

Figure 15. PET-crystallization half times vs. PET percent at 130°C

of PET. The same behavior was observed by DLI and can be explained as a consequence of the dilution effect, resulting in a need for each component to segregate from the mixture in order to crystallize.

The final PBT crystallinity decreases with the increase of PET content, while a reverse effect occurs with the PET crystallinity, which increases markedly with the presence of PBT. The experimental evidence from DSC shows a similar increase of crystallinity; the same effect is seen in the density measurements. The apparent degree of crystallinity of the 20/80 blend is higher than that of the 0/100 sample, which is a clear indication of the increase of PET crystallinity, especially if we consider that there is DSC, WAXS, and IR evidence that PBT does not crystallize in this sample. The apparent degrees of crystallinity for the 40/60 and 50/50 blends are also higher than the expected values and confirm the increase in crystallinity of PET since the IR and DSC evidence shows that the crystallinity of PBT decreases.

The microscopic examination shows for PBT small spherulites which become immediately volume filling, yielding a circularly symmetrical SALS pattern. For PET we can observe volume-filling spherulites with distinct maltese crosses (Figure 5). When the PBT percentage increases,

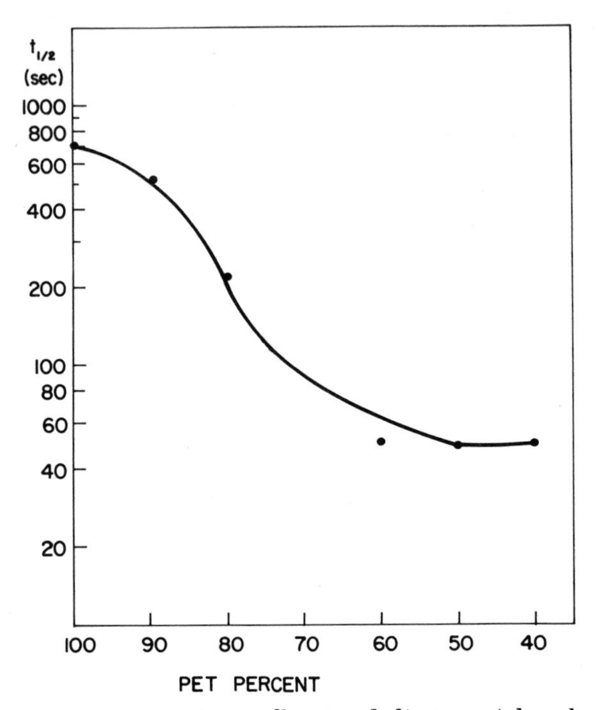

Figure 16. PET-crystallization half times (plotted logarithmically) vs. PET percent at 110°C

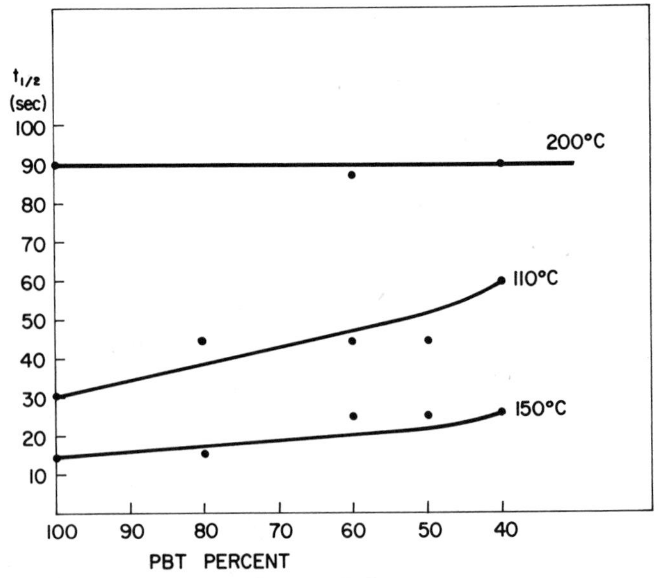

Figure 17. PBT-crystallization half times vs. PBT percent at 200°, 150°, and 110°C

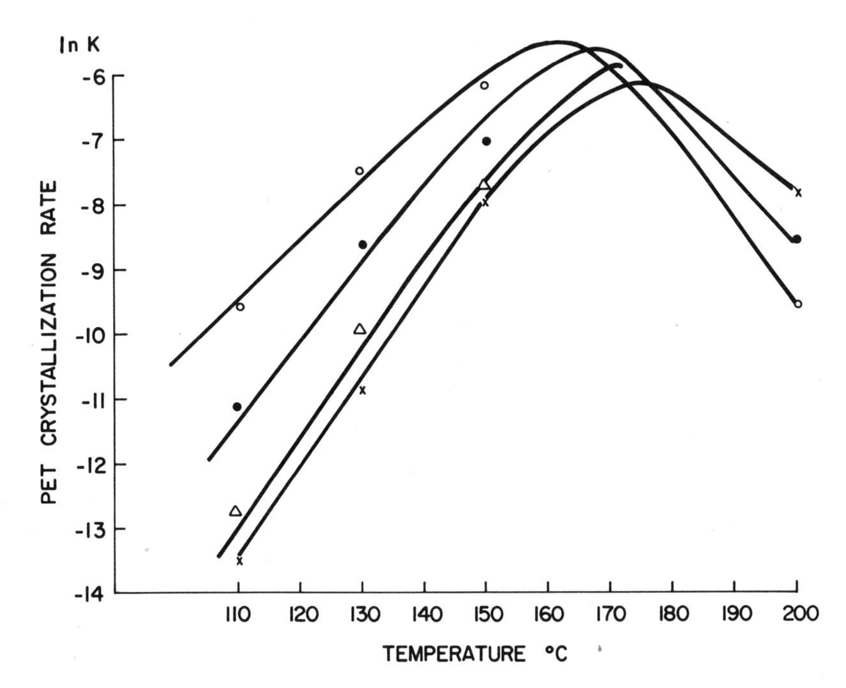

Figure 18. PET-crystallization rate vs. temperature for the 0/100 (×), 10/90 (△), 20/80 (●), and 40/60 (○) (PBT/PET) blends

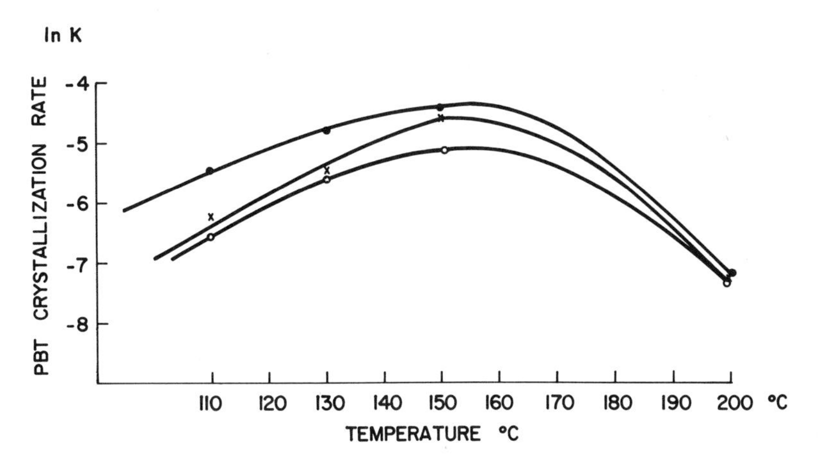

Figure 19. PBT-crystallization rate vs. temperature for the 100/0 (●), 80/20 (×), and 60/40 (○) (PBT/PET) blends

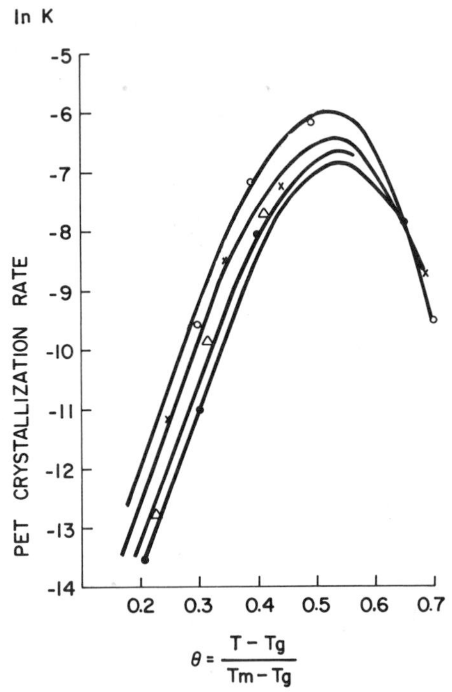

Figure 20. PET-crystallization rate vs. reduced temperature for the 0/100 (●), 10/90 (△), 20/80 (×), and 40/60 (○) (PBT/PET) blends

waviness of the maltese cross appears until the extinction pattern becomes lost in the 40/60 blend (Figure 6).

Crystallization at 150°C. The crystallization at 150°C was carried out by annealing the glassy samples that had been quenched from the melt. It can be seen how the PET-crystallization half-times decrease with increasing PBT concentration, which decreases the T_g of the blend. Similarly the crystallization half times for PBT increase with increasing PET, which increases the T_g of the blends. The change in the crystallization half-times of each component in the blends can be explained in terms of the changing T_g. The crystallization rate is controlled by T_c-T_g in the diffusion-controlled region; the crystallization temperature remains constant but the T_g, as we have shown, changes with blend composition. Therefore, T_c-T_g changes with blend composition and this affects the crystallization rate of each component. The crystallization rates of the PET component increase with increasing PBT composition because the T_g of the blends is lowered; similarly the crystallization rate of the PBT component decreases with increasing amounts of PET, which causes an increase of the T_g.

Because of the closeness of the annealing temperature (150°C) to the temperature of maximum rate of crystallization, both PBT and PET show at this temperature their lowest crystallization half-times. Since the distance from the changing T_g is quite high, the changes in crystallization rate are not as strong as those observed at lower annealing temperatures.

The PET crystallinity remains fairly constant with the presence of PET in the blends (Figure 22). There are two opposite effects at this temperature that probably compensate each other. These are the dilution effect by the PBT and the increase of mobility with the lowering of the T_g. The results from the apparent crystallinities obtained through density are further evidence of this behavior. The apparent degrees of crystallinity are very close to the values expected from the density of the pure polymers, showing no drastic changes in the partial crystallinities.

Microscopic examination of these samples show for the PET-rich blends the presence of very small spherulites. The PBT-rich blends show very low birefringence, low turbidity, and no organized structures. The SALS patterns are circularly symmetrical.

Crystallization at 130° and 110°C. The most profound changes in the crystallization behavior with blend composition are seen at the 130° and the 110°C crystallization temperatures. The PET-crystallization half-

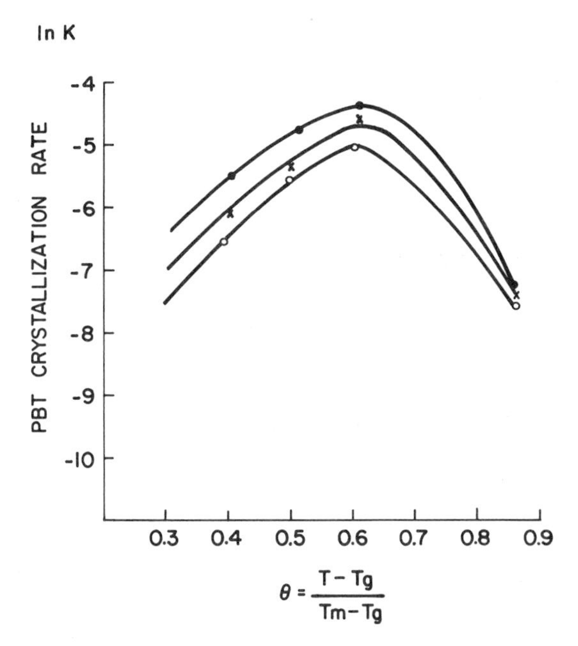

Figure 21. PBT-crystallization rate vs. reduced temperature for the 100/0 (●), 80/20 (×), and 60/40 (○) (PBT/PET) blends

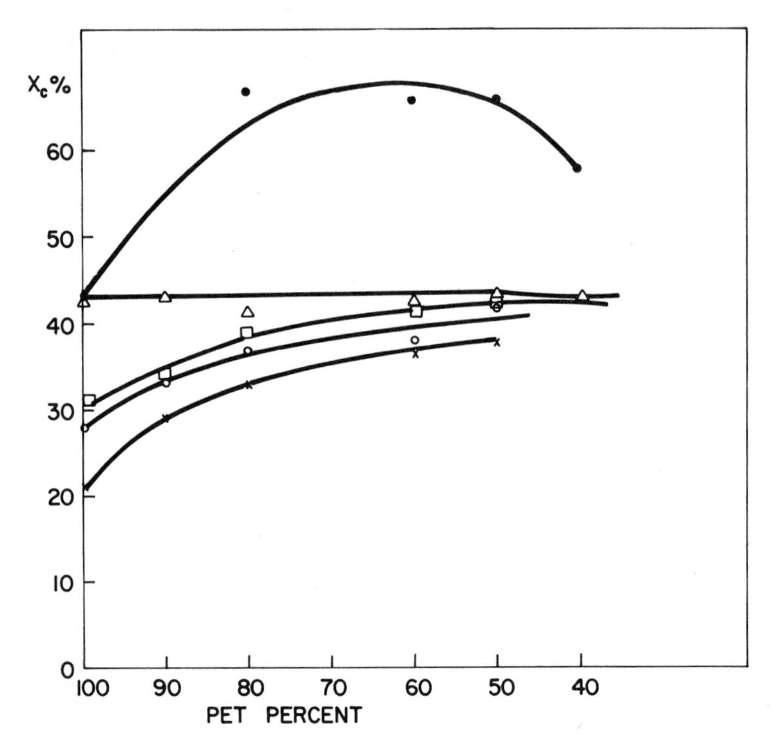

Figure 22. PET ultimate degree of crystallinity vs. PET percent at 200° (●), 150° (△), 130° (□), 110° (○), and 90°C (×)

times decrease drastically with PBT percentage, while those of PBT increase with PET content. This reflects the effect of the changing T_g. The effect is large for PET since it is a slowly crystallizing polymer, and this annealing temperature is quite close to its T_g.

The final crystallinity for PBT is lower than for the sample annealed at 150°C and decreases with increasing PET percentage in the blends. At the same time the PET crystallinity increases with the presence of PBT quite markedly. Both phenomena can be accounted for as a consequence of the change in T_g in the blends.

The interpretation of the effect of the difference between T_c and T_g on the ultimate degree of crystallinity depends on the approach taken to explain the nature of the T_g. In terms of a free-volume theory, one expects always to obtain the same ultimate degree of crystallinity regardless of the annealing temperature; the only difference will be that the crystallization will proceed at a higher rate at high annealing temperatures.

However, if the T_g is explained according to the Gibbs–DiMarzio second-order transition theory, it may be expected that not only the crystallization rate depends on the annealing temperature, but also the ultimate degree of crystallinity achieved will similarly vary. At higher annealing temperatures, a large number of configurations are available which allows for individual rearrangements into different configurations that will facilitate the achievement of higher ultimate degrees of crystallinity. We also must consider that as crystallization proceeds, the T_g will increase because of the introduction of crystallites, and the T_g will become closer to the crystallization temperature. Therefore, crystallization will stop sooner at lower crystallization temperatures, yielding lower ultimate crystallinities. The change of T_g in the blends with blend composition changes T_c-T_g, influencing therefore the ultimate crystallinities.

Microscopic examination of the samples crystallized at 130°C shows very low turbidity and birefringence for the PBT samples; the turbidity in the blends increased, and small spherulites were present for PET. The samples crystallized at 110°C again showed small spherulites for PET, and no organized structures were observed in the blends of intermediate composition although their turbidity was quite high; with samples of very high PBT composition, the turbidity was lost.

Crystallization at 90°C. The crystallization at 90°C is very slow and since the samples were not crystallized long enough to allow for a definite determination of the crystallization half-times, these have not been plotted. The plotted degrees of crystallinity for PBT and PET are those achieved after 30 minutes at the crystallization temperature.

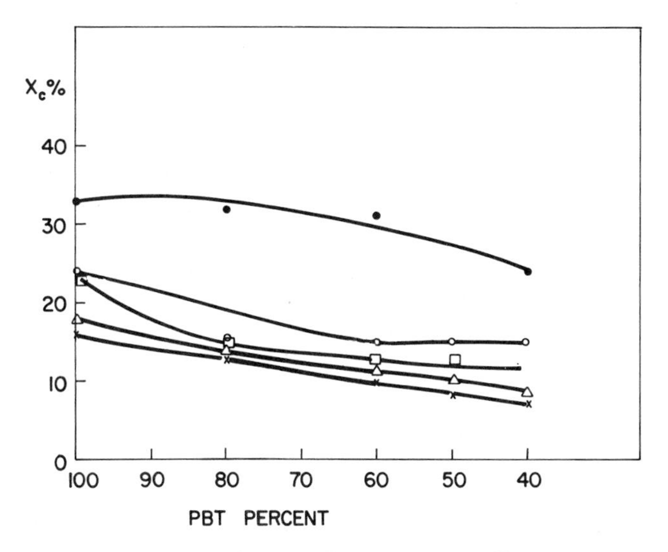

Figure 23. PBT ultimate degree of crystallinity vs. PBT percent at 200° (●), 150° (○), 130° (□), 110° (△), and 90°C (×)

Although PET has not completed its crystallization, PBT and the blends have done so because of their higher rate. As in previous cases, we can observe how the PBT crystallinity decreases with PET content in the PBT-rich blends.

The Avrami analysis was performed on the crystallization data. The DLI measurements provided high Avrami exponents in the blends, but the analysis on the DLI data is extremely inaccurate because of all the difficulties inherent to the method. The IR measurements show Avrami exponents that range from 2.5 to 1.5 for PET and PBT. These studies were made on the diffusion-controlled region. It is our belief that the Avrami analysis strongly depends on the method used to follow the crystallization and always has to be accompanied with direct observations on the morphology.

Overall Crystallization Behavior. If instead of plotting the crystallization half times vs. blend composition we plot the crystallization rates for each component in the blends vs. temperature of crystallization, we obtain a series of curves which show a maximum. The right side of these curves is nucleation controlled while the left side is diffusion controlled (Figures 18 and 19).

For the PET component we can observe how increasing amounts of PBT shift the value of the maximum towards lower temperatures, increasing at the same time the crystallization rate. A similar behavior is observed by Boon and Azcue (*29*) when looking at the crystallization of polystyrene with a diluent; an increase in crystallization rate and a shift of the maximum towards lower temperatures was observed. Similarly for the PBT component we observed a decrease in the crystallization rate and a shift of the maximum towards higher temperatures.

Both effects can be accounted for in terms of the change in the T_g of the blends caused by the introduction of the second component. In the PET case, the introduction of PBT in the blends lowers the T_g, causing the increase in crystallization rate and the shift of the maximum in crystallization rate towards lower temperatures. Similarly for the PBT component, the addition of PET increases the T_g which causes a decrease of the rate and a shift in the maximum towards higher temperatures. One can plot the crystallization rate vs. a reduced temperature that would account for the shift of T_g. This reduced temperature is defined as:

$$\theta = \frac{T - T_g}{T_m - T_g} \tag{5}$$

When doing so, we can observe how the maximum occurs at the same reduced temperature for the polymer and the blends, correcting, therefore, the shift in the crystallization rate curves (Figures 20 and 21).

We can therefore conclude that the changes in the crystallization behavior of the blends is attributable to the change in T_g with blend composition. We know that each component crystallizes separately and according to its own crystalline structure since DSC has shown the presence of only one T_g in the blends which depends upon blend composition. We also have shown that the crystallization behavior of each component strongly depends upon the value of that T_g. Therefore, we must conclude that there is compatibility between both polymers in their amorphous state. Although when they crystallize, there is a phase separation with two different crystalline structures. Their crystallization behavior suggests that they do not crystallize from two segregated phases, but that in fact there is an ultimate mixing with only one amorphous phase from which the molecules crystallize in a manner dictated by the particular characteristics of that amorphous phase.

Comparison between the Crystallization from the Melt or the Glass. The crystallization studies at low temperatures can either be performed on samples cooled from the melt to the crystallization temperature or on quenched samples which are annealed to the crystallization temperature. Each method would be indicated for the particular region where equilibration to the crystallization temperature would be faster without having induced crystallinity in the amorphous polymer. In general, crystallization from the melt is better at temperatures above that which gives the maximum in growth rate, and crystallization from the glass is better for temperatures lower than the maximum.

In our case since we have a very fast crystallizing polymer (PBT), we wanted to see if there was any difference in the crystallization of the blends depending on how the study was performed. Two sets of studies were done, in which the samples were crystallized at 110°C in one of them from the glass and in the other from the melt. The crystallization half-times and ultimate crystallinities obtained for each component were plotted in Figures 24, 25, 26, and 27.

We can observe how for the PET component the crystallization behavior is sensibly the same, some differences appearing only for the blends of intermediate composition where the crystallization rate is higher. The PBT-crystallization behavior is markedly different. We observe a much faster crystallization behavior when coming from the melt than when annealing the glass; the crystallinities obtained are also much higher and the samples present a different physical appearance. Those crystallized from the melt are very turbid while those from the glass present no turbidity, and practically no birefringent structures are present.

We must therefore conclude that for slowly crystallizing polymers both methods can be used, and the growth rates and morphologies ob-

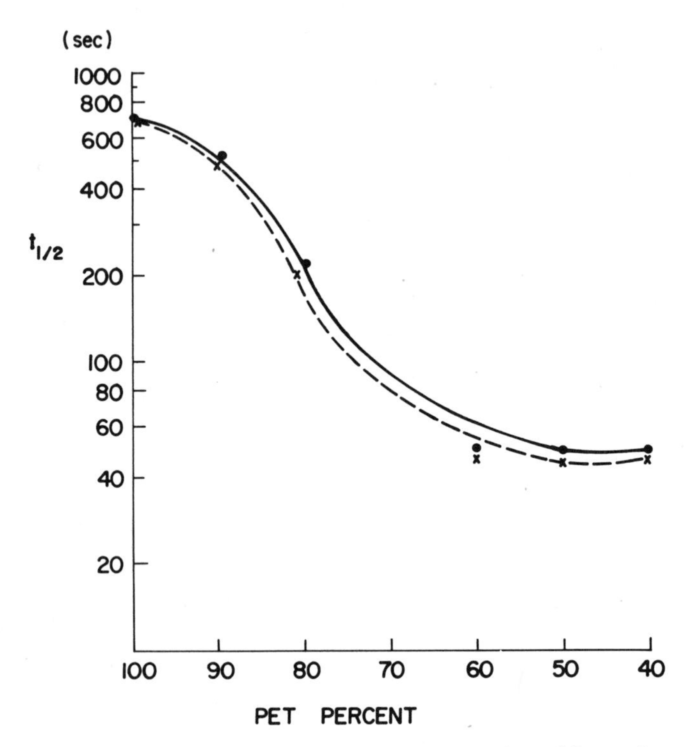

Figure 24. PET-crystallization half times (plotted logarithmically) vs. PET percent. Samples crystallized from glass (●) or melt (×) at 110°C.

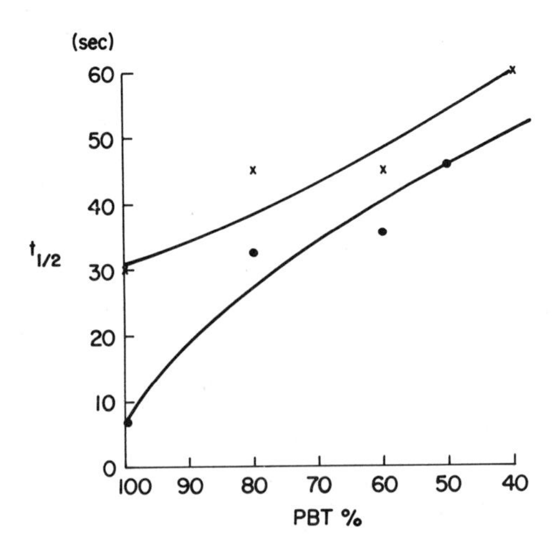

Figure 25. PBT-crystallization half times vs. PBT percent. Samples crystallized from glass (×) or melt (●) at 110°C.

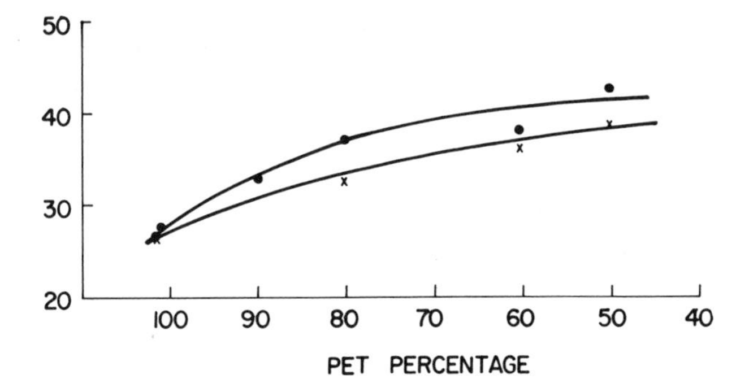

Figure 26. PET ultimate degree of crystallinity vs. PET percent in samples crystallized from glass (×) or melt (●) at 110°C

tained will be the same and will only depend upon the crystallization temperature. Since the maximum in nucleation rate occurs at lower temperatures than the maximum in growth rate, in quenched samples which are annealed above the first one, very high nucleation densities are achieved which yield samples with a large number of spherulites. These do not achieve as large a radius as those in samples crystallized from the melt at the same temperature. Identical behavior was observed by Van Antwerpen (*10*) on samples of PET of different molecular weights.

In those polymers, like PBT, which have a very high maximum crystallization rate, different growth rates and morphologies are obtained for the two modes of crystallization. When these samples are crystallized

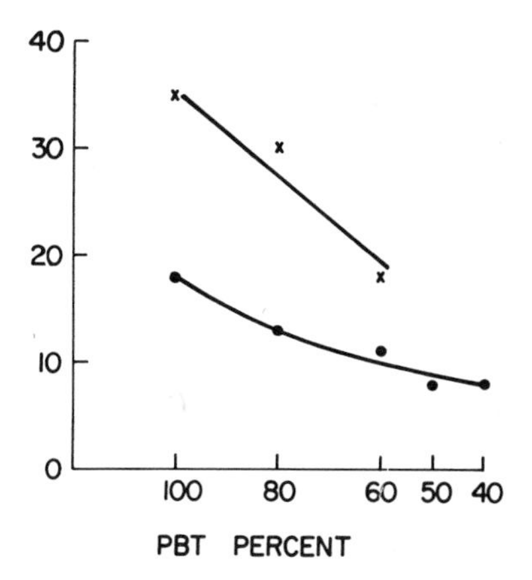

Figure 27. PBT ultimate degree of crystallinity vs. PBT percent in samples crystallized from glass (●) or melt (×) at 110°C

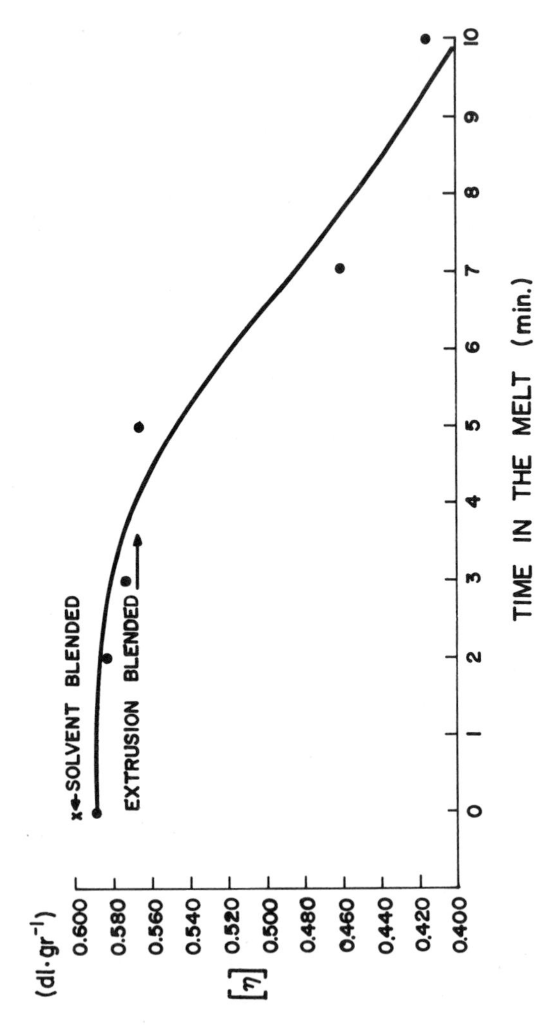

Figure 28. Intrinsic viscosity vs. time in the melt for the 50/50 blend

from the melt at low temperatures and upon going through the maximum, a substantial nucleation and crystallization occurs. Therefore, superstructures like spherulites are formed and they continue their crystallization at the desired temperature. However, if the same samples are quenched from the melt to the amorphous state, although a higher number of nuclei may be formed, they do not grow since the cooling step is too fast. Therefore, when annealed at the crystallization temperature, those nuclei grow at the rate dictated by that temperature. Because of this high number and characteristics of the crystallization at that temperature, no supersructures are seen.

Trans-Esterification Studies. It has been suggested that the compatibility in these blends arises from the fact that ester interchange may occur between both polymers in the blend during the melting process, forming a copolymer of both species, which would show only one T_g. Although the arguments in favor of compatibility based on the crystallization behavior of each component cannot be dismissed in terms of the ester-interchange reaction occurring, we have looked to what extent it might affect our results. Flory (*30, 31, 32*) has shown how this reaction is quite fast in the presence of a catalyst. In our case, the polymers have

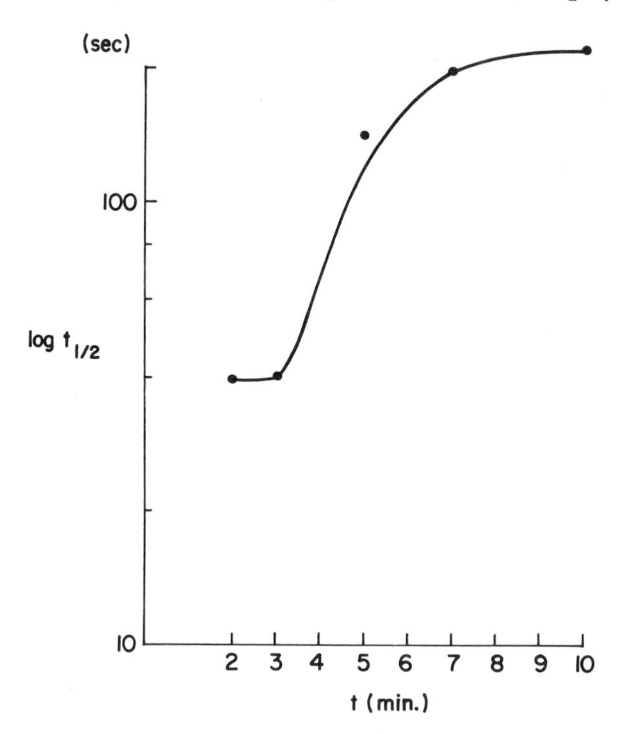

Figure 29. PBT-crystallization half times vs. time in the melt for the 50/50 blend crystallized at 90°C

been statistical and no catalyst is present, but since they are kept at high temperatures (280°C) for two minutes, it is important to determine how this affects the crystallization behavior.

It is expected that if trans esterification occurs, the intrinsic viscosity of the samples will decrease and the rates of crystallization and ultimate degrees of crystallinity will also decrease. Therefore a series of studies on the 50/50 blend which was kept in the melt for 2, 3, 5, 7, and 10 minutes were performed. The crystallization was followed by DSC and IR. Simultaneous measurements of the intrinsic viscosity were made on a 40:60 mixture of 1,1,2,2,-tetrachloroethane phenol. The results are plotted in Figures 28, 29, and 30.

It is possible to observe how the intrinsic viscosity remains quite constant for up to five minutes in the melt, after which there is a sharp decrease. Similar results are observed with the crystallization half-times which remain constant for up to three minutes in the melt; after that there is a sharp increase.

This slowing down of the crystallization also was observed by DSC. Although both effects, the drop in intrinsic viscosity and slowing of the

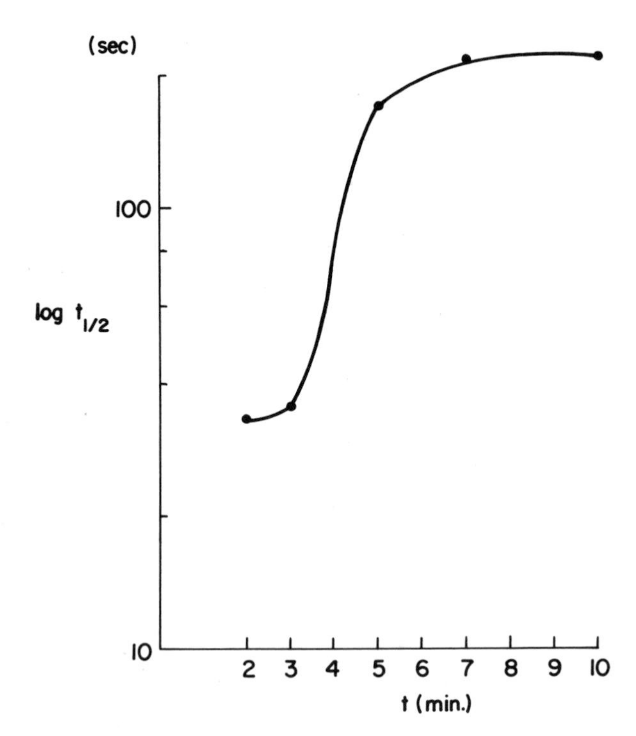

Figure 30. PET-crystallization half times vs. time in the melt for the 50/50 blend crystallized at 90°C

crystallization, might be attributable to other degradation processes (33) besides trans esterification, the fact that they do not occur until after at least three minutes makes us feel comfortable knowing that no significant trans esterification has occurred during the preparation of the samples, which were kept in the melt for exactly two minutes. Therefore, we do not expect the crystallization behavior to be affected by these preparation conditions. Further information was obtained from samples which were solvent blended instead of extrusion blended. They showed no significant difference in behavior from those samples that had been extrusion blended when both sets of samples were kept for two minutes in the melt. To further inquire on the possible effect of the processing conditions on the crystallization behavior, a series of studies were performed on samples that had been under different conditions. The processed samples were extruded at different rates, with residence times varying from 50 to 20 sec and stabilizer concentration ranging from 0 to 0.5%. No significant changes in the crystallization behavior of these samples were found.

Conclusions

Blends of polybutylene terephthalate and polyethylene terephthalate are believed to be compatible in the amorphous phase as judged from (a) the existence of a single glass-transition temperature intermediate between those of the pure components and (b) the observation that the crystallization kinetics of the blend may be understood on the basis of this intermediate T_g. While trans esterification occurs in the melt, it is possible to make T_g and crystallization kinetics measurements under conditions where it is not significant. When the melted blend crystallizes, crystals of each of the components form, as judged from x-ray diffraction, IR absorption, and DSC. There is no evidence for cocrystallization. There is a slight mutual melting point depression.

The individual crystalline components exhibit characteristic IR bands. By calibrating the intensities of these against densities using the pure components, it is possible to use these for the measurement of degrees of crystallinity with respect to each of the components. The change of these with time for samples cooled from the melt to a particular crystallization temperature and for samples annealed from the quenched glass has been followed to characterize the crystallization kinetics. These are explainable in terms of classical nucleation and growth theory, providing one takes into account the degree of supercooling with respect to each of the components as well as a diffusion process dependent upon the single T_g of the blend. Crystallization also was followed by observation of density and depolarized light transmission, but their interpretations for the blend was only qualitative.

Morphology changes were observed by optical microscopy and small-angle light scattering. The pure components exhibit spherulitic structures, each with different orientation of the optic axis with respect to the spherulite radius. Spherulites become disordered and larger with the introduction of small amounts of the second component. Larger amounts of the second component result in a loss of spherulitic order.

Acknowledgment

This work was principally supported by a grant from the General Electric Co. The studies also were partly supported by grant #DMR75-05004 from the National Science Foundation, grants from the Army Research Office and from the Materials Research Laboratory of the University of Massachusetts. One of us (A.E.) appreciates the travel support from the International Commission for Cultural Exchange between the United States and Spain. We would like to express also our appreciation to E. Balizer who obtained the calorimetry measurements.

Literature Cited

1. Ong, C. J., Ph.D. Thesis, University of Massachusetts, Amherst (1974).
2. Ong, C. J., Price, F. P., *J. Polym. Sci., Polym. Symp.* (1977).
3. Khambatta, F. H., Ph.D. Thesis, University of Massachusetts, Amserst, MA (1976).
4. Khambatta, F. H., Russell, T., Warner, R., Stein, R. S., *J. Polym. Sci., Polym. Phys. Ed.* (1976) **14**, 1391.
5. Wenig, W., Karasz, F. E., MacKnight, W. J., *J. Appl. Phys.* (1975) **46,** 4194.
6. Warner, F. P., MacKnight, W. J., Stein, R. S., *J. Polym. Sci., Polym. Phys. Ed.* (1977) **15,** 2113.
7. Misra, A., Stein, R. S., *Bull. Am. Phys. Soc., Ser. II* (1975) **20**, 391.
8. Wasiak, A., Stein, R. S., *Polym. Prepr., Am. Chem. Soc., Div. Polym. Chem.* (1975) **16**(2), 643.
9. Keller, A., *J. Polym. Sci.* (1955) **17,** 291.
10. Van Antwerpen, F., Van Krevelen, D. W., *J. Polym. Sci., Polym. Phys. Ed.* (1972) **10**, 2423.
11. Misra, A., Ph.D. Thesis, University of Massachusetts, Amherst, MA (1976).
12. Mencik, Z., *J. Polym. Sci., Polym. Phys. Ed.* (1975) **13,** 2173.
13. Daubeny, R., Bunn, C. W., Brown, C. J., *Proc. R. Soc. London, Ser. A* (1954) **226**, 531.
14. Kilian, H. G., Halboth, H., Jenckel, E., *Kolloid-Z.* (1960) **172,** 166.
15. Stein, R. S., Warner, F. P., Escala, A., Balizer, E., Russell, T., Koberstein, J., *Am. Chem. Soc., Div. Org. Coat. Plast. Chem., Prepr.* (1977) **37**(1), 7.
16. Stein, R. S., Khambatta, F. H., Warner, F. P., Russell, T., Escala, A., Balizer, E., *J. Polym. Sci., Polym. Symp.*, in press.
17. Magill, J. H., *Polymer* (1962) **3,** 35.
18. Cobbs, W. H., Burton, R. L., *J. Polym. Sci.* (1953) **10,** 275.
19. D'Esposito, L., Koenig, J. L., *J. Polym. Sci., Polym. Phys. Ed.* (1976) **14,** 1731.

20. Bahl, S., Cornell, D., Boerio, F., *J. Polym. Sci., Polym. Lett. Ed.* (1974) **12**, 131.
21. Staumbaugh, B., Koenig, J. L., Lando, J. B., *J. Polym. Sci., Polym. Lett. Ed.* (1977) **15**, 299.
22. Bunn, C. W., Brown, C. J., *Proc. R. Soc. London, Ser. A* (1954) **226**, 531.
23. Kokoudu, M., Sakakibara, Y., Tadokoro, H., *Macromolecules* (1976) **9**, 226.
24. Alter, U., Bonart, R., *Colloid Polym. Sci.* (1976) **254**, 348.
25. Boye, C. A., Overton, J. R., *Bull. Am. Phys. Soc., Ser. II* (1974) **10**, 352.
26. Slagowski, E. L., Chang, E. P., *J. Appl. Phys.*, unpublished data.
27. Mandelkern, L., "Crystallization of Polymers," Chap. 8, McGraw-Hill, New York (1964), 233.
28. Hoffman, J. D., Weeks, J. J., *J. Chem. Phys.* (1962) **37**(8), 1723.
29. Boon, J., Azcue, J. M., *J. Polym. Sci., Part A* (1968) **A2**(6), 885.
30. Flory, P. J., *J. Am. Chem. Soc.* (1940) **62**(9), 1057.
31. Ibid., 2255.
32. Flory, P. J., *J. Am. Chem. Soc.* (1952) **64**, 2205.
33. Peebles, L. H., Huffman, M. W., Ablett, C. T., *J. Polym. Sci., Part A-1* (1969) **7**, 479.

RECEIVED June 6, 1978.

25

Characterization of Injection-Molded Impact Polypropylene Blends

H. K. ASAR, M. B. RHODES[1], and R. SALOVEY

Department of Chemical Engineering, University of Southern California, Los Angeles, CA 90007

The impact properties of an injection-molded blend of polypropylene (PP) and ethylene–propylene–diene terpolymer (EPDM) are sensitive to blend composition, processing variables, and testing conditions. Instrumented impact testing coupled with morphological characterization of the fractured surfaces elucidate the mechanism of fracture upon impact. Observations show that: (1) local variations in EPDM concentration and domain sizes can result in a twofold difference in the total energy absorbed during impact; (2) deformation and orientation characteristics of the EPDM phase can result in a reduced compressive modulus on impact as well as a yielding during the initiation phase; and (3) impact velocity variation over the range 3.43–1.30 m/sec results in a transition from brittle to ductile failure.

The toughening of brittle glassy polystyrene by the inclusion of polybutadiene elastomer has led to the development of high-impact polystyrene and has opened a new field of technology (*1*). Typically, a bulk polymerization is conducted with the elastomeric component dissolved in the monomer of the glassy polymer (*2*). Resultant polymer composites are morphologically complex and an understanding of the mechanism of rubber toughening requires a detailed characterization of the structure. By relating the structure or morphology to the physical

[1] On leave from the University of Massachusetts.

0-8412-0457-8/79/33-176-489$07.25/0

properties it is then possible to ascertain the important factors. For high-impact strength, it is apparently necessary to have an elastomer dispersed in a continuous rigid matrix. The rubbery dispersed phase must be sufficiently incompatible with the matrix to maintain phase separation, yet adequately compatible to provide for strong adhesion between the phases and a fine dispersion of the rubber into micron-sized particles (*3*).

Toughened polypropylene may be prepared by block copolymerization in which ethylene monomer is added during the final stages of the polymerization of propylene (*4*). Thus, some polypropylene chains would contain an end block of rubbery ethylene–propylene copolymer. Alternatively, a blend of an elastomeric copolymer of ethylene and propylene (EPR or EPDM) with isotactic polypropylene (PP) can produce an impact-resistant polymer (*5*).

The rheological properties of blends of linear polyethylene and PP are sensitive to composition and temperature (*6*). Tensile bars of injection-molded PP evidence a skin-core morphology characterized by an outer quenched nonspherulitic skin, an oriented-row-nucleated spherulitic shear zone, and a randomly nucleated spherulitic core (*7, 8, 9*). It was observed that structural variations depended on processing conditions. Similarly, the morphology of injection-molded ethylene–propylene copolymers reflects the processing conditions (*10*). Moreover, samples differing in morphology exhibit variations in physical properties, specifically impact strength (*10*). It is expected that the morphology and properties of blends of EPDM and PP depend on the manner of preparation and are sensitive to processing conditions.

Preliminary to a study of the relations among the morphology, processing variables, and physical properties of polymer blends, it is essential to develop sensitive techniques for correlating blend behavior. In this chapter, we illustrate the value of combining instrumented impact measurements with a morphological examination of the resultant fracture surfaces in elucidating the toughening mechanism in injection-molded blends of PP/EPDM.

Experimental Details

Materials. The polymers used are commercial isotactic polypropylene (PP, Hercules, Profax 6523) and ethylene–propylene–diene terpolymer (EPDM, duPont, Nordel Hydrocarbon Rubber). Physical properties of the two polymers are listed below.

Material	*Density (g/cm^3)*	*Melt Flow Index (g/10 min)*	*Mooney Visc (ML-4 at 250°F)*	*Wt Avg Mol Wt*
PP	.902	4	—	325,000
EPDM	.850	—	70	261,000

Sample Preparation. A master batch blend containing 30% EPDM was prepared using a 2.5-in. twin screw extruder and adding 0.2% by weight of Ethyl 330 antioxidant. This master batch was mixed with the polypropylene to prepare other compositions of interest. The test samples were molded using an Arburg 200μ injection molding machine with barrel heaters set at 200°C, the melt temperature of the polymer mixture. Variation in the melt temperature produced no significant effect on the impact properties of the blends. The back pressure and injection pressure were held at 1400 and 1200 psi, respectively, unless otherwise indicated. The cycle time was maintained at 35 seconds and the injection speed was set at 3. The mold temperature was varied between 20° and 87°C.

Test and Characterization Methods. IMPACT TESTING. The injection-molded bars were notched according to ASTM D-256, using a milling machine. The notch had an included angle of 45°, a tip radius of 0.01 in., and a notch depth of 0.1 in. Impact test data were collected by two techniques using a Tinius Olsen Plastic Impact Tester and the Izod configuration. In one series of tests, an operator read the data directly off the instrument scale while in the second series, the entire test was instrumented through the use of the DYNATUP system (Effects Technology).

Noninstrumented Izod Testing. The impact strength, reported in joules per meter of notch, was measured at temperatures between 25° and −30°C. The samples were cooled to the test temperature using a Lauda Kryostat and equilibrated in the Kryostat for at least one and one half hours prior to testing. Five to seven seconds were required to remove a sample from the Kryostat, mount, and perform the Izod impact test. Ten samples of each composition were tested at each temperature.

Instrumented Izod Testing. Samples of injection-molded polymer blends were tested on the DYNATUP Model ETI-300 system. (All instrumented Izod testing was performed at the Effects Technology Laboratories in Santa Barbara, California, with the cooperation of Donald R. Ireland and Lee Wogulis.) A diagram of the instrumented Izod equipment is shown in Figure 1. Attached to the Izod pendulum is an instrumented tup in which a semiconductor strain gauge monitors the compression loading during the time of contact with the specimen. The tup velocity is recorded through the use of fiber optics with a controlled light beam and photo sensor. The microprocessor is the integrating unit of the assembly, processing the signals from the tup and the fiber optics velocometer unit, providing both a graphical and tabulated form of impact data. The unit has great flexibility because of both program and data storage as well as operator control through the keyboard.

Figure 2 illustrates an idealized instrumented impact curve associated with these impact studies. Load is plotted against time and the area under the curve represents energy. The fracture process is divided into two stages in which the first is identified as initiation and the second as crack propagation. The initiation part of the process occasionally demonstrates yielding as evidenced by a change in the initial slope. The maximum load corresponds to the fracture load and marks the beginning of the propagation step. This part of the response may be very short, in which case the load rapidly decays, or it may return more gradually to the initial load level. The propagation time indicates the manner in which energy propagates through the sample.

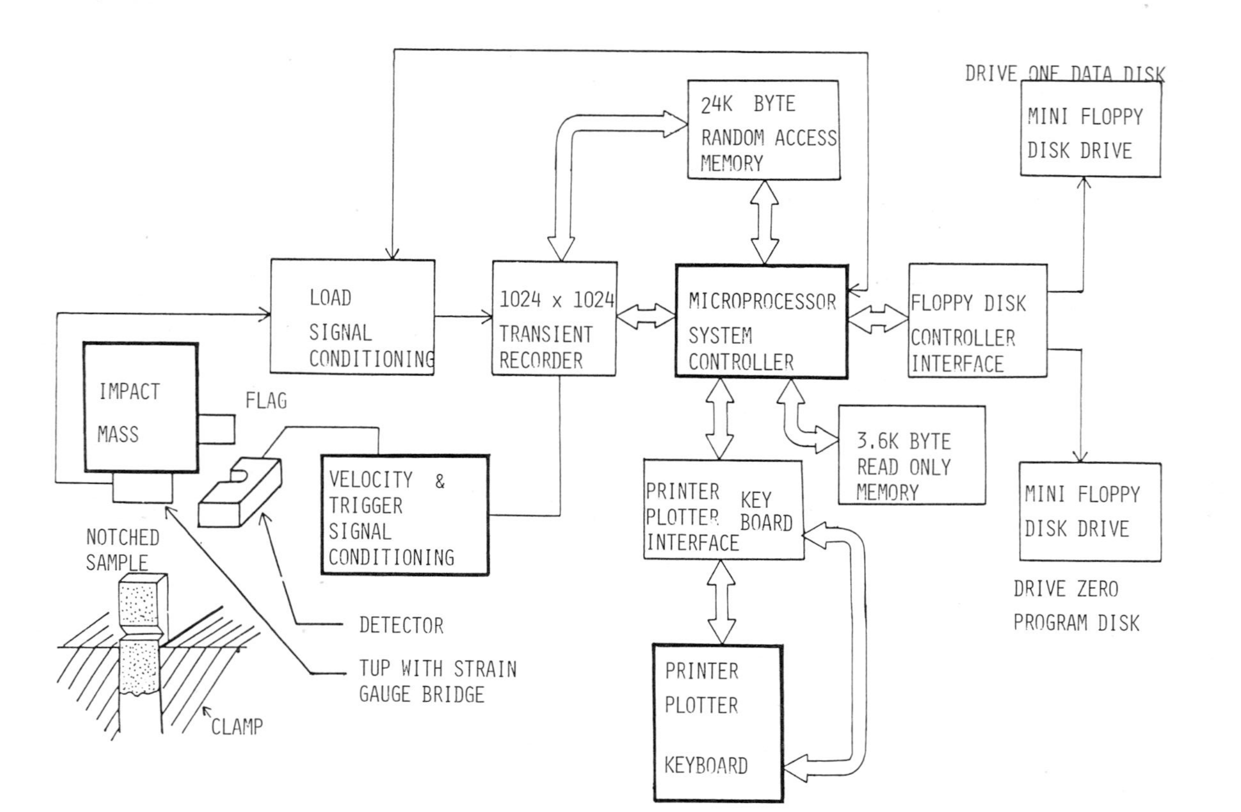

Figure 1. Block diagram—Dynatup model ETI-300 instrumented impact system

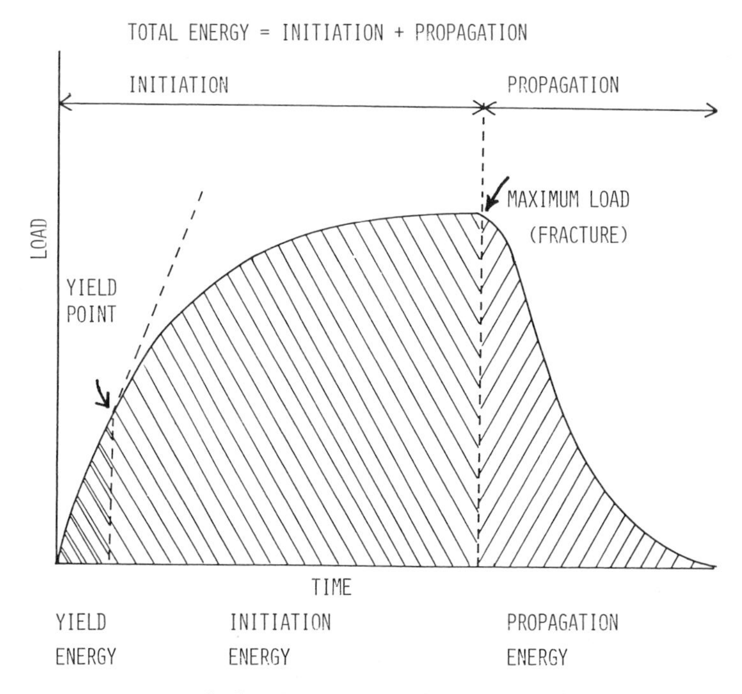

Figure 2. Idealized instrumented impact graphical output

Examples of the on-line computer output are illustrated in Figures 3 and 4. Figure 3a shows a typical load energy curve with its accompanying numerical data in Figure 4a. This initial output is then supplemented, after a curve fitting analysis, by the plot illustrated in Figure 3b and by the data tabulated in Figure 4b. Data shown in Figure 4a records the initial energy of the impact, the total time of sample-tup contact during the impact process, as well as the time associated with the initiation phase of the process. Also with this data output comes the assignment of the energy absorption for the initiation and propagation phases. The curve fitting from Figure 3b is used to give the data shown in Figure 4b. This includes a slope value (an initial modulus), a yield load value (if applicable), and a final analysis of the energy absorption into yielding, initiation, and propagation. The time parameter also is identified with these processes. The deflection data, useful in many engineering impact tests, has an ambiguous interpretation in these polymer blend investigations. The computed quantities of most significance in this research are the values associated with yielding during the initiation step of the fracture and the values that permit a comparison between initiation and propagation energies.

MICROSCOPY. Three levels of magnification were required to characterize adequately the morphology of the molded polymer samples and of fracture surfaces after the impact test. A low power binocular microscope covering the magnification range from 2 to 12 times permitted initial evaluation of the fracture surface after which specific areas and

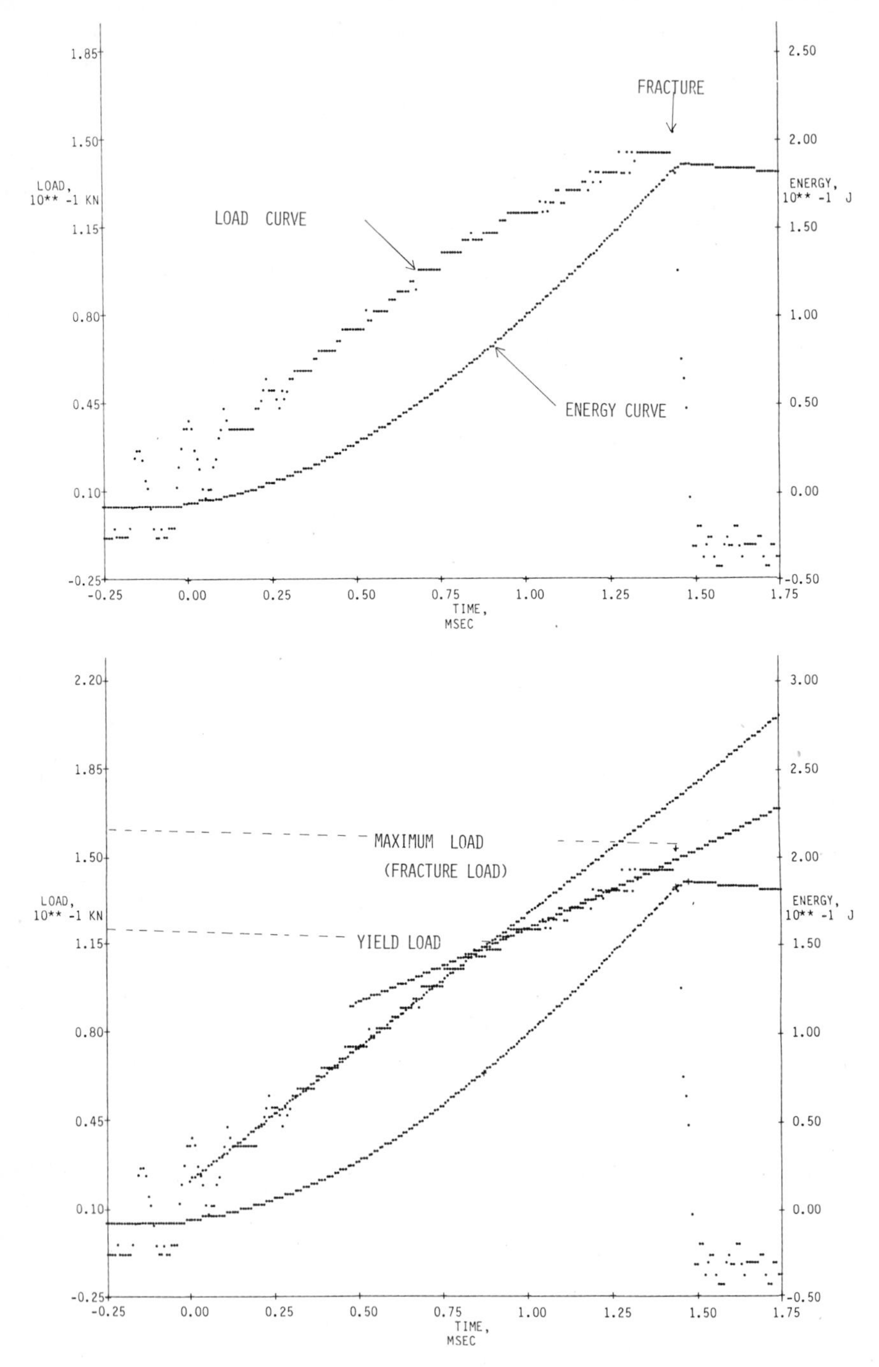

Figure 3. Typical instrumented impact graphical output. (a, top*) Load/energy data; (b,* bottom*) computer analysis.*

TEST	TEMP	IMPACT VELOCITY	ENERGY	TIME,10** 0 MSEC		LOAD,10**-4KN	ENERGY,10** -2 J		
	C	M/S	J	INIA	TOTAL	MAX	INIA	PROP	TOTAL
G-1	20.0	1.43	3.9	1.43	1.47	1523	18.1	0.3	18.5

TEST	TEMP	IMPACT VELOCITY	ENERGY	SLOPE	TIME,10** 0 MSEC			LOAD,10**-4KN			DEFLECTION,10** -2MM			ENERGY,10** -2 J			
	C	M/S	J	10** -1N/MM	YIELD	INIA	TOTAL	YIELD	MAX	FRAC	YIELD	INIA	TOTAL	YIELD	INIA	PROP	TOTAL
G-1	20.0	1.43	3.90	747.57	0.87	1.43	1.47	1130	1523	1523	123	203	208	7.7	18.1	0.3	18.5

Figure 4. Typical instrumented impact numerical output. (a, top*) Numerical data; (b,* bottom*) computer analysis.*

samples were selected for a detailed examination with the scanning electron microscope (SEM). Fracture surfaces were metal coated and photographed in a Cambridge Stereoscan S4-10 at magnifications ranging from 200 to 20,000 times. In addition to the fracture surface study, the internal morphology was investigated using a Zeiss Photomicroscope with polarization optics. In this latter approach, samples were microtomed parallel and perpendicular to the mold flow direction using an American Optical rotary microtome. Section thickness varied between 5 to 15 microns and the magnification from 60 to 300 times.

DIFFERENTIAL SCANNING CALORIMETRY. A Perkin–Elmer Model DSC-1B calorimeter was used to examine crystallinity by measuring areas under the fusion curve as a function of elastomer composition and processing variables. Areas of endotherms were calibrated against an indium standard and the crystallinity calculated using a value of −138 J/g for a 100% crystalline polypropylene polymer (*11*).

X-RAY ANALYSIS. Wide angle x-ray patterns were photographically recorded with 30-minute exposures from selected areas of the injection-molded samples.

Results and Discussion

Impact Testing (Noninstrumented). Because the Izod impact is a failure test, some scatter of data is expected. Reasons for this scatter include the presence of weak domains, mechanical anisotropy, and any variability in test conditions. Extreme care was taken to maintain constant test conditions. To increase the reliability of the results, 10 test pieces of each sample were tested and the average was calculated.

Typical data are presented in Table I for samples of varying composition from 0 to 30% EPDM. All data in the table refer to samples injection molded at 1200 psi with one of three mold temperatures, 20°, 50°, or 87°C and under test temperatures that ranged from −30° up to 25°C.

Figure 5 is a plot of the 20°C mold temperature data showing impact strengths measured in joules per meter of notch as a function of test temperature for different blend compositions. The sensitivity of the impact strength to test temperature increases with EPDM content, and the temperature range in which a rapid change in impact strength takes place appears to shift to lower temperatures as the EPDM content increases from 20 to 30%. Figure 6 shows impact strength plotted against percent EPDM for two test temperatures. As expected, high EPDM content results in a tougher material, and lower test temperatures result in reduced impact strength; also, the difference in impact strength at room temperature and at −30°C decreases as the EPDM content decreases.

Figure 7 shows the impact strength plotted against mold temperature for various testing temperatures. Impact strength goes through a minimum, one which appears to lessen as the testing temperature decreases. The effect of secondary crystallization can be observed by comparing

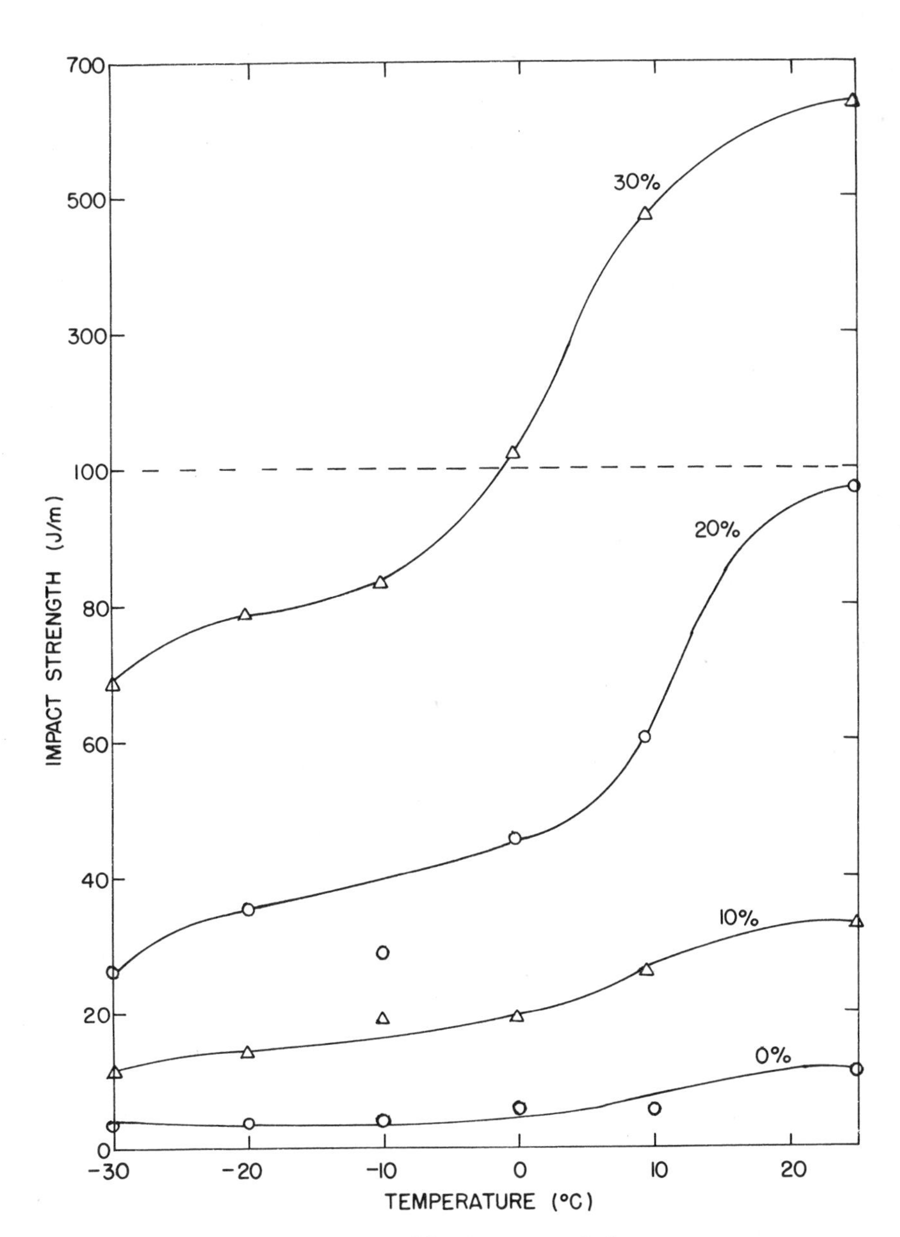

Figure 5. Impact strength of PP blends of specified EPDM contents, injection molded at a mold temperature of 20°C as á function of impact test temperature

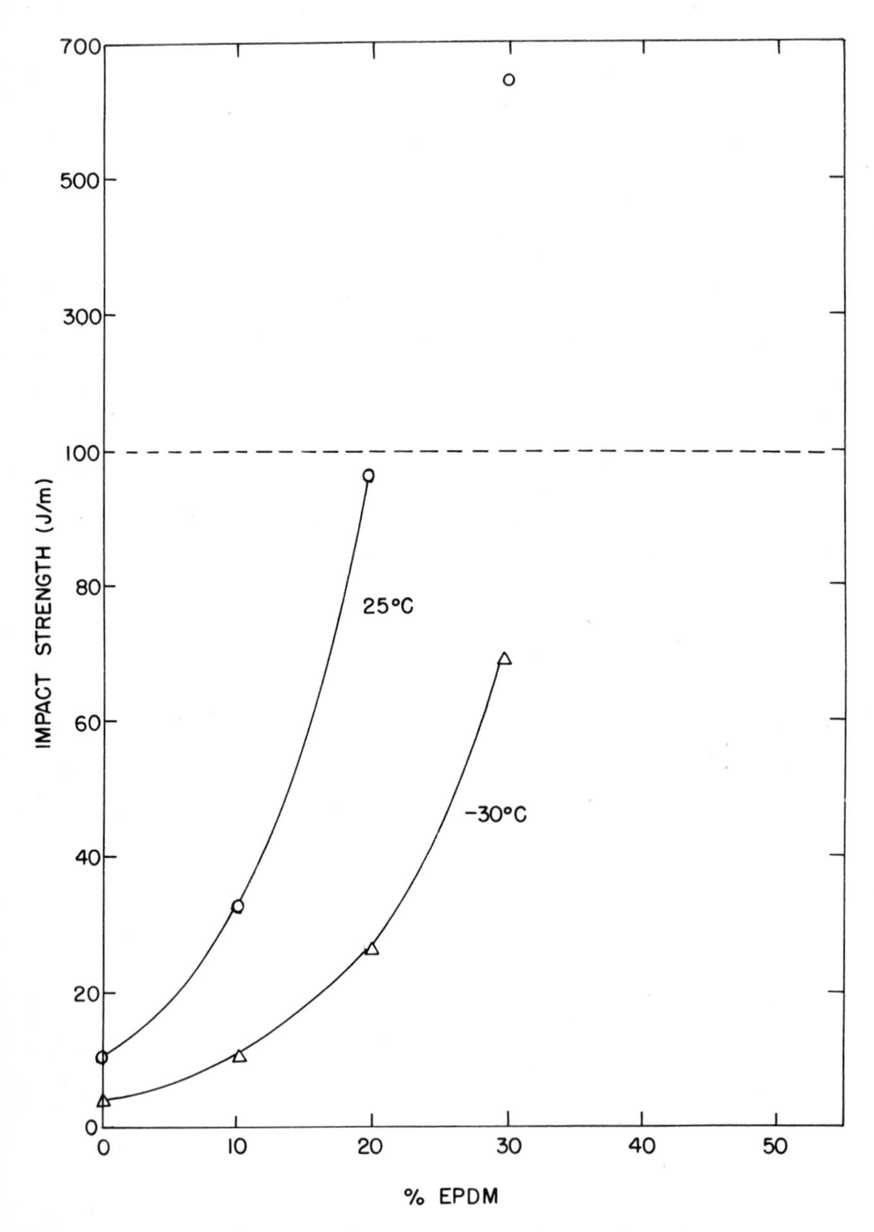

Figure 6. Dependence of the impact strength at 25° and −30°C of PP blends, on EPDM content

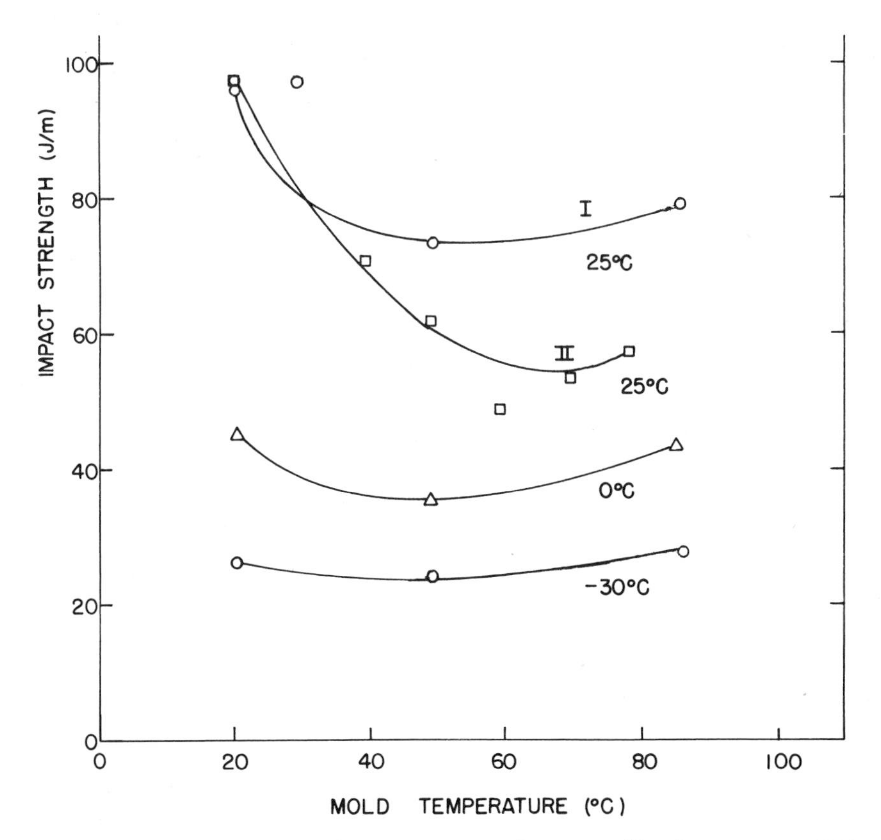

Figure 7. Dependence of the impact strength of a PP blend containing 20% EPDM on mold temperature at test temperatures of −30°, 0°, and 25°C. Curve I refers to testing two days after molding, while Curve II corresponds to seven days.

Curves I and II at 25°C. Samples corresponding to Curve I were tested two days after molding while those of Curve II were tested seven days after molding.

Morphology. It is well documented that injection-molded polypropylene is characterized by at least three principal zones identified as the skin, shear, and core regions (*7, 8, 9, 10, 12*). The relative dimensions of these regions are affected by processing conditions, and frequently the demarcation between these zones, especially between the shear region and the core, is not very obvious. It has been observed that the skin region is oriented but nonspherulitic; the shear region is spherulitic, frequently characterized by row nucleation; and the core region is spherulitic, having spherulites of larger size than those found within the shear region. Figure 8a illustrates a typical skin and shear zone morphology while Figure 8b shows the spherulitic core region.

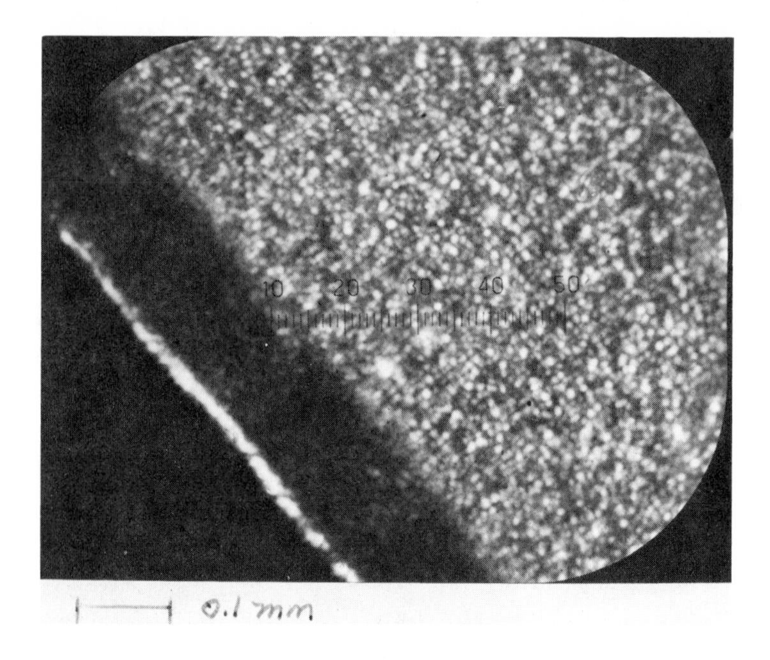

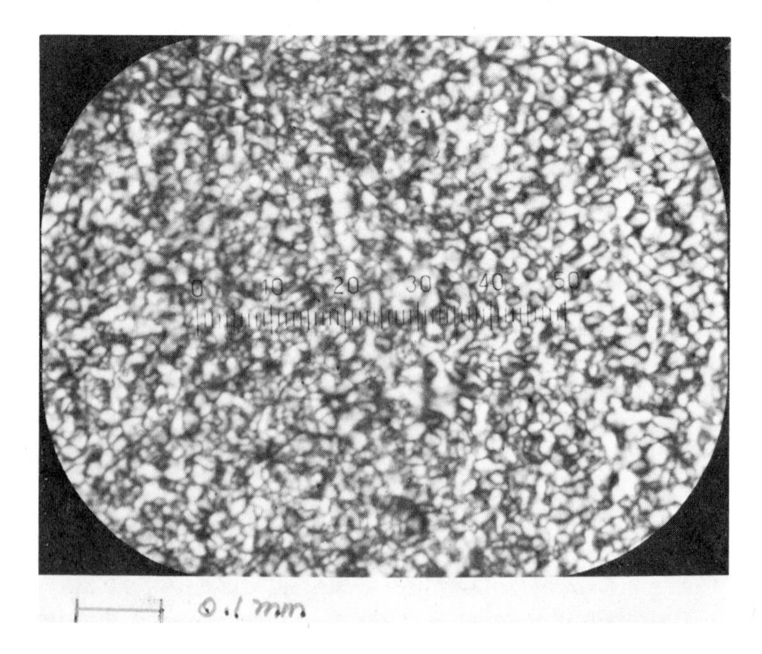

Figure 8. Skin/core morphology of injection-molded PP—polarizing optics. (a, top*) Skin and shear zone; (b,* bottom*) spherulitic core.*

Variations in the EPDM content as well as in the processing conditions result in characteristic textural changes both within and between these three regions. Many of the following observations also have been reported by the previously mentioned investigators:

(1) Spherulite size shows a general increase in going from the shear region towards the core.

(2) Samples with smaller spherulites are typical of the lower mold temperatures and show high impact strengths.

(3) Increasing EPDM content results in irregular spherulitic texture, smaller spherulite size, and loss of sharpness in the spherulite boundaries.

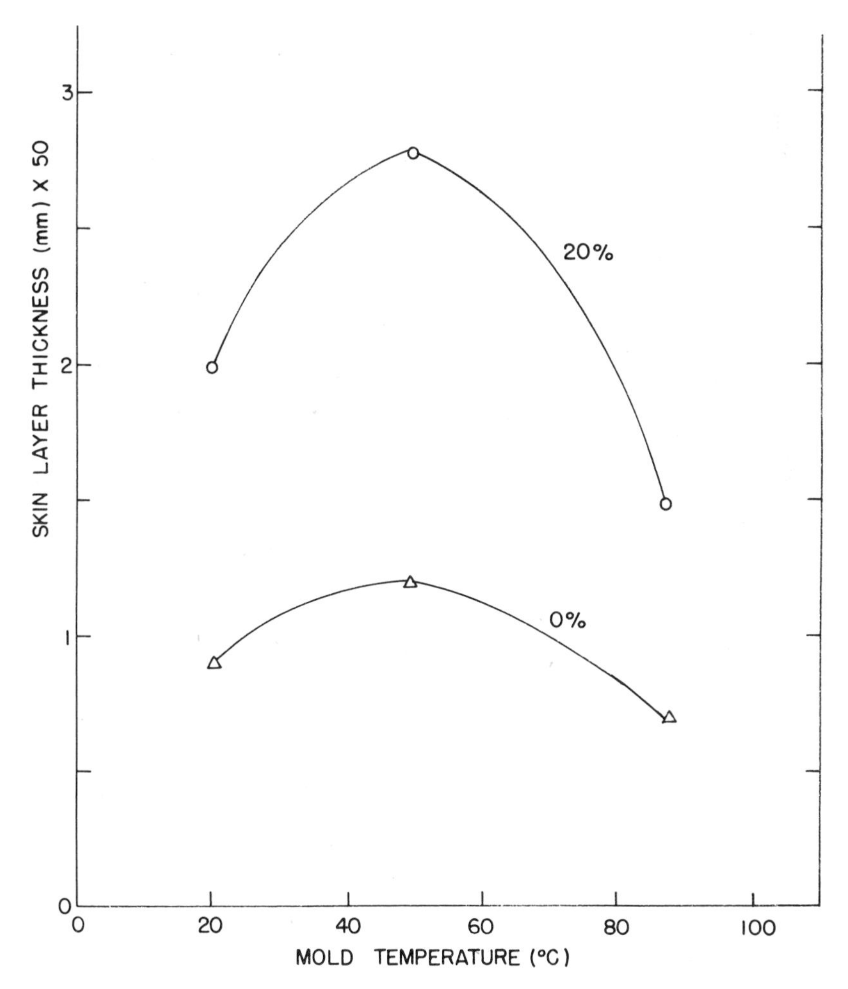

Figure 9. Skin layer thickness of injection-molded PP and PP blend with 20% EPDM at various mold temperatures

(4) The skin region is oriented along the direction of flow, as are the EPDM domains contained within the skin region.

(5) The thickness of the skin layer varied as a function of mold temperature, exhibiting a maximum at the 50°C mold temperature. This is shown in the data plotted in Figure 9 for pure PP and for a PP blend containing 20% EPDM.

Many investigators have studied the relationship between injection molding conditions and the mechanical properties of the polymer (*13, 14, 15*). Factors such as polymer molecular weight, degree of supercooling, and melt shear conditions significantly contribute to the observed results (*16*). Current observations are consistent with these reports.

Crystalline orientation and the degree of crystallinity were not only influenced by the molding conditions but by the EPDM composition as well. Figure 10 shows the calorimetric crystallinity for varying EPDM

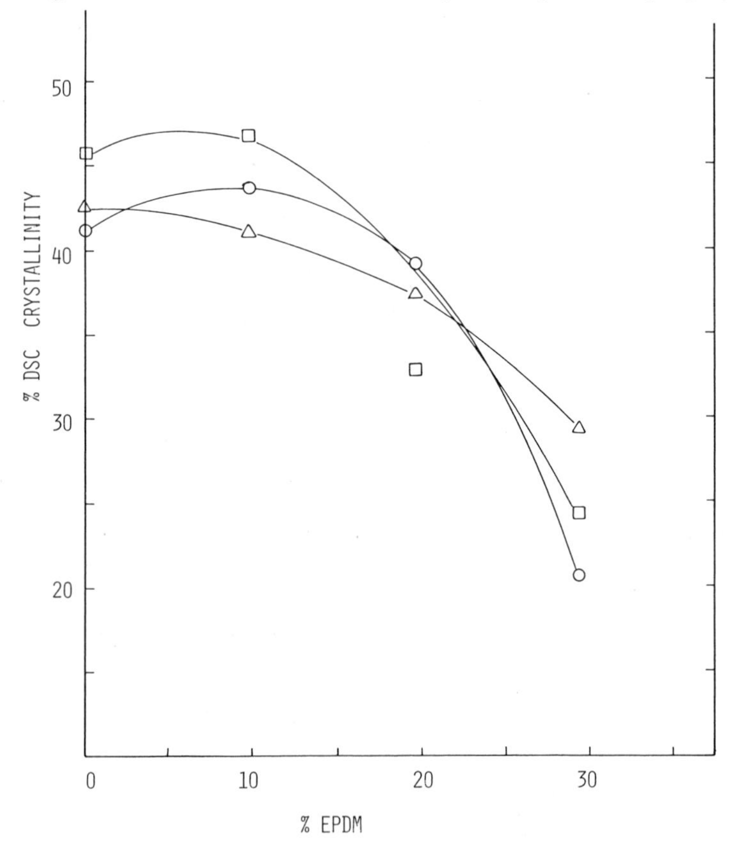

Figure 10. Calorimetric crystallinity of PP blends with various EPDM contents at specified molding temperatures and normalized to constant weight of PP. Mold temperature: (□) 87°C; (△) 50°C; (○) 25°C.

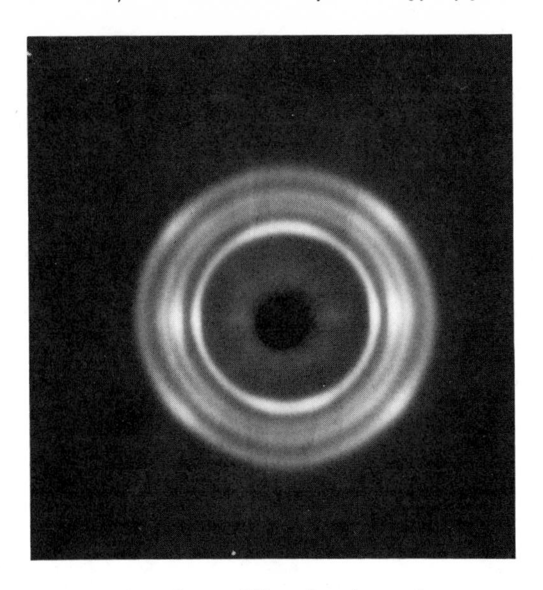

Figure 11. X-ray diffraction from the core region of injection-molded PP containing 30% EPDM showing (a + c) *axis orientation*

contents and mold temperatures. The initial increase may be kinetic in origin as a reult of increased nucleation sites (*8*). The x-ray diffraction patterns almost always showed increased orientation within the skin and shear zones as compared with the core region. Samples routinely demonstrated the $(a + c)$ axis orientation in both the core and shear regions (*9*). Figure 11 illustrates that even when the EPDM composition reached 30%, samples showed this orientation although the orientation distribution was broad.

Instrumented Izod Impact Tests. Interpreting the nature of the strengthening process operating in these blends was difficult because of the scatter in data obtained in the Izod tests. However, the use of instrumented impact testing offered a more quantitative analysis of toughening by monitoring the sample-tup interaction as a function of time. Such an approach provided a quantitative description of the fracture process and related fracture to morphology and processing conditions. For these studies, samples were either pure polypropylene or 20% EPDM.

Impact tests made with the DYNATUP system indicate that the impact process is not restricted to a simple mechanism. Although means are available to isolate specific characteristics of the fracture process, this information is of little value unless it is specifically correlated with sample morphology. When the detailed test data are related to morphology, an understanding of the complex mechanisms operating during fracture is possible.

Irrespective of the specific characteristics of the fracture process, whether related to the load or energy features of the fracture, samples showing the most consistent behavior had been prepared at the two extremes in the processing variables, either at low mold temperatures and high injection pressures or at high mold temperatures and low injection pressures. It is suggested that these two processing variables independently influence the impact behavior. Any trends observed operated towards either of these extremes, although sufficient data does not exist at this time to establish a more detailed correlation between injection-molding conditions and polymer-impact strength. Fracture surface morphology is correlated with impact test conditions and load-energy characteristics. The following examples will illustrate the effect of impact velocity on the sample response and the relationship of yielding and energy absorption on the fracture morphology.

Impact Velocity. A series of tests were run on a sample containing 20% EPDM molded at 87°C with an injection pressure of 600 psi in which the tup velocity at impact was varied from 0.59 to 3.43 m/sec. An impact velocity of 0.59 m/sec was insufficient to fracture the notched bars. A velocity of 1.10 m/sec resulted in a very ductile fracture. Figure 12 shows a low magnification of the fracture surface where irregular

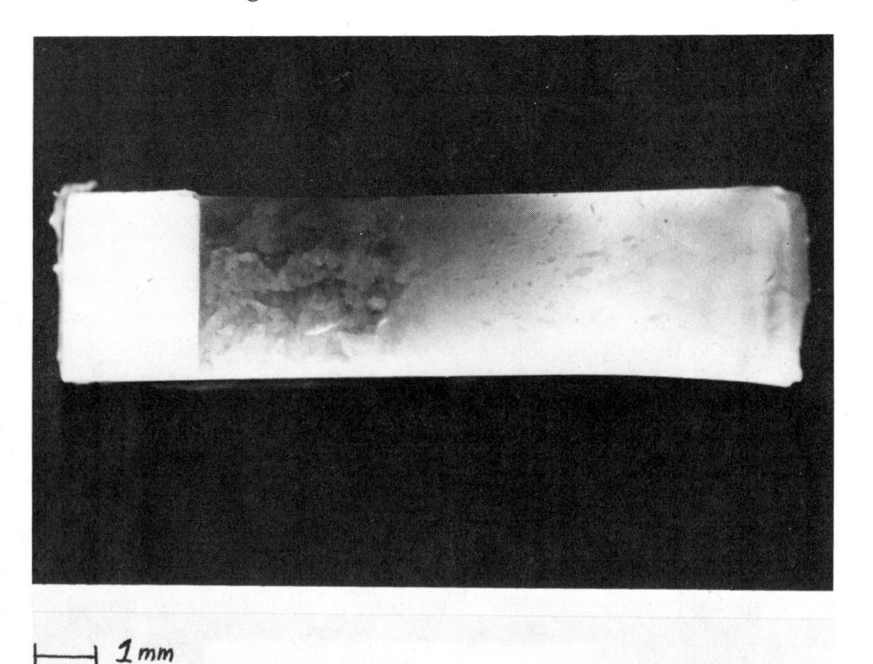

Figure 12. Fracture surface of injection-molded PP containing 20% EPDM, impact velocity of 1.10 m/sec. The notch is to the left.

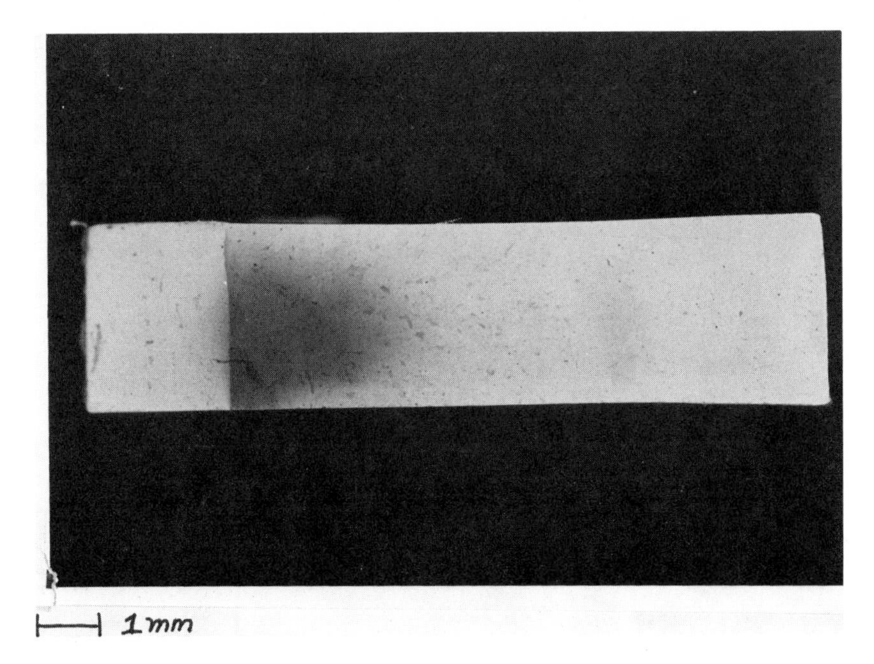

Figure 13. Fracture surface of injection-molded PP containing 20% EPDM, impact velocity of 1.42 m/sec. The notch is to the left.

disruption of the surface is apparent and a major portion of the surface is seen to be stress whitened. The stress-whitened area is dark because of crazing and the formation of microvoids. However, when the tup velocity is raised to 1.42 m/sec, the fracture surface is more characteristic of a brittle process (Figure 13). The stress-whitened area seen at the notch tip is characteristic of the blends and is not observed in fractured surfaces of pure polypropylene bars.

Three load energy curves are presented in Figures 14, 15, and 16 for impact velocities of 1.10, 1.42, and 3.43 m/sec. The total energy of absorption during the fracture process for these three different impact velocities is 77.3×10^{-2}, 58.1×10^{-2}, and 20.2×10^{-2} J, respectively, a trend which is consistent with the appearance of the fracture surfaces. In the highest velocity impact, only 6% of the 20.2×10^{-2} J (the total absorbed energy) is associated with the propagation step. At high velocity, the brittle fracture without yielding is an effect analogous to that resulting from a lowered test temperature. One observes no yielding during initiation, and the propagation energy and impact strength are both very low. At 1.42 m/sec, the fracture is basically brittle, there is some yielding during crack initiation, and the propagation energy has increased to 24% of the total 58.1×10^{-2} J. At 1.10 m/sec, almost 60% of the total energy was required for the propagation process. The ratio of

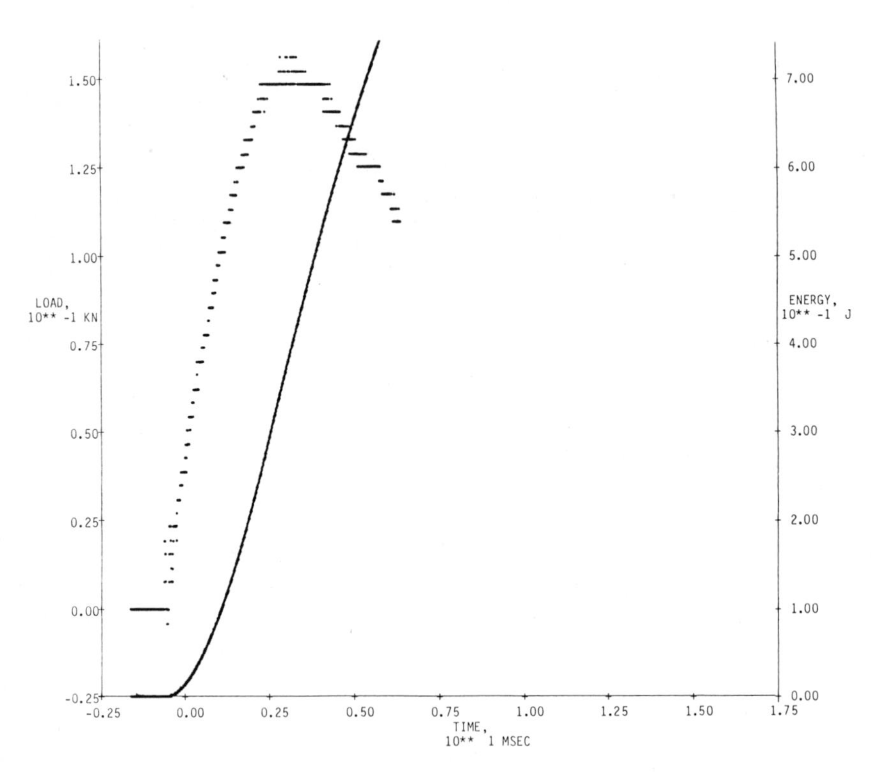

Figure 14. Instrumented impact load and energy as a function of time for an impact velocity of 1.10 m/sec.

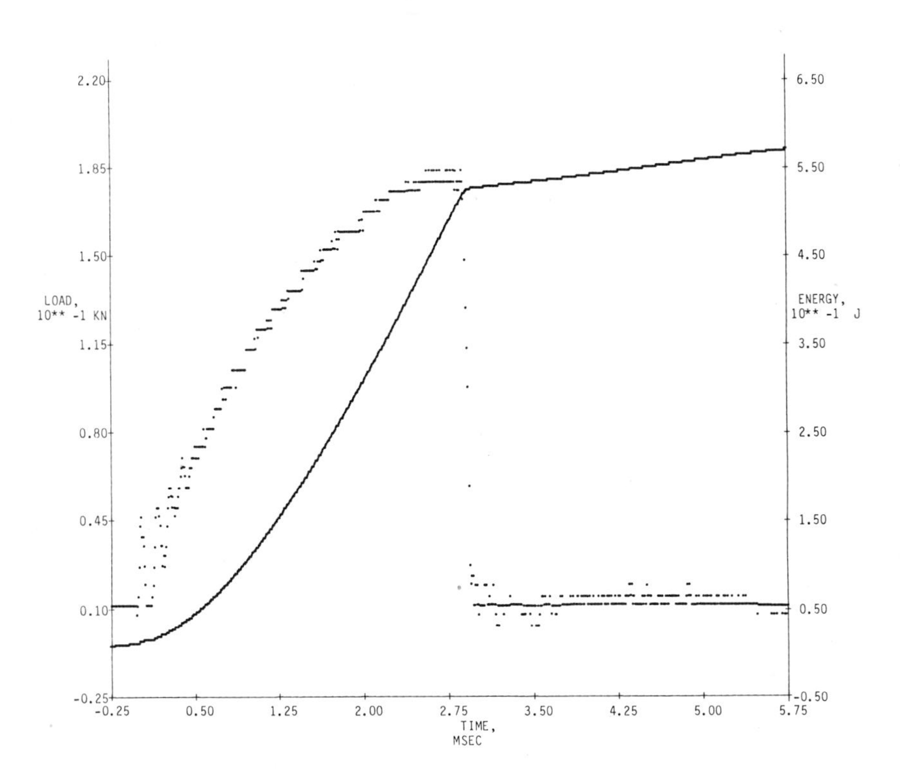

Figure 15. Instrumented impact load and energy as a function of time for an impact velocity of 1.42 m/sec

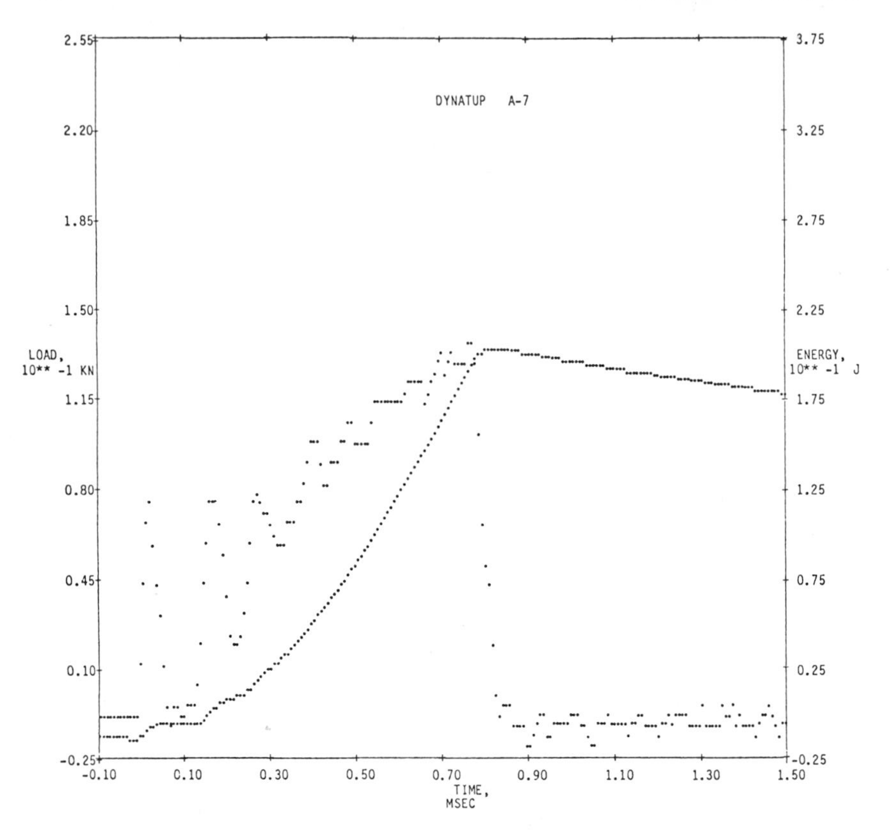

Figure 16. Instrumented impact load and energy as a function of time for an impact velocity of 3.43 m/sec

propagation to initiation energy is identified as a "ductility index" and is a function of the impact velocity for the polypropylene blends. For these three velocities, 1.10, 1.42, and 3.43 m/sec, the ductility index values are 1.34, 0.32, and 0.06, respectively. A series such as this emphasizes the need to document carefully the impact test conditions if valid interpretations are to be extracted from the test data.

YIELDING AND ENERGY ABSORPTION. The sensitivity of the instrumented impact measurements enables us to detect a variability in the impact response for samples of identical composition and processing history. This is illustrated with PP blends containing 20% EPDM and processed at a mold temperature of 87°C and an injection pressure of 1200 psi. Using duplicate samples, the total impact energy absorbed is similar (17.3 and 18.9 $\times$ 10^{-2} J), yet one sample gave a reduced modulus, as calculated from the initial slope of the load/time curve and also demonstrated a yielding during crack initiation (Figures 17 and 18).

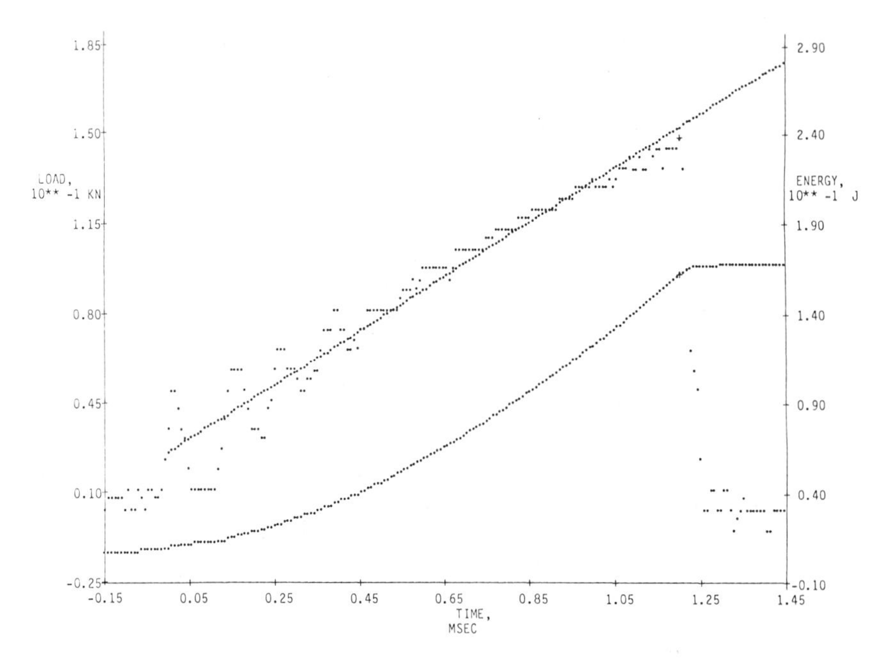

Figure 17. Instrumented impact for a PP blend containing 20% EPDM

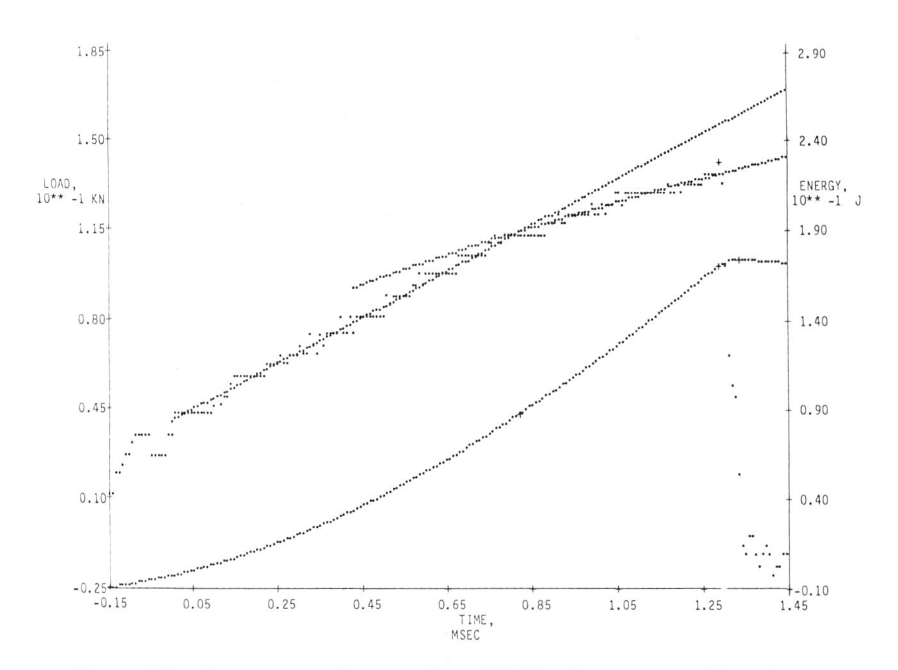

Figure 18. Instrumented impact for a PP blend containing 20% EPDM

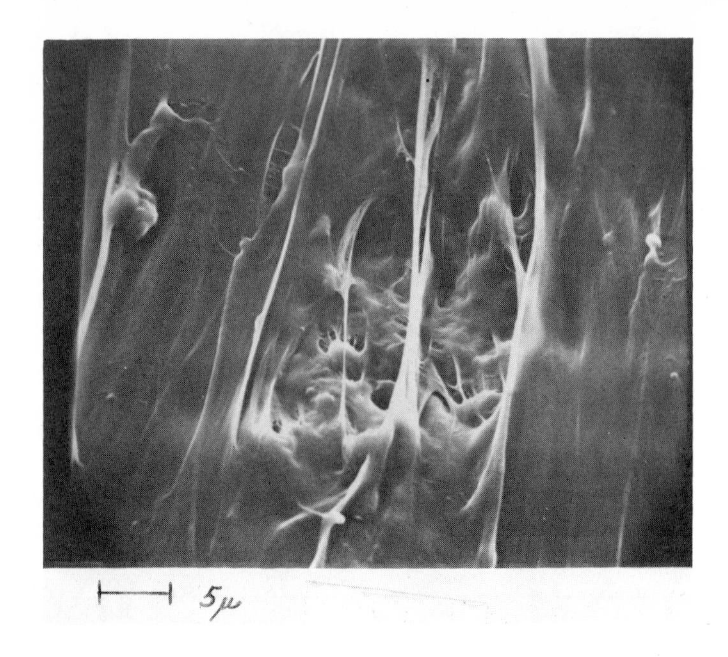

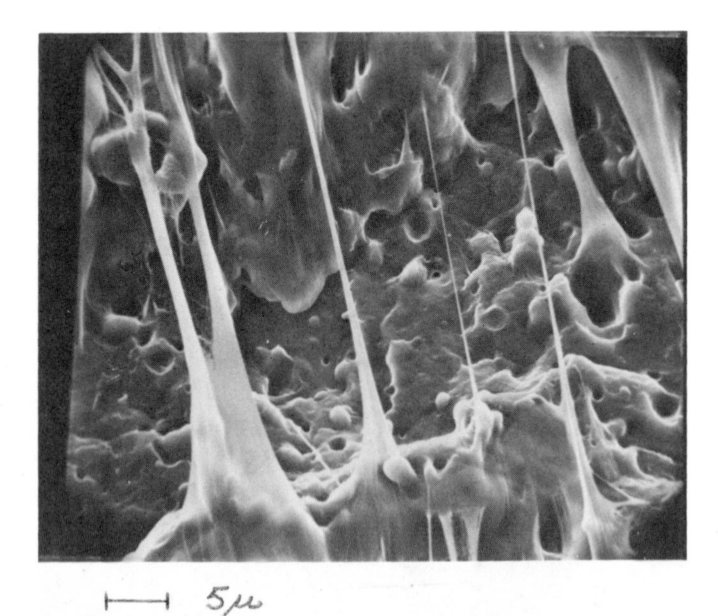

Figure 19. Scanning electron micrographs from the fracture surface of a PP blend with 20% EPDM. (a, top*) Corresponding to Figure 17; (b,* bottom*) corresponding to Figure 18.*

Investigation using SEM of the fracture surface of the region just adjacent to the stress-whitened area shows the corresponding morphologies in Figure 19. The sample that does not yield (Figure 19a) shows a much smoother fracture surface, whereas the sample that yielded (Figure 19b) shows extensive drawing and a very high concentration of drawn material. The extreme drawing evidenced in Figure 19b is associated with the yielding process during the initiation phase of the process.

Another pair of identical samples with diverse impact behavior produce the instrumented impact curves shown in Figures 20 and 21. These refer to PP blends with 20% EPDM molded at 50°C and 600 psi injection pressure. For this case, the total energy absorbed by the two samples is different (15.9 vs. 27.7 × 10^{2-} J), as is the response to yielding. For the sample corresponding to Figure 20 yielding is evident during initiation, but there is very little propagation energy. In Figure 21 there is no apparent yielding. However, both the initiation and propagation energies are appreciably larger, and the total impact energy is almost doubled. The morphological relationship for this behavior is shown in scanning

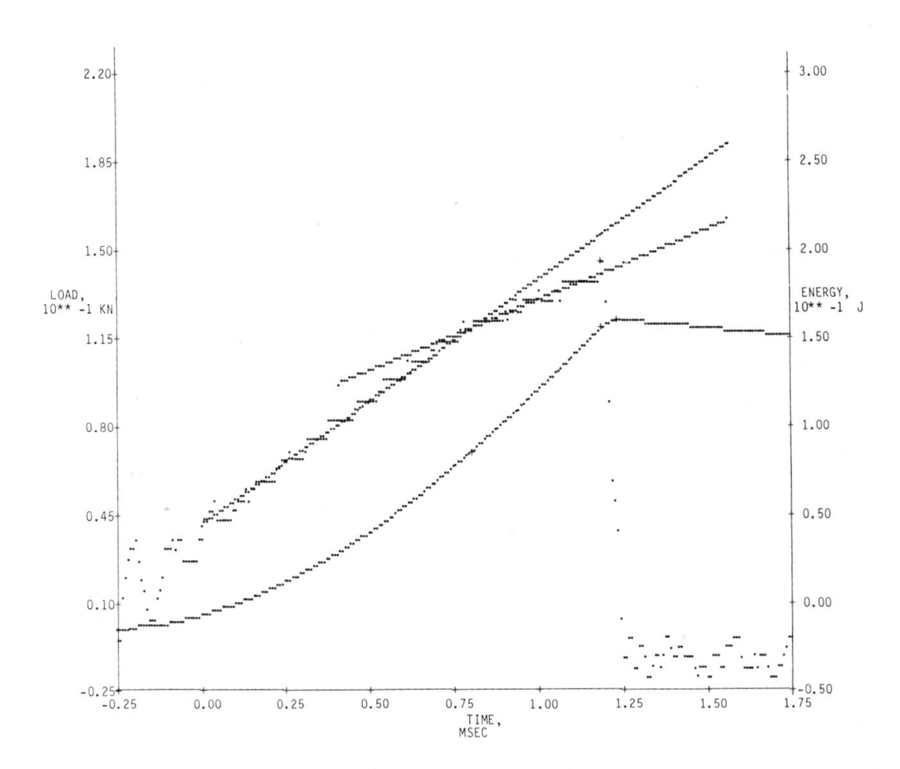

Figure 20. Instrumented impact for a PP blend with 20% EPDM

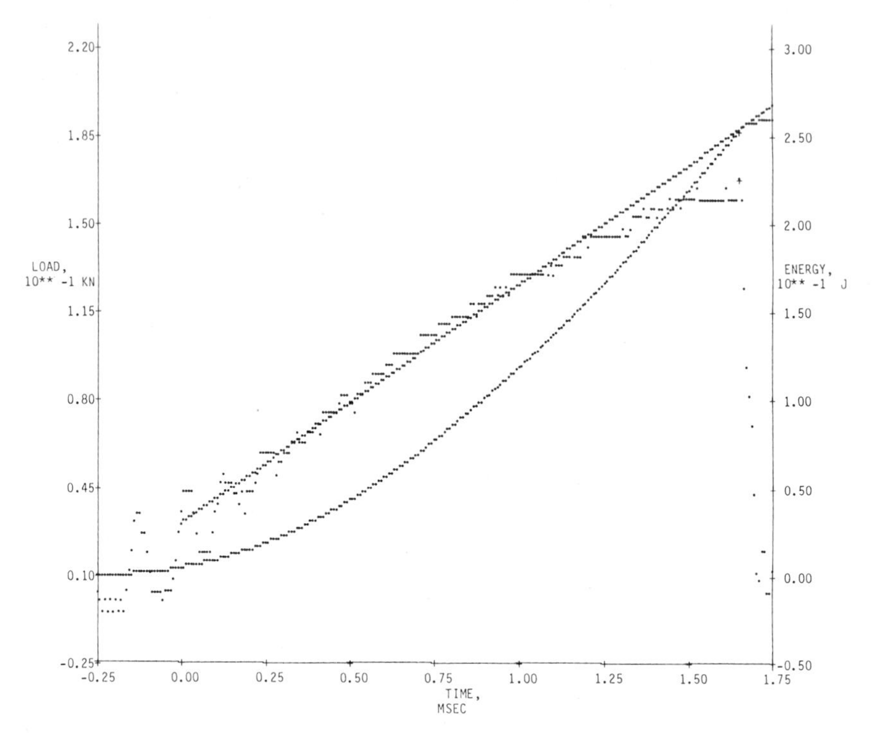

Figure 21. Instrumented impact for a PP blend with 20% EPDM

electron micrographs of the fracture surfaces in the core region adjacent to the stress-whitened area (Figure 22). The size distribution and number of EPDM domains is considerably different in the two samples. The sample that showed greater energy absorption (Figure 22b) is characterized by a larger number of domains and a more extensive distribution of domain sizes. Such characteristics of the EPDM domains favors increased total energy absorption with a finite percent of this energy associated with the propagation phase.

Conclusions

The impact properties of an injection-molded blend of PP and EPDM are sensitive to blend composition, processing variables, and test conditions. At all temperatures tested, the impact strength of blends of PP with up to 30% EPDM increase with EPDM content and decrease with test temperature. The temperature dependence of the impact strength increased with the fraction of EPDM. The impact strength appears to

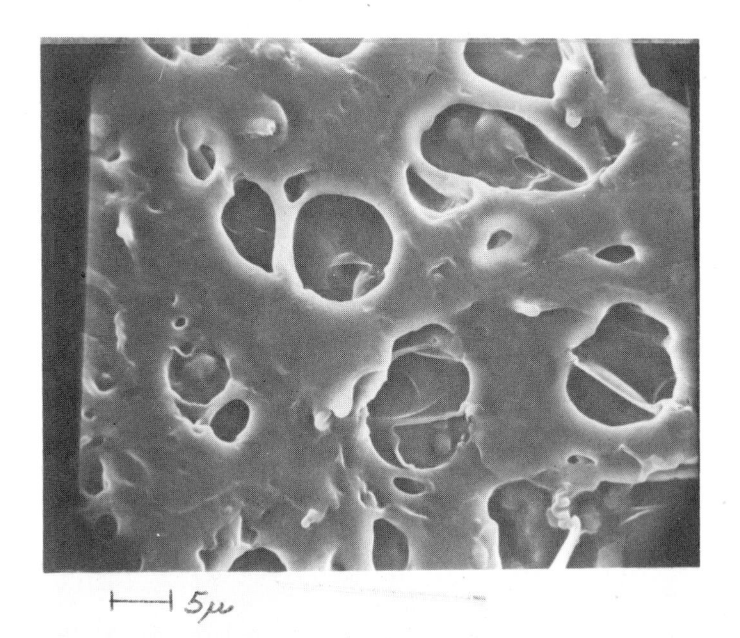

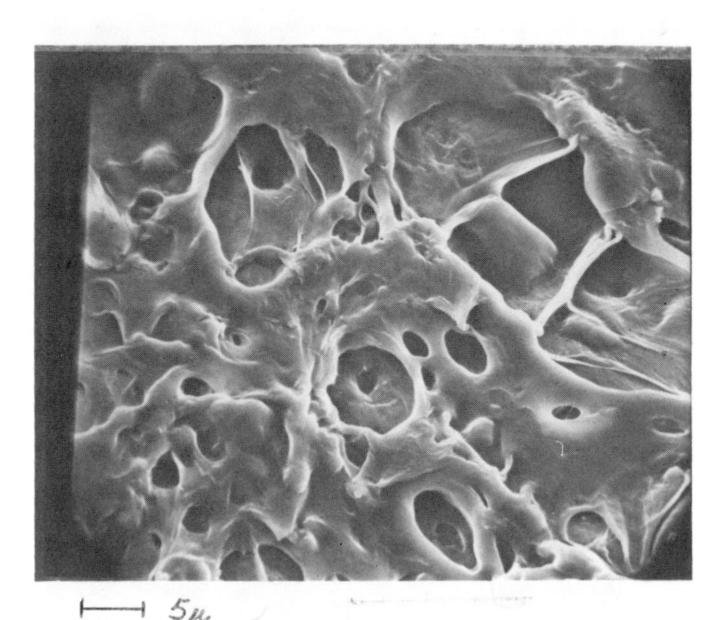

Figure 22. Scanning electron micrographs from the fracture surface of a PP blend with 20% EPDM. (a, top*) Corresponding to Figure 20; (b,* bottom*) corresponding to Figure 21.*

exhibit a minimum with the mold temperature of injection molding. Typical Izod impact testing is usually not sensitive nor reproducible enough to permit more quantitative correlations. The impact energy values listed in Table I illustrate these limitations. However, morphological evaluation of the bulk sample as well as the fracture surface can provide the unifying link in elucidating the fundamental basis of impact behavior.

Initial consideration of the morphological characteristics with the impact properties refer to the skin, shear, and core regions of the injection-molded samples. The extent and orientation of these regions vary with the EPDM content and processing conditions. EPDM domains in the skin region are oriented along the direction of flow. The skin layer thickness of PP/EPDM blends showed a pronounced maximum with mold temperature. Increasing fractions of EPDM produced an increasingly irregular spherulitic texture in shear and core zones. As the fraction of EPDM blended with PP increases, an increase in calorimetric crystallinity, followed by a marked decrease, is noted.

Although the processing variables appear to have correlations with morphology, results of impact tests fail to produce a definitive description of the morphological response to impact. However, this problem is shown to be substantially eliminated through the use of instrumented impact testing (*17*). Such tests measure the time dependence of the impact load borne by the sample and indicate that impact fracture is a

Table I. Impact Strength for PP/EPDM Blends (J/m of Notch)

		Test Temperature					
		25°C		*0°C*		*−30°C*	
Sample % PP/% EPDM	*Mold Temp (°C)*	*Mean*	*Standard Deviation*	*Mean*	*Standard Deviation*	*Mean*	*Standard Deviation*
100/0	20	11.2	3.9	6.2	1.7	3.5	1.4
100/0	50	14.0	5.3	5.6	1.5	2.5	0.9
100/0	87	11.3	4.5	4.3	0.6	1.9	0.1
90/10	20	33.1	7.8	19.0	2.4	12.0	4.2
90/10	50	37.1	14.6	21.9	6.6	13.5	3.6
90/10	87	32.1	6.1	20.6	3.2	12.8	3.8
80/20	20	97.2	19.2	45.7	7.1	26.6	7.8
80/20	50	74.6	16.3	35.5	3.5	25.1	5.7
80/20	87	80.0	19.6	44.4	11.3	28.4	8.8
70/30	20	Deformation of sample, no fracture		128.8	34.7	69.4	17.0
70/30	50			123.2	14.7	64.0	13.2
70/30	87			104.6	8.7	58.7	7.4

complex process consisting of initiation and propagation stages. Since sample deflection is linear with time, the initial slope of the load/time curve gives the sample modulus. Yielding is often revealed during crack initiation, and the initiation stage usually terminates in a maximum or fracture load. Energies corresponding to the various stages give important details of the impact process. For instance, with a change in the impact velocity from 3.43 to 1.10 m/sec, the impact behavior changed markedly with the propagation energy increasing from 24 to 60% of the total energy. Examination of the corresponding fracture surfaces by optical microscopy reveals a marked increase in stress whitening because of crazing and microvoid formation. A blend of PP with 20% EPDM evidenced a transition from brittle to ductile impact failure with decreasing velocity of impact, the total absorbed energy increasing from 0.202 to 0.773 J as the impact velocity decreased from 3.43 to 1.10 m/sec. Fluctuations in local blend composition and domain size distribution are inferred from variations in load/energy/time curves from instrumented impact studies on duplicate injection-molded samples. An examination of scanning electron micrographs of corresponding fracture surfaces confirms the morphological variability and when combined with impact studies provides a powerful tool for explaining the fundamental structural basis of impact behavior.

Microscopy, both optical and scanning electron, together with instrumented impact testing provides a method for the critical examination of any polyblend system. The use of these two somewhat diverse experimental approaches aids in assigning the appropriate significance to the characteristics important in the fabrication of polyblend systems. The approach provides a very sensitive technique for distinguishing subtle variations in the impact behavior and correlating them with morphology. Although the polymer samples reported here are unsuitable for quantitatively evaluating absolute fracture mechanisms, they are capable of contrasting and comparing characteristics of the fracture process. The distinction between ductile and brittle fracture as presented by various investigators has a corresponding identification by these tests, first by the load energy curve and its accompanying data and secondly by the microscopy giving a one-on-one correspondence (*3, 18*). The relative importance of the two steps in the fracture process, initiation and propagation, is likewise easily correlated from the impact data. Then the various factors affecting impact strength also can be monitored through the assignment of the specific morphologies associated with the well-documented impact test (*19*). Thus, through the use of these methods there exists a means to describe a fracture mechanism and to test its applicability and eventually permit enhanced control of the toughening process for polymers.

Acknowledgment

This research was supported by the Los Angeles Rubber Group, the duPont Company, and the Materials Research Laboratory at the University of Massachusetts. Technical assistance was kindly provided by Janis Levin of the University of Massachusetts and by Donald R. Ireland and Lee Wogulis of Effects Technology.

Literature Cited

1. Amos, J. L., *Polym. Eng. Sci.* (1974) **14**, 1.
2. Freeguard, G. F., *Br. Polym. J.* (1974) **6**, 205.
3. Bucknall, C. B., "Toughened Plastics," Applied Science Publishers, London, 1977.
4. Heggs, T. C., in "Block Copolymers," D. C. Allport, W. H. Janes, Eds., p. 105, Applied Science Publishers, London, 1973.
5. Thamm, R. C., *Rubber Chem. Technol.* (1977) **50**, 24.
6. Noel, O. F., Carley, J. F., *Polym. Eng. Sci.* (1975) **15**, 117.
7. Kantz, M. R., Newman, H. D., Jr., Stigale, F. H., *J. Appl. Polym. Sci.* (1972) **16**, 1249.
8. Fitchmun, D. R., Mencik, Z., *J. Polym. Sci., Polym. Phys. Ed.* (1973) **11**, 951.
9. Mencik, Z., Fitchum, D. R., *J. Polym. Sci., Polym. Phys. Ed.* (1973) **11**, 973.
10. Henke, S. J., Smith, C. E., Abbott, R. F., *Polym. Eng. Sci.* (1975) **15**, 79.
11. Miller, R. L., in "Polymer Handbook," J. Brandrup, E. H. Immergut, Eds., p. III-10, John Wiley, New York, 1975.
12. Dragaun, H., Hubeny, H., Muschik, H., *J. Polym. Sci., Polym. Phys. Ed.* (1977) **15**, 1779.
13. Jackson, G. B., Ballman, R. L., *SPE J.* (1960) **16**, 1147.
14. Han, C. D. Villamizar, C. A., Kim, Y. W., *J. Appl. Polym. Sci.* (1977) **21**, 353.
15. Djurner, K., Kubat, J., Rigdahl, M., *Polymer* (1977) **18**, 1068.
16. Reinshagen, J. H., Dunlap, R. W., *J. Appl. Polym. Sci.* (1975) **19**, 1037.
17. Server, W. L., Ireland, D. R., Special Technical Publication **563**, p. 74, ASTM, Philadelphia, 1974.
18. Ward, I. M., "Mechanical Properties of Solid Polymers," Chapter 12, Wiley-Interscience, New York, 1971.
19. Nielsen, L. E., "Mechanical Properties of Polymers and Composites," Chapter V, Vol. 2, Marcel Dekker, New York, 1974.

RECEIVED April 14, 1978.

26

Segmental Orientation, Physical Properties, and Morphology of Poly-ϵ-Caprolactone Blends

DOUGLAS S. HUBBELL and STUART L. COOPER

Department of Chemical Engineering, University of Wisconsin, Madison, WI 53706

The compatibility, mechanical properties, and segmental orientation characteristics of poly-ϵ-caprolactone (PCL) blended with poly(vinyl chloride) (PVC) and nitrocellulose (NC) are described in this study. In PVC blends, the amorphous components were compatible from 0–100% PCL concentration, while in the NC system compatibility was achieved in the range 50–100% PCL. Above 50% PCL concentration, PCL crystallinity was present in both blend systems. Differential IR dichroism was used to follow the dynamic strain-induced orientation of the constituent chains in the blends. It was found for amorphous compatible blends that the PCL oriented in essentially the same manner as NC and the isotactic segments of PVC. Syndiotactic PVC segments showed higher orientation functions, implying a microcrystalline PVC phase.

The objectives of the present study were to analyze polymer blends of poly-ϵ-caprolactone with poly(vinyl chloride) and nitrocellulose. The research included determination of compatibility and characterization of the morphological, mechanical, and orientation properties of the polycaprolactone blends.

Experimental

Poly-ϵ-caprolactone designated as PCL-700 was supplied by J. V. Koleske of the Union Carbide Corp. This polymer has been used in several other blend studies (*1–6*). The polymers blended with PCL were poly(vinyl chloride) (PVC) and nitrocellulose (NC). The poly-

0-8412-0457-8/79/33-176-517$05.00/0

(vinyl chloride) used was Union Carbide's QYTQ-387. The nitrocellulose, supplied by Hercules Inc., was designated as RS ½ sec. The NC is reported to have 11.8–12.2% nitrogen (*7*), which corresponds to 2.25 ± 0.06 nitro groups per anhydroglucose ring. The molecular weights, densities, and solubility parameters for PCL, NC, and PVC appear in Table I. The solubility parameters, when possible, have been calculated by the group contribution methods of Small (*8*) and Hoy (*9*). A small difference between solubility parameters of components in a mixture is usually considered a necessary but not sufficient condition for compatibility. A full discussion of the limitations of the solubility parameter approach to compatibility is given in the review of polymer blend technology by Paul and Barlow in Chapter 17.

All samples for this study were prepared by spin casting from polymer solutions (*9, 10*). In this technique, a solution of polymer and solvent is forced against the walls of a spinning cylinder. The solvent is gradually evaporated under a slight vacuum and also by heat if desired, with the polymer precipitating onto a sheet of aluminum or paper lining the wall. With this method, it has been demonstrated that unoriented films with thicknesses as small as 4 μ can be made. Tetrahydrofuran (THF) was used as the spin-casting solvent. Films were cast at room temperature from approximately 15 mL of a 1% solution.

All samples were dried in a constant stream of air for several hours and then dried for at least four days under vacuum at room temperature. After drying, the films were aged in desiccators at room temperature for at least two weeks. This allows the PCL crystallinity to approach its equilibrium value (*6*). Throughout this chapter the blends are designated by the weight percent of the polymer blended with polycaprolactone. Thus 25% PVC is a 25/75 blend of PVC/PCL.

In favorable situations, segmental orientation in polymer blends can be followed using the technique of IR dichroism. The results can be expressed in terms of the Hermans orientation function shown in Equation 1.

$$f = (3\,\overline{\cos^2\theta} - 1)/2 \tag{1}$$

The orientation function varies from unity for perfect axial alignment of the chain backbone to $-\frac{1}{2}$ for perfect transverse orientation, with $f = 0$ for random orientation. Commonly, dichroism data are in the form of absorptions for plane-polarized light perpendicular and parallel

Table I. Materials Studied

	Polymers		
	PCL	*PVC*	*NC*
M_n	13,000	35,000	45,000
M_w	24,000	72,000	—
Density (g/cm³ at 20°C)	1.149[a]	1.39	1.58–1.65
δ(cal/cm³)$^{1/2}$ Hoy	9.43	9.47	9.95
δ(cal/cm³)$^{1/2}$ Small	9.34	9.55	—

[a] 1.094 and 1.187 are the densities for amorphous and crystalline PCL.

to the direction of stretch ($A_\perp$ and $A_{||}$, respectively). The orientation function is related to the ratio of these absorptions ($D = A_{||}/A_\perp$) by Equation 2, where D_0 is the dichroic ratio for perfect orientation in the

$$f = \frac{(D_0 + 2)(D - 1)}{(D_0 - 1)(D + 2)} \tag{2}$$

stretch direction, equal to ($2 \cot^2 \alpha$). α is the angle between the transition moment vector and the local chain axis.

In this study, dynamic experiments were performed using a differential method of dichroism. The sample was strained in the common beam of a Perkin–Elmer model 180 double-beam spectrophotometer, with the transmitted radiation split into two beams polarized parallel to and perpendicular to the direction of stretch. The recorded output was the difference between the two absorptions. The orientation function was calculated using Equation 3, where A_0 is the unpolarized absorption of

$$f = \frac{D_0 + 2}{D_0 - 1} \cdot \frac{A_{||} - A_\perp}{3A_0(l/l_0)^{1/2}} \tag{3}$$

the undeformed sample and l and l_0 are the stretched and unstretched sample lengths. The term $A_0(l/l_0)^{-1/2}$ equals the unpolarized absorption of any strain level, assuming constant volume deformation.

The IR apparatus and dichroism methods used in this work have been previously described (*11, 12*). In the differential dichroism experiment, the samples were stretched from both ends simultaneously at a true strain rate of approximately 30% per min. A motorized stretching jig was used which fits inside the instrument in the path of the common beam at a 45° angle. Wire grid polarizers were set at 45° in the reference and sample beams, and the instantaneous dichroic difference $\Delta A = A_{||} - A_\perp$ was recorded with the instrument operating in constant, wave-number mode.

The dynamic elastic modulus (E'), the loss modulus (E''), and the tan δ were measured simultaneously by the Rheovibron dynamic viscoelastometer model DDV-II (Toyo Measuring Instruments Co., Ltd.). Measurements were made at a frequency of 110 Hz, starting from −140°C and heating at 1°–2°C per minute until the samples became too soft to be tested. The readings were taken every 4°–8°C except in the transition zone when the readings were taken every 2°C. The sample chamber was kept dry by a stream of moisture-free nitrogen.

Differential scanning calorimetery (DSC) was used to measure heat capacity as a function of temperature. The DSC used in this study was a Perkin–Elmer model DSC-2. Liquid nitrogen was used as a heat sink and helium was used as the purge gas. Samples were usually about 30 mg, and a heating rate of 20°C/min was used for measuring T_gs and T_ms.

Results and Discussion

Dynamic Mechanical Properties. The dynamic mechanical properties of the PVC/PCL and NC/PCL blend systems are shown in Figures

1 and 2. Although the T_gs found by this method are somewhat less precise than those found by the DSC, it has been shown (*13, 14*) that mechanical testing can detect segmental motion on a smaller scale than thermal analysis. Thus, multiple phases can be more easily detected.

Figure 1 shows that each homopolymer exhibits two relaxations in the E'' curve. The higher temperature peak corresponds to a glass transition, while the lower temperature peak can be attributed to a secondary

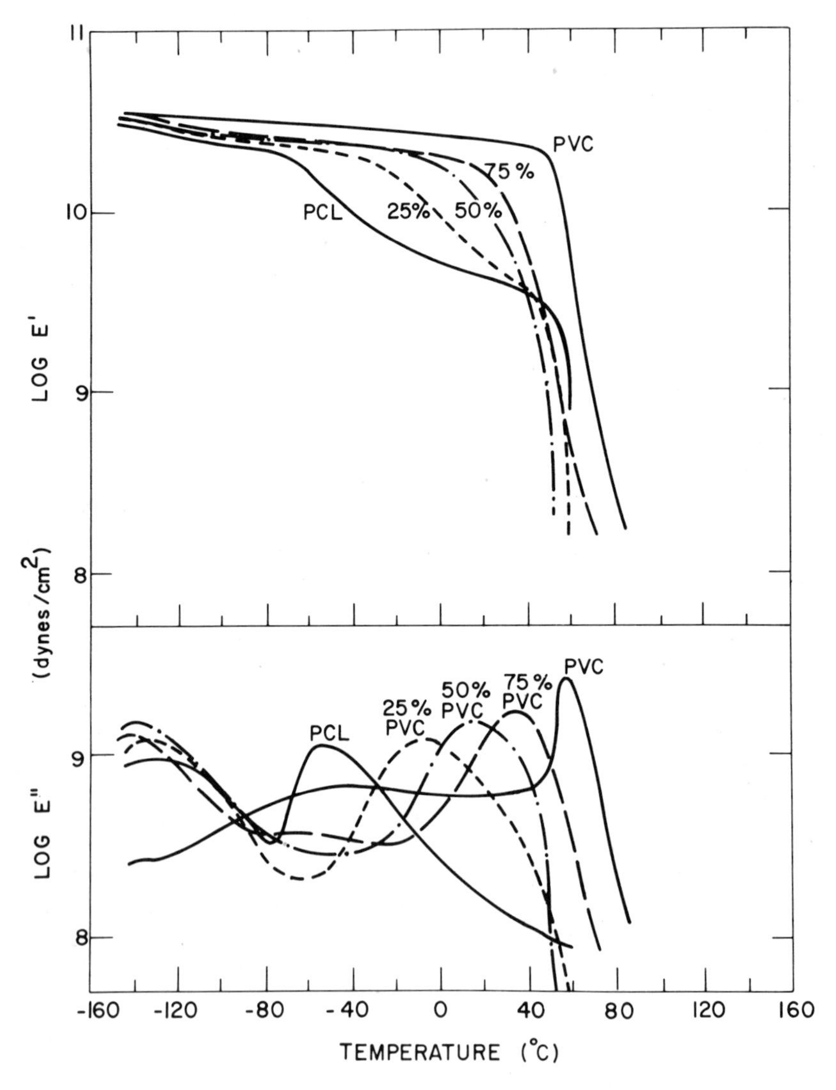

Journal of Applied Polymer Science

Figure 1. Temperature dependence of E′ *and* E″ *for PVC/PCL* (1)

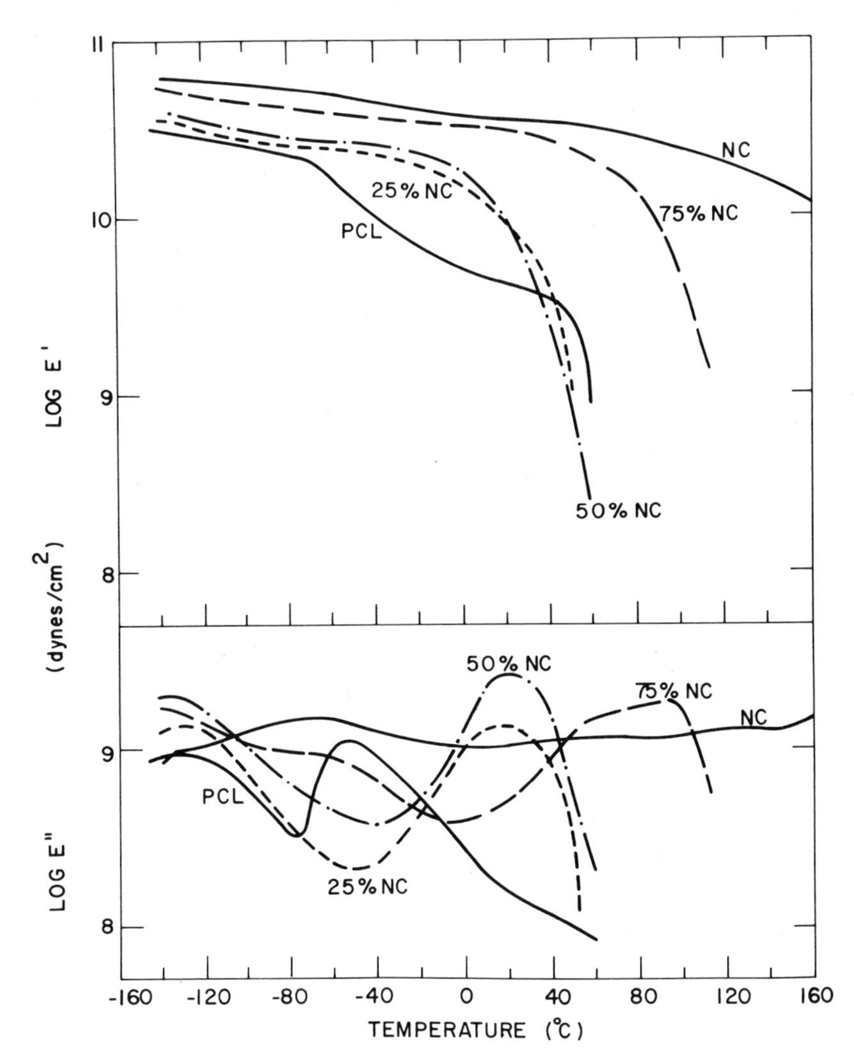

Figure 2. Temperature dependence of E′ *and* E″ *for NC/PCL (1)*

relaxation (*2, 6*). In the blends, only one glass-transition peak is evident in the E'' curves, though the secondary relaxations of the homopolymers are also evident to some extent. The sharp declines in the E' and E'' curves for 100% PCL, 25 and 50% PVC in the neighborhood of 50°–60°C can be attributed to the melting of the crystalline PCL. The dynamic mechanical testing results for NC/PCL blends are presented in Figure 2. The E' curves for 25 and 50% NC seem quite similar with the samples becoming very soft in the region of the PCL melting point. The 75 and

100% NC samples maintain high elastic moduli to much higher temperatures. The E'' curves for the blends show the PCL methylene relaxation at lower temperatures and prominent peaks above 0°C. The NC sample seems to lack a relaxation peak which would correspond to a T_g but does show a broad peak centered at about −68°C. This peak also appears to be present in the 50 and 75% NC samples.

The lack of an obvious NC T_g peak makes analysis of this blend system somewhat ambiguous. The major transition of the 25% NC blend as shown by the E'' curve is the melting of the PCL at 18°C. The T_g of the amorphous 50% NC blend is in a similar temperature range at 23°C. The symmetry of the E'' peak for the 50% NC blend suggests a compatible blend since only one relaxation is present.

In the 75%-NC-blend curve, the interpretation is quite different. The broad relaxation seems to have components at 63° and 95°C. Possibly the upper component of the E'' relaxation represents the T_g of a nearly pure NC phase. The peak at 63°C would then represent a second phase containing both NC and PCL. However, both peaks could represent NC-rich microphases of differing compositions.

In random copolymers, there is a high degree of interaction between the different repeat units because they are held together chemically. The glass transition of a random copolymer consequently reflects the composition of the copolymer. Thus, the T_g of random copolymers can usually be represented as a monotonic function of composition by one of the various copolymer equations. Similarly, if blends of polymers are mixed at the segmental level, only one T_g should be evident, and a copolymer equation should be applicable.

The Gordon–Taylor copolymer equation (Equation 4) has been applied to the T_g data in this study to determine if the blends are single-

$$T_{g12} = T_{g1} + kW_2(T_{g2} - T_{g12})/W_1 \quad (4)$$

phase systems or not. These data include the results for samples annealed at room temperature and also for the same samples which have been heated above the melting point of PCL and then quenched rapidly in the DSC to 150 K. For a well-mixed system, the plot of T_{g12} vs. $[(T_{g2} - T_{g12})W_2/W_1]$, where W_1 and W_2 are the weight fractions of polymers 1 and 2, respectively in the amorphous phase, will yield a straight line with a slope of k and an ordinate intercept of T_{g1}.

The T_gs for the PVC/PCL and NC/PCL systems have been plotted vs. $[(T_{g2} - T_{g12})W_2/W_1]$ in Figure 3. For the quenched samples, W_1 and W_2 are the weight fractions of PCL and PVC or NC, respectively. Straight lines have been fitted by least-squares analysis to this data. For the unquenched samples, W_1 is the weight fraction of PCL in the amor-

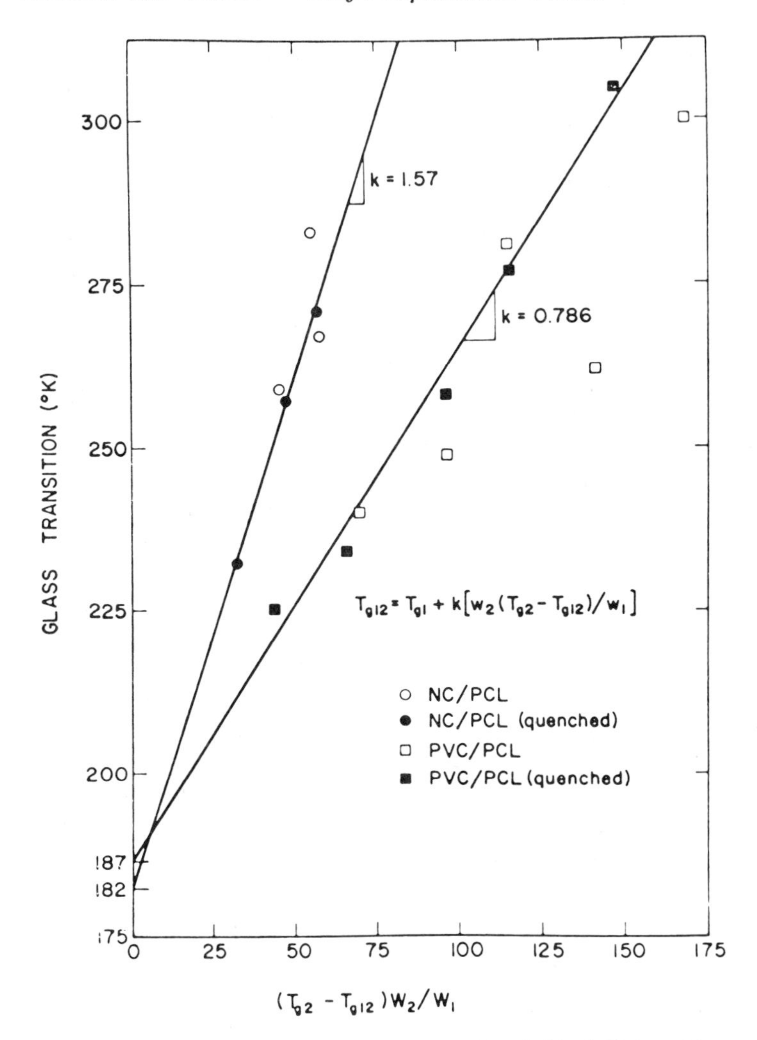

Journal of Applied Polymer Science

Figure 3. T_g data for PVC/PCL and NC/PCL blends plotted according to the Gordon–Taylor equation. The lines are fit to the quenched data (1).

phous phase. The percent crystallinity of PCL was calculated from DSC heat of fusion measurements. The variable W_2 is the weight fraction of PVC or NC in the amorphous phase, assuming that all of the PVC and NC was noncrystalline.

Figure 3 shows that the Gordon–Taylor equation fits the quenched data for the PVC/PCL blends quite well. The thermograms for the blends of 25, 35, and 50% PVC display PCL heats of fusion, and consequently, the unquenched T_gs for these samples lie above the curve because the crystallinity reduces the concentration of PCL in the amorphous phase.

The NC/PCL data present a different situation. Blends of 25 and 40% NC showed the opaqueness and a DSC endotherm, which indicate the presence of crystallinity. Consequently, melting and quenching produced lower T_gs as expected. Pure NC and 50% NC did not show either of these features. Yet, the T_g dropped from 338 to 328 K for NC and from 283 to 271 K for 50% NC after the heating–quenching cycle. Blends of 65 and 75% NC showed high initial T_gs but failed to show any transition after quenching.

The Gordon–Taylor equation fits the quenched points reasonably well for 25, 40, and 50% NC. Thus, NC/PCL blends appear to be compatible in the range 0–50% NC. The unquenched 50% NC does not fit onto the curve even though it did not contain crystalline PCL. Consequently, it would seem that microheterogeneity is present in coarse enough form to allow detection of one of the phases and that the segregation was eliminated by the heating. This trend is continued by the data for 65 and 75% NC. Obviously, large-scale phase separation has produced nearly pure nitrocellulose domains in the 65–75% NC region.

IR Dichroism. Two types of IR-dichroism experiments were used in this study to follow segmental orientation. First, dynamic differential dichroism was used to follow chain orientation while the sample was elongated at a constant strain rate. This experiment was performed with different IR peaks which allowed a comparison of the molecular orientations for each blend constituent. Second, a cyclic experiment was used where the film was strained to a predetermined elongation, relaxed at the same strain rate until the stress was reduced to zero, and then elongated to a higher level of strain, and so forth.

The orientation of the two components of a compatible (*1*), amorphous 50% NC blend is shown in Figure 4. The films have been strained to 25, 50, and 75% in successive cycles, with each strain increment followed by a relaxation to zero stress. The orientation of the PCL and NC have been followed by using the carbonyl (1,728 cm^{-1}) and NO_2 (1,660 cm^{-1}) stretching peaks, respectively.

Figure 4 shows that the orientation of the two chains follow similar paths. The similarity suggests that the local environment is very much the same for both blend constituents. In contrast, a phase-separated binary mixture of polymer segments will exhibit a different orientation for each component. For example, high-density polyethylene and isotactic polypropylene form multiphase systems when blended together (*16*). When the incompatible blends are deformed, the component which constitutes the continuous phase always orients to a higher degree than

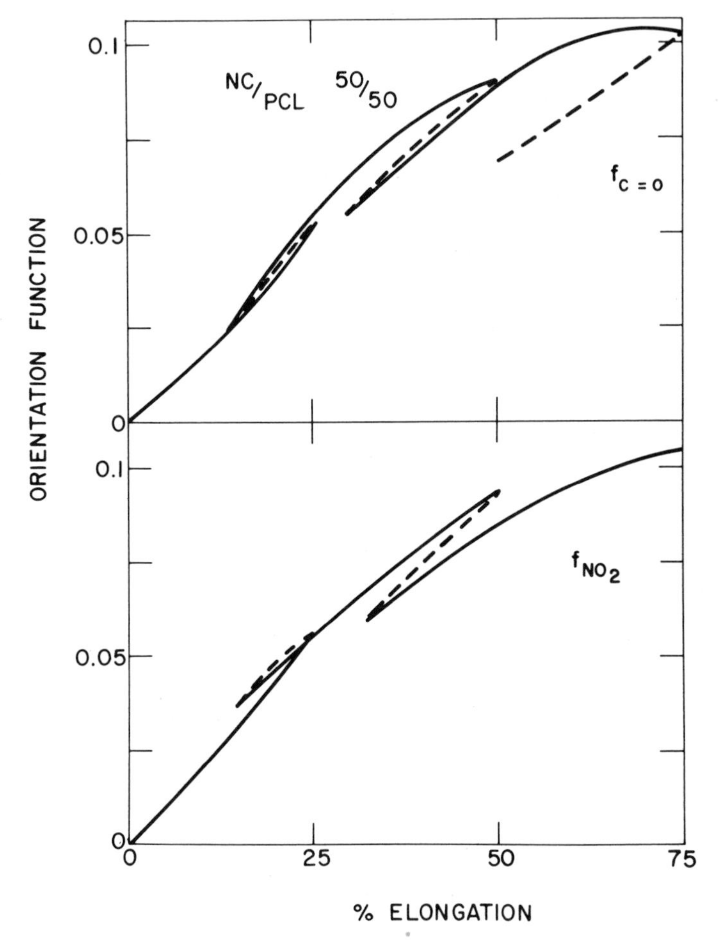

Journal of Polymer Science, Polymer Physics Edition

Figure 4. Carbonyl and NO_2 orientation function vs. elongation curves for 50% NC in a cyclic IR dichroism experiment (19)

the component in the dispersed phase, regardless of which polymer dominates the composition.

The orientation functions for a 75/25 PVC/PCL blend are presented in Figure 5, which shows that the PCL chains in the 75% PVC blend orient in essentially the same way as the isotactic segments and the other folded-chain PVC segments represented by the 693 cm^{-1} peak. This peak is shifted from 691 cm^{-1} for pure PVC. The shape and magnitude of the orientation functions for the carbonyl and 693-cm^{-1} C–Cl peak are similar to the orientation functions shown for the amorphous 50% NC blend.

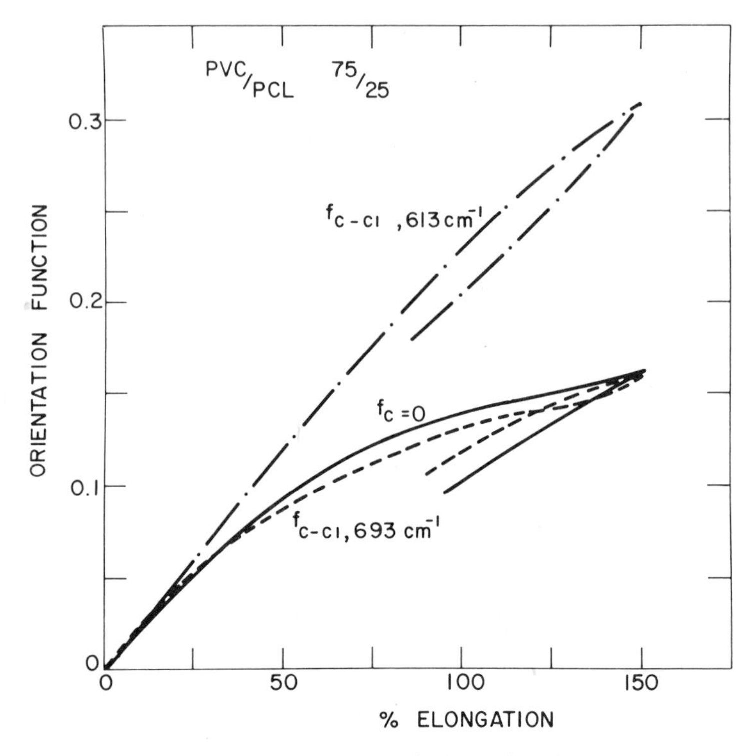

Journal of Polymer Science, Polymer Physics Edition

Figure 5. Carbonyl and C–Cl orientation functions vs. elongation curves for 75% PVC (19)

However, the syndiotactic segments of extended chains (613 cm^{-1}) are obviously situated in a different local environment. These more rigid segments form into microcrystalline phases which orient as units to much higher degrees than the amorphous isotactic sequences. The C–Cl peak at 613 cm^{-1} arises mainly from seqeunces of four or more trans conformations in syndiotactic repeat units. With syndiotactic sequences of four or more repeat units, crystallinity can form (*17, 18*). This ease of crystallization has been ascribed to the strong dipole–dipole interaction between the C–Cl bonds.

Based on the results of two amorphous compatible blend systems of 50/50 NC/PCL and 75/25 PVC/PCL, the following conclusion can be drawn. Compatible amorphous blend constituents show identical segmental orientation behavior, indicating good mixing at the molecular level. Thus, an amorphous compatible blend can exhibit the characteristics of a single homopolymer not only in its glass-transition behavior and mechanical properties but also in the uniform way in which the polymer chains orient.

There are however, hypothetical polymer blend situations which may not conform to this model. First, one might have two incompatible polymers which form an interpenetrating network such that the phases effectively act mechanically in parallel. Thus their orientation functions would be related to their respective moduli and could possibly be quite similar. Second, if one considers an uncrosslinked network as effectively crosslinked by entanglements, one would expect that the orientation function of such a chain will vary inversely with the number of segments between crosslink points. Consequently, compatible chains of two different types will only have the same orientation function provided that the number of such segments is equal. Thus a compatible blend of a polymer with stiff chains and one of flexible chains will yield higher orientation functions for the stiff-chain component. Intrachain stiffness is indeed a factor, as brought out by the high orientation of the syndiotactic PVC sequences reported in this work.

Conclusions

Blends of PCL with PVC were shown to exhibit only one glass-transition temperature (T_g), which for blends quenched from the melt could be represented as a function of composition by the Gordon–Taylor T_g equation. By similar criterion, the PCL blends with NC were shown to be compatible for the composition range 0–50% NC by weight. With higher concentrations of NC, both thermal and mechanical testing indicated that multiple amorphous phases were present, even though the films were clear.

Differential IR dichroism was used to follow the dynamic strain-induced orientation of the constituent chains in PVC/PCL and NC/PCL blends. It was found for amorphous compatible blends that PCL oriented in essentially the same manner as NC and the isotactic segments of PVC. Syndiotactic PVC segments showed much higher orientation functions, which implied the existence of a microcrystalline PVC phase.

Literature Cited

1. Hubbell, D. S., Cooper, S. L., *J. Appl. Polym. Sci.* (1977) **21**, 3035.
2. Ong, C., Ph.D. thesis, University of Massachusetts, Amherst (1973).
3. Khambatta, F. B., Stein, R. S., *Polym. Prepr. Am. Chem. Soc., Div. Polym. Chem.* (1974) **15**, 260.
4. Khambatta, F. B., Warner, F., Russell, T., Stein, R. S., *J. Polym. Sci., Polym. Phys. Ed.* (1976) **14**, 1391.
5. Olabisi, O., *Macromolecules* (1975) **8**, 316.
6. Koleske, J. V., Lundberg, R. D., *J. Polym. Sci., Part A-2* (1969) **7**, 795.
7. "Hercules Nitrocellulose, Chemical and Physical Properties," Hercules, Inc.,
8. Small, P. A., *J. Appl. Chem.* (1953) **3**, 71.

9. Hoy, K. L., *Paint Technol.* (1970) **42,** 76.
10. Shen, M., Mehra, U., Niinomi, M., Koberstein, J. T., Cooper, S. L., *J. Appl. Phys.* (1974) **45,** 4182.
11. Seymour, R. W., Allegrezza, A. E., Jr., Cooper, S. L., *Macromolecules* (1973) **6,** 896.
12. Allegrezza, A. E., Jr., Seymour, R. W., Ng, H. N., Cooper, S. L., *Polymer* (1974) **15,** 433.
13. Stoelting, J., Karasz, F. E., MacKnight, W. J., *Polym. Eng. Sci.* (1970) **10,** 133.
14. MacKnight, W. J., Stoelting, J., Karasz, F. E., "Multicomponent Polymer Systems," N. Platzer, Ed., ADV. CHEM. SER. (1971) **99,** 29.
15. Brode, G. L., Koleske, J. V., *J. Macromol. Sci., Chem.* (1972) **A6,** 1109.
16. Asada, T., Fukao, T., Tanaka, A., Onogi, S., *J. Soc. Mater. Sci., Kyoto* (1968) **17,** 59.
17. Koleske, J. V., Wartman, L. H., "Poly(vinyl Chloride)," p. 33–38, Gordon and Breach, New York, 1969.
18. Nakajima, A., Hamada, H., Hayashi, S., *J. Macromol. Chem.* (1966) **95,** 40.
19. Hubble, D. S., Cooper, S. L., *J. Polym. Sci., Polym. Phys. Ed.* (1977) **15,** 1143.

RECEIVED April 14, 1978.

Brillouin Scattering from Polymer Blends

G. D. PATTERSON

Bell Laboratories, Murray Hill, NJ 07974

Brillouin scattering measures the velocity and attenuation of hypersonic thermal acoustic phonons. A theory of Brillouin scattering from polymer blends is presented and illustrated qualitatively by several examples. The study of blend compatibility is illustrated for the system PMMA-PVF$_2$. The detection of inhomogeneous additives is shown for commercial PVC film and cellulose acetate, and simultaneous measurements on separated phases are presented for Mylar film. The main purpose of the paper is to stimulate further work in a potentially promising field.

Brillouin scattering measures the spectrum attributable to the interaction of light with thermal acoustic phonons (*1*). The scattered light is shifted in frequency with a splitting given by

$$\Delta\omega = qV(q) \tag{1}$$

where $q = (4\pi n/\lambda) \sin \theta/2$ is the magnitude of the scattering vector for light of incident vacuum wavelength λ traveling in a medium of refractive index n and scattered through an angle θ in the scattering plane, and $V(q)$ is the velocity of the phonons with wavevector magnitude q. The shifted peaks have a linewidth given by

$$\Gamma = \alpha V/2\pi \tag{2}$$

where α is the phonon attenuation coefficient and Γ is the halfwidth at halfheight measured in Hertz.

Brillouin scattering has now been studied in a large number of bulk amorphous polymers (*2–13*) and the behavior of $\Delta\omega$ and Γ have been determined as a function of temperature. If the incident light is polarized

0-8412-0457-8/79/33-176-529$05.00/0

vertically with respect to the scattering plane and the scattered light is observed through an analyzer set vertical, then the Brillouin peaks attributable to the longitudinal phonons are observed. If the analyzer is set horizontal with respect to the scattering plane then the I_{HV} spectrum displays the Brillouin peaks attributable to transverse phonons. A spectrum of bisphenol A polycarbonate showing both longitudinal and transverse Brillouin peaks is presented in Figure 1. There is also a strong central peak which will not be discussed in this chapter.

At high temperatures, the longitudinal phonon velocity is given by Equation 3 (*13*):

$$V_l = (\gamma/\rho\beta_T)^{1/2} \tag{3}$$

where $\gamma = C_p/C_v$ is the ratio of specific heats, ρ is the density, and β_T is the isothermal compressibility. The linewidth is (*13*):

$$\Gamma_l = \frac{q^2}{2\rho}\left[\eta_v + \frac{4}{3}\eta_s + \frac{\kappa(\gamma - 1)}{C_p}\right] \tag{4}$$

where η_v is the volume viscosity, η_s is the shear viscosity, and κ is the thermal conductivity.

As the fluid is cooled, the relaxation times of the system become comparable with $(\Delta\omega_l)^{-1}$, and the viscoelastic nature of polymer liquids must be taken into account. The longitudinal velocity becomes (*13*):

$$V_l = \left(\frac{\gamma(q)M'(q)}{\rho}\right)^{1/2} \tag{5}$$

where $M(q) = K(q) + 4/3G(q)$ is the longitudinal modulus, $K(q)$ is the modulus of compression, $G(q)$ is the shear modulus, and $M'(q)$ is the real part of the complex modulus. In the viscoelastic region, transverse phonons can propagate with a velocity given by (*13*):

$$V_t = \left(\frac{G_\infty}{\rho} - \frac{1}{(2\tau_s q)^2}\right)^{1/2} \tag{6}$$

where G_∞ is the limiting shear modulus and τ_s is the average shear relaxation time. For small values of $q\tau$, the transverse phonons will be overdamped and only a central peak will be observed. The longitudinal linewidth Γ_l can still be represented as in Equation 4, but all the quantities in the brackets must be evaluated at finite q since $q\tau$ will be significant. The value of Γ_l will go through a maximum when $V_l q\tau$ equals one. The transverse linewidth is given by (*13*):

$$\Gamma_t = 1/2\tau_s \tag{7}$$

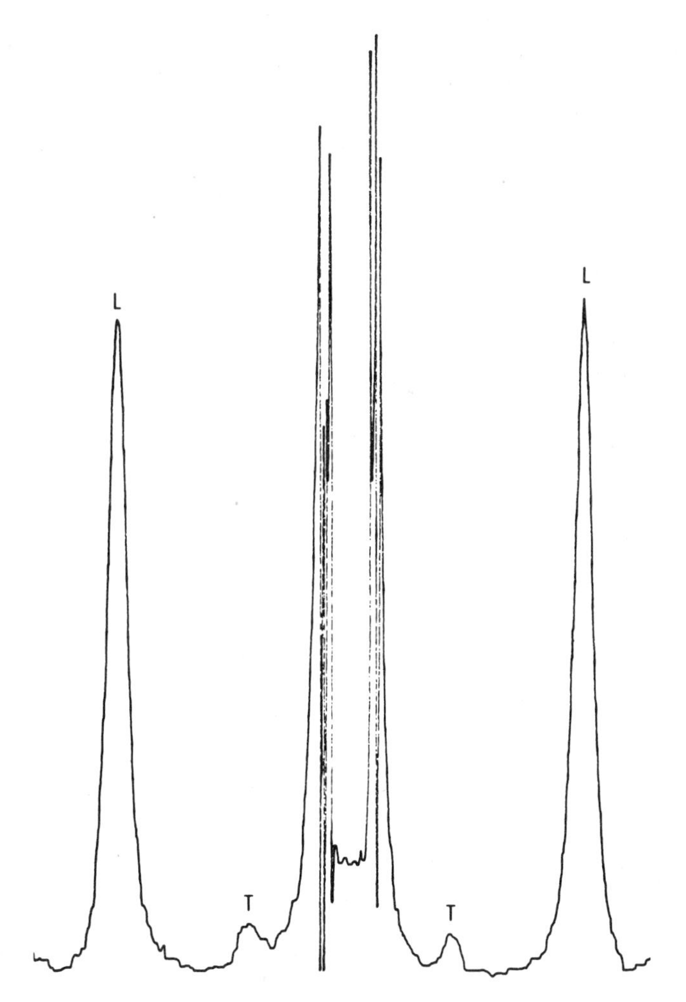

Figure 1. Rayleigh–Brillouin spectrum of bisphenol A polycarbonate showing both longitudinal (L) and transverse (T) Brillouin peaks

As the fluid is cooled further towards the glass transition the phonon velocities reach their limiting frequency values (*13*):

$$V_l = \left(\frac{M_\infty}{\rho}\right)^{1/2}$$

$$V_t = \left(\frac{G_\infty}{\rho}\right)^{1/2} \tag{8}$$

where M_∞ and G_∞ are the infinite frequency values of the longitudinal and shear modulus. The longitudinal and transverse linewidths are predicted to be negligible since the relaxation times that determine them become very long. However, the observed longitudinal linewidths are much larger than predicted and in fact are in the range 100–400 MHz near T_g. The reason for the excess linewidths is inhomogeneous phonon attenuation attributable to regions of different average density. This effect is usually ignored in the simple linear theories of Brillouin scattering.

In the present chapter we consider the effect of a mixture of species on the Brillouin spectrum. The theory will be described and examples will be presented which illustrate the theory.

Theory

The linear theory of Brillouin scattering from a viscoelastic medium has been presented by Rytov (*14*). A summary and discussion of this theory are given in Ref. *13*. In a real experiment, light scattering is observed from a finite scattering volume and at a finite value of q. As a result, the predicted spectrum must be averaged over the scattering volume with a characteristic length given by $2\pi/q$.

$$I_1(q,\omega) \propto \int_V P(\rho,M) I_1(q,\omega,\rho,M)\, d\rho\, dM \tag{9}$$

where $P(\rho,M)$ is the probability that a volume of radius $2\pi/q$ has a density ρ and modulus M, $I_1(q,\omega,\rho,M)$ is the spectrum predicted for such a material, and the integration is carried out over the scattering volume.

For a homogeneous, low-viscosity fluid the probability $P(\rho,M)$ is essentially a delta function centered at $<\rho>$ and $<M>$ and the linear result is recovered. However, when $V_1 q\tau$ is much greater than one, there should be a thermodynamic distribution of density and modulus for a region of size $2\pi/q$.

$$P(\rho,M) \propto \exp\left(-\frac{(\rho - <\rho>)^2}{<\rho^2> - <\rho>^2}\right) \exp\left(-\frac{(M - <M>)^2}{<M^2> - <M>^2}\right) \tag{10}$$

Thus, for $V_1 q\tau >> 1$, the observed spectrum should be of the form:

$$I_1(q,\omega) \propto \exp\left(-\frac{\left(\omega^2 - q^2\left\langle\frac{M_\infty}{\rho}\right\rangle\right)^2}{2q^4\left(\left\langle\left(\frac{M_\infty}{d}\right)^2\right\rangle - \left\langle\frac{M_\infty}{\rho}\right\rangle^2\right)}\right) \tag{11}$$

since

$$I_1(q,\omega,\rho,M_\infty) = \delta\left(\omega^2 - q^2\left(\frac{M_\infty}{\rho}\right)\right) \quad (12)$$

near T_g.

The above result seems to adequately describe the observed Brillouin scattering of a homogeneous single-component-polymer fluid near its glass transition (*13*). If a binary mixture is truly homogeneous, then Equation 11 also should be valid for this case. However, if the two components phase separate, then the distribution function given by Equation 10 is invalid and further analysis is required.

If the microphase regions are small relative to $2\pi/q$, then some additional broadening will be observed attributable to greater fluctuations in in M_∞/ρ. If the moduli of the two phases differ significantly, then as the size of the regions grows, the Brillouin peaks will broaden considerably and will have non-Gaussian shapes. Eventually two sets of Brillouin peaks will be observed, reflecting phase-separated regions with size large relative to $2\pi/q$ but still smaller than the scattering volume. For large phase-separated regions, the observed spectrum will depend strongly on the part of the sample that is probed.

If the two polymers are compatible at high temperatures, Brillouin studies of the quenched glassy materials can reveal how homogeneous the mixtures remain on cooling. Examples of this type of study will be presented. If there are well-defined, phase-separated regions, then the behavior of both phases can be studied simultaneously. This case also will be presented.

Experimental

Typical Brillouin splittings are in the range 10^8–10^{10} Hz. Brillouin linewidths are in the range 16^6–10^9 Hz. Thus, a very high resolution instrument is required. Brillouin scattering is a weak effect so that an intense source of light is needed and a sensitive means of detection. All of these criteria have been met, and Brillouin scattering is now a routine tool of experimental physics (*13*).

The light source is an argon ion laser operated in a single-frequency mode. More than one watt of power can typically be obtained with a laser linewidth of approximately 10 MHz. The incident-beam polarization can be continually adjusted with respect to the scattering plane.

The frequency distribution of the scattered light is analyzed with a Fabry–Perot interferometer. A detailed discussion of this instrument is given in Ref. *12, 13,* and *15*. For the present chapter, it is most important to note the high contrast that can be obtained by operating the interferometer in the multiple pass mode. The Brillouin peaks in a polymer-blend sample or in a semicrystalline polymer sample can be more than six orders of magnitude smaller than the strong central peak. Contrasts of greater than 10^7 are routinely available now for a three-pass system. Thus Brillouin scattering from polymer blends is quite feasible.

The scattered light is detected with a photomultiplier tube and photon-counting electronics. The digital spectrum is then recorded with a multichannel analyzer and analyzed with a computer. A typical experimental arrangement is shown in Figure 2.

Translucent bulk samples are not suitable for study, but polymer films are quite acceptable (*16*). With films care must be taken to properly define the true scattering angle. The use of index-matching fluids is not recommended because they are likely to alter the polymer sample.

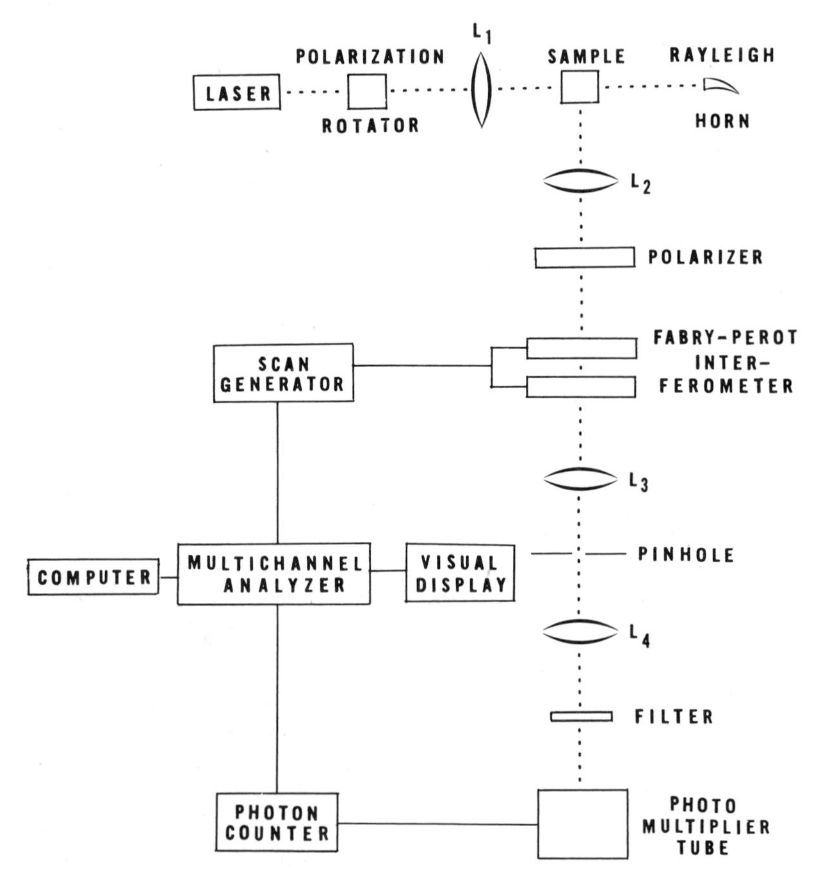

Figure 2. Light-scattering spectrometer

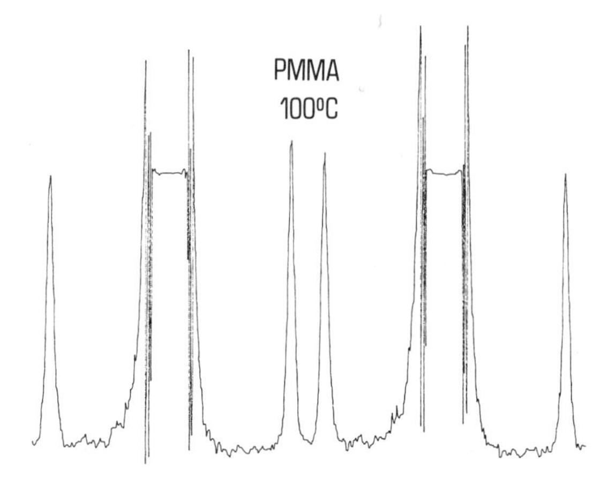

Figure 3. Brillouin spectrum of a film of pure PMMA at 100°C showing two Fabry–Perot orders

Examples and Discussion

Polymer blends of poly(methyl methacrylate) (PMMA) and poly(vinylidene fluoride) (PVF_2) have been shown to be compatible above the melting point of PVF_2 (*17*). Clear films can be prepared by rapid quenching from the melt. The Brillouin spectrum of a film of pure PMMA near its glass transition is shown in Figure 3. The longitudinal peaks are sharp and well resolved. The corresponding spectrum of a blend of 75% by weight PVF_2 with PMMA is shown in Figure 4. The

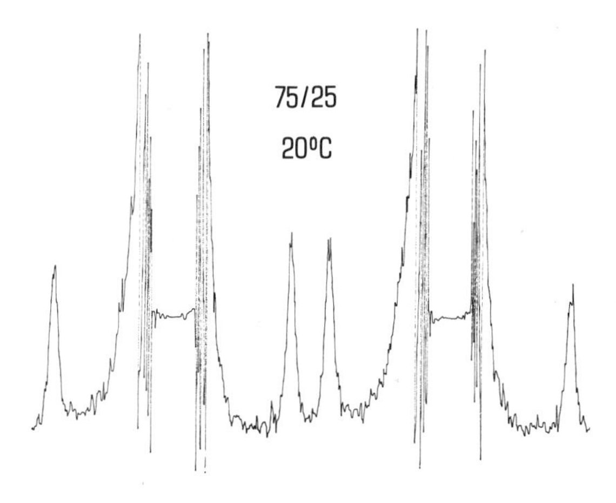

Figure 4. Brillouin spectrum of a quenched film of 75% PVF_2 25% PMMA by weight at 20°C

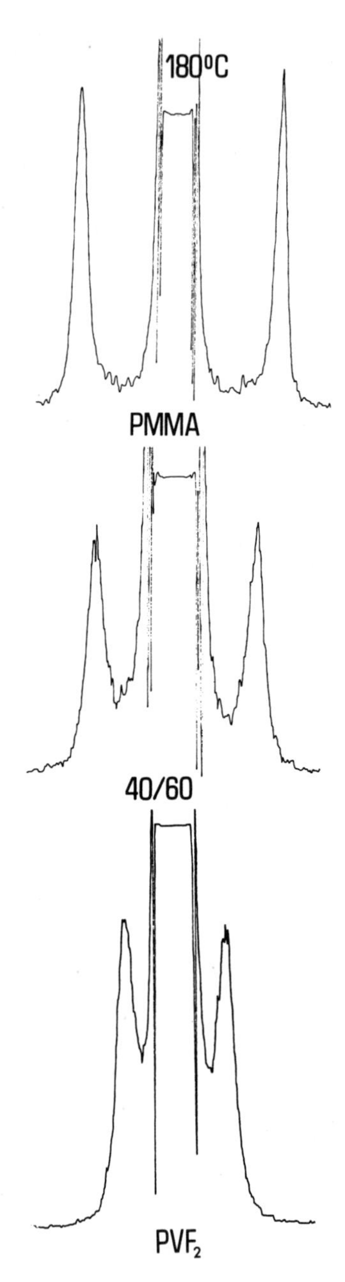

Figure 5. Brillouin spectra of PMMA, a 40% PVF_2–60% PMMA mixture, and pure PVF_2 at 180°C

quenched film appears to be homogeneous based on the sharp peaks. Measurements of the Brillouin splitting as a function of temperature have been carried out for several PVF_2–PMMA blends (*18*). The quenched films appear amorphous and a single glass transition is observed. Above T_g, the PVF_2 crystallizes from the mixture and the Brillouin splitting is greater than would be predicted for the amorphous blend. Finally the PVF_2 melts and the spectrum is again characteristic of a homogeneous amorphous phase. The spectra of PMMA, a 40% PVF_2 blend, and pure PVF_2 at 180°C are shown in Figure 5. The linewidths at 180°C are determined mostly by the dynamic viscosity contributions given in Equation 4.

Many commercial polymer films are blends of different polymers and other additives. The Brillouin spectrum (*16*) of Vynalloy PVC film is shown in Figure 6. The small peaks at higher Brillouin splitting are attributed to the presence of styrene–acrylonitrile copolymer that is blended into the film. These peaks are absent in the spectrum of pure PVC. The spectrum (*16*) of a commercial cellulose acetate film is presented in Figure 7. The lower frequency shoulder is probably caused by the inhomogeneous addition of a plasticizer. Again, pure cellulose-acetate films do not display the above feature. The existence of mechanical inhomogeneities is easily detected using Brillouin scattering.

A Brillouin spectrum of commercial Mylar film is shown in Figure 8. The longitudinal (L) and transverse (T) Brillouin peaks are easily seen.

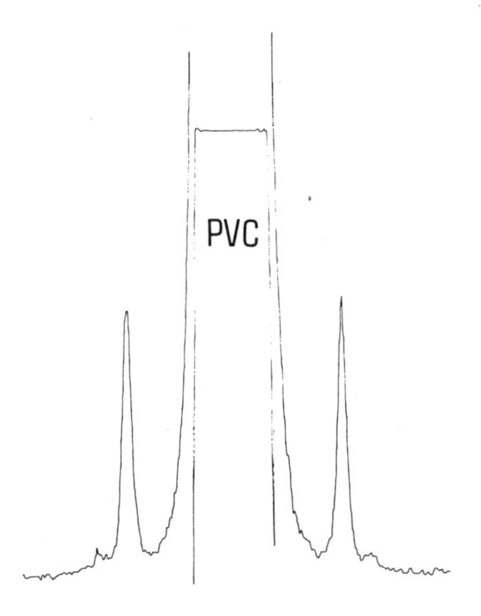

Figure 6. Brillouin spectrum of Vynalloy PVC film

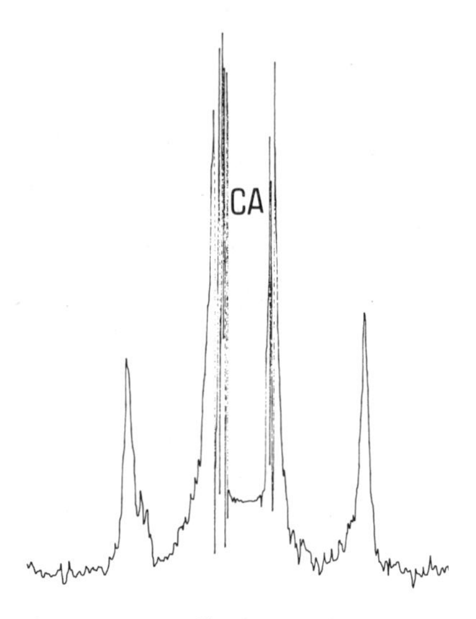

Figure 7. Brillouin spectrum of a commercial cellulose-acetate film

These are the only peaks seen in quenched amorphous poly(ethylene terephthalate) (PET) films (*19*). The third set of peaks was a mystery until the Brillouin spectrum was examined using a larger free-spectral range. The result is shown in Figure 9. The transverse peaks are now hidden in the wings of the central peak. The third set of peaks has a Brillouin splitting almost twice that of the normal longitudinal peaks. This feature is attributable to the presence of crystals of the cyclic trimer of PET on the surface of the Mylar film. The Brillouin splitting of the crystals is plotted as a function of temperature in Figure 10.

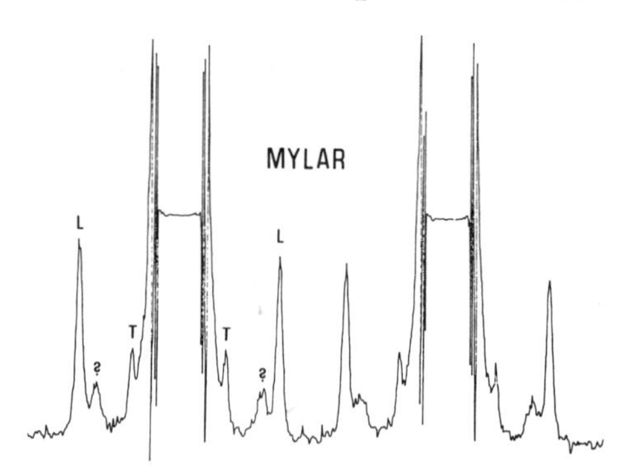

Figure 8. Brillouin spectrum of commercial Mylar film showing longitudinal (L), transverse (T), and unknown (?) Brillouin peaks

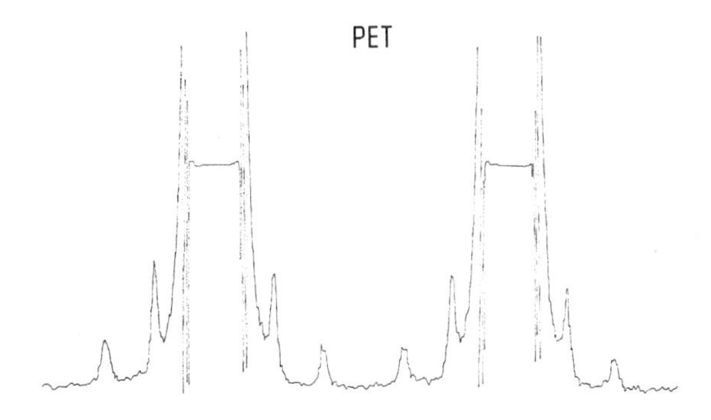

Figure 9. Brillouin spectrum of commercial Mylar film at larger free-spectral range than in Figure 8

Commercial Mylar film is partially crystalline and mechanically anisotropic. The longitudinal Brillouin splitting is observed to be greater than that for quenched amorphous PET and to be different for films oriented parallel and perpendicular to the film-roll edge. Longitudinal Brillouin splittings for Mylar and for amorphous PET are plotted vs. temperature in Figure 11. The apparent value of T_g also is elevated for the Mylar film from 70° to 80°C.

Measurement of the Brillouin spectrum of polymer films is now straightforward using multiple-pass Fabry–Perot interferometry. The use of Brillouin scattering to study polymer blends as films should be a very fruitful area for further study.

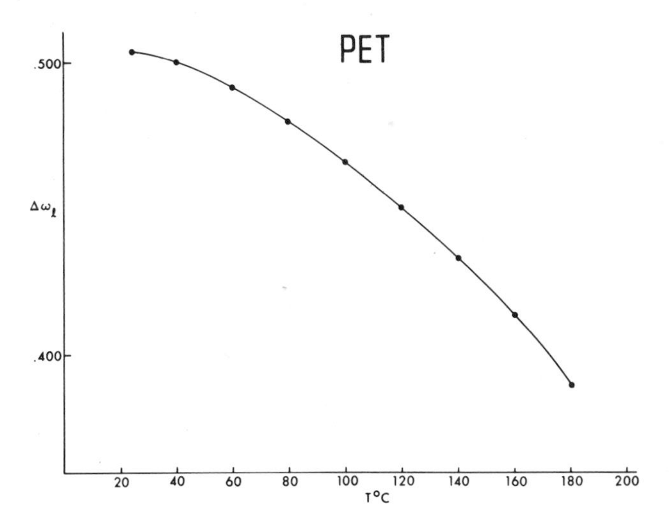

Figure 10. Brillouin splitting $\Delta\omega_l$ vs. temperature for cyclic trimer PET crystals on the surface of Mylar film

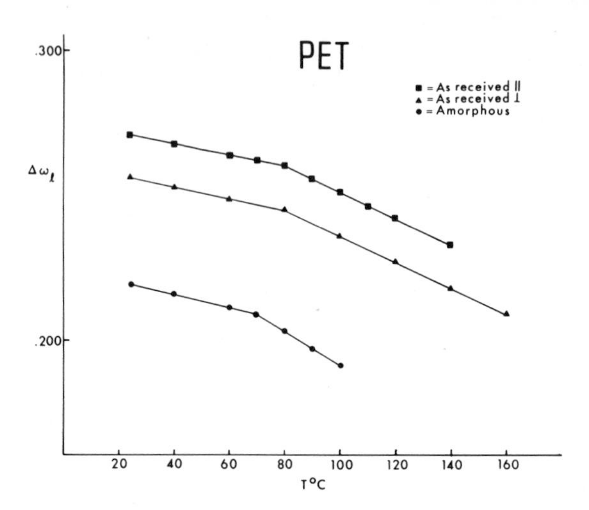

Figure 11. Brillouin splittings $\Delta\omega_l$ vs. temperature for amorphous PET, Mylar film oriented parallel to the film-roll edge and perpendicular to the film edge

Literature Cited

1. Brillouin, L., *Ann. Phys. (Paris)* (1922) **17,** 88.
2. Peticolas, W. L., Stegeman, G. I. A., Stoidreff, B. P., *Phys. Rev. Lett.* (1967) **18,** 1130.
3. Friedman, E. A., Ritger, A. J., Andrews, R. D., *J. Appl. Phys.* (1969) **40,** 4243.
4. Romberger, A. B., Eastman, D. P., Hunt, J. L., *J. Chem. Phys.* (1969) **51,** 3723.
5. Wang, C. H., Huang, Y. Y., *J. Chem. Phys.* (1976) **64,** 4847.
6. Coakley, R. W., Mitchell, R. S., Stevens, J. R., Hunt, J. L., *J. Appl. Phys.* (1976) **47,** 4271.
7. Brody, E. M., Lubell, C. J., Beatty, C. L., *J. Polym. Sci., Polym. Phys. Ed.* (1975) **13,** 295.
8. Jackson, D. A., Pentecost, H. T. A., Powles, J. G., *Mol. Phys.* (1972) **23,** 425.
9. Lindsay, S. M., Hartley, A. J., Shepherd, I. W., *Polymer* (1976) **17,** 501.
10. Mitchell, R. S., Guillet, J. E., *J. Polym. Sci., Polym. Phys. Ed.* (1974) **12,** 713.
11. Patterson, G. D., *J. Polym. Sci., Polym. Phys. Ed.* (1976) **14,** 741.
12. Patterson, G. D., *J. Macromol. Sci., Phys.* (1977) **B13,** 647.
13. Patterson, G. D., "Methods of Experimental Physics: Polymer Physics," R. Fava, Ed., Academic, New York, 1979.
14. Rytov, S. M., *Sov. Phys. JETP, Engl. Transl.* (1970) **31,** 1163.
15. Patterson, G. D., ADV. IN CHEM. SER. (1974) **174.**
16. Patterson, G. D., *J. Polym. Sci., Polym. Phys. Ed.* (1976) **14,** 143.
17. Nishi, T., Wang, T. T., *Macromolecules* (1975) **8,** 909.
18. Patterson, G. D., Nishi, T., Wang, T. T., *Macromolecules* (1976) **9,** 603.
19. Patterson, G. D., *J. Polym. Sci., Polym. Phys. Ed.* (1976) **14,** 1909.

RECEIVED April 14, 1978.

28

Domain Flow in Two-Phase Polymer Systems

A Study of the Flow Properties of a Styrene–Methylmethacrylate Diblock Copolymer and SBS Triblock Copolymers

J. LYNGAAE-JØRGENSEN, NARASAIAH ALLE, and F. L. MARTEN

Instituttet for Kemiindustri, The Technical University of Denmark, DK-2800 Lyngby, Denmark

Based on thermodynamic considerations, criteria for the existence of domains in the melt in simple shear fields are developed. Above a critical shear stress, experimental data for the investigated block copolymers form a master curve when reduced viscosity is plotted against reduced shear rate. Furthermore the zero shear viscosity corresponding to data above a critical shear stress follow the WLF equation for temperatures in a range: $T_g + 100°C$. This temperature dependence is characteristic of homopolymers. The experimental evidence indicates that domains exist in the melt below a critical value of shear stress. Above a critical shear stress the last traces of the domains are destroyed and a melt where the single polymer molecules constitute the flow units is formed in simple shear flow fields.

A number of investigations concerning the morphology of AB and ABA block copolymers have shown that these systems normally exhibit phase separation if the blocks consist of sufficiently long chains. One block type often is found to constitute a dispersed phase in a continuous matrix of the second block type.

A thermodynamic treatment of phase separation in block copolymers has been given by Krause (*1, 2*). As is the case with blends of homopolymers, phase separation in block copolymers is caused by a positive free energy of mixing. Meier (*3*) has presented a treatment of micro-

0-8412-0457-8/79/33-176-541$05.00/0

phase separation and domain formation and obtains a relation between domain size and chain length of the block. Furthermore, Meier gives a basis for estimation of critical block chain length for phase separation and domain morphology.

Previous investigations (*4, 5, 6, 7, 8*) have shown that block copolymers exhibit unusual melt rheological properties such as a very high viscosity, elasticity, and non-Newtonian behavior even at very low shear rates which are all attributed to the multiphase structure resulting from the incompatibility between the two copolymer units in the melt state.

The purpose of this research project was to investigate whether single polymer molecules could constitute the flow units (in the following referred to as a monomolecular melt) in melts of a diblock copolymer and a triblock copolymer, respectively. If a monomolecular melt does exist, under what conditions will it be formed?

Experimental

Sample Materials. A styrene–butadiene–styrene (SBS) triblock copolymer from Phillips Petroleum named Solprene 414C was studied in the investigation.

The material and information on molecular weight data were kindly supplied by R. H. Burr, Phillips Petroleum Co.: $\overline{M}_w = 129{,}000$, $\overline{M}_n = 107{,}000$, $\overline{M}_w/\overline{M}_n = 1.21$, styrene content 40% by weight. (Our own GPC measurements showed styrene content 39% and $\overline{M}_w/\overline{M}_n = 1.38$ without correction for band spreading.) SBS and SB materials caused problems because of thermal (and possibly mechanical) degradation.

A diblock copolymer model material with blocks of methylmethacrylate (approximately 25%) and styrene was prepared since this system should be thermally stable. The diblock copolymer was prepared using a technique described by Rempp et al. (*9, 10*) with slight modifications. The following amounts were used: methylmethacrylate 50 g (0.5 mol), styrene 150 g (1.44 mol), solvent THF 1,000 mL, and *n*-butyllithium $2.5 \cdot 10^{-3}$ mol (anionic catalyst). Reaction temperature, −55°C.

The monomers were vacuum distilled three times and dried with CaH_2. THF was refluxed over $CuCl_2$ (5 hr), refluxed five times over CaH_2 (8 hr), distilled, and finally refluxed over $LiAlH_4$. All treatments were carried out under nitrogen. Before the monomers were added, a drop of styrene was added to the solvent which was titrated with *n*-butyl lithium until the weakly red, styrylic anion color was stable.

The solution was diluted with cyclohexane, a solvent for polystyrene and a nonsolvent for polymethylmethacrylate, and the product thereafter precipitated by addition to methanol; THF/cyclohexane/methanol = 1:2:10 (by volume). The product was finally dried to constant weight.

Characterization of Molecular Structure. The molecular weight distribution (MWD) was determined before and after processing using a GPC with two detectors. The method is described by Runyon et al. (*11*).

The GPC apparatus was Waters Assoc. Model 200. The column combination was 10^6, $2 \cdot 10^4$, 10^4, 10^3 Å polystyrene gel. The GPC instrument was run under the following conditions: flow rate: 1 mL/min; injection volume: 2 mL; sample concentration: $2.5 \cdot 10^{-3}$ g/mL. THF was used as the solvent.

The columns were calibrated with polystyrene standards from Pressure Chemical Co. and from National Bureau of Standards under the above mentioned conditions except for the injection volume which was 0.5 mL.

The calibration found for PS standards was used directly in the calculations since intrinsic viscosity–molecular weight relations for PS and PMMA samples in THF do not deviate significantly, $[\eta] = 1.17 \cdot 10^{-2} \cdot \overline{M}_w^{0.717}$ for PS in THF (*12*) and $[\eta] = 1.28 \cdot 10^{-2} \cdot \overline{M}_w^{0.69}$ [cm³/g] for PMMA in THF (*13*) (the last relation is based on very few points with quite large scatter).

The styrene content was determined to 74% by weight with 5% of the total mass as pure polystyrene, $\overline{M}_w = 91{,}000$, $\overline{M}_n = 50{,}000$, $\overline{M}_w/\overline{M}_n = 1.82$. Except for a low-molecular-weight tailing of pure polystyrene (5%), no deviation from the average composition was observed. No significant changes in MWD were observed after processing except for the two highest temperatures, where $\overline{M}_w$ dropped to $\sim$ 80,000.

Rheometry. Melt-flow properties at low shear rates in a steady state rotational mode were determined by using a Rheometrics Mechanical Spectrometer on a cone and plate set up (DDC-1) and (DDC-2) between 160° and 250°C. The details of the geometry were: disks, diameter = 25 and 50 mm, respectively; cone angle, 0.040 rad; gap, 0.050 mm. After loading the material, measurements were performed in succession starting with low and thereafter increasing shear rates. Additional measurements on the triblock copolymer were performed on a Brabender Plastograph and a Rheometric's rheometer in a bicone geometry as covered by Ref. *14*.

Theoretical (Basic Hypothesis). The purpose of this part is to derive a criterion for the transition from a two-phase melt state to a monomolecular melt state. The basic idea behind the derivation is as follows. The total change in free energy (ΔG_T) by removing one domain from a melt at constant shear stress (τ) would consist of two contributions, $\Delta G_T = \Delta G_{mix} + \Delta G_2$. One contribution ΔG_{mix} corresponds to the change in the free energy of mixing. Since the systems considered are originally two-phase systems (for $\tau = 0$), ΔG_{mix} is always positive.

The action of a domain in a polymer melt (at constant shear stress) can be shown to be equivalent to the action of a giant crosslink in a rubber. Removing of one "crosslink" is accompanied by a negative "free-energy change" (ΔG_2).

The conditions where the last domain vanishes is found as follows. A condition for equilibrium between a two-phase structure and a homogeneous-melt structure is that the chemical potential of a repeat unit in the domain $\mu_A{}^D$ and in the monomolecular melt state μ_A are the same:

$$\mu_A{}^D = \mu_A \text{ or } \mu_A - \mu_A{}^D = 0 \qquad (1)$$

The chemical potential of a repeat unit in a domain at shear stress τ can be written as the identity:

$$\mu_A{}^D = \mu_A{}^o + \mu_A{}^D(-\mu_A{}^o)$$

where $\mu_A{}^0$ is the chemical potential of a repeat unit of pure amorphous polymer A.

$$\mu_A - \mu_A{}^D = \mu_A - \mu_A{}^o - (\mu_A{}^D - \mu_A{}^o) = 0$$

where $\mu_A - \mu_A{}^0$ is equal to the positive change in chemical potential on mixing and $-(\mu_A{}^D - \mu_A{}^0)$ is equal to the negative change in chemical potential realized when a domain acting as a giant crosslink is removed.

That is,

$$(\mu_A - \mu_A{}^o)_{mix} = \mu_A{}^D - \mu_A{}^o. \quad (2)$$

$(\mu_A - \mu_A{}^0)_{mix}$ is found by partial differentiation with respect to the number of polymer repeat units found in domains (U_A), from the expression for the Gibbs free energy change on domain destruction found by Krause (2).

$$(\mu_A - \mu_A{}^o)_{mix} = \left(\frac{\partial \Delta G_{mix}}{\partial U_A}\right)_{T,P,U_B}$$

S. Krause's thermodynamic treatment of the Gibbs free-energy change of mixing on domain destruction for the whole system of N_c copolymer molecules occupying volume V gives

$$\Delta G_{mix}/kT = (V/V_r)v_A v_B \chi_{AB}(1 = 2/z) + N_c \ln(v_A{}^{v_A} v_B{}^{v_B})$$
$$- 2\,Nc\,(m-1)(\Delta S_{dis}/R) + N_c\,ln(m-1) \quad (3)$$

where V is the total volume of the system, V_r is the volume of a lattice site, V_A and V_B are the volume fractions of monomer A and B in the copolymer molecule, respectively, z is the coordination number of the lattice, χ_{AB} is the interaction parameter between A units and B units, m is the number of blocks in the block copolymer molecule, $\Delta S_{dis}/R$ is the disorientation entropy gain on fusion per segment of polymer which is again related as $\Delta S_{dis}/R = \ln\,[(z-1)/e]$, where e is the base of the natural logarithms, k is the Boltzman constant, and T is the absolute temperature. On differentiating Equation 3 with respect to polymer repeat units in a domain (U_A),

$$(\mu_A - \mu_A^o)_{mix} = \left(\frac{\partial \Delta G_m}{\partial U_A}\right)_{T,P,U_B} =$$

$$\frac{V_A}{\overline{\overline{V}}} RT \left\{ \chi_{AB}\left(1 - \frac{2}{z}\right) v_B^2 + \frac{1}{\overline{X}_n}\left[v_B \ln \frac{v_A}{v_B} + \ln(v_A^{v_A} \cdot v_B^{v_B})\right] - \frac{2}{\overline{X}_n}(m-1)\frac{\Delta S_{dis}}{R} + \frac{1}{\overline{X}_n}\ln(m-1) \right\} \quad (4)$$

for a copolymer with molar volumes V_A, V_B and mole fractions n_A and n_B, $V = n_A V_B + n_B V_B$, v_A is the volume fraction of repeat units found in domains, $v_B = 1 - v_A$ and X_n is the degree of polymerization.

As shown in a later publication,

$$(\mu_A{}^D - \mu_A{}^o) = \frac{\tau^2}{Q \cdot T} \quad (5)$$

where τ is the shear stress, T absolute temperature, and Q is a constant depending on the molecular structure of the material (calculated as 0.166 (erg mol/cm^6 K) for the PS-PMMA block copolymer).

$$Q = \frac{2a^2 \cdot \rho \cdot R}{\overline{V}\,\overline{M}_c \cdot H_2}$$

where a is approximately equal to the ratio between the second and the first normal stress difference (approximately 0.1), ρ is the melt density, R is the gas constant, V is the average molar volume of the repeat units, M_c is the critical molecular weight where the exponent a in the equation $\eta = K \cdot M_w{}^a$ changes from 1 to 3.5, and H is the heterogeneity of the sample: $H = M_w/M_n$.

Substituting Equations 4 and 5 into Equation 2 and simplifying, the following expression is found:

$$\frac{\tau_{cr}{}^{*2}}{T} = A'T + B' \quad (6)$$

$$\text{for } \chi_{AB} = \alpha + \frac{\beta}{T}, \quad (7)$$

where α and β are constants, $\tau_{cr}{}^*$ is the shear stress where the last domain disappear, and

$$A' = Q\frac{V_A}{\overline{V}}\left\{ Rv_B^2\left(1-\frac{2}{z}\right)\alpha + R/\overline{X}_n\left[\ln v_A^{v_A}v_B^{v_B} + v_B \ln\frac{v_A}{v_B}\right]\right.$$

$$\left. -\,(2R/\overline{X}_n)\,(m-1)\,(\Delta S_{dis}/R) + \frac{R}{\overline{X}_n}\ln(m-1)\right\}$$

$$B' = Q\frac{V_A}{\overline{V}}\,Rv_B^2\left(1-\frac{2}{z}\right)\cdot\beta$$

Since A' and B' are constants for a given copolymer and the possible temperature interval is relatively limited, a transition is predicted at an approximately constant shear stress. Melt transitions have in fact been reported at approximately constant shear stresses for styrene–butadiene–styrene triblock copolymers (*4*). However, this behavior was certainly not observed for the styrene methylmethacrylate diblock copolymer.

A possible explanation for the observed behavior is given as follows. The exact movement of a domain in a shear flow field is difficult to describe if the domain is a viscoelastic liquid because the domain will deform. However, whether a domain is "rigid" or deformable it has to rotate in order to constitute a flow unit. The force holding the domain together is the "chemical potential force" opposing mixing of A and B segments. However, if flow processes occur inside a domain it may break up before the conditions derived above are reached. The following hypothesis could serve as a first approximate rationalization of the observed behavior.

The probability P that a part of a polymer molecule belonging to a domain will be "torn" out by flow processes is assumed to be proportional to the ratio between the energy which is actually transmitted by the melt to that which is theoretically necessary for destruction of a domain. P is furthermore inversely proportional to the "domain viscosity."

That is

$$P\alpha\,\frac{(\mu_u^D - \mu_A)\,dU_A}{(\mu_A^D - \mu_A)^{\text{theoretical}}dU_A}\cdot\frac{1}{\eta_D} = \frac{\tau_{cr}^2}{\tau_{rc}^{*2}}\frac{1}{\eta_D} = \text{constant} = K$$

This leads to a prediction of the form

$$\tau_{cr}^2 = K\tau_{cr}^{*2}\cdot\eta_D \text{ for } K\cdot\eta_D \leq 1 \qquad (8a)$$

$$\tau_{cr} = \tau_{cr}^* \qquad \text{for } K\cdot\eta_D \geq 1 \qquad (8b)$$

where τ_{cr}^* is the theoretical value of critical shear stress and η_D is the domain viscosity.

Results

SM Copolymer. Figure 1 shows double logarithmic delineations of shear stress against shear rate with temperatures between 160° and 250°C as discrete variables. All of the curves show a break at some critical shear stress τ_{cr} or shear rate γ_{cr}. Corresponding to the parts of the curves representing shear rates higher than γ_{cr}, Ferry's equation (15) $(1/\eta) = (1/\eta_o) + b\tau$ has been used to define a zero shear viscosity. Such plots gave straight lines. The zero shear viscosities ($\eta_o{}^*$) found by this procedure are given in Table I together with τ_{cr} values.

Since the data for the styrene–methylmethacrylate copolymer show that τ_{cr} and γ_{cr} occur (Figure 1) very close to the Newtonian range, this empirical procedure should give reliable values of zero shear viscosity.

SBS Polymers. As mentioned earlier, processing of SBS samples caused degradation. This can be seen in Figures 2a and 2b, showing the molecular-weight distribution of Solprene 414C before and after processing on a Brabender plastograph.

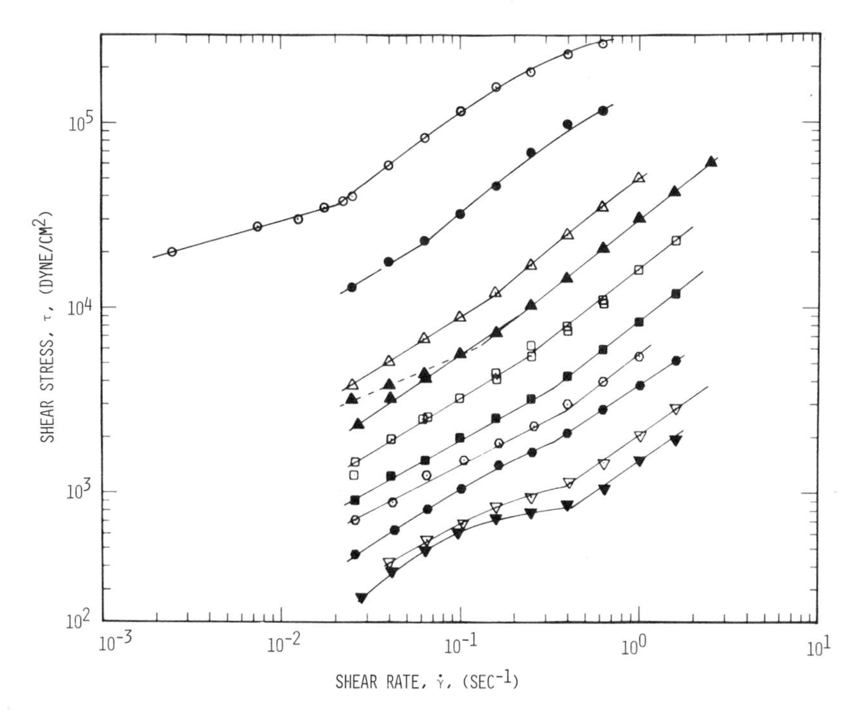

Figure 1. Shows a double logarithmic delineation of shear stress against shear rate with temperature as discrete variable; (○) 160°, (●) 170°, (△) 180°, (▲) 190°, (□) 200°, (■) 210°, (⬡) 220°. (⬢) 230°, (▽) 240°, (▼) 250°C. Material: S–M diblock polymer.

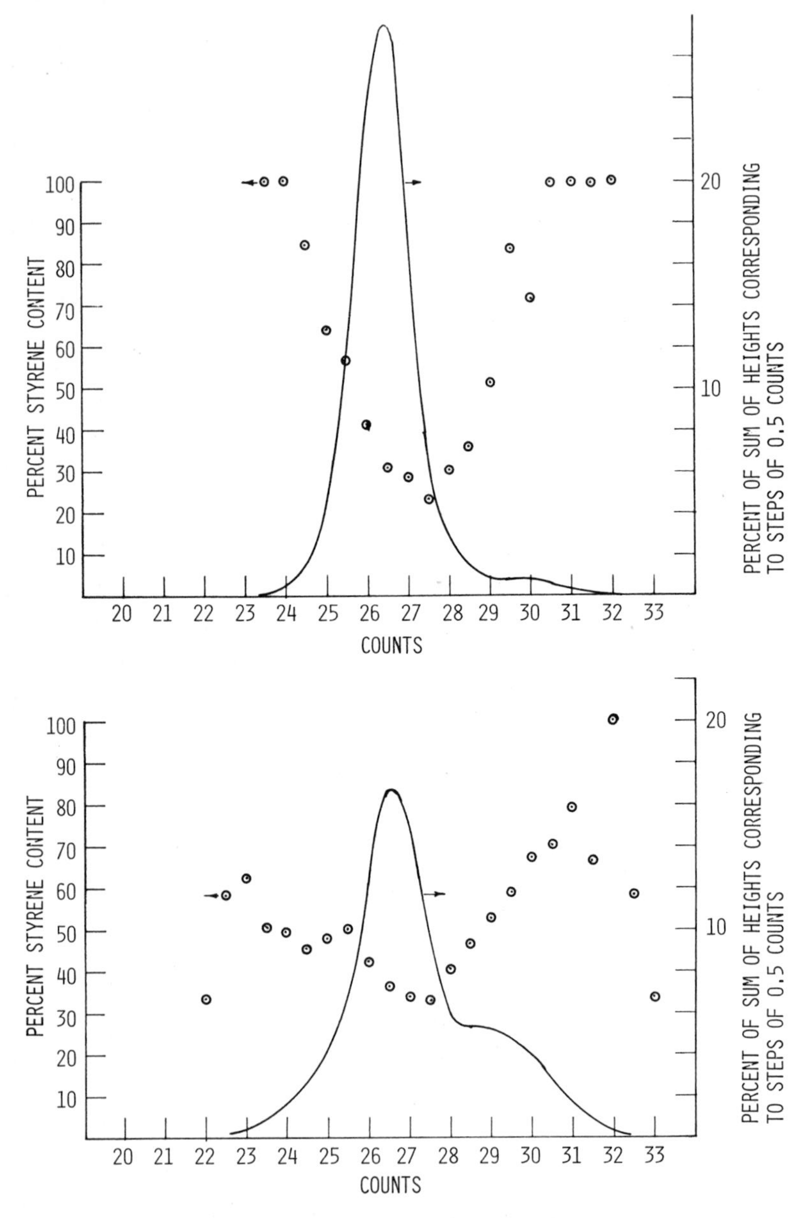

Figure 2. Shows the influence of degradation under processing for SBS copolymer 414C. (a) The molecular weight distribution (MWD) before processing. (b) The MWD after 41 min treatment in a Brabender Plastograph, Walzenkneter 30; temperature ~ 180°C and rate of rotation: 110 rotations per minute (with 1% Ionox added).

Table I. Data for the Styrene–Methylmethacrylate Diblock Copolymers

Temperature (°C)	*Zero Shear Viscosity* $\eta_0^* \times 10^{-4}$ *(poise)*	*Critical Shear Stress* $\tau_{cr} \times 10^{-4}$ *(dyn/cm²)*
160	222.20	3.6
170	38.46	2.5
180	8.33	1.08
190	4.17	0.6
		0.85
200	2.18	0.6
210	1.13	0.37
220	0.61	0.28
230	0.33	0.19
240	0.19	0.112
250	0.11	0.087

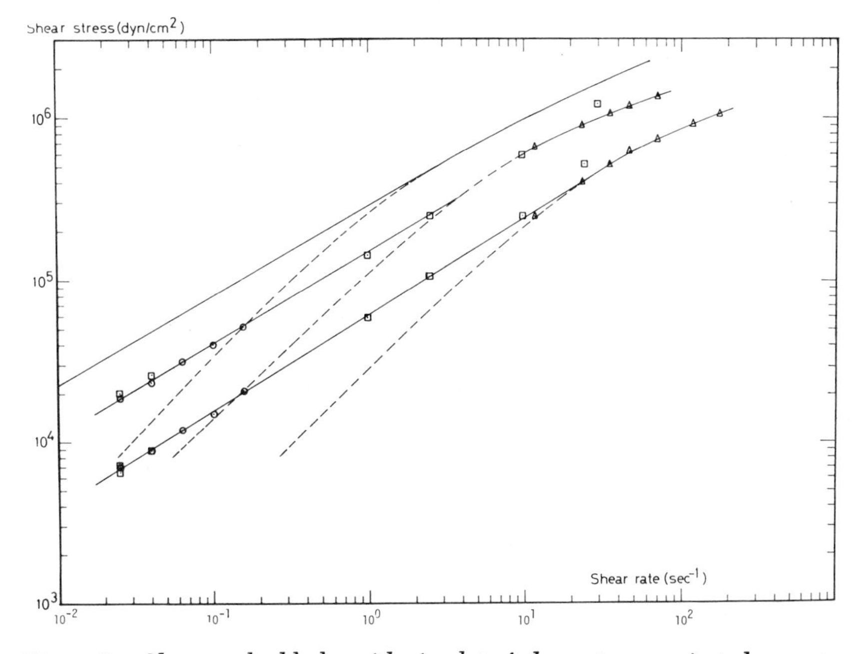

Figure 3. Shows a double logarithmic plot of shear stress against shear rate. The far left curve is taken from Ref. 4 for a SBS copolymer with 39% Styrene (see Table II) at 150°C. The two other curves represent SBS copolymer 414C at 160° and 200°C, respectively. (○) Cone and plate data, (□) bicone data, (△) Brabender data. (– –) Flow curve corresponding to the master curve for homogeneous melts shown on Figure 7.

Since this was the case, all data from Brabender measurements were measured as quickly as possible. When a constant reading could not be achieved, the slowly changing torque was extrapolated to zero time. The correction was normally less than 10%. The data for Solprene 414C are shown in a double logarithmic plot of shear stress against shear rate on Figure 3.

The following relations were found by calibration:

$$\tau = 1200\, M$$
$$\dot{\gamma} = 1.2 \cdot N$$

For Brabender measurements with a mixing chamber, Walzenkneter 30, M is the measured torque, and N is the number of rotations per minute (τ in dyn/cm^2 and M in g · m).

$$\tau = 120 \cdot M$$
$$\dot{\gamma} = 10\, N'$$

For Rheometrics measurements with a bicone geometry, (N' radians per sec, τ is in dyn/cm^2 and M is in g · cm).

Discussion

Styrene–methylmethacrylate Diblock Copolymer. Referring to Meier's (3) work, we expect that our model system consists of rather small spherical domains of PMMA blocks in a PS matrix (domain diameter around 500 Å). The method of precipitation should point to the same conclusion.

The experimentally observed data may be explained by a hypothesis mentioned under theoretical part (b). A delineation of log τ_{cr} against log η_D should thus constitute a straight line with slope 0.5.

Since the critical conditions (τ_{cr}, γ_{cr}) were reached in areas where the melt was close to the first Newtonian range, η_D should be proportional to the zero shear viscosity of the monomolecular melt. (As a first approximation, $\eta_D = (M_{PMMA}/M)^{3.5}\ \eta_o$, where M_{PMMA} is the molecular weight of a PMMA block and M is the molecular weight of a polymer molecule.)

Thus a delineation of log τ_{cr} against log η_o should be a straight line with slope 0.5 and intercept $\frac{1}{2}\log[(\tau_{cr}{}^*)^2 \cdot K \cdot B]$ for log $\eta_o = 0$. $\tau_{cr}{}^*$ is the theoretical shear stress calculated by use of Equation 6, K is the constant in Equation 8a, and $B \cong (M_{PMMA}/M)^{3.5}$. Furthermore, (log $\tau_{cr}{}^*$, log $1/KB$) should be a point on the line where the slope changes from 0.5 to 0.

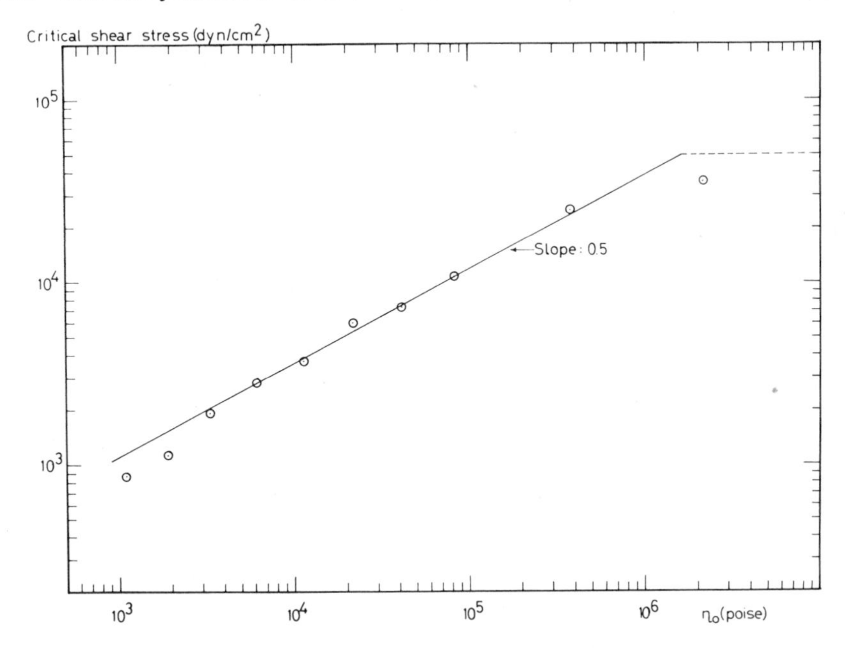

Figure 4. Shows a double logarithmic delineation of critical shear stress against the zero shear viscosity η_o for the S–M diblock copolymer*

Figure 4 shows a delineation of log τ_{cr} against log η_o. The slope is 0.5 within experimental uncertainty and the intercept $\frac{1}{2}\log[\tau_{cr}^*)^2 \cdot K \cdot B]$ $= 1.556$ from which $[(\tau_{cr}^*)^2 \cdot K \cdot B]$ is 1294 (cgs units) at log $\eta_o = 0$.

Calculation of τ^*_{cr}. The values recommended by S. Krause $\Delta S_{dis}/R = 1$, $z = 8$, and $v_B = 0.756$, $v_A = 0.244$, $\chi_n = 500$ was used in Equation 6.

The interaction parameter was estimated as follows:

$$\chi_{AB} = \alpha + \frac{\beta}{T}$$

$$\frac{\beta}{T} = \frac{\overline{V}}{RT}(\delta_A - \delta_B)^2 \rightarrow \beta = \frac{\overline{V}}{R}(\delta_A - \delta_B)^2$$

where δ_A and δ_B are the solubility parameters of repeat units of polymethylmethacrylate and polystyrene, respectively, and $\alpha \simeq 0$.

In the evaluation of Q: (cgs units) $a = 0{,}1$, $\rho = 1$, $R = 8.317 \cdot 10^7$, $\overline{V} = 102$, $\overline{M}_c = 25{,}000$, $H = 1.96$. Since PS repeat units and PMMA repeat units have very different polar and hydrogen-bonding character, the work of Charles Hansen (*16, 17*) was used in the calculation of β. That is, contributions from dispersion forces, polar forces, and hydrogen-bonding forces were added according to the equation:

$$(\delta_A - \delta_B)^2 = (\delta_A - \delta_B)_d^2 + (\delta_A - \delta_B)_p^2 + (\delta_A - \delta_B)^{h2} =$$
$$(8.6–9.2)^2 + (3.0–4.0)^2 + (2.0–4.2)^2 = 9.2$$

and $\beta = 469.2$. With these values inserted in Equation 6, $\tau_{cr}{}^* = 5 \cdot 10^4$ dyn/cm² for $T = 200°C$ (and $\tau_{cr}{}^*$ is predicted to change very slowly with temperature).

With the value of $\tau_{cr}{}^*$ known, $K \cdot B$ can be estimated either from the intercept $\frac{1}{2}\log((\tau_{cr}{}^*)^2 \cdot K \cdot B)$ in the double logarithmic delineation of shear stress against zero shear viscosity or as $KB = 1/\eta_o{}^*$, where $\eta_o{}^*$ is the value of zero shear viscosity corresponding to $\tau_{cr}{}^*$ (read from the straight line on Figure 4).

That is, $K \cdot B = 5.2 \cdot 10^{-7}$ poise^{-1} from the intercept and $K \cdot B = 6.25 \cdot 10^{-7}$ poise^{-1} from the point where $\tau_{cr}{}^* = 5 \cdot 10^4$ dyn/cm² crosses the straight line $\log \tau_{cr} - \log \eta_o$. Thus the hypothesis seems to be reasonable, giving approximately the same values for $K \cdot B$ and the correct slope.

Two criteria were used to examine the hypothesis that above a critical shear stress the melt was in a homogeneous melt state. The first

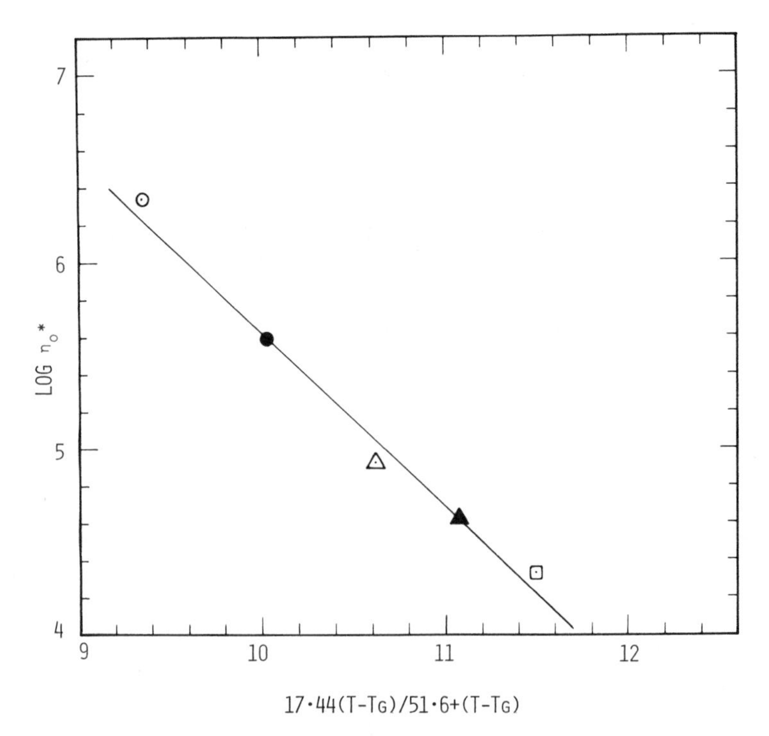

Figure 5. Shows a plot of the logarithm of the zero shear viscosity against 17.44 (T − T$_g$)/51.6 + (T − T$_g$). (○) 160°, (●) 170°, (△) 180°, (▲) 190°, and (□) 200°C. Material: S–M diblock copolymer.

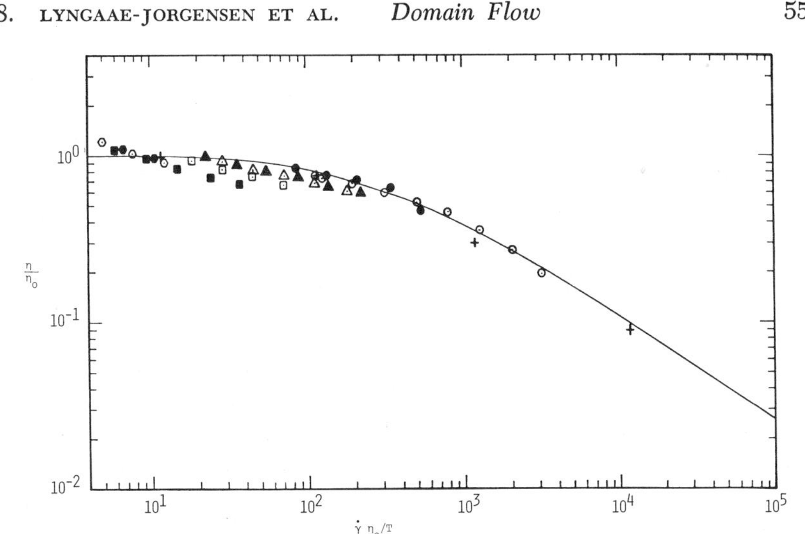

Figure 6. Shows a double logarithmic delineation of $\eta/\eta_o{}^*$ against $\eta_o{}^*\gamma/T$ for the S–M diblock copolymer. (○) 160°, (●) 170°, (△) 180°, (▲) 190°, (□) 200°, (■) 210°, (⬡) 220, and (⬢) 230°C. Full-drawn curve is a master curve for polystyrene samples with $M_w/M_n \approx 2$ and (+) represents Utracki's (12) data for polystyrene samples with a heterogeneity of 2.3.

criterion was to see whether the zero shear viscosities ($\eta_o{}^*$) showed a temperature dependence as expected for monomolecular melts. The second was to investigate if a master curve of the type η/η_o against $\lambda\gamma$ (where λ is a relaxation time) of the same form as the one shown for PS samples is applicable to the diblock melt data.

(a) Since the data are close to the glass-transition temperature of PS and PMMA (100° and 104°C, respectively), our data should follow the William–Landel–Ferry (*18*) (WLF) equation with $T_g \simeq 100°C$. Figure 5 shows a delineation of log $\eta_o{}^*$ against $[17.44\ (T - T_g)]/[51.6 + (T - T_g)]$. The experimentally determined slope of this curve is -0.95. This slope is not significantly different from the expected value of -1.

(b) A delineation of log η/η_o against log $(\eta_o/T)\gamma$ is a master curve for PS samples with $M_w/M_n \approx 2$. If the block copolymer is in a monomolecular melt state, it must comply with the same master curve. This is so as shown on Figure 6.

The fact that the melt of the block copolymer above a critical shear stress shows the same structure–viscosity relations as found for linear melts in simple shear flow is taken as an indication of a monomolecular melt state.

SBS-Triblock Copolymers (40% Styrene). The following values were used in calculation of $\tau_{cr}{}^*$ for Solprene 414C:

$$\frac{R}{\Delta S_{dis}} = 1, z = 8, v_B = 0.6, v_A = 0.4, \chi_n = 1760, \beta = \frac{\overline{V}}{R} (\delta_A - \delta_B)^2$$

where δ_A and δ_B are the solubility parameters of repeat units of polystyrene and polybutadiene, respectively,

$$\delta_A = 9.1 \left(\frac{\text{cal}}{\text{cm}^3}\right)^{\frac{1}{2}}, \delta_B = 8.5 \left(\frac{\text{cal}}{\text{cm}^3}\right)^{\frac{1}{2}}, \text{and } \overline{V} = 39 \frac{\text{cm}^3}{\text{mol}}$$

$\beta = 7.0$, $\alpha \cong 0$, $a = 0.1$, $H = 1.3$, $\overline{M}_c = \overline{M}_c^{PB} \cdot n_{PB} + \overline{M}_c^{PS} \cdot n_{PS} =$ 9400 g/mol for $\overline{M}_c^{PB} = 5600$ g/mol, $\overline{M}_c^{PS} = 31{,}200$ g/mol, and mole fractions $n_{PB} = 0.852$ and $n_{PS} = 0.148$, respectively. This gives $Q =$ 2.685 [(erg mol)/(cm^6 · K)] and $\tau_{cr}^* = 5.2 \cdot 10^5$ dyn/cm^2 for $T =$ 200°C.

From Figure 3 it is seen that this prediction seems to be correct for both our own data and those reported by Holden, Bishop, and Legge (D) for a SBS copolymer with 40% styrene and measured with a bicone geometry.

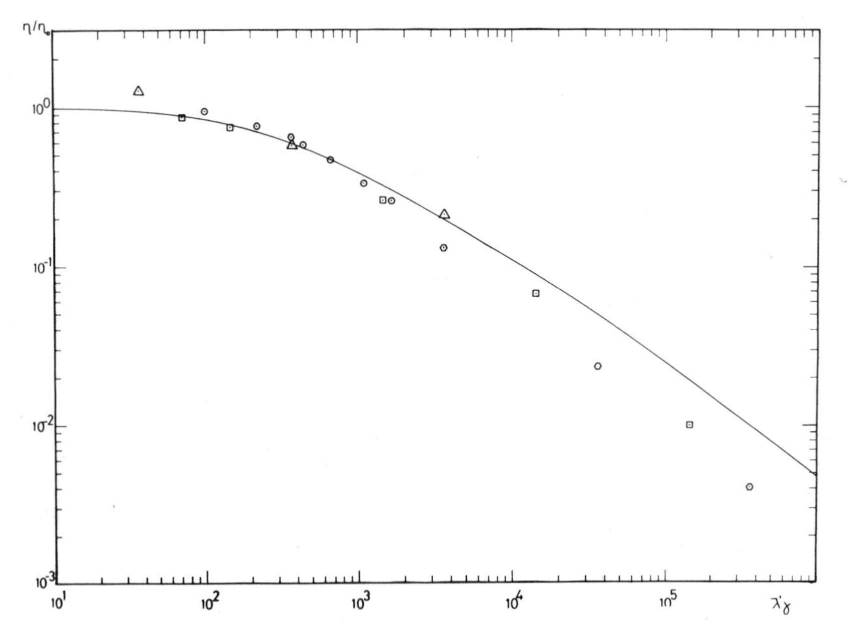

Figure 7. Shows a double logarithmic plot of η/η_o^* against f · $\eta_o^*\gamma$/T for SBS copolymers. The full-drawn curve is the same master curve as shown on Figure 4. (○) SBS–414C at 200°C, (□) SBS (14S–74B–14S) at 175°C, (⬡) SBS (20S–110B–20S) at 175°C, and (△) SBS (10S–52B–10S) at 175°C. The η data for the last three samples has been taken from Ref. 4. The η_o^* values used are given in Table II.

Table II. SBS Copolymers

Sample	*Temperature (°C)*	*Zero Shear Viscosity* $\eta_o^* \times 10^{-5}$ *(poise)*
16S–52B–16S (39%) (*4*)	150	3.3
Solprene 414C (40%S)	200	0.264
10S–52B–10S (27.8%S) (*4*)	175	1.0
14S–74B–14S (27.5%S) (*4*)	175	39
20S–110B–20S (26.7%S) (*4*)	175	1000

A delineation of log η/η_o against log $[(\eta_o/T)\dot{\gamma} \cdot f]$ should be the polystyrene master curve with $M_w/M_n = 2$ for SBS melts above the critical shear stress (f is a theoretical factor correcting for differences in structure such as different heterogeneity (to be given in a later publication)).

Figure 7 shows that for four different SBS polymers with styrene content 27% and 40%, a delineation of logη/η_o against log $[(\eta_o/T)\dot{\gamma} \cdot f]$ is in fact a master curve: f (40% styrene) = 0.196, f (27% styrene) = 0.164. The zero shear viscosities for the SBS samples in Figure 7 are given in Table II. The fact that the SBS master curve deviates from the PS master curve at high values of $(\eta_o/T)\dot{\gamma} \cdot f$ may be caused by nonisothermal conditions in the capillary.

Conclusion

The experimental evidence indicates that below a critical value of shear stress, domains exist in the melt. Above a critical value of shear stress, the domains are destroyed and a monomolecular melt phase is formed under shear flow conditions.

The critical shear stress seems to be predictable using the theoretical expression given in Equation 6 in this chapter, at least when the viscosity of the domains exceeds a critical value.

The predicted critical shear stress is approximately constant (varies slowly with temperature).

List of Symbols

a = ratio between the second and first normal stress difference
A = polymer originally in domains
A' = constant in Equation 6
b = empirical constant in Ferry's equation
B = the ratio between the domain viscosity η_D and the zero shear viscosity η_o^*
B' = constant in Equation 6

e = base of natural logarithm
f = correction factor for differences in structure
ΔG_{mix} = free-energy change on mixing
ΔG_2 = free-energy change attributable to removal of one domain in a melt experiencing shear flow
ΔG_T = total free energy change
H = heterogeneity index
k = Boltzman constant
K = empirical constant in Equation 8a
m = number of blocks
MMA = methylmethacrylate
M_{PMMA} = molecular weight of polymethylmethacrylate
M_{PS} = molecular weight of polystyrene
M_{PB} = molecular weight of polybutadiene
M = molecular weight of a copolymer molecule
M = torque
MWD = molecular weight distribution
$\overline{M}_w$ = weight average molecular weight
$\overline{M}_n$ = number average molecular weight
$\overline{M}_c$ = critical molecular weight
n = mole fraction
n_A = mole fraction of polymer A
n_B = mole fraction of polymer B
N = number of rotations per minute
N' = number of radians per second
N_c = number of copolymer molecules
P = probability
PS = polystyrene
PB = polybutadiene
Q = empirical constant in Equation 5
R = gas constant
S = styrene
SBS = styrene–butadiene–triblock copolymer
SM = styrene–methylmethacrylate diblock copolymer
$\Delta S_{dis}/R$ = disorientation entropy
T = absolute temperature
T_g = glass-transition temperature
THF = tetrahydrofuran
U_A = number of repeat units of polymer A
U_B = number of repeat units of polymer B
v_A = volume fraction of polymer A
v_B = volume fraction of polymer B
V_A = molar volume of monomer A

V_B = molar volume of monomer B
V = total volume
$\overline{V}$ = average molar volume of the repeat units
V_r = average volume of a lattice site ($V_r = \overline{V}$)
$\overline{X}_n$ = degree of polymerization
z = coordination number
α = empirical constant in Equation 7
β = empirical constant in Equation 7
δ_B, δ_A = solubility parameters of repeat units in domains and in continuous phase, respectively
$\dot{\gamma}$ = shear rate
$\dot{\gamma}_{cr}$ = critical shear rate
η = melt viscosity
η_o = zero shear viscosity
η_o^* = estimated zero shear viscosity ($\eta_o^* = \eta_o$)
η_D = domain viscosity
μ_A^0 = chemical potential of a repeat unit of pure polymer A
μ_A = chemical potential of a repeat unit of polymer A in the homogeneous melt
μ_A^D = chemical potential of a repeat unit in domains
ρ = melt density
τ = shear stress
τ_{cr} = critical shear stress
τ_{cr}^* = theoretical critical shear stress
χ_{AB} = interaction parameter

Acknowledgment

Narasaiah Alle wishes to express his gratitude to the Danish International Development Agency (DANIDA) for financial support of this research project.

Literature Cited

1. Krause, S., *J. Polym. Sci., Polym. Phys. Ed.* (1969) **7**, 249.
2. Krause, S., *Macromolecules* (1970) **3**, 84.
3. Meier, D. J., *J. Polym. Sci., Polym. Symp.* (1969) **C26**, 81.
4. Holden, G., Bishop, E. T., Legge, N. R., *J. Polym. Sci., Polym. Symp.* (1969) **C26**, 37.
5. Leblanc, J. L., *Rheol. Acta* (1976) **15**, 654.
6. Chung, C. I., Giale, J. C., *J. Polym. Sci., Polym. Phys. Ed.* (1976) **14**, 1149.
7. Kraus, G., Naylor, F. E., Rollman, K. W., *J. Polym. Sci., Polym. Phys. Ed.* (1971) **9**, 1839.
8. Arnold, K. R., Meier, D. J., *J. Appl. Polym. Sci.* (1970) **14**, 427.
9. Rempp, P., Loucheux, M. H., *Bull. Soc. Chim. Fr.* (1958) 1497.
10. Freyss, D., Leng, M., Rempp, P., *Bull. Soc. Chim. Fr.* (1964) **2**, 221.

11. Runyon, J. R., Barnes, P. E., Rud, J. F., Tung, L. H., *J. Appl. Polym. Sci.* (1969) **13**, 2359.
12. Kolinský, M., Janca, J., *J. Polym. Sci., Polym. Chem. Ed.* (1974) **12,** 1181.
13. Rudin, A., Hoegy, H. L. W., *J. Polym. Sci., Polym. Chem. Ed.* (1972) **10,** 217.
14. Lyngaae-Jørgensen, J., *Polym. Eng. Sci.* (1974) **14,** 342.
15. Ferry, J. D., *J. Am. Chem. Soc.* (1942) **6,** 1330.
16. Hansen, C. M., *J. Paint Technol.* (1967) **39,** 104.
17. Hansen, C. M., "The Three Dimensional Solubility Parameter and Solvent Diffusion Coefficients," Thesis, Techn. University of Denmark (1967).
18. Aklonis, J. J., MacKnight, W. J., M. Shen, "Introduction to Polymer Viscoelasticity," p. 49–51, Wiley–Interscience, New York, 1972.
19. Utracki, L. A., Baker Djian, Z., Kamal, M. R., *J. Appl. Polym. Sci.* (1975) **19,** 481.

RECEIVED April 14, 1978.

29

Polymer Compatibilization: Blends of Polyarylethers with Styrenic Interpolymers

OLAGOKE OLABISI and A. G. FARNHAM

Research and Development Department, Chemicals and Plastics,
Union Carbide Corp., Bound Brook, NJ 08805

The polymer compatibilization concept is to modify advantageously the mechanical properties of a normally incompatible mixture by the addition of a suitable compatibilizing agent consisting usually of a block or graft copolymer. These interpolymers act as interfacial bridges between the components in the mixture. In this work, the 'compatibilization concept' is invoked, but pendant chemical groups are the compatibilizing agents rather than the more usual copolymers. Nitrile and/or ester groups have been attached conveniently to the backbone of a series of polyarylethers to compatibilize the ethers with αmethyl styrene/methyl methacrylate/acrylonitrile interpolymers. The choice of the appropriate pendant chemical groups was made on the basis of the observation that the styrene interpolymers are miscible with polymers that contain these specific groups.

A miscible polymer–polymer blend almost always yields a physical-property spectrum superior to the individual components, and this allows the development of a new set of products with significant savings in capital investment. Partly for this reason and partly because new and commercially viable polymers are becoming harder to 'come by,' the plastic industry has expended a sizeable sum towards identifying miscible high-performance polymer mixtures. Indeed, the more recent renewed experimental and theoretical programs have resulted in an increased number of known miscible blends. On the commercial scene, however, successful miscible polymer–polymer blends are still rather few and are limited to

0-8412-0457-8/79/33-176-559$06.75/0

polystyrene-poly(2,6-dimethyl 1,4-phenylene oxide), elastomer blends, miscible polymeric coatings, and blends of PVC with its permanent plasticizers.

On the other hand, some mechanically compatible blends as well as some dispersed two-phase systems have made respectable inroads into the commercial scene. Many of these are blends of low-impact resins with high-impact elastomeric polymers; examples are polystyrene/rubber, poly(styrene-co-acrylonitrile)/rubber, poly(methyl methacrylate)/rubber, poly(ethylene propylene)/propylene rubber, and bis-A polycarbonate/ABS as well as blends of polyvinyl chloride with ABS or PMMA or chlorinated polyethylene.

Another concept which has been exploited is the 'compatibilization' technique (*1–7*). The idea is to advantageously modify the mechanical properties of a normally incompatible mixture by the addition of suitable compatibilizing agents. Use of such agents is exemplified by the blend of poly(ethyl acrylate) with polystyrene-grafted poly(ethyl acrylate) (*1, 2*); the mixture of polyvinyl chloride with polybutadiene that has been separately grafted onto polystyrene, poly(methyl methacrylate), and the copolymer of the two (*3*); the mixture of polystyrene, polybutadiene, and polystyrene-grafted onto polybutadiene; and the mixture of polyvinyl chloride, cellulose, and poly(ethyl acrylate) grafted onto cellulose (*4*). A similar concept has been used in developing reinforced and filled polymer composites where glass fibers are mixed with polyester epoxy resin in the presence of the compatibilizing graft copolymer of glass with polyester or glass with epoxy resin. In the field of adhesives, improved adhesive bond is achieved between, for example, PVC floor tiles and concrete, wood, or gypsum by using a compatibilizing agent composed of graft copolymer of PMMA with natural rubber or epoxy resin (*5, 6*).

The common factor in all of these examples is that each system takes advantage of the compatibility of the compatibilizing agent with one or more of the components of the mixture. The agents are composed of a block or graft copolymer of one polymer molecule to another, and the generally accepted explanation is that the agents act as interfacial bridges which couple the components in the mixture. The resulting blend is usually immiscible but marginally compatible, and the degree of phase separation is reduced considerably.

In the recent miscibility studies with α-methyl styrene/acrylonitrile copolymer and a α-methyl styrene/methyl methacrylate/acrylonitrile terpolymer (*8*), it was found that almost all miscible second components contain amides, imides, nitriles, or esters, each of which contains lone-pair electrons capable of donor–acceptor complexation—a state which was earlier speculated to contribute somewhat to the formation of compatible polymer mixtures (*9*). It was decided, therefore, that introduction of these type of groups onto the backbone of a series of otherwise insoluble, high impact, high T_g polyarylethers might yield compatible blends of superior properties. That is, the 'compatibilization concept' is invoked, but now pendant chemical groups are the compatibilizing agents as opposed to the block and graft copolymers previously used. It may be more appropriate to refer to these pendant chemical groups as 'internal' compatibilizing agents and to the others as 'external' compatibilizing agents.

Experimental

Synthesis. Five different polyaryl ethers were made from the condensation product, resulting from the reaction of phenol and levulinic acid, commonly referred to as diphenolic acid, and one or more of the following monomers: bisphenol A, dichlorodiphenyl sulfone, 2,6-dichlorobenzonitrile, and 4,4′-difluorobenzophenone. The resulting polymers were subsequently methylated such that the common monomer becomes (**1**):

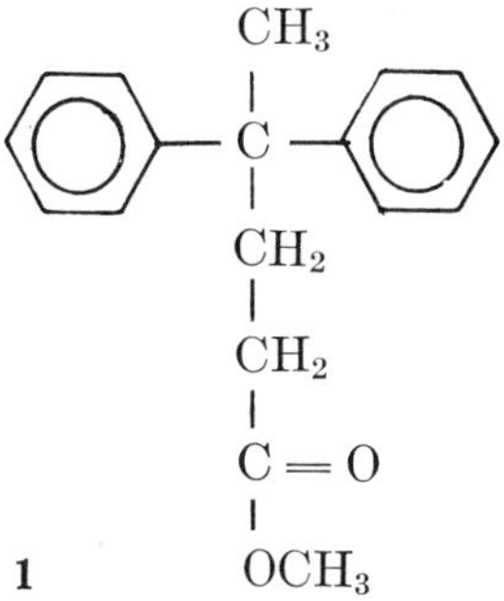

CH_3 | C | CH_2 | CH_2 | C=O | OCH_3 **1**

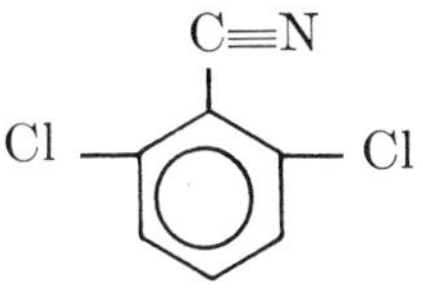

C≡N Cl Cl **2**

Six more polyaryl ethers were made from 2,6-dichlorobenzonitrile (**2**) and one of the following monomers: 1,1′-bis(4-hydroxy-3,5-dimethyl phenyl) cyclohexane; 2,2′-bis(4-hydroxyphenyl)-2-phenyl ethane; 1,3-bis(4-hydroxyphenyl)-1-ethyl cyclohexane; 2-(4-hydroxyphenyl)-2-[3-(4-hydroxyphenyl)-4-methyl cyclohexyl] propane; 2,2′-bis(4-hydroxy-3,5-dimethyl phenyl) propane; and bisphenol A.

Because the procedure used in making these polymers is similar (*10*), it is sufficient to describe one, namely, the synthesis of the terpolyether from bisphenol A, diphenolic acid, and dichlorodiphenyl sulfone, whose polymerization scheme is:

$$NaO\text{-}C_6H_4\text{-}C(CH_3)_2\text{-}C_6H_4\text{-}ONa + NaO\text{-}C_6H_4\text{-}C(CH_3)(CH_2CH_2COONa)\text{-}C_6H_4\text{-}ONa +$$

$$2\ Cl\text{-}C_6H_4\text{-}SO_2\text{-}C_6H_4\text{-}Cl \xrightarrow{DMSO}$$

$$\text{-}(O\text{-}C_6H_4\text{-}C(CH_3)_2\text{-}C_6H_4\text{-}O)_{0.5}\text{-}(C_6H_4\text{-}SO_2\text{-}C_6H_4)_1\text{-}(O\text{-}C_6H_4\text{-}C(CH_3)(CH_2CH_2COONa)\text{-}C_6H_4\text{-}O)_{0.5}\text{-} \xrightarrow{CH_3Cl}$$

$$\text{-}(O\text{-}C_6H_4\text{-}C(CH_3)_2\text{-}C_6H_4\text{-}O)_{0.5}\text{-}(C_6H_4\text{-}SO_2\text{-}C_6H_4)_1\text{-}(O\text{-}C_6H_4\text{-}C(CH_3)(CH_2CH_2COOCH_3)\text{-}C_6H_4\text{-}O)_{0.5}\text{-} + NaCl$$

The following were placed in a 500-cm^3 four-neck reaction flask fitted with stirrer, thermometer, nitrogen sparge tube, helices packed column, and Dean–Stark moisture trap with condenser: 25.68 g bisphenol A (.1125 mol), 32.21 g diphenolic acid (.1125 mol), 112.5 cm^3 Me_2SO, and 200 cm^3 toluene. The solution was warmed to about 50°C, air was displaced by nitrogen sparge, and 22.51 g NaOH (.5625 mol as 50.62% solution) was added. The mixture was stirred and heated to reflux with water separation and removal until no more water was evident in the reflux distillate. Toluene was distilled off to a pot temperature of about 160°C, and a solution of 64.65 g 4,4′-dichlorodiphenyl sulfone (.255 mol) in 110 cm^3 toluene was added. Toluene was removed to give a reaction temperature of 160°–165°C and maintained for about 2 hr. The mixture was cooled to 120°–125°C and methyl chloride bubbled in at this temperature to saturation until the mixture became uniform and the pH had dropped to give an acidic test with bromo cresol purple indicator (pH ~ 5). The polymer solution was diluted with 75 cm^3 chlorobenzene and filtered to remove salt. The clear, filtered solution was precipitated into methanol using a Waring blender, and the polymer fluff was filtered and washed with fresh methanol and dried under vacuum. Yield: 100.8 g (calc 107.6), RV_{CHCl_3} = 0.33. IR analysis of a film showed a deep absorption band for carbonyl ($\sim 5.75\ \mu$) but no band for free OH or carbonyl (at $\sim 3\ \mu$), indicating adequate conversion of sodium carboxylate to methyl ester.

NMR Characterization. The structures of the polyarylethers A–K (Table I) were confirmed through their NMR spectra. As an example of the NMR data, observe the spectrum in Figure 1 for sample K, the condensation product of disodium salt of diphenolic acid and 2,6-dichlorobezonitrile which was subsequently methylated as discussed above.

The spectrum is separated into two regions, namely, a low-field quartet plus doublet of lines and a high-field singlet with one doublet. The low-field doublet of lines, denoted by *p*, constitutes those hydrogens meta in the nitrile group, H_p. This doublet would appear as a singlet were it not for the proton para to the nitrile group, H_s. The interaction between the two magnetic nuclei is responsible for the splitting of the nuclear-spin energy levels. The quartet can be considered to be made up of a pair of doublets q and r, and such a quartet is very characteristic of para-disubstituted benzenes where the substituents are different. Proton H_s is not readily observable but its presence can be ascertained from the spectrum integral shown in triplicate in Figure 1.

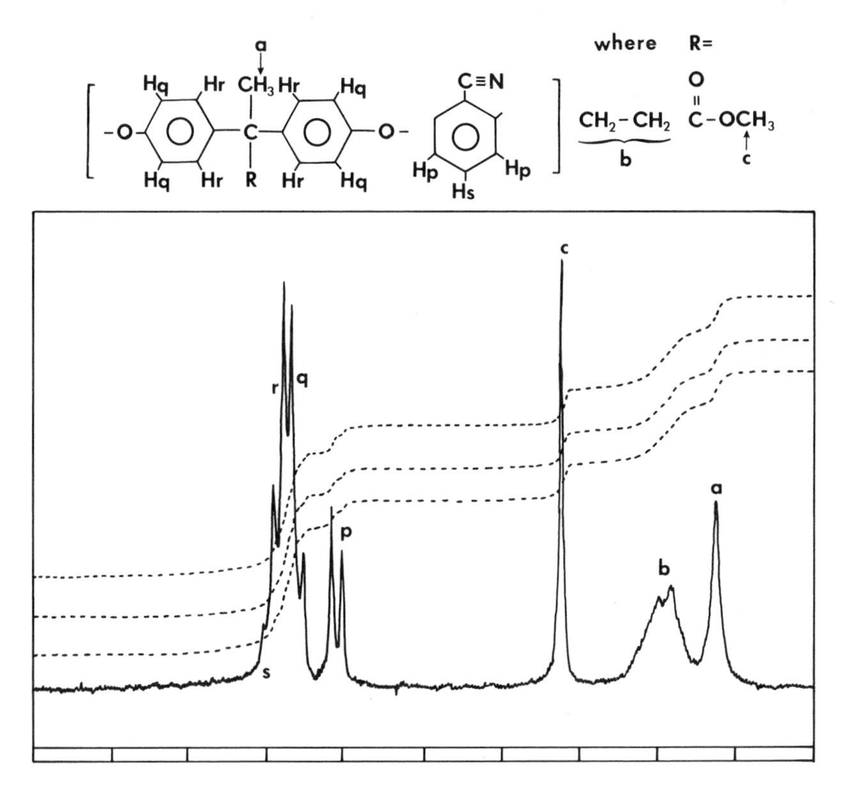

Figure 1. NMR spectra of polymer K

Table I. Polyarylether Structures

Polymer Designation	*Polymer Repeat Unit*	$\frac{\eta sp}{C}$ *(RV)*	T_g *(°C)*
Polysulfone	$[-O-C_6H_4-C(CH_3)_2-C_6H_4-O-C_6H_4-SO_2-C_6H_4-]$	0.48	184
A	$[-O-C_6H_4-C(CH_3)_2-C_6H_4-O-]_{n/2}[-C_6H_4-SO_2-C_6H_4-]_n[-O-C_6H_4-C(CH_3)(CH_2CH_2COOCH_3)-C_6H_4-O-]_{n/2}$	0.33	154
B	$[-C_6H_4-SO_2-C_6H_4-]_n[-O-C_6H_4-C(CH_3)(CH_2CH_2COOCH_3)-C_6H_4-O-]_n$	0.48	152
C	$[-O-C_6H_4-C(CH_3)(CH_2CH_2COOCH_3)-C_6H_4-O-C_6H_4-C(=O)-C_6H_4-]$	0.76	123
D	$[-O-C_6H_4-C(CH_3)_2-C_6H_4-O-]_{n/2}[-C_6H_3(CN)-]_n[-O-C_6H_4-C(CH_3)(CH_2CH_2COOCH_3)-C_6H_4-O-]_{n/2}$	0.52	149

Table I. Continued

Polymer Designation	*Polymer Repeat Unit*	$\frac{\eta sp}{C}$ *(RV)*	T_g *(°C)*
E		0.61	245
F		0.74	185
G		0.77	198
H		0.76	212
I		0.94	280
J		0.69	170

Table I. Continued

Polymer Designation	*Polymer Repeat Unit*	$\frac{\eta sp}{C}$ *(RV)*	T_g *(°C)*
K	$-[O-C_6H_4-C(CH_3)(CH_2CH_2COOCH_3)-C_6H_4-O-C_6H_3(CN)]-$	0.42	135

The high field singlet 'c' is attributable to the carbomethoxy methyl hydrogens, and singlet 'a' is attributable to the three equivalent methyl hydrogens designated as such on the structure in Figure 1. Peak b, attributable to the $—CH_2—CH_2—$ group, appears as a doublet because the methylene groups are not equivalent and they couple with each other, yielding the rather broadened doublet resonance.

The relative chemical shifts of the peaks a, b, and c are as expected for aliphatic hydrogens, and the integrals are in the approximate ratio 1:1.3:1, in excellent agreement with theory. Lastly, from the integrals, the ratio of the aromatic to the aliphatic protons is estimated to be 1:1.18; this should be compared with the theoretical value of 1:1.1 calculated from the fact that the monomer contains 11 aromatic and 10 aliphatic protons.

Mechanical Characterization. The polymer–polymer blends were made by using a Two-Roll Mill or a Brabender Plasticorder maintained at a temperature of ~ 250°C. Samples for the Instron tester and dynamic mechanical testing were prepared by compression molding. The heat distortion temperature (HDT), izod, and flexural properties were obtained from 125-mil-thick samples, the dynamic mechanical testing was done on 20-mil-thick samples, and all other data were obtained from 15-mil-thick samples. The mechanical property data were obtained at 25°C and the dynamic mechanical testing was conducted with a torsion pendulum equipment similar in design to that reported by Nielsen (*11*). Sample dimensions for the torsion pendulum were chosen to give a nominal frequency of 1 Hz in the solid state. The glass-transition temperatures were obtained from resilience–temperature measurements.

In using the resilience data for locating the T_g, one subjects a polymeric sample to a stress which is just enough to give one-percent strain at a given temperature T. On releasing the stress, the sample completely recovers (100 percent resilience) if $T < T_g$. As the temperature increases, the percent of recovered strain decreases, reaching a minimum at $T \approx T_g$. At temperatures just above the T_g the resilience increases again, reaching a maximum at a value usually less than 100 percent. At still higher temperatures, the resilience monotonically decreases to zero unless the material crosslinks.

All of the T_g data reported in Table II were obtained with the resilience method; a few selected results were doubly checked with the values obtained from the torsion-pendulum data. Note the identical value of the T_g for αMS/AN/MMA measured by the resilience method and that obtained with the torsion-pendulum data on Figure 12.

Results and Discussion

The terpolymer of α-methyl styrene/methyl methacrylate/acrylonitrile (60/20/20) is miscible in all proportion with poly(methyl methacrylate) (8). Figure 2 illustrates the dependence of T_g and HDT on

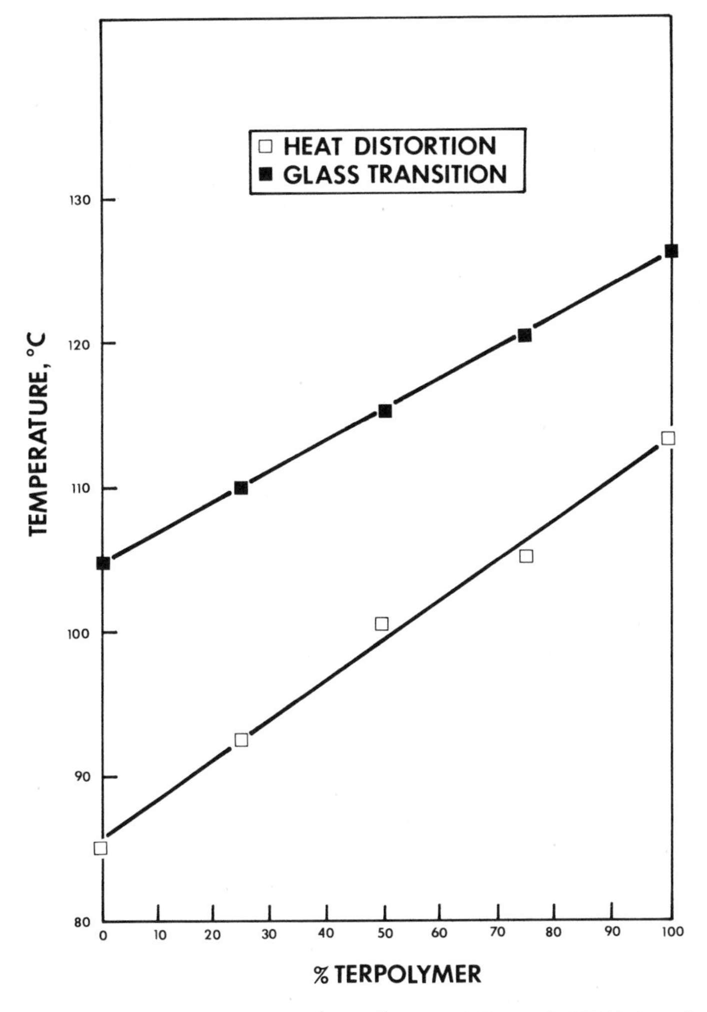

Figure 2. Composition dependence of T_g and HDT for the terpolymer/PMMA blends

the weight-percent terpolymer in the PMMA/terpolymer blends. Each data point corresponds to the single, distinct T_g or HDT for the blend. Such single, distinct T_gs intermediate between the pure component T_gs (as shown in Figure 2) are usually considered a necessary but not a sufficient condition for the establishment of polymer–polymer miscibility. It indicates that the two component polymers are miscible up to a scale corresponding to that responsible for the cooperative main-chain movement associated with the glass-transition temperature. Figure 3 illustrates a significant attribute of a miscible blend, namely, that the mechanical properties of the blends at intermediate compositions are superior to those of the two constituent polymers. This point is particularly more evident in the flexural properties.

More or less similar behavior has been observed (8) in the blends of the copolymer or the terpolymer with the following: bis-A polycarbonate, polyvinyl chloride, poly(ethyl methacrylate), and a terpolymer made from methyl methacrylate, *N,N′*-dimethyl acrylamide, and *N*-phenyl-maleimide. Because of this unique miscibility characteristic of the α-methyl styrene interpolymers, an attempt was made at compatibilizing polyarylethers with the interpolymers by attaching pendant chemical groups known to exist in systems with which the interpolymers are miscible.

Table I illustrates the structures of the modified polyarylethers, their glass-transition temperatures, and their reduced specific viscosities (RV) measured in chloroform at 25°C at a concentration of 0.2 g/100 mL.

Table II. Mechanical Properties

	Structure	*Blends*
	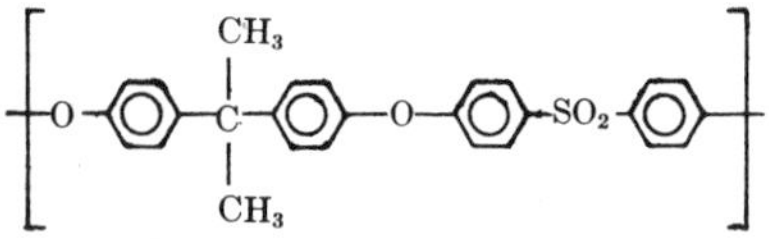 $-[-O-C_6H_4-C(CH_3)_2-C_6H_4-O-C_6H_4-SO_2-C_6H_4-]-$ Polysulfone	50% αMS/AN/MAA 50% αMS/AN Pure PSF
A	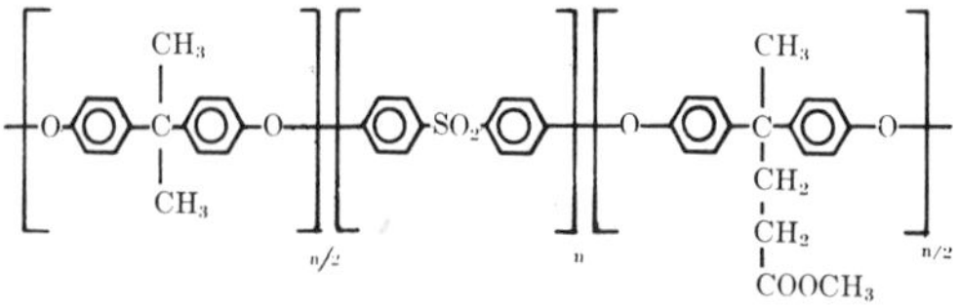 $-[-O-C_6H_4-C(CH_3)_2-C_6H_4-O-]_{n/2}-[-C_6H_4-SO_2-C_6H_4-]_n-[-O-C_6H_4-C(CH_3)(CH_2CH_2COOCH_3)-C_6H_4-O-]_{n/2}-$	50% αMS/AN/MAA 50% αMS/AN Pure A

Table II contains the mechanical properties and the glass-transition temperatures of the styrenic interpolymers, the various polyarylethers, and their 50/50 blends.

The blends of polysulfone with the α-methyl styrene polymers are immiscible, as evidenced by the double glass-transition temperatures in Table II. To improve the miscibility characteristics, polysulfone was modified in two ways. First, 25% of the bisphenol A was replaced by monomer I which contains a pendant ester group and, when no improvement resulted, the whole 50% of the bisphenol A was replaced. Again, the blends remain immiscible as evidenced from Figures 4 and 5 and from Table II. Further, the presence of the pendant ester group in polymer C does not improve the miscibility picture even though one would expect a favorable contribution from the carbonyl group on account of the miscibility of polycarbonate with the α-methyl styrene polymers.

A pendant nitrile group was also tried as a compatibilizing agent in polymers E–J with no significant success. Combination of the pendant nitrile with the pendant ester groups in polymer D also fails to improve the miscibility picture. In all of these cases, two minima are exhibited in the resilience–temperature curves, examples of which appear in Figures 4, 5, 6, 7, and 8 for the 50/50 blends of polymers A, B, D, F, and H with the styrenic interpolymers. The slight shifts in T_g which appear in some cases indicate the 'slight' affinity of the constituent polymers, but the consistent cloudiness is a further proof of the two-phase state of the

of Polyarylether Blends

1% Sec Modulus (psi)	*Yield Strength (psi)*	*Yield Elong.*	*Strength @ Break*	*Elong. @ Break*	*Pendulum Impact (ft lb/in³)*	T_g (°C)	*Plaque Appearance*
314,000	None	—	8,900	3.5	3.6	126,171	Cloudy
350,000	"	—	8,500	2.5	2.3	114,164	"
360,000	10,200	6.0	7,400	50.0	110.0	185	Transparent
400,000	None	—	8,200	2.0	3.0	126,154	Cloudy
410,000	"	—	5,300	1.5	1.5	110,142	"
357,000	"	—	8,100	2.5	7.5	154	Transparent

Table II.

	Structure	*Blends*
B	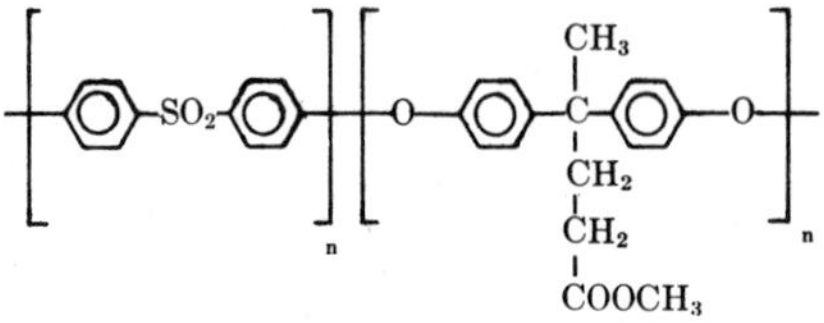 SO_2; O; CH_3; C; CH_2; CH_2; $COOCH_3$; O; n; n	50% αMS/AN/MAA 50% αMS/AN Pure B
C	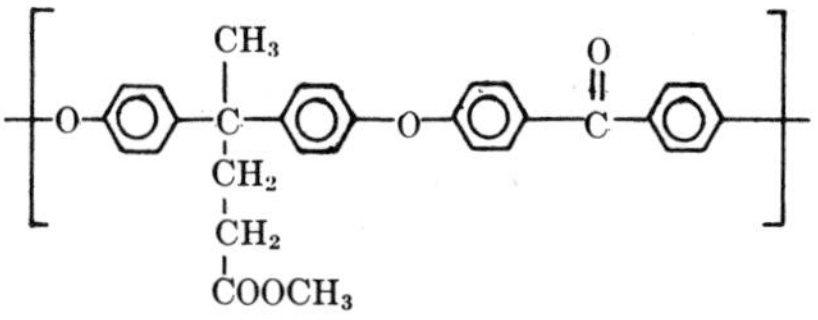 O; CH_3; C; CH_2; CH_2; $COOCH_3$; O; O; C	50% αMS/AN/MAA 50% αMS/AN Pure C
D	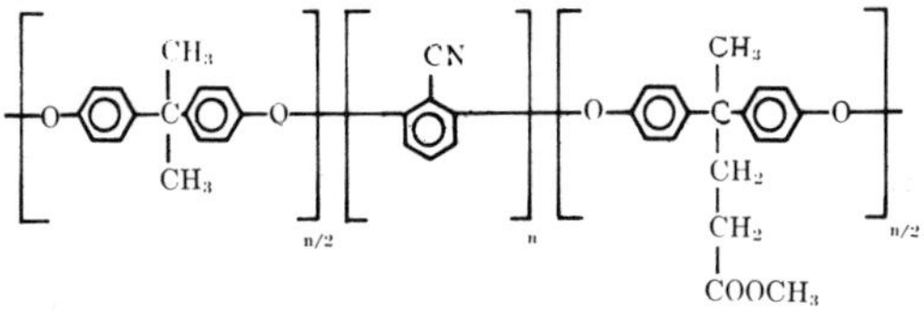 O; CH_3; C; CH_3; O; n/2; CN; n; O; CH_3; C; CH_2; CH_2; $COOCH_3$; O; n/2	50% αMS/AN/MAA 50% αMS/AN Pure D
E	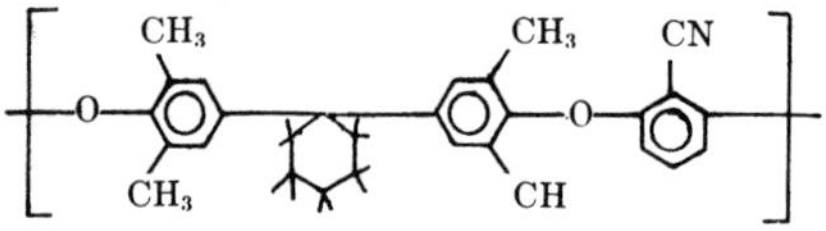 O; CH_3; CH_3; CH_3; CH; O; CN	50% αMS/AN/MAA 50% αMS/AN Pure E
F	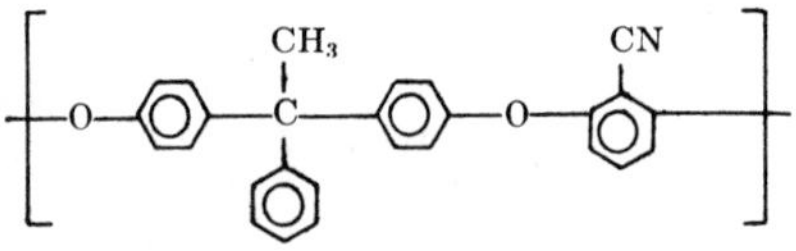 O; CH_3; C; O; CN	50% αMS/AN/MAA 50% αMS/AN Pure F
G	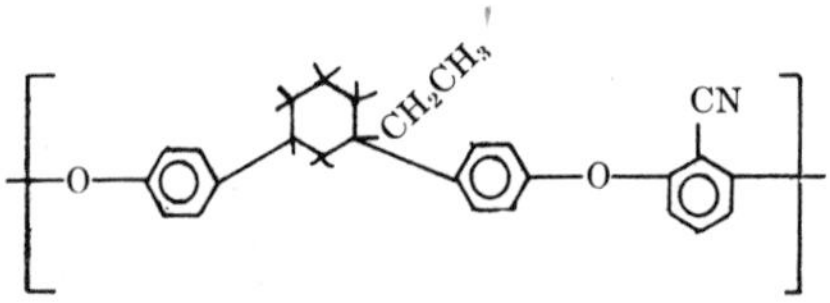 O; CH_2CH_3; O; CN	50% αMS/AN/MAA 50% αMS/AN Pure G

Continued

1% Sec Modulus (psi)	Yield Strength (psi)	Yield Elong.	Strength @ Break	Elong. @ Break	Pendulum Impact (ft lb/in³)	T_g (°C)	Plaque Appearance
384,000	"	—	11,300	4.5	3.7	127,141	Cloudy
335,000	"	—	8,300	2.5	2.1	110,136	"
336,000	"	—	9,300	3.0	45.0	152	Transparent
389,000	"	—	7,000	1.5	8.0	126	Cloudy
360,000	"	—	3,700	3.0	3.5	115	"
373,000	8,600	2.5	7,000	5.0	67.0	123	Transparent
338,000	None	—	10,900	4.5	4.5	126,140	Cloudy
353,000	"	—	9,000	3.0	3.8	114,134	"
320,000	"	—	10,800	6.0	87.0	149	Transparent
404,000	"	—	10,200	3.0	1.0	126,223	Cloudy
411,000	"	—	7,700	2.0	3.0	112,185	"
320,000	"	—	10,600	5.0	9.4	245	Transparent
320,000	"	—	10,300	4.5	3.9	120,160	Cloudy
288,000	"	—	6,400	2.5	3.0	115,179	"
347,000	15,000	6.5	12,600	10.0	51.0	185	Transparent
292,000	None	—	9,600	4.0	3.9	118,193	Cloudy
332,000	"	—	4,300	1.5	2.1	116,188	"
252,000	12,390	11.0	12,380	49.0	137.0	198	Transparent

Table II.

	Structure	*Blends*
H	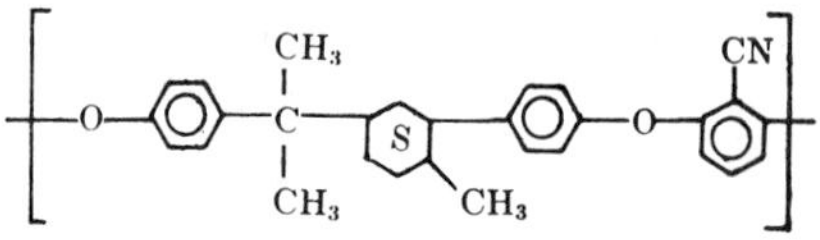 CH_3, C, CH_3, S, CH_3, CN, O	50% αMS/AN/MAA 50% αMS/AN Pure H
I	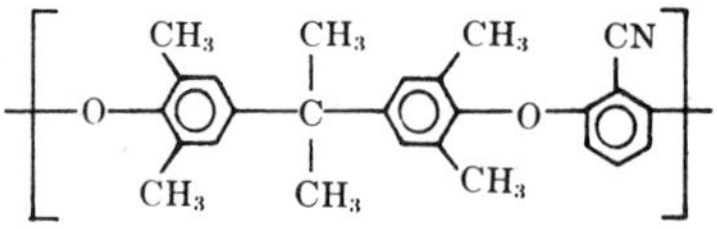 CH_3, CH_3, CH_3, CH_3, CH_3, CH_3, C, CN, O	50% αMS/AN/MAA 50% αMS/AN Pure I
J	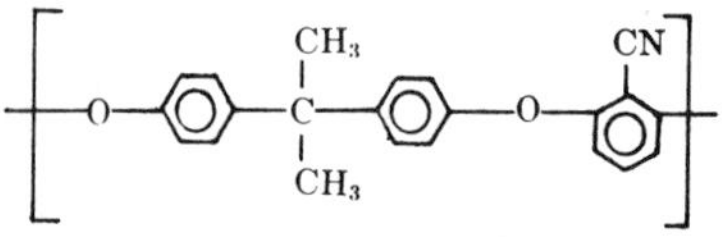 CH_3, C, CH_3, CN, O	50% αMS/AN/MAA 50% αMS/AN Pure J
K	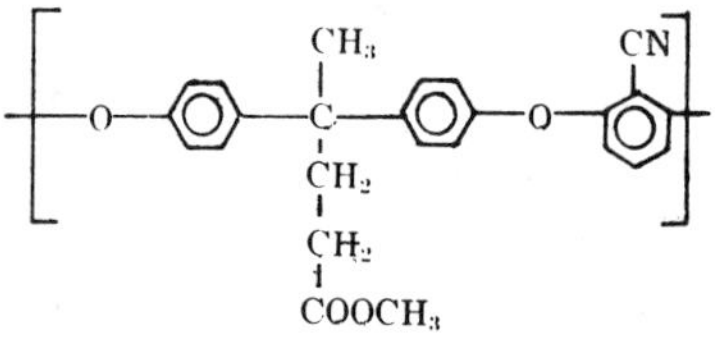 CH_3, C, CH_2, CH_2, $COOCH_3$, CN, O	50% αMS/AN/MAA 50% αMS/AN Pure K
	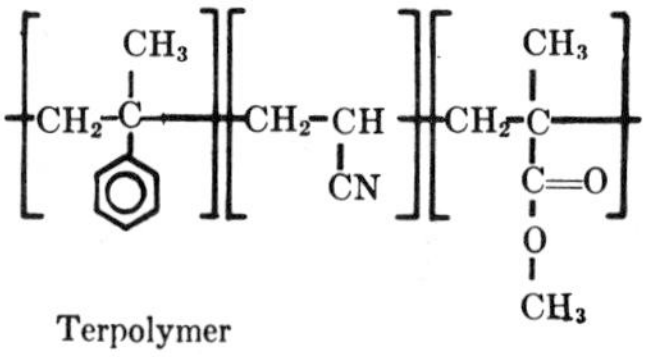 Terpolymer: CH_3, CH_2–C, CH_2–CH, CN, CH_2–C, CH_3, C=O, O, CH_3; 600 αMS, 200 AN, 200 MMA	Pure αMS/AN/MMA
	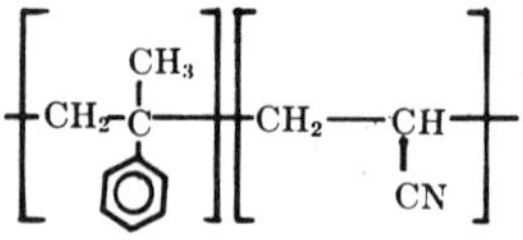 Copolymer: CH_3, CH_2–C, CH_2—CH, CN; 690 αMS, 310 AN	Pure αMS/AN

Continued

1% Sec Modulus (psi)	*Yield Strength (psi)*	*Yield Elong.*	*Strength @ Break*	*Elong. @ Break*	*Pendulum Impact (ft lb/in³)*	T_g *(°C)*	*Plaque Appearance*
400,000	"	—	11,200	4.0	1.0	121,186	Cloudy
347,000	"	—	3,900	1.5	1.0	113,195	Very Hazy
332,000	"	—	11,800	6.0	8.0	212	Transparent
303,000	"	—	8,900	3.5	4.3	126,203	Cloudy
320,000	"	—	4,600	1.5	1.7	114,162	Very Hazy
293,000	12,100	8.5	10,760	15.0	15.5	234	Transparent
304,000	"	—	12,000	7.0	7.5	126,146	Cloudy
320,000	"	—	9,600	3.5	4.0	111,148	"
330,000	13,000	5.8	12,790	26.0	185.0	161	Transparent
355,000	"	—	8,100	2.5	3.0	130	Cloudy
410,000	"	—	8,700	3.0	3.0	120	Hazy
360,000	"	—	11,000	4.5	60.0	128	Transparent
450,000	None	—	7,600	2.0	6.0	126	Transparent
437,000	None	—	8,300	2.0	3.0	108	Transparent

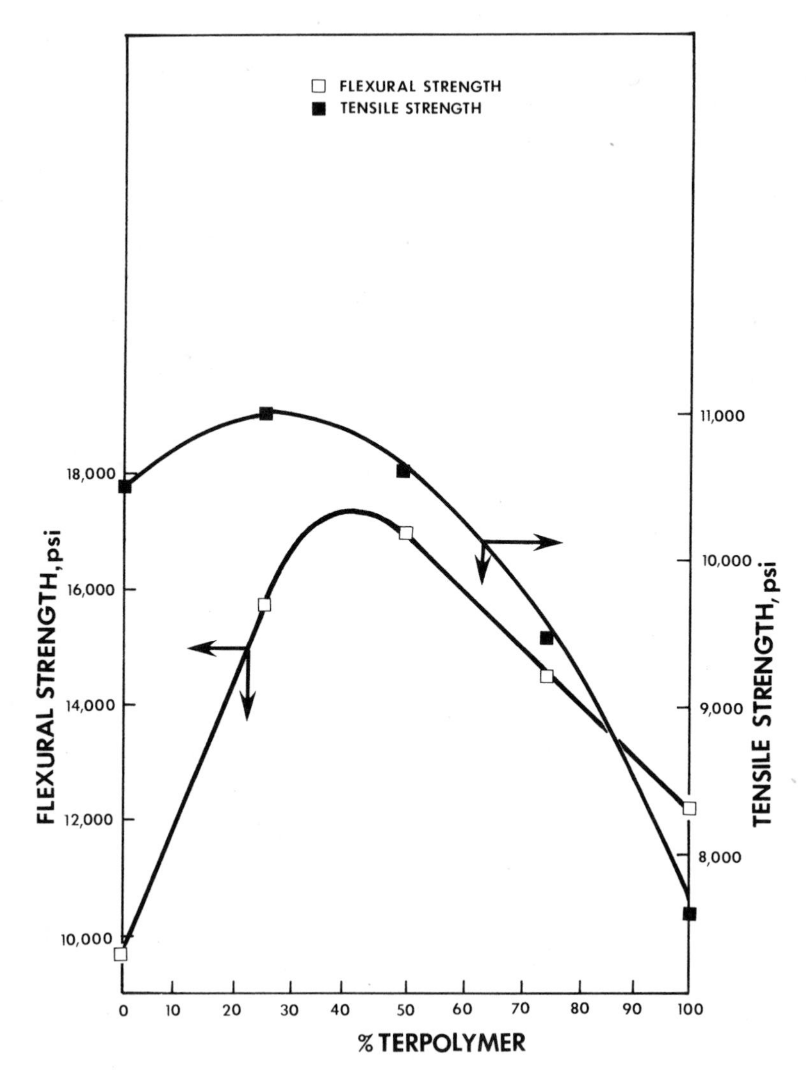

Figure 3. Composition dependence of the flexural and tensile strengths for the terpolymer/blends

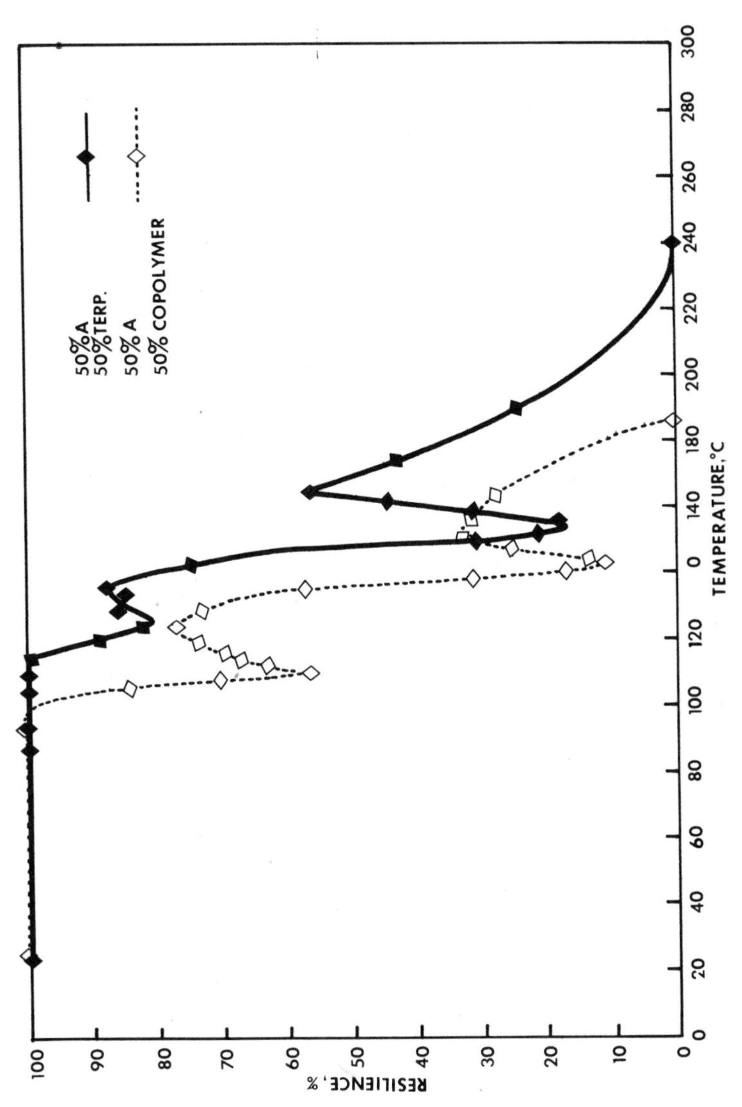

Figure 4. Resilence–temperature curves for the blends of polymer A with the styrenic interpolymers

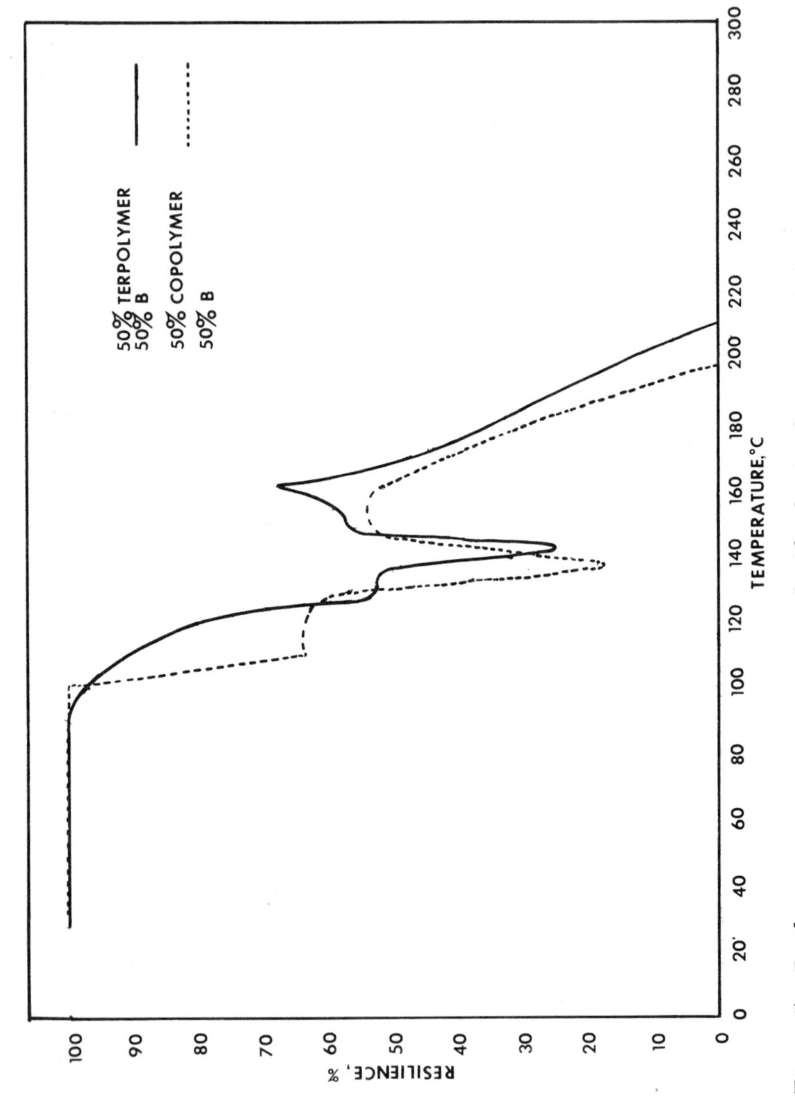

Figure 5. Resilience–temperature curves for the blends of polymer B with the styrenic interpolymers

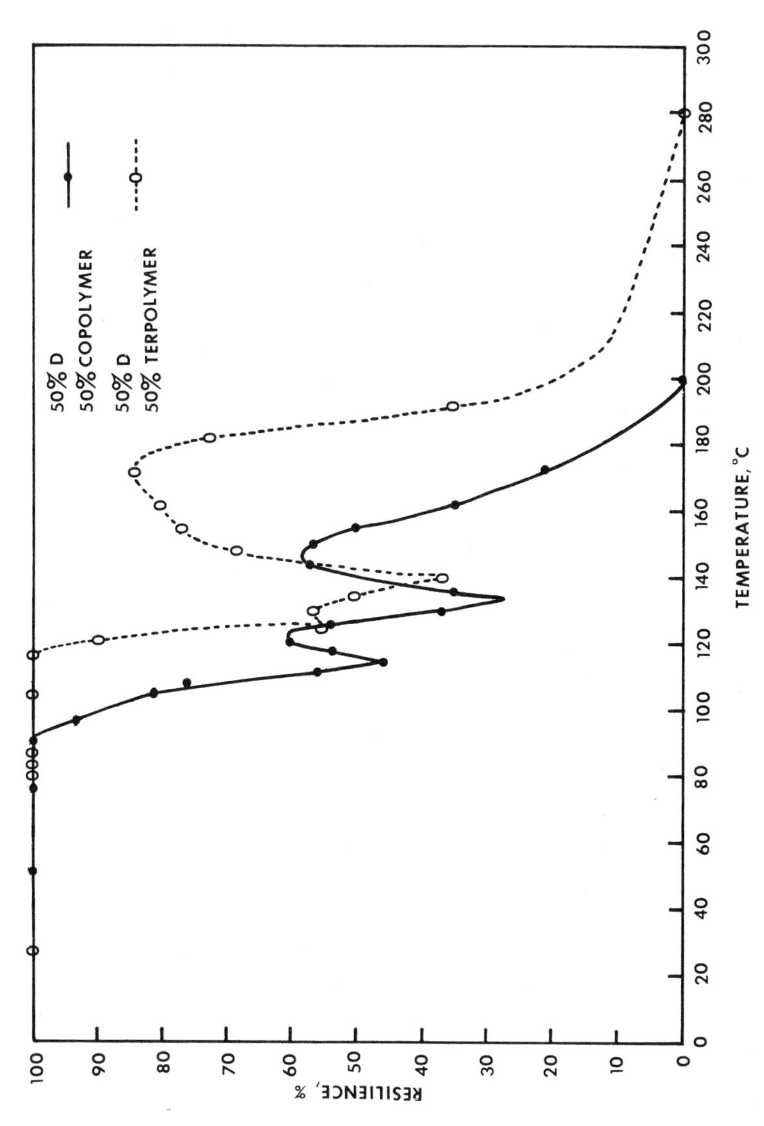

Figure 6. Resilience–temperature curves for the blends of polymer D with the styrenic interpolymers

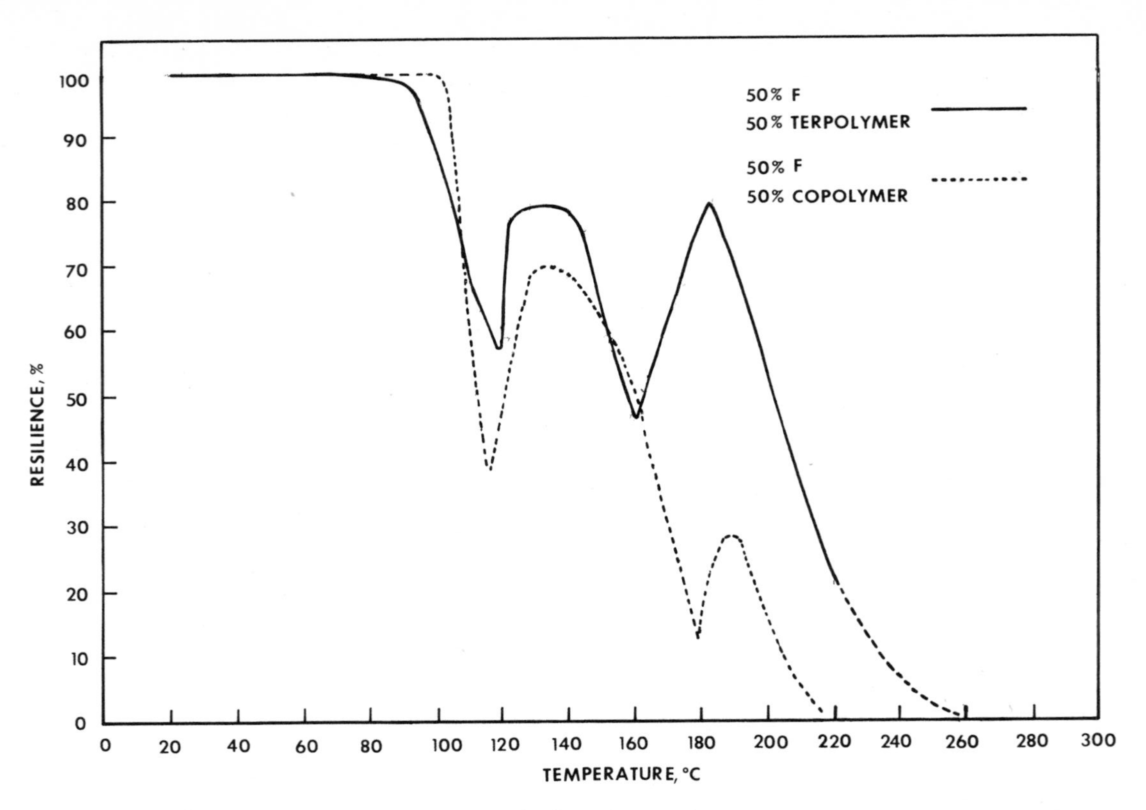

Figure 7. Resilience–temperature curves for the blends of polymer F with the styrenic interpolymers

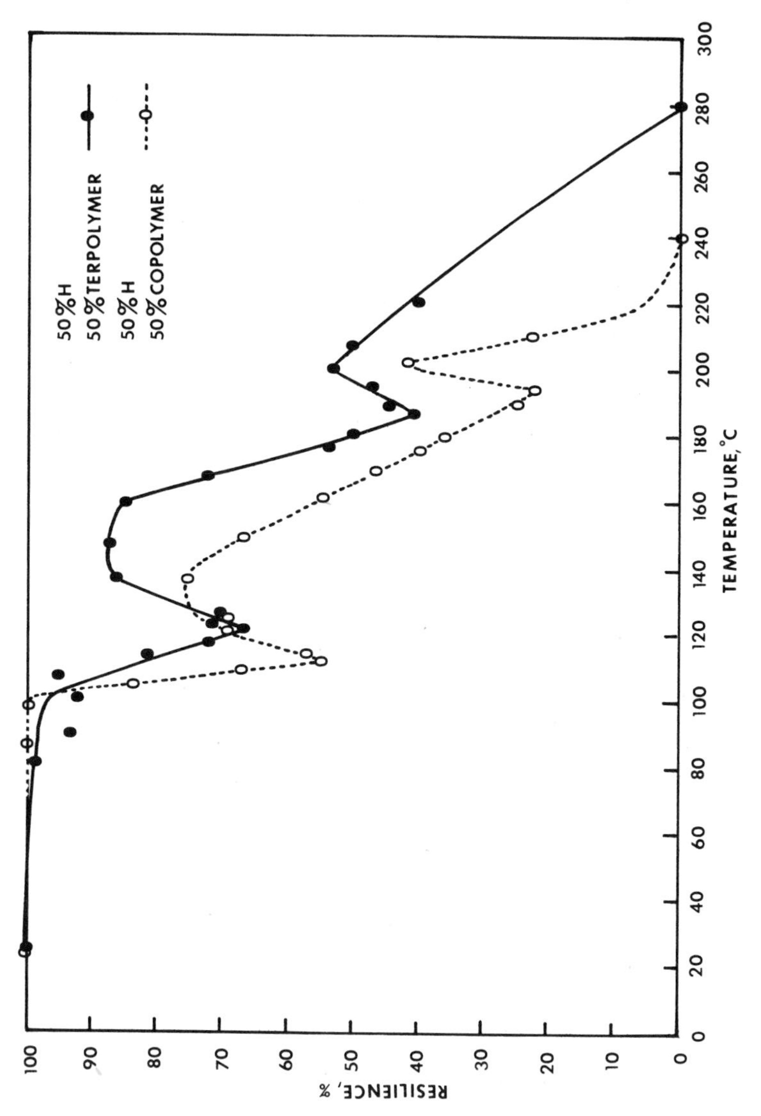

Figure 8. Resilience–temperature curves for the blends of polymer H with the styrenic interpolymers

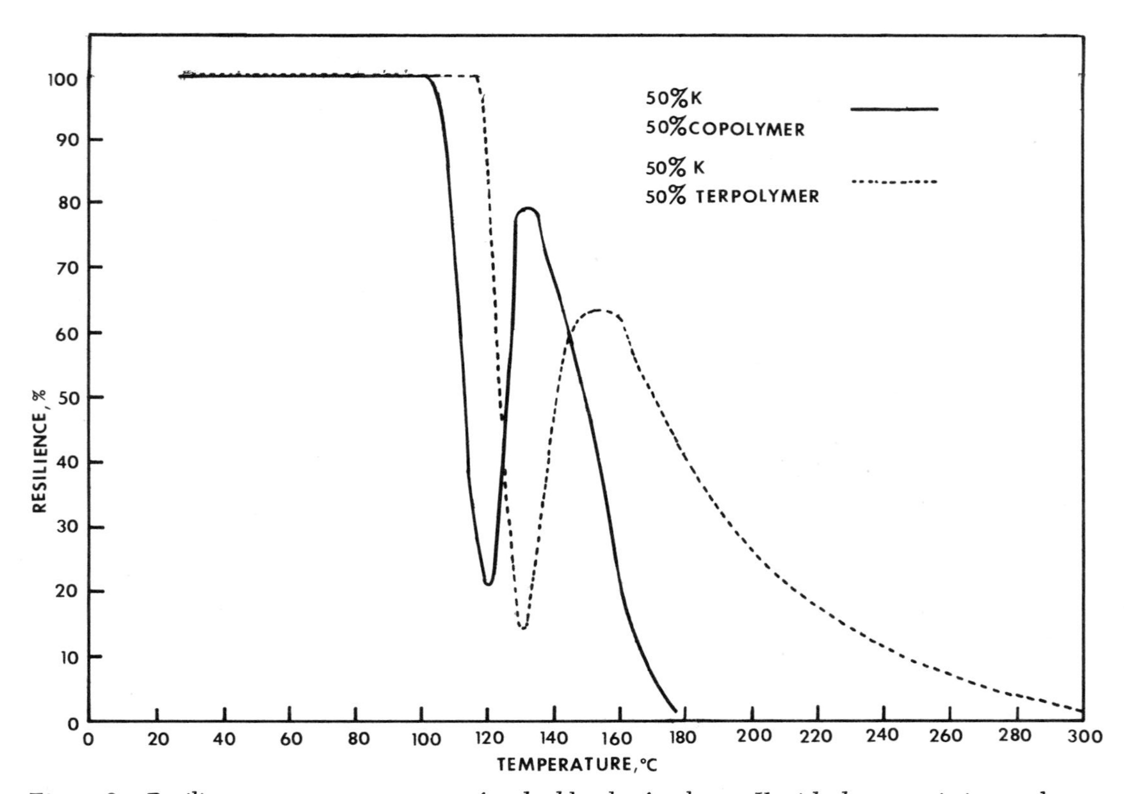

Figure 9. Resilience–temperature curves for the blends of polymer K with the styrenic interpolymers

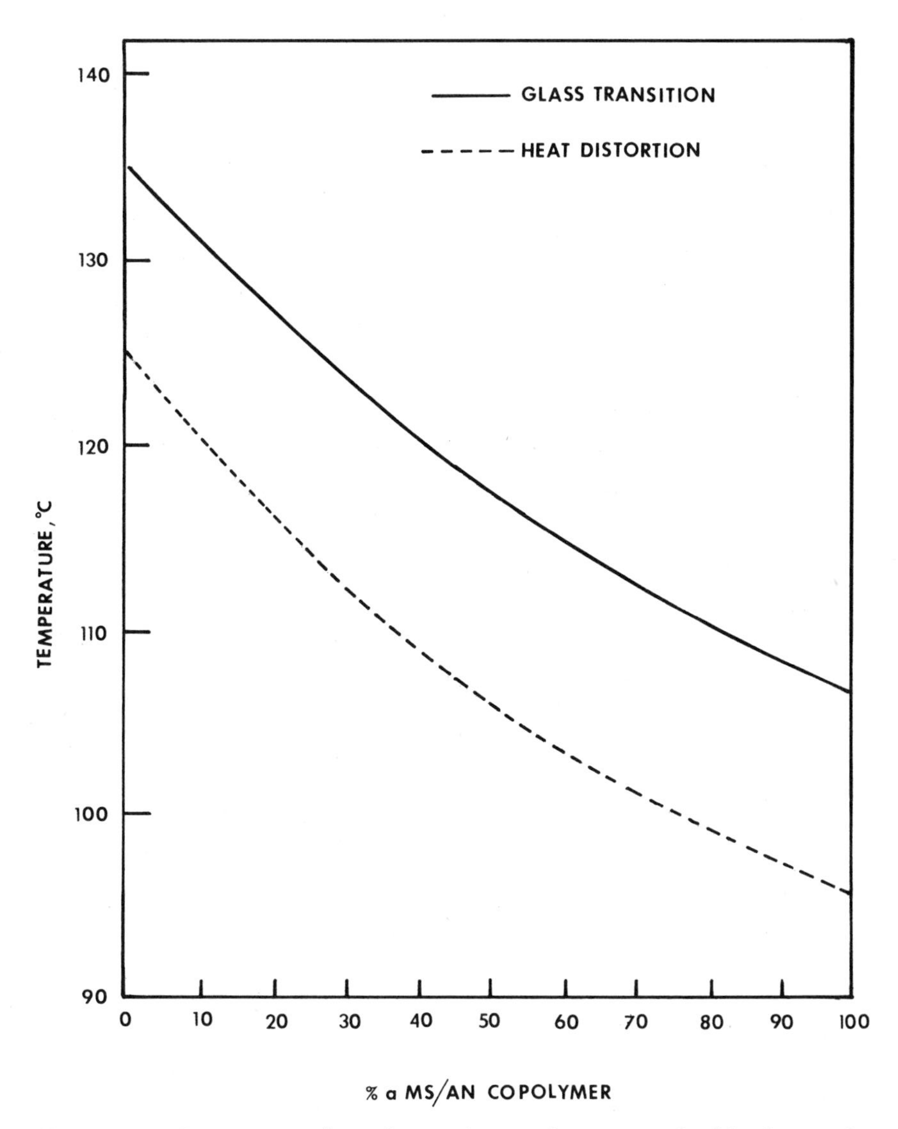

Figure 10. Composition dependence of T_g and HDT for the blends of polymer K with αMS/AN copolymer

blends. Nonetheless, the pretty good mechanical properties indicate that the blends are mechanically compatible even though they are thermodynamically immiscible.

The best results are obtained with polyarylether designated as K, and Figure 9 represents the resilience–temperature curve showing the single T_g for the 50/50 mixture of the polymer K with α-methyl styrene and co- and terpolymers. The dependence of the T_g and HDT on the percent polyarylether in the αMS/AN blends appears in Figure 10. Even though the T_gs and HDTs of the blends are intermediate between the pure component values, the blends are believed to be two-phase owing to their consistent opacity. Additional revealing evidence of the possible immiscibility of these mixtures is embodied in the temperature dependence of dynamic mechanical data illustrated in Figures 11 and 12. Note

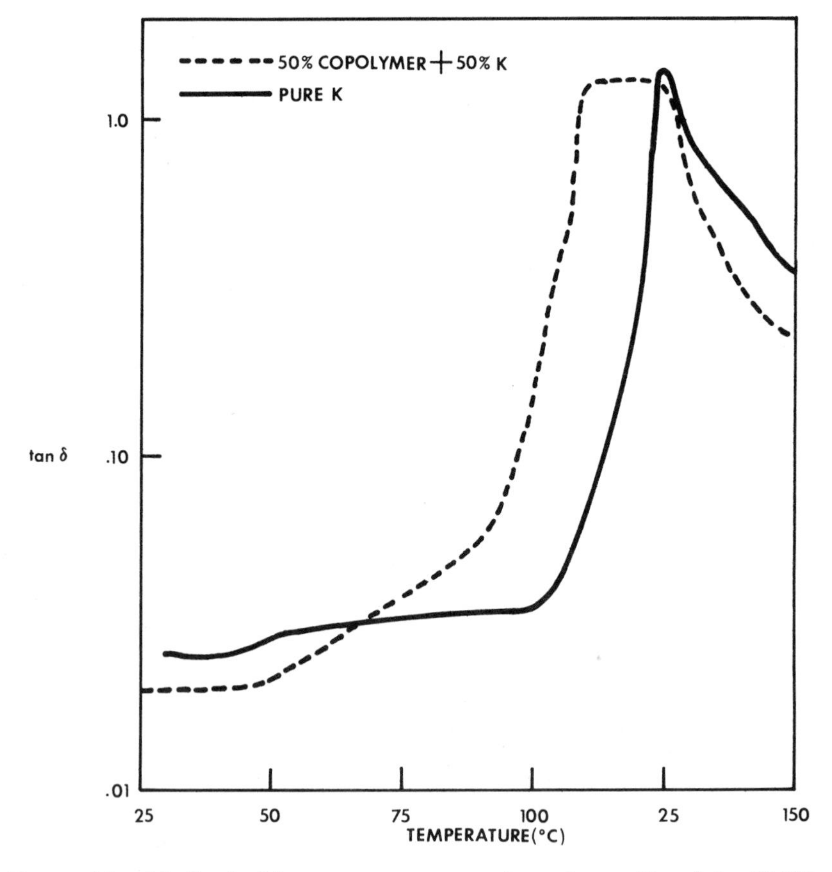

Figure 11. Mechanical loss vs. temperature for polymer K and its 50/50 blend with αMS/AN copolymer

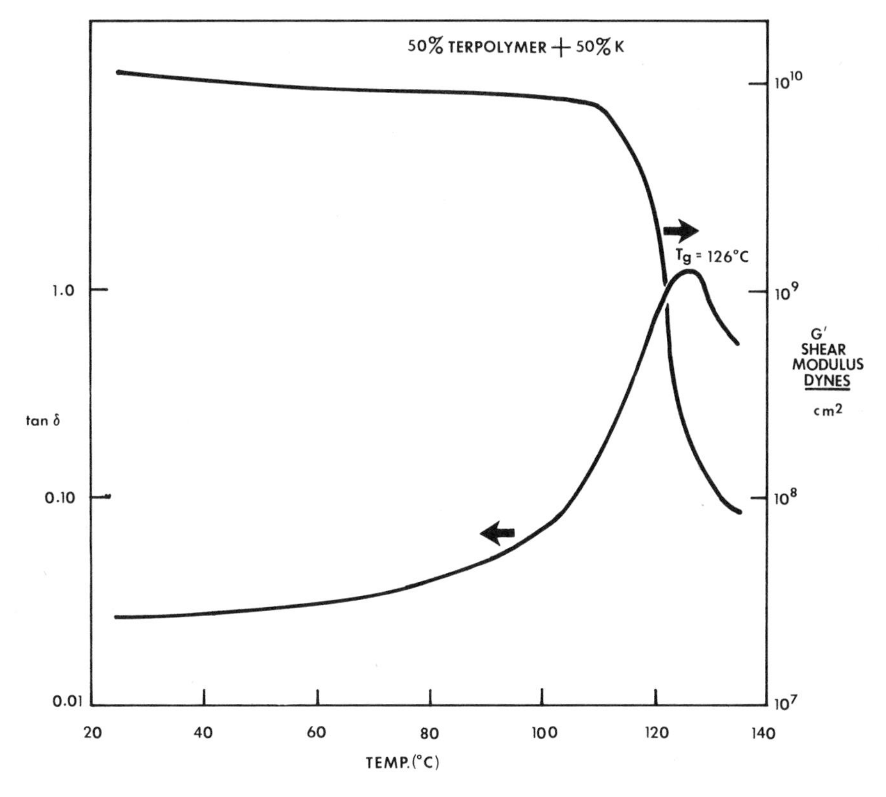

Figure 12. Shear modulus and mechanical loss vs. temperature for the 50/50 blend of polymer K with αMS/AN/MMMA terpolymer

the very broad transition curves which appear to average the values of the separate component glass transitions. Although the broadening of the tan δ can be interpreted as an artifact of the instrumentation, a miscible blend would ordinarily give a sharp transition curve similar in behavior to that of a pure polymer. That is, the behavior of the best system of this study may be that of a marginally miscible blend.

The composition dependence of the flexural strength of the αMS/AN-copolymer blend with polyarylether K appears in Figure 13. As the composition of the copolymer increases, the strength first increases, reaches a maximum, and then decreases. It actually exhibits a minimum at about ~ 80% αMS/AN. This behavior can only substantiate earlier suggestions regarding the possible immiscibility of these systems. All of the other mechanical properties indicate that mixtures with polyarylether K may not be miscible but are mechanically compatible. Finally, it is interesting to note that at least one of the pendant chemical groups present on K exists on either of the α-methyl styrene interpolymers. It

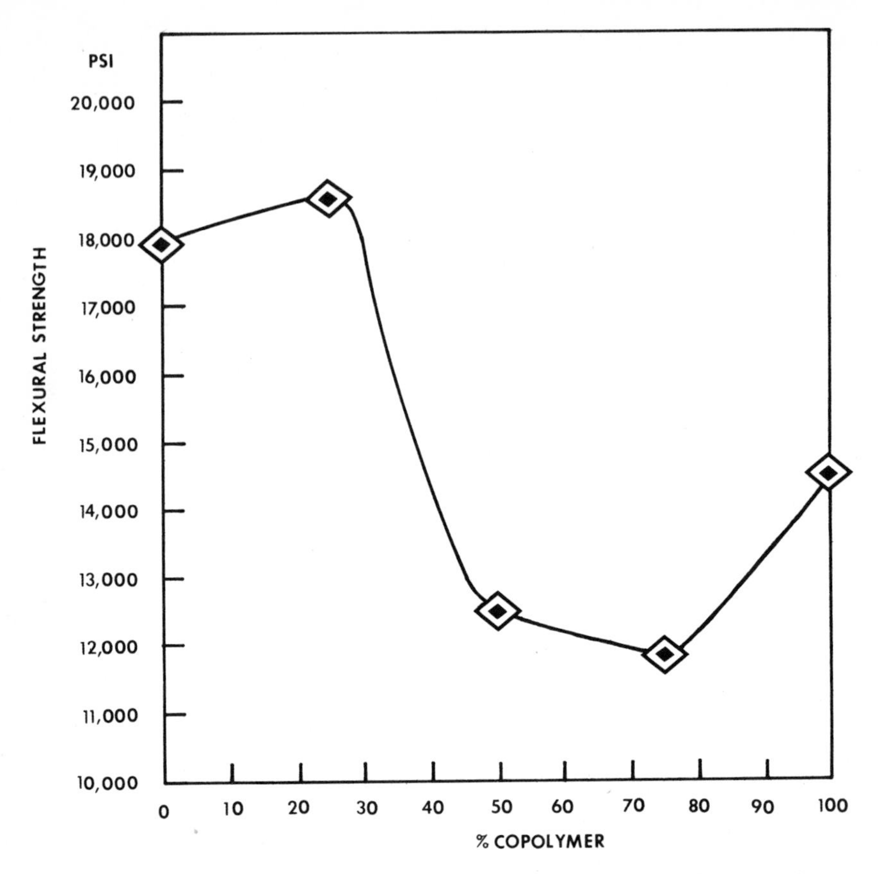

Figure 13. Composition dependence of the flexural and tensile strengths for the blend of polymer K with αMS/AN copolymer

may well be that the old concept of 'like dissolves like' suffices to explain the compatibility characteristics of polyarylether K; but then, the same concept fails when applied to polymers A–J.

Conclusion

α-Methyl styrene/methyl methacrylate/acrylonitrile terpolymer and α-methyl styrene/acrylonitrile copolymer are miscible with poly(methyl methacrylate) and some other polymers containing amides, imides, nitriles, or esters, each of which contains lone-pair electrons capable of donor–acceptor complex formation. By attaching this type of chemical groups onto the backbone of an otherwise immiscible polyarylether, it was possible to produce compatible mixtures of the styrenic interpolymers with the polyarylether.

The resulting mixtures exhibit single T_gs intermediate between the component T_gs, but the transition curves are broad and the plaques are opaque to light transmission. Nonetheless, the rather good mechanical properties indicate that the blends are mechanically compatible even though they are thermodynamically immiscible. That is, compatibilization has been achieved via the use of pendant chemical groups as internal compatibilizing agents.

Acknowledgment

We thank the Union Carbide Corp. for permission to publish this work. Acknowledgment also goes to J. J. Bohan and L. B. Conte for their technical assistance; to A. D. Hammerich for the NMR measurements and to L. M. Robeson for the dynamic mechanical data.

Literature Cited

1. Molau, G. E., *J. Polym. Sci., Part A* (1965) **3**, 1267.
2. Ibid., 4235.
3. Himei, S., Takine, M., Akita, K., Kanegafuchi Chemical Industry Co., Japanese Patent **866** (1967); *Chem. Abstr.* (1967) **67**, 22418.
4. Amicon Corp., French Patent **1,539,053** (1968).
5. Bevan, A. R., Bloomfield, G. F., *Adhes. Age* (1964) **7**(2), 36.
6. Ibid., **7**(3), 34.
7. Gayyord, N. G., "Copolymers, Polyblends and Composites," ADV. CHEM. SER. (1975) **142**, 76.
8. Olabisi, O., unpublished data.
9. Olabisi, O., *Macromolecules* (1975) **8**, 316.
10. Farnham, A. G., Johnson, R. N., Union Carbide Corporation, U.S. Patent **3,332,909** (1968); U.S. Patent **4,108,837** (1978).
11. Nielson, L. E., *Rev. Sci. Inst.* (1951) **22**, 690.

RECEIVED April 14, 1978.

30

Effect of Branch Length of Neoprene-g-Poly(tetrahydrofuran) Copolymers on Properties

J. LEHMANN[1] and P. DREYFUSS

Institute of Polymer Science, University of Akron, Akron, OH 44325

Neoprene-g-poly(tetrahydrofuran) copolymers with branches of different length were synthesized. Solubility, NMR, and GPC indicated that pure graft copolymer was produced in all cases. The copolymers exhibited one T_g which was between the T_gs of the homopolymers and a crystalline melting temperature slightly lower than the T_m of polytetrahydrofuran. Polarizing optical microscopy on thin films revealed spherulitic domains in a nonspherulitic matrix. Tensile strength tests of uncured specimens indicated that breaking-elongation decreases and ultimate tensile strength increases as the length of the polytetrahydrofuran branches increases. Permanent set increased with branch length and was thermally reversible. Adhesive characteristics in standard neoprene recipes were very similar to those of neoprene or polytetrahydrofuran with plastic or metal adheronds.

When certain allyl halides react with a suitable inorganic salt in the presence of a heterocyclic monomer, polymerization ensues (*1*). If the halide is a polymeric halide, graft copolymer results. The objectives of the present work are: (1) to synthesize and characterize such graft copolymers and (2) to test some of their mechanical properties and relate them to the content of branch which has been grafted from the polymer.

[1] Current address: Electrochemical Industries (Frutarom), P.O. Box 1929, Haifa 31000, Israel.

0-8412-0457-8/79/33-176-587$05.00/0

As the polymeric halide for our initial investigation we have chosen polychloroprene (Neoprene), which has the following structure:

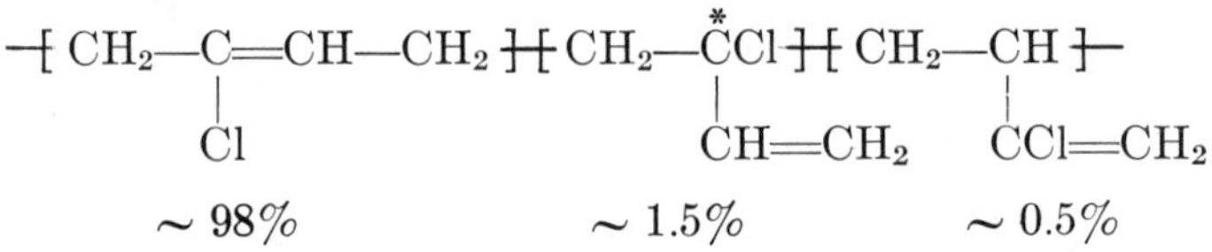

$$-\!\!\left[CH_2-\underset{\substack{|\\ Cl}}{C}=CH-CH_2 \right]\!\!\left[CH_2-\underset{\substack{|\\ CH=CH_2}}{\overset{*}{C}Cl} \right]\!\!\left[CH_2-\underset{\substack{|\\ CCl=CH_2}}{CH} \right]\!\!-$$

$$\sim 98\% \qquad\qquad \sim 1.5\% \qquad\qquad \sim 0.5\%$$

The tertiary allylic chlorides labeled above with an asterisk are the grafting sites (*1*). Tetrahydrofuran (THF) was chosen as the heterocyclic monomer because its chemistry is so well known that it is the model heterocycle of choice in oxonium-ion polymerizations (*2*).

This chapter describes the synthesis of the selected graft copolymers and some of their characterization by solubility, NMR, IR, GPC, DTA, optical microscopy, stress–strain properties, and work of adhesion. Correlations with polytetrahydrofuran (PTHF) content and branch length are reported. These correlations are based on calculated numbers derived from gravimetrically determined conversions and the number of active halogens on the backbone assuming 100% reactivity of the halogens. Studies with model, small organic halides suggest that this is a reasonable first approximation (*3, 4*). Attempts to characterize the branches experimentally have been unsuccessful so far.

Experimental

Materials. THF monomer was refluxed under nitrogen first over lithium aluminum hydride for 48 hr, then over sodium–potassium alloy. The distilled monomer was used within 24–48 hr. Methylene chloride (CH_2Cl_2) was refluxed over calcium hydride overnight under nitrogen and then distilled under nitrogen. Silver hexafluorophosphate ($AgPF_6$) (Alfa Inorganics or Ozark Mahoning) was used as received. Triphenylphosphine was recrystallized from ether and dried in a vacuum oven at room temperature.

Polychloroprene (Neoprene W or AC Soft, E. I. duPont de Nemours and Co.) was dissolved in toluene, precipitated in methanol, and dried in a vacuum oven at room temperature. This procedure was repeated three times. The polymer was used immediately after the last precipitation. Neoprene W and Neoprene AC Soft have $\overline{M}_n$s of 180,000 and 230,000, respectively, and allylic chloride contents of 1.45 and 0.77 mol%, respectively. The $\overline{M}_n$s were determined by osmometry, and mol% allylic chloride are based on an IR measurement of the 1,2-addition-product band of polychloroprene at 921 cm^{-1} (*5*). PTHF used for DTA and stress–strain measurements had $\overline{M}_n = 95{,}000$.

The compounding ingredients for preparation of adhesion test specimens, magnesia (MgO), zinc oxide, Zalba Special (hindered phenol antioxidant from E. I. duPont de Nemours and Co.), and Bakelite-brand *t*-butylphenolic resin CK-1634 (Union Carbide Corp.) were used as received. Aluminum, chrome, and brass substrates were prepared as previously described (*6*). Poly-2,6-dimethyl-1,4-phenylene oxide (PPO, General Electric Co.) sheets were molded between preheated Ferrotype plates backed by stainless steel plates at 290°C in a Pasadena Press using 30,000 lb on a 5 in. ram. A cycle with 1.5 min preheat and 2.2 min at full pressure followed by 5 min in a cold press at the same pressure was most suitable. Polyethylene terephthalate (Mylar, E. I. duPont de Nemours and Co.) was secured to an aluminum plate using a polyester resin (Adhesive 46971, E. I. duPont de Nemours and Co.) and pressing for 1 hr at about 125°C and 20,000 lb per 5-in. ram force.

Polymerizations. Graft copolymerizations were carried out in bottles or round-bottom flasks under a dry nitrogen atmosphere in a dry box or under a dry nitrogen blanket. The backbone was first dissolved in the monomer by letting a mixture of purified monomer, polymer, and in some cases CH_2Cl_2 stand under nitrogen for 24–48 hr with occasional shaking. $AgPF_6$ solution in CH_2Cl_2 was added with shaking. The reactions were terminated by addition of either concentrated ammonium hydroxide or triphenylphosphine under nitrogen. When termination was completed, an antioxidant (a 1% solution of 1,2-dihydro-2,2,4-trimethylquinoline in THF or CH_2Cl_2) was added. The copolymers were then isolated by evaporating volatiles and drying in vacuum. The copolymerization details are given in Table I and the resulting graft copolymer compositions are shown in Table II. PTHF for DTA and stress–strain measurements was similarly prepared except that allyl chloride was used instead of neoprene.

Characterization. Extractions were carried out by stirring the polymers with several changes of the desired solvent until no more material would dissolve. IR spectra were obtained on a Perkin–Elmer 521 Grating Infrared Spectrometer. Polymer films were cast on a NaCl crystal from a suitable solvent (usually CCl_4). Compositions were determined from ^1H NMR spectra using CCl_4 solutions and a Varian T-60 NMR spectrometer. Gel-permeation chromatograms of dilute polymer solutions in THF at 37°C were obtained on a Waters Associates Ana-Prep Chromatogram. Details of the GPC analysis have been described (*7, 8*).

Number-average molecular weights were measured using a Mechrolab 500 Series Membrane Osmometer and toluene at 38°C. DTA measurements were made using a duPont 900 Instrument with a DSC cell and a heating rate of 20°C/min. Stress–strain measurements and adhesion tests were made with a table-model Instron Model TNN. Microdumbbells for tensile tests were cut from films cast from CH_2Cl_2. Optical microscopy studies were made using a Leitz Orthoplan microscope with and without crossed polaroids.

Samples for 180° peel tests were prepared according to standard neoprene recipes (*9*). Neoprene AC or a graft copolymer prepared from it was mixed on an open mill at room temperature for 20–25 min with the following compounding ingredients in succession:

Material	Parts
Neoprene AC (or neoprene-g-PTHF)	100
Magnesia	4
Zalba special	2
Zinc oxide	5

Table I. Graft

No.	Neoprene (g)	Neoprene Active Cl (mmol)[b]	Monomer (mL)
	PTHF from Neoprene W		
1	0.80	0.18	22
2	1.53	0.25	20
3	0.79	0.13	20
4	1.56	0.26	20
5	0.79	0.13	20
6	1.56	0.26	20
	PTHF from Neoprene AC Soft		
7	19.38	1.7	150
8	20.10	1.7	150

[a] All polymerizations were run at room temperature.
[b] By IR analysis.
[c] Terminated with NH_4OH. Other polymerizations were terminated with 1–2 parts Ph_3P based on the $AgPF_6$. This is equivalent to 2–3 parts Ph_3P based on active Cl.

The milled stock (111 parts) was dispersed in CH_2Cl_2 and a solution of *t*-butylphenolic resin (40 parts) in CH_2Cl_2 was stirred in. Additional CH_2Cl_2 was added so that the final slurry had 20% solids. The solvent was then evaporated and a thin film was produced. This film was pressed against the prepared substrate and covered with a sheet of finely-woven cotton cloth. The final sandwich was first pressed in a mold for 30 min at room temperature and 25,000 lb on a 5 in. ram and then cured for 40 min at 150°C and 30,000 lb on a 5 in. ram. The thickness of the elastomer layer in the resulting cloth–elastomer–substrate sandwich was ~ 0.4 mm. Peeling experiments were carried out on strips of cloth-backed elastomer layer after trimming them to a uniform width on the substrate of 2 cm.

Table II. Graft Copolymer Compositions

No.	Composition[a] Backbone (%)	Composition[a] Branch (%)	Branch[b] $\overline{M}_n \times 10^{-3}$
1	78	22	1.75
2	68	32	2.68
3	40	60	8.75
4	14	86	36.4
5	13	87	42.9
6	10	90	52.2
7	86	14	1.10
8	31	69	25.7

[a] By ^1H NMR.
[b] Calculated from conversion assuming all the allylic chlorines of the backbone have reacted with $AgPF_6$ to give active sites for the graft copolymerization.

Copolymerization Details[a]

CH_2Cl_2 (mL)	$AgPF_6$ (mmol)	Pzn. Time (hr)	Conversion (%)
	PTHF from Neoprene W		
10	0.2	17[c]	1.2[d]
—	0.4	2.5	4.0
11	0.2	5.5	6.7
—	0.4	3.2[c]	46.5
11	0.2	30[c]	28
—	0.3	69.5	78
	PTHF from Neoprene AC Soft		
80	4.8	3.25	2.5
80	3.0	48	30

[d] The conversion was low for the time polymerized because of an impurity in the neoprene. The polymer is included in the table because it was used for some of the characterizations described below.

The cloth-backed elastomer layer was peeled off a short distance, bent back through 180°, and then stripped off at 1 cm/min. The peel force P per unit width of the detaching layer was calculated from the time-average force observed (*10*).

Results and Discussion

Solubility and Composition. The copolymers are quite soluble and continue to be soluble if antioxidant is added; otherwise some gelatin occurs. The usual solvents for neoprene and PTHF (CH_2Cl_2, $CHCl_3$, CCl_4, toluene, THF, etc.) are also solvents for the graft copolymers. But there are some unexpected features about the solubility. Benzene dissolves both the backbone and the branch quite readily. Yet at least one graft copolymer, which dissolved readily and completely in CH_2Cl_2 and ethyl acetate, swelled but did not dissolve in benzene. Also some of the copolymers were not completely soluble in toluene and THF. Still all the neoprene-g-PTHF copolymers were 100% soluble in ethyl acetate, a nonsolvent for neoprene. As shown in Table III, the ethyl-acetate extracts had in each case the same composition by ^1H NMR as the crude products. Thus it appears that the backbone was pulled into its nonsolvent by the PTHF branches. This means that no unreacted backbone remained.

Molecular-Weight Distribution. The solid line in Figure 1 shows a typical gel-permeation-chromatogram trace for an unfractionated neoprene-g-PTHF (Polymer 1, Table I) that was soluble in THF. The

Table III. Comparison of Composition of Different Samples of Neoprene-g-PTHF

Sample No.	Composition Determined by: ^1H NMR on Crude Product	^1H NMR on Ethyl Acetate Extract	Weight
2 neoprene (%)	68	67	68
PTHF (%)	32	33	32
3 neoprene (%)	38	38	40
PTHF (%)	62	62	60
6 neoprene (%)	10	10	10
PTHF (%)	90	90	90
7 neoprene (%)	87	87	86
PTHF (%)	13	13	14

polymer is characterized by a unimolecular weight distribution and has a high-molecular-weight tail. There is no evidence of homopolytetrahydrofuran, which should appear at high count, 50–60, at this low molecular weight. Compared with the original untreated backbone (dashed curve in Figure 1), the maximum is shifted to higher count (lower molecular weight); but compared with backbone exposed to typical reaction conditions using nonpolymerizable-heterocycle 2-methyltetrahydrofuran in-

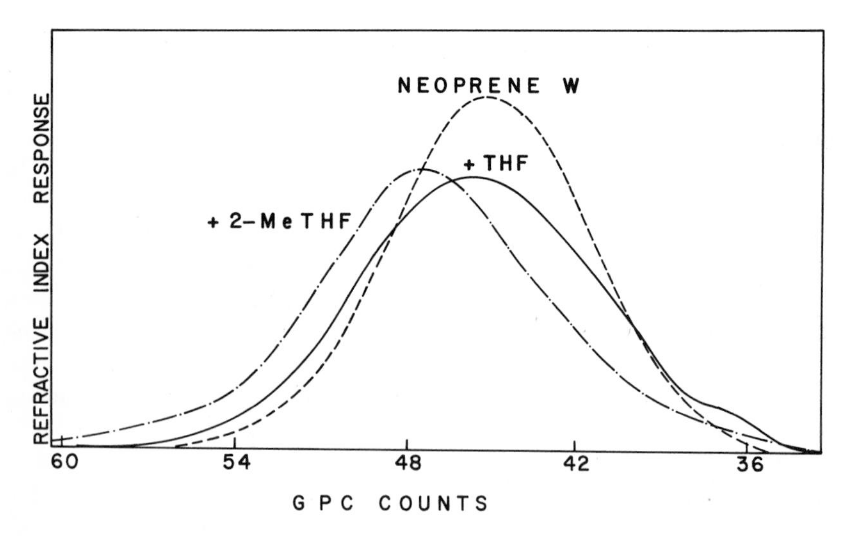

Figure 1. Gel-permeation chromatograms of unfractionated Neoprene W-g-PTHF. Polymer 1, Table I (——); Neoprene W after reaction with $AgPF_6$ in 2-methyltetrahydrofuran (· — · —); Neoprene W before exposure to grafting reaction conditions (– – –).

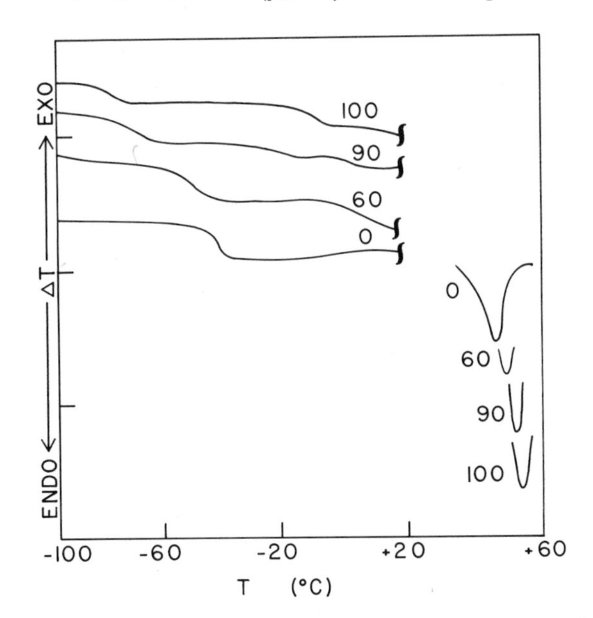

Figure 2. DTA thermograms of Neoprene W-g-PTHFs. Scan rate = 20°C/min. The numbers on the curves indicate the percent PTHF in the polymer.

stead of THF, the maximum is shifted to lower count (higher molecular weight). The data indicate that neoprene is significantly degraded under reaction conditions and that, as would be expected, the molecular size of the graft copolymer is increased compared with the treated backbone.

Differential Thermal Analysis. The unfractionated graft copolymers showed transitions between those of the homopolymers when examined by DTA. Two transitions were found for each copolymer examined. As shown in Figures 2 and 3, the T_gs decreased and the T_ms increased as the percent by weight of PTHF increased.

Morphology. The DTA results suggest that only one phase might be present in the graft copolymers. However, optical micrographs of a thin film of graft copolymer taken between crossed polaroids show two phases, one spherulitic and the other nonspherulitic. The size of the spherulitic regions varied with the PTHF content. Typical photomicrographs are shown in Figure 4.

Stress–Strain Properties. Figure 5 shows the engineering stress δ in MPa plotted vs. elongation E for solution-cast uncured neoprene-g-PTHF copolymers of different compositions. Curves for Neoprene W and PTHF are also plotted for comparison. The curves show that the elongation at break decreases with increasing PTHF content in the graft copolymer, the ultimate tensile strength increases with increasing PTHF in the graft

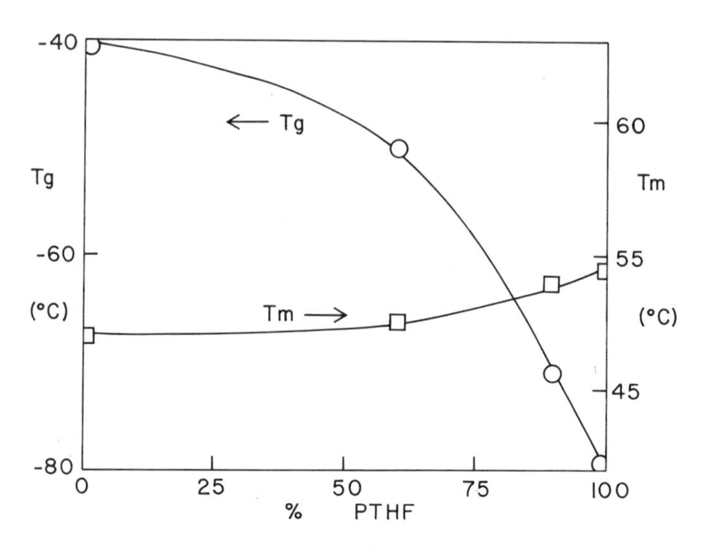

Figure 3. Variation of T_g and T_m of Neoprene F-g-PTHF with percent PTHF

copolymer, and elongation and ultimate tensile strength of the graft copolymers lie between those of the homopolymers. Permanent set was calculated after 100% elongation and is given in Table IV. Permanent set increased with increasing PTHF content in the graft copolymer. However, if the samples were heated gently, the samples returned to their original shape and the permanent set was lost presumably because of melting of the crystallites formed on stretching.

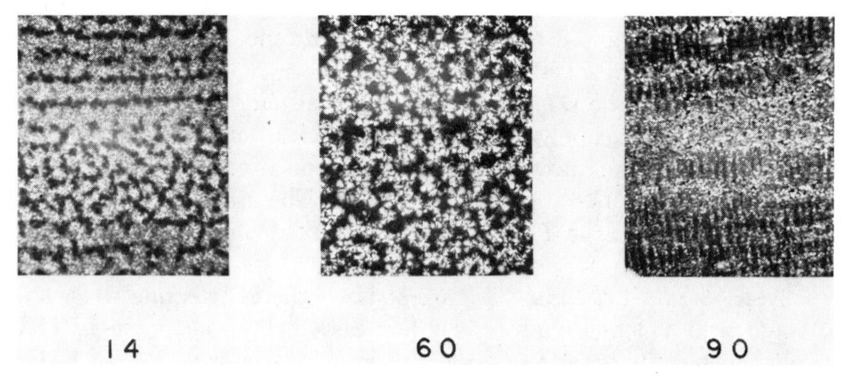

Figure 4. Optical micrographs of neoprene-g-PTHF with 90, 60, and 14% PTHF, samples 6, 3, and 7 in Table I, respectively. Crossed polaroids at 250×. The numbers below the micrographs give the % PTHF in the graft copolymers.

Table IV. Permanent Set vs. Graft Copolymer Composition

PTHF (%)	*Permanent Set after 16 hr (%)*
100	15
90	50
60	20
14	9.5
0	5

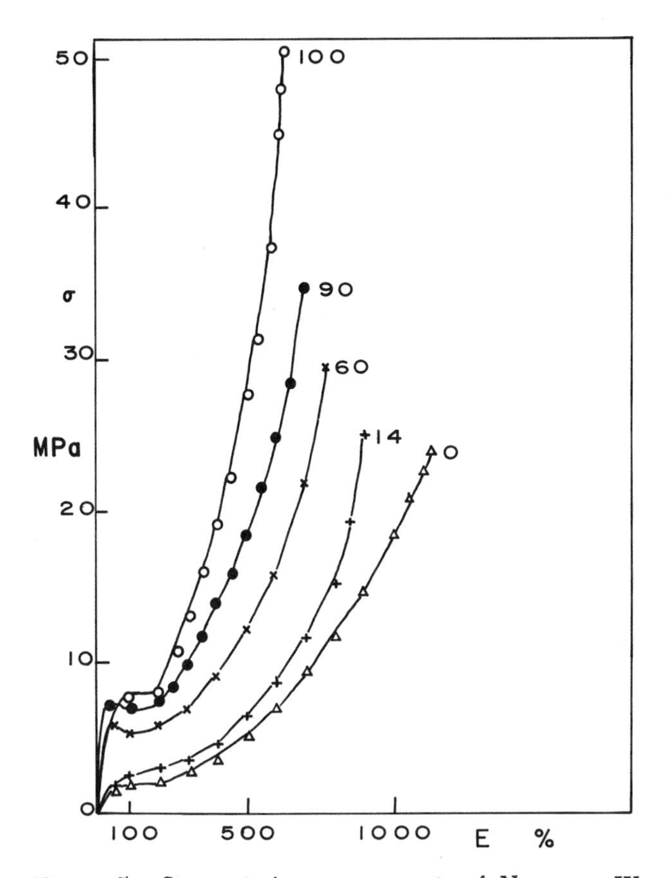

Figure 5. Stress–strain measurements of Neoprene W (△), PTHF (○), and neoprene-g-PTHF with varying PTHF content: 90% (·), 60% (×), 14% (+), samples 6, 3, and 7 in Table I, respectively.

Table V. Adhesion of Neoprene AC Soft-g-PTHF Copolymers, Neoprene AC Soft, and PTHF to Various Substrates

		Peeling Force on Substrate Listed (N/m)[a]				
Sample	*PTHF (%)*	*PPO*	*Mylar*	*Aluminum*	*Chrome*	*Brass*
Neoprene[b]	0	490	560	1220	170	480
7[c]	14	370	760	840	220	290
8[c]	69	310	780	850	300	280
PTHF	100	630[d]	400[d]	530[d]	120[d]	520[d]

[a] 180° peel tests at 1 cm/min crosshead speed.
[b] AC Soft.
[c] See Tables I and II.
[d] All samples except chrome failed in a mixed cohesive–adhesive mode. Tests were run 1–2 days after preparation of the test specimens. Since PTHF may crystallize slowly even in the presence of the compounding diluents, there could be an important effect of aging.

Adhesion Properties. The adhesion properties of two graft copolymers (numbers 7 and 8 in Table I) were studied using several rigid substrates and were compared with the adhesive properties of the homopolymers from which they were derived. The data are given in Table V. Although the peeling force measured at 180°C and 1 cm/min varied considerably with the substrate, it was not very sensitive to the composition of the adhesive. As the PTHF content increased, a small decrease in peeling force was observed with PPO, aluminum, and brass substrates. A small increase was observed with Mylar and chrome substrates.

Summary. Solubility, ^{1}H NMR, DTA, and GPC all support the conclusion that pure graft copolymer is obtained by the present method of synthesis. The only purification necessary is removal of the silver salts. Properties of the neoprene-g-PTHF copolymers depend on composition and branch length and generally lie between those of the backbone and branch.

Acknowledgment

This work was supported by the U.S. Army Research Office. We thank the E. I. duPont de Nemours and Co. for supplying the samples of Neoprene W, Neoprene AC Soft, and some of the reagents used in the adhesion testing.

Literature Cited

1. Dreyfuss, P., Kennedy, J. P., *J. Polym. Sci., Polym. Symp.* (1976) **56,** 129.
2. Dreyfuss, P., Dreyfuss, M. P., *Adv. Polym. Sci.* (1967) **4,** 528.
3. Lee, K., Dreyfuss, P., *ACS Symp. Ser.* (1977) **59,** 24.
4. Quirk, R., Lee, D. P., Dreyfuss, P., unpublished data.

5. Maynard, J. T., Mochel, W. E., *J. Polym. Sci.* (1954) **13**, 251.
6. Cagle, C. V., "Adhesive Bonding: Techniques and Applications," Chap. 5, McGraw Hill, New York, 1968.
7. Fetters, L. J., Morton, M., *Macromolecules* (1974) **7**, 552.
8. Kennedy, J. P., Smith, R. R., "Recent Advances in Polymer Blends, Grafts, and Blocks," L. H. Sperling, Ed., p. 303, Plenum, New York, 1974.
9. Elastomers Chemicals Dept., Elastomers Laboratory, "Neoprene Solvent Adhesives," E. I. duPont de Nemours and Co., Inc.
10. Ahagon, A., Gent, A. N., *J. Polym. Sci., Polym. Phys. Ed.* (1975) **13**, 1285.

RECEIVED April 14, 1978.

31

Evaluation of the Polymer–Polymer Interaction Parameter from the Equation of State Thermodynamic Theory

RYONG-JOON ROE

Department of Materials Science and Metallurgical Engineering,
University of Cincinnati, Cincinnati, OH 45221

The strength of interaction between dissimilar components greatly influence the morphology of microdomains in polymer blends and block copolymers. In the theories of polymer–polymer interfaces in multicomponent systems, the interaction between components usually is expressed in terms of a single parameter independent of concentration. The validity of such an approximation is examined here in the light of the equation-of-state thermodynamic theory of polymer solutions and mixtures developed by Flory and co-workers. Using the parameters evaluated by these workers, we have computed the free energy of mixing for the pairs, polyethylene/polyisobutylene, polyisobutylene/polystyrene, and natural rubber/polystyrene. The results show that the polymer–polymer interaction parameter Λ *only moderately depends on concentration and temperature.*

The morphology and properties of polymer–polymer blends and block copolymers depend to a large measure on the strength of interaction between the two, chemically different components in the system. The degree of separation into distinct phases, the thickness of the interface between them, and the size and shape of the block copolymer domains are some of the important features which are governed largely by the polymer–polymer interaction.

There are now available a number of theoretical treatments (*1–11*) dealing with the problem of polymer compatibility and the polymer–polymer interfaces either in polymer blends or in block copolymers.

0-8412-0457-8/79/33-176-599$05.00/0

These theories differ from each other greatly in the way they evaluate the entropic effect of domain formation, i.e., the restrictions to the polymer chain conformations imposed by the presence of the interface. However, most of them share a common approximation in which the energetic interaction between two dissimilar polymer segments is represented by a single parameter, usually denoted by χ, which is assumed to be a function of temperature only for a given polymer–polymer pair. The χ parameter was originally introduced in the polymer solution thermodynamics to represent the van Laar heat of interaction. It was soon realized, both experimentally and theoretically, that the value of χ is not a constant for a given polymer–solvent pair but depends on the concentration (and to some extent on the molecular weight) of the polymer, and also that a sizeable contribution of an entropic nature to χ has to be admitted in order to account for its temperature dependence.

When attempting to use these polymer interface theories (*1–11*) in planning and interpreting experiments, we encounter two problems associated with the interaction parameter in polymer–polymer systems. First we have to find a way of determining the values of the interaction parameters for the polymer pairs of interest. There is as yet no general experimental procedure which allows us evaluation of the χ parameter. Without reliable values of the χ parameter, comparative tests of the competing theories can only be made qualitatively. Secondly, to be able to use these theories with more confidence we have to have some idea about the dependence of the χ parameter on concentration, temperature, etc.

The equation-of-state thermodynamics, recently developed by Flory and his co-workers (*12, 13, 14*), has achieved a considerable improvement in treating the thermodynamics of polymer solutions over the classical Flory–Huggins type solution theory. Whereas in the latter the volume change on mixing is ignored, the newer theory takes into account the equation-of-state contribution, or the contribution by the change in free volume on mixing, to the free energy of mixing. In this work we evaluate the polymer–polymer interaction parameters for a few polymer pairs of interest by closely following the method of calculation suggested by the Flory equation-of-state thermodynamics and using the values of molecular parameters of component polymers evaluated by them. The results are examined to gain insight into the dependence of the interaction parameter on various physical parameters of the system such as the concentration and temperature.

Evaluation of the Polymer–Polymer Interaction Parameter

The strength of polymer–polymer interaction is represented here by a new parameter Λ instead of the usual χ. The "polymer–polymer interaction parameter" Λ is defined by the following equation:

$$\Delta G_M = RT\left(\frac{1}{V_1}\phi_1 \ln \phi_1 + \frac{1}{V_2}\phi_2 \ln \phi_2\right) + \Lambda \phi_1 \phi_2 \tag{1}$$

where ΔG_M is the free energy of mixing per unit volume of the mixture, V_1 and V_2 are the molar volumes of the polymers 1 and 2, and ϕ_1 and ϕ_2 are the volume fractions of the two polymers in the mixture. The first term represents the usual combinatorial entropy of mixing and therefore the term containing Λ includes all other contributions to the free energy of mixing. Λ thus includes the enthalpic interaction between the two components but also the effect of entropy changes unaccounted for by the simple combinatorial term.

It has the dimension of energy per unit volume. When both Λ and χ are independent of concentration, Λ is equal to $\chi RT/V_2$. When χ varies with concentration the relation between Λ and χ is a little more complex because χ in polymer solution theories is usually defined in terms of the excess chemical potential rather than the excess free energy of mixing given in Equation 1. The polymer–polymer interaction parameters appearing in various theories of interfaces (*1–11*) are usually defined with regard to the local free energy of mixing, and therefore the definition of Λ given in Equation 1 is more appropriate than the one given in terms of the excess chemical potential.

There are other advantages of the above definition of Λ, one of them being the fact that the values of χ depend on the assumed volume of a polymer segment (usually taken equal to the volume of the solvent molecule V_2), while Λ is defined per unit volume of the mixture. Another advantage is that the values of Λ can be related approximately to the solubility parameters of the components δ_1 and δ_2 by:

$$\Lambda = (\delta_1 - \delta_2)^2 \tag{2}$$

In the Flory equation-of-state theory, the thermodynamic properties of a pure component liquid (either monomeric or polymeric) are represented completely by means of three characteristic constants, v_1^*, T_1^*, and p_1^*, which can be evaluated from the *P–V–T* and other thermodynamic properties of the liquid. The equation-of-state thermodynamics then recognizes three types of contributions to the free energy of mixing two liquid components. The first is the one attributable to the combinatorial entropy of mixing as expressed by the first term in Equation 1. The second is the change in enthalpy, $\Delta H_{M,\,\text{contact}}$, arising from creation of 1, 2 nearest neighbor contacts on mixing. The third is the free volume (or the equation-of-state) contribution resulting from the change in volume on mixing. This latter contribution contains both enthalpic and entropic components, $\Delta H_{M, fv}$ and $-T\Delta S_{M, fv}$. The quantities, expressed per unit volume of the mixture, are given by:

$$\Delta H_{M, contact} = X_{12}\, \theta_2\, \phi_1/\tilde{v}^2 \tag{3}$$

$$\Delta H_{M, fv} = \sum_{i=1,2} \phi_i p_i^* \left(\tilde{v}_i^{-1} - \tilde{v}^{-1}\right)/\tilde{v} \tag{4}$$

$$-T\Delta S_{M, fv} = \sum_{i=1,2} 3\tilde{v}^{-1}\, \phi_i\, p_i^*\, \tilde{T}_i \times \ln\left[(\tilde{v}_i^{1/3} - 1)/(\tilde{v}^{1/3} - 1)\right] \tag{5}$$

where $\tilde{v}_1 = v_1/v_1^*$, $\tilde{T}_1 = T_1/T_1^*$, and the unsubscripted quantities refer to the mixture.

To evaluate the expressions 3, 4, and 5 one needs, in addition to the characteristic parameters v_i^*, p_i^*, T_i^* for each component, the value of X_{12} denoting the energy change on contact of 1, 2 polymer segments and the surface fraction θ_2 defined by:

$$\theta_2 = \frac{\phi_2}{\phi_1\,(s_1/s_2) + \phi_2} \tag{6}$$

where s_1/s_2 denotes the ratio of surface areas of the two types of segments (occupying equal volumes). The derivation of the expressions 3, 4, and 5 and more precise meaning of the various symbols can be found in the original papers (*12, 13, 14*) by Flory and co-workers.

In view of 3, 4, and 5, Λ defined in Equation 1 can be expressed as:

$$\Lambda = \Lambda_{h, contact} + \Lambda_{h, fv} + \Lambda_{s, fv} \tag{7}$$

where

$$\Lambda_{h, contact} = \Delta H_{M, contact}/\phi_1\, \phi_2 \tag{8}$$

$$\Lambda_{h, fv} = \Delta H_{M, fv}/\phi_1\, \phi_2 \tag{9}$$

and

$$\Lambda_{s, fv} = -T\Delta S_{M, fv}/\phi_1\, \phi_2 \tag{10}$$

Results and Discussion

Three polymer pairs, polyethylene(PE)/polyisobutylene(PIB), PIB/polystyrene(PS), and natural rubber(NR)/PS, are considered. The characteristic parameters, v^*, p^*, T^*, for each of these polymers have been evaluated by Flory and co-workers and are used here as given.

The first pair, PE/PIB, already has been examined by Flory, Eichinger, and Orwoll (*15*) to ascertain their possible mutual compatibility.

For this pair X_{12} and s_1/s_2 were given as 0.20 ± 0.10 cal/cm^3 and 1/0.72, respectively, the former determined as a result of their examination of a series of mixtures of PIB with *n*-alkanes.

For the PIB/PS pair, we take X_{12} equal to ca. 40 J/cm^3 (9.6 cal/cm^3), since four systems consisting of an aromatic and an aliphatic component were found to give similar values of X_{12}: 42 J/cm^3 for PIB/benzene (*16*), 42 J/cm^3 for cyclohexane/benzene (*17*), 40 J/cm^3 for *n*-heptane/benzene (*17*), and 42 J/cm^3 for cyclohexane/PS (*18*). The ratio s_1/s_2 can be determined by examination of the molecular models of respective polymer segments. To avoid introducing any additional parameters which could be construed as adjustable, we obtained the s_1/s_2 ratio of 1.24 for PIB/PS by multiplying the s_1/s_2 ratios 0.62 for PIB/cyclohexane (*19*) and 2.0 for cyclohexane/PS (*18*), which were determined in the previous studies of these polymer–solvent systems.

The value of X_{12} for benzene/natural rubber was given (*13*) as 1.40 cal/cm^3. In the absence of any other data on similar systems, we have taken X_{12} for NR/PS to be also 1.40 cal/cm^3, on the assumption that the interaction of NR segments with any aromatic component is likely to be similar. The value of s_1/s_2 for NR/PS is taken as 1.9, equaling the product of the previously reported values of the ratio: 0.90 for NR/benzene (*13*), 1/0.58 for benzene/PIB (*16*), and 1.24 for PIB/PS.

The values of the polymer–polymer interaction parameter Λ, calculated by use of the X_{12} and s_1/s_2 values discussed above for the three polymer pairs, are tabulated in Table I. To show the concentration dependence, these values are given for each pair at three different concentrations, corresponding to the weight fraction w_1 of the first component equal to 0.1, 0.5, and 0.9.

For PE/PIB, both aliphatic hydrocarbons, the exchange energy parameter X_{12} is very small, and the largest contribution to Λ arises from the free volume entropy effect. The reduced volume $\tilde{v} = v/v^*$ changes from 1.182 for PE and 1.149 for PIB to the intermediate value of 1.166 for the mixture at $w_1 = 0.5$. The net excess volume of mixing predicted is very small (and positive). However, the changes in enthalpy and entropy of each component arising from their respective changes in the free volume on mixing (i.e., the individual terms in expressions 4 and 5) fail to cancel each other completely, and the net difference, especially the entropy effect represented by $\Lambda_{s,\,fv}$, remains positive and finite. Its magnitude, although fairly small in absolute terms, is still larger than the contact enthalpy term $\Lambda_{h,\,contact}$. Even if X_{12} were zero, the PE/PIB pair would have been incompatible because of the free volume effect.

For PIB/PS, the large X_{12} value gives rise to a repulsion and a consequent small expansion of free volume on mixing. The $\tilde{v}$ values for the two components are fairly similar ($\tilde{v} = 1.149$ and 1.153 for PIB and PS, respectively). Although the $\tilde{v}$ value of 1.154 for the mixture at $w_1 =$

0.5 indicates expansion of the free volume on mixing, the resulting changes in enthalpy and entropy, as represented by $\Lambda_{h, fv}$ and $\Lambda_{s, fv}$, happen to cancel each other almost completely. The Λ value for PIB/PS therefore arises solely from the contact term $\Lambda_{h, contact}$.

The NR/PS is of more interest here because of its similarity to polybutadiene/PS and polyisoprene/PS which are the pairs most widely studied in block copolymer systems. No excess volume of mixing is predicted for this pair, the v values, 1.172 for NR and 1.152 for PS, changing to 1.162 for the mixture at $w_1 = 0.5$. Although $\Lambda_{h, fv}$ and $\Lambda_{s, fv}$ fail to cancel each other completely, the net sum of the two is still only about 10% of the total Λ, indicating that here too the contact term is the most important effect giving rise to the incompatibility of the pair.

In describing the polymer–polymer interaction parameters for pairs which are normally incompatible as those discussed in this work, it therefore appears that the free volume (or the equation-of-state) contribution can usually be neglected except when the pair is on the verge of compatibility. This is in a marked contrast to polymer–solvent systems where the free volume effect is usually very large. From this follows that the dependence of Λ for polymer–polymer pairs on concentration or on temperature would also be fairly small. Table I shows only a modest variation of Λ with a change in w_1 from 0.1 to 0.9. The relatively more pronounced variation of Λ for NR/PS as compared with other two pairs still arises mostly from the contact enthalpy term and reflects the larger s_1/s_2 ratio for this system.

The temperature dependence of Λ is summarized in Table II. It shows that Λ can be taken for practical purposes as being independent of temperature. No similar calculation was made for NR/PS because the characteristic parameters for NR at temperatures other than 25°C are not available. It is however very likely that Λ for NR/PS would also

Table I. Polymer–Polymer Interaction Parameter Λ and Its Components[a]

	w_1	Λ	$\Lambda_{h, contact}$	$\Lambda_{h, fv}$	$\Lambda_{s, fv}$
PE/PIB at 25°C	0.1	0.33	0.14	−0.03	0.22
	0.5	0.30	0.12	−0.03	0.21
	0.9	0.27	0.11	−0.03	0.20
PIB/PS at 25°C	0.1	7.02	7.02	1.20	−1.20
	0.5	6.49	6.48	1.10	−1.09
	0.9	5.96	5.95	1.00	−0.99
NR/PS at 25°C	0.1	1.03	0.95	0.14	−0.06
	0.5	0.78	0.70	0.11	−0.03
	0.9	0.64	0.56	0.09	−0.01

[a] Λ in cal/cm^3.

Table II. Temperature Dependence of Λ

	T (°C)	Λ	$\Lambda_{h,contact}$	$\Lambda_{s,fv}$	$\Lambda_{h,fv}$
PE/PIB at $w_1 = 0.5$	25	0.30	0.12	−0.03	0.21
	100	0.28	0.11	−0.04	0.20
	150	0.26	0.11	−0.04	0.19
PIB/PS at $w_1 = 0.5$	0	6.53	6.52	1.02	−1.01
	25	6.39	6.38	1.10	−1.08
	100	6.00	5.99	1.31	−1.30

be fairly independent of temperature, in view of the dominant contribution to Λ by the contact enthalpy term as discussed above.

The fact that Λ is positive and fairly independent of temperature for these three pairs of polymers means that as the temperature is raised, their degree of incompatibility will diminish and the interfacial boundary between phases will become more diffuse. In the case of block copolymer systems the constraint imposed by the requirement that the block joints be at the domain boundaries provides additional free energy term to make the mixing more favorable. Thus, when the block lengths are fairly moderate, the first term in Equation 1 will become sufficiently large and negative to render the two blocks thermodynamically miscible. Evidence of such homogenization of a styrene–butadiene–styrene block copolymers at increased temperature has been reported from rheological studies (*20, 21*).

There are currently known (*22*) several polymer pairs, such as PS/poly(vinyl methyl ether) (*23, 24*), which appear to be truly compatible thermodynamically. In most of these the compatibility probably arises from specific interactions other than dispersion forces which make Λ for the pair negative. The increase in temperature generally will weaken the specific interaction, and the increasingly unfavorable free volume effect (*25*) will eventually make Λ positive. Thus these compatible pairs will in general exhibit a lower critical solution temperature phenomenon on heating. In contrast, the three polymer pairs discussed in this work are, at room temperature, all below their upper critical solution temperature. The thermodynamic properties of block copolymers formed from these polymer pairs can therefore be discussed in terms of a constant polymer–polymer interaction parameter Λ without referring to the possible complications arising from the free volume effect. In this vein it seems unlikely that a styrene/α-methylstyrene block copolymer (*26*) will exhibit enhanced domain segregation on raising the temperature.

It is gratifying to see that the value of Λ for NR/PS given in Table I is in excellent agreement with 0.8 cal/cm^3 which was estimated by Meier (*4*) for polybutadiene/PS from the solubility parameter difference. It is, however, doubtful that the present approach of using the equation-of-state data can ever be a practical means of evaluating Λ. A reliable value of X_{12} for a polymer–polymer pair can be estimated only when data on a number of similar polymer–solvent systems are available. There is also a theoretical difficulty arising from the unsymmetric definition of X_{12}, so that $X_{21} \neq X_{12}$. In the treatment of polymer–solvent systems, an additional entropic term containing a parameter Q_{12} was often introduced to achieve good agreement between theory and experiment, but it is difficult to estimate the error arising from omission of this term for polymer–polymer systems.

Because of these uncertainties, equations 1, 2, 3, 4, and 5 may not be relied upon as a means of quantitative evaluation of Λ until more data for other polymer–solvent systems become available. The equation-of-state thermodynamics is, however, useful in its ability to give us insight into the physical factors and their relative magnitudes which contribute to the polymer–polymer interaction parameter. The results in this work clearly show that the dependence of Λ on concentration and temperature is moderate. This gives a justification as a good approximation to the use of a constant polymer–polymer interaction parameter in the polymer interface theories where the polymer concentration encompasses the whole range $w_1 = 0$ to 1 across the phase boundary.

Acknowledgment

This research was supported by the Office of Naval Research.

Literature Cited

1. Roe, R. J., *J. Chem. Phys.* (1974) **60,** 4192.
2. Roe, R. J., *J. Chem. Phys.* (1975) **62,** 490.
3. Meier, D. J., *J. Polym. Sci.* (1969) **26**(C), 81.
4. Meier, D. J., "Block and Graft Copolymers," p. 105, Burke and Weiss, Ed., Syracuse University, Syracust, NY, 1973.
5. Meier, D. J., *Polym. Prepr. Am. Chem. Soc., Div. Polym. Chem.* (1974) **15,** 171.
6. Helfand, E., Tagami, Y., *J. Chem. Phys.* (1971) **56,** 3592.
7. Helfand, E., Wasserman, Z. R., *Macromolecules* (1976) **9,** 879.
8. Nose, T., *Polym. J.* (1975) **8,** 96.
9. Krause, S., *Macromolecules* (1970) **3,** 84.
10. Boehm, R. E., Krigbaum, W. R., *J. Polym. Sci.* (1976) **54**C, 153.
11. Leary, D. F., Willimas, M. C., *J. Polym. Sci., Polym. Phys. Ed.* (1973) **11,** 345.
12. Flory, P. J., *J. Am. Chem. Soc.* (1965) **87,** 1833.
13. Eichinger, B. E., Flory, P. J., *Trans. Faraday Soc.* (1968) **64,** 2035.
14. Flory, P. J., *Discuss. Faraday Soc.* (1970) **49,** 7.
15. Flory, P. J., Eichinger, B. E., Orwoll, R. A., *Macromolecules* (1968) **1,** 287.
16. Eichinger, B. E., Flory, P. J., *Trans. Faraday Soc.* (1968) **64,** 2053.
17. Abe, A., Flory, P. J., *J. Am. Chem. Soc.* (1965) **87,** 1838.
18. Hocker, H., Shih, H., Flory, P. J., *Trans. Faraday Soc.* (1971) **67,** 2277.
19. Eichinger, B. E., Flory, P. J., *Trans. Faraday Soc.* (1968) **64,** 2061.

20. Chung, C. I., Gale, J. C., *J. Polym. Sci.* (1976) **14**(A-2), 1149.
21. Gouinlock, E. V., Porter, R. S., *Polym. Prepr. Am. Chem. Soc., Div. Polym. Chem.* (1977) **18**(1), 245.
22. Krause, S., *J. Macromol. Sci., Rev. Macromol. Chem.* (1972) **7**, 251.
23. Nishi, T., Kwei, T. K., *Polymer* (1975) **16**, 285.
24. Robard, A., Patterson, D., *Macromolecules* (1977) **10**, 1021.
25. McMaster, L. P., *Macromolecules* (1973) **6**, 760.
26. Robeson, L. M., Matzner, M., Fetters, L. J., McGrath, J. E., "Recent Advances in Polymer Blends, Grafts and Blocks," p. 281, L. H. Sperling, Ed., Plenum, New York, 1974.

RECEIVED December 14, 1978.

32

A Proposed Generalized Nomenclature Scheme for Multipolymer and Multimonomer Systems

L. H. SPERLING and E. M. CORWIN

Materials Research Center, Coxe Laboratory # 32, Lehigh University, Bethlehem, PA 18015

The number of two-polymer, multipolymer, and multimonomer systems reported in the scientific and patent literature continues to rise without an adequate nomenclature to describe the several materials. This chapter is divided into three parts. (1) A proposed nomenclature system which uses a short list of elements (polymers or polymer reaction products). These elements are reacted together in specific ways by binary operations which join the two polymers to form blends, grafts, blocks, crosslinked systems, or more complex combinations. (2) The relationship between the proposed nomenclature and the mathematics of ring theory (a form of the "new math") is discussed. (3) A few experimental examples now in the literature are mentioned to show how the new nomenclature scheme already has been used to discover new multipolymer systems.

A good nomenclature scheme should accomplish two tasks: (1) accurately name all of the existing known compositions and (2) provide a method for systematic predictions concerning missing, unsynthesized, or unrecognized materials included within the nomenclature scheme (*1*).

Let us now turn to the stateof the art of nomenclature in multicomponent polymers (2–7). Many simple materials already have precise names. For example, poly(butadiene-b-styrene) is represented by the structure in Equation 1, where A stands for the butadiene mer and B represents the styrene mer in block copolymer arrangement, as indicated by the small -b-.

0-8412-0457-8/79/33-176-609$05.50/0

$$A{-}A{-}A{-}\ldots{-}A{-}B{-}B{-}B{-}\ldots{-}B \tag{1}$$

$$\begin{array}{l} A{-}A{-}A{-}A{-}A{-}A{-}\ldots{-}A \\ \qquad\qquad\quad | \\ \qquad\qquad\quad B{-}B{-}B{-}B{-}\ldots{-}B \end{array} \tag{2}$$

Similarly, poly(butadiene-g-styrene) represents the graft copolymer in Equation 2, where it is understood that the symbol -g- means graft copolymer, the first indicated species forms the backbone, and the second the side chain. Similarly, the symbol -co- stands for a random copolymer, -a- for an alternating copolymer, and occasionally the symbol -cl- is used for crosslinked systems.

Materials such as blends (*8*), interpenetrating polymer networks (IPNs) (*9, 10*), AB-crosslinked copolymers (*11*), and many other combinations have no existing systematic nomenclature. The situation is further complicated by the fact that the time sequence of polymerization, grafting, and/or crosslinking each produce materials possessing different morphologies and often widely different mechanical or physical properties (*7*). An examination of the patent literature (*12*) reveals complex combinations of up to five polymers. An exact description of each requires a long paragraph, and the identification of the isomeric possibilities presents a serious challenge even to the expert. The need for a more comprehensive nomenclature has now become imperative.

Because of the above considerations, Sperling and co-workers (*12, 13*) have evolved a tentative nomenclature scheme for polymer blends, isomeric graft copolymers, and IPNs, and have gradually broadened and clarified the system. However, a certain level of achievement has been attained, and even though the proposed system still has faults, it will be presented below.

A few final points are in order.

(1) From a notation point of view, the system to be presented reads from left to right, corresponding to ordinary chemical notation (*2, 3, 4*). Some of the earlier papers had notation which read from right to left, following standard mathematical notation. However, since the following is primarily intended for people working in the chemical arts, the system now reads in the sequence ordinarily used by chemists.

(2) The nomenclature system has been submitted to the Nomenclature Committee of the Polymer Division of the American Chemical Society (*14*) for evaluation. The authors invite the readers to help, if they see possible improvements, errors, inconsistencies, etc. A nomenclature system has value only if it accurately serves its purpose and is in a form acceptable to "the workers in the field."

Table I. Basic Polymeric Elements

Symbol	*Meaning*
P	linear polymer
R	random copolymer
A	alternating copolymer
M	mechanical or physical blend
G	graft copolymer
C	crosslinked copolymer
B	block copolymer
S	starblock copolymer
I	interpenetrating polymer network
U	unknown reaction mixtures

Proposed Nomenclature for Multipolymer Systems

The number of two-polymer and multipolymer combinations reported in the scientific and patent literature continues to rise without an adequate nomenclature to describe the several materials. More than 200 distinct topological methods of organization are already known (*7, 12*) and new methods are being reported frequently. In addition to the usual molecular specifications, the time sequence of synthesis is important in many cases and needs to be preserved in the nomenclature for a full comprehension of the final product.

The following nomenclature scheme uses a short list of elements in Table I (polymers and polymer-reaction products) which are reacted together in specific ways by binary operations (a joining of two elements), Table II. A series of numerical subscripts are used on the elements to allow an arbitrarily large number of different polymers to be designated conveniently.

Table II. Binary Operations and Associated Reactions

Symbol	*Reaction Induced*
O_R	random copolymer formation
O_A	alternating copolymer formation
O_M	mechanical or physical blending[a]
O_G	graft copolymer formation
O_C	crosslinked network formation
O_B	block copolymer formation
O_S	starblock copolymer formation
O_I	interpenetrating polymer network
O_U	unknown reaction mixtures

[a] A blend indicates a mixture of two or more molecular species without chemical bonding between them and can be induced by mechanical blending, physical means such as coprecipitation, or chemical means such as degrafting.

This nomenclature primarily answers three questions about the chemical entity: (1) which polymers are combined, (2) principal modes of combination, and (3) the time sequence of reaction. Other items of information, such as weight proportions, molecular weights, morphology, tacticity, etc. can be included as ancillary items but will not be discussed in detail below.

While the nomenclature scheme was originally derived to describe polymer blends, blocks, grafts, and interpenetrating polymer networks, it was felt important to generalize the scheme by including random and alternating copolymers and starblock materials although the first two are formed through direct combination of monomers herein. Through the addition of further operations, additional types of materials can easily be included.

As with other nomenclature schemes, the objective will be to name the most important product(s) of a reaction. In some cases, several isomers are possible, but listing each one all the time becomes cumbersome and perhaps misleading.

Subscripts. The subscripts, 1, 2, 3, . . . *i*, *j*, . . . will indicate a numbering of the polymers such as in Equations 3 and 4. When more

$$P_1 \quad \text{Polymer 1} \tag{3}$$

$$P_j \quad \text{Polymer } j \tag{4}$$

than one subscript appears, the first appearing subscript indicates the first formed polymer, and the second subscript, the second formed polymer as shown in Table III. In R, A, and M materials, as defined in Table

Table III. Illustration of the Use of Subscripts

P_1	A homopolymer composed of monomer 1 mers.
G_{12}	A graft copolymer having polymer 1 as the backbone and polymer 2 as the side chain.
C_{11}	A crosslinked polymer composed of a network of polymer 1 chains.
R_{27}	A random copolymer composed of monomers 2 and 7. (In R and A materials, the time element can be replaced by the composition importance order.)
U_{12}	Unknown reaction mixture of polymer 1 and polymer 2. Specific examples: mechano-chemical blends which contain unknown proportions of grafts, blocks, and homopolymers or solution graft copolymers which contain much homopolymer. Various isomers can be formed.
M_{21}	Mechanical or chemical blend, with no time sequence. First subscript listed may have the higher weight percentage, etc.

I, the time sequence is unimportant. The first listed subscript should be the most important.

Reaction Examples. The binary operation symbol used between any two elements reacts them together in the required manner. (The reader is reminded that all combinations, mathematical or chemical, are binary operations. The advent of modern computers has focused much attention on this often neglected fact.) The first formed element appears on the left and the reaction-time sequence (when required) proceeds from left to right.

Some examples illustrate the use of this notation. A graft copolymer having polymer 1 as the backbone and polymer 2 as the side chain, synthesized in that order (Equation 5); a crosslinked copolymer made from polymers *i* and *j* (Equation 6) (an AB-crosslinked copolymer); an alternating copolymer of polymers 3 and 5 (Equation 7). (For random and alternating copolymers, one really considers mixtures of mers rather than large chain portions as in block copolymers.) Equation 8 represents a physical blend of graft copolymers G_{13} and G_{23}, where polymers 1 and 2 were blended first, followed by the polymerization of polymer 3 with grafting onto polymers 1 and 2. On the other hand, in Equation 9, the symbol $G_{1(2,3)}$ indicates both polymers grafted onto polymer 1. (Alternately, the symbol $U_{1(2,3)}$ might be called for since the exact modes of grafting cannot be specified, or which product(s) will dominate). Figure 1 illustrates some simple structures.

$$P_1 \, O_G \, P_2 = G_{12} \tag{5}$$

$$P_i \, O_C \, P_j = C_{ij} \tag{6}$$

$$P_3 \, O_A \, P_5 = A_{35} \tag{7}$$

$$(P_1 \, O_M \, P_2) \, O_G \, P_3 = G_{13} \, O_M \, G_{23} \tag{8}$$

$$P_1 \, O_G \, (P_2 \, O_M \, P_3) = G_{12} \, O_M \, G_{13} \, O_M \, G_{1(2,3)} \tag{9}$$

The above examples use the abstract element symbols P_1, G_{12}, etc. Briefly let us show how this system will operate with real monomers, polymers, and multipolymer combinations, using chemical notation instead of the elements but retaining the binary operation notation.

More than one possible notation appears possible.

(1) Direction substitution of element notation (Equation 10) representing the grafting of polystyrene side chains onto polybutadiene backbone. An AB-crosslinked copolymer of Bamford and Eastmond (14a)

$$\text{polybutadiene } O_G \text{ polystyrene} \quad (10)$$

$$\text{poly(vinyl trichloroacetate) } O_C \text{ polystyrene} \quad (11)$$

is shown in Equation 11. Here, two polymers are caused to form one network through intermolecular covalent bonds.

An IPN of two different polymers is shown in Equation 12, where the poly(ethyl acrylate) network is formed first, then swollen with styrene and crosslinker, and polymerized in situ (*16*).

$$\text{poly(ethyl acrylate) } O_C \text{ poly(ethyl acrylate)} \\ O_I \text{ polystyrene } O_C \text{ polystyrene} \quad (12)$$

$$\text{polyethylene } O_M \text{ polypropylene} \quad (13)$$

Mechanical blends, usually not considered as single chemical entities, can easily be represented by Equation 13. This particular combination was presented by Deanin and Sansone as part of the present symposium (*17*).

(2) Retention of abstract elemental symbolism, but equating elements and polymers separately. This notation method is more suitable for more complex reaction mixtures. The equivalent of Equation 10 can be written as in Equation 14.

$$G_{12}; \quad P_1 = \text{polybutadiene} \\ P_2 = \text{polystyrene} \quad (14)$$

More complex examples of this second scheme will await the introduction of vertical notation, below.

Vertical Notation

For very complex notation, part of the structure may be shown vertically. Two-dimensional nomenclature, and particularly structures, are common throughout chemistry. Examples:

Structure	Nomenclature
P_1	P_1
	O_G
P_2	$P_2\,O_C\,(P_4\,O_A\,P_6)$
	O_G
A_{46}	$P_3\,O_B\,P_5$
P_3, P_2	
A_{46}	
P_5	(15)

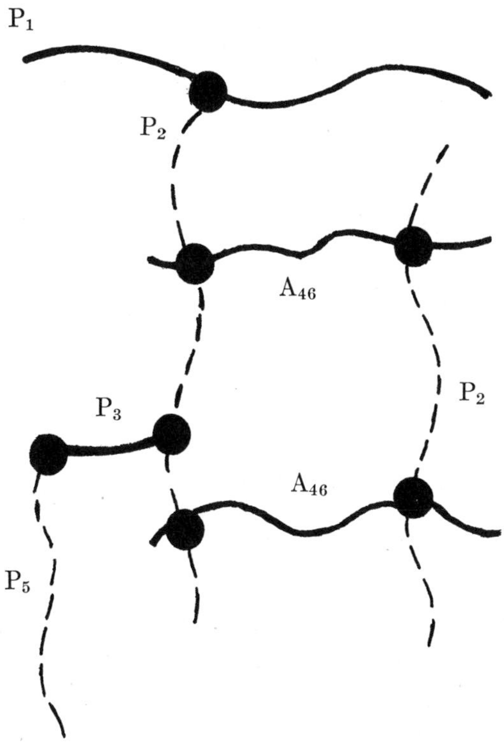

Let us return to the second method of naming materials, started above, with the aid of vertical as well as horizontal representation. The compositions in Equation 15 might have been composed of those in Equation 16. Again, it should be understood in the present context that P_4 and P_6 appear in the monomeric or single mer form.

P_1 = polybutadiene
P_2 = poly(vinyl alcohol)
P_3 = poly(ethylene terephthalate)
P_4 = polystyrene
P_5 = nylon 66
P_6 = poly(maleic acid anhydride)

(16)

One further example might be the thermoplastic elastomer, triblock copolymer of polystyrene, polybutadiene, and polystyrene, in that order. The polybutadiene in this case has been grafted with poly(methyl methacrylate). The proposed nomenclature is shown in Equation 17. For emphasis, the notation is read from top to bottom, and from left to right.

$$\begin{array}{ll} P_1 & P_1 = \text{polystyrene} \\ O_B & P_2 = \text{polybutadiene} \\ P_2\ O_G\ P_3 & P_3 = \text{poly(methyl methacrylate)} \\ O_B & \\ P_1 & \end{array} \quad (17)$$

Tables. The value of the proposed system lies principally in its capability of depicting very complex multipolymer combinations. These tables, unlimited in size, join combinations of elements systematically. Tables IV and V provide examples. Each binary operation has a table and all of the elements can appear in any table. The rows are reacted with the columns in that order. For example, in Table V, P_1 at the left of the row is reacted with P_2 at the top of the column to synthesize G_{12}.

A comment is in order about the complex subscripting in the lower portions of Table V. If it is presumed that the reaction site occurs at the end of that molecular portion most recently reacted, then taking P_2 at the left of the row and reacting it with G_{12} at the top of the column produces G_{221}. The first subscript 2 comes from the P_2 of the row. The second subscript 2 comes from the growing end of the side chain of the G_{12}, yielding the final structure as in Equation 17a. However, looking

Table IV. Mechanical and Physical Polymer Blends[a]

O_M	P_1	P_2	M_{12}	G_{12} . . .
P_1	$2P_1$	M_{12}	$2P_1\ O_M\ P_2 \cong M_{12}$	$P_1\ O_M\ G_{12}$
P_2	M_{21}	$2P_2$	$P_1\ O_M\ 2P_2 \cong M_{12}$	$P_2\ O_M\ G_{12}$
M_{12}	$2P_1\ O_M\ P_2 \cong M_{12}$	$P_1\ O_M\ 2P_2 \cong M_{12}$	$2M_{12}$	$M_{12}\ O_M\ G_{12}$
G_{12}	$G_{12}\ O_M\ P_1$	$G_{12}\ O_M\ M_2$	$G_{12}\ O_M\ M_{12}$	$2G_{12}$
. . .				

[a] Note $2P_1\ O_M\ P_2$ and $P_1\ O_M\ 2P_2$ were taken to approximate M_{12}, as indicated.

Table V. Graft and Branch Copolymers

O_G	P_1	P_2	M_{12}	G_{12} . . .
P_1	G_{11}	G_{12}	G_{11} O_M G_{21}	G_{112}
P_2	G_{21}	G_{22}	G_{21} O_M G_{22}	G_{212}
M_{12}	G_{11} O_M G_{21}	P_{12} O_M G_{22}	G_{21} O_M G_{22} G_{11} O_M G_{12}	G_{112} O_M G_{212}
G_{12}	G_{121}	G_{122}	G_{121} O_M G_{212}	G_{2121}
.				
.				
.				

ahead to the discussion leading up to Equation 29, the result G_{212} is attained. This has the geometric structure as in Equation 17b. This latter is preferred and used in Table V.

Obviously, in more complex cases, one may not know which product will predominate, or a mixture of isomers may be generated. However, by specifying the reaction rules a series of tables can be generated, each yielding the chemical isomers in the equivalent table positions.

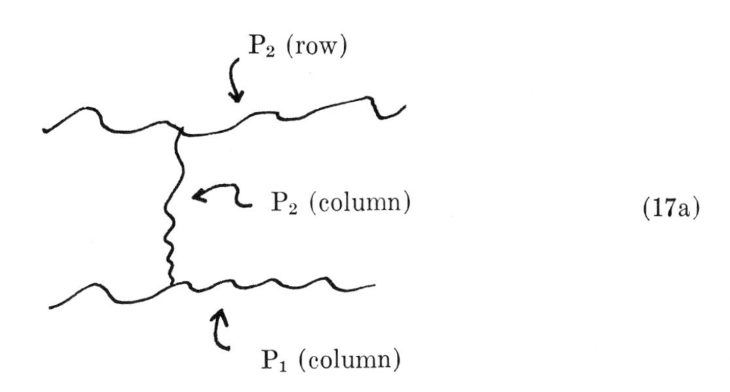

(17a)

P_2 (row)

P_1 (column)

P_2 (column)

(17b)

Simultaneous Mixing or Reacting. One of the advantages of the proposed nomenclature is the preservation of the time sequence of polymer reactions. Examples include grafting reactions and interpenetrating polymer networks. In some cases, like the blends or random copolymerization, the time sequence has no meaning.

In other cases, it may be important to indicate that the mixing or reactions were carried out simultaneously. This will be indicated, where necessary, by brackets. *See* example in Equation 18. This indicates a simultaneous interpenetrating network, where C_{11} was formed by one

$$[C_{11}\ O_I\ C_{22}] \tag{18}$$

$$C_{11}\ O_I\ C_{22} \tag{19}$$

reaction (say addition), C_{22} was found independently by another reaction (say condensation), and the two networks are simultaneously polymerized. Lacking brackets (Equation 19), a sequential synthesis of an interpenetrating polymer network is indicated. In both cases, this special mode of interpenetration is indicated by the binary operation O_I.

Coefficients and Molecular Weights. Where proportions are known, coefficients can be used (Equation 20). This indicates five moles of P_2

$$P_1\ O_R\ 5P_2 \tag{20}$$

per mole of P_I. Molecular weights, where known, may be indicated by parentheses following the element: $P_1(5 \times 10^4$ g/mol) O_G $P_2(3 \times 10^5$ g/mol).

Multiple Mathematical Ring Structure of the Nomenclature Scheme

In the previous papers of this series, the concepts of mathematical group theory and then ring theory were brought to bear on the nomenclature and structure of multicomponent polymer materials (*12, 18, 19*). In each case, no proof of the existence of either a mathematical group or ring was offered, but rather only the notions or concepts were applied.

The objective of this portion of the chapter will be to present such a proof, with certain physical restrictions. With the additions of a zero, additive inverses, and coefficients, and simple restrictions regarding reactivity sites, a series of formal rings will be shown to exist. Table IV will be shown to form an "additive" group held in common with the different "multiplication" modes of joining to form a series of rings. Several theorems of a general nature are applied to the formal ring, yielding insight into synthetic methods. The chemical literature relating to the several multipolymer combinations (*20, 28*) is broad and recently has been reviewed (*18, 29*).

$$R = (P_1, P_2, P_3, \ldots, P_n) \tag{21}$$

A generating set R of n linear polymer elements is defined (Equation 21), where n represents the number of different kinds of polymer chains required. Each polymer has an additive inverse which will be written P_1^{-1}, P_2^{-1}, etc. By definition the inverses subtract, or take away. Rings also require a zero which adds nothing to a system and which, in the multiplicative sense, removes everything.

These elements are combined by any of the binary operations shown in Table II. While O_M forms the addition mode, all others constitute modes of "multiplication." Simple examples of modes of combination are shown in Figure 1.

Thus the series of rings will be shown to each contain the O_M binary operation, combined with a different one of the multiplicative binary operations. For brevity, the combinations generated by O_G will be examined in detail. The extension to the other tables is obvious in many cases.

Several mathematical texts were used. The principal text referred to is McCoy (*30*), but books by Baumslag and Chander (*31*), Goldhaber and Ehrich (*32*), and Cotton (*33*) will be referenced also. For convenience of the reader, the page numbers of the various theorems and statements will be given in the text. As stated previously, the notation will read from left to right, in the normal chemical sense, rather than right to left as frequently used by mathematicians.

Six Requirements of a Ring. The following discussion is based on establishing that Laws One through Six of McCoy (*30*, p. 23) hold for the binary operations O_M and O_G. A small finite segment of the "addition" and "multiplication" tables generated by O_M and O_G are shown in Table IV and V, respectively.

Law One: $a + b = b + a$ (commutative law of addition). In the synthesis of polymer blends, the polymers (elements) may be added in any order to get the same chemical mixture. Thus, the elements are interchangeable with respect to position, and may be written in any order.

$$P_1 \, O_M \, P_2 = M_{12} = M_{21} = P_2 \, O_M \, P_1 \tag{22}$$

Law Two: $(a + b) + c = a + (b + c)$ (associative law of addition). Blending a mixture of two polymers with a third yields the same ternary chemical mixture as blending the mixture of the second and third with the first.

$$(P_1 \, O_M \, P_2) \, O_M \, P_3 = P_1 \, O_M \, (P_2 O_M \, P_3) = M_{123} \tag{23}$$

Law Three: There exists an element 0 of R (set of all elements) such that $a + 0 = a$ for every element of R (existence of a zero). Under the binary operation O_M, it means that nothing was added. $P_1 \, O_M \, 0 = P_I$,

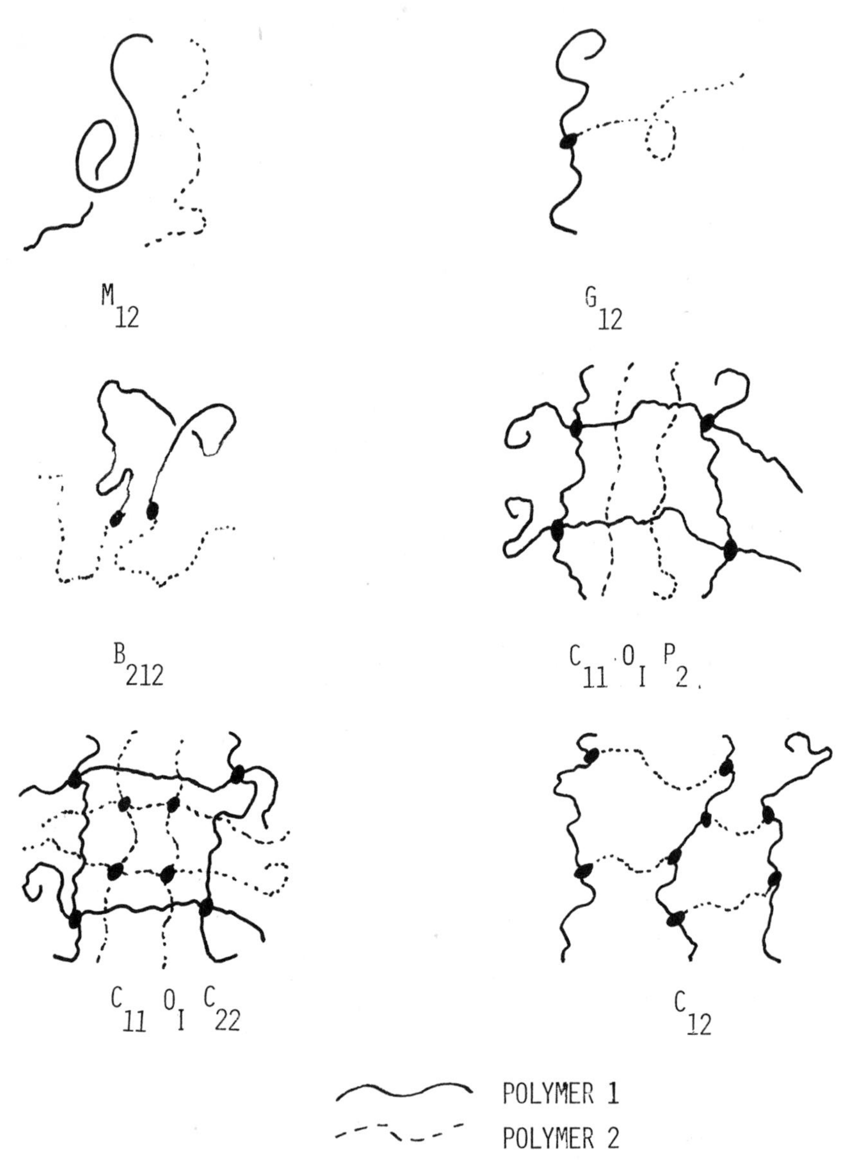

Figure 1. Simple two-polymer combinations

etc. Under the binary operation O_G, generating Table V, it follows immediately that multiplication by zero removes the element (Equation 24). The mathematics illustrated in Equation 24 is convenient for pouring out the pot, unzipping a chain polymerization, etc.

$$0\ O_G\ P_1 = 0 \tag{24}$$

Law Four: If $a \in R$, there exists $x \in R$ such that $a + x = 0$ (existence of additive inverses). The inverse elements combine with the positive elements as follows in Equation 25. This leads to the removal mechanism in the additive table.

$$P_1 \, O_M \, P_1^{-1} = 0 \tag{25}$$

The appearance of an inverse element in the final combination of positive and inverse elements means that the product cannot be made in the laboratory. Examples include those in Equations 26 and 27. The

$$P_1 \, O_M \, P_2^{-1} = M_{1-2} \tag{26}$$

$$P_2 \, O_G \, P_1^{-1} = G_{2-1} \tag{27}$$

following two laws are more difficult to apply exactly without the introduction of some arbitrary rules regarding reactive sites in chemically combining polymers.

Law Five: $(ab)c = a(bc)$ (associative law of multiplication). The two rules are adopted to eliminate ambiguous cases. Rule 1: complex elements must be broken down into their fundamental counterparts and then reacted; and rule 2: a reaction in a series of reactions joins the new polymer to the next element on the right. Thus we have Equations 28, 29, 30, 31, and 32. Finally, in conformity with Law Five, we have Equations 33 and 34. The chemical structures associated with Equations 28, 29, and 30 are shown in Figure 2.

$$P_1 \, O_G \, P_2 = G_{12} \tag{28}$$

$$P_1 \, O_G \, P_2 \, O_G \, P_3 = G_{123} \tag{29}$$

$$P_1 \, O_G \, P_2 \, O_G \, P_3 \, O_G \, P_4 = G_{1234} \tag{30}$$

$$G_{12} \, O_G \, G_{34} = P_1 \, O_G \, P_2 \, O_G \, P_3 \, O_G \, P_4 \tag{31}$$

$$G_{12} \, O_G \, G_{34} = G_{1234} \tag{32}$$

$$(P_1 \, O_G \, P_2) \, O_G \, P_3 = P_1 \, O_G \, (P_2 \, O_G \, P_3) \tag{33}$$

$$G_{123} = G_{123} \tag{34}$$

From a chemical point of view, it may not be known if P_3 in Equation 19 will react with P_1 or P_2 or both. Further tables can be set up, giving the isomeric structures of the multipolymer grafts. From a mathematical or chemical point of view, each element in the new tables will be

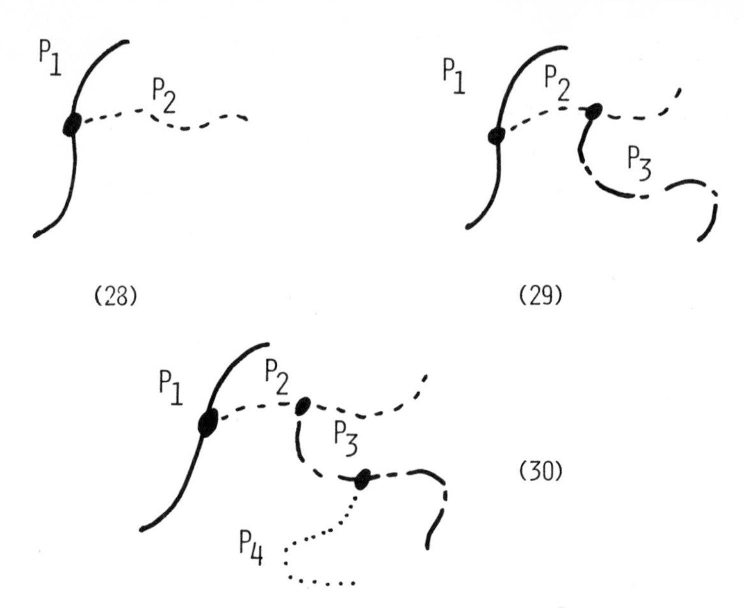

Figure 2. Products of sequential grafting reactions

mathematically or chemically isomorphic with the formal ring element. A series of functions could be generated to reach each of the isomers, showing the relationship among them. By isomorphic, it is not meant that the isomorphism is between the rings or tables, but rather that there is a natural one-to-one and onto mapping, or a natural bijection of the elements.

Law Six: $a(b + c) = ab + ac$, $(b + c)a = ba + ca$ (distributive laws). The simple graft copolymer interpretation would be as in Equations 35 and 36. The structures corresponding to Equations 35 and 36 are shown in Figure 3.

$$P_1\, O_G\, (P_2\, O_M\, P_3) = G_{12}\, O_M\, G_{13} \tag{35}$$

$$(P_1\, O_M\, P_2)\, O_G\, P_3 = G_{13}\, O_M\, G_{23} \tag{36}$$

Without further rules, one does not know whether P_2 and P_3 both react with P_1 to form Equation 37. (*See* Figure 3.) The use of the Rule 2 (*above*) denies Equation 37. However, the structure can be reached through the construction of alternate isomeric reaction tables, similar to that suggested above. It should be noted that the elements of Equations 35 and 36 are homomorphic with the element in Equation 37. The term homomorphic is used because there is not a one-to-one correspondence between the elements of Equations 35 and 36 and Equation 37.

$$P_1\,O_G\,(P_2\,O_M\,P_3) = G_{1(23)} \quad (37)$$

The reader will note that coefficients have been introduced into Table IV and elsewhere as needed. According to the rearrangement theorem, Cotton (33, p. 6), "Each row and each column in the group addition table lists each of the group elements once and only once. From this, it follows that no two rows can be identical nor can any two columns be identical. Thus, each row and each column is a rearranged list of the

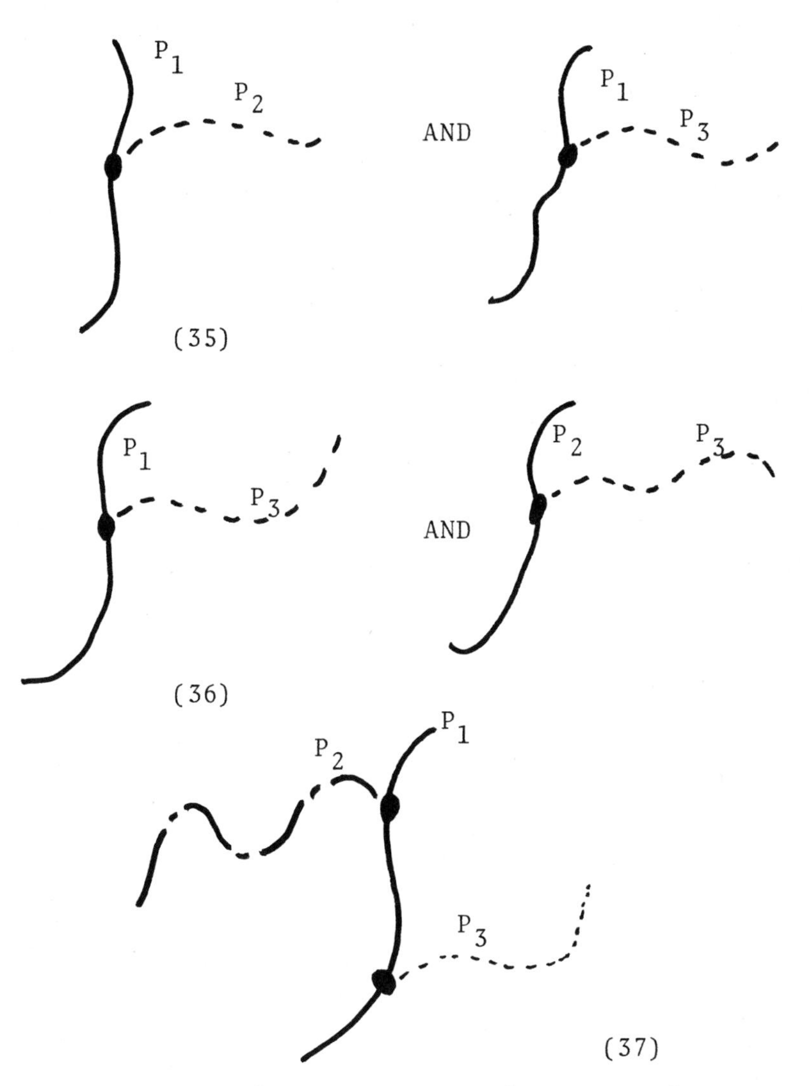

Figure 3. Graft copolymer structures. Numbers refer to the equation numbers in the text.

group elements." (Table IV forms a mathematical group, as shown below.)

$$P_1\ O_M\ P_1 = 2P_1 \tag{38}$$

In other words, $2P_1$ atnd not P_1, should be used. The reader should note that while the ring is finitely generated by $R = (P_1, P_2, P_3, \ldots, P_n)$, the number of elements is infinite. In particular, this means that an infinite column or row length would be required to duplicate all the elements.

The above discussion shows that with (i) a zero, (ii) inverse additive elements, and (iii) coefficients, the set defined by R and the binary operations O_M and O_G form a ring. According to Laws Five and Six, only one chemical isomer is permitted. However the "legal" isomer element is either isomorphic or homomorphic with the other chemical possibilities, which may be formed into alternate tables according to simple chemical rules.

Rings Involving O_C and O_B Operations. Let us now examine the tables generated by the O_C and O_B binary operations. Each element in these two tables is isomorphic, i.e., has a bijection, with its corresponding element in the O_G-generated table. Laws One, Two, Three, and Four apply only to the O_M-generated table. The applicability of Laws Five and Six needs to be established.

Law Five:

(a) Crosslinked Copolymers. The structure (Equation 39) indicates an AB-crosslinked copolymer (*15, 34*) generated by three polymers. From a topological point of view, both sides of the Equation 39 yield the same structure.

$$(P_1\ O_C\ P_2)\ O_C\ P_3 = P_1\ O_C\ (P_2\ O_C\ P_3) \tag{39}$$

(b) Block Copolymers. The structure (Equation 40) also forms the same structure on both sides. Rules 1 and 2 may be used as necessary.

$$(P_1\ O_B\ P_2)\ O_B\ P_3 = P_1\ O_B\ (P_2\ O_B\ P_3) \tag{40}$$

Law Six:

(a) Crosslinked Copolymers. The structure (Equation 41) forms an interesting kind of interpenetrating polymer network (IPN) (*35, 36*) where each network is an AB-crosslinked copolymer (*15, 34*). The structure (Equation 42) is also an IPN composed of AB crosslinked copolymers.

$$P_1\ O_C\ (P_2\ O_M\ P_3) = P_1\ O_C\ P_2\ O_M\ P_1\ O_C\ P_3 \tag{41}$$

$$(P_1\ O_M\ P_2)\ O_C\ P_3 = P_1\ O_C\ P_3\ O_M\ P_2\ O_C\ P_3 \tag{42}$$

(b) Block Copolymers. The structures (Equations 43 and 44) are also well defined.

$$P_1\ O_B\ (P_2\ O_M\ P_3) = P_1\ O_B\ P_2\ O_M\ P_1\ O_B\ P_3 \qquad (43)$$

$$(P_1\ O_M\ P_2)\ O_B\ P_3 = P_2\ O_B\ P_3\ O_M\ P_1\ O_B\ P_3 \qquad (44)$$

Mixed Elements. Clearly, it would be advantageous from a nomenclature point of view to have elements composed by combinations of operations via O_G, O_C, and O_B. These can be easily introduced but are also subject to reactivity restrictions, leading to a third rule. Rule 3: binary element combinations formed by operations not part of the ring under consideration behave as single elements and cannot be separated into more simple elements. Togther with Rules 1 and 2, this Rule 3 eliminates ambiguous reactions in most of the cases of interest. There may be more complex combinations that Rules 1, 2, and 3 do not completely define. In each case, isomorphic or homomorphic elements may exist but can be reached as per discussion above.

Use of Theorems. The establishment of the present nomenclature and associated structures as a series of related rings suggests the search for theorems that can be adoptable to the present case or that can generate new ones. The following two theorems are self evident for the chemist.

McCoy (*30*, p. 33) theorem: the zero of a ring R, whose existence is asserted by Law Three, is unique.

McCoy (*30*, p. 34) theorem: if a, b, and c are elements of a ring, R, the following are true:

$$\text{(i) if } a + c = b + c, \text{ then } a = b;$$

$$\text{(ii) if } c + a = c + b, \text{ then } a = b.$$

An example of (ii) is shown in Equations 45 and 46.

$$P_3\ O_M\ G_{12} = P_3\ O_M\ (P_1\ O_G\ P_2) \qquad (45)$$

then

$$G_{12} = P_1\ O_G\ P_2 \qquad (46)$$

And a corollary: the additive inverse of an element of a ring, R, whose existence is asserted by Law Four, is unique.

McCoy (*30*, p. 36) theorem: if a and b are elements of a ring, R, the equation $a + x = b$ has in R the unique solution $x = b - a$.

Two examples will be given.
(1) Let $a = P_1$ and $b = M_{12}$, then $x = P_2$
(2) Let $a = P_2^{-1}$ and $b = P_1$, then we have:

$$P_1^{-1} O_M x = P_1 \tag{47}$$

$$x = P_1 O_M P_1 = 2 P_1. \tag{48}$$

This last shows the importance of coefficients in the system.

McCoy (*30, p. 38*) theorem: for each element a of a ring, R, we have $a \cdot 0 = 0 \cdot a = 0$. In chemical terms, the element a was removed.

McCoy (*30,* p. 39) let R be a ring and S a nonempty subset of the set R. Then S is a subring if and only if the following conditions hold:

(i) S is closed under the operations of addition and multiplication defined on R; and

(ii) if $a \in S$, then $-a \in S$.

The simplest example in the present case is shown in Equations 49 and 50. The subring generated by P_1, i.e., all positive and negative integer multiples of P_1 and zero, fulfills the necessary requirements.

$$R = (P_1, P_2, P_3, \ldots P_n) \tag{49}$$

$$S = (P_1) \tag{50}$$

A Mathematical Group

Laws Two, Three, and Four form the basis for a group, McCoy (*30,* p. 135). Since all three apply to the elements under the binary operation O_M, it follows that the polymer blends constitute an additive group. In fact, O_M generates a free abelian group on (P_1, P_2, P_3, . . . , P_n) since its elements (plus inverses and zero) are commutative under O_M.

Examination of Table IV reveals an interesting set of subgroups. An example is shown in Table VI. Composed of the zero, any element, and its inverse, Table VI is a finitely generated (but infinite) subgroup

Table VI. A Finitely Generated Group of P_1 Combinations

O_M	O	P_1	P_1^{-1} . . .
O	O	P_1	P_1^{-1}
P_1	P_1	$2P_1$	O
P_1^{-1}	P_1^{-1}	O	$2P_1^{-1}$
.			
.			
.			

of the elements in Table IV. This statement can be generalized. From Goldhaber and Ehrlich (*32*, p. 110), "Every subgroup of a finitely generated abelian group is finitely generated."

Because Law One is followed ($P_2 O_M P_1 = P_1 O_M P_2$), a corollary can be written: the elements of the table generated by O_M-binary operations are symmetric about the diagonal.

Summary of Ring Notation. In summary, the addition of a zero, additive inverses, and coefficients allows for Laws One through Six to be followed. The polymer blend, graft, and IPN nomenclature scheme forms three rings, and the fact that the binary operation O_M on $R = (P_1, P_2, P_3, \ldots, P_n)$ constitutes a group puts the system on an improved mathematical ground. Similar relationships can be developed for the other operations.

Several theorems were examined and adopted or shown to hold for the present ring system. Such theorems show new insight into laboratory related problems and deserve further study.

The value of the present nomenclature is twofold.

(1) It does provide a nomenclature, where none existed before. The nomenclature, however, remains incompleted and requires further consideration.

(2) For the research polymer chemist, the scheme has the function of being able to systematically suggest new or unrecognized polymer combinations or new reactions.

The chemical nomenclature scheme makes no mention of morphology although this subject is of paramount importance in governing properties (*7, 8, 33, 34, 35*). As a nomenclature, it only speaks to how the various polymer chains are combined in space.

Mathematical Functions and Applications

In the proceeding, the notions of elements and binary operations were developed. An important part of the mathematics of ring theory involves the concept of functions.

Upon examination of Tables IV and V, a striking symmetry becomes apparent. Elements on either side of the diagonal running from the upper left to the lower right are clearly related. For example, the qualitative relationship between G_{12} and G_{21} is obvious, but we now observe that they are quantitatively related in Table V, lying in symmetric positions across the diagonal. In fact, a function, γ, can be defined which will cause the element to be moved to its corresponding position across the diagonal and to physically adopt its new structure (*12, 13*). *See* Figure 4. In general, two functions can be defined. In addition to γ, a function of β can be defined as one which moves an element from a position in one

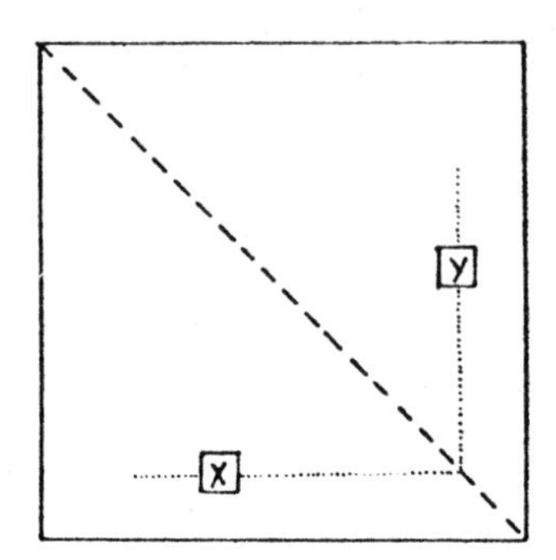

Figure 4. The function γ moves an element to its symmetric position across the diagonal. In this case, elements X *and* Y *are interchanged.*

table to its corresponding position in another table and simultaneously requires the element to undergo the necessary chemical transformation.

The function β may be subscripted for clarity, indicating, as necessary, the tables from which and to which the element is being moved. Some examples are in order.

$$\gamma\, G_{12} = G_{21} \tag{51}$$

$$\beta_{GC}\, G_{12} = C_{12} \tag{52}$$

$$\beta_{IM}\, C_{11}\, O_I\, C_{22} = P_1\, O_M\, P_2 \tag{53}$$

This last creates a kind of chemical blend and was recently accomplished (9) to produce a novel morphology.

One can also consider the transformation of one polymer species into another by post-polymerization reactions. For example, if polymer I is polystyrene and polymer II is sulfonated polystyrene, one can use the function δ for the transformation:

$$\delta\, P_1 = P_2 \qquad \begin{array}{l} P_1 = \text{polystyrene} \\ P_2 = \text{sulfonated polystyrene} \end{array} \tag{54}$$

For a crosslinked random copolymer of styrene and ethyl vinyl benzene (assuming only the styrene reacts), we could write:

$$\delta_{23} \begin{pmatrix} P_1 \\ O_R \\ P_2 \end{pmatrix} O_C \begin{pmatrix} P_1 \\ O_R \\ P_2 \end{pmatrix} = \begin{pmatrix} P_1 \\ O_R \\ P_3 \end{pmatrix} O_C \begin{pmatrix} P_1 \\ O_R \\ P_3 \end{pmatrix} \tag{55}$$

where P_1 = ethyl vinyl benzene, P_2 = styrene, P_3 = sulfonated styrene, and where the function δ_{23} represents the selective element reacted.

In general, reactions can be found to move the elements from one table to another under the function of β, or within the table under the function γ. Again, while the nomenclature and the mathematical ring notation do not speak about morphology or physical properties, the methods outlined above can be used to reach novel materials.

Literature Cited

1. Parkes, G. D., "Mellor's Modern Inorganic Chemistry," Chap. 8, Longmans, Green, and Co., 1951.
2. "IUPAC Nomenclature of Regular Single-Stranded Organic Polymers," *Pure Appl. Chem.* (1976) **48,** 373.
3. "Basic Denitions of Terms Relating to Polymers," pp. 479–491, IUPAC, Butterworths, 1974.
4. "List of Standard Abbreviations (Symbols) for Synthetic Polymers and Polymer Materials," pp. 475–576, IUPAC, Butterworths, 1974.
5. Ceresa, R. J., "Block and Graft Copolymers," Butterworths, 1962.
6. Burlant, W. J., Hoffman, A. S., "Block and Graft Polymers," Reinhold, 1960.
7. Manson, J. A., Sperling, L. H., "Polymer Blends and Composites," Plenum, 1976.
8. Bucknall, C. B., "Toughened Plastics," Appl. Sci. London, 1977.
9. Neubauer, E. A., Thomas, D. A., Sperling, L. H., *Polymer* (1978) **19,** 188.
10. Klempner, D., Frisch, K. C., *J. Elastoplast.* (1973) **5,** 196.
11. Eastwood, G. C., Phillips, D. G., "Polymer Alloys," D. Klempner, K. C. Frisch, Eds., Plenum, 1977.
12. Sperling, L. H., Ferguson, K. B., *Macromolecules* (1975) **8,** 691.
13. Sperling, L. H., Ferguson, K. B., Manson, J. A., Corwin, E. M., Siegfried, D. L., *Macromolecules* (1976) **9,** 743.
14. Donaruma, G., Vice President for Academic Affairs, New Mexico Institute of Mining & Technology, Socorro, NM, 87801, Chairman.
15. Bamford, C. H., Eastmond, G. C., "Recent Advances in Polymer Blends, Grafts, and Blocks," L. H. Sperling, Ed., Plenum, New York, 1974.
16. Huelck, V., Thomas, D. A., Sperling, L. H., *Macromolecules* (1972) **5,** 340, 348.
17. Deanin, R. D., Sansone, M. F., *Polym. Prepr., Am. Chem. Soc., Div. Polym. Chem.* (1978) **19**(1), 211.
18. Sperling, L. H., "Recent Advances in Polymer Blends, Grafts, and Blocks," L. H. Sperling, Ed., Plenum, New York, 1974.
19. Sperling, L. H., "Toughness and Brittleness of Plastics," R. D. Deanin, A. M. Crugnola, Eds., ADV. CHEM. SER. (1976) **154,** Chap. 14.
20. Aggarwal, S. L., *Polymer* (1976) **17,** 838.
21. Imken, R. L., Paul, D. R., Barlow, J. W., *Polym. Eng. Sci.* (1976) **16,** 593.
22. Ceresa, R. J., "Block and Graft Copolymers," Chap. 1, Butterworths, London, 1962.
23. Burlant, W. J., Hoffman, A. S., "Block and Graft Polymers," Chap. 1, Reinhold, New York, 1960.
24. Ceresa, R. J., Ed., "Block and Graft Copolymerization," Vol. 1, Wiley, New York, 1973.
25. Burke, J. J., Weiss, V., Eds., "Block and Graft Copolymers," Syracuse, NY, 1973.
26. Molau, G. E., Ed., "Colloidal and Morphological Behavior of Block and Graft Copolymers," Plenum, New York, 1971.
27. Platzer, N., Ed., ADV. CHEM. SER. (1971) **99.**
28. Platzer, N., Ed., ADV. CHEM. SER. (1975) **142.**

29. Sperling, L. H., *Macromol. Rev.* (1977) **12**, 141.
30. McCoy, N. H., "Fundamentals of Abstract Algebra," Allyn and Bacon, Boston, 1972.
31. Baumslag, B., Chandler, B., "Theory and Problems of Group Theory," McGraw–Hill, 1968.
32. Goldhaber, K., Ehrlich, G., "Algebra," Macmillan, London, 1970.
33. Cotton, F. A., "Chemical Applications of Group Theory," 2nd ed., Wiley–Interscience, New York, 1971.
34. Bamford, C. H., Eastmond, G. C., Whittle, D., *Polymer* (1975) **16**, 377.
35. Sperling, L. H., "Encyclopedia of Polymer Science and Technology," Supplement No. 1, N. Bikales, Ed., p. 288, Wiley, 1976.
36. Kim, S. C., Klempner, D., Frisch, K. C., Radigan, W., Frisch, H. L., *Macromolecules* (1976) **9**, 258.

RECEIVED June 6, 1978. The authors are pleased to acknowledge the partial support of the Air Force Office of Scientific Research, Grant No. AFOSR-76-2945A and B.

INDEX

INDEX

D

N

O

P

Q

R

The text of this book is set in 10 point Caledonia with two points of leading. The chapter numerals are set in 30 point Garamond; the chapter titles are set in 18 point Garamond Bold.

The book is printed offset on Text White Opaque 50-pound. The cover is Joanna Book Binding blue linen.

Jacket design by Esther Agonis. Editing and production by Saundra Goss.

The book was composed by Service Composition Co., Baltimore, MD, printed and bound by The Maple Press Co., York, PA.